GEOCHIMICA ET COSMOCHIMICA ACTA

Supplement 14

PROCEEDINGS
OF THE
ELEVENTH LUNAR AND PLANETARY SCIENCE
CONFERENCE

Houston, Texas, March 17–21, 1980

COVER ILLUSTRATIONS

The ceaseless search for money with which to fund research has formed an important part of the professional scientist's lifestyle for millenia. One of the earliest known formal proposals in support of research in the lunar and planetary sciences was submitted to the Vatican (Pope Clement XI) in 1711 by L. Marsigli and E. Manfredi who were hoping to establish an astronomical observatory at Bologna.

Being as aware as their modern counterparts that a classy proposal which explains itself well is more likely to be funded than is a less elegant proposal, Marsigli and Manfredi commissioned a series of eight oil paintings to illustrate the instruments which would be domiciled in the new observatory and the various heavenly bodies which would be studied there. Two artists collaborated to produce the paintings: D. Creti, a popular landscape painter, painted the backgrounds and a miniaturist, R. Manzini, inserted the planets, the sun, and a comet into Creti's compositions. The histories of Manfredi's and Marsigli's successful application for funds and of the Institute of the Sciences of Bologna which they founded are discussed in the paper (Volume I, pages xiiiff.) by Silvio Bedini, Keeper of the Rare Books, Smithsonian Institution, who discovered and identified the eight paintings in the Vatican collection.

All eight of the paintings will be found in Bedini's article. Three illuminate the covers of these Proceedings:

Volume 1. The planet Saturn, shown with astronomers who are working with portions of an aerial telescope.

Volume 2. The planet Jupiter with three of its moons. Note the long telescope on a standard in the right foreground.

Volume 3. Earth's moon being viewed with a medium-sized telescope.

These paintings are reproduced courtesy of the Vatican Monumenti, Musei e Gallerie Pontificie.

GEOCHIMICA ET COSMOCHIMICA ACTA

Journal of The Geochemical Society and The Meteoritical Society

Supplement 14

PROCEEDINGS
OF THE
ELEVENTH LUNAR AND PLANETARY
SCIENCE CONFERENCE

Houston, Texas, March 17–21, 1980

Compiled by the
Lunar and Planetary Institute
Houston, Texas

Volume 3
Physical Processes

Pergamon Press

New York • Oxford • Toronto • Sydney • Frankfurt • Paris

Pergamon Press Offices:

U.S.A.	Pergamon Press Inc., Maxwell House, Fairview Park, Elmsford, New York 10523, U.S.A.
U.K.	Pergamon Press Ltd., Headington Hill Hall, Oxford OX30BW, England
CANADA	Pergamon of Canada Ltd., 150 Consumers Road, Willowdale, Ontario M2J 1P9, Canada
AUSTRALIA	Pergamon Press (Aust) Pty. Ltd., P.O. Box 544, Potts Point, NSW 2011, Australia
FRANCE	Pergamon Press SARL, 24 rue des Ecoles, 75240 Paris, Cedex 05, France
FEDERAL REPUBLIC OF GERMANY	Pergamon Press GmbH, 6242 Kronberg/Taunus, Pferdstrasse 1, Federal Republic of Germany

First edition 1980

Type set by Precision Typographers,
printed by Publishers Production International,
and bound by Arnold's Bindery
in the United States of America

Library of Congress Cataloging in Publication Data

Lunar and Planetary Science Conference, 11th Houston,
 Tex., 1980.
  Proceedings of the Eleventh Lunar and Planetary
Science Conference, Houston, Texas March 17–21, 1980.

  (Geochimica et cosmochimica acta: Supplement; 14)
  Includes bibliographies and indexes.
  CONTENTS: v. 1. Igneous processes and remote sen-
sing.—v. 2. Meteorite and regolith studies.—
v. 3. Physical processes.
  1. Lunar geology—Congresses. 2. Planets—Geology—
Congresses. 3. Solar system—Congresses. I. Lunar
and Planetary Institute. II. Series.
QB592.L84 1980   559.9   80–24270
ISBN 0-08-026314-3 (set)

R. J. Pike, U.S. Geological Survey, Menlo Park, California 94025

A. M. Reid, Geology Branch, NASA Johnson Space Center, Houston, Texas 77058

G. Ryder, Northrop Services, Incorporated, Houston, Texas 77034

R. B. Schaal, Lockheed Engineering and Management Services Company, Incorporated, Houston, Texas 77058

R. N. Schock, Lawrence Livermore Laboratory, Livermore, California 94520

C. H. Simonds, Northrop Services, Incorporated, Houston, Texas 77034

L. J. Srnka, Exxon Production Research Company, Houston, Texas 77001

D. Stöffler, Lunar and Planetary Institute, Houston, Texas 77058

E. Stolper, Division of Geological and Planetary Sciences, California Institute of Technology, Pasadena, California 91125

G. J. Taylor, Department of Geology and Institute of Meteoritics, University of New Mexico, Albuquerque, New Mexico 87131

D. Walker, Hoffman Laboratory, Harvard University, Cambridge, Massachusetts 02138

R. J. Williams, Geochemistry Branch, NASA Johnson Space Center, Houston, Texas 77058

A. Zaikowski, Department of Energy, Bendix Field Engineering Corporation, Grand Junction, Colorado 81501

TECHNICAL EDITOR

P. R. Criswell, Lunar and Planetary Institute, Houston, Texas 77058

EDITORIAL STAFF

K. Hrametz   B. Crowley
R. Edwards   M. Dickey
K. Christianson
Lunar and Planetary Institute, Houston, Texas 77058

CONTENTS

Volume Three—Physical Processes

GEOPHYSICAL INVESTIGATIONS
Laboratory Measurements

Planetary Measurements

Structural Implications and Models

IMPACT PROCESSES

Experimental Studies

Crater Morphology and Frequency

Models Theory

VOLCANIC—TECTONIC PROCESSES

PLANETARY ATMOSPHERES

Proc. Lunar Planet. Sci. Conf. 11th (1980), p. 1777–1788.
Printed in the United States of America

Thermal diffusivity of two Apollo 11 samples, 10020,44 and 10065,23: Effect of petrofabrics on the thermal conductivity of porous lunar rocks under vacuum

K. Horai[1] and J. L. Winkler, Jr.[2]

[1] Lamont-Doherty Geological Observatory of Columbia University, Palisades, New York, 10964
[2] Lockheed Electronics Corporation, Houston, Texas, 77058

Abstract—A remeasurement of the thermal diffusivity on two Apollo 11 rock samples (basalt, 10020,44; and, breccia, 10065,23), in the temperature range between 80 and 460 K, under both atmospheric and vacuum conditions, revealed that porosity is not the only major parameter determining the porous lunar material's thermal conductivity. A comparison with previous thermal diffusivity data on Apollo 17 rock samples shows the importance of petrofabrics. Effects of interstitial gaseous pressure on the porous lunar material's thermal conductivity will be totally different depending on the abundance of the sample's microcracks. For a sample abundant in microcracks, adsorption and desorption of gas molecules on the surfaces of mineral grains will decisively control the sample's intergranular thermal contact. Interstitial gaseous pressure being reduced, the thermal conductivity will be lowered drastically because the absence of an adsorption layer tends to isolate thermally the sample's mineral grains. The intergranular connectivity also influences the thermal conductivity. If a rock sample's petrographic texture is such that the mineral grains form a well-connected network, leaving the voids geometrically isolated, the increase in the thermal conductivity can be large enough to offset the effect of porosity.

INTRODUCTION

Thermal diffusivity measurements on four Apollo 11 rock samples were made immediately after the return of Apollo 11 mission (Horai *et al.*, 1970) using the modified Angstrom method (Kanamori *et al.*, 1968; 1969) in the temperature range between 150 and 450 K under 1 atm of air. The measurements indicated that the thermal diffusivity as a function of temperature is nearly identical for the crystalline rock samples (10020,44 and 10057,76) and for the breccias (10046,51 and 10065,23), but the crystalline rocks have a distinctively higher thermal diffusivity than the breccias. The above study did not include, however, a measurement under a vacuum condition, pertinent to a discussion of the lunar surface. To refine the previous result, the thermal diffusivity was remeasured on two of the above samples using a more systematic procedure.

SAMPLE

A description of the samples selected for the present study can be found in Table 1. The same method as employed previously in our laboratory was used to determine the porosity (Horai and Winkler, 1976).

EXPERIMENTAL

As with the previous measurements, the modified Angstrom method was used for the determination of thermal diffusivity. Temperature on one side of the parallelepiped sample was varied periodically with an amplitude of several degrees Kelvin, and the sample thermal diffusivity was determined from the amplitude decay and the phase lag of the temperature wave propagating through the sample, by measuring the surface temperature on the other side of the sample where a total reflection of the temperature wave was assumed.

The measurement apparatus used in the present study has been described before (Horai and Winkler, 1976). The measurement on sample 12002,85 indicated that the thermal diffusivity determined with this apparatus depended strongly on the frequency of the temperature wave (Horai and Winkler, 1975). It was necessary, therefore, to make measurements at four different temperature wave frequencies ranging from 16.7 to 33.3 mHz, hoping that their average would be close to the sample's thermal diffusivity. For four Apollo 17 rock samples the thermal diffusivity determination was less sensitive to the frequency so that the average of measurements using two different frequencies, of either 3.0, 5.0, or 10.00 mHz, sufficed to give a reliable thermal diffusivity value (Horai and Winkler, 1976).

The present study revealed that the two Apollo 11 samples' thermal diffusivity was nearly independent of the temperature wave frequency. As Fig. 1 shows, each sample's thermal diffusivity varies little for a frequency changing by a factor of five. Subsequently, a fixed frequency, 8.0 mHz for sample 10020,44 and 3.0 mHz for sample 10065,23, was used for all measurements with changing temperature and pressure conditions, assuming that the thermal diffusivity value obtained at this frequency does not differ significantly from the sample's intrinsic thermal diffusivity.

The apparent dependency of the thermal diffusivity on the temperature wave frequency is less pronounced as the sample thickness increases. The thickness of the samples studied in our laboratory was 0.33 cm for sample 12002,85, 0.58 to 0.77 cm for Apollo 17 samples, and 0.97 to 0.98 cm for Apollo 11 samples. The temperature wave's higher order Fourier components decay more rapidly than those of the lower orders as the wave travels through the sample. Probably, sample 12002,85 was not thick enough for the higher order Fourier components to decay completely. As the modified Angstrom technique is developed on the theory of amplitude decay and phase lag of a purely sinusoidal temperature wave, any persisting higher order Fourier component, resulting in a wave form distortion, interferes with the thermal diffusivity determination. Fourier analysis of input and output temperature wave forms may have been an appropriate method of data analysis that should be employed for the measurement of sample 12002,85.

Table 1. Sample description.

Sample	Dimensions (cm × cm × cm)	Mass, m (g)	Bulk density, ρ (g/ml)	Intrinsic density, ρ_o (g/ml)	Porosity, p (%)
10020,44	0.98 × 0.98 × 2.03	5.942	3.060	3.251	5.9
10065,23	0.97 × 1.00 × 1.97	4.483	2.367	3.115	24.0

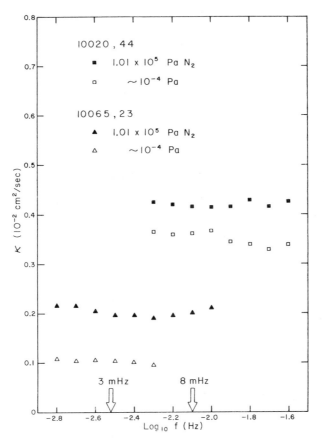

Fig. 1. Dependency of lunar sample thermal diffusivity κ determined by the modified Angstrom method on the frequency f of temperature wave.

MEASUREMENT RESULT

Figure 2 summarizes the measurement's result. For both samples 10020,44 and 10065,23, the thermal diffusivity was determined at a 20 K interval from 80 to 460 K. The vacuum chamber containing the thermal diffusivity measurement apparatus was first evacuated and then filled with N_2 gas. As the samples were desiccated in a vacuum oven before installed in the measurement apparatus and were contacted with the dry N_2 gas after subjected to a high vacuum, they are believed to be free from the effect of moisture. The gaseous pressure was maintained at either 1.01×10^5 Pa (1 atm) or 1.0×10^{-4} Pa (ca 10^{-6} torr) during the measurement. As the void spaces in a porous lunar rock are likely to be interconnected, the sample's environmental gaseous pressure can be assumed to be equal to the sample's interstitial gaseous pressure. The error bounds of thermal diffusivity data points in Fig. 2, estimated from the scattering of individual determinations, are $\pm 10\%$ for sample 10020,44 and $\pm 6\%$ for sample 10065,23.

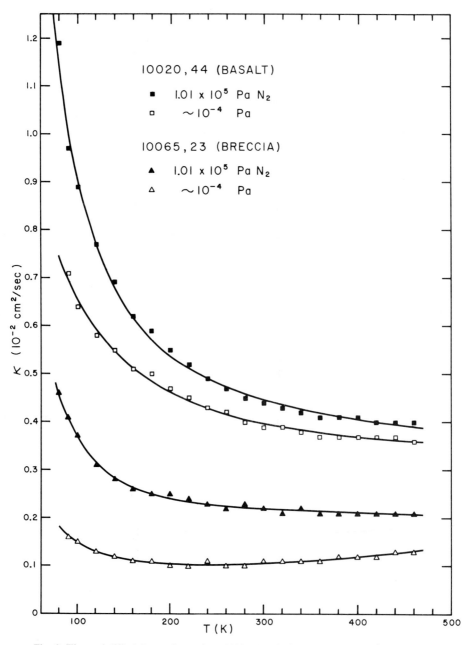

Fig. 2. Thermal diffusivity κ of samples 10020,44 and 10065,23 as a function of temperature T with interstitial gaseous pressure of N_2 1.01×10^5 Pa and about 10^{-4} Pa.

Table 2. Thermal diffusivity κ (cm²/s) as a function of temperature T (K): $\kappa = A + B/T + C/T^2 + DT^2$.

Sample	Condition	A $(10^{-2}cm^2/s)$	B $(cm^2\ K/s)$	C $(10^2cm^2\ K^2/s)$	D $(10^{-8}cm^2/s\ K^2)$
10020,44	1.01×10^5 Pa N₂	0.333	0.252	0.323	−0.063
	~10^{-4} Pa N₂	0.242	0.452	−0.040	0.100
10065,23	1.01×10^5 Pa N₂	0.214	−0.036	0.188	−0.036
	~10^{-4} Pa N₂	0.062	0.042	0.040	0.276

Under both atmospheric and vacuum conditions, the thermal diffusivity, κ, as a function of temperature, T, is adequately represented by the formula $\kappa = A + B/T + C/T^2 + DT^2$ as illustrated in Fig. 2. To fit the equation to the data, the coefficients were determined by the least-square criterion as listed in Table 2.

EFFECT OF ADSORBED GAS MOLECULES ON THE THERMAL CONDUCTIVITY OF POROUS ROCKS

For both the crystalline rock sample (10020,44) and the breccia (10065,23), the sample's thermal diffusivity decreases as the interstitial gaseous pressure is lowered from 1 atm to high vacuum. Primarily, this is because the interstitial gas's effective thermal conductivity becomes extremely low under vacuum conditions. The data presented in this paper suggest, however, that a porous rock sample's thermal conductivity under vacuum is controlled more subtly by the gas molecule's interaction with the sample's intergranular thermal contact. The values of sample thermal diffusivity necessary for the following discussion are summarized in Table 3.

The table shows that the sample 10020,44's thermal diffusivity at 200 K decreases by 14 percent, from 5.37×10^{-3} cm²/s to 4.62×10^{-3} cm/s, as the

Table 3. Thermal diffusivity of Apollo 11 and 17 rock samples at 200 K under atmospheric and vacuum conditions.

Sample	Porosity, p (%)	Thermal diffusivity, κ $(10^{-3}\ cm^2/s)$		Decrease in thermal diffusivity, $\Delta\kappa/\kappa$ (%)
		1.01×10^5 Pa N₂	10^{-4} Pa N₂	
10020,44 (basalt)	5.9	5.37	4.62	14
10065,23 (breccia)	24.0	2.42	1.04	57
		1.01×10^5 Pa Air	10^{-4} Pa Air	
70215,18 (basalt)	5.3	5.61	2.68	52
70017,77 (basalt)	22.7	5.51	2.74	50

interstitial gaseous pressure of N_2 is decreased from 1.01×10^5 Pa to 1.0×10^{-4} Pa. It is contrasted with the data on the sample 70215,18 reported previously (Horai and Winkler, 1976), showing a 52 percent decrease of the sample's thermal diffusivity, from 5.61×10^{-3} cm^2/s to 2.68×10^{-3} cm^2/s, with the lowering pressure of air from 1 atm (1.01×10^5 Pa) to 10^{-6} torr (ca 1.0×10^{-4} Pa) at 200 K. Since the sample's porosity is about the same (5.9 percent for samples 10020,44 and 5.3 percent for sample 70215,18), the comparison of the data indicates that the volumetric fraction of the void spaces in the sample is not the only decisive factor controlling the lunar rock sample's thermal conduction under vacuum.

We infer that the geometrical configuration, namely the size and shape distribution, of the void spaces is not the same for samples 10020,44 and 70215,18. Probably, the void spaces of the sample 10020,44 are mostly spherical pores, whereas microcracks are abundant in the sample 70215,18. The effect of rarefied gas on the intergranular thermal contact will then be totally different for these two samples. As we indicated previously (Horai and Winkler, 1976), a porous rock sample's intergranular thermal contact can be influenced strongly by an adsorption of gas molecules. Figure 3 is a model we propose for mechanisms of thermal conduction that differ according to the sample type. For 10020,44 type samples, molecules of the interstitial gas are adsorbed on the surface of the spherical pores. As the gaseous pressure is reduced, the gas molecules are desorbed. Since, however, the sample's intergranular thermal contact is little affected by the desorption in the spherical pore, the sample's thermal conductivity is decreased as the effective thermal conductivity of the interstitial gas diminishes with the reduced gaseous pressure.

The situation is completely different if the adsorption occurs in microcracks as in sample 70215,18. According to the theory of surface chemistry (for example, Aveyard and Haydon, 1973), neutral gas such as N_2 can form a multilayer of adsorbed molecules at a temperature even higher than the gas's critical temperature (126.3 K for N_2). The thickness of the layer can be more than ten molecular diameters. As the diameter of the N_2 molecule is 0.4 nm, an adsorption multilayer of 100 molecular diameter thickness can fill a microcrack of 0.04 μm wide. The effect of adsorption will be particularly noticeable near the crack's edge where the adsorptive potential exercising on the adsorbate is at least twice as much as is expected from a half-space adsorbent. It is known that the molecular structure of an adsorption layer is similar to that of a liquid. Evidently, the material's thermal conductivity increases substantially as it is transformed from a gas to the liquid state. The thermal contact between the mineral grains separated by a microcrack will be improved greatly if the crack is filled with the adsorption layer.

For the 70215,18 type samples, the effective thermal conductive cross section between the mineral grains diminishes rapidly as the adsorbed gas molecules are removed. The thermal conductivity decreases drastically under vacuum conditions, as the sample's mineral grains separated by microcracks tend to be thermally isolated by the desorption of gas molecules filling the cracks. Accordingly, the thermal conductivity, hence the diffusivity, of sample 70215,18 is substantially lower than that of 10020,44 under vacuum. Under atmospheric conditions, both

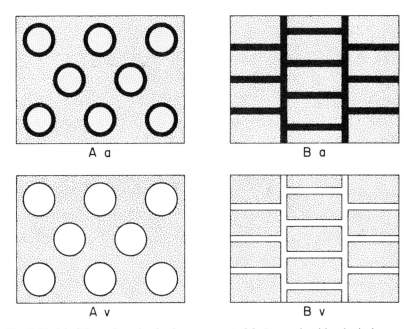

Fig. 3. Model of thermal conduction in porous material. A, sample with spherical pores: solid, continuous phase; gas, isolated phase. The sample's intergranular thermal conductive cross section is not altered appreciably by adsorption (a) and desorption (v) of gas. B, sample abundant in microcracks: solid, isolated phase; gas, continuous phase. Adsorption of gas (a) increases the sample's intergranular thermal contact drastically. Gas molecules being desorbed under vacuum condition (v), mineral grains are isolated thermally.

samples show about the same thermal conductivity. The intergranular thermal contact of sample 70215,18 with all its microcracks filled with adsorbed gas molecules, is comparable to that of sample 10020,44 in which mineral grains are not separated by microcracks. The macroscopic sample's thermal conductive cross section will then be determined by the sample's porosity. The porosity of sample 70215,18, with a volumetric fraction of macropores and macrocracks* that are too wide to be filled by the adsorption layer, is about the same as that of sample 10020,44. The micropores and microcracks, however numerous, will not contribute much to the sample's porosity.

Effect of the rock's texture on the sample's thermal conductivity

A rock's petrographic texture has a strong control on the sample's intergranular thermal contact. This is apparent in the data presented in Fig. 2. Under both atmospheric and vacuum conditions, the thermal diffusivity of sample 10065,23

---

*Pores (and cracks) are classified according to size as follows (Aveyard and Haydon, 1974): micropores, <1.5 nm; transitional pores or mesopores, 1.5 ~ 100 nm; macropores, >100 nm.

is definitely lower than that of sample 10020,44. As indicated in Table 1, the two samples have a markedly different porosity; 5.9 percent for 10020,44 and 24 percent for 10065,23. The sample 10065,23's larger porosity, however, may not solely be responsible for the sample's lower diffusivity. Sample 70017,77, one of the four Apollo 17 samples studied previously (Horai and Winkler, 1976), has a porosity of 22.7 percent, but its thermal diffusivity at 200 K is 5.51×10^{-3} cm^2/s and 2.74×10^{-3} cm^2/s, respectively, under atmospheric and vacuum conditions. More likely, the lower thermal diffusivity of samples 10065,23 is attributed to its loose petrographic texture. Even a visual inspection will show the weak intergranular cohesion of sample 10065,23. Probably, the sample 10065,23's intergranular voids are generally so wide that the gas molecules adsorbed on the mineral grains' surface do not improve appreciably the sample's intergranular thermal contact. This does not preclude, however, a possibility that the sample contains a substantial number of micropores and microcracks. The decrease of sample 10065,23's thermal diffusivity at 200 K, from 2.42×10^{-3} cm^2/s to 1.04×10^{-3} cm^2/s, with decreasing gaseous pressure of N_2 from 1.01×10^5 Pa to 1.0×10^{-4} Pa is too large to be explained by the absence of the gaseous phase alone.

In spite of its large porosity, sample 70017,77 has a thermal diffusivity that is comparable, both under atmospheric and vacuum conditions, to the less porous sample 70215,18's thermal diffusivity. The sample 70017,77 is basalt, hence its intergranular cohesion is much stronger than that of breccia. A 50 percent decrease of thermal diffusivity at 200 K with the decrease of interstitial gaseous pressure from 1 atm to high vacuum suggests the microcracks are abundant in this sample also. With the microcracks filled with the adsorbed gas molecules, the sample's thermal conductivity becomes equal to that of sample 10020,44 and 70215,18. It is possible that the sample 70017,77's intergranular connectivity is good enough to compensate the decrease of thermal conductivity due to its larger porosity.

CONCLUDING REMARKS

A rather speculative interpretation of the data given above needs to be elaborated by further experiments. It is necessary to study the sample's petrofabric in detail after the thermal property measurements were completed. As the lunar rock samples are undestroyable, they are not suitable for this purpose. Studies on homogeneous terrestrial rock samples are desirable, since petrographic thin sections are easily made from a part of the samples on which thermal conductivity or diffusivity has been measured.

In the Appendix, a result of theoretical calculation on the thermal conductivity of two phase material is presented. This will supplement the qualitative argument developed in the text.

Acknowledgments—Constructive criticisms given by T. Takahashi, J. S. Weaver, W. Durham, and an anonymous Proceedings review were most helpful. Financial support was provided by NASA grant NGR-33-008-177. Lamont Doherty Geological Observatory Contribution No. 3059.

REFERENCES

Aveyard R. and Haydon D. A. (1973) *An Introduction to the Principle of Surface Chemistry*. Cambridge Univ. Press, London. 323 pp.

Hashin Z. and Shtrikman S. (1962) A variational approach to the theory of the effective magnetic permeability of multiphase materials. *J. Appl. Phys.* **33**, 3125–3131.

Hemingway B. S., Robie R. A., and Wilson W. H. (1973) Specific heats of lunar soils, basalt, and breccias from the Apollo 14, 15 and 16 landing sites, between 90 and 350°K. *Proc. Lunar Sci. Conf. 4th*, p. 2481–2487.

Horai K. and Baldridge S. (1972) Thermal conductivity of nineteen igneous rocks. *Phys. Earth Planet. Inter.* **5**, 151–166.

Horai K., Simmons G., Kanamori H., and Wones D. (1970) Thermal diffusivity, conductivity, and thermal inertia of Apollo 11 lunar material. *Proc. Apollo 11 Lunar Sci. Conf.*, p. 2243–2249.

Horai K. and Winkler J. L. (1975) Thermal diffusivity of lunar rock sample 12002,85. *Proc. Lunar Sci. Conf. 6th*, p. 3207–3215.

Horai K. and Winkler J. L. (1976) Thermal diffusivity of four Apollo 17 rock samples. *Proc. Lunar Sci. Conf. 7th*, p. 3183–3204.

Kanamori H., Fujii N., and Mizutani H. (1968) Thermal diffusivity measurement of rock-forming minerals from 300° to 1100° K. *J. Geophys. Res.* **73**, 595–605.

Kanamori H., Mizutani H., and Fujii N. (1969) Method of thermal diffusivity measurements. *J. Phys. Earth* **17**, 43–53.

Kawada K. (1964) Variation of thermal conductivity of rocks. *Bull. Earthquake Res. Inst. Univ. Tokyo* **42**, 631–647.

Powell R. L. and Childs G. E. (1972) Thermal conductivity. In *American Institute of Physics Handbook*, 3rd ed., section 4g. McGraw-Hill, N.Y.

Robertson E. G. and Peck D. L. (1974) Thermal conductivity of vesicular basalt from Hawaii. *J. Geophys. Res.* **79**, 4875–4888.

Robie R. A. and Hemingway B. S. (1971) Specific heats of lunar breccia (10021) and olivine dolerite (12018) between 90° and 350° Kelvin. *Proc. Lunar Sci. Conf. 2nd*, p. 2361–2365.

Robie R. A. Hemingway B. S., and Witson W. H. (1970) Specific heats of lunar surface materials from 90° and 350°K. *Proc. Apollo 11 Lunar Sci. Conf.*, p. 2361–2367.

APPENDIX

Thermal conductivity of two-phase material according to Hashin and Shtrikman's theory

Theoretical bounds for the thermal conductivity of two-phase material have been given by Hashin and Shtrikman (1962). For material consisting of gaseous and solid phases, they are

$$k_U = k_s \left[1 + \frac{v}{(1-v)/3 - 1(1-r)} \right] \tag{A1}$$

$$k_L = k_s \left[r + \frac{r(1-v)}{v/3 + r/(1-r)} \right] \tag{A2}$$

$$r = k_g/k_s \tag{A3}$$

where k_s and k_g are the thermal conductivities of solid and gas, respectively, and v is the volumetric fraction of the gaseous phase in the material. For a porous solid rock sample with its void spaces filled with gas, v is equal to the sample's porosity, p. For a macroscopically isotropic and homogeneous two-phase material in which both solid and gaseous phases are completely randomly admixed, arithmetic mean of the bounds

$$k = (k_U + k_L)/2 \tag{A4}$$

represents satisfactorily the material's thermal conductivity.

Figure A1 is an illustration of k/k_s as a function of p for various r. The figure shows that the thermal conductivity of porous rock sample with its void spaces filled with gas decreases as the porosity increases. Naturally, the decrease of thermal conductivity is more rapid as the interstitial gas is less conductive. At 200 K, the thermal conductivity of N_2 gas is 18.7 mW/m K and that of air 18.3 mW/m K (Powell and Childs, 1972). At ordinary temperature, the thermal conductivity of basaltic material ranges from 1.8 to 2.3 W/m K (Robertson and Peck, 1974; Horai and Baldridge, 1972). At 200 K, the range will be from 2.0 to 2.5 W/m K as the thermal conductivity of basalt

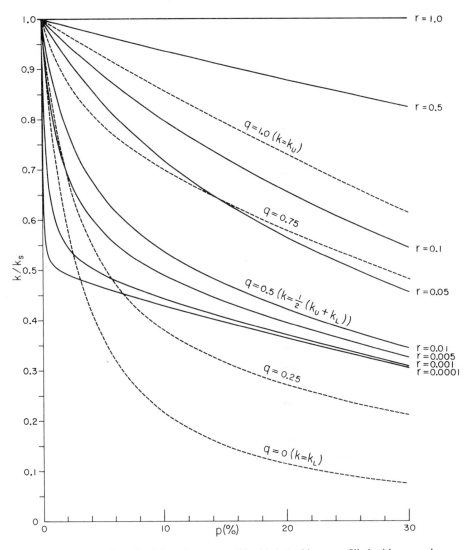

Fig. A1. Thermal conductivity of porous solid with its void spaces filled with gas, calculated according to Hashin and Shtrikman's theory. Average thermal conductivity k of the theoretical upper and lower bounds, k_U and k_L, normalized by k_s is given as a function of porosity p for various $r = k_g/k_s$, where k_g and k_s are the thermal conductivities of solid and gaseous phases, respectively. For $r = 0.01$, $k/k_s = [q\,k_U + (1 - q)k_L]/k_s$ with $q = 0.25$ and 0.75 is also illustrated.

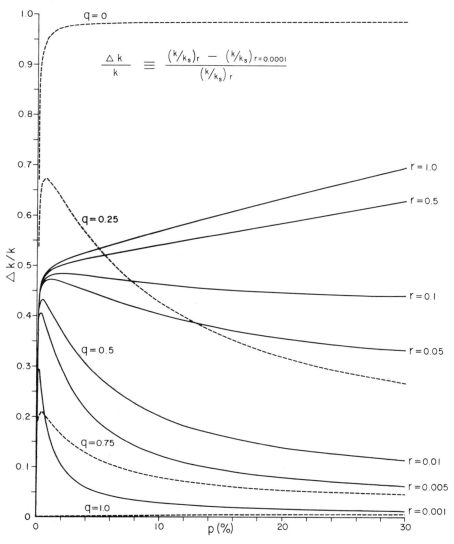

Fig. A2. A plot of $\Delta k/k = [(k/k_s)_r - (k/k_s)_{r=0.0001}]/(k/k_s)_r$ as a function of p for various r, indicating the effect of diminishing effective thermal conductivity of interstitial gas on the thermal conductivity of porous solid.

increases with decreasing temperature (Kawada, 1964). Therefore, $0.005 < r < 0.01$ will be a reasonable estimate for a porous basaltic lunar rock sample with its void spaces filled with gaseous N_2 or air at 200 K. The figure shows that for $r = 0.01$, the decrease of sample thermal conductivity is from 0.62 k_s to 0.39 k_s as the sample's porosity increases from 6 to 24 percent.

The theoretical bounds given by (A1) and (A2) are the limits that the thermal conductivity of two-phase material can take: k_U will be assumed if solid constitutes a continuous phase and gas an isolated phase; and, conversely, k_L if solid and gas are isolated and continuous phases, respectively. A porous lunar rock sample's void spaces are interconnected as well as its mineral grains. Accordingly, both solid and gaseous phases are continuous.

Table A1. Estimation of the sample's solid phase connectivity q from the variation of thermal conductivity.

| Sample | Porosity p (%) | Effect of porosity | | | | Effect of reduced gaseous pressure | |
| | | Under 1 atm | | Under vacuum | | | |
		$(k/k_s)_a$ (%)	q	$(k/k_s)_v$ (%)	q	$\Delta k/k$ (%)	q
10020,44	5.9	47	0.27	40	0.44	14	0.67
10065,23	24.0	17	0.14	7	0.11	57	0.09
70215,18	5.3	49	0.27	23	0.25	52	0.24
70017,77	22.7	39	0.49	20	0.28	50	0.12

In Fig. A1, k_U/k_s and k_L/k_s are illustrated as a function of p for r = 0.01. A wide range around the mean $(k_U + k_L)/2k_s$ shows that the porous lunar rock sample's thermal conductivity can vary according to the sample's texture as much as the limits indicate. For a sample in which both solid and gaseous phases are equally well connected, the thermal conductivity will be given by (A4). If, however, the sample's texture is such that either of the sample's solid or gaseous phases is better connected than the other, the thermal conductivity will increase or decrease depending on the connectivity of the phases. The sample's solid phase connectivity can be expressed numerically by q if we put

$$k = q\,k_U + (1 - q)k_L.$$

In Fig. A1, besides the curves for q = 1.0, 0.5, and 0, those for q = 0.75 and 0.25 are also illustrated for r = 0.01. It is easily seen from the figure that the change of thermal conductivity can be large enough to compensate the effect of porosity.

Figure A2 is a plot of $\Delta k/k = [(k/k_s)_r - (k/k_s)_{r=0.0001}]/(k/k_s)_r$ as a function of p, illustrating the effect of reduced interstitial gaseous pressure on the porous material's thermal conductivity. Comparing Figs. A1 and A2 with the data presented in Table 3, the sample's solid phase connectivity can be estimated. Assuming k_s = 2.0 W/m K (r = 0.009), the sample's solid phase thermal diffusivity is calculated as $\kappa_s = k_s/\rho_0\,c$, where ρ_0 is the intrinsic density of the sample (Table 1) and c = 0.562 J/g K is the lunar material's specific heat at 200 K (Hemingway *et al.*, 1973; Robie and Hemingway, 1971; Robie *et al.*, 1970). Then, the decrease in thermal conductivity due to increasing porosity p is obtained from the thermal diffusivity data as $k/k_s = (1 - p)\,\kappa/\kappa_s$. The effect of reduced interstitial gaseous pressure on thermal conductivity is derived simply from the diffusivity: $\Delta k/k = \Delta \kappa/\kappa$.

The result summarized in Table A1 shows some conflicts which probably originated from the assumed k_s that may be different according to the samples. For example, a notable discrepancy is seen between the values of q estimated from k/k_s for sample 10020,44 under atmospheric and vacuum conditions. Another example is the value of q obtained from $\Delta k/k$ which appears to be too large for sample 10020,44 and too small for sample 70017,77. The result is otherwise consistent with the discussion developed in the text. 1) Sample 10020,44's larger q than sample 10065,23's indicates that the basalt sample's mineral grain connectivity is generally better than the breccia's. In fact, the breccia sample 10065,23's q is lower than any of the three basalt samples'. 2) For sample 10065,23, 70215,18, and 70017,77, the values of q estimated under atmospheric condition are larger than those obtained under vacuum condition, testifying the presence of microcracks in these samples. The sample's apparent mineral grain connectivity increases as the sample's microcracks are filled with adsorbed gas molecules. Under the condition of 1 atm, sample 70215,18's apparent mineral grain connectivity is equal to sample 10020,44's mineral grain connectivity. 3) Sample 70017,77's large q value under atmospheric condition indicates that the sample's mineral grain connectivity is large enough to offset the effect of porosity. The sharp decrease of q under vacuum condition shows that the sample contains numerous microcracks.

Proc. Lunar Planet. Sci. Conf. 11th (1980), p. 1789–1799.
Printed in the United States of America

Magnetic classification of Antarctic meteorites

Takesi Nagata

National Institute of Polar Research, Tokyo 173, Japan

Abstract—Thirty one chondrites (including 14 Antarctic chondrites) and 18 Antarctic achondrites are magnetically classified into 8 chemically different groups, namely, E-, H-, L-, LL- and C- chondrites, ureilites, diogenites and basaltic achondrites (eucrites and howardites). The principle of magnetic classification is based on the content of metallic NiFe represented by the saturation magnetization (I_s) as a parameter and the relative content of Ni in the NiFe metals represented by a ratio of the saturation magnetization of kamacite to the total I_s value ($I_s(\alpha)/I_s$) as the other parameter.

On an $I_s(\alpha)/I_s$ versus I_s diagram, the eight chemical groups of chondrites and achondrites are confined to respective domains well separated from one another.

1. INTRODUCTION

Magnetic analyses of extraterrestrial materials such as lunar rocks and meteorites have frequently revealed their unique usefulness in detecting and characterizing the ferromagnetic constituents in those materials such as native iron and magnetite. For example, the abundance and characteristics of native iron in an eucritic unique achondrite ALHA 77005 recently found in Antarctica is reasonably well determined by measuring its magnetic hysteresis curve and thermomagnetic curves, though no metal was observed in this achondrite by ordinary petrographic analyses (McSween *et al.*, 1979a, b). As discussed later in this report, the average abundance of native iron in ALHA 77005 is about 0.042 wt%, which can be quantitatively estimated by a magnetic analysis without any difficulty.

Since the magnetic analyses can give rise to an estimate of the bulk abundance of ferromagnetic constituents and their average chemical composition, those magnetic techniques have been often applied on describing and characterizing metallic components in lunar materials and meteorites (e.g., Nagata *et al.*, 1971, 1974; Gus'kova, 1972). One of the valuable applications of the magnetic techniques in studies of meteorites may be a magnetic classification of stony meteorites. Since the content of metallic iron in chondrites decreases in the sequence of their chemically classified groups of enstatite chondrite (E), olivine bronzite chondrite (H), olivine hypersthene chondrite (L), olivine pigeonite chondrite (LL) and carbonaceous chondrite (C) (Urey and Craig, 1953; Mason 1962a) and the content of metallic iron in achondrites except ureilites is distinctly smaller than that in chondrites, the chemical classification of stony meteorites could be performed by evaluating the content of metallic iron with the aid of measuring the saturation magnetization (I_s). On the other hand, it has been pointed out by Prior (1920) that the ratio of Ni content to Fe content (Ni/Fe) in the metallic component in chondrites increases in the sequence of E-chondrite, H-chondrite, L-chondrite and LL-chondrite. In reality, the metallic component in E-chondrite is almost kamacite, that in H-chondrite is kamacite plus a small amount of plessite, that in L-chondrite is kamacite plus a larger amount of plessite and/or taenite, and that in LL-chondrite is mostly taenite. The contents of kamacite, plessite and taenite in meteorites can be evaluated by their thermomagnetic analysis. On the basis of a combination of the Urey-Craig-Mason law in regard to the abundance of metallic nickel-iron and the Prior rule in regard to Ni/Fe in the metallic component, a magnetic classification scheme of stony meteorites has been proposed with reasonable success (Nagata and Sugiura, 1976; Nagata, 1979a, b, c). In the proposed magnetic classification scheme, however, achondrites (except ureilites) are classified into a single group, in which the saturation magnetization (I_s) is very small (<0.6 emu/gm). In the present study, an attempt is made to distinguish magnetically pyroxene-plagioclase achondrites (eucrite and howardite) from the diogenite group.

Up to the present time, 6 irons, 1 stony-iron, 14 chondrites and 18 achondrites of the Antarctic collection of meteorites have been magnetically analyzed. Iron meteorites and stony-iron meteorites can be easily distinguished from stony meteorites without any precise mineralogical, chemical or magnetic examination. For example, the density of examined 6 Antarctic iron meteorites ranges from 6.868 to 7.836, while that of an Antarctic pallasite (Yamato-74044) is 5.083. These values of density of irons and stony-irons are distinctly larger than the density of any stony meteorite. In the present note, therefore, the magnetic properties of Antarctic chondrites and achondrites only will be dealt with.

2. MAGNETIC CLASSIFICATION OF ANTARCTIC CHONDRITES

A group of Antarctic chondrites which had been examined magnetically have been petrographically and chemically classified into 1 E-chondrite, 4 H-chondrites, 4 L-chondrites, 3 LL-chondrites and 2 C-chondrites, as listed in Table 1. A magnetic classification of these 14 Antarctic chondrites together with other 17 chondrites (4 H-chondrites, 6 L-chondrites, 1 LL-chondrite and 6 C-chondrites) has already been reported (Nagata, 1979a, b, c). Therefore, only a brief summary of a magnetic classification of Antarctic chondrites will be outlined in the present note.

As shown in Table 1, the saturation magnetization (I_s) is the largest for E-chondrites and the smallest for LL-chondrites, and generally,

$$I_s(E) > I_s(H) > I_s(L) > I_s(LL). \tag{1}$$

Since the ferromagnetic constituents of E-, H-, L- and LL-chondrites are FeNi alloys, inequality relation (1) may approximately represent the well-known rela-

Table 1. Saturation magnetization (I_s) and forromagnetic constituents of Antarctic chondrites.

Chondrites	Classification	I_s (emu-gm)	$\dfrac{I_s(\alpha)}{I_s}$	$\dfrac{I_s(\alpha + \gamma)}{I_s}$	$\dfrac{I_s(\gamma)}{I_s}$	$\dfrac{I_s(Mt)}{I_s}$
Yamato 691	$E_{3,4}$	48.0	0.97	0	0.03	0
Yamato 74031	H_5	33.5	0.95	0.05	0	0
Yamato 74647	H_5	27.9	0.94	0.06	0	0
MB 7602	H_6	27.4	0.88	0.10	0.02	0
Yamato 694	H_6	32.3	0.94	0.06	0	0
Yamato 74191	L_4	6.8	0.79	0.21	0	0
Yamato 7304	L_5	16.6	0.90	0	0.10	0
Yamato 74362	L_6	8.1	0.81	0.19	0	0
ALHA 7609	L_6	8.4	0.65	0.35	0	0
ALHA 7604	LL_3	5.2	0.60	0.40	0	0
Yamato 74442	LL_4	6.0	0.45	0.35	0.20	0
Yamato 74646	$LL_{5,6}$	3.2	0.19	0.07	0.74	0
Yamato 74662	C_2	0.81	0.05	0	0.85	0.10
Yamato 693	C_3	10.8	0	0	0	1.00

Remarks: MB = Mt. Boldr   ALHA = Allan Hills

tionship between FeO content and the content of metallic iron plus Fe in FeS in these chondrites, found by Urey and Craig (1953) and re-examined by Mason (1962a,b). As for C-chondrites, I_s of Yamato-6903 is 10.8 emu/gm whereas I_s of Yamato 74662 is only 0.81 emu/gm. As far as the I_s-value is concerned, therefore, C-chondrites may not be able to be distinguished from L- and LL-chondrites.

In Table 1, ratios of saturation magnetizations of α-phase (kamacite), ($\alpha + \gamma$)-phase (plessite) and γ-phase (taenite) to the total saturation magnetizations (I_s), namely, $I_s(\alpha)/I_s$, $I_s(\alpha + \gamma)/I_s$, and $I_s(\gamma)/I_s$, of these 14 Antarctic chondrites are given, where $I_s(Mt)/I_s$ in the cases of carbonaceous chondrites presents ratio of saturation magnetization of magnetite component to I_s. The saturation magnetizations of each phase, $I_s(\alpha)$, $I_s(\alpha + \gamma)$, $I_s(\gamma)$ and $I_s(Mt)$, are estimated by analyzing the thermomagnetic curve on the basis of thermomagnetic characteristics of each phase. The thermomagnetic curves of the α-phase are characterized by their thermally irreversible but reproducible transformation between α and γ phases. Those of ($\alpha + \gamma$)-phase are characterized by an irreversible transformation from the martensitic plessite structure of ($\alpha + \gamma$)-phase to γ-phase by heating in vacuum, while those of γ-phase are identified by their thermally reversible magnetization with characteristic low Curie point. The thermomagnetic curve of γ-phases of around 60 wt% in Ni content, however, can hardly be distinguished from that of magnetite, because their Curie points are very close to each other. In Table 1, the ratio of $I_s(\alpha)/I_s$ of E-chondrite, Yamato 691, is close to unity, while that of H-, L- and LL-chondrites ranges respectively 0.88~0.95, 0.65~0.90 and 0.19~0.60. We may conclude thus,

$$[I_s(\alpha)/I_s]_E \gtrsim [I_s(\alpha)/I_s]_H \gtrsim [I_s(\alpha)/I_s]_L > [I_s(\alpha)/I_s]_{LL}. \qquad (2)$$

The inequality relationship given by (2) in regard to $I_s(\alpha)/I_s$ corresponds to the well-known chemical composition of metallic component in chondrites, found by Prior (1920), that ratio of Ni-content to Fe-content in metal in chondrites increases in a sequence order as $E \rightarrow H \rightarrow L \rightarrow LL$. If the Ni-content is less than 7 wt% in all metallic grains, all metals are in α-phase. If the Ni-content somewhat exceeds 7 wt%, being 7~20%, metals consist of α- and $(\alpha + \gamma)$-phases. When the Ni-content increases further, metals comprise α-, $(\alpha + \gamma)$- and γ-phases, or α- and γ-phases and γ-phase becomes more abundant than α-phase with an increase of Ni content. If Ni becomes very rich, occupying 50 wt% or more of metals, entire metals become γ-phase. In this sense, ratio $I_s(\alpha)/I_s$ could be a reasonable measure for evaluating the poorness of Ni-content in metallic components in chondrites. The ratio $I_s(\alpha)/I_s$ can characterize the group of carbonaceous chondrites also, well separated from the other chondrite groups, because the major ferromagnetic constituents in carbonaceous chondrites are either magnetites or taenites, which contain no α-phase. Hence,

$$[I_s(\alpha)/I_s]_{LL} > [I_s(\alpha)/I_s]_C. \tag{2'}$$

If both I_s and $I_s(\alpha)/I_s$ are adopted as two parameters to characterize the abundance and composition of metallic component, then, the five chemical groups of chondrite can be well separately represented.

In Fig. 1, $I_s(\alpha)/I_s$ is plotted against I_s for 31 chondrites including 14 Antarctic chondrites, where the Antarctic chondrites are plotted with either full symbols

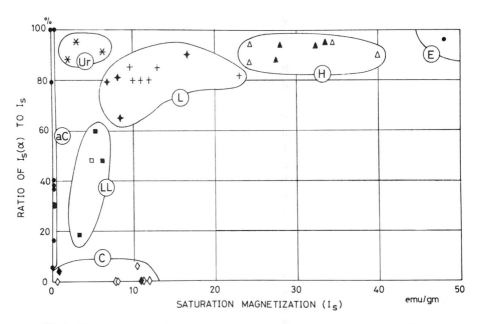

Fig. 1. Magnetic classification of chondrites on an $I_s(\alpha)/I_s$ versus I_s diagram. Filled in symbols indicate Antarctic meteorites.

or thick symbols to be distinguished from other chondrites which are plotted with either hollow symbols or thin symbols. In the $I_s(\alpha)/I_s$ versus I_s diagram in the figure, E-, H-, L-, LL- and C-chondrites give well defined groups. In the same diagram, $I_s(\alpha)/I_s$ and I_s values of diogenites, eucrites, and howardites are plotted as a single group of achondrite, while those of ureilites are grouped in a separate domain.

3. MAGNETIC CLASSIFICATION OF ANTARCTIC ACHONDRITES

In previous works (Nagata and Sugiura, 1976; Nagata, 1979a,b), diogenites, eucrites and howardites have been placed in a single domain in the $I_s(\alpha)/I_s$ versus I_s diagram for a classification of stony meteorites, where the achondrite domain is characterized by its small value of I_s ranging from 0.04 to 0.55 emu/gm.

Recent collections of Antarctic meteorites have produced a number of achondrites for magnetic examination. Eighteen Antarctic achondrites which have been magnetically examined to date are chemically and petrographically classified into 8 diogenites, 5 eucrites, 1 howardite, 3 ureilites and a unique achondrite whose parent planet looks likely to be similar to that of the Shergottite (McSween *et al.*, 1979a,b).

The magnetically measurable parameters of achondrites are the saturation magnetization (I_s) which can give the content of native metallic iron, the phase composition of native metallic iron in terms of α-, $(\alpha + \gamma)$- and γ-phases and the paramagnetic susceptibility (χ_p) which is approximately proportional to the Fe^{2+} content. Since a distinctly low content of nickel in eucrites compared with Ni-content in diogenites has been reported (Duke 1965) and that the ratio FeO/(FeO + MgO) in diogenites is much lower than that in eucrites and howardites has been reported (e.g., Mason, 1962b), it may be possible to distinguish basaltic achondrites (eucrites and howardites) magnetically from hypersthene achondrites (diogenites).

Figures 2 and 3 illustrate typical examples of thermomagnetic curves (applied magnetic field H_{ex} = 10k.Oe) in 10^{-5} Torr atmosphere for an eucrite and a diogenite respectively. The thermomagnetic curve of the Yamato-74159 eucrite in Fig. 2 consists of a paramagnetic magnetization ($\chi_p H_{ex}$) and a ferromagnetization of kamacite phase of less than 3 wt% in Ni content. The thermomagnetic curve of the Yamato-74136 diogenite in Fig. 3 consists of $\chi_p H_{ex}$, ferromagnetic contribution of a kamacite phase of less than 3 wt% Ni-content and ferromagnetic contribution of a plessite phase which has an irreversible $(\alpha + \gamma) \rightarrow \gamma$ transformation temperature at 580°C in the first run heating curve. In the furst run cooling curve of this diogenite, a small ferromagnetic moment of the transformed taenite phase is superposed upon the ferromagnetic moment of kamacite phase at temperatures below about 200°C. These results indicate that Ni content in the plessite phase is 25~30 wt%. The heating and cooling thermomagnetic curves in a second run of these two achondrites are almost the same as the respective first run cooling curves.

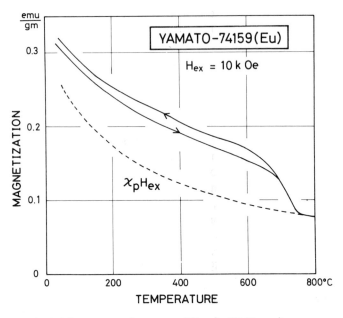

Fig. 2. Thermomagnetic curves of Yamato-74159 eucrite.

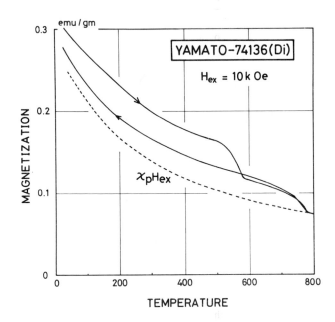

Fig. 3. Thermomagnetic curves of Yamato-74136 diogenite.

Analyzing these thermomagnetic curves, the paramagnetic susceptibility dependent on temperature T, $\chi_p(T)$, and the total saturation magnetization (I_s), and the saturation magnetization of α- and $(\alpha + \gamma)$-phases are estimated for each achondrite sample. For all the 18 achondrites examined, it is experimentally confirmed that $I_s = I_s(\alpha) + I_s(\alpha + \gamma)$, the metallic component in achondrites containing no appreciable amount of γ-phase in their original state.

In Table 2, I_s, $I_s(\alpha)/I_s$ and $\chi_p(T = 300°K)$ of 18 Antarctic achondrites are summarized, where $1 - I_s(\alpha)/I_s = I_s(\alpha + \gamma)/I_s$. In regard to the I_s-values, it may be concluded that I_s of achondrites except ureilites is smaller than 0.53 emu/gm, which means that the content of native metallic iron in these achondrites is smaller than about 0.27 wt%. I_s-values of some diogenites and eucrites range from 0.04 to 0.08 emu/gm, which correspond to the metallic iron content of $0.02 \sim 0.04$ wt%. It is known, on the other hand, that ureilites are the only achondrites with an appreciable amount of nickel iron. In fact, the I_s value of examined 3 Antarctic ureilites ranges from 2.2 to 6.4 emu/gm, which correspond to the content of kamacite metal of $1.1 \sim 3.2$ wt%. It may be interesting to note that one of the examined ureilites, Yamato 74659, contains 3.02% of carbon (Gibson and Yanai, 1979).

A remarkable difference between a group of diogenites and a group of pyroxene-plagioclase achondrites (i.e., eucrites and howardites) can be observed in their values of $I_s(\alpha)/I_s$ given in Table 2. Here a unique achondrite, ALHA-77005, is eucritic and chemically belongs to the group of pyroxene-plagioclase achon-

Table 2. Magnetic parameters of Antarctic achondrites.

Achondrites	Classification	I_s (emu/gm)	$I_s(\alpha)/I_s$	χ_p(300 K) (emu/Oe/gm)
Yamato 692	Diogenite	0.19	0.40	2.8×10^{-5}
Yamato 74013	Diogenite	0.17	0.37	3.0×10^{-5}
Yamato 74037	Diogenite	0.22	0.38	2.68×10^{-5}
Yamato 74097	Diogenite	0.32	0.29	2.40×10^{-5}
Yamato 74136	Diogenite	0.038	0.40	2.47×10^{-5}
Yamato 74648	Diogenite	0.20	0.29	2.35×10^{-5}
Yamato 75032	Diogenite	0.042	0.08	2.65×10^{-5}
ALHA 77256	Diogenite	0.16	0.17	2.50×10^{-5}
Yamato 74123	Ureilite	6.40	0.91	—
Yamato 74659	Urelite	2.23	0.88	—
ALHA 77257	Ureilite	3.14	0.95	—
Yamato 74159	Eucrite	0.061	1.00	3.09×10^{-5}
Yamato 74450	Eucrite	0.22	1.00	2.94×10^{-5}
ALHA 7605	Eucrite	0.076	1.00	3.23×10^{-5}
ALHA 77302	Eucrite	0.012	1.00	3.00×10^{-5}
ALHA 78040	Eucrite	0.083	0.79	2.83×10^{-5}
Yamato 7307	Howardite	0.53	1.00	3.3×10^{-5}
ALHA 77005	Eucritic unique achondrite	0.085	1.00	2.45×10^{-5}

drites. For diogenites $I_s(\alpha)/I_s \leqq 0.40$, whereas $I_s(\alpha)/I_s \geqq 0.79$ for the group of eucrites and howardites. In the cases of diogenites, the ferromagnetic component in addition to α-phase (kamacite) is mostly $(\alpha + \gamma)$-phase (plessite). It may thus be concluded that the bulk Ni content in the metallic component in eucrites and howardites is much smaller than that in diogenites.

Duke (1965) determined Ni content in metallic nickel iron in 7 eucrites and 1 howardite petrologically. The nickel irons in these eucrites are all kamacites of less than 1% in Ni content, while Ni content in nickel iron metals in a howardite is about 3.5%. If these eucrites and the howardite are thermomagnetically analyzed, they all should give $I_s(\alpha)/I_s = 1.00$. In ALHA-78040 eucrite, $I_s(\alpha)/I_s = 0.79$ and $I_s(\alpha + \gamma)/I_s = 0.21$ and a transformation temperature from $(\alpha + \gamma)$-phase to γ-phase in the first heating process $(H^*_{(\alpha+\gamma)\to\gamma})$ is 575°C. The plessite phase of $H^*_{(\alpha+\gamma)\to\gamma} = 575°C$ indicates that the Ni content in the plessite phase is about 30 wt%, while $I_s(\alpha + \gamma)$ for 30% Ni plessite is estimated to be about 150 emu/gm (Nagata, 1979a). The Ni content in kamacite phase in this eucrite is estimated to be much less than 3 wt% from the thermomagnetic characteristics of α-phase. Then, the average Ni content in metallic nickel iron in this eucrite amounts to about 8 wt%. This amount of Ni content is unusually high as the metallic nickel iron component in eucrites. However, the unusually high content of nickel in ALHA-78040 eucrite is still distinctly lower than that in the group of diogenites listed in Table 2. Thus, eucrites and howardites can be distinctly separated from diogenites in terms of $I_s(\alpha)/I_s$. Since the I_s-values of ureilites are much larger than those of diogenites, eucrites and howardites, the three groups of achondrites, i.e., hypersthene achondrite group (diogenites), olivine-pigeonite achondrite group (ureilites) and pyroxene-plagioclase achondrite group (eucrites and howardites) can be plotted within three separate domains on an $I_s(\alpha)/I_s$ versus I_s diagram, as illustrated in Fig. 4.

If $I_s(\alpha)/I_s$ and I_s values of both chondrites and achondrites are plotted on a single $I_s(\alpha)/I_s$ versus I_s diagram, the ureilite domain is located between the L-chondrite domain and the domain of general achondrites except ureilites, as illustrated in Fig. 1. Combining Fig. 4 with Fig. 1, all different chemical groups of stony meteorites so far examined, i.e., E-, H-, L-, LL- and C-chondrites and diogenites, ureilites and pyroxene-plagioclase achondrites (eucrites and howardites) are well separately grouped on an $I_s(\alpha)/I_s$ versus I_s diagram. It is hoped further that another relatively abundant Ca-poor achondrite, aubrite, also will be magnetically examined in the future.

4. PARAMAGNETIC SUSCEPTIBILITY OF ACHONDRITES

As shown in Figs. 2 and 3, for example, the paramagnetic component occupies a considerably large portion of the total magnetization of achondrites when they are magnetized in a sufficiently strong magnetic field, because the content of ferromagnetic metallic nickel-iron is extremely small in achondrites in comparison with that in the ordinary chondrites. From both the magnetization curve at a certain temperature and the thermomagnetic curve in a sufficiently strong mag-

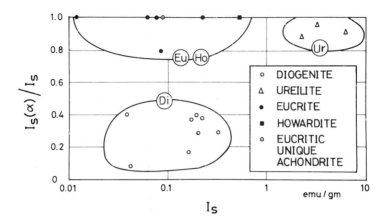

Fig. 4. Magnetic classification of Antarctic achondrites on an $I_s(\alpha)/I_s$ versus I_s diagram.

netic field, therefore, the paramagnetic susceptibility of achondrites can be determined with a considerably good accuracy as a function of temperature. In Table 2, the paramagnetic susceptibility at temperature $T = 300°K$ thus determined is given for 15 achondrites (except for 3 ureilites).

The paramagnetism of stony meteorites is due to the presence of Fe^{2+} and Mn^{2+} in their silicate minerals. Since the content of Mn^{2+} is very small compared with that of Fe^{2+} in stony meteorites and the content of Fe^{2+} considerably varies in various stony meteorites, the paramagnetic susceptibility could be a reasonably good indicator of the content of Fe^{2+}.

When the content of Fe^{2+} in weight in non-magnetic matrix is noted by $m(Fe^{2+})$, the paramagnetic susceptibility (χ_p) at temperature T is given by

$$\chi_p(T) = \frac{6.45 \times 10^{-2}}{T} m(Fe^{2+}). \quad (emu/gm/Oe) \qquad (3)$$

When the content of FeO in weight is noted by m(FeO), χ_p at $T = 300°K$ is numerically expressed as

$$\chi_p(300°K) = 1.67 \times 10^{-4} m(FeO). \quad (emu/gm) \qquad (4)$$

$\chi_p(300)$ values of diogenites, eucrites and howardites, given in Table 2, range from 2.35×10^{-5} to 3.3×10^{-5} emu/gm, which correspond to the FeO content of 14.1 $\sim$ 19.8 wt%. These amounts of FeO content are reasonable values for hypersthene achondrites and pyroxene-plagioclase achondrites.

It appears in Table 2 that $\chi_p(300°K)$ values of the pyroxene-plagioclase achondrite group are a little larger than those of the hypersthene achondrite group, but the observed difference between the two groups may not be regarded significantly large. However, the paramagnetic susceptibility which is approximately proportional to the content of FeO should be able to be a reasonably good parameter to magnetically distinguish enstatite achondrites (aubrites) from the other achondrites, because the FeO content in aubrites is distinctly smaller than that in any other achondrite (e.g., Mason 1962b).

5. CONCLUDING REMARKS

The magnetic classification scheme proposed here is based on two characteristic magnetic properties of stony meteorites: One is the saturation magnetization (I_s) which is approximately proportional to the content of metallic nicked iron or magnetite and the other is a ratio of the saturation magnetization of kamacite component alone to the total saturation magnetization ($I_s(\alpha)/I_s$) which is an indicator of Ni content in the metals. The saturation magnetization of stony meteorites at room temperature can be easily measured with the aid of a magnetic balance or a vibration magnetometer for a small sample piece of 20–100 mg in weight in a magnetic field larger than 10kOe.

The saturation magnetization of kamacite component can be definitely determined in the heating and cooling thermomagnetic curves because of its characteristic thermal change behavior that a phase transition from γ to α in the cooling process takes place at a temperature ($H^*_{\gamma \to \alpha}$) lower than a phase transition temperature from α to γ ($H^*_{\alpha \to \gamma}$) in the heating process if the Ni content in kamacite is in a range of 3–7 wt%, and that its thermographic curves are almost identical to those of pure metallic iron if its Ni content is less than 3 wt%.

The magnetization of kamacite component can thus be clearly distinguished from that of other ferromagnetic constituents such as plessite, taenite and magnetite. The thermomagnetic curves of stony meteorites also can be measured by a magnetic balance or a vibration magnetometer, without any difficulty, in a magnetic field larger than 10 kOe, though meteorite samples must be kept in a low pressure atmosphere less than 10^{-4} Torr during the whole heating processes.

Since the plots of I_s and $I_s(\alpha)/I_s$ of various stony meteorites on a $I_s(\alpha)/I_s$ versus I_s diagram are well separately grouped in accordance with their chemical compositions, and since magnetic measuring and analysis techniques do not require any specific skill, a magnetic classification scheme for stony meteorites could be proposed as a practical method for identifying the chemically classified groups of stony meteorites.

Acknowledgment—The author's thanks are due to M. Funaki for his assistance in magnetic measurements of a number of Antarctic stony meteorites.

REFERENCES

Duke M. B. (1965) Metallic irons in besaltic achondrites. *J. Geophys. Res.* **70**, 1523–1527.

Gibson E. K. Jr., and Yanai K. (1979) Total carbon and sulfur abundances in Antarctic meteorites. *Memoirs of the National Institute of Polar Research, Special Issue No. 15* (T. Nagata, ed.), p. 189–195. Nat'l. Inst. Polar Research, Tokyo.

Gus'kova Ye. G. (1972) *The Magnetic Properties of Meteorites.* Nauka, Leningrad. 107 pp.

Mason B. (1962a) The classification of the chondritic meteorites. *Amer. Museum Novitates.* **No. 2085**, 1–20.

Mason B. (1962b) *Meteorites*. Wiley, N.Y. 274 pp.

McSween H. Y. Jr., Taylor L. A., and Stolper E. M. (1979a) Allan Hills 77005; a new meteorite type found in Antarctica. *Science* **204**, 1201–1203.

McSween H. Y. Jr., Stopler E. M., Taylor J. A., O'Kelley G. D., Eldridge J. S., Biswas S., Ngo H. T., and Lipschutz M. E. (1979b) Peterogenic relationship between Allan Hills 77005 and other achondrites. *Earth Planet. Sci. Lett.* **45**, 275–284.

Nagata T. (1979a) Magnetic classification of Antarctic stony meteorites (III). *Memoirs of the National Institute of Polar Research, Special Issue No. 12* (T. Nagata, ed.), p. 223–237. Nat'l. Inst. Polar Research, Tokyo.

Nagata T. (1979b) Meteorite magnetism and the early solar system magnetic field. *Phys. Earth. Planet. Inter.* **20**, 324–341.

Nagata T. (1979c) Magnetic classification of Stony Meteorites (IV). *Memoirs of the National Institute of Polar Research, Special Issue No. 15* (T. Nagata, ed.). p. 273–279. Nat'l. Inst. Polar Research, Tokyo.

Nagata T., Fisher R. M., and Schwerer F. C. (1971) Lunar rock magnetism. *The Moon* **4**, 160–186.

Nagata T., Fisher R. M., and Schwerer F. C. (1974) Some characteristic magnetic properties of lunar materials. *The Moon* **9**, 63–77.

Nagata T. and Sugiura N. (1976) Magnetic characteristic of some Yamato meteorites-magnetic classification of stone meteorites. *Memoirs of the National Institute of Polar Research, Ser. C. No. 10* (T. Nagata, ed.), p. 30–58. Nat'l Inst. Polar Research, Tokyo.

Prior G. T. (1920) The classification of meteorites. *Mineral. Mag.* **19**, 51–63.

Urey H. C. and Craig H. (1953) The composition of the stone meteorites and the origin of meteorites. *Geochim. Cosmochim. Acta.* **4**, 36–82.

Proc. Lunar Planet. Sci. Conf. 11th (1980), p. 1801–1813.
Printed in the United States of America

Comparisons of magnetic paleointensity methods using a lunar sample

N. Sugiura and D. W. Strangway

Department of Geology, University of Toronto, Toronto, Ontario, M5S 1A1, Canada

Abstract—Lunar sample paleointensity determinations have been made using several techniques—Shaw's method, the ARM method, the IRM method and Thellier's method. We have applied these to a single sample of lunar breccia sample 60255, and show that the first three can give false results which appear to be acceptable. High values are indicated and are interpreted as the result of acquiring a very stable viscous remanent magnetization. This remanence is hard to remove by alternating fields but is readily removed by heating to 300°C.

Thellier's method, on the other hand, gave a value of about 500nt which appears to be of lunar origin. When samples which have been analyzed using Thellier's method are examined, there is little evidence left of a lunar ancient magnetic field with a strong time dependence.

INTRODUCTION

Ancient lunar magnetic field intensities have been estimated using various methods such as Thellier's method, the anhysteretic remanent magnetization method (ARM), Shaw's method, and the isothermal remanent magnetization method (IRM) (e.g., Cisowski and Fuller, 1977). The first three methods have a built-in criterion by which NRM is identified as a thermoremanent magnetization (TRM). Shaw's method (Shaw, 1974) has been shown to be as reliable as Thellier's method (Kono, 1978) in some cases. The ARM method (Stephenson and Collinson, 1974) is considered to be less reliable because the ratio of TRM to ARM, which has to be assumed to be a constant (f = 1.34) has been shown to change with grain size by a factor ~ 10 (Levi and Merrill, 1976). The IRM method (Cisowski *et al.*, 1975) does not have a criterion which identifies the natural remanent magnetization (NRM) as TRM, and it is, therefore, the least reliable method. In the present study, these four methods are examined using a lunar breccia sample (60255).

SAMPLE DESCRIPTION AND EXPERIMENTS

60255 is a breccia with a significant amount of white clast material in a glassy matrix. The exact age of formation is not known, but the age is considered to be

~3.9 b.y. like other lunar samples at the Apollo 16 landing site. Two samples, 60255,26, and 60255,19, were used for the experiments. The former is composed of small chips. The latter was a large (~12 gm) sample which was sawed into five pieces. Sub-sample 1 is rich in glass, sub-samples 2, 3 and 4 contain a medium amount of glass and sub-sample 5 contains the least amount of glass. Two generations of vesicular glass splash on the surface have been noted (Warner, 1972). Magnetic properties of this breccia have previously been studied by Nagata *et al.*, (1973). Although sample 60255 was cited as one of the samples which shows a characteristic behavior of textural remanence (Brecher, 1976), it appears to be isotropic and lacks evidence of post-metamorphism shock effects (Butler, 1972).

   Remanent magnetization was measured using a superconducting magnetometer (Develco, Inc.). The samples were first heated to a temperature of ~100°C under vacuum and then they were vacuum sealed in quartz tubes with Ti metal as an oxygen getter for the heating steps (Taylor, 1979) up to ~550°C. The titanium was then separated for the rest of the heating steps. The ambient magnetic field during thermal demagnetization was less than 10nt. Samples were stored in a low field room ($\leq$100 nt) for two weeks before the experiments.

ARM method

The ARM method is based on the fact that TRM and ARM induced in a small, direct field have similar coercivity spectra (e.g., Levi and Merrill, 1976). Paleointensity is calculated from the slope of the linear portion of the NRM vs. ARM plot (Fig. 1) at different steps of demagnetization by alternating fields using $h_p = f \cdot \dfrac{\Delta NRM}{\Delta ARM} h_A$; where h_p is the paleofield and h_A is the field used to apply the ARM. The factor f may depend on the grain size (Levi and Merrill, 1976), but it is usual to choose f = 1.34. The NRM of 60255 hardly changes direction during alternating field (AF) demagnetization (Figs. 1 and 3). The ARM and the NRM have similar coercivity spectra in the range from 10 to 100 mt (100 to 1000 Oe). The ARM method, using f = 1.34, gives a relatively large paleointensity—2800 nt.

Shaw's Method

In Shaw's method, the coercivity spectra of NRM and TRM are compared by AF demagnetization. To make sure that the coercivity spectrum was not changed by the heating which created the laboratory TRM, coercivity spectra of ARM before and after the heating are compared (Fig. 2). The paleointensity itself is determined from the slope of the linear portion of the NRM vs. TRM plot (Fig. 2) by $h_p = \dfrac{\Delta NRM}{\Delta TRM} h_A$. Shaw's method also gives a relatively large paleointensity of 3200 nt for sample 60255,26.

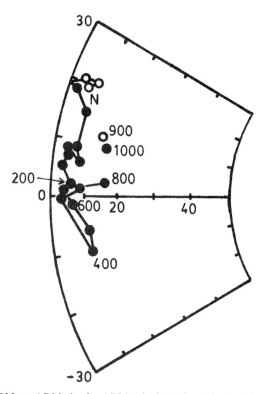

Fig. 1. (a) NRM vs. ARM plot for ARM paleointensity estimate. The numbers adjacent to points indicate the peak alternating field.

IRM method

The ratio of NRM to saturation isothermal remanence (SIRM) gives a rough estimate of the paleointensity. NRM and SIRM of a small chip of 60255,26 are 2.95×10^{-5} emu/gm and 2.41×10^{-2} emu/gm, respectively. Using the ratio due to Cisowski and Fuller (1977), the IRM method gives a large paleointensity of 6000 nt.

Thellier's method

In this method, the acquisition of partial thermoremanent magnetism (pTRM) on laboratory heating and the loss of NRM by stepwise thermal demagnetization are compared. In the case of sample 60255, thermal demagnetization of the NRM [Fig. 3(b)] revealed that the NRM which seemed to be unidirectional during AF demagnetization, is composed of three components. The reason that AF demag-

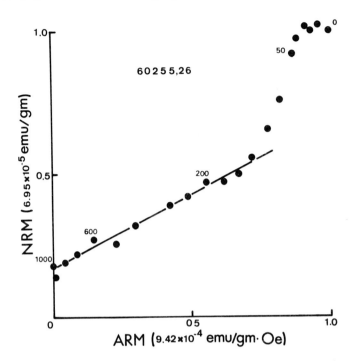

Fig. 1. (b) Direction changes on alternating field demagnetization of the sample 60255,26 on which the ARM method was used.

netization of 60255,26 [Fig. 1(b)] does not show much directional change is either (1) high coercivity component is small (≪20%) in this sample or (2) the direction of the low coercivity component is accidentally similar to that of the high coercivity component. In the case of sub-samples 60255,19-1 and 4, only <40% of NRM is erased by 100 mt AF demagnetization. The largest component of the NRM is erased by heating to 200°C (Fig. 3) and is believed to be of viscous origin. The medium temperature (200–500°C) component of the NRM is stronger in sub-samples 2, 3 than in sub-sample 5. The high temperature component (≥500 ~ 600°C) is evident in sub-sample 5, but is obscure (probably masked by spurious magnetization) in sub-samples 2, 3. This high temperature component is the most likely candidate for the magnetization acquired when the breccia was formed. The difference with regard to the intensity of the medium temperature component between sub-samples 2, 3 and 5 can be explained by (1) Sub-samples 2, 3 contain more glass than sub-sample 5. Since glass is often enriched in fine metallic particles, there may be more low blocking temperature material in these samples. (Sub-samples 2, 3 are considerably more viscous than sub-sample 5); and/or (2) The breccia was inhomogeneously remagnetized when the vesicular glass splash covered one of the surfaces of the rock. Although the relative position of the glass splash and the sub-samples are not known, it is conceivable that sub-samples 2, 3 were reheated to a higher temperature than sub-sample 5, and thus acquired more medium temperature component and/or (3) low to intermediate pressure

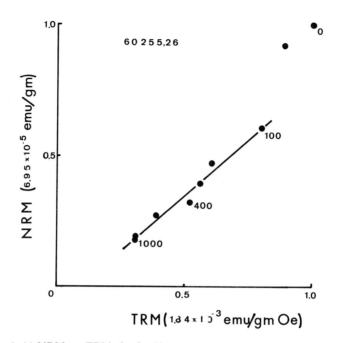

Fig. 2. (a) NRM vs. TRM plot for Shaw's method of paleointensity estimate.

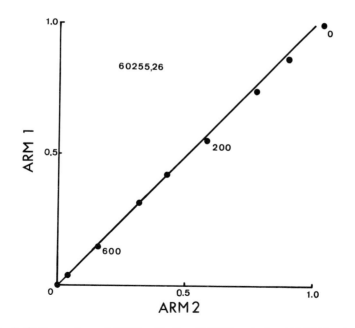

Fig. 2. (b) Comparison of ARM's of before (ARM1) and after (ARM2) heating.

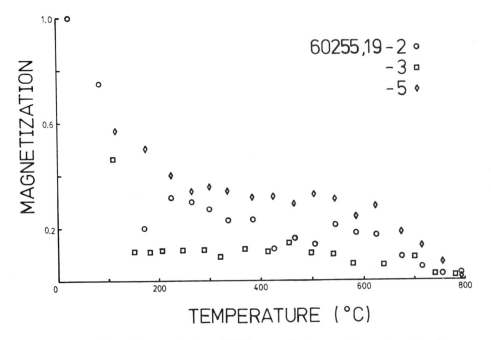

Fig. 3.(a). Thermal demagnetization of NRM, normalized at room temperature intensity.

shock remanence, although the sample lacks evidence of strong shock effects.

The NRM vs. pTRM plot (Fig. 4) of the medium and high temperature component is very scattered. The main source of the noise is the viscous remanence picked up after the heating experiment during various exposures to the earth's field. Because the NRM direction changes as a result of thermal-demagnetization, vectorial difference of the NRM between each heating step should be plotted on the ordinate to determine the paleointensity. This was not done because the experimental error was too large to get accurate paleointensity value. Roughly speaking, however, a paleointensity of the order of 500 nt can be estimated from both high and medium temperature components of both sub-samples 2 and 5.

ADDITIONAL EXPERIMENTS

pTRM—To confirm that the largest component of the NRM is of viscous origin, a partial thermoremanence ($25 < T < 190°C$, H = 0.05 mt) was AF demagnetized and compared with the ARM (Fig. 5). Theoretically (Néel, 1949) assuming single domain particles, this pTRM should have magnetic properties similar to the viscous remanence acquired in 10 years in the geomagnetic field. The pTRM is extremely stable upon AF demagnetization. The pTRM and the ARM have similar coercivity spectra at high fields ($50 \lesssim H \lesssim 100$ mt). The apparent paleoin-

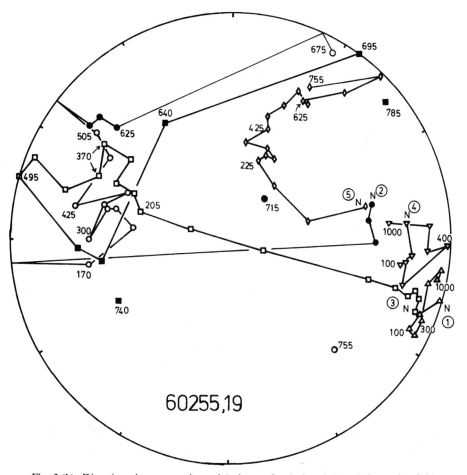

60255,19

Fig. 3.(b). Direction changes on thermal (sub-samples 2, 3 and 5) and alternating field (sub-samples 1 and 4) demagnetization of 60255,19. The numbers adjacent to points indicate either temperature (°C) or the peak alternating field (Oe).

tensity determined from the comparison of the pTRM and the ARM is about 6000 nt. Thus, although the small chips of the breccia may be inhomogeneous, it is more likely that the ARM method will give a false paleointensity, if the sample can acquire a strong viscous remanence which is stable to AF demagnetization.

Thermomagnetic Analysis—Although the result of Thellier's experiment on 60255 is one of the most successful ones we have obtained to date, sub-sample 2 showed a sharp decrease of pTRM intensity at high temperature [at 715°C (Fig. 3 (a)]. Such a decrease of pTRM at high temperatures has been obtained in many samples under various conditions (Pearce *et al.*, 1976, Sugiura *et al.*, 1978, Sugiura *et al.*, 1979). To clarify the cause of the decrease of the pTRM at high temperature, thermomagnetic curves of sample 60255,26 were measured under

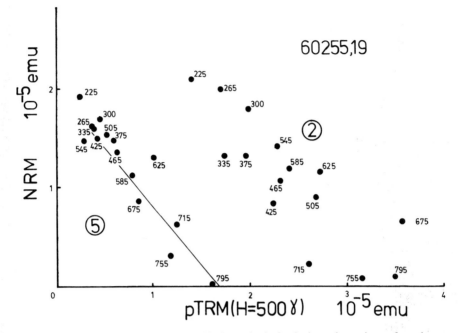

Fig. 4. NRM vs. pTRM plot for Thellier's method of paleointensity estimate for subsamples 2 (right) and 5 (left). The solid line is a least squares fit (from 500° to 795°) with a slope of 590 nt.

various conditions (Fig. 6). The samples were prepared as follows: Sample A was sealed in a quartz tube under vacuum (10^{-3} torr). Samples B, C and E were sealed in quartz tubes under vacuum with Ti metal, then heated to 400, 530 and 560°C, respectively. Then, the Ti metal was separated for the higher temperature steps (for detailed procedures, see Taylor, 1979). Sample F was sealed in a quartz tube under vacuum with Ti metal which was not separated during the experiment. Thus the atmosphere in the sealed quartz tube is more reducing in the order from A to F. Thermomagnetic curves of these samples were measured with a vibrating sample magnetometer (PAR). The saturation magnetization as a function of temperature [Js(T)] curve of sample D was measured under high vacuum ($\sim 10^{-5}$ torr) with a magnetic balance. Samples A through E show (1) a decrease of Js (saturation magnetization) after heating and (2) a decrease of the iron-nickel ($\gamma \rightarrow \alpha$) phase transition temperature (650~550°C→550~450°C; these are not Curie temperatures because repeated heating-cooling curves show the characteristic thermal hysteresis of $\alpha \rightarrow \gamma \cdot \gamma \rightarrow \alpha$ transitions). These results can be explained by the selective oxidation of iron in small iron-nickel alloy [superparamagnetic~single domain size grains which are abundant in lunar glass (Housley *et al.*, 1973)]. By an oxidizing reaction such as $2Fe + SiO_2 + O_2 \rightarrow Fe_2SiO_4$, ferromagnetic iron converts to paramagnetic olivine, thus decreasing the saturation magnetization.

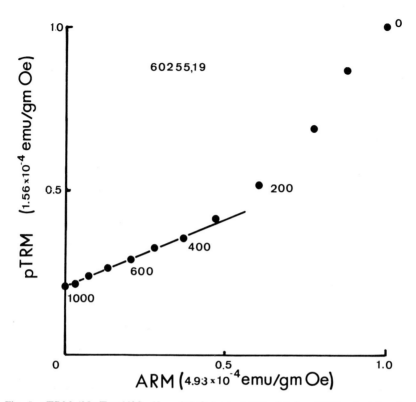

Fig. 5. pTRM (25≤T≤190°C, H = 0.5 Oe) vs. ARM plot for ARM paleointensity estimate.

It would appear that the metal is also enriched in nickel, thus decreasing the $\gamma \rightarrow \alpha$ transition temperature. Disappearance of single domain grains by this oxidation, which starts at 600°C (see heating curves A and B) explains the decrease of pTRM at high temperature. The decrease of the saturation magnetization is quite remarkable since there is almost no decrease of ARM intensity (induced with maximum 100 mt alternating field) [Fig. 2 (b)]. This suggests that there is a very high content of superparamagnetic and single domain particles present which do not contribute to the ARM. Sample F was reduced during the measurement resulting in the creation of metallic iron and an increase of Js at room temperature. This reduction started at 600°C (heating curve F). Heating in vacuum (sample D) looks better than most of the sealed quartz methods (samples A, B, C) for avoiding oxidation, but each successive heating oxidizes the sample to a certain extent, and so they are not optimum for Thellier's method. The sealed quartz method, with Ti, separated after heating to 560°C (sample E), appears to be the best heating approach for Thellier's method at least for the case of sample 60255. The separation temperature for the Ti metal must be lower than 600°C to avoid reduction. It should be noted that 60255 contains a small amount of rust (Butler,

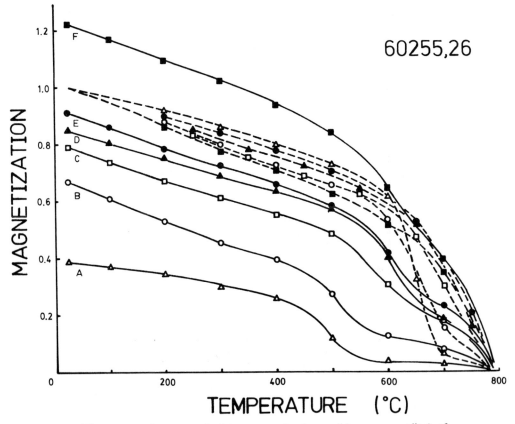

Fig. 6. Thermomagnetic curves (broken curves; heating, solid curves; cooling) of 60255,26 under different conditions. All curves are normalized to initial magnetization at room temperature. Applied field is 800 mt.

1972), so it may be that this procedure will differ for different samples.

Sample 60255, 19-5, probably contained a smaller amount of super-paramagnetic and small single domain particles and/or rust than 60255, 19-2, giving a better paleointensity estimate by Thellier's method.

DISCUSSION AND CONCLUSIONS

It was shown that the ARM method (2800 nt), Shaw's method (3200 nt) and the IRM method (6000 nt) could give false paleointensities. This is due to the presence of a strong, stable (against AF demagnetization) viscous remanence in sample 60255. If a very large alternating field, enough to demagnetize all the NRM, is available for AF demagnetization, and if the primary NRM of lunar origin is comparable in intensity with the viscous remanence, we were able to recognize that the NRM was made of two components and could avoid determining a false paleointensity. But since microscopic coercive forces in lunar samples are distributed up to 1000 mt (Dunlop *et al.*, 1973), and the primary NRM/VRM could be ~0.1 (60255,19 sub-sample 3), we are very likely to generally get a false paleointensity from the ARM method. Since the most stable component of NRM which is carried by SD grains seems to be most easily affected by heat treatment, we are even more likely to get a false result from Shaw's method than from the ARM method. It is not a surprise that the IRM method gave a very large paleointensity for 60255, because the NRM is largely composed of viscous remanence acquired in the geomagnetic field. Lunar samples generally contain very fine metallic grains and acquire viscous remanence quite easily. Although fine grains are more abundant in fines and breccias, igneous rocks also have some fine metallic grains (e.g., Schwerer and Nagata, 1976) and can acquire viscous remanence. A fine-grained vesicular basalt (10017) has a NRM which behaves just like the NRM of the breccia 60255 (Stephenson *et al.*, 1977, Hoffman *et al.*, 1979). It seems probable that the NRM is largely composed of VRM. Remarkably, linear segments are found in NRM vs. ARM plots and NRM vs. TRM plots for 10017, and large paleointensities 0.07 mt and 0.093 mt from ARM method (Stephenson *et al.*, 1977, Hoffman *et al.*, 1979) and 0.071 mt from Shaw's method (Hoffman *et al.*, 1977) were obtained. The NRM, however, was completely destroyed by heating to 300°C (Hoffman *et al.*, 1979). In addition, the NRM in another chip of 10017 is less stable than TRM (Sugiura *et al.*, 1978). All these facts can be consistently explained if we assume that the NRM is entirely made of viscous remanence which is stable to AF demagnetization, just like the breccia 60255.

Thus, we have to critically examine the paleointensities which were obtained from IRM, ARM and Shaw's method which may be contaminated by viscous effects. Stability of the NRM against a thermal demagnetization up to 200~300°C appears to be a necessary condition to make reliable paleointensity determinations. Unfortunately, the stability of the NRM against thermal demagnetization is not known for many samples on which paleointensities were estimated by ARM method. If we omit these estimates, then only five published paleointensi-

Table 1.

	paleointensity Oe	method	age b.y.	references
15498	0.021–0.05	Thellier	<3.3	Gose *et al.* (1973)
62235	1.2	Thellier	3.9	Collinson *et al.* (1973)
	1.4	ARM		Stephenson *et al.* (1974)
70019	0.025	Thellier	0.003	Sugiura *et al.* (1979)
70215	0.02–0.075	Thellier	3.84	Stephenson *et al.* (1974)
	0.06	ARM		
60255	0.005	Thellier	3.9	this study

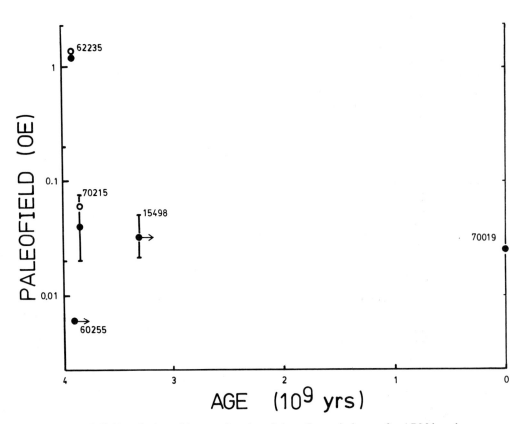

Fig. 7. Reliable paleointensities as a function of time. Open circles are for ARM based values where Thellier determinations are also available.

ties are considered to be reliable (Table 1). We need more data to study the origin and variation of the lunar magnetic field with time (Fig. 7).

Thellier's method was successful on sub-sample 5, but not as successful on sub-sample 2. These sub-samples are only ~1 cm apart from each other, but the latter contains considerably more glass (hence more small single domain grains) and is magnetically more viscous than the former. The reasons for the less successful result on sub-sample 2 are: (1) It picks up large viscous remanence, resulting in high noise to signal ratio. (2) As shown by thermomagnetic analysis, small single domain grains are susceptible to physicochemical changes (oxidation to ferrous iron, change of transition temperature and spontaneous magnetization, change of grain size). Some part of the NRM may be lost by these physicochemical changes. Partial TRM capacity also will change with these physicochemical changes.

Although the sealed quartz tube technique certainly minimizes the physicochemical changes, to get a good result from Thellier's method, we need a good sample like sub-sample 5 whose NRM seems to be carried by relatively large SD grains.

Acknowledgments—We thank L. A. Taylor for suggesting the sample preparation technique for paleointensity determination by Thellier's method, and the National Science and Engineering Research Council for financial support.

REFERENCES

Brecher A. (1976) Textural remanence: A new model of lunar rock magnetism. *Earth Planet. Sci. Lett.* **29**, 131–145.

Butler P. Jr. (1972) Thin Section Description. 60255. In *Apollo 16 Lunar Sample Information Catalog,* p. 102–103. MSC 03210, NASA Johnson Space Center, Houston.

Cisowski S. M. and Fuller M. D. (1977) On the intensity of ancient lunar fields. *Proc. Lunar Sci. Conf. 8th,* p. 725–750.

Cisowski S. M., Fuller M. D., Wu Y. M., Rose M. F., and Wasilewski P. J. (1975) Magnetic effects of shock and their implications for magnetism of lunar samples. *Proc. Lunar Sci. Conf. 6th,* p. 3123–3141.

Collinson D. W., Stephenson A., and Runcorn S. K. (1973) Magnetic properties of Apollo 15 and 16 rocks. *Proc. Lunar Sci. Conf. 4th,* p. 2963–2976.

Dunlop D. J., Gose W. A., Pearce G. W., and Strangway D. W. (1973) Magnetic properties and granulometry of metallic iron in lunar breccia 14313. *Proc. Lunar Sci. Conf. 4th,* p. 2977–2990.

Gose W. A., Strangway D. W., and Pearce G. W. (1973) A determination of the intensity of the ancient lunar magnetic field. *The Moon* **5**, 196–201.

Hoffman K. A., Baker J. R., and Banerjee S. K. (1979) Combining paleo-intensity methods: A dual-valued determination on lunar sample 10017, 135. *Phys. Earth Planet. Inter.* **26**, 317–323.

Housley R. M., Grant R. W., and Paton N. E. (1973) Origin and characteristics of excess Fe metal in lunar glass welded aggregates. *Proc. Lunar Sci. Conf. 4th,* p. 2737–2749.

Kono M. (1978) Reliability of paleointensity methods using alternating field demagnetization and anhysteretic remanence. *Geophys. J. Roy. Astron. Soc.* **54**, 241–261.

Levi S. and Merrill R. T. (1976) A comparison of ARM and TRM in magnetite. *Earth Planet. Sci. Lett.* **32**, 171–184.

Nagata T., Fisher R. M., Schwerer F. C., Fuller M. D., and Dunn J. R. (1973) Magnetic properties and natural remanent magnetization of Apollo 15 and 16 lunar materials. *Proc. Lunar Sci. Conf. 4th,* p. 3019–3043.

Néel L. (1949) Théorie du trainage magnétique des ferromagnétiques en grains fins avec applications aux terres cuites. *Ann. Geophys.* **5**, 99–136.

Pearce G. W., Hoye G. S., Strangway D. W., Walker B. M., and Taylor L. A. (1976) Some complexities in the determination of lunar paleointensities. *Proc. Lunar Sci. Conf. 7th,* p. 3271–3297.

Schwerer F. C. and Nagata T. (1976) Ferromagnetic-superparamagnetic granulometry of lunar surface materials. *Proc. Lunar Sci. Conf. 7th,* p. 759–778.

Shaw J. (1974) A new method of determining the magnitude of the paleomagnetic field, application to five historic lavas and five archaeological samples. *Geophys. J. Roy. Astron. Soc.* **39**, 133–141.

Stephenson A. and Collinson D. W. (1974) Lunar magnetic field paleointensities determined by an anhysteretic remanent magnetization method. *Earth Planet. Sci. Lett.* **23**, 220–228.

Stephenson A., Collinson D. W., and Runcorn S. K. (1974) Lunar magnetic field paleointensity determinations on Apollo 11, 16 and 17 rocks. *Proc. Lunar Sci. Conf. 5th,* p. 2859–2871.

Stephenson A., Runcorn S. K., and Collinson D. W. (1977) Paleointensity estimates from lunar samples 10017 and 10020. *Proc. Lunar Sci. Conf. 8th,* p. 579–687.

Sugiura N., Strangway D. W., and Pearce G. W. (1978) Heating experiments and paleointensity determinations. *Proc. Lunar Planet. Sci. Conf. 9th,* p. 3151–3163.

Sugiura N., Wu Y. M., Pearce G. W., Strangway D. W., and Taylor L. A. (1979) A new magnetic paleointensity value "for a young lunar glass." *Proc. Lunar Planet. Sci. Conf. 10th,* p. 2189–2197.

Taylor L. A. (1979) An effective sample preparation technique for paleointensity determinations at elevated temperatures (abstract). In *Lunar and Planetary Science X,* p. 1209–1211. Lunar and Planetary Institute, Houston.

Warner J. L. (1972) 60255 In *Apollo 16 Lunar Sample Information Catalog,* p. 101. MSC 03210, NASA Johnson Space Center, Houston.

Proc. Lunar Planet. Sci. Conf. 11th (1980), p. 1815–1823.
Printed in the United States of America

Compressive strength, seismic Q, and elastic modulus

B. R. Tittmann[1], V. A. Clark[1], and T. W. Spencer[2]

[1]Rockwell International Science Center Thousand Oaks, California 91360
[2]Texas A & M University, College Station, Texas 77843

Abstract—Small amounts of water have previously been found to have a dramatic effect on some of the mechanical properties of rocks. This suggests the importance of controlling the environment in which these measurements are made. Since the effect of moisture is so large, studying the variation of these mechanical properties with partial pressure of water and other volatiles can provide important clues for understanding "room dry" rocks. Thus measurements are presented for Q and velocity which are compared with reported values for uniaxial compressive strength for quartzitic types of rocks.

The compressive strength and Q are both seen to be affected by water vapor in a similar way suggesting a relationship between these mechanical properties. Since volatiles are known to affect the surface energy in different ways, Q and velocity were measured for a variety of volatiles and were found to be reduced by benzene, hexane, ethyl alcohol, methyl alcohol, and water with increasing severity and in qualitative agreement with the reduction in surface energy. This result suggested that volatile assisted crack growth is the mechanism by which both the strength and seismic dissipation are affected. Calculations based on a theory by Spetzler and Getting show that contributions to seismic Q from crack growth are to be expected and are especially significant at elevated temperatures and high tectonic stresses.

INTRODUCTION

The high mechanical strength of lunar rock, the isostatically uncompensated lunar surface features, the high seismic Q and the low near surface elastic velocity values are considered by many investigators manifestations of a unique mechanical state arising from a long term absence of water. Many mechanisms such as breaking of cold welds, water assisted crack growth, stress induced diffusion of H_2O films, boundary lubrication, or dislocation motion have been suggested to account for the role of water in the mechanical properties. It is well known that the presence of water affects most mechanical properties but the mechanisms for the various properties may or may not be related. To test and guide the development of models we have carried out experiments to document some of the mechanical properties.

COMPRESSIVE STRENGTH

The presence of water has long been known to aid in crack growth in silicates (Martin, 1972; Wiederhorn, 1969). Rocks have a much higher compressive strength in a vacuum. Mizutani *et al.* (1977) found the ultimate failure strength of Ralston intrusive (mafic latite) was 2.18 kbar to 4.43 kbar when it was measured in wet and dry environments respectively. Martin (1972) measured the rate of growth of cracks in single crystal quartz as a function of stress level, temperature, and water pressure. He found the time for crack growth to be proportional to the log of the water pressure, and obtained an activation energy of 25 Kcals for water assisted crack growth. Calculations based on Martin's model and data show crack growth of 0.2 mm in 100 sec with 645 bars of applied stress and 405 mb water pressure at 153°C.

More recently, Spetzler and Getting (1980) in a general theory of attenuation for seismic waves based on thermally activated processes show that at elevated temperatures and high tectonic stresses cracking due to stress corrosion can have a significant contribution to seismic Q.

In uniaxial compression experiments, Colback and Wiid (1965) found that the compressive strength of two quartzite rock types depended on the relative humidity. Their data is replotted as strength versus relative humidity in Fig. 1. The shape of the curve is characterized by a very sharp decrease in strength with P/P_0. To test the possible relationship between seismic Q and strength we have carried out Q measurements as a function of humidity on quartzite rock types and these are discussed below.

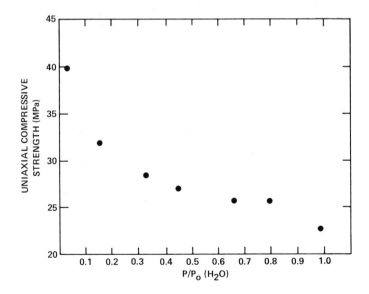

Fig. 1. Relationship between uniaxial compressive strength and the partial pressure of water for quartzitic sandstone specimens. From Colback and Wiid (1965).

SEISMIC Q AND MODULUS

Several investigators have found that small amounts of water drastically affect the seismic dissipation (Q^{-1}) of lunar and terrestrial rocks (Tittmann *et al.*, 1975; Pandit and Tozer, 1970; Gordon and Davis, 1968). Thus, water in trace amounts seems to play an important role in the attenuation of seismic waves in room-dry rocks, and investigation of the variation of Q^{-1} with the amount of water present should provide useful clues in understanding the attenuation mechanisms. In addition, small amounts of water have been found to lower the modulus in some rocks (Wyllie *et al.*, 1962). The modulus of the dry rock is used in theoretical modeling of the velocities of water-saturated rocks (Toksöz *et al.*, 1976). Since small amounts of water can have large effects on Q^{-1}, and sometimes on the velocity, it is insufficient to refer to the state of a rock as "dry." The relative humidity of the atmosphere during measurements must also be specified.

1. Experimental procedure

Measurements using the vibrating bar method were performed in both torsional and extensional modes (Nowick and Berry, 1972). A detailed description of the experimental procedure is given in Clark (1980). The samples were cylinders 5 inches long and .561 inches in diameter. Frequencies were in the range of 5 kHz to 20 kHz. Strain amplitudes were less than 10^{-7} (calibrated with a capacitive pickup, ADE Microsense). Magnetic transducers (Airpax) were used both for source and receiver. Coupling to the rock bar was provided by tabs of Armco iron, glued on with Duco. Experiments were performed in air and in a vacuum chamber.

The chamber was usually evacuated overnight. Water was then introduced into the chamber from a sample bottle which had previously been degassed by freezing with liquid nitrogen and pumping. The benzene was supplied water and air free. Pressure was measured with a Wallace and Tiernan gauge which could be read to 0.1 torr (0.1 mm).

Typically a waiting time of 45–60 minutes was sufficient for equilibrium to be achieved. The equilibration time varied strongly with the permeability of rocks. For example, with very porous Berea sandstone equilibrium was achieved instantaneously, whereas with very tight Tennessee Sandstone 30–45 minutes were necessary.

In this experiment the pressure measurement was the main source of error; this was the result of accumulation of air in the system through small leaks. Thus, the measured pressures could be slightly higher than the actual vapor pressure. When the system was shut off from the pump, a rise in pressure amounting to less than .05 torr/hr usually occurred. Part of this increase was due to outgassing from the walls of the system. Runs were usually completed in approximately ten hours, and thus the maximum error in pressure measurement would be less than 5%.

A voltage controlled oscillator, driven by a ramp was automatically swept through a preset frequency range. The output from the receiver was amplified and fed into the y-axis of an x-y recorder; the ramp controlled the x-axis. The Q^{-1} and velocity were calculated from the width of the resonant frequency (Nowick and Berry, 1972). The error was about 5% for Q^{-1} measurement. The resonant frequency measured had to be corrected for the effect of the iron end tabs (Clark, 1980); the error was about 2% for velocity.

2. Results

Table 1 contains the sample descriptions for the two rock types used, Sioux quartzite and Coconino Sandstone and Figs. 2 and 3 contain the Q data. Q_s, the attenuation in the torsional mode, is identical to the Q of shear waves propagating

Table 1. Survey of sample characteristics.

Sample	Analysis	Description	Porosity %	Permeability md	Bulk density g/cm^3	Grain density g/cm^3	Period	Grain size
Sioux Quartzite	99% quartz 1% hematite, zircon, magnetite, rutile, and amphibole	Flat bedded orthoquartzite. Silica bonding along concavoconvex grain boundaries.	<.4	Very low	2.63	2.65	Precambian	Fine to med.
Coconino sandstone	90% quartz 9% rock fragments 1% clay and iron oxides	Flat bedded. Well sorted. Silica cemented sandstone with abundant enhedral quartz overgrowths	8.9	1.4	2.38	2.65	Permian	Fine

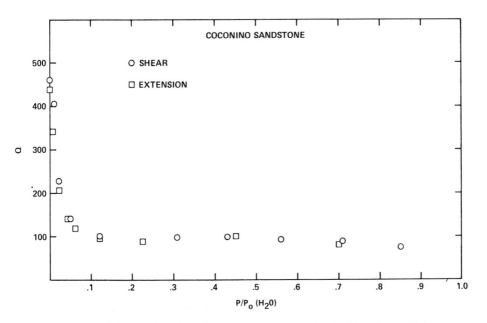

Fig. 2. Q_E and Q_S as a function of water vapor pressure for the Coconino sandstone.

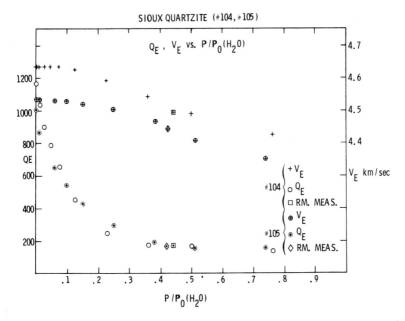

Fig. 3. Q and velocity for two samples of Sioux quartzite as a function of the partial pressure of water vapor. The measurements taken under room conditions are plotted at the appropriate relative humidity.

Table 2. Surface-energy decrease of quartz in various saturated vapors.*

Saturated vapor	Surface-energy decrease $(\gamma v - \gamma f)$ ergs/cm^2
water	244
n-propylalcohol	110
acetone	85
benzene	52

* Boyd and Livingston (1942)

in an infinite isotropic medium. Q_E is the attenuation in the extensional (bar) mode. Q_P, the Q of compressional waves propagating in an infinite isotropic medium, can be calculated from Q_E and Q_S by using complex moduli (Tittmann *et al.*, 1977). For the Coconino sandstone Q_E and Q_S are plotted versus the relative partial pressure of water vapor in Fig. 2. A relative partial pressure of .5 corresponds to a relative humidity of 50%. For the Sioux quartzite Q_E and v_E are plotted versus relative partial pressure for two samples. The shapes of the curves for Q are similar to the curves of uniaxial compressive strength versus partial pressure of water.

The water uptake on the quartzite sample was measured at $P/P_o = 0.93$ with a precision micro balance to be about 0.05 mg of H_2O per gm of sample. This corresponds to a saturation of 3%. In the Coconino sandstone the corresponding

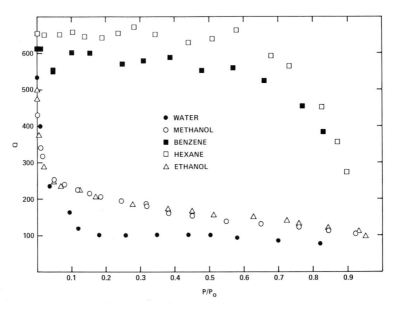

Fig. 4. Q of Coconino sandstone vs. relative partial pressure.

value was 2.0 mg of H_2O per gm of sample. Could motion of water molecules at highly strained points such as crack tips cause attenuation? Such a mechanism would be consistent with the observation that the Sioux quartzite exhibits large changes in Q as a function of relative humidity but adsorbs little water.

EFFECT OF VOLATILES ON Q AND VELOCITY

Table 2 presents data by Boyd and Livingston (1942) on how the surface energy of quartz decreases in various saturated vapors. The results imply that the stress necessary to open up a crack tip in quartz is lowered most effectively by H_2O, somewhat less effectively by n-propylalcohol, and least by benzene. Tittmann *et al.* (1976) had shown that a variety of volatiles could have large effects on the Q of a basalt when the rock was exposed to the volatile at its equilibrium vapor pressure. Figures 4 and 5 show Q and velocity measurements on the Coconino sandstone as a function of relative partial pressure and large differences are seen in the effects of different volatiles, namely water, methanol, alcohol, benzene and hexane. In qualitative agreement with the trends in reduction of surface energy, water effects the Q and velocity the most and then in order of diminishing effects methanol, ethyl alcohol, benzene, and hexane. One difficulty in using the alcohol and many other polar molecules is removing trace amounts of water present. The alcohols used in these experiments were dried with molecular sieves prior to use.

DISCUSSION

Walsh (1966) and Gordon and Davis (1968) noted a large difference in the attenuation between single crystals and monomineralic rocks made up of the same

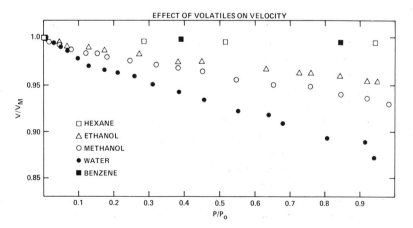

Fig. 5. Velocity in Coconio sandstone vs. relative partial pressure.

crystals; they attributed the large attenuation observed in the rocks to the presence of grain boundaries and cracks. Walsh (1966) proposed frictional sliding on cracks. Gordon and Davis (1968) proposed that some form of "interface damping" takes place. The low Q^{-1} values obtained by Tittmann *et al.* (1975) after evacuation show that a large part of the attenuation in rocks under room conditions is due to the presence of adsorbed water in the cracks and pores. The results of Fig. 4 are not easily explained on the basis of lubrication of grain boundaries and cracks, since comparable coverage of benzene and hexane should be able to lubricate the contacts at least qualitatively as well as the alcohols. Walsh's (1966) model predicts explicitly how the Q depends on the coefficient of friction. Bowden and Hanwell (1966) measured the increase in the value of the coefficient of friction in a variety of nonmetals and found the values to increase from about 0.3 at room conditions to 0.7–1.0 under vacuum. For this range of values Walsh's (1966) model predicts that the ratio Q_p/Q_S increases as the coefficient of friction increases. The data in Fig. 2 show a constant ratio of Q_E/Q_S, however, between room conditions and vacuum in apparent contradiction to the model.

The similarity of the dependence on humidity of compressive strength and Q (see Fig. 1, 2, and 3) suggests a possible relationship between these two parameters. The work of Spetzler and Getting (1980) predicts a contribution to attenuation from stress corrosion cracking. For their study, a solid is modeled by parallel and series configuration of dashpots and springs. The contribution of stress and temperature activated processes to the long term dissipative behavior of this system is analyzed. Data from brittle rock deformation experiments suggest that one such process, stress corrosion cracking, may make a significant contribution to Q, especially for long period oscillations under significant tectonic stress. For the extension of the calculations to other conditions, for example, to high pressures, experimental data on creep, strength, and Q are needed.

Tittmann *et al.* (1980) have proposed a model for seismic attenuation based on the deformational effects primarily driven by the perturbations at crack tips, asperities, and other zones of high stress concentrations. In these zones, the small strains created by the passage of the seismic wave exert strong forces causing diffusion or shear displacement of thin films of adsorbed water. At crack tips the opening or closing under compressional strains causes cyclic "hopping" of the volatile molecules in and out of the crack tip. At grain boundaries and crack asperities, shearing strains cause a shearing of the volatile film against interlayer forces and compressive strains cause a squeezing of the film out of the contact area. The sudden breaking of the volatile bonds to the grain surfaces or other volatiles produces phonon energy thus taking energy from the passing waves. Readsorption of the volatiles is out of phase with the wave so that the energy is not returned coherently. The process of bond making and bond breaking occurs in the interior of the rock on all those surfaces exposed to the atmosphere. This process is not yet documented on a quantitative basis to allow critical comparisons between theory and experiment. A theoretical companion paper (Richardson and Tittmann, 1980) discusses some of the results in terms of a phenomenological theory.

Acknowledgment—This material is based in part upon work supported by the National Science Foundation under grant No. EAR 7906 709.

REFERENCES

Bowden F. P. and Hanwell A. E. (1966) The friction of clean crystal surfaces. *Proc. Roy. Soc. London* **A295**, 233–243.

Boyd G. E. and Livingston H. K. (1942) Adsorption and energy changes at crystalline solid surfaces. *J. Amer. Chem. Soc.* **64**, 2383.

Clark V. A. (1980) Effects of volatiles on attenuation and velocity of elastic waves in sedimentary rocks. Ph.D. dissertation, Texas A&M Univ., College Station.

Colback P. S. and Wiid B. L. (1965) The influence of moisture content on compressive strength of rocks. In *Proc. 3rd National Symposium on Rock Mechanics*, p. 65–83. Toronto, Canada.

Gordon R. B. and Davis L. A. (1968) Velocity and attenuation of seismic waves in imperfectly elastic rock. *J. Geophys. Res.* **73**, 3917–3935.

Martin R. J. (1972) Time dependent crack growth in quartz and its application to the creep of rocks. *J. Geophys. Res.* **77**, 1406–1419.

Mizutani H., Spetzler H., Getting I., Martin R. T., and Soga N. (1977) The effect of outgassing upon the closure of cracks and the strength of lunar analogues. *Proc. Lunar Sci. Conf. 8th*, p. 1235–1248.

Nowick A. S. and Berry B. S. (1972) Mechanical vibration formulas. In *Anelastic Relaxation in Crystalline Solids*, p. 621–631. Academic, N.Y.

Pandit B. I. and Tozer D. C. (1970) Anomalous propagation of elastic energy within the moon. *Nature* **226**, 335.

Richardson J. M. and Tittmann B. R. (1980) Phenomenological theory of attentuation and propagation velocity of elastic waves in rocks. *Proc. Lunar Planet. Sci. Conf. 11th*. This volume.

Spetzler H. and Getting I. (1980) The contribution of stress corrosion cracking to Q (abstract). In *Lunar and Planetary Science XI*, p. 1070–1072. Lunar and Planetary Institute, Houston.

Tittmann B. R., Ahlberg L., and Curnow J. (1976) Internal friction and velocity measurements. *Proc. Lunar Sci. Conf. 7th*, p. 3123–3132.

Tittmann B. R., Ahlberg L., Nadler H., Curnow J., Smith T., and Cohen E. R. (1977) Internal friction quality factor Q under confining pressure. *Proc. Lunar Sci. Conf. 8th*, p. 1209–1224.

Tittmann B. R., Clark V. A., Richardson J. M., and Spencer T. W. (1980) Possible mechanism of seismic attenuation in rocks containing small amounts of volatiles. *J. Geophys. Res.* In press.

Tittmann B. R., Curnow J. M., and Houseley R. M. (1975) Internal friction quality factor Q-3100 achieved in lunar rock 70215, 85. *Proc. Lunar Sci. Conf. 6th*, p. 3217–3226.

Toksöz M. N., Cheng C. H., and Timur A. (1976) Velocities of seismic waves in porous rocks. *Geophys.* **41**, 621–645.

Walsh J. B. (1966) Seismic wave attenuation in rock due to friction. *J. Geophys. Res.* **71**, 2591–2599.

Wiederhorn S. M. (1967). Influence of water vapor on crack propagation in soda lime glass. *J. Amer. Ceram. Soc.* **50**, 407–414.

Wyllie M. R., Gregory A. R., and Gardner G. H. (1962). Studies of elastic wave attenuation in porous media. *Geophys.* **27**, 569–589.

Proc. Lunar Planet. Sci. Conf. 11th (1980), p. 1825–1835.
Printed in the United States of America

The contribution of activated processes to Q

H. A. Spetzler, Ivan C. Getting, and Peter L. Swanson

Department of Geological Sciences, Cooperative Institute for Research in Environmental
Sciences, University of Colorado/NOAA, Boulder, Colorado 80309

Abstract—The possible role of activated processes in seismic attenuation is investigated. In this study, a solid is modeled by a parallel and series configuration of dashpots and springs. The contribution of stress and temperature activated processes to the long term dissipative behavior of this system is analysed. Data from brittle rock deformation experiments suggest that one such process, stress corrosion cracking, may make a significant contribution to the attenuation factor, Q, especially for long period oscillations under significant tectonic stress.

INTRODUCTION

Mechanisms responsible for the attenuation of seismic waves in planetary interiors have received considerable attention (Anderson and Hart, 1978; Hart *et al.*, 1977; Liu and Archambeau, 1975; Jackson and Anderson, 1970). In this treatment, we develop a simple model of inelastic behavior in solids and consider how activated processes, such as stress corrosion cracking, may contribute to seismic attenuation. The behavior of the model with stress corrosion cracking as the dominant absorption mechanism is examined and the determination of the model parameters is discussed.

The absence of significant amounts of moisture on moon is thought by many to account for its high value of Q (Tittman *et al.*, 1980). The transient stresses at cracking sites in rocks due to the passage of seismic waves are not thought to be sufficient to produce significant crack growth associated attenuation (Tittman *et al.*, 1980). We examine, however, the superposition of these small seismic wave associated stresses with much larger relatively static stresses such as those associated with tectonics or lithostatic overburden.

We consider the behavior of a solid which is modeled by the configuration depicted in Fig. 1a and 1b. Consider first the response of the system in Fig. 1b to a step function stress of magnitude S_0, shown in Fig. 1c. The instantaneous strain is determined by the unrelaxed modulus, $M_U = M_2$. The strain associated with the relaxed modulus, $M_R = \dfrac{M_1 M_2}{M_1 + M_2}$, is reached after several relaxation

1825

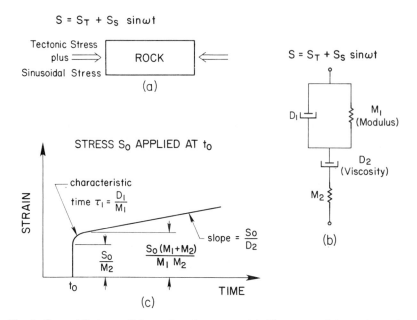

Fig. 1. General features of the rock and stress model. The stress of the rock consists of a constant component, S_T, plus a sinusoidal component, shown in (a). The dashpot and spring configuration used to model the mechanical behavior is shown in (b). The response of this system to a constant stress of magnitude, S_0, is shown in (c).

times, $\tau_1 = D_1/M_1$ of the $M_1 - D_1$ combination. The displacement of D_2 is responsible for the long term creep properties of the solid. After several relaxation times, τ_1, the constant stress on the system is supported by M_1, M_2 and D_2, D_1 then bears an insignificant portion of the load.

If a sinusoidal (seismic) stress, $S_S \sin \omega t$, is now superimposed on a constant (tectonic) stress, S_T, energy from the seismic wave is lost in D_2 by modulation of the previously steady state creep. This energy loss is associated with the oscillatory excursions of D_2. If the seismic stress period is near τ_1, energy is also lost in D_1 (see Fig. 2).

For a real solid, the values of D_1, D_2, M_1 and M_2 are such that at low temperatures and seismic frequencies, attentuation is largely due to D_1 and if a single mechanism is involved, the attenuation peaks when the seismic period is equal to τ_1. Attenuation due to the modulation of creep (D_2) is directly proportional to the period of the seismic wave and increases as the tectonic stress increases. Since D_1 does not support the tectonic stress, attenuation due to D_1 at first glance should be independent of tectonic stress. In a real solid however, this is not the case since a steady state stress modifies the physical properties of the solid by controlling such parameters as dislocation densities, and grain orientations and sizes. Thus the values of D_1 as well as D_2 must be dependent upon the background tectonic stress.

Absorption during activated creep

Much attention has been given (Anderson and Minster, 1979; Minster and Anderson, 1980) to attenuation of seismic waves involving short term relaxation, such as would be due to the D_1–M_1 system in Fig. 1. Here we will consider only the contribution to attenuation due to modulation of steady state deformation (D_2). We are borrowing the general formulation from stress corrosion theory (Charles and Hillig, 1962; Wiederhorn, 1969), but are not restricting ourselves to that mechanism. We are considering an activated process where the rate of the reaction, r, depends exponentially upon the temperature, T, such that

$$r = r_0 \exp\left[-A^*/RT\right] \tag{1}$$

where r_0 is the pre-exponential term, A^* the activation energy and R the gas constant. In the absence of a temperature gradient or a deviatoric stress, an equilibrium will be reached where the activated process results in equal reaction rates in the forward and reverse directions. For example, at a given temperature an equilibrium concentration of vacancies or dislocations will exist. As many are created as are destroyed. Figure 3b shows how we perceive the effect of deviatoric stress, σ, upon a system that is otherwise in equilibrium. The stress reduces the energy barrier and drives the system in the forward direction. The stress and temperature activated reaction rate may then be written as

$$r = r_0\{\exp[-(A^* - \sigma V^*)/RT] - \exp[-A^*/RT]\} \tag{2}$$

where A^* is the activation energy, which is a function of the hydrostatic component of the stress, P. It should in general be written as $A^* + PV^*$, where V^* is the appropriate activation volume; i.e., as the hydrostatic pressure increases so does the barrier height. σ is the deviatoric stress at the site within the solid where the reaction takes place. It is in general much greater than the regional deviatoric stress, S, yet proportional to it; i.e., stress concentrations occur at

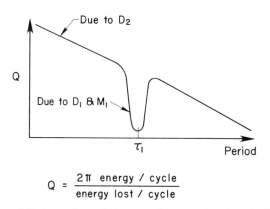

$$Q = \frac{2\pi \ \text{energy / cycle}}{\text{energy lost / cycle}}$$

Fig. 2. Behavior of Q, the reciprocal of attenuation, as a function of seismic period for the system depicted in Fig. 1.

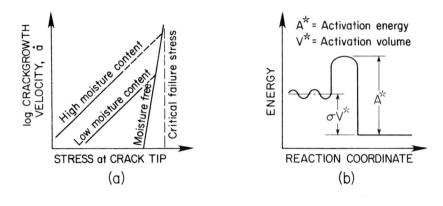

$$\dot{a} = a_0 \, P_{H_2O} \left\{ \exp\left[-\frac{(A^*-\sigma V^*)}{RT}\right] - \exp\left[-\frac{A^*}{RT}\right]\right\} + \sum_{i=1}^{n} \left\{a_{0i} \exp\left[-\frac{A_i^*-\sigma V_i^*}{RT}\right] - \exp\left[-\frac{A_i^*}{RT}\right]\right\}$$

Fig. 3. Crack growth and activated processes. (Note—in the text, crack velocity is denoted by v whereas the symbol à appears in this figure.) (a) shows the moisture and stress dependence of crack velocity. The energy diagram for a stress activated process is shown in (b). The equation gives a general expression for crack velocity as due to water induced stress corrosion cracking plus other activated processes.

grain boundaries, vacancies, dislocations, etc. V^* is an activation volume characteristic of the particular stress activated process. To avoid the difficulty of having to determine a stress concentration factor, $k = \dfrac{\sigma}{S}$, we define an ultimate deviatoric stress σ_u such that

$$\sigma_u = \frac{A^*}{V^*}. \tag{3}$$

From Fig. 3b it is clear that σ_u is the stress at which the barrier height is reduced to zero.

In terms of σ_u, Eq. (2) becomes

$$r = r_0 \left\{ \exp\left[-A^*\left[1 - \frac{\sigma}{\sigma_u}\right]\middle/RT\right] - \exp[-A^*/RT]\right\}. \tag{4}$$

In analogy to σ_u we define an ultimate applied stress $S_u = \dfrac{\sigma_u}{k}$ at which the barrier is reduced to zero.

The strain rate, $\dot{\epsilon}$, resulting from a stress activated reaction is proportional to the reaction rate. For a tectonic stress, S_T, with a superimposed seismic stress, $S_S \sin \omega t$, the resulting strain rate is

$$\dot{\epsilon} = B \left\{ \exp\left[-A^*\left(1 - \frac{S_T + S_S \sin \omega t}{S_u}\right)\middle/RT\right] - \exp[-A^*/RT]\right\} \tag{5}$$

where B is the proportionality constant between strain rate and reaction rate.

To calculate Q, the inverse of attenuation, for the seismic wave we recall a definition of Q

$$Q = \frac{2\pi \times \text{energy/cycle}}{\text{energy lost/cycle}}. \tag{6}$$

For a sinusoidal stress wave, $S_S \sin \omega t$, the energy stored per cycle per unit volume is

$$E = \frac{S_S^2}{2M} \tag{7}$$

where M is the appropriate modulus of the material. The energy lost per cycle per unit volume, E_1, is obtained by integrating the strain rate times the stress over one period, τ,

$$E_1 = \int_0^\tau (S_T + S_S \sin \omega t) B \left\{ \exp\left[-A^* \left(1 - \frac{S_1 S_S \sin \omega t}{S_u} \right) \middle/ RT \right] - \exp[-A^*/RT] \right\} dt$$

$$- \int_0^\tau S_T B \left\{ \exp\left[-A^* \left(1 - \frac{S_T}{S_u} \right) \middle/ RT \right] - \exp[-A^*/RT] \right\} dt. \tag{8}$$

The first term is the energy expended per unit volume during one cycle due to tectonic stress and the superimposed seismic stress. The second term is the energy expended per unit volume during the time of one cycle due to tectonic stress alone. Integration of (8) and substitution into (6) yields

$$Q = \frac{2\pi RT S_u \exp\left[A^* \left(1 - \frac{S_T}{S_u} \right) \middle/ RT \right]}{M B \tau A^* \left[1 + \frac{A^* S_T}{2RT S_u} \right]}. \tag{9}$$

Since $\frac{A^* S_S}{RT S_u} \ll 1$ the integration of (8) was performed after expanding $\exp[A^* S_S \sin \omega t / RT S_u]$ into a series. It is interesting to note that Q is independent of the amplitude of the seismic stress (as long as $\frac{A^* S_S}{RT S_u} \ll 1$), decreases as the tectonic stress increases, and is inversely proportional to the period of the seismic wave. To obtain numerical results from (9), the activation energy A^* and the activation volume V^* (or its equivalent $\frac{A^*}{k S_u}$) as well as B, the proportionality constant between the reaction rate and the strain rate must be determined experimentally.

Determination of model input parameters

The prevalent role of stress corrosion cracking in the time dependent mechanical behavior of geologic materials (Anderson and Grew, 1977) and the strong dependence of Q on moisture (Tittman *et al.*, 1979) suggests that stress corrosion cracking may possibly be a mechanism of seismic energy absorption. In the past, stress corrosion cracking has been extensively studied by material scientists. Only recently, however, have there been attempts to directly study stress corrosion by geo-scientists.

Determination of the thermodynamic parameters describing stress corrosion cracking, A^* and V^*, for simple homogeneous systems is accomplished through the use of various fracture mechanics techniques. A fracture is propagated slowly (controllably $\sim 10^{-2}$ to 10^{-9} meters/second) under a regulated environment in a specimen with a geometry of known stress intensity factor, K_I (mode I—opening or tensile mode). Subcritical fracture data is typically displayed in a plot of the logarithm of the crack velocity, v, versus the stress intensity factor (see Fig. 3a). The shift in the K_I, v diagram to higher crack velocities with increased temperature allows a determination of the activation energy, A^*, for the mechanism producing subcritical crack growth. The slope of the K_I, v curve determines the activation volume, V^*, when a crack tip radius is assumed.

Estimates of A^* have been given for single crystal quartz by Martin (1972) and Scholz (1972) as 108 ± 3 kJ/mole and 96 to 150 kJ/mole, respectively. Atkinson (1979) has reported a value of 69.5 ± 1.7 kJ/mole for a natural quartz polycrystal, Arkansas Novaculite. Measurements of A^* in polyphase aggregates have not yet been made. Swanson and Spetzler (1979) have shown that the fracture path (e.g., intergranular vs. transgranular) in Westerly granite double torsion specimens is dependent on both the amount of moisture in the environment and rate of crack growth. Consequently it is not certain if A^* measured in this process will correspond to single crystal values. The measured slope of the K_I, log v curve obtained for the granite immersed in water at room temperature using the double torsion technique (Evans, 1972; see Fig. 4 and Appendix) was $5.3 \times 10^{-5} \mathrm{m}^{5/2}/$ joule. Discrepancies with the same measurement by Atkinson and Rawlings (1979) using the same technique illustrates the current difficulties in directly applying fracture mechanics techniques to rocks. (A brief description of the double torsion technique is given in the appendix).

For water induced stress corrosion cracking, the parameter B in Eq. (5) may be separated into two components: the partial pressure of water at the crack tip, $P_{H_2O}^n$, with $n \sim 1$ (Soga et al., 1979a), and a new proportionality constant, b.

$$B = P_{H_2O}^n \cdot b \tag{12}$$

The value of b may then be determined from experimental stress-strain data in which the partial pressure of water is known. From Mizutani et al. (1977) and Spetzler et al. (1980) we estimate a strain rate of $10^{-8} \mathrm{s}^{-1}$ for Ralston intrusive at room temperature (300K) and 100% relative humidity, $P_{H_2O} = 3.9 \times 10^3$Pa (29 mmHg), at an axial stress of 400 MPa (4kbars). Neglecting the back reaction term, $\exp[-A^*/RT]$, these values give

$$b = 2.6 \times 10^{-12} \exp[A^*(1 - 4 \times 10^8/S_u)/300R] \tag{13}$$

and Q becomes

$$Q = \frac{3.9 \times 10^{11} \exp[(A^*/R)(1 - S_T/S_u)/T - (1 - 4 \times 10^8/S_u)/300)]}{P_{H_2O}TM(A^*/RTS_u)(1 + S_TA^*/2RTS_u)}. \tag{14}$$

DOUBLE TORSION SPECIMEN

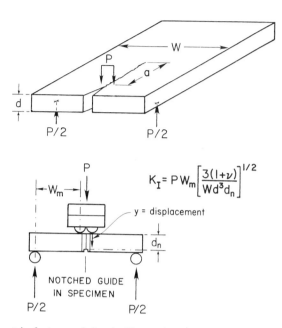

$$K_I = P W_m \left[\frac{3(1+\nu)}{Wd^3 d_n}\right]^{1/2}$$

y = displacement

NOTCHED GUIDE
IN SPECIMEN

Fig. 4. Geometric features of the double torsion specimen used for "single crack" propagation measurements. The displacement, y, is measured from the unstressed position.

DISCUSSION

To check the applicability of this absorption mechanism we use the following experimentally determined values for Ralston intrusive as used by Spetzler *et al.* (1980): $A^* = 1.1 \times 10^5$ J/mole (26 kcal/mole), $S_u = 1.23$ GPa (12.3 kbars) and $M = 100$ GPa (1 megabar). Taking only one moisture content, $P = 3.9 \times 10^3$ Pa (29 mmHg) which corresponds to saturation at 300K, we calculate the values of the product $Q\tau$ in Table 1.

At tidal periods for the earth-moon system of $\tau \sim 10^5$s, Q becomes important (~65) at 400K. For seismic waves of 1 second period Q is 10^4 at 500K with zero tectonic stress and 600 at 500K with a tectonic stress of 100 MPa (1 kbar).

Further perspective on the contribution of the stress corrosion mechanism to Q may be acquired by comparing the values of Q from Eq. (14) with the corresponding strain rates predicted by Eq. (5). In Fig. 5 we have plotted log $(Q\tau)$ vs. temperature for two partial pressures of water and two tectonic stresses (descending curves). The corresponding strain rates are plotted for two different tectonic stresses of 0.1 and 1.0 MPa (1 and 10 bars). To illustrate the point of the data presented in Fig. 5 we assume that a Q of 1000 at a frequency of 1 Hz was determined for a crustal depth corresponding to a temperature of 500K. It is

Table 1. The quantity Q times period ($Q\tau$) at various temperatures and tectonic stress levels.

Tectonic stress (MPa)	0	1	10	100
Temperature (K)				
300	3.0×10^{11}	3.0×10^{11}	1.8×10^{11}	3.0×10^{9}
400	6.5×10^{6}	6.3×10^{6}	4.4×10^{6}	1.8×10^{5}
500	1.0×10^{4}	9.0×10^{3}	8.0×10^{3}	6.0×10^{2}
600	1.6×10^{2}	1.5×10^{2}	1.2×10^{2}	1.4×10^{1}

clear from the $Q\tau$ curves for vapor saturation that the appropriate Q would be about 10 for zero tectonic stress and about 1 for a tectonic stress of 100 MPa. Clearly the partial pressure of water would have to be lower than the vapor pressure. At 500K, a tectonic stress of 100 MPa and a partial pressure of 3.9×10^{3}Pa, Q is about 1000. The strain rate under those conditions, however, would be about $2 \times 10^{-6}\text{s}^{-1}$, certainly not a geological strain rate. For a geologically

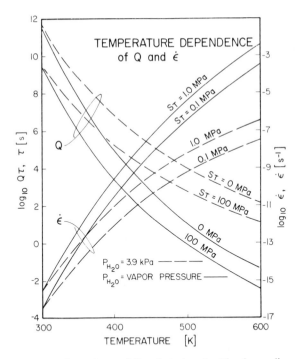

Fig. 5. The temperature dependence of Q and strain rate. The descending curves show the temperature dependence of Q, expressed as log $Q\tau$, where τ is the seismic period. Curves for two tectonic stresses, S_T, and two partial pressures of water conditions are shown. The vapor pressure of water is 3.9 kPa just above room temperature. For the solid curves, the partial pressure of water is equal to the vapor pressure at temperature. The ascending curves show the creep rate which would result from the tectonic stress levels indicated (right hand axis) for the same moisture levels.

more reasonable strain rate of $10^{-11}s^{-1}$ a tectonic stress lower than 0.1 MPa is required. The implications must be: 1. The partial pressure of H_2O must be about $3.9 \times 10^3 Pa$ or lower. If it is $3.9 \times 10^3 Pa$ then the tectonic stress must be below 0.1 MPa. 2. If the tectonic stress is on the order of 1 MPa, and a geologic strain rate of $\sim 10^{-11}s^{-1}$ or lower is observed, then the partial pressure of water must be considerably below $3.9 \times 10^3 Pa$ and the absorption mechanism for the Q value of 500 is not due to stress corrosion.

The values above apply only to room pressure conditions. Extrapolation to high confining pressures requires the inclusion of a term that increases the activation energy in response to the confining pressure. A^* is replaced by $A^* + PV^*$ as described above. Typical values for V^* are in the range from $(2-10) \times 10^{-6}$ $m^3/mole$. Further modification is required because of the effect of confining pressure on crack structure and through it on S_u and b. To extend these calculations to high pressures, experimental data on creep, strength, and Q, such as those of Kohlstedt *et al.* (1976), Soga *et al.* (1979b), Brodsky and Spetzler (1979) and Tittman *et al.* (1979) are needed.

Acknowledgments—This work was supported by National Aeronautics and Space Administration grant NSG 7584. We thank Hitoshi Mizutani and Kei Kurita for their technical advice.

REFERENCES

Anderson D. L. and Hart R. S. (1978) The Q of the Earth. *J. Geophys. Res.* **83**, 5869–5882.

Anderson D. L. and Minster J. B. (1979) The physics of creep and attenuation in the mantle. *Proc. XVII General Assembly of the Int'l Union Geod. Geophys.,* Canberra, Australia. In press.

Anderson O. L. and Grew P. C. (1977) Stress corrosion theory of crack propagation with applications to geophysics. *Rev. Geophys. Space Phys.* **15**, 77–104.

Atkinson B. K. (1979) Stress corrosion and the rate-dependent tensile failure of a fine-grained quartz rock. *Tectonophys.* In press.

Atkinson B. K. and Rawlings R. D. (1979) Acoustic emission during subcritical tensile cracking of gabbro and granite (abstract). *EOS, (Trans. Amer. Geophys. Union)* **60**, 740.

Brodsky N. and Spetzler H. A. (1979) Time dependent deformation of a basalt at low differential stress. *Proc. Lunar Sci. Conf. 10th,* p. 2155–2163.

Charles R. J. and Hillig W. B. (1962) The kinetics of glass failure by stress corrosion. *Proc. Symposium sur la Resistance Mecanique due Verre et les Moyens de l'Ameliorer,* Union Scientifique Continentale du Verre, Charleroi, Belgium, 511–527.

Evans A. G. (1972) A method for evaluating the time-dependent failure characteristics of brittle materials—and its application to polycrystalline alumina. *J. Mater. Sci.* **7**, 1137–1146.

Hart R. S., Anderson D. L., and Kanamori H. (1977) The effect of attenuation on gross Earth models. *J. Geophys. Res.* **82**, 1647–1654.

Jackson D. D. and Anderson D. L. (1970) Physical mechanisms of seismic wave attenuation. *Rev. Geophys. Space Phys.* **8**, 1–63.

Kim K. and Stout M. D. (1978) Determination of effective elastic modulus by compliance calibration for measurement of fracture toughness of rock. *Proc. U.S. 19th National Symposium on Rock Mechanics,* p. 203–209. Reno, Nevada.

Kohlstedt D. L., Goetze C., and Durham W. B. (1976) Experimental deformation of single crystal olivine with application to flow in the mantle. In *The Physics and Chemistry of Minerals and Rocks* (S. K. Runcorn, ed.), p. 35–49. Wiley, London.

Liu H. P. and Archambeau C. B. (1975) The effect of anelasticity on periods of the Earth's free oscillations (toroidal modes). *Geophys. J. Roy. Astron. Soc*. **43**, 795–814.

Martin R. J. III (1972) Time-dependent crack growth in quartz and its application to the creep of rocks. *J. Geophys. Res*. **77**, 1406–1419.

Minster J. B. and D. L. Anderson (1980) Dislocations and nonelastic processes in the mantle. *J. Geophys. Res*. In press.

Mizutani H., Spetzler H. A., Getting I. C., Martin R. J. III, and Soga N. (1977) The effect of outgassing upon the closure of cracks and the strength of lunar analogues. *Proc. Lunar Sci. Conf. 8th*, p. 1235–1248.

Scholz C. H. (1972) Static fatigue of quartz. *J. Geophys. Res*. **77**, 2104–2114.

Soga N., Okamoto T., Hanada T., and Kunugi M. (1979a) Chemical reaction between water vapor and stressed glass. *J. Amer. Ceram. Soc*. **62**, 309–310.

Soga N., Spetzler H. A., and Mizutani H. (1979b) Comparison of single crack propagation in lunar analogue glass and the failure strength of rocks. *Proc. Lunar Sci. Conf. 10th*, p. 2165–2173.

Spetzler H. A., Sondergeld C., and Getting I. C. (1980) The influence of strain rate and moisture on rock failure. Submitted to *Proc. XVII General Assembly of the Int'l Union Geod. Geophys*., Canberra, Australia. In press.

Swanson P. L. (1980) Observations of the fracture process in Westerly granite double torsion specimens (abstract). *EOS (Trans. Amer. Geophys. Union)* **61**, 372.

Swanson P. L. and Spetzler H. A. (1979) Strength of rock as a function of single crack growth parameters and crack morphology (abstract). *EOS (Trans. Amer. Geophys. Union)* **60**, 940.

Tittman B. R., Nadler H., Clark V., and Coombe L. (1979) Seismic Q and velocity at depth. *Proc. Lunar Sci. Conf. 10th*, p. 2131–2145.

Tittman B. R., Clark V., Arora A., and Spencer T. W. (1980) Compressive strength, seismic Q, elastic modulus, and acoustic emission studies of rock. In *Lunar and Planetary Science XI*, p. 1155–1157. Lunar and Planetary Institute, Houston.

Wiederhorn S. M. (1969) Fracture of ceramics. In *Mechanical and Thermal Properties of Ceramics*. Nat. Bur. Standards Publ. 303, p. 217–241. Washington, D.C.

Williams D. P. and Evans A. G. (1973) A simple method for studying slow crack growth. *J. Test. Eval*. **1**, 264–270.

APPENDIX

The major advantage of the double torsion specimen (Evans, 1972) over other more standard fracture geometries is the crack length independence of K_I making it well-suited to opaque materials and harsh environments. The expression for K_I as given by Williams and Evans (1973) is

$$K_I = PW_m \left[\frac{3(1 + \nu)}{Wd^3 d_n} \right]^{\frac{1}{4}}$$

where ν is Poisson's ratio and the remainder of the terms are identified in Fig. 4. A linear relation between the specimen compliance (y/p—see Fig. 4) and crack length, a, (Evans, 1972; Williams and Evans, 1973; Kim and Stout, 1978) results in a crack velocity which is easily measured. With the relaxation method the sample is rapidly loaded to a preselected fraction of K_{IC} (the fracture toughness) and then the displacement of the loading points, y, is fixed. The crack velocity at various loads, P, is determined from the ensuing force decay by

$$v = \frac{v'}{P^2} \left(\frac{dP}{dt} \right)$$

where v' is a constant (Evans, 1972).

A detailed study of the assumptions used in applying the double torsion theory to experiments on rocks has revealed major violations (Swanson, 1980). These difficulties, encountered with the fine-grained Westerly granite, can be overcome with appropriate modifications to the technique. Acoustic emission location studies show that there is an extraneous contribution to dP/dt in the equation above from time dependent frictional resistance along the crack walls (see Fig. 4). Double exposure holographic interferometry shows that the "elastic" portions of the specimen are also susceptible to anomalous behavior when in the presence of moisture. Adjustment of the grain size to plate thickness ratio and isolation of the "elastic" portions of the specimen from the control environment should eliminate these difficulties.

Proc. Lunar Planet. Sci. Conf. 11th (1980), p. 1837–1846.
Printed in the United States of America

Phenomenological theory of attenuation and propagation velocity in rocks

J. M. Richardson and B. R. Tittmann

Rockwell International Science Center, Thousand Oaks, California 91360

Abstract—A phenomenological theory of velocity and attenuation of elastic waves in rocks is proposed and applied to two rock types (i.e., Coconino sandstone and Sioux quartzite). The theory is based upon the two main assumptions: (a) that the macroscopic behavior can be modelled as a superposition of linear dissipative processes and (b) that these processes involve thermal activation with flat distribution of activation energies over a range whose end points depend on the partial pressure of the volatile to which the rock is exposed.

1. INTRODUCTION

The purpose of this short note is to present a phenomenological theory of the attenuation and propagation velocity of elastic waves in rocks in the linear regime. The theory does not attempt to provide a complete prediction of the macroscopic behavior of elastic waves starting with a particular assumed statistical model of the microstructure. It pursues the less ambitious objective of deriving a partial description of macroscopic behavior based on certain general assumptions concerning the nature of the microstructure. Two of these assumptions are (a) that the complex modulus is represented as a superposition of elementary nonresonant relaxation processes, and (b) that these involve thermal activation with a broad distribution of activation energies. This latter property is almost inevitable in permanently disordered systems if the disorder directly affects the activation energies. We also allow the possibility of additional sets of relaxation processes involving very short relaxation times.

The assumption of a broad distribution of activation energies has been utilized in other fields of physics. For example, Nowick and Berry (1972) consider such an assumption in their phenomenological description of anelastic relaxation. Also, Fröhlich (1958) considers an identical assumption in his phenomenological theory of dielectric relaxation.

The theory is compared with experimental measurements on the dependence of attenuation and velocity on moisture uptake in Coconino sandstone and Sioux quartzite. Even though these rocks are very different from each other, the theory works well for both of them. Similar investigations of a number of other rocks are currently underway.

2. GENERALITIES

If we make the assumption that the macroscopic loss is dominated by elastic dissipative processes (as contrasted with inertial dissipative processes), it follows that the attenuation and propagation velocity are determined by the appropriate complex modulus (e.g., the shear modulus in the case of torsional modes, Young's modulus in the case of extensional modes, etc.). Denoting the modulus by $\lambda = \lambda(\omega)$ at the frequency ω, we can express Q and the propagation velocity v in terms of the following expressions

$$Q^{-1} = \text{Arg } \lambda \simeq \frac{\lambda''}{\lambda'} \tag{2.1}$$

and

$$v = \left(\frac{\lambda'}{\rho}\right)^{1/2} \tag{2.2}$$

where λ' and λ'' are the real and imaginary parts of λ, respectively. If the attenuation and dispersion are sufficiently small we can rewrite (2.1) and (2.2) in the forms

$$Q^{-1} \simeq \frac{\lambda''}{\lambda_0'} \tag{2.3}$$

$$\frac{v_0 - v}{v_0} \simeq \frac{1}{2} \frac{\lambda_0' - \lambda'}{\lambda_0'} \tag{2.4}$$

where λ_0 and v_0 are the modulus and propagation velocity of perfectly dry rock and where, using our earlier convention, λ_0' is the real part of λ_0. In writing the above expressions we have assumed that at the frequency ω the implicit time factor is $\exp(i\omega t)$.

We now assume that the complex modulus can be represented in terms of superposition of elementary relaxation processes, i.e.,

$$\lambda = \lambda^0 - \gamma \int_0^\infty d\tau P(\tau) \frac{1}{1 + i\omega\tau} \tag{2.5}$$

where λ^0 is a reference modulus existing in the absence of dissipative processes, τ is a relaxation time, and $P(\tau)$ is a probability density representing the distribution of relaxation times. The constant coefficient γ reflects the density of elementary relaxation processes and their strengths of interaction with the gross elastic field. Of course, one would expect each elementary process to have its own γ with a value different from those ascribed to other elementary processes. However, if the values of γ and τ are statistically independent, one obtains (2.5) but with γ now interpreted as the average coefficient.

There is nothing new in Eq. (2.5). The idea of a superposition of linear dissipative processes has been developed in many areas of physical science (e.g., rheology, dielectric relaxation, etc.). However, in the next section we will consider a particular form of $P(\tau)$ that is almost inevitable for certain classes of disordered materials.

3. BROAD DISTRIBUTION OF ACTIVATION ENERGIES

In view of the above discussion it is reasonable to consider the following assumptions:

Assumption I.
For each elementary dissipative process the relaxation time is given by

$$\tau = \tau^0 \exp(E/kT) \tag{3.1}$$

where τ^0 is a characteristic relaxation time, E is the activation energy, and kT is the thermal energy (k is the Boltzmann constant and T is the absolute temperature).

Assumption II.
The activation energies are uniformly distributed in the interval $[E_{min}, E_{max}]$.

Assumption III.
The quantities τ^0 and $E_{max} - E_{min}$ are independent of the partial pressure of the volatile (e.g., H_2O) while E_{max} and E_{min} individually depend on this partial pressure.

These three assumptions imply that $P(\tau)$ is now given by

$$\gamma P(\tau) = \Gamma \frac{1}{\tau}, \tau_{min} \leq \tau \leq \tau_{max}$$
$$= 0, \text{ otherwise}, \tag{3.2}$$

where, obviously,

$$\tau_{min} = \tau^0 \exp(E_{min}/kT) \tag{3.3}$$

$$\tau_{max} = \tau^0 \exp(E_{max}/kT). \tag{3.4}$$

The coefficients γ and Γ are related by integrating (3.2) with the result

$$\gamma = \Gamma \log(\tau_{max}/\tau_{min})$$
$$= \Gamma(E_{max} - E_{min})/kT. \tag{3.5}$$

Assumption III clearly implies that Γ is also independent of volatile partial pressure if γ is; however, this last assumption may not be valid. Before proceeding further, we will make one further assumption, namely:

Assumption IV.
We assume that the signal frequency ω is much larger than the minimum relaxation frequency $1/\tau_{\text{max}}$, or equivalently,

$$\omega \tau_{\text{max}} \gg 1. \qquad (3.6)$$

Substituting (3.2) into (2.5) and introducing (3.6), we obtain the main result

$$\lambda - \lambda^0 = -\Gamma\phi(\omega\tau_{\text{min}}) \qquad (3.7)$$

where, letting $x = \omega\tau_{\text{min}}$,

$$\phi(x) = \log\left(1 + \frac{1}{ix}\right)$$
$$= \frac{1}{2}\log\left(1 + \frac{1}{x^2}\right) + i \tan^{-1} x - \frac{1}{2}\pi i, \qquad (3.8)$$

a special case of the results of Nowick and Berry (1972) and Fröhlich (1958). A plot of the Re$\phi(x)$ vs. Im $\phi(x)$ is presented in Fig. 1. Equation (3.7) gives the dependence of the complex modulus on the frequency ω under the assumption that λ^0, Γ, and τ_{min} are frequency-independent. Since our experiments thus far have been performed at a fixed frequency with varying partial pressures of volatiles we must direct our attention to the dependence of λ^0, Γ, and τ_{min} on the latter. Although Γ will certainly depend on the volatile partial pressure to some extent, a preliminary analysis indicates that this dependence will not be very significant and thus we will defer its consideration to a later time. Concerning the

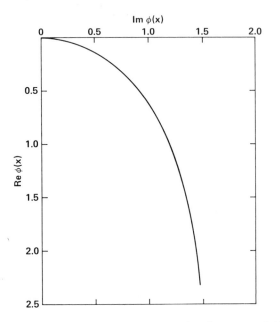

Fig. 1. The real part of the function $\phi(x)$ versus the imaginary part of $\phi(x)$.

dependence of λ^0 and τ_{min} upon the volatile partial pressure we make the following additional assumptions:

Assumption V.
We assume that λ^0 is real and depends upon the volatile partial pressure in accordance with the Brunauer-Emmett-Teller (1938) isotherm. In explicit mathematical form we assume

$$\lambda^0 - \lambda_0^0 = \xi\beta(z) \tag{3.9}$$

where λ_0^0 is the dry reference modulus, ξ is a constant (usually positive), and z is the relative partial pressure given by

$$z = \frac{p}{p^0}, \tag{3.10}$$

where, in turn, p is the partial pressure of the volatile and p^0 is at saturation value at the ambient temperature. The function $\beta(z)$ embraces the BET isotherm and is given by

$$\beta(z) = \frac{z}{(1 - z)[1 + (c - 1)z]} \tag{3.10a}$$

where

$$c = \exp[(Q_1 - Q_v)/kT] \tag{3.10b}$$

in which Q_1 is the heat of adsorption of a molecule in the first adsorbed layer and Q_v is the heat of vaporization.

Assumption VI.
We further assume that in Eq. (3.3) the dependence of τ^0 on volatile partial pressure is insignificant compared with the dependence of the activation energy E_{min} and that the latter quantity depends linearly on this pressure in the following manner

$$E_{min} = E_{min,o} - \eta z \tag{3.11}$$

where η is a constant (usually positive).

With these assumptions we obtain

$$\lambda = \lambda_0^0 + \xi\beta(z) - \Gamma\phi(z) \tag{3.12}$$

where ϕ is now written as a function of $z = p/p^0$. The earlier dependent variable $x = \omega\tau_{min}$ is given in terms of z by the expression

$$x = x_o \exp(-\eta z/kT) \tag{3.13}$$

where x_o, the dry value of x, is given by

$$x_o = \omega\tau^0 \exp(E_{min,o}/kT). \tag{3.14}$$

It is convenient to introduce a new variable

$$y = -\log x = -\log x_o + \frac{\eta z}{kT} \tag{3.15}$$

which is linear in z but with a translation and scale change. Expressed into terms of y we obtain

$$\phi(y) = \log[1 - i \exp(y)] \tag{3.16}$$

from which we infer that

$$\text{Re } \phi(y) = \frac{1}{2} \log[1 + \exp(2y)] \tag{3.17}$$

$$\text{Im } \phi(y) = \tan^{-1} \exp(y) \tag{3.18}$$

Plots of $\text{Re}\phi(y)$ and $-\text{Im}\phi(y)$ as functions of y are given in Fig. 2.

The relations between the quantities Q^{-1} and $(v_0 - v)/v_0$ and the complex modulus λ are now readily obtained. First we obtain the real and imaginary parts given by

$$\lambda' = \lambda_0^0 - \xi\beta(z) - \Gamma \text{ Re } \phi(z) \tag{3.19}$$

$$\lambda'' = -\Gamma \text{ Im } \phi(z). \tag{3.20}$$

We also obtain for the dry value of λ'

$$\lambda_0' = \lambda_0^0 - \Gamma \text{ Re } \phi(o). \tag{3.21}$$

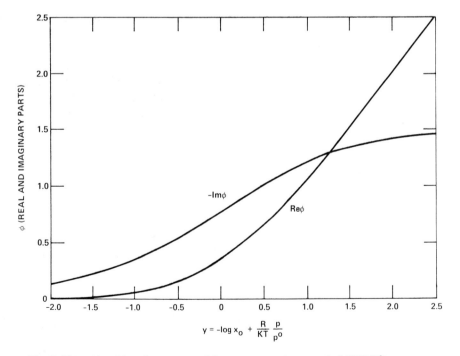

Fig. 2. The real and imaginary parts of ϕ versus $y = -\log x_0 + (\eta/kT)(P/P^0)$.

Equations (2.3) and (2.4) now take the form

$$Q^{-1} = -\frac{\Gamma}{\lambda_0'}\, \text{Im}\, \phi(z)$$

$$\simeq -\frac{\Gamma}{\lambda_0^0}\, \text{Im}\, \phi(z) \tag{3.22}$$

$$\frac{v_0 - v}{v_0} = \frac{\xi\beta(z)}{2\lambda_0'} + \frac{\Gamma}{2\lambda_0'}[\text{Re}\, \phi(z) - \text{Re}\, \phi(o)]$$

$$\simeq \frac{\xi\beta(z)}{2\lambda_0^0} + \frac{\Gamma}{2\lambda_0^0}[\text{Re}\, \phi(z) - \text{R}\, \phi(o)] \tag{3.23}$$

In deriving the last lines of Eqs. (3.22) and (3.23) we have neglected $\Gamma\, \text{Re}\phi(o)$ in the relation between λ_0^1 and λ_0^0 [see (3.21)]. In some experimental determinations of $(v_0 - v)/v_0$ the velocity v_0 refers not to the perfectly dry state but to the small but nonvanishing value of z, which we will denote by z_0. In such a case (3.2.3) must be rewritten in the form

$$\frac{v_0 - v}{v_0} \simeq \frac{\xi[\beta(z) - \beta(z_0)]}{2\lambda_0^0} + \frac{\Gamma}{2\lambda_0^0}[\text{Re}\, \phi(z) - \text{Re}\, \phi(z_0)] \tag{3.24}$$

where we have again used the approximation of neglecting $\Gamma\text{Re}\phi(o)$ in the denominator.

4. COMPARISON WITH EXPERIMENT

We turn now to a comparison of the phenomenological theory outlined in the previous section with the experimental results of Tittmann *et al.* (1980) on velocity and attenuation. This somewhat preliminary comparison deals with Coconino sandstone and Sioux quartzite exposed to varying partial pressures of H_2O. We consider only a limited range of the latter quantity since presumably another regime sets in above a critical value.

In the theory there are four adjustable parameters (or combinations of parameters), namely Γ/λ_0^0, log x_0, η/kT, and $\xi/2\lambda_0^0$ which were defined and discussed in the last section. Our procedure is, first, to find the best values of the first three parameters to give a best fit of the theoretical expression for Q^{-1} [Eq. (3.22)] to the experimental values of Q^{-1}, both regarded as functions of $z = p/p^0$ the relative partial pressure of H_2O. Roughly speaking the procedure is to determine the multiplier Γ/λ_0^0 from the early saturation level of the experimental Q^{-1} and then

Table 1. Parameter values giving best agreement with experiment.

Rock type	Γ/λ_0^0	log x_0	η/kT	$\xi/2\lambda_0^0$
Coconino sandstone	6.57×10^{-3}	1.02	30	0.17
Sioux quartzite	5.0×10^{-3}	1.71	6.7	0.01

the other parameters log x_o and R/kT are determined to bring the flex point and horizontal scale of the theoretical S-shaped Q^{-1} curve into consistency with the experimental data. The second step is to determine the fourth parameter $\xi/2\lambda_0^0$ (keeping the first three parameters fixed) to give a best fit of the theoretical expression for $(v_o - v)/v_o$ [Eq. (3.24)] to the experimetal data. In the case of

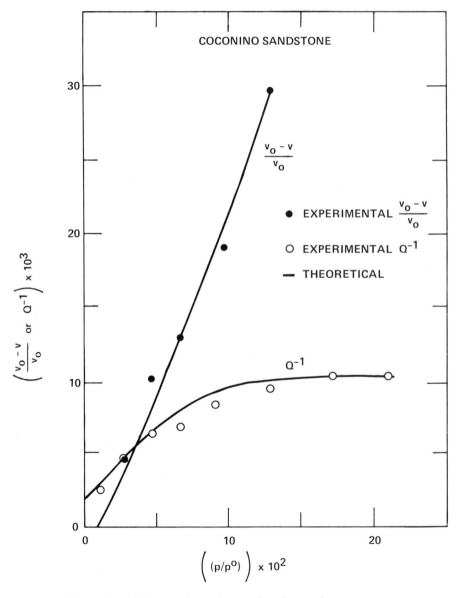

Fig. 3. Comparison of theory and experiment—Coconino sandstone.

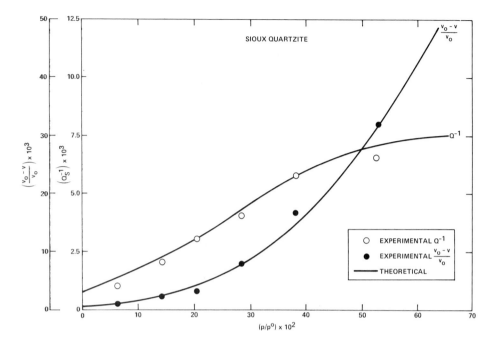

Fig. 4. Comparison of theory and experiment—Sioux quartzite.

Coconino sandstone the comparison with experiment was limited to relatively small values of $z = p/p^0$ and we consequently made the approximation $(z) \simeq z$ in which case the constant c is irrelevant. However, in the case of Sioux quartzite, this was not the case and the correct expression for $\beta(z)$ was used. Here we chose the value $c = 0.1$ from an inspection of the shape of the experimental curve. The best values of the first four parameters are summarized in Table 1 for the two rock types considered. In Figs. 3 and 4 we compare the theoretical curves (with the best parameter values) for Q^{-1} and $(v_0 - v)/v_0$ vs. $z = p/p^0$ with the experimental points.

In our opinion, the comparison of theory and experiment is surprisingly good considering the complexity of the physical systems under consideration. The comparison would be substantially improved if a more sophisticated method of parameter estimation were used. It is to be strongly emphasized that the comparison has a significance that goes beyond that of mere curve-fitting. The reader is reminded that the frequency-dependent part of the complex modulus, based on Assumptions I–IV enumerated in the last section, involves a universal function $\phi(\omega\tau_{min})$ which, when regarded as a function of $z = p/p^0$, is subject only to translation and to horizontal and vertical scale changes without change of shape. Thus the comparison of theory and experiment for Q^{-1} is significant to this degree. The comparison for $(v_0 - v)/v_0$ is slightly less significant because of an additive term linear in $\beta(z)$ with the adjustable coefficients $\xi/2\lambda_0^0$ and c.

A few final remarks are in order concerning the phenomenological nature of the present theory. The successful comparison with experiment does not lend support to particular underlying mechanisms, but only to a certain class of possible mechanisms that are consistent with the assumptions enumerated in the last section. To be more specific, there are apparently two sub-classes of mechanisms present: one with very short relaxation times giving an almost instantaneous response relative to the signal frequency and a second sub-class with a distribution of relaxation times straddling the reciprocal signal frequency in certain intervals of $z = p/p^0$. Possible mechanisms in the first sub-class must then conform to Assumption V. A plausible hypothesis is that the dominant mechanism here entails modulus reduction due to intragranular cracks caused by a stress corrosion process involving H_2O. The second sub-class of possible mechanisms must similarly conform to the remaining assumptions. A number of plausible mechanisms can be postulated, but the further narrowing of the sub-class will require more detailed microscopic observations.

Clearly further work remains to be done. An important future experimental objective is the separation of the effects of the fast-relaxing mechanisms using the frequency dependence, thus allowing a simpler, less heavily parameterized, explanation of the effects of the remaining slower-relaxing mechanisms. Further theoretical work needs to be done on the weakening of the overly idealized assumptions used here and on the more careful treatment of the way in which the elementary processes are to be combined to yield macroscopic behavior.

Acknowledgments—This work was supported in part by NSF Contract No. EAR-79-067-79.

REFERENCES

Brunauer S., Emmett P. H., and Teller E. (1938) Adsorption of gases in multimolecular layers. *J. Amer. Chem. Soc.* **60**, 309–319.

Fröhlich H. (1958) *Theory of dielectrics.* Oxford Univ. Press. 192 pp.

Nowick A. and Berry B. S. (1972) Anelastic relaxation in crystalline solids, Academic, N.Y. 677 pp.

Tittmann B. R., Clark V. A., and Spencer T. W. (1980) Compressive strength, seismic Q, and elastic modulus. *Proc. Lunar Planet. Sci. Conf. 11th.* This volume.

Proc. Lunar Planet. Sci. Conf. 11th (1980), p. 1847–1853.
Printed in the United States of America

Shallow moonquakes: How they compare with earthquakes

Yosio Nakamura

The University of Texas, Marine Science Institute, Galveston Geophysics Laboratory,
Galveston, Texas 77550

Abstract—Of three types of moonquakes strong enough to be detectable at large distances—deep moonquakes, meteoroid impacts and shallow moonquakes—only shallow moonquakes are similar in nature to earthquakes. A comparison of various characteristics of moonquakes with those of earthquakes indeed shows a remarkable similarity between shallow moonquakes and intraplate earthquakes: (1) their occurrences are not controlled by tides; (2) they appear to occur in locations where there is evidence of structural weaknesses; (3) the relative abundances of small and large quakes (b-values) are similar, suggesting similar mechanisms; and (4) even the levels of activity may be close. The shallow moonquakes may be quite comparable in nature to intraplate earthquakes, and they may be of similar origin.

INTRODUCTION

The present-day moon is often regarded as a planet of extreme quiescence. The surface of the moon is nearly completely devoid of any evidence of tectonic activity during the last three billion years. The observed seismicity also appears to be very low compared with the earth (Latham *et al.*, 1972). Studies based on abundant Apollo lunar rock samples have left an impression on many of us that all significant activity in the moon occurred only during the first one and a half billion years following its creation. Yet, the measured heat-flow values are comparable to those of the earth (Langseth *et al.*, 1976), indicating that a large amount of potential tectonic energy is still stored in the lunar interior. The heat flow values certainly do not suggest that the moon is a cold, dead planet.

I have therefore re-examined various properties of moonquakes to see how they compare with earthquakes and to consider their significance in terms of the present-day tectonics of the lunar interior. This short paper presents some comparisons of moonquakes with earthquakes, and shows how I conclude that the present-day dynamics of the lunar interior may be quite comparable to the tectonics of the earth if plate boundaries are excluded.

MOONQUAKES

There are three types of moonquakes large enough to be observable at great distances. The most numerous are deep moonquakes. About two thousand of them were detected annually during the Apollo lunar seismic experiment. They occur at depths about halfway to the center of the moon. The most striking characteristic of deep moonquakes is the regularity in time of their occurrence, showing a clear correlation with the tidal periodicity of the moon. No such quakes have been identified in the earth; a reason may be that terrestrial seismographs are not sufficiently sensitive to detect them even if they exist in the earth. Deep moonquakes are very small. Despite their large numbers, the total energy released by them is quite insignificant in comparison with that of earthquakes (Lammlein *et al.*, 1974). Deep moonquakes appear to represent merely a process of storage and release of tidal energy without a significant release of tectonic energy (Nakamura, 1978; Koyama and Nakamura, 1980).

The next most abundant are moonquakes caused by impacts of meteoroids. About three hundred of them were observed yearly by the Apollo seismic network. They are obviously of external origin, and are in no way comparable to earthquakes.

The third type of moonquakes is the shallow moonquakes. They are the rarest—only four to five of them were detected yearly by the Apollo seismic network. However, they represent the most energetic sources in the moon, and account for most of the seismic energy released in the moon. They occur at depths generally shallower than about 100 km (Nakamura *et al.*, 1979), and appear to be the only moonquakes that may be related to earthquakes in their origin.

SHALLOW MOONQUAKES AND EARTHQUAKES

Tidal correlation

Shallow moonquakes occur quite randomly in time, as seen in Fig. 1. Unlike deep moonquakes, no clear correlation with the tidal cycle is observed. A strong tidal coupling would appear as periodicities of a month or seven and a half months. No such periodicities are observed for shallow moonquakes. Tidal periodicity of earthquake occurrence has been a subject of study by several investigators in the past in the hope of finding evidence for tidal triggering of earthquakes. The results, however, have been questionable at best (e.g., Knopoff, 1964). Tectonic quakes perhaps do not generally show any clear evidence of being triggered by tidal forces. In this respect, shallow moonquakes are similar to earthquakes.

Epicentral distribution

Though we detected only 28 shallow moonquakes, their epicenters appear to be distributed randomly on the lunar surface (Nakamura *et al.*, 1979). This is in contrast to the distribution of earthquakes, the overwhelming majority of which are concentrated within narrow seismic belts along plate boundaries. The relative

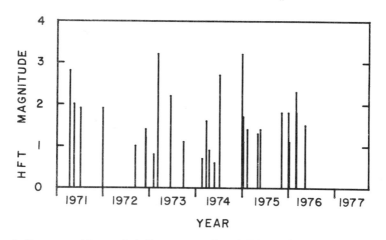

Fig. 1. Occurrence history of shallow moonquakes. HFT magnitude is defined by log (SPZ envelope amplitude in DU reduced to 60° distance). It is estimated to be approximately 1½ less than the body-wave magnitude as estimated by Goins (see text). HFT (high-frequency teleseismic) is the term previously used for this group of moonquakes. The period before the installation of Apollo 14 seismic station (February, 1971) has been excluded because shallow moonquakes could not be identified on Apollo 12 station records, which lacked a short-period component.

movements of plates are believed to be the direct cause of these earthquakes. On the other hand, there is no indication of current lunar plate movements, and little evidence of seismic belts on the moon. One might, therefore, conclude that shallow moonquakes are quite dissimilar to earthquakes.

However, there are earthquakes that occur away from plate boundaries on the earth. While not numerous, some of them can be quite large. Some similarities can be recognized between the distributions of these intraplate earthquakes and of shallow moonquakes.

Sites of intraplate seismicity can be identified as zones of pre-existing weaknesses in lithospheric plates on the earth (Sykes, 1978). Although epicenters of shallow moonquakes are not very accurately determined, their distribution seems to be correlated with the distribution of ancient impact basins (Nakamura *et al.,* 1979), which may also represent zones of weakness.

Observations of quakes associated with ancient impact structures are not confined to the moon. For example, Leblanc *et al.* (1973) report a series of St. Lawrence Valley earthquakes in the Canadian shield associated with an impact structure of middle Ordovician to late Devonian age, i.e., some 300 to 400 million years old. They suggest that these earthquakes manifest yielding of the weakened crust under the impact crater to the post-glacial strain field acting over a broader region. The lunar quakes may represent yielding to a similarly broad strain field due to some other cause such as cooling of the lunar interior.

Anomalies in electrical conductivity are found associated with impact basins on the moon (Schubert *et al.,* 1974; Sonett *et al.,* 1974; Dyal and Daily, 1979).

Such anomalies represent differing physical properties between impact basins and surrounding areas, and may suggest the presence of different thermal conductivity and temperature regimes. It is reasonable to expect that concentrations of tectonic stresses in such heterogeneous regions would cause moonquakes to occur there. The association of intraplate seismicity with anomalous distribution of electrical conductivity is also observed on the earth (e.g., Lilley, 1975).

b-value

One way to characterize a given population of earthquakes is to compare the relative abundance of large and small earthquakes. This relationship is normally plotted as a magnitude-frequency diagram. Figure 2 shows this relationship for the shallow moonquakes. The relative abundance is usually expressed by the slope of such a curve, called its "b-value." The b-value of shallow moonquakes, as determined by the maximum likelihood estimate of Aki (1965), is 0.55. (Events smaller than HFT magnitude 1.0 have been excluded from this determination because it is likely that many small events were not detected.) In comparison, deep moonquakes give b-values generally in excess of 1.5 (Lammlein et al., 1974), while most earthquakes give values close to 1.0 (Matuzawa, 1964). The low b-value for the shallow moonquakes means that large moonquakes are proportionately more abundant than in normal earthquakes.

Interestingly, low b-values of 0.5 to 0.6 are also found for earthquakes occurring in continental interiors (Matuzawa, 1964; Lammlein et al., 1971). Although the physical significance of b-values is not well understood, the common and signif-

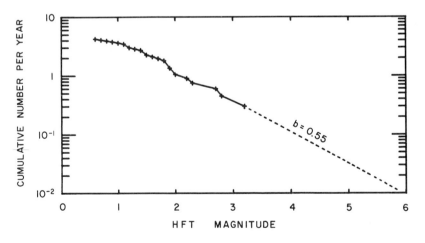

Fig. 2. Cumulative frequency-magnitude distribution of detected shallow moonquakes. The ordinate of each point represents the number of events N observed per year to have magnitude M equal to or greater than the value represented by the abscissa. The b-value is defined by $-d(\log N)/dm$, i.e., the slope of the distribution curve. The dashed line is an extrapolation of the slope representing $b = 0.55$.

icantly lower than normal b-values for shallow moonquakes and intraplate earthquakes suggest a possibility of similar processes that cause these two groups of
quakes.

Energy release

The level of shallow moonquake activity, i.e., the average amount of seismic
energy released by all shallow moonquakes, is not easy to estimate because of
the short duration of observation, as will be explained below. Values such as
2×10^{17} ergs/year given by Goins *et al*. (1980), I believe, are a gross underestimation because of this.

Let us consider a population of quakes, with a set of observed quakes representing a sample from the population. If large quakes are proportionately abundant, as in the case of shallow moonquakes, most of the energy released by the
entire population is attributable to the largest quake in the population. The difficulty in estimation arises when the size of the sample (observations) is so small
that one cannot reasonably estimate the magnitude of the largest quake in the
population. Then, the energy release estimated from the largest observed quakes
is quite likely to be well below the true rate of energy release over a long term.

The latest estimate (Neal Goins, pers. comm., 1979) assigns a body-wave magnitude of 4.8 to the largest observed shallow moonquake. If this were the largest
ever expected, the average energy release rate would be about 2×10^{17} ergs/year
for the entire moon. However, there is no reason to believe that we have observed, in the short duration of our observation, the largest shallow moonquakes,
which are expected to occur much less frequently than once every eight years.
The magnitude-frequency relation of Fig. 2, in fact, shows no indication of the
line deviating from the linear trend at large magnitude. This means that larger
shallow moonquakes are indeed expected in a longer period of observation.

Thus, it is quite reasonable to expect that the magnitude-frequency relation of
Fig. 2 can be extrapolated further to larger magnitude, following the dashed line
of Fig. 2. How much extrapolation is justifiable cannot be determined from the
presently available data. However, if we extrapolate this relation to HFT magnitudes of 4.0, 5.0 and 6.0 (equivalent to body-wave magnitudes of roughly 5½
to 7½), for example, we obtain estimated energy release rates of 10^{18}, 3×10^{19}
and 10^{21} ergs/year, respectively, for the entire moon. Shallow moonquakes of
such magnitudes are expected only once in every 9, 32 and 115 years, respectively.

These estimated long-term energy release rates are still very small fractions of
the average annual energy release by all the earthquakes, which is estimated to
be 10^{24} to 10^{25} ergs/year (Stacy, 1977). Of this amount, intraplate earthquakes
account for about 0.2% (Gutenberg and Richter, 1949); i.e., about 10^{22} ergs/year.

To compare these values with the above estimates for the moon may not be a
simple matter. Besides the simple difference in size of these two planets, the
process of how planetary bodies of different sizes envolve need be considered,
which is beyond the scope of this paper. As a rough comparison, if one takes

into account only the volume ratio of 49.2 to 1 and surface ratio of 13.5 to 1 between the earth and the moon, the energy release rate of intraplate earthquakes is roughly equivalent to 10^{20} to 10^{21} ergs/year for either volume or surface equivalent to the moon. This is quite comparable to the above estimates for the shallow moonquakes.

This comparison, of course, is valid only under the assumption that the magnitude-frequency relation of Fig. 2 can be extrapolated to larger magnitude. To obtain an energy release rate comparable to intraplate earthquakes, one needs to extrapolate the shallow moonquake frequency curve to HFT magnitude of 5½ to 6, or equivalent body-wave magnitude of 7 to 7½. The expected frequency of occurrence of such moonquakes is once in 50 to 100 years. Such intervals are also comparable to those of rare, large intraplate earthquakes.

The largest expected shallow moonquakes may have magnitudes even greater than those used in the above examples. Since the lunar lithosphere is thicker than the terrestrial lithosphere, the former may be able to accumulate larger stress than the latter. As a consequence, much larger amounts of energy may be stored in the lunar lithosphere than in the terrestrial lithosphere for infrequent release by rare, large moonquakes.

CONCLUDING REMARKS

The above comparisons of moonquakes with earthquakes thus show that shallow moonquakes are quite similar in many respects to intraplate earthquakes. To summarize the comparisons: (1) their occurrences are apparently not tidally controlled; (2) their locations in relation to structural weaknesses and other geological features are similar; (3) the b-values are about the same, suggesting similar nature of source mechanisms; and (4) even the estimated long-term energy release rates are about the same.

These similarities certainly are not solid evidence that shallow moonquakes are caused in exactly the same way as intraplate earthquakes. Rather, they may be viewed only as circumstantial evidence. However, there are enough similarities that we can reasonably accept shallow moonquakes as the lunar equivalent of intraplate earthquakes.

There are some important implications of this observation. First, for the moon: Even though shallow moonquakes appear to be associated with ancient impact basins (Nakamura *et al.*, 1979), the estimated amount of energy released by them is far greater than expected from simple isostatic adjustments of original disturbances caused by impacts. The energy of shallow moonquakes, therefore, must be supplied by other, contemporary tectonic sources.

Second, for the earth: Recently, there has been increased interest in interpreting intraplate seismicity in terms of plate tectonics (e.g., Sykes, 1978). To the contrary, shallow moonquakes have shown us that it is possible for a significant number of tectonic quakes to occur without apparent plate movement. The lunar data thus suggest that the plate tectonic interpretation of intraplate seismicity may not be a valid one.

Acknowledgments—I would like to thank several staff members of the Galveston Geophysics Laboratory, including Drs. Cliff Frohlich, Abou-Bakr Ibrahim and Junji Koyama, for stimulating discussions during the course of this study, and especially Drs. Gary Latham, Jim Dorman and Cliff Frohlich for reviewing the manuscript and offering constructive comments. This research was supported by NASA Grant NSG-7418. University of Texas Marine Science Institute Contribution No. 436, Galveston Geophysics Laboratory.

REFERENCES

Aki K. (1965) Maximum likelihood estimate of b in the formula $\log N = a - bM$ and its confidence limits. *Bull. Earthquake Res. Inst. Univ. Tokyo* **43**, 237–239.

Dyal P. and Daily W. D. (1979) Electrical conductivity anomalies associated with circular lunar maria. *Proc. Lunar Planet. Sci. Conf. 10th*, p. 2291–2297.

Goins N. R., Dainty A. M., and Toksöz M. N. (1980) Seismic energy release of the moon (abstract). In *Lunar and Planetary Science XI*, p. 336–338. Lunar and Planetary Institute, Houston.

Gutenberg B. and Richter C. F. (1949) *Seismicity of the Earth*. Princeton Univ., Princeton, New Jersey. 273 pp.

Koyama J. and Nakamura Y. (1980) Focal mechanism of deep moonquakes. *Proc. Lunar Planet. Sci. Conf. 11th*. This volume.

Knopoff L. (1964) Earth tides as a triggering mechanism for earthquakes. *Bull. Seismol. Soc. Amer.* **54**, 1865–1870.

Lammlein D. R., Latham G. V., Dorman J., Nakamura Y., and Ewing M. (1974) Lunar seismicity, structure, and tectonics. *Rev. Geophys. Space Phys.* **12**, 1–21.

Lammlein D. R., Sbar M. L., and Dorman J. (1971) A microearthquake reconnaissance of southeastern Missouri and western Tennessee. *Bull. Seismol. Soc. Amer.* **61**, 1705–1716.

Langseth M. G., Keihm S. J., and Peters K. (1976) Revised lunar heat-flow values. *Proc. Lunar Sci. Conf. 7th*, p. 3143–3171.

Latham G., Ewing M., Dorman J., Lammlein D., Press F., Toksöz N., Sutton G., Duennebier F., and Nakamura Y. (1972) Moonquakes and lunar tectonism results from Apollo passive seismic experiment. *Proc. Lunar Sci. Conf. 3rd*, p. 2519–2526.

Leblanc G., Stevens A. E., and Wetmiller R. J. (1973) A microearthquake survey of the St. Lawrence Valley near La Malbaie, Quebec. *Can. J. Earth Sci.* **10**, 42–53.

Lilley F. E. M. (1975) Electrical conductivity anomalies and continental seismicity in Australia. *Nature* **257**, 381–382.

Matuzawa T. (1964) *Study of Earthquakes*. Uno Shoten, Tokyo. 213 pp.

Nakamura Y. (1978) A_1 moonquakes: Source distribution and mechanism. *Proc. Lunar Planet. Sci. Conf. 9th*, p. 3589–3607.

Nakamura Y., Latham G. V., Dorman H. J., Ibrahim A. K., Koyama J., and Horvath P. (1979) Shallow moonquakes: Depth, distribution and implications as to the present state of the lunar interior. *Proc. Lunar Planet. Sci. Conf. 10th*, p. 2299–2309.

Schubert G., Smith B. F., Sonett C. P., Colburn D. S., and Schwartz K. (1974) Polarized magnetic field fluctuations at the Apollo 15 site: Possible regional influence on lunar induction. *Science* **183**, 1194–1197.

Sonett C. P., Smith B. F., Schubert G., Colburn D. S., and Schwartz K. (1974) Polarized electromagnetic response of the moon. *Proc. Lunar Sci. Conf. 5th*, p. 3073–3089.

Stacey F. D. (1977) *Physics of the Earth*. 2nd ed. Wiley, N.Y. 414 pp.

Sykes L. R. (1978) Intraplate seismicity, reactivation of preexisting zones of weakness, alkaline magnetism, and other tectonism post-dating continental fragmentation. *Rev. Geophys. Space Phys.* **16**, 621–688.

Proc. Lunar Planet. Sci. Conf. 11th (1980), p. 1855–1865.
Printed in the United States of America

Focal mechanism of deep moonquakes

Junji Koyama* and Yosio Nakamura

Marine Science Institute, Geophysics Laboratory, University of Texas, Galveston, Texas 77550

Abstract—To elucidate the focal mechanism of deep moonquakes, we analyzed S-wave polarizations of deep moonquake signals from the A_1 source region. At station 12, where the available data are of the highest quality, the variation of polarization angle with the anomalistic phase of the moon indicates agreement with that expected from the focal mechanism model of Nakamura. Focal mechanism solutions were derived for eight A_1 moonquakes assuming that moonquakes are, like earthquakes, caused by a shear fracture on a fault plane. The mechanism solutions generally indicate a nearly horizontal or almost vertical faulting. The slip directions estimated from the solutions are different from one another, again suggesting variation of focal mechanisms as a function of the tidal phase of the moon.

INTRODUCTION

Two distinct types of moonquakes which have been identified so far are deep moonquakes (Latham *et al.*, 1971) and shallow moonquakes (Nakamura *et al.*, 1974). Deep moonquakes, which are numerous but small, exhibit several characteristic features: First, they are concentrated mainly at depths between 800 and 1000 km, roughly halfway to the center of the moon. Second, they occur repeatedly in each of a number of distinct source regions, producing groups of seismic signals with nearly identical waveforms. Third, they are closely synchronized with the periodic tides raised on the moon by the earth and the sun. Shallow moonquakes, in contrast, are sparse but relatively large. They are believed to occur at depths shallower than 100 km and show no apparent periodicity in their activity (Nakamura *et al.*, 1979).

The most active group of deep moonquakes, designated A_1, includes 187 events detected between 1969 and 1977, among which are many of the largest deep moonquakes ever observed. Some of the A_1 events during 1972 to 1974 showed reversed polarity of seismic signals (Lammlein, 1977; Nakamura, 1978). Toksöz *et al.* (1977) and Cheng and Toksöz (1978) compared the calculated tidal stresses in a moon model with the occurrence of A_1 moonquakes, and concluded that the presence of a constant ambient tectonic stress is necessary to explain the observed reversal in the polarities of A_1 moonquake signals.

---

* Present address: Geophysical Institute, Faculty of Science, Tohoku University, Sendai 980, Japan

On the other hand, Nakamura (1978) analyzed the variation of amplitude ratios of P and S waves generated by A_1 moonquakes and showed that the mechanism at the source is not just a reversal of motion as suggested by the Toksöz model, but a rotation of slip vector controlled by the shifting tidal stress field. He consequently suggested that deep moonquakes are the manifestation of mere storage and release of tidal energy within tectonically-inactive lunar interior.

In order to learn more about the focal mechanism of deep moonquakes, we looked for further clues in the properties of S-waves generated by deep moonquakes. We examined polarizations of S waves because they are directly related to the focal mechanism, and are an independent source of information relative to both P wave polarities and amplitude ratios of P and S waves previously analyzed (Lammlein, 1977; Chen and Toksöz, 1978; Nakamura, 1978).

METHOD OF ANALYSIS

Focal mechanisms of earthquakes are usually studied by analyzing polarities of P-wave first motion and/or directions of S-wave initial motions (Honda, 1962). However, the complexity of lunar seismic signals due to intensive scattering and the paucity of seismic stations on the moon preclude the application of these conventional analysis techniques. The method used in this study is to evaluate the directions of particle motion of S waves in a statistical manner.

Let the observed seismic signal around the S-wave onset time be given as a set of discrete time series x_i and y_i on two horizontal orthogonal coordinates (x, y) of the instrument. (The Apollo lunar seismometers use left-handed coordinate system with the positive z-axis pointing upward. The orientations of horizontal axes are given in the caption for Table 1). The signals can also be expressed in another set of time series X_i and Y_i on a horizontal coordinate system (X, Y) which is rotated by an angle α with respect to (x, y). The correlation coefficient r between the components X_i and Y_i is defined by

$$r^2 = E(X \cdot Y)^2 / E(X^2)E(Y^2), \tag{1}$$

where $E(X)$ denotes the mathematical expectation of X. It is assumed that the means of X_i and Y_i are zero; i.e., dc components have been removed from the signal. The correlation coefficient r vanishes when X and Y axes are the major and minor axes of the orbital motions of S waves. Then, the angle α is given as a solution of $E(X \cdot Y) = 0$, i.e.,

$$\tan \alpha = \{ D \pm \sqrt{D^2 + S_{xy}^2} \} / S_{xy}, \tag{2}$$
$$\text{where } S_{xy} = E(x \cdot y), D = (S_{yy} - S_{xx})/2, S_{yy} = E(y^2), \text{ and } S_{xx} = E(x^2).$$

The plus sign in the numerator corresponds to a rotation in which the X axis coincides with the major axis and the Y axis with the minor axis of the orbital motions of S waves. The use of minus sign places the Y axis on the major, and the X axis on the minor axis. We adopt the former convention in the analysis that follows. Then the angle α becomes the direction of the major axis measured clockwise from the x axis. The ratio of $E(X^2)$ to $E(Y^2)$ gives a measure of ellipticity of the particle orbital motions. We will instead define a linearity coefficient f of the particle motion by

$$f = 1 - \sqrt{E(X^2)/E(Y^2)}, \tag{3}$$

which leads to

$$f = 1 - \sqrt{(S_{yy}\gamma^2 + 2S_{xy}\gamma + S_{xx})/(S_{xx}\gamma^2 - 2S_{xy}\gamma + S_{yy})}, \tag{4}$$

where $\gamma = (D - \sqrt{D^2 + S_{xy}^2})/S_{xy}$.

DIRECTION OF S-WAVE ORBITAL MOTIONS

The present analysis was based on A_1 deep moonquakes detected through 1977, the end of the passive seismic experiment of the Apollo project. First, for each seismogram of these moonquakes, a 7.2 second data secton (48 data points) containing the initial S-wave arrival was selected for each of the two horizontal components as indicated by boxed sections in the sample seismograms of Fig. 1. Seismic signals recorded at stations 12 and 15 generally show well defined S-wave

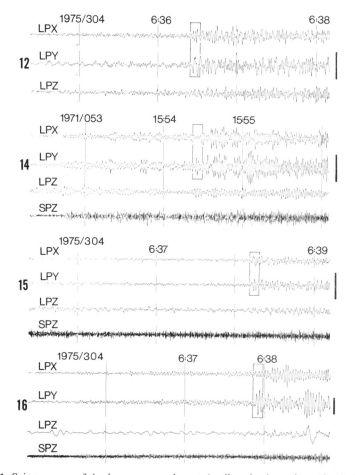

Fig. 1. Seismograms of A_1 deep moonquakes at Apollo seismic stations 12, 14, 15 and 16. LPZ and LPY are long-period horizontal components, and LPZ and SPZ are long- and short-period vertical components, respectively. Signals around S-wave arrivals used for the analysis are indicated in the figure. Except for those of station 14, all seismograms were recorded in the flat-mode response of the seismometers. Vertical bars indicate amplitude of 20 DU, where DU (digital unit) is the unit of signal digitization performed on the moon.

Table 1. Directions of S-wave orbital motions and linearity coefficients. The directions of orbital motions are measured clockwise from the positive x direction of each instrument. The positive x directions are due south, due north, due north, and N25.5°W for stations 12, 14, 15, and 16, respectively.

			Station							
			12		14		15		16	
Year	Day	Time[1]	α	f	α	f	α	f	α	f
1970	9	0203	116°	0.67						
	35	2041	111	.60						
	38	133	104	.55						
	63	424	124	.59						
	64	2344	106	.56						
	91	953	108	.65						
	116	1431	124	.65						
	145	1214	111	.56						
	171	032	134	.54						
	173	1158	115	.56						
	201	1144	114	.65						
	226	1827	125	.57						
	252	1212	117	.52						
	280	549	129	.61						
	284	938	123	.57						
	307	1716	126	.55						
	361	2036	130	.58						
	363	1428	111	.62						
	365	1543	114	.63						
1971	28	1500	115	.61						
	51	1508	138	.60						
	53	1554	129	.63	88°	0.68				
	56	1217	93	.50	110	.53				
	80	1631	130	.61	95	.74				
	82	2115	128	.53	92	.54				
	85	0027			92	.52				
	107	613			88	.50				
	137	537	141	.69	91	.53				
	160	1018	138	.61						
	163	640	120	.65	90	.84				
	187	1655	118	.54	86	.60				
	216	716	133	.60	90	.55				
	218	722	124	.54	89	.82				
	245	2314	127	.57	85	.54	89°	0.70		
	273	2034			92	.53				
1972	17	414	148	.62			88	.53		
	44	2126			91	.72	96	.62		
	164	2349							80°	0.65
1973	20	007							85	.85
	50	1257							82	.72
	60	717	105	.51					100	.62

[1]approximate arrival time of s-wave at station 12

Table 1. (*Continued*)

Year	Day	Time	Station 12 α	12 f	14 α	14 f	15 α	15 f	16 α	16 f
1973	88	640			91	.86	123	.56		
	116	1058							86	.73
	127	1908	112	.54	80	.55				
	156	1112	114	.53	84	.67			85	.84
	201	1903	113	.58						
	229	1712	90	.63						
	241	611			89	.56				
	253	1100							76	.67
	270	1747			87°	0.67			101°	0.60
	273	414			94	.81	101°	0.55		
	303	102			92	.76	121	.50		
	321	731	120°	0.67	91	.55			84	.85
	330	236			91	.67				
1974	124	2213					108	.65	81	.85
	127	1949							85	.80
	151	1150	153	.56	91	.59	112	.52	85	.88
	178	2011			89	.57			77	.79
	343	2349			91	.77	119	.65	80	.71
1975	86	1849	105	.52			60	.63	88	.84
	113	1155	119	.54			89	.52	84	.87
	140	1730	118	.57					89	.81
	168	2230	106	.58					88	.85
	250	952	107	.66					87	.74
	276	624	119	.52					88	.72
	278	747	107	.52					93	.56
	304	636	116	.52					86	.71
	331	612							87	.75
1976	20	930	126	.54					91	.66
	76	006			87	.54			94	.71
	102	1604							66	.65
	157	052			94	.74				
	184	122			92	.81			90	.76
	210	1755			90	.84			87	.70
	238	1427							87	.72
	264	1653							83	.72
	266	2019							89	.71
	293	123							88	.76
	294	1937							84	.65
	297	136							87	.67
	320	1500							82	.69
	324	2241							80	.56
	351	1318			91	.78			87	.63
1977	12	700							95	.54
	63	1221			103	.56			85	.67
	65	1019			89	.57			87	.61
	91	1739							88	.63

Table 1. (Continued)

| | | | \multicolumn{8}{c}{Station} | | | | | | | |
| | | | 12 | | 14 | | 15 | | 16 | |
Year	Day	Time	α	f	α	f	α	f	α	f
1977	92	1548	116	.60					80	.78
	118	1330							85	.83
	147	1206							83	.68
	149	1137					90	.70	88	.80
	174	222			96	.67				
	175	1449			85	.75			88	.82
	176	2332							90	.74
	202	1139							86	.83
	229	1651			90	.93			87	.80
	254	1047			88	.68			84	.79
	256	2309			89	.78	97	.56	85	.82

arrivals, while those at stations 14 and 16 are not so well defined but of a ringing nature caused presumably by surface layering at these sites. The time window included 3 to 5 cycles of seismic waves. For each selected section, the angle α of the particle orbital motions of S waves as defined by Eq. (2) and the linearity coefficient f as defined by Eq. (4) were calculated. Table 1 lists the computed values. A blank space indicates either that there are no data, signal amplitudes are too small for the analysis, or that the linearity coefficient is less than 0.5. The linearity coefficient, f, greater than 0.5 means that more than 80% of the wave energy is in the direction of the major axis. Events for which reliable directions could not be determined for any station are not included in the table. The data cover a full period of the six-year tidal periodicity for stations 12 and 14, and successively shorter periods for the later missions, being reduced to 5 years for station 16.

Figure 2 shows directions of S-wave orbital motions at station 12 plotted against the anomalistic phase of the moon at the time of occurrence of each A_1 event. No systematic difference can be found between the flat-mode and peaked-mode data, though the latter data scatters more.

The model by Toksöz et al. (1977) involves a simple reversal of slip directions, and thus predicts two discrete values of S-wave polarization angles differing by 180° from each other for all the A_1 moonquakes. Since these two polarization angles would give exactly the same angle α of the S-wave orbital motion, there should be a single constant value of α in Fig. 2. The model by Nakamura (1978), on the other hand, predicts strong dependence of S-wave polarization on the tidal phase of the moon as indicated by the solid curves in Fig. 2. The agreement between the observation and the theory is encouraging. The necessary adjustment of the parameters may indicate that the true coordinates of the A_1 source region are slightly east of the assumed coordinates (14.4°S, 34.0°W), or that local, lateral velocity heterogeneities are affecting the direction of arrivals of seismic rays.

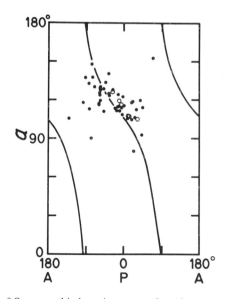

Fig. 2. Directions of S-wave orbital motions at station 12 measured from the positive x coordinate of the instrument. They are plotted against anomalistic phase of the moon at the time of occurrence of each event. A and P represent apogee and perigee, respectively. Open circles indicate those obtained from seismograms in flat-mode response, and solid circles in peaked-mode response. Solid curves in the figure show theoretical directions of S-wave orbital motions derived from the model of focal mechanism for A_1 moonquakes by Nakamura (1978). The adjusted parameters are $\theta = 35°$, $\psi = 100°$ and $A_z = 200°$, where θ is the angle between the normal to the fault plane and the line connecting the focus with the station, ψ is the azimuth of the slip direction at perigee relative to the projection of the source-station line on the fault plane and A_z is the azimuth of the epicenter looking from the station. The corresponding parameters of the original model by Nakamura (1978) are $\theta = 28°$, $\psi = 85°$ and $A_z = 222°$.

For stations 14, 15, and 16 (Fig. 3), the agreement between the observations and the theory is poor. The nearly constant directions of S-wave polarization at stations 14 and 16, if taken by themselves, may suggest a simple reversal of slip directions as indicated by the model of Toksöz *et al.* (1977). However, the constant polarization angle at station 14 is clearly inconsistent with those at station 12, which is very close to station 14. Most of the directions at stations 14 and 16 are about 90°, which simply means the initial S-wave amplitudes on the y component are always much larger than those on x components at these stations.

There are some uncertainties about the response of horizontal-component seismometers at these stations. It has long been noted that the y component always shows larger amplitudes than the x component regardless of the source type and source azimuth. There is no indication that actual instrumental gains are different. The discrepancy may be caused by local structural heterogeneities and/or by the way the seismometers are installed on the lunar surface. Since no one knows for sure what causes this discrepancy and especially what would be its effect on initial arrivals such as those used in the present study, we decided not to

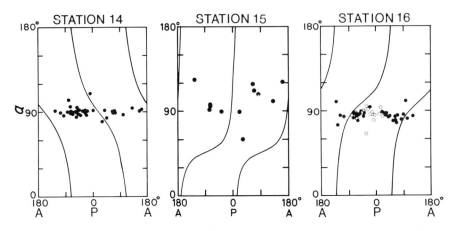

Fig. 3. Directions of S-wave orbital motions at stations 14, 15 and 16. See the caption for Fig. 2 for further explanation.

"correct" for the amplitude differences. Besides, an inspection of sample seismograms in Fig. 1 clearly shows that the difference in the initial S-wave amplitude between x and y components are much more than those of the remainder of the seismograms.

Additionally, for station 14, there is an uncertainty in determining when the shear wave actually arrives. The seismometers at this station could not be operated stably in the flat (wide-band) mode. As a result, there is a possibility that the phase we used is not a real, pure S arrival. Nevertheless, the concentration of the observed polarizations at these stations near 90°, where the theoretical curves have inflections, is encouraging. The large scatter of the data at station 15 is most likely to be due to very small signal amplitudes normally recorded at this station.

FOCAL MECHANISM SOLUTIONS OF A_1 MOONQUAKES

If we assume that moonquakes, like earthquakes, are caused by a shear fracture on a fault plane, we can determine the orientation of the equivalent force system to cause a shear fracture from three observations of polarization angles (Hirasawa, 1966). Three determinations of polarization angles have been made for each of eight events out of 187 A_1 moonquakes. Figure 4 shows the focal mechanism solutions derived for these events. Three solutions exist for most events; however, some of them can be rejected by taking into account the relative polarity relations among seismograms at different stations as given by Nakamura (1978). Since the polarization angles determined in this study do not contain information as to the polarity of S-wave initial motions, it is impossible to tell which of the quadrants represents compression, or rarefaction, for P-wave initial motions.

For those events for which station 14 data are not used, at least one acceptable solution shows a focal mechanism in which one of the nodal planes is nearly vertical and the other horizontal. A typical example is the one for the event of 1975, day 86 in Fig. 4. Although three observations of polarization angles are just enough to determine the orientation of the force system, and despite the fact that polarization angles at some stations may not be as reliable as others and that some of the data may be contaminated by scattered signals, the solutions are fairly close to those postulated by Nakamura (1978) from amplitude ratios of P and S waves. Nakamura (1978) chose the horizontal nodal plane to be the fault plane from the analysis of the relative hypocenter distribution of A_1 moonquakes.

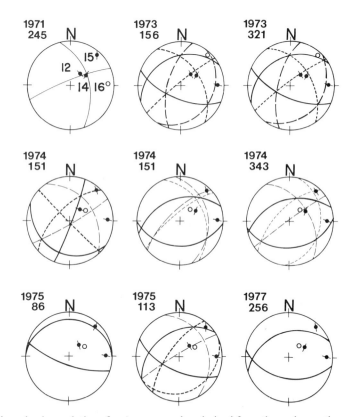

Fig. 4. Focal mechanism solutions for A_1 moonquakes derived from three observations of polarization angles of S-waves. Thick curves indicate sets of two nodal planes of P waves, the fault plane and auxiliary plane, that satisfy the relative polarity relations among seismograms at different stations given in Nakamura (1978). Thin curves indicate those which do not satisfy the polarity relation. They are plotted on the upper hemisphere using equal area net. Solid dots indicate stations used for the solution, each with a line indicating the S-wave polarization angle. Open circles simply indicate locations of other stations not used for the solution. Whether a given quadrant is compressional or dilatational cannot be determined in this solution as explained in the text.

DISCUSSION AND CONCLUSIONS

Two models for the focal mechanism of deep moonquakes have been proposed by Toksöz *et al.* (1977) and Nakamura (1978). The former postulates a combination of the cyclic tidal stresses and a constant ambient tectonics stress to produce moonquakes when tidal unloading allows slippage to occur in one of two opposite directions. The latter postulates a slippage on a horizontal fault plane in a direction that rotates with the tidal phase of the moon. Although both of the models assume a priori a shear fracture as the focal mechanism, these two models lead to quite different concepts of the focal process of deep moonquakes; one is a fracture involving tectonic stress as in earthquakes, and the other is a fracture due to fatigue by cyclic tidal loading.

To clarify this situation, we have analyzed S-wave orbital motions generated by A_1 deep moonquakes. The observed directions of S-wave orbital motions at station 12 indicate time-dependent characteristics of focal mechanisms for A_1 moonquakes in agreement with the results by Nakamura (1978). Although the directions of S-wave orbital motions at other stations may be less reliable than those at station 12, the polarizations of S waves generally agree with those expected for a shear fracture on a fault.

Focal mechanism solutions were derived for eight A_1 events, assuming that moonquakes, like earthquakes, are caused by shear fracture. Most of the solutions show a mechanism of nearly horizontal or almost vertical faulting, which is in agreement with the one postulated by Nakamura (1978) from the amplitude ratio of P and S waves. Because of the paucity of observations, the mechanism solutions are not so constrained as to be unique; however, the mechanisms are different from one another, again suggesting temporal variation of the slip direction on the fault. The variation sometimes results in a reversal of polarity of A_1 seismograms at certain stations.

Our current thinking on the focal mechanism of deep moonquakes may be summarized as follows. Toksöz *et al.* (1977) first postulated a mechanism model just to explain the polarity reversal of seismograms at certain stations. Later, Nakamura (1978) showed that a different model was needed to explain several other aspects of the observed data. These two models are not necessarily mutually exclusive in all respects. In fact, many features of the models are common, including an existence of nearly horizontal fault planes and the dominant role played by the tidal stress. Nakamura's observation indicates that it is not necessary to have accumulated tectonic stress for the deep moonquakes to occur there. A small amount of tectonic stress may be present, but it cannot play a dominant role without contradicting the observed amplitude variations. Thus we have a model in which the tidal effects play a major role both in supplying the necessary strain energy and some mechanisms to initiate moonquakes. The study reported here gives a support to such a model by showing the observed shear wave polarizations are also in agreement.

The maximum shear stresses induced by tides are less than 1 bar at all depths of a perfectly elastic moon with no small scale heterogeneities (Toksöz *et al.,*

1977; Lammlein, 1977). The magnitude of shear stress in such a moon is dependent on the shear modulus profile in its interior. However, the shear stress reaches a maximum around the depths of deep moonquake sources irrespective of assumed lunar structure. Toksöz *et al.* (1977) estimated the magnitude of ambient tectonic stress to be about 0.5 bar for the polarity reversal of seismograms of A_1 moonquakes. These stress levels are far below the fracture strength of most materials. Fatigue cracking is more likely as the focal process of deep moonquakes under low-stress, cyclic tidal loading. Though calculated tidal stresses based on elastic moon models do not show a sign reversal in any stress components, fatigue cracks are non-linear processes and it is conceivable that polarity reversals in A_1 moonquake seismograms do represent such a process. Some suggestions come from experimental studies of acoustic emissions. Acoustic emissions are observed for rock samples under uniaxial load not only during loading but also during unloading of the sample (Khair, 1977). The rotation of slip directions may be a result of similar non-linear processes under varying tidal stresses.

Acknowledgments—We would like to thank several staff members of the Galveston Geophysics Laboratory, including Drs. Cliff Frohlich, James Dorman and Gary Latham for stimulating discussions during the course of this study and for reviewing the manuscript. This research was supported by NASA Grant NSG-7418, and is University of Texas Marine Science Institute Contribution No. 437, Galveston Geophysics Laboratory.

REFERENCES

Cheng C. H. and Toksöz M. N. (1978) Tidal stresses in the moon. *J. Geophys. Res.* **83,** 845–855.
Hirasawa T. (1966) A least squares method for the focal mechanism determination from S wave data; Part 1. *Bull. Earthq. Res. Inst., Tokyo Univ.,* **44,** 901–918.
Honda H. (1962) Earthquake mechanism and seismic waves. *Geophys. Notes, Tokyo Univ.* **15** Suppl., 1–97.
Khair A. W. (1977) A study of acoustic emission during laboratory tests on Tennessee sandstone. In *Proc. 1st Conf. Acoustic Emission/Microseismic Activity* (H. R. Hardy and F. W. Leighton, eds.), p. 57–86. Trans Tech Publ., Clausthal, Germany.
Lammlein D. R. (1977) Lunar seismicity and tectonics. *Phys. Earth Planet. Inter.* **14,** 224–273.
Latham G., Ewing M., Press F., Sutton G., Dorman J., Nakamura Y., Toksöz N., Duennebier F., and Lammlein D. (1971) Passive seismic experiment. NASA SP-272, p. 133–161.
Nakamura Y. (1978) A_1 moonquakes: Source distribution and mechanism. *Proc. Lunar Planet. Sci. Conf. 9th,* p. 3589–3607.
Nakamura Y., Dorman J., Duennebier F., Ewing M., Lammlein D., and Latham G. (1974) High-frequency lunar teleseismic events. *Proc. Lunar Sci. Conf. 5th,* p. 2883–2890.
Nakamura Y., Latham G., Dorman J., Ibrahim A., Koyama J., and Horvath P. (1979) Shallow moonquakes: Depth, distribution and implications as to the present state of the lunar interior. *Proc. Lunar Planet. Sci. Conf. 10th,* p. 2299–2309.
Toksöz M. N., Goins N. R., and Cheng C. H. (1977) Moonquakes: Mechanisms and relation to tidal stresses. *Science* **196,** 979–981.

Proc. Lunar Planet. Sci. Conf. 11th (1980), p. 1867–1877.
Printed in the United States of America

Lunar polar wandering

S. K. Runcorn

Institute of Lunar & Planetary Sciences, School of Physics, The University, Newcastle upon Tyne, NE 1 7RU, England

Abstract—The modelling of lunar magnetic anomalies, observed by Apollo 15 and 16 subsatellite magnetometers, gives the direction of magnetization of areas of the lunar crust. On the hypothesis of an ancient lunar magnetic field generated by a core dynamo, pole positions are calculated. Polar wandering of 90° in the early moon is found and explained by changes in the moment of inertia due to early impact.

INTRODUCTION

Lunar palaeomagnetism has been studied under a handicap which was never experienced in similar studies on terrestrial rocks, the orientation of the Apollo samples when they were in bed rock is unknown. Consequently the subject has proceeded along different lines: emphasis being placed on palaeointensity studies, on the origin of the magnetizing field, and on the nature of the magnetizing process (Fuller, 1974). By contrast it was the varying palaeomagnetic directions of terrestrial samples of different geological age which excited controversy. They were initially interpreted in terms of a polar wandering curve (Creer *et al.*, 1954), and when different curves were found for rocks from different continents (Runcorn, 1956; Irving, 1956) they became the first quantitative evidence for continental drift. Thus the debate in lunar palaeomagnetism has hitherto not concerned the mechanics of the moon but the processes generating magnetic fields. However, in this paper I explore whether some recently determined lunar palaeomagnetic directions provide evidence for early reorientation of the moon with respect to its axis of rotation.

LUNAR MAGNETIC ANOMALIES

Recently a breakthrough has occurred through success in modelling the magnetic anomalies mapped by the subsatellites of the Apollo 15 and 16 missions in terms of sources of permanent magnetization at the lunar surface. Thirty-five such anomalies on the far side of the moon have been fitted by 35 dipoles (Hood *et al.*, 1978) and recently the Reiner γ anomaly has been modelled (Hood *et al.*, 1979). It has been shown that these anomalies, if modelled by a dipole, must be placed at a depth of about 50–70 km below the lunar surface, or if modelled by uniformly magnetized thin plates their horizontal dimensions must be of the order of 100–200 km. In fact, these models are equivalent for the magnetic field of a thin disc of radius a, observed at a height of h, well represented by a dipole, oriented in the same direction as the magnetization of the disc but placed $[\sqrt{a^2 + h^2} - h]$ below its centre.

The importance of these results is that if the magnetization of the sources is not to be greater than that of the returned samples, the sources must be uniformly magnetized on a horizontal scale of 100 km and in depth at least 1–10 km. The physical or chemical changes involved in the processes of magnetization are obscure. It has been plausibly argued that, as large meteorite impacts have occurred, especially in the early history of the highland crust, and as cometary impacts must have been frequent, the magnetizations might well have been produced by the physical processes, involving much energy, momentarily accompanying these impacts. We may reasonably choose to be agnostic about the nature of the magnetizing process, but it is a fact that both the crystalline rocks, i.e., the basalts of the mare basins, the anorthosites of the highlands, and the high grade breccias, deposited as layers of debris on the surface, would have acquired a thermoremanent magnetization (TRM) during their initial slow cooling at the surface if a steady ambient magnetic field had been present at that time. The impact process, involving transient stress, might momentarily have magnetized the crust in the direction of an amplified solar wind field, or other such transient field associated with the plasma generated by the explosion, but it is not plausible that this field could be uniform over horizontal distances of 100–200 km, and it is certain that it could not have been maintained unchanged over a sufficient period to produce a TRM. If it is to be maintained that the magnetization is due to physical or chemical processes attending the impact, the presence of a uniform background field is necessary. In this case, as in TRM, the directions of magnetization of the sources preserve the direction of this ambient field at the time of their magnetization.

In spite of the handicap of the unknown orientation of the lunar rocks, arguments have been presented that this magnetizing field was of internal origin. It was inferred to have been a dipole field generated by dynamo action in a lunar core. This hypothesis is consistent with the observed absence of a present day dipole moment arising from the magnetization of the lunar crust (Runcorn, 1975). It is also a simple way to explain the exponential decay of the palaeointensities with age (Collinson et al., 1977). There is insufficient data on the palaeomagnetic directions to allow a decisive test of this hypothesis. In this paper we will assume it to be true and follow up the consequences with a view to testing the hypothesis.

PALAEOMAGNETISM AND POLE POSITIONS

In the fluid metallic core of any planet, it was shown (Runcorn, 1954) that the Coriolis force is by many orders of magnitude a dominant term in the magnetohydrodynamic equations. It follows that the mean field is axially symmetrical and other general considerations, especially if the core is small, lead to the expectation of an axial dipole field (Runcorn, 1954). On this hypothesis the pole or axis of rotation can be calculated from the mean palaeomagnetic direction at a single site.

From the directions of crustal magnetization determined by Hood *et al.* (1978) from 35 far side anomalies, I calculated the corresponding pole positions and found they were grouped around antipodal mean positions along the east-west equatorial axis of the moon (Runcorn, 1978). The groupings can be explained on the hypothesis that the early moon possessed a dipole field generated in a core, if 90° of polar wandering since the date of magnetization is assumed to have occurred. The fact that the two clusters of poles are 180° apart may be explained either by reversals of the early lunar dynamo or by assuming that some of the sources are limited areas of breccia or igneous bodies which became magnetized along the lunar field to higher intensities than the surrounding crust, and that other sources are craters or regions of the crust which were demagnetized by impact. The age of these anomalies is not known with certainty, but is probably either that of the differentiation of the highlands or that of the great impacts and their ejecta—the Wasserburg cataclysm, between 4.4 and 4.1 b.y. ago. However, an alternative interpretation of these far side anomalies, which was originally put forward by Hood *et al.* (1978), is that they are directions of magnetization randomly directed in the present lunar equatorial plane. This was tested and rejected on a statistical argument (Runcorn, 1978) but the anomalies are too small for the determinations of magnetizations to be very accurate.

The Reiner γ anomaly, however, is larger and well determined, and from the successive satellite passages the contouring of the anomaly can be done in some detail. From earlier data Hood *et al.* (1979) obtain a direction of magnetization N 335°E and pointing upwards making an angle of 65° to the vertical. This corresponds to a pole position 58°N 171°E. However, a later set of data gives a direction of magnetization N 310°E, 60° to the vertical, corresponding to a pole position 35°N 180°E. The origin of the Reiner γ feature has raised a lively debate, and Schultz & Srnka (1980) suggest that it and other such swirls could be a relatively recent comet impact. If so, could the magnetization be implanted by the solar wind magnetic field, be compressed, and the field amplified by the comet as they suggest? A strong argument against this suggestion is that the resultant magnetization would be expected to be horizontal, yet it has a large angle with the horizontal. Further, the swirl deposit itself, whether it is from the comet, or is debris from the nearby crater or other breccia, is only 1 m deep, as determined from the shadows it casts. It is unreasonable to suppose that the magnetic anomaly arises from its magnetization alone as the intensity would be very much greater than could be explained with a reasonable iron content. Consequently, one interprets the surface feature as in some way a chemical deposit arising from physico-chemical processes down to a depth of 10 km or so, such as might be produced by degassing of volatiles through cracks produced by the focussing of shock waves from an antipodal impact as El Baz *et al.*, (1980) suggests. Whatever the merits of such speculations, it remains a fact that the magnetized region has uniformity of direction of magnetization on the horizontal scale of 100 km and the vertical scale of 10 km and it is reasonable to make the hypothesis that it records the existence of a lunar field at a time probably considerably younger than that field preserved in the far side anomalies. The latitude of the pole position fits this assumption of an intermediate age and suggests that this pole is on a curve of successive movement of the pole.

MECHANICS OF POLAR WANDERING

The mechanics of this phenomenon of lunar polar wandering was discussed by Gold (1955) and by Runcorn (1955). Euler showed that a body would rotate with stability about either its axes of maximum or of minimum moment of inertia. The impacts forming the great basins and their later flooding with lava could obviously have made considerably changes in the positions of these axes. It is therefore interesting that the pole positions determined above place the maria in orientations which are suggestive that they have a relationship to the process of polar wandering. The pole corresponding to the far side anomalies is such that the maria can be enclosed between lines of longitude 90° apart (see Fig. 1). The Reiner γ pole places the equator near to the great circular maria, Mare Orientale, Mare Imbrium, Mare Serenitatis, Mare Crisium and Mare Smythii (see Fig. 2). These basins are all formed about 4.1 b.y. ago and are the last of the great collisions although Mare Smythii may be older. Therefore the bodies which caused these basins must have been in the ecliptic, supposing that poles of the ecliptic and axis of rotation have always been close. Some support for this conclusion can be adduced from the fact that the debris flung out from the Mare Orientale impact has a strong NW-SE orientation and this would result from a body coming at a low angle from this palaeoequator. Similarly, it has always been noticed that strong N-S grooves characterise the Imbrium debris south of the basin, again to be expected from a body coming from this palaeoequator, which runs in a north east direction from Orientale and is roughly east west near Imbrium. This tendency for debris to be ejected at right angles to the impacting body's approach is, of course, derived from model experiments (Gault & Wedekind, 1978) but some physical arguments suggest that the extrapolation required to relate these laboratory experiments to the early moon may not be unreasonable. The low latitudes of these impacts require that the impacting bodies were moving very close to the ecliptic. This is inconceivable if the bodies came from outside the earth-moon system but might be reasonable were they orbiting around the moon or the earth. It is therefore tentatively suggested that they were moons, similar to Deismos and Phobos, which collided with the moon in its retreat from the earth due to tidal friction. It has been noticed for a century that the circular maria lie on a great circle (Malcuit *et al.*, 1975) and attempts to understand the implication of this are timely.

In an interesting paper Melosh (1975) finds that the axis of maximum moment

Fig. 1. Moon oriented with respect to pole derived from far side anomalies.

of inertia of a simplified model of the moon—a uniform sphere with additional surface masses as required by the mascon gravity anomalies—lies within 5° or 10° of the present axis of rotation and the axis of minimum moment of inertia lies in the direction towards the earth. Whether or not such a simple model is realistic, the exercise suggests that before the emplacement of the mascons, i.e., before the flooding of the circular maria basins with lava, the axis of maximum moment of inertia was different from the present. The emplacement of the mascons, as dated by the mare basalts, occurred over 3.2 b.y. ago, although some later sub-

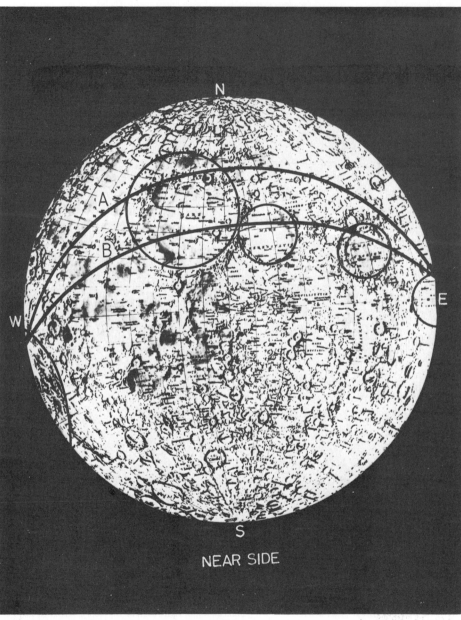

(b)

Fig. 2. Palaeoequators corresponding to poles derived from the Reiner γ anomaly: (a) palaeoequators with great circular mare (near side), (b) palaeoequators with earlier basins now incorporated in irregular mare (near side), (c) palaeoequators (far side). Palaeoequator (a) based on revised determination of crustal direction of magnetization by Hood *et al.* in this volume. Palaeoequator (b) based on earlier data of Hood *et al.* (1979).

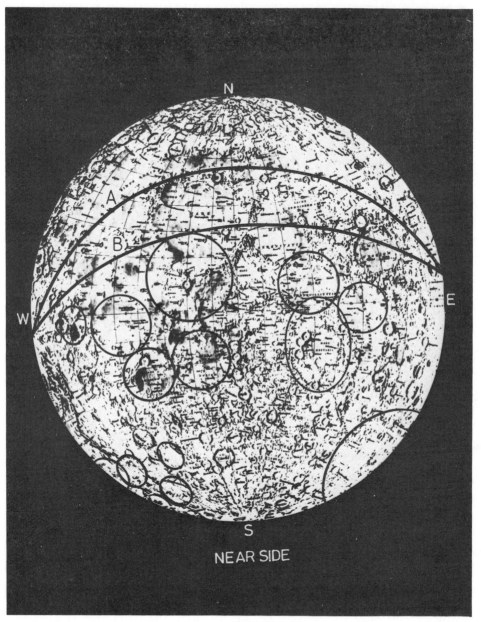

(a)

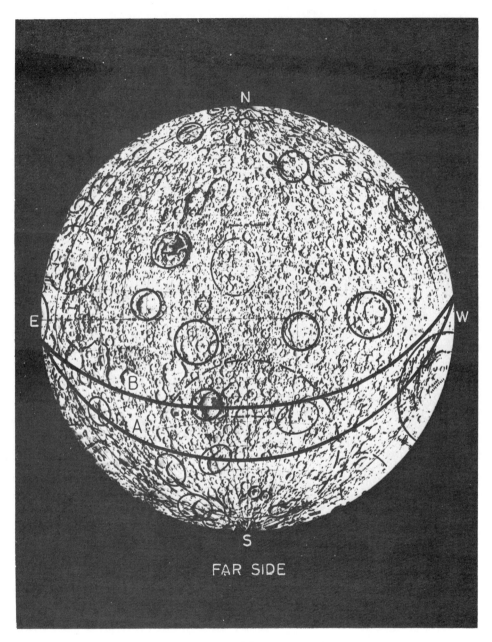

(c)

sidence, reducing the size of the uncompensated masses, probably occurred (Runcorn, 1974) more recently. Thus, if the age of the Reiner γ anomaly was between 3.2 b.y. and 4.0 b.y. ago then the events would be interpreted as follows. The moon rotated about the Reiner γ pole, the circular basins were produced by impacting bodies, but due to the weakness of the lithosphere, isostatic equilibrium was soon reached by the upwarping of the mantle, partly filling the holes in the highland crustal shell created by the impact (Runcorn, 1974). The resulting depressions, being compensated, did not greatly alter the principal axes of inertia. Later on when the flooding of these basins by lava created the mascons, which were supported by the strength of the lithosphere (by that time much thicker), polar wandering commenced bringing the axis to its present position. On this simple model, of course, an additional property of the lunar interior has to be assumed: below the lithosphere it must behave like a fluid of high viscostiy, so that the equatorial bulge due to rotation can readjust to the changing position of the axis of rotation. In other words, solid state creep must be the dominant mechanical property in the long term.

However, the present differences between the moments of inertia are much greater than are expected on the hydrostatic theory, and much greater than can be explained on this simple model of a uniform moon plus the mascons (Runcorn, 1967). The classical explanation was always that this nonhydrostatic figure arose from early distortion of the moon which was retained by the finite strength of its interior. This hypothesis of course would preclude polar wandering. But I suggested the alternative hypothesis that the non-hydrostatic shape arises from a second degree harmonic convection current (Runcorn, 1962; 1967). Of the arguments I presented in favour of convection, the most important is that the present ellipticity of the surface of the maria is not that of the present second harmonic component of the gravitational potential. Consequently, the process causing the nonhydrostatic figure must have been occurring after the flooding of the maria (Runcorn, 1977): this is possible if the explanation is convection, but impossible if it is finite strength. Convection described by a first harmonic, or any other apart from the second, makes no contribution to the moment of inertia tensor. On the theory of this paper the present pattern of convection would have become established after the axis of rotation had come to its present position.

It may be asked: why did the second harmonic convection arrange itself with the uprising current along the earth-moon direction and the downgoing current at right angles in the equatorial plane, so that the axis of maximum moment of inertia of the flow was aligned along the axis of rotation, and the axis of minimum moment of inertia in the earth direction, like that of the lithosphere containing the mascons? The answer must be that this configuration is one of minimum energy and the convection naturally picks out this sub-term of the second degree harmonic.

It remains to suggest how the moon came to rotate about the Reiner γ pole prior to the final great basin forming impacts about 4 b.y. ago when its original rotation was about the pole of the far side anomalies which we date as rather older. It is reasonable to suppose that this earlier reorientation was connected with the excavation of the older basins, which are now the irregular maria.

CONCLUSION

From the subsatellites launched during the Apollo 15 and 16 missions, three component magnetic maps have been made of parts of the lunar surface, from which the directions of magnetization of regions of the crust have been determined. The understanding of the mechanism responsible for the enhanced levels of magnetization intensities in such regions has not progressed far, but it is a reasonable hypothesis to suppose that their directions are those of an early lunar field. No mechanism for generating this field, except that of a core dynamo, has been put forward in a form easy to test. But on the core hypothesis, the poles of rotation of the moon in its early history have been found and relate in interesting ways to the maria basins. The redistribution of mass caused by the early bombardment and the later filling of the large basins by lava, cannot have failed to change the principal axes of inertia. Thus a convincing explanation of the polar wandering inferred from palaeomagnetism is given, provided that solid state creep is assumed to occur below the lunar lithosphere. This account of the early mechanics of the moon seems reasonable and so in turn gives some support to the interpretation of lunar palaeomagnetism, which we have long given i.e., that it records the existence of an early lunar magnetic field generated in a small iron core. At any rate, the conclusions reached here open up new ways of testing this hypothesis for there is preserved, yet not fully interpreted, much information about the early mechanical history of the moon. This account of polar wandering also gives support to the convection theory of the nonhydrostatic figure of the moon, for which I have long argued.

REFERENCES

Collinson D. W., Stephenson A., and Runcorn S. K. (1977) Intensity and origin of the ancient lunar magnetic field. *Phil. Trans. Roy. Soc. London* **A285,** 241–247.

Creer K. M., Irving E., and Runcorn S. K. (1954) The direction of the geomagnetic field in remote epochs in Great Britain. *J. Geomag. Geoelectr.* **4,** 163–168.

El Baz F., Lin L. P., Hood L. L., and Runcorn S. K. (1980) Magnetic anomalies antipodal to large impact basins (abstract). In *Lunar and Planetary Science XI,* p. 626–627. Lunar and Planetary Institute, Houston.

Fuller M. (1974) Lunar magnetism. *Rev. Geophys. Space Phys.* **12,** 23.

Gault D. E. and Wedekind J. A. (1978) Experimental studies of oblique impact. *Proc. Lunar Planet Sci. Conf. 9th,* p. 3843–3875.

Gold T. (1955) Instability of the earth's axis of rotation. *Nature* **175,** 526–529.

Hood L. L., Coleman P. S., and Wilhelms D. F. (1979) The moon: Sources of the crustal magnetic anomalies. *Science* **204,** 53–57.

Hood L. L., Russell C. T., and Coleman P. J. (1978) The magnetization of the lunar crust as deduced from orbital surveys. *Proc. Lunar Planet. Sci. Conf. 9th,* p. 3057–3078.

Irving E. (1956) Paleomagnetic and paleoclimatic aspects of polar wandering *Geophis. Pura. Appl.* **33,** 23–41.

Malcuit R. I., Byerly G. R., Vogel T. A., and Stoeckley T. R. (1975) The great-circle pattern of large circular maria: Product of an earth-moon encounter. *The Moon,* **12,** 55–62.

Melosh H. J. (1975) Mascons and the moon's orientation. *Earth Planet. Sci. Lett.* **25,** 322–326.

Runcorn S. K. (1954) The earth's core (abstract). *EOS (Trans. Amer. Geophys. Union)* **35,** 49–63.

Runcorn S. K. (1955) Rock magnetism—geophysical aspects. *Adv. in Phys. Suppl. to Phil. Mag.* **4,** 244–291.

Runcorn S. K. (1956) Palaeomagnetic comparisons between Europe and North America. *Proc. Geol. Assoc. Can.,* **8,** 77–85.

Runcorn S. K. (1962) Convection in the moon. *Nature* **195,** 1150–1151.

Runcorn S. K. (1967) Convection in the moon and the existence of a lunar core. *Proc. Trans. Roy. Soc. London* **A296,** 270–284.

Runcorn S. K. (1974) On the origin of mascons and moonquakes. *Proc. Lunar Sci. Conf. 5th,* p. 3115–3126.

Runcorn S. K. (1975) On the interpretation of lunar magnetism. *Phys. Earth Planet. Inter.* **10,** 327–335.

Runcorn S. K. (1977) Interpretation of lunar potential fields. *Phil. Trans. Roy. Soc. London* **A285,** 507–516.

Runcorn, S. K. (1978) The origin of lunar palaeomagnetism. *Nature* **275,** 430–432.

Schultz P. H. and Srnka L. (1980) Cometary collisions on the moon and Mercury. *Nature* **284,** 22–26.

Proc. Lunar Planet. Sci. Conf. 11th (1980), p. 1879–1896.
Printed in the United States of America

Bulk magnetization properties of the Fra Mauro and Reiner Gamma Formations

L. L. Hood

Lunar and Planetary Laboratory, University of Arizona, Tucson, Arizona 85721

Abstract—Higher-resolution contour maps of the vector components of the lunar crustal magnetic field within the selenographic range 25°W to 62.5°W and 2.5°N to 10°N are used to set probable limits on the magnetization properties of two surface geologic units. As previously reported, the mapped magnetic anomalies are generally correlated with exposed segments of the Fra Mauro Formation while the largest single anomaly occurs over the Reiner Gamma Formation on western Oceanus Procellarum. An interative forward modeling procedure is used to estimate bulk directions of magnetization and lower limits on the intensities of magnetization within these units assuming that they are the sources of the associated anomalies.

The lower bounds on mean magnetization intensities within the studied part of the Fra Mauro Formation (maximum assumed thickness 1 km) are in the range $6 - 27 \times 10^{-5}$ e.m.u./g. These values are not inconsistent with laboratory data for the returned breccias and soils. The dipole moment per unit area for Reiner Gamma and vicinity is found to range from about 43 to 189 G-cm. Although the thickness of the Reiner Gamma unit is not precisely known, a probable maximum thickness of $\sim$100 m, estimated from geological considerations, yields minimum mean intensities of $1.5 - 6 \times 10^{-3}$ e.m.u./g. The bulk direction of magnetization within the Reiner Gamma source layer (horizontal scale $\sim$100 km) is essentially uniform to the degree determinable with present data. However, the Fra Mauro Formation is inferred to be magnetized very inhomogeneously in both direction and intensity on horizontal scales > 100 km. The degree to which these results are likely to be dependent on the choice of source model geometry is discussed.

I. INTRODUCTION

Associations of orbital magnetic anomalies with surface geologic units provide a useful source of quantitative information on the nature of lunar crustal magnetization. Previous geologic-magnetic studies using Apollo 15 and 16 subsatellite magnetometer data include a correlative analysis of magnetic field measurements vs. surface geology and relative age across a part of the far-side highlands (Russell *et al.*, 1977), a modeling study of the Van de Graaff-Aitken anomalies on the south-central far side (Hood *et al.*, 1978), and an analysis of magnetic anomalies on the central near side with emphasis on the geologic origin of certain correlated surface units (Hood *et al.*, 1979b). Complementary efforts using electron-reflectance measurements include the correlative analyses of Lin *et al.* (1977) across the near-side maria and Anderson and Wilhelms (1979) across the central far side beneath the Apollo 16 subsatellite orbit track. Finally, the Rima Sirsalis anomaly

(Anderson *et al.*, 1977) has been subjected to a quantitative modeling analysis using both the subsatellite magnetometer data and electron-reflectance data by Srnka *et al.* (1979).

Among the 2179 usable orbits of Apollo 15 subsatellite magnetometer data and the 426 orbits of Apollo 16 subsatellite measurements, orbits 198–270 of the Apollo 16 subsatellite mission are characterized by successive low-altitude passes across a geologically well-studied portion of the central near side. These passes occur in the lunar wake or shadow region when the moon was in the solar wind so that magnetic field variations of external origin are relatively minimal. The magnetometer data obtained during these intervals are therefore particularly suitable for both correlative and modeling studies.

We have previously reported the construction from the Apollo 16 wake measurements of the first low-altitude magnetic anomaly map of a portion of the lunar near side (Hood *et al.*, 1979a,b). Magnetic anomalies were found to exist primarily over areas dominated geologically by the Fra Mauro and Cayley Formations but were relatively absent over the western maria and over the craters Copernicus, Kepler, and Reiner and their encircling ejecta. The largest single anomaly was correlated with a conspicuous light-colored marking on western Oceanus Procellarum known as Reiner Gamma.

The associations of mapped anomalies on the lunar near side with the Fra Mauro and Cayley Formations have received qualitative support from the study of electron-reflection maxima across the far side by Anderson and Wilhelms (1979) in which the identified magnetized regions occur peripheral to large impact basins in areas where deposits of basin ejecta are observed or inferred. It has also been shown that the global distribution of Reiner Gamma-like deposits is generally correlated with the occurrence of regions of strong magnetic anomalies directly detected at high altitudes with the subsatellite magnetometers (Hood *et al.*, 1979b,c; Hood *et al.*, 1980) and indirectly by the electron-reflection technique (Lin, 1979; Lin *et al.*, 1980).

Since the magnetic anomalies associated with Reiner Gamma and with the Fra Mauro Formation appear to be at least partly representative of a large fraction of the moon, detailed studies of the available low-altitude magnetometer data can lead to results of general utility. In this paper, the modeling approach of Hood *et al.*, (1978a,b) is applied to calculate bulk directions of magnetization and lower bounds on mean intensities of magnetization within these surface geologic units assuming that they are the sources of the associated magnetic anomalies. The inferred directional properties of the magnetization are then used to further constrain theories of the origin of the magnetizing field(s) and the calculated magnetization intensities are qualitatively compared to available returned sample data.

The magnetic anomaly map used for this study is a higher-resolution section of an improved version of the near-side map published in Hood *et al.* 1979b). The methods employed in its construction and the resulting correlations of magnetic field maxima with surface geology will first be described.

II. DATA PROCESSING AND FIELD MAPPING

The previously published contour maps of the crustal magnetic field on the central near side possessed a maximum resolution of 1.2° or ~36 km. During the past year, refinements in data processing and selection have resulted in the production of a second set of maps of this region with less coverage but with greater accuracy and resolution. These processing refinements are summarized below.

The Apollo subsatellite telemetry system operated in both a "stored" and a "real-time" data transmission mode depending on whether direct earth-based communication with the spacecraft was possible. The temporal resolution of the measurements in the stored mode was relatively low and consisted of weighted 24-second averages computed on board the subsatellite. In the interest of consistency, the initial processing of the much higher-resolution real-time measurements included a weighted-average computation which effectively converted these measurements to the same resolution as the stored-mode data. Since the subsatellite orbit period was about 120 minutes, a 24-second data point separation corresponds to 1.2° or about 36 km along the orbit track. As long as the subsatellite altitude above the lunar surface exceeded this length (as it usually did), then the stored-mode resolution was adequate to resolve all fields of crustal origin. However, the Apollo 16 subsatellite magnetometer measurements across the lunar near side were often obtained at altitudes less than 36 km (minimum altitude ~11 km within the region selected for mapping). For this reason, significant errors in resolution were introduced into both single-orbit plots and two-dimensional maps of this data. To overcome this difficulty, the original real-time measurements have been reprocessed to produce 5-second weighted averages with a corresponding five-fold increase in spatial resolution (to 7 km along the orbit track). Fig. 1 is a comparison of the radial magnetic field component measured across the near side during orbits 214 to 222 prior to and after the reprocessing was carried out. Repeating patterns of field fluctuations at the 5-second resolution on successive orbits verify that additional information has indeed been extracted.

Higher-resolution contour maps of the Apollo 16 subsatellite measurements have been constructed for several selenographic areas within the region of available coverage on the near side (for a coverage map, see Fig. 2 of Hood *et al.*, 1979b). As before, the data were two-dimensionally filtered across adjacent orbits to minimize non-repeating, short-wavelength contributions. The method adopted for this purpose has been described in further detail by Hood *et al.* (1980). Shown in Fig. 2 is one magnetic anomaly map for the the longitude range 25°W to 65°W which resulted when the higher-resolution data was passed through a two-dimensional filter with a minimum passed wavelength of 0.625° (compared to 1.25° on earlier maps). Only Apollo 16 subsatellite orbits 220 to 252 were selected to further minimize external noise contributions. From comparisons of single-orbit plots of this data, an estimate for the noise level remaining on the contour maps of Fig. 2 can be made. This level for the radial field component map is approximately 0.5γ ($1\gamma = 10^{-5}$ G = 1 nT) while that for the east, north, and total field maps is approximately 1γ. However, due to the larger amplitudes of the Reiner Gamma anomaly (total field maximum $> 30\gamma$ on some single-orbit plots), the noise levels for this portion of the field map may be as large as 2 to 3γ.

Correlations with surface geology

The largest single anomaly within the limited region of coverage of Fig. 2 (at 58.3°W, 7.8°N) again correlates precisely with the location of Reiner Gamma on western Oceanus Procellarum as earlier reported. However, several new details are noteworthy. First, the sharp anomaly peak is more fully resolved on this map as evidenced by its amplitude of 22γ compared to 11γ on earlier maps. Second, as can be seen in Fig. 3, the contour lines of the total field maximum exhibit an

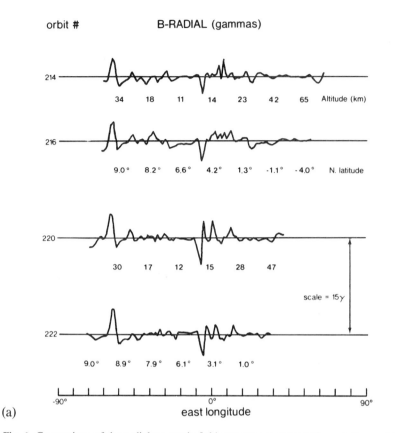

orbit #   B-RADIAL (gammas)

Fig. 1. Comparison of the radial magnetic field component at (a) 1.2° resolution and (b) 0.3° resolution for a series of Apollo 16 subsatellite orbit segments. The quadratic trend was removed from the measured field and the residuals were plotted against longitude. Latitude and altitude above the lunar surface are indicated numerically below alternate data segments.

elongation which is parallel to the main axis of the Reiner Gamma albedo feature and which makes an angle of about 20° with the subsatellite orbit track. The elongation of contour lines suggests that the source may also be elongated in this direction, consistent with the possibility that a surficial layer roughly coincident with the albedo feature is the main anomaly source. It should be noted however that high-resolution coverage of magnetometer data is restricted to a narrow latitudinal band which only marginally includes the largest concentration of swirls. Other swirls to the northeast and southwest are not covered. Lin (1979) has interpreted electron-reflectance data to conclude that a broader distribution of anomaly sources may be present in the Reiner Gamma region. Thus the possibility of other anomaly sources which do not correlate with albedo features can not be excluded.

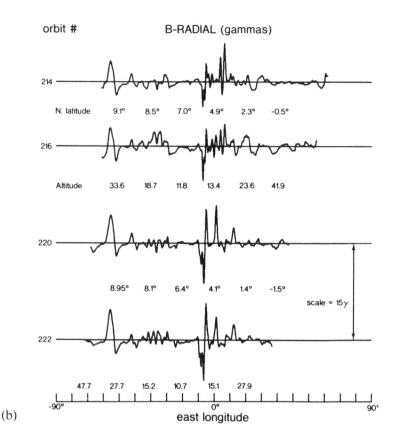

orbit #   B-RADIAL (gammas)

(b)

east longitude

We have previously reported general associations of medium-amplitude anomalies with areas dominated geologically by the Fra Mauro and Cayley Formations (Hood *et al.*, 1979a,b) but the correlations evident on the higher-resolution maps are more pronounced and increase the probability that basin ejecta materials themselves are sources of some observed anomalies. The group of weaker anomalies near 35°W in Fig. 2 are generally associated with remnants of the Fra Mauro Formation that have survived inundation by mare basalt. The largest total field peak (6γ) is located at 35°W, 5°N on Fra Mauro terrain near the crater Encke. Several elongations of contours or magnetic "lineaments" are present in this region which are parallel to major axes of Fra Mauro deposition, i.e., radial and subradial to the Imbrium basin. These are most easily discerned on the radial field map which contains less distortion from external noise contributions. Figure 4 is a superposition of a portion of this map onto a geologic map of the Kepler region. The several field maxima occur primarily over the roughly textured Fra Mauro unit while the crustal field is relatively weak over the Copernican crater Kepler and over the Imbrian crater Encke.

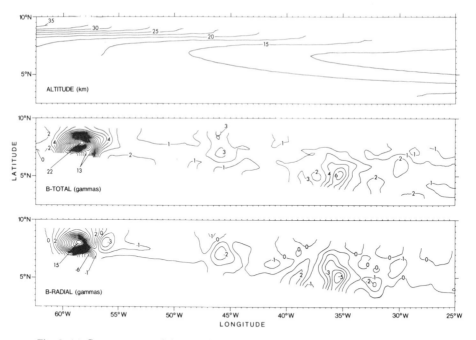

Fig. 2. (a) Contour maps of the Apollo 16 subsatellite altitude, the field magnitude and radial field component measured by the subsatellite magnetometer across a section of the lunar near side. The contour interval on the altitude plot is 2.5 km. The contour interval on the field plots is 0.5γ.

III. MODELS OF CRUSTAL MAGNETIZATION

Mathematical inversions of potential field data are inherently non-unique. Nevertheless, if geological or geophysical constraints on the range of source model geometries are available, useful limits on the bulk magnetization properties of crustal geologic units can be estimated.

The geological constraints for the present analysis come from the correlations of anomaly maxima with surface geology discussed above. It is generally acknowledged that the Fra Mauro and Cayley Formations represent primary and/ or secondary basin ejecta emplaced at 3.8 to 4.0 b.y. ago (Taylor, 1975). Although the origin of Reiner Gamma and other similar markings found elsewhere on the moon is currently the subject of some debate, available photogeologic data were interpreted by Hood *et al*. (1979b) to suggest that Reiner Gamma may consist of unusually strongly magnetized deposits of ejecta from secondary craters of the nearby primary impact crater Cavalerius. The observed correlations of both the Fra Mauro and Reiner Gamma Formations with magnetic anomalies were therefore considered to be generally consistent with the hypothesis that variably magnetized ejecta materials produced during meteoroid impacts (but magnetized in an unspecified manner) are major sources of the magnetic anomalies detected

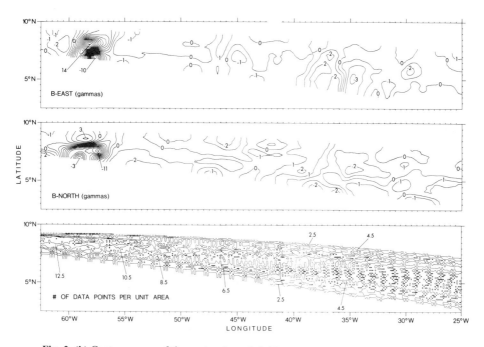

Fig. 2. (b) Contour maps of the east and north field components and the number of data points per $0.625° \times 0.625°$ unit of area. The contour interval on the bottom panel is two data points.

from orbit. The ejecta deposit hypothesis was first proposed by Strangway *et al.* (1973) on the basis of returned sample and surface magnetic field investigations. Thus, a source model consisting of a surface layer of material coherently magnetized over a substantial area is indicated.

An alternate possible explanation for Reiner Gamma and similar albedo markings found on the lunar far side is that these features represent residues of recent cometary impacts (Schultz and Srnka, 1980). There are a number of arguments against the cometary impact hypothesis; however, they are beyond the scope of this paper and the reader is referred to the discussion in Hood (1980). Other suggested explanations which pre-date the magnetics data have been summarized by Schultz (1976). In any case, the superposition of Reiner Gamma on mare basalt flows, materials that are elsewhere typically poorly magnetized, and the absence of a measurable associated gravity anomaly are consistent with the possibility that the source of the observed magnetic anomaly is a surficial layer of material magnetized nearly uniformly over a substantial area.

For these reasons, it was assumed in the work reported here that the sources of the magnetic anomalies mapped in Fig. 2 are surface layers of material with thicknesses much less than the subsatellite altitude. The continuous and constantly changing surface density of magnetization was represented, as a first approximation, by a distribution of uniformly magnetized circular plates separated by unmagnetized material. Plate locations, radii, dipole moments per unit area,

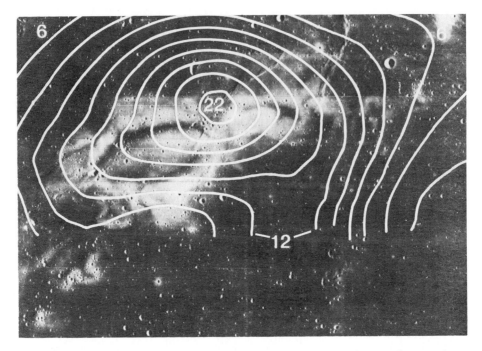

Fig. 3. Superposition of the total field maximum of the large magnetic anomaly mapped in Fig. 2 onto a section of Lunar Orbiter IV frame 157. Note the elongation of contour lines parallel to the main axis of the Reiner Gamma albedo feature. The orbit tracks make angles with the direction of the elongation of approximately 30°.

and bulk directions of magnetization were determined in such a manner as to minimize the r.m.s. deviations of the model field components from the observed field components. In addition, available estimates for maximum thickness of the geologic units associated with the observed anomalies were used to obtain approximate lower bounds on mean intensities of magnetization.

A modeling procedure which is in keeping with the assumptions stated above has been described by Hood *et al.* (1978). This procedure was modified slightly for the present study. Initial estimates for plate locations and radii were first made based respectively on the locations of total field maxima and on horizontal field gradients measured outward from these maxima. The method of Talwani (1965) was used to calculate the model magnetic field at individual points along the available orbit tracks for a range of dipole moments per unit area and for a range of directions of magnetization. The trial directions of magnetization were equally spaced at intervals of about 4° within a cone of angular radius 24° about a chosen initial value. The r.m.s. deviation of the model field components from the observed field components was computed at each step to obtain the minimum-variance fit. If the "best-fitting" direction of magnetization fell along the edge of the sampling cone, the initial chosen direction was reset accordingly and the step was repeated. If the selected dipole moment per unit area occurred at the maximum or minimum of the sampled range, a similar procedure was followed.

When the set of magnetization parameters corresponding to the minimum variance between modeled and observed magnetic field values had been selected, the resulting model field was again calculated at individual data points along the available orbit tracks and passed through the same two-dimensional filtering routines used to construct the observed field maps. A direct comparison between contour maps of the model magnetic field and the observed magnetic field was then possible. Based on this comparison, adjustments to the assumed plate locations and radii were made and the entire procedure was repeated. The process was continued until the r.m.s. deviation became comparable to estimated noise levels or the r.m.s. deviation asymptotically approached a contant value.

Although the procedure used to obtain surface distributions of magnetization consistent with available data was simple, care was taken to examine the significance of the final model by determining the range of model parameters which

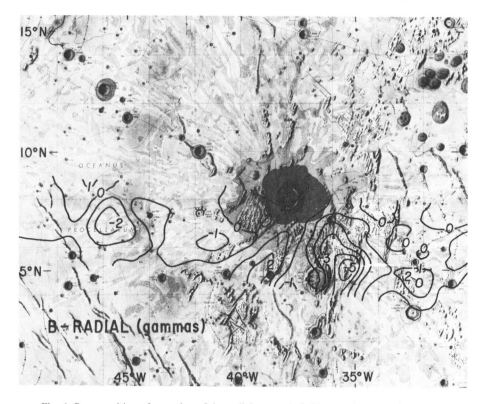

Fig. 4. Superposition of a portion of the radial magnetic field anomaly map (Fig. 2) onto a geologic map of the Kepler region (Hackman, 1962). The largest maxima are correlated with exposed segments of the Fra Mauro Formation (a roughly textured unit of primary Imbrium Basin ejecta) while the craters Kepler and Encke and areas dominated by mare basalt are magnetically relatively weak.

provide equally or nearly equally small r.m.s. values. Assuming that the surface-layer source model is valid, the set of magnetization models which satisfy the observational constraints could be identified.

Application to the Fra Mauro Formation

As discussed in Section II, the radial field maxima located near 35°W, 5°N of Fig. 2 are correlated with segments of primary Imbrium basin ejecta which have survived inundation by mare basalt. Two main anomaly complexes, centered at approximately 36°W,5°N and 39°W,5°N, are sufficiently distinct to allow separate modeling. The largest peak (maximum amplitude -5γ) was modeled using three separate plates to roughly match the oblong shape and variable surface density of magnetization of the source. The final model parameters are listed as plates 1–3 in Table 1. The smaller elongated maximum at 39°W,5°N was modeled using four adjacent plates with final model parameters listed as plates 4–7 in Table 1. Contained in Table 1 are the longitude and latitude of the plate centers in degrees, their radii in kilometers, the directions of magnetization specified by angles θ and ϕ, and the dipole moments per unit area in Gauss-cm. The angle θ is the polar angle measured from the local outward radius vector at the longitude and latitude of the plate center and ϕ is the azimuth measured counterclockwise from east about the zenith (e.g., $\theta = 90°$ $\phi = 90°$ is due north). It should also be mentioned that the dipole moment per unit area given in the last column is numerically equal to the magnetization intensity in gammas ($1\gamma = 10^{-5}$ G.) if the thickness of the plate is 1 km.

Contour plots of the magnetic field components produced by the entire distribution of magnetization listed in Table 1 are presented in Fig. 5 for comparison to Fig. 2. The Pythagorean mean of the r.m.s. deviations of the model field components from the observed field components was 1.03γ which is of the order of the estimated noise level. The directions of magnetization for each set of adjacent plates (1–3 and 4–7) were assumed to be uniform during the model-fitting analysis. The resulting reduction of the r.m.s. to a value comparable to the noise level supports the approximate validity of this assumption.

Table 1.

#	W. Long	N. Lat.	Rad. (km)	θ	ϕ	G-cm
1.	35.1°	5.2°	20	160°	298°	38
2.	35.9°	6.0°	20	160°	298°	22
3.	35.7°	4.4°	10	160°	298°	82
4.	38.4°	5.9°	13	27°	123°	18
5.	38.8°	5.2°	13	27°	123°	34
6.	39.1°	4.8°	13	27°	123°	20
7.	38.0°	6.3°	8	27°	123°	25

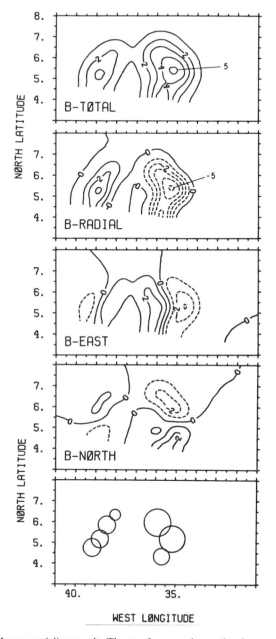

Fig. 5. Fra Mauro modeling result. The top four panels are the three vector field components and the field magnitude (contour interval 1γ) that would be produced at sub-satellite altitudes by the magnetization model listed in Table 1. The seven surface plates used to represent the major anomaly sources are sketched in the bottom panel.

The model parameters of Table 1 are also indicated schematically in the bottom panel of Fig. 5. The locations of plates 1–3 correspond approximately to the large segment of Fra Mauro terrain located just east and northeast of Encke (see Fig. 4). However, the same Fra Mauro unit extends further to the northeast but evidently has no strong radial component of magnetization. The locations of plates 4–7 nearly correspond to a second segment of Fra Mauro terrain elongated in a direction radial to the Imbrium basin (Fig. 4). The slight westward offset of the inferred plate locations from the visible Fra Mauro unit may indicate partial burial of the anomaly source by a thin layer of mare basalt.

Since the maximum thickness of the Fra Mauro layer in the Kepler region is about 1 km, the dipole moments per unit area given in the last column of Table 1 imply a lower bound on the mean intensities of magnetization within this unit (and within the boundaries of Fig. 2) in the range $18 - 82\gamma$. Assuming a mass density of 3 g/cm^3, these values correspond to a range of $6 - 27 \times 10^{-5}$ e.m.u./ g. Because the magnetization intensity is a function of the inferred plate radius which is in turn only weakly determined by the observed field gradients, it is suggested that these lower bounds be regarded as accurate only to within a factor of two. Also, the thickness of the magnetized layer may be less than 1 km and may vary with horizontal position. Nevertheless, this range of magnetization intensity appears to compare favorably with measured values for breccias and soils returned from the Apollo landing sites (e.g., Strangway et al., 1973; Fuller, 1974). More detailed comparisons between the Apollo 14 breccia magnetization properties and those inferred here on the basis of orbital data could be instructive.

The direction of magnetization for the main Fra Mauro source (plates 1–3) is radially inward and inclined by approximately 20° toward the southeast while that of the secondary source (plates 4–7) is radially outward and inclined about 27° toward the northwest. Because these directions of magnetization are nearly opposite to one another, two interpretations are possible in principle. One is that the two anomalies are caused by a uniformly magnetized layer whose thickness varies so that one region is thicker than average and the other is thinner. The second interpretation is that the two anomalies are caused by separate exposures of material magnetized in nearly opposite directions. The surface plate distribution obtained from the analysis could equally well represent either model. Although it is not possible to uniquely determine the actual source geometry from the magnetometer data alone, geological considerations can again be used to indicate the most probable model.

The Fra Mauro Formation was initially deposited as a layer of variable thickness peripheral to the Imbrium basin. Subsequent flooding by mare basalt largely inundated low-lying areas so that only the thickest layers of ejecta or layers of ejecta lying on elevated terrain remain exposed. Comparison of Fig. 2 to surface geologic maps shows that areas dominated by mare basalt (particularly the young mare basalt of western Oceanus Procellarum) are only weakly magnetized relative to other areas. Therefore, the association of the two anomalies studied here with distinct exposures of Fra Mauro terrain separated by mare basalt (Fig. 4) can not be easily interpreted as due to alternately thinner and thicker segments of a uniformly magnetized layer. Instead, the most probable interpretation is that both unflooded segments of Fra Mauro terrain are relatively thick and are separated by poorly magnetized material. Adopting this interpretation, it follows that the two exposures, separated by a distance >100 km, are magnetized in very different directions.

It should be stressed that the inferred large-scale inhomogeneity of magnetization of the Fra Mauro Formation is a preliminary result which must be tested in additional modeling efforts elsewhere. Fortunately, additional Apollo 16 subsatellite coverage at slightly higher altitudes is available over Fra Mauro terrain near the 0° longitude meridian. Qualitative examinations of available maps of this region support the possibility of variable directions of magnetization but firm conclusions must await a more detailed analysis. This will be the subject of future work.

Application to the Reiner Gamma Formation

The magnetic anomaly associated with Reiner Gamma has previously been modeled using the lower-resolution (~1.2°) subsatellite magnetometer data (Hood *et al.*, 1979a,b). The anomaly source was represented in the latter references by a thirteen-sided polygonal plate whose shape was chosen to approximate that of the visible albedo marking. The derived dipole moment per unit area was 208 ± 90 G-cm and the inferred direction of magnetization (assumed uniform) had local angular coordinates of $\theta = 65°$, $\phi = 115°$.

For the present analysis, no assumption is made about the lateral shape of the anomaly source and the higher-resolution (~0.3°) data set is used to obtain the minimum-variance fit. The source is represented by three separate plates whose locations are initially estimated to account for the elongated shape of the total field maximum. The bulk direction and surface density of magnetization for each plate are left as free parameters to allow, at least to first order, for the possibility of lateral variations in magnetization properties across the source. As will be seen, the major difference from previous analyses of the resulting model parameters is that the bulk magnetization vector has a stronger eastward component. This is a direct consequence of the increased resolution since the north component of the magnetic field is now weaker relative to the east component (see Fig. 2) rather than stronger as was the case in the original lower-resolution data set (see the maps contained in Hood *et al.*, 1979b). Despite the increased amplitude of the anomaly as mapped in higher resolution (22γ compared to 11γ on earlier maps), the inferred maximum dipole moment per unit area is slightly less than was determined from the lower-resolution data. This is due to the relaxation of our earlier assumption that the lateral shape of the anomaly source is defined by the visible albedo marking.

The final model parameters are numerically listed in Table 2 and the model

Table 2.

#	W. Long	N. Lat	Rad. (Km)	θ	φ	G-cm
1.	58.1°	7.95°	17	54°	129°	146°
2.	59.4°	7.5 °	35	61°	130°	102°
3.	57.5°	7.95°	44	66°	140°	43°

magnetic field components are contoured in Fig. 6 in the same format as Fig. 5. The Pythagorean mean of the r.m.s. deviations from observed values of the model field components asymptotically approached a constant value of $2.5 - 2.7\gamma$ which is of the order of the estimated noise level for this data (Section II). The mean r.m.s. for the model parameters listed in Table 2 was 2.57γ. As indicated in the bottom panel of Fig. 6, plate 1 is located near the total field maximum and is characterized by the largest dipole moment per unit area of 146γ. The remaining two plates have much larger radii and correspondingly smaller surface densities of magnetization. However, it is evident that plate 3 overlaps plate 1 so that the actual maximum surface density of magnetization for the Reiner Gamma source infered from this analysis is $146 + 43 = 189\gamma$.

Assuming that the Reiner Gamma albedo feature is due to the existence of a deposit on the mare basalt surface, it is possible to estimate a probable thickness of the order of 10 m for the layer from the absence of visible shadows at low angles of the sun's illumination (Hood *et al.*, 1979a). Conservatively, an upper limit of 100 m for the mean deposit thickness may be assumed yielding corresponding minimum mean magnetization intensities of $1.5 - 6 \times 10^{-3}$ e.m.u./g. These values are significantly larger than those of the most strongly magnetized returned breccias (e.g., Strangway *et al.*, 1973). Thus, one is faced with two alternatives: Either the original assumption that a surficial deposit is responsible for the anomaly is in error so that the thickness is underestimated or the magnetization level is indeed excessive. In the absence of additional geological or geophysical constraints, neither possibility can be completely excluded.

The directions of magnetization listed in Table 2 are inclined with respect to one another by angles of as much as 25°. The significance of this difference in directions was tested by repeating the analysis with the directions of magnetization of all three plates held in parallel. If the r.m.s. deviation were to increase significantly, then the inferred non-uniformity of the direction of magnetization across the source could be regarded as real. However, a nearly equally good fit with a mean r.m.s. of 2.61γ was obtained. The directions of magnetization for the three plates had local angular coordinates of $\theta = 63°$, $\phi = 135°$. It is concluded that the direction of magnetization within the main Reiner Gamma source is not sufficiently different from uniform to be detectable with present data. The same is not true of the intensity of magnetization (or, alternatively, the thickness of the source layer), however. When a similar test was applied with the dipole moments per unit area held constant, a significantly larger mean r.m.s. was obtained.

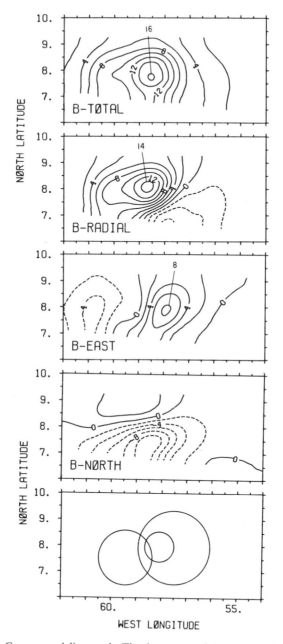

Fig. 6. Reiner Gamma modeling result. The three vector field components and the field magnitude (contour interval 2γ) are computed from the magnetization model listed in Table 2. The distribution of three surface plates used to represent the main Reiner Gamma source is indicated in the bottom panel.

IV. DISCUSSION

In the work reported here, bulk magnetization properties within two lunar surface geologic units have been inferred using low-altitude, high-resolution Apollo 16 subsatellite magnetometer data. On the basis of correlations of mapped anomalies with relatively surficial units on the central near side, a surface plate model with thickness much less than the subsatellite altitude was adopted and was used to represent the sources of the largest anomalies. The results strongly suggest that directional coherence of the surface density of magnetization can occur over horizontal scales up to 100 km. Tentative evidence for a lack of directional coherence on scales > 100 km was found in the case of the Fra Mauro Formation.

Any interpretation of these inferred magnetization properties must first consider the geologic origin of source bodies. Although the origin of the Reiner Gamma Formation continues to be debated, there is general agreement that the Fra Mauro Formation is Imbrium basin ejecta emplaced at about 3.9 b.y. ago (Taylor, 1975). Acquisition of shock remanence and/or thermoremanence by this ejecta in the presence of a strong magnetic field (or fields) at the time of deposition is the most probable explanation for its apparent magnetization. By extension, virtually every other large basin-producing meteoroid impact on the moon during its early history may have resulted in other similar variably magnetized ejecta materials. The results reported by Anderson and Wilhelms (1979) are consistent with this possibility. The ubiquitous detection of small and medium-amplitude magnetic anomalies from lunar orbit across the highlands is probably best understood on this basis.

Assuming that the directions of magnetization preserved in Fra Mauro ejecta materials are an indicator of the orientation of the magnetizing field at the time of impact, the inferred directional properties can be used to constrain the identity and origin of this field. The coherence with which the studied exposures are evidently magnetized is supportive of such an application and indicates that at least these regions have not experienced major alterations of their bulk magnetization properties by subsequent bombardments. The present availability of only three directions of magnetization on the lunar near side precludes a statistical analysis of directional properties similar to that applied to far-side data by Hood *et al.* (1979a,b). Nevertheless, when geological considerations are added to the modeling results, several possible deductions appear to be worthy of discussion.

First, it is quite likely that the several exposures of the Fra Mauro Formation formed nearly contemporaneously, i.e., within a brief interval following the Imbrium impact. Bulk magnetization characteristics were probably acquired during roughly the same interval although exact simultaneity in the acquisition of these properties may have been prevented by different cooling rates, etc. The inferred properties of the present magnetization of this unit suggest the absence of coherence in either intensity or direction on horizontal scales greater than 100 km. This can be explained either by very rapid fluctuations in the orientation of the large-scale magnetizing field or by a relatively constant magnetizing field whose spatial scale-size was not greater than 100 km.

A scale-size of only 100 km for the coherence of the observed magnetization is most easily understood if relatively local processes either greatly distorted any existing large-scale magnetizing field or themselves generated short-lived magnetic fields of the required intensities. A number of such mechanisms have been suggested (see the reviews by Fuller, 1974; Dyal *et al.*, 1974) but in view of the association of mapped anomalies with ejecta materials, processes associated with large meteoroid impacts would appear to be least ad hoc. A local mechanism for generating strong transient magnetic fields during hypervelocity meteorid impacts has been proposed on theoretical grounds by Srnka (1977). However, experimental efforts have not yet succeeded in confirming this process (Srnka *et al.*, 1979).

The lack of complete randomness of the set of magnetization vectors on the south-central far side has been used to argue against an entirely local origin of the lunar magnetizing field(s) (Hood *et al.*, 1978). Additional work to determine a statistically significant number of magnetization vectors on the near side will further assist in distinguishing between locally generated magnetizing fields and large-scale magnetizing fields distorted by local processes. Until this work is complete, direct applications of the inferred directions of magnetization to derive more general constraints on lunar history (e.g., Runcorn, 1979) should be considered with caution.

Finally, it should be emphasized that a simple form of observational evidence for local magnetic field generation during hypervelocity impacts would be provided by the identification of a clearly young ($\ll 3$ b.y. old) but strongly magnetized geologic unit. Photogeologic evidence discussed in Hood *et al.* (1979b) do not seem to allow the use of Reiner Gamma for this purpose. Preservation of its relatively high albedo could be accounted for by magnetic deflection of the solar wind ion bombardment (Hood and Schubert, 1980). However, other Reiner Gamma-like swirls exist on the lunar far side in areas of strong crustal magnetization. Some of these, particularly those which may be associated with the late Copernican crater Goddard A, located just north of Mare Marginis, are strong candidates for further study in this context. Until such evidence is found, it will remain difficult to rule out the possible role of an early strong large-scale magnetizing field in the lunar environment.

Acknowledgments—The higher-resolution magnetic anomaly map contained in this paper was produced in collaboration with P. J. Coleman Jr. and C. T. Russell of the University of California at Los Angeles. Computational assistance was provided by R. Warniers at UCLA and T. Trebisky at the University of Arizona. Constructive criticisms of an earlier version of this manuscript by C. T. Russell are especially appreciated and useful discussions with P. J. Coleman Jr., L. J. Srnka, C. P. Sonett, F. Herbert, R. P. Lin, P. Schultz, S. K. Runcorn, and D. E. Wilhelms are gratefully acknowledged. Supported by the National Aeronautics and Space Administration under grants NSG 7020 (at the University of Arizona) and NGR 05-007-351 (at the University of California, Los Angeles).

REFERENCES

Anderson K. A., Lin R. P., McGuire R. E., McCoy J. E., Russell C. T., and Coleman P. J. Jr. (1977) Linear magnetization feature associated with Rima Sirsalis. *Earth Planet. Sci. Lett.* **34**, 141–151.

Anderson K. A. and Wilhelms D. E. (1979) Correlation of lunar far-side magnetized regions with ringed impact basins. *Earth Planet. Sci. Lett.* **46**, 107–112.

Dyal P., Parkin C. W., and Daily W. D. (1974) Magnetism and the interior of the moon. *Rev. Geophys. Space Phys.* **12**, 568–591.

Fuller M. (1974) Lunar magnetism. *Rev. Geophys. Space Phys.* **12**, 23–70.

Hackman R. J. (1962) Geologic map and sections of the Kepler region of the moon. U.S. Geol. Survey Map I-355.

Hood L. L., Coleman P. J. Jr., Russell C. T., and Wilhelms D. E. (1979c) Lunar magnetic anomalies detected by the Apollo substatellite magnetometers. *Phys. Earth Planet. Inter.* **20**, 291–311.

Hood L. L., Coleman P. J. Jr., and Wilhelms D. E. (1979a) The moon: Sources of the crustal magnetic anomalies. *Science* **204**, 53–57.

Hood L. L., Coleman P. J. Jr., and Wilhelms D. E. (1979b) Lunar nearside magnetic anomalies. *Proc Lunar Planet. Sci. Conf. 10th,* p. 2235–2257.

Hood L. L., Russell C. T., and Coleman P. J. Jr. (1978) The magnetization of the lunar crust as deduced from orbital surveys. *Proc. Lunar Planet. Sci. Conf. 9th,* p. 3057–3078.

Hood L. L., Russell C. T., and Coleman P. J. Jr. (1980) Contour maps of lunar remanent magnetic fields. *J. Geophys. Res.* In press.

Hood L. L. and Schubert G. (1980) Lunar magnetic anomalies and surface optical properties. *Science* **208**, 49–51.

Lin. R. P. (1979) High spatial resolution measurements of surface magnetic fields of the lunar frontside. *Proc. Lunar Planet. Sci. Conf. 10th,* p. 2259–2264.

Lin R. P., Anderson K. A., Russell C. T., Boyce J. M., Masursky H., Wilhelms D. E., and Stuart-Alexander D. E. (1977) Correlations of lunar surface remanent magnetic fields to surface geologic age (abstract). In *Lunar Science VIII, Supp. A,* p. 14–15. The Lunar Science Institute, Houston.

Lin R. P., El-Baz F., Hood L. L., Runcorn S. K., and Schultz P. H. (1980) Magnetic anomalies antipodal to large impact basins (abstract). In *Lunar and Planetary Science XI,* p. 626–627. Lunar and Planetary Institute, Houston.

Runcorn S. K. (1979) An iron core in the moon generating an early magnetic field? *Proc. Lunar Planet. Sci. Conf. 10th,* p. 2325–2333.

Russell C. T., Weiss H., Coleman P. J., Soderblom L. A., Stuart-Alexander D. E., and Wilhelms D. E. (1977) Geologic-magnetic correlations on the moon: Apollo subsatellite results. *Proc. Lunar Sci. Conf. 8th,* p. 1171–1185.

Schultz P. (1976) *Moon Morphology.* Univ. Texas Press, Austin. 626 pp.

Schultz P. H. and Srnka L. J. (1980) Cometary collisions on the moon and Mercury. *Nature* **284**, 22–26.

Srnka L. J. (1977) Spontaneous magnetic field generation in hypervelocity impacts. *Proc. Lunar Sci. Conf. 8th,* p. 785–792.

Srnka L. J., Hoyt J. L., Harvey J. V. S., and McCoy J. E. (1979) A study of the Rima Sirsalis lunar magnetic anomaly. *Phys. Earth Planet. Inter.* **20**, 281–290.

Srnka L. J., Martelli G., Newton G., Cisowski S. M., Fuller M. D., and Schaal R. B. (1979) Magnetic field and shock effects and remanent magnetization in a hypervelocity impact experiment. *Earth Planet. Sci. Lett.* **42**, 127–137.

Strangway D. W., Gose W., Pearce G., and McConnell R. K. (1973) Lunar magnetic anomalies and the Cayley Formation. *Nature* **246**, 112–114.

Talwani M. (1965) Computation with the help of a digital computer of magnetic anomalies caused by bodies of arbitrary shape. *Geophys.* **30**, 797–817.

Taylor S. R. (1975) *Lunar Science: A Post-Apollo View.* Pergamon, N.Y. 372 pp.

Proc. Lunar Planet. Sci. Conf. 11th (1980), p. 1897–1906.
Printed in the United States of America

On the search for an intrinsic magnetic field at Venus*

C. T. Russell, R. C. Elphic, J. G. Luhmann, and J. A. Slavin

Institute of Geophysics and Planetary Physics, University of California
Los Angeles, California 90024

Abstract—Magnetic field observations obtained by the Pioneer Venus orbiter at low altitude are now available for two sets of orbits in the Venus wake. Data from these 130 orbits are examined for possible surface correlated features or any intrinsic magnetic moment. No surface correlated magnetic fields are observed, but our threshold for the detectability of such fields at Venus is about an order of magnitude greater than at the moon. A surface feature of 10° extent would have to create an anomaly of at least 5γ at 200 km to be detected in the Pioneer Venus data. Using measurements averaged in 72 $10° \times 10°$ bins we obtain a planetary magnetic dipole moment of $0.87 \pm 3.00 \times 10^{21}$ Gauss-cm^3. Thus the upper limit of the present day Venus moment is less than 4×10^{-5} of the terrestrial moment.

INTRODUCTION

Until seismic measurements are undertaken on the surface of Venus, we have only gravity and magnetic field data from which to infer the interior structure of Venus. Inferring interior structure from magnetic field measurements is made all the more difficult because we are uncertain about the physics of the dynamo mechanism and the energy source driving this mechanism, even for the earth. Nevertheless, there exists one scaling law which orders planetary moments moderately well (Busse, 1976), and this scaling law predicts that if Venus has a dynamo then it should be large enough to be readily detected by orbiting vehicles. If Venus does not have a detectable magnetic field, as appears to be the case, while its nearby 'twin' planet, the earth, has a strong dynamo, we should be able to learn more either about the interior of Venus or the dynamo mechanism or possibly both.

The earliest Venus missions showed that Venus possessed at most a weak planetary magnetic field. Venera 4 showed only small changes in the magnetic field direction as it descended to 200 km. Dolginov *et al.* (1968) initially interpreted these data to constrain Venus' magnetic moment to a value of less than 3×10^{22} Gauss-cm^3. They later lowered this estimate to less than 10^{22} Gauss-cm^3 (Dolginov *et al.*, 1969). These limits were questioned and it was suggested that

*IGPP Publication No. 2036

the planetary moment could be as high as 6.5×10^{22} Gauss-cm³ (Russell, 1976). However, it was later revealed that Venera 4 encountered Venus north of the ecliptic plane and the vector variations in the magnetic field were consistent with a draped solar wind field rather than an extended intrinsic field (Dolginov *et al.*, 1978).

The first Venus orbiters, Venera 9 and 10, were placed into orbit in October 1975 with 48 hour orbits and periapsis altitudes of 1500 km. The paucity of tracking coverage and lack of knowledge of the orientation of the measurement platform when the spacecraft were far from the planet permitted some differences in the interpretation of the magnetometer data. Dolginov *et al.* (1978) interpreted the measurements to indicate an intrinsic moment of about 2×10^{22} Gauss-cm³. Yeroshenko (1979) interpreted the observations in terms of induced fields and concluded that 2×10^{22} Gauss-cm³ was an upper limit on the intrinsic magnetic moment.

The orbit of the Pioneer Venus orbiter has great advantages over previous missions for studying the planetary magnetic field. The spacecraft has a very low periapsis altitude, close to 140 km on the nightside, and is in near polar orbit. On the nightside of Venus, in the Venus wake, the effects of the solar wind interaction with the ionosphere are minimal (Russell *et al.*, 1979a), so we use this region for our planetary field search. On the other hand, the near wake region is not a vacuum and there are local current systems in the night ionosphere in addition to the currents associated with the draped interplanetary magnetic field. The strength of the magnetic field on the nightside of the planet due to external sources is much less than that in front of the planet but it is significant and determines the 'noise-level' of our analysis. On the other hand, given sufficient data these external sources should average to zero. In this study we assume that this is true and do not attempt to correct for external sources either on a case by case or statistical basis.

The data to be used in this study were obtained from the fluxgate magnetometer on the Pioneer Venus orbiter. The details of the instrumentation have been described by Russell *et al.* (1980a); the data processing appropriate to this study by Russell *et al.* (1980b). A preliminary survey of the observations made by this instrument has been given by Russell *et al.* (1979b); a more detailed account of the solar wind interaction has been given by Slavin *et al.* (1980) and of the dayside ionopause and ionosphere by Elphic *et al.* (1980).

Our initial survey of any possible intrinsic Venus magnetic field used data from only 42 orbits during the satellite's first traversal of the Venus solar wind wake (orbits 50–91) (Russell *et al.*, 1980b). These data were averaged in $10° \times 10°$ bins. Only 37 of these bins had data from 5 or more orbits and were used in subsequent analysis. In the present study we have used data from the full first wake passage (orbits 33 to 110) and all the data available from the second wake passage (orbits 282–344). Thus, in this paper we will be using three times the amount of data available in our initial study to continue our search for an intrinsic magnetic field at Venus due either to internal dynamo action or to remanence in the Venus crust.

DATA AVAILABILITY

Before proceeding with the analysis of the data it is instructive to examine the availability and suitability of data for the mapping study, the planetary relationship of data from the first and second wake passages and future data availability. Data from all planetary longitudes would be available if we were to include data from all solar zenith angles. However, in the dayside ionosphere the thermal pressure of the plasma is much greater than we would expect the magnetic pressure of any planetary field to be. Hence, any planetary magnetic field on the dayside would be highly distorted by ionospheric plasma motions, and we restrict our mapping to the nightside of the planet where the ionospheric plasma pressure is less and where any planetary field penetrating the ionospheric plasma would be dragged by the tailward convection of the plasma. The magnetic field convected against the planet by the solar wind drapes across the dayside ionosphere and closes behind the planet. To minimize the effect on our mapping studies of this "draped field" we have restricted our mapping volume to altitudes below 600 km and to solar zenith angles greater than 120°. This choice is somewhat arbitrary and we will investigate the effects of this choice below by examining the dependence of our results on solar zenith angle and altitude.

During each wake passage, we cross about 110° of planetary longitude while the spacecraft remains in our arbitrary mapping volume. Due to the fact that Venus rotates in a retrograde sense with a period of 243 days while its orbital period is 225 days, the orbit plane which is fixed in inertial space traverses a 30° interval of new longitudes and 80° of previously covered longitudes in the course of each subsequent wake passage. These new longitudes occur at the beginning of the wake passage and occur at lower planetary longitudes, i.e., to the west of the earlier coverage.

When the Pioneer Venus orbiter began its second wake passage two events transpired which prevented the acquisition of data. First, Venus passed through superior conjunction. Thus Venus was as far away from the earth as it can get and the telemetry signal had to pass through the solar corona. Second, Saturn was in the same part of the sky as Venus and Pioneer 11 was encountering Saturn. The net result was that Pioneer Venus was not tracked for almost a full month. This month included all new planetary longitudes. Thus, all the data obtained during the second wake passage was over planetary longitudes covered in the first wake traversal. While useful for providing redundant coverage, the second wake passage did not increase the region of the planetary surface explored magnetically. As shown in Fig. 1, Beta Regio is the only famous Venus landmark covered in our survey.

The third wake passage which at this writing has just begun will cover the missing 30° of the second wake passage and add yet another 30° interval of uncovered longitudes. Thus we will eventually have 170° of the surface covered. We will not get a fourth wake passage at low altitudes because of constraints imposed by the on-board gas supply. Periapsis altitude cannot be maintained after

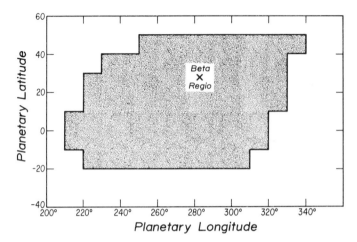

Fig. 1. Planetary longitude and latitude surveyed during the first two wake passages of Pioneer Venus.

the current wake passage since the remaining gas will be necessary to maintain spin axis orientation. Periapsis altitude will rise rapidly at first and then more slowly reaching about 2000 km in the year 1986. Then periapsis will decrease until the satellite is lost to entry into the Venus atmosphere sometime in 1992.

DATA PROCESSING

We simply repeated the analysis that we performed on the initial wake observations (Russell *et al.*, 1980b), sorting the data into bins 10° in longitude by 10° in latitude and averaging these data by component according to the number of data points in each bin and calculating the standard deviation and probable error of the mean for each bin. We have calculated the probable error of the mean by assuming that the degrees of freedom are set by the number of different orbits contributing to each bin, not the number of original data points. There were 72 bins with data from 5 or more orbits. The maximum number of orbits contributing to a bin was 18. The average magnetic field in each of these bins was then inverted using the average latitude, longitude and altitude of the bin to find the planet-centered dipole moment that would give the observed average field. We would expect that if the deduced moments were in fact due to a randomly oriented external source of "noise" rather than an internal field then increasing the number of independent samples would lead to a decrease in the inferred moment. Figure 2 shows the magnitude of the inverted moment as a function of the number of different orbits contributing to a bin. There is much scatter in these moments. Nevertheless the trend indicated by the overlapping medians is clear. The moment decreases as the number of orbits increases. Since the number of orbits covers a range of about 4 we would expect the derived moment to vary by about a factor of 2. The trend in the medians is in accord with our expectations.

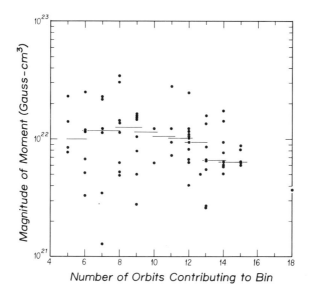

Fig. 2. Magnetic moment versus number of orbits contributing to average. Horizontal bars are overlapped medians.

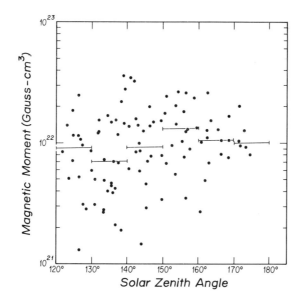

Fig. 3. Magnetic moment versus solar zenith angle. Two wake passages have been plotted separately. Horizontal bars show medians.

We can use these same data to test our arbitrary choice of mapping volume.
Figure 3 shows the magnitude of the derived moment as a function of solar zenith
angle. Here we have plotted data from each wake passage separately because
data over the same planetary longitude is generally obtained at two quite different
solar zenith angles in subsequent wake passages. The medians show no obvious
change in the moment or the deviation about the median with solar zenith angle.
The apparent decrease in scatter with increasing solar zenith angle is due simply
to the decrease in the number of measurements at high solar zenith angles. There
is nothing in these data that indicates there is any advantage to further restricting the range of solar zenith angles considered.

Figure 4 shows the derived moments versus altitude. If the field were constant
with altitude the derived moment would increase about 18% from 200 km to
550 km. The moments appear to increase about 50% from low to high altitudes
but the statistical significance of variation is low. Again there is no obvious advantage to decreasing the size of the region of space used in our mapping study.
Whether our volume can be expanded somewhat awaits further study.

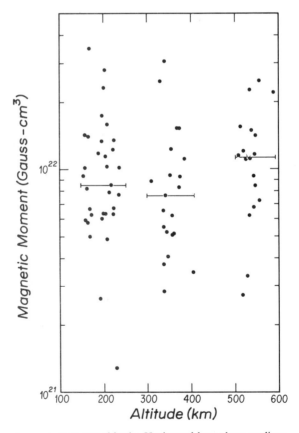

Fig. 4. Magnetic moment versus altitude. Horizontal bars show medians.

Table 1. Location, Average Field and Probable Error of Mean for Most Non-Random Bins.

Longitude	Latitude	Orbits	BR/σ	BE/σ	BN/σ	Angle between fields on two passages
235.4	−11.7	7	6.45/3.59	2.37/2.2	6.92/1.52	40.4°
235.6	34.1	8	−4.88/2.00	7.48/4.26	8.48/2.72	44.5°
273.8	43.6	9	−0.86/0.69	−4.84/2.28	−2.03/1.64	51.5°
294.9	−13.3	7	−4.86/1.66	−5.44/2.14	−4.65/2.89	N/A
294.8	−4.5	9	−6.70/3.11	−3.15/2.60	−3.42/2.27	N/A
295.1	15.2	14	−3.95/1.83	−0.90/4.23	5.52/2.19	84.5°
305.0	14.6	8	−9.70/2.30	−12.16/2.59	6.04/3.75	N/A
304.6	25.5	11	−6.69/2.58	−4.51/3.19	9.94/2.59	105.1°
304.8	34.7	12	−5.85/2.68	−2.49/2.00	8.77/2.97	22.7°
323.2	15.4	5	−2.04/0.92	5.02/4.37	−2.76/2.19	N/A

SURFACE CORRELATED FIELDS

We have not performed a careful examination of all successive passes using high resolution wake data, searching for repeated features. We have done this for limited regions and found no such features (Russell *et al.*, 1980b). We plan to perform a more complete survey at a later date. For now we will concentrate on searching for large scale surface correlated fields using the data in the $10° \times 10°$ bins. We examined the field in each of the 72 bins in local planetary coordinates of radial, east, and north components. We then compared the average of each component with our estimate of the probable error of the mean. We took as our estimate the standard deviation of the component divided by the square root of the number of orbits contributing to that bin. Of the 216 components, 94 or 43.5% were greater than one probable error of the mean from zero, 26 or 12% greater than two, 6 or 2.8% greater than 3 and 3 or 1.4% greater than 4. If the distribution of observed fields was due to a random Gaussian distribution of fields from external sources with zero mean and if we have chosen a good estimate of the probable error of the mean then the observed distribution is decidedly non-random because we would only expect 32%, 4.6%, 0.3% and 0.01% of the data to be more than 1, 2, 3 and 4 times the probable error of the mean. On the other hand, if the actual probable error of the mean is only 40% greater than we assumed it to be then the observed distribution of moments should be expected to arise at random.

To test further the randomness of the data we have listed in Table 1 the most non-random fields observed. The table gives the average longitude and latitude of the observations, the number of orbits in the average, and the components of the field and our estimate of the probable error of the mean. Two of the bins have components over four standard deviations of the mean away from zero: 235°E, −12°N and 305°E and 15°N. This fact would argue strongly that the fields in these bins were not due to a random source if we could demonstrate that our estimate of the probable error in the mean were correct. In fact external conditions could persist on average 2 orbits or even more. Thus, we do not feel this table presents conclusive evidence of crustal magnetism. As a further check we compared the average direction of the field obtained on the two wake passages whenever there was redundant data. We see that three of the nine entries include data from only one wake passage (the first) and that the field directions differ by moderately large angles from one year to the next, ranging from 23° to 105°. Again this argues against a crustal origin for these fields.

THE PLANETARY MAGNETIC MOMENT

When we invert the magnetic fields in our 72 bins, we obtain 72 estimates of the planetary magnetic moment. If we assume that each of these values is an independent estimate of the moment then the resultant moment and probable error of the mean are:

M_x: $0.32 \pm 0.81 \times 10^{21}$ Gauss-cm^3
M_y: $0.55 \pm 0.74 \times 10^{21}$ Gauss-cm^3
M_z: $-0.59 \pm 1.02 \times 10^{21}$ Gauss-cm^3
$|M|$: $0.87 \pm 1.50 \times 10^{21}$ Gauss-cm^3

On the other hand, it is unlikely that these 72 estimates are independent since the same external conditions prevail on each orbital pass across all latitudes (at roughly a constant longitude). Thus a better estimate of the degrees of freedom is the number of longitudinal bins. Since the data from the second wake passage is clearly independent of the first we add the number of longitudinal bins in each wake passage. This number totals to 18. Thus we must double the error bars and the total moment and its uncertainty becomes $0.87 \pm 3.00 \times 10^{21}$ Gauss-cm^3. We note that our limit on the Venus magnetic moment must be considered to be 3×10^{21} Gauss-cm^3, not 0.87×10^{21} Gauss-cm^3. This magnetic moment corresponds to an equatorial surface field of slightly over 1 gamma (nT).

DISCUSSION AND CONCLUSIONS

We have searched the Pioneer Venus orbiter magnetic field data for convincing evidence that Venus has intrinsic magnetic fields of either crustal or dynamo origin and have found none. Our analysis indicates that any surface correlated field must have an average magnitude (averaged over 10° at 200 km altitude) of less than about 5 gammas. On the other hand, we note that on the moon which has extensive crustal magnetism the average field at 100 km is less than 1γ (Russell *et al.*, 1975). Thus we are far from establishing that the Venus crust is less magnetized than that of the moon. Furthermore, we have surveyed but a fraction of the surface of Venus and have not overflown either Aphrodite Terra or Ishtar Terra.

The weakness of the dipole moment is quite surprising in view of our expectations based on the planetary dynamo scaling law of Busse (1976), and the conclusions of Dolginov *et al.* (1978). The apparent explanation in the former case is that the energy available to drive a dynamo in the Venus interior is not sufficient to maintain one, and in the second case that Dolginov *et al.* have mistakenly identified magnetic fields of external origin to be of internal origin. Since the energy source for the earth's magnetic dynamo is the subject of debate we can only speculate why it is absent at Venus. Stevenson (1979) has postulated that the very small difference in size between Venus and the earth may be sufficient to lower the pressure in the center of Venus and prevent the formation of a solid core. The latent heat released in the freezing of a solid inner core has been postulated to be one of the driving energy sources for the terrestrial dynamo (Gubbins *et al.*, 1979).

We plan to continue our search for a intrinsic magnetic vield at Venus as additional data become available but do not expect our limits on the Venus moment to change markedly from those given herein since we only expect to increase our low altitude data base by about 50%. We also will continue our search for crustal magnetism as additional surface coverage becomes available.

Acknowledgments—This research was supported by the National Aeronautics and Space Administration under contract NAS2-9491.

REFERENCES

Busse F. H. (1976) Generation of planetary magnetism by convection. *Phys. Earth Planet. Inter.* **12**, 350–358.

Dolginov Sh. Sh., Yeroshenko E. G., and Zhuzgov L. N. (1968) Magnetic field investigation with spacecraft Venera-4. *Kosmich. Issled.* **6**, 562.

Dolginov Sh. Sh., Yeroshenko Ye. G., and Davis L. (1969) On the nature of the magnetic field near Venus. *Kosmich. Issled.* **7**, 747–752.

Dolginov Sh. Sh., Zhuzgov L. N., Sharova V. A., and Buzin V. B. (1978) Magnetic field and magnetosphere of the planet Venus. *Kosmich. Issled.* **16**, 827–863.

Elphic R. C., Russell C. T., and Slavin J. A (1980) Observations of the dayside ionopause and ionosphere of Venus. *J. Geophys. Res.* In press.

Gubbins D., Masters T. G., and Jacobs J. A. (1979) Thermal evolution of the earth's core. *Geophys. J. Roy. Astron. Soc.* **59**, 57.

Russell C. T. (1976) The magnetic moment of Venus: Venera-4 measurements reinterpreted. *Geophys. Res. Lett.* **3**, 125–129.

Russell C. T., Coleman P. J. Jr., Fleming B. K., Hilburn L. D., Ioannidis G., Lichtenstein B. R., and Schubert G. (1975) The fine scale lunar magnetic field. *Proc. Lunar Sci. Conf. 6th.* p. 2955–2969.

Russell C. T., Elphic R. C., and Slavin J. A. (1979a) Initial Pioneer Venus magnetic field results: Nightside observations. *Science* **205**, 114–116.

Russell C. T., Elphic R. C., and Slavin J. A. (1979b) Initial Pioneer Venus magnetometer observations. *Proc. Lunar Planet. Sci. Conf. 10th,* p. 2277–2290.

Russell C. T., Snare R. C., Means J. D., and Elphic R. C. (1980a) Pioneer Venus Orbiter fluxgate magnetometer. *ISEE Trans. Geosci. Remote Sensing* **GE-18**, 32–35.

Russell C. T., Elphic R. C., and Slavin J. A. (1980b) Limits on the possible intrinsic magnetic field of Venus. *J. Geophys. Res.* In press.

Slavin J. A., Elphic R. C., Russell C. T., Scarf F. L., Wolfe J. H., Mihalov J. D., Intriligator D. S., Brace L. H., Taylor H. A. Jr., and Daniell R. E. Jr. (1980) The solar wind interaction with Venus: Pioneer Venus observations of bow shock location and structure. *J. Geophys. Res.* In press.

Stevenson D. J. (1979) Comparative thermal evolutions and magnetism of the earth and Venus (abstract). *EOS (Trans. Amer. Geophys. Union)* **60**, 305.

Yerohenko E. G. (1979) Unipolar induction effects in the magnetic tail of Venus. *Kosmich. Issled.* **17**, 93–105.

Proc. Lunar Planet. Sci. Conf. 11th (1980), p. 1907–1913.
Printed in the United States of America

The core and the magnetic field of Uranus

M. Torbett and R. Smoluchowski

Departments of Astronomy and Physics, University of Texas, Austin, Texas 78712

Abstract—The presence of a magnetic field on Uranus can be explained in a natural way using the recent three-layer model of this planet because, in contrast to earlier models, the core is metallic, mostly liquid and gravitational differentiation produces enough heat in it to drive a hydromagnetic dynamo.

The existence of a magnetic field on Uranus is suggested by the tentative observation of 0.5 Mhz pulses from the planet (Brown, 1976) which are analogous to the Jovian decametric radiation. Attempts to account for this field, however, were hampered by the lack of suitable conditions within the planet for the generation of magnetic fields in the models so far proposed. Analysis of a recent three-layer model, though, leads to the conclusion that the electrical conductivity, physical state, and thermal gradients are such that a thermally driven hydromagnetic dynamo is likely to be operating in the core.

The generation of a magnetic field is highly sensitive to the internal structure of a planet. A satisfactory model for the Uranian field is complicated by the lack of more detailed knowledge of composition, equations of state, and by the non-uniqueness of the models of the interior. For instance, the two-layer model proposed by Podolak and Cameron (1974) could account for the field only if the planet possessed a metallic hydrogen or a metallically conductive molecular hydrogen layer as suggested by the lower transition pressure to metallic hydrogen of ~2 Mbar (Hawke *et al.*, 1978; Torbett and Smoluchowski, 1979). This layer is marginally possible at the core boundary pressures yet the thinness of this shell makes it unlikely that a field could be generated there. Furthermore, in this model the core turns out to be solid. The ionic conductivity of liquid H_2O is difficult to estimate because while initially below 1 Mbar it increases (Nellis, pers. comm., 1979) as a result of dissociation and formation of ions, at higher pressures it should decrease because of the usual drop in ionic mobility with increasing pressure. The activation volume which controls this effect has not yet been measured. The question of pressure-metallization of H_2O appears to be still unsettled (Vereschagin *et al.*, 1975). It is thus very likely that the conductivity in the ice layer will be too low to permit the operation of a dynamo. A dynamo operating at the core-mantle boundary would require convective velocities far too large to be

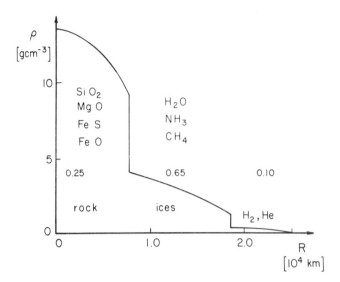

Fig. 1. Three-layer model of Uranus according to Hubbard and MacFarlane (1980).

compatible with the observational upper limits of the surface heat flux (Loewenstein *et al.*, 1977). A precessionally driven dynamo might lead to higher velocities but its presence is very unlikely (Smoluchowski, 1979).

The new three-layer model of Hubbard and MacFarlane (1979), which is based on better equations of state, seems to provide an improved fit to the observed gravitational moments, the optical oblateness, and the chemical constraints than previous models. For this reason, an analysis to check whether this structure was suitable for the generation of a field was undertaken. The model, shown in Fig. 1, consists of a "rocky" core, an "ice" mantle, and a H_2-He envelope comprising 25, 65 and 10 percent of the planet by mass respectively. The existence of a metallically conductive layer is excluded in this structure since the pressure at the base of the H_2-He envelope is well below 2 Mbar. For reasons similar to those given above, the conductivity of the ice layer is probably too low for the operation of a dynamo within the limitations imposed on the convective velocities by the low surface heat flux. If a layer of metallic water is present at the core boundary it is too thin to sustain a dynamo. It remains then to analyze the planet's core which contains ~3.5 earth masses of rock and has a radius of 7700 km (R_c/R_u ~ 0.3). The core is considered to consist of solar proportions of the nonvolatiles and can be roughly approximated by a mixture of 38, 25, 25, 12 percent by mass of MgO, SiO_2, FeS, FeO respectively. To investigate the physical state of the core, the melting temperatures of these compounds were calculated in the Debye approximation (Zharkov and Kalinin, 1971).

$$T_m = T_o \left(\frac{\rho_o}{\rho}\right)^{2/3} \left(\frac{\theta}{\theta_o}\right)^2$$

where T_o, ρ_o, θ_o are the melting point, density, and Debye temperature at standard conditions. The variation of the Debye temperature with density is given by

$$\theta = \theta_0 \left\{ \exp \int_{\rho_0}^{\rho} \frac{\gamma}{\rho} \, d\rho \right\}$$

where γ, the Grüneisen gamma, were taken from Zharkov and Kalinin (1971). The density dependence of γ for FeS was assumed to be similar to that for FeO. The equations of state for these compounds were taken from Zharkov *et al.* (1974) and the results are shown in Fig. 2. For an isothermal core at 7000°K, as assumed in the three-layer model, the iron compounds are liquid throughout the core, MgO appears solid everywhere, while SiO$_2$ seems to be solid in the inner

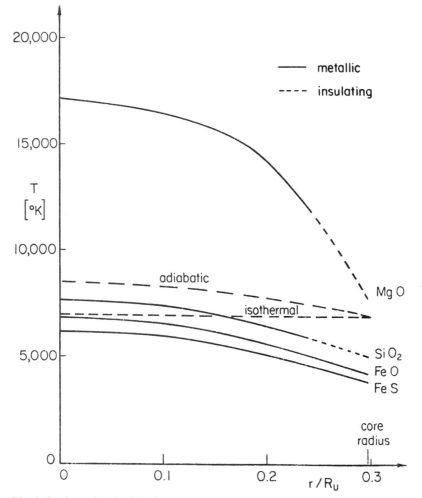

Fig. 2. Isothermal and adiabatic temperature gradients in the core of Uranus and the melting temperatures of the constituents. R_u is the radius of the planet.

regions of the core. In this simple view, then, roughly half of the core would be liquid. If, however, an adiabatic temperature gradient, given by

$$\ln \frac{T_l}{T_{lo}} = \int_{l_o}^{l} \frac{g\alpha}{C_p} \, dl$$

where g = acceleration of gravity, α = thermal expansivity and C_p = specific heat is extended from 7000°K at the core boundary to the center of the planet, the result is as shown in Fig. 2. Within the limits of uncertainty all materials except MgO appear liquid. Actually, the core is expected to be rather more liquid than this estimate suggests in view of the fact that the compounds in the core will not retain their individual identity but will form composite materials and eutectics which generally melt at considerably lower temperatures than the constituent compounds. Hence, it is assumed that the bulk of the core is indeed in a liquid state. The same is true if, as discussed below, the core becomes metallic.

To investigate the electrical conductivity of the core, Herzfeld's simple but surprisingly successful molar refractivity criterion for metallic conduction (Ross, 1972) was applied to the core compounds individually. This condition, stated briefly, is that the electrons are freed for metallic conduction when the volume available to the bound electrons becomes less than that occupied by the volume of the molecule. The results are shown in Fig. 3, where metallic conductivity is

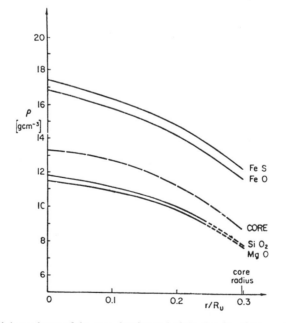

Fig. 3. Radial dependence of the core density and of the density of its constituents. Full line—metallic, dashed line—nonmetallic. R_u is the radius of the planet.

indicated by solid lines. MgO and SiO_2 are found to be metallic up to about 85% of the radius of the core, FeO is metallic throughout the core and FeS is known to become metallic even in the kilobar range of pressures (Stiller *et al.*, 1974). Thus, it is concluded that the bulk of the core possesses metallic conductivity.

The total primordial thermal energy of the planet derived from virialized gravitational energy and early radiogenic heating as estimated from Hubbard's evolutionary cooling sequences is $>10^{40}$ ergs. It can be shown that this heat was originally sufficient to keep the core temperature in excess of that required for it to be entirely liquid in the metallic state. Furthermore, the temperature gradient produced convection and, in this manner, served to keep the core well-stirred and homogeneous as well as generating an early magnetic field. However, as the core cooled and the temperature distribution tended to become subadiabatic and convection became less effective, the originally homogeneous core started undergoing gravitational fractionation due to the density disparity of the constituents. A diffusion controlled "rainout" of the iron compounds towards the center ensued. This process evolves considerable heat and, in analogy to the situation in the H-He system (Salpeter, 1973), leads to a self-regulating mechanism where the rate of segregation and heat production is limited by convective re-mixing. In order to estimate the energy available from this mechanism, the difference in gravitational potential energy between a core with uniformly distributed materials and a totally segregated core was calculated and a value of $\Delta\phi \sim 10^{39}$ ergs obtained. This energy distributed, say, over the age of the solar system corresponds to an average surface heat flux of $\sim$40 ergs $cm^{-2}s^{-1}$, which is well compatible with the observational upper limit of $\sim 10^2$ ergs $cm^{-2}s^{-1}$. Thus it seems reasonable to conclude that, at present, the temperature distribution within the core is not isothermal but slightly superadiabatic. Unfortunately, a lack of knowledge of the phase diagrams of the multicomponent alloys at the pressures and temperatures encountered in the core together with a lack of information about mutual solubilities makes a more quantitative estimate of the heat generation rate impossible. It should be noted, however, that with increasing pressures the miscibility gap of silicate and sulfide liquids increases (Huang and Williams, 1980), which will tend to promote segregation.

Thus, with a liquid and metallically conductive core possessing an internal energy source a thermally convective dynamo seems likely. Using a mixing-length theory in the convenient formulation of J. Scalo (pers. comm.) the convective velocity is given by

$$U = \left(\frac{gQ(\nabla - \nabla s)}{2H_p}\right)^{1/2} l$$

where $Q \equiv -\left(\dfrac{d \ln \rho}{d \ln T}\right)p$, H_p is pressure scale height and l is the mixing length. Assuming a typical superadiabaticity of $\nabla - \nabla_s \sim 10^{-6}$, which is a lower bound for convection being able to carry the bulk of the flux, a thermal expansion coefficient $\alpha \sim 10^{-6}\text{deg}^{-1}$, and a mixing length $l = 10^{-3}r_c$, one obtains for the convective velocity a value of

$$U \sim 10^{-1} \text{ cm s}^{-1}.$$

The mixing length was chosen in order to obtain a heat flux compatible with observations. The heat flux carried by this convection is given by (Scalo, l.c.)

$$\phi = \frac{\rho C_p U \, l T (\nabla - \nabla_s)}{2H_p}.$$

For the present conditions this yields a heat flux of $\phi \sim 10^2$ ergs cm^{-2}s^{-1} at the core boundary or $\phi \sim 10$ erg cm^{-2}s^{-1} at the planet's surface which is well within the observational limits.

It remains to check whether this convection in the core can sustain a hydromagnetic dynamo. The necessary condition for the operation of such a dynamo is that the magnetic Reynolds number

$$R_m = \mu\sigma UL$$

be greater than about 10 where μ = permeability, σ = conductivity, U = convective velocity, and L = size of the core. In view of the highly heterogeneous nature of the core, it seems reasonable to assume a value of $\sigma \sim 10^{-5}$ emu for the electrical conductivity, which is typical of low conductivity metals. With these values for σ and U the above condition for magnetic field generation is seen to be satisfied by at least an order of magnitude.

Thus, the existence of a magnetic field on Uranus can be understood as being a natural result of gravitational differentiation in a liquid, metallic core. The presence of a heat source in the core implies that the central temperature is $\sim 8400°$K rather than 7000°K in the original model. This difference will not substantially affect the model's other results since a comparable flux was already taken into account and since the thermal pressure was taken to be negligible originally. Finally, Busse's (1976) semi-empirical rule applied to the Uranian core leads to an estimate for the field strength of $\sim 10^{-1}$ gauss at the planet's surface.

Acknowledgments—Supported by NASA Grant NSG-7505 supplement 1.

REFERENCES

Brown L. W. (1976) Possible radio emission from Uranus at 0.5 MHz. *Astrophys. J.* **207**, L209-L212.
Busse F. H. (1976) Generation of planetary magnetism by convection. *Phys. Earth Planet. Inter.* **12**, 350–358.
Hawke R. S., Burgess T. J., Duerre D. E., Huebel J. G., Keeler R. N., Klapper H., and Wallace W. C. (1978) Observation of electrical conductivity of isentropically compressed hydrogen at megabar pressures. *Phys. Rev. Lett.* **41**, 994–997.

Huang W. L. and Williams R. J. (1980) Melting relations of portion of the system Fe-S-Si-O to 32 KB with implication to the nature of mantle-core boundary (abstract). In *Lunar and Planetary Science XI*, p. 486–488. Lunar and Planetary Institute, Houston.

Hubbard W. B. and MacFarlane J. J. (1979) Structure and evolution of Uranus and Neptune. *J. Geophys. Res.* **85**, 225–234.

Hubbard W. B. (1978) Comparative thermal evolution of Uranus and Neptune. *Icarus* **35**, 177–181.

Loewenstein R. F., Harper D. A., Moseley S. H., Telesco C. M., Thromson H. A. Jr., Hildebrand R. H., Whitcomb S. E., Winston R., and Steining R. F. (1977) Far infrared and submilimeter observations of the planets. *Icarus* **31**, 315–324.

Podolak M. and Cameron A. G. W. (1974) Models of the giant planets. *Icarus* **22**, 123–148.

Ross M. (1972) On the Herzfeld theory of metallization; application to rare gases, alkali halides and diatomic molecules. *J. Chem. Phys.* **56**, 4651–4653.

Salpeter E. E. (1973) On convection and gravitational layering in Jupiter and in stars of low mass. *Astrophys. J.* **181**, L83-L86.

Smoluchowski R. (1979) Origin of the magnetic fields in the giant planets. *Phys. Earth Planet. Inter.* **20**, 247–254.

Stiller H., Siepold U., and Vollstaedt H. (1974) Electrical, thermoelectrical and elastic properties of transition metal chalcogenides under critical pressure-temperature conditions. *Fiz. Svoistva Gorn. Prod. Miner. Vys. Temp. Mater. Vses. Soveshch.* **4**, 198–200.

Torbett M. and Smoluchowski R. (1979) The structure and the magnetic field of Uranus. *Geophys. Res. Lett.* **6**, 675.

Vereshchagin L. F., Yakovliev E. N., and Timofeev Yu. A. (1975) Transition of H_2O into the conducting state at static pressures P $\approx$ 1 Mbar. *J. Exp. Theor. Phys. Lett.* **21**, 304–305.

Zharkov V. N. and Kalinin V. A. (1971) *Equations of State for Solids at High Pressures and Temperatures.* Plenum, N.Y. 380 pp.

Zharkov V. N., Trubitsyn V. P., Tsarevskij I. A., and Makalkin A. B. (1974) The equations of state of cosmochemical materials and the structure of the major planets. *Phys. Solid Earth* **10**, 610–617. (In English).

Proc. Lunar Planet. Sci. Conf. 11th (1980), p. 1915–1929.
Printed in the United States of America

On the early global melting of the terrestrial planets

Charles J. Hostetler and Michael J. Drake

Department of Planetary Sciences and Lunar and Planetary Laboratory, University of Arizona,
Tucson, Arizona 85721

Abstract—It is generally agreed that the moon underwent a global differentiation event approximately contemporaneously with its formation. The heating mechanism(s) responsible for melting the moon cannot be identified unequivocally. In the absence of this information, a chronology is constructed in order to examine the time intervals during which various heating mechanisms may have operated, either singly or in concert. The thermal input to each terrestrial planet relative to the moon is calculated for each plausible combination of heating mechanisms. It is concluded that unless a combination of electrical induction and tidal heating contributed the major portion of thermal input to the moon, it is probable that all terrestrial planets underwent early global differentiation. If induction and tides did contribute most of the thermal input into the moon, it is possible that the ancient southern cratered highlands of Mars could consist of primordial, albeit weathered, Mars material. The possibility exists that a sample return from Mars could constrain the timescale for accretion of the terrestrial planets and the heating mechanisms responsible for the early global melting of the moon and planets.

INTRODUCTION

In order to decipher the crustal records of the terrestrial planets, we must investigate their early thermal histories. The conceptual formalism used in constructing early thermal histories is based largely on our understanding of the lunar thermal history as deduced from returned lunar samples and remote sensing devices. The mare/highlands physical and chemical dichotomy and an early separation of chemical and isotopic complementary reservoirs, (seen preserved in mare basalts which erupted much later than the emplacement of the highlands anorthosites) seem to point to a moon which underwent an early global differentiation or "magma ocean" event* (see, for example, Taylor, 1975). Although we have as yet no returned samples from the other terrestrial planets, it is reasonable to ask if these other bodies also underwent global differentiation events similar to the moon.

A definitive answer to this question requires knowledge of physical conditions in the early solar system and a knowledge of the mechanisms responsible for

---

*By "global" we refer to igneous processes responsible for large scale chemical differentiation. These processes may encompass superposition of large melt pools, magma oceans, or complete melting of the planet.

causing the lunar magma ocean. We clearly do not possess the latter. However, in this work we consider all of the mechanisms known to us which might have been responsible for the melting of the moon, and estimate their effects on all of the terrestrial planets relative to their effects on the moon. We also establish a chronology of the early solar system in order to evaluate the time periods during which each of the mechanisms might have been active. With the relative effects and timescales, we can examine the consequences for the early thermal histories of the terrestrial planets of combinations of mechanisms which might have operated to produce a lunar magma ocean. Although we cannot obtain unique solutions using this approach we can deduce the probability of planetary magma oceans on a scale comparable to the lunar magma ocean.

ASSUMPTIONS CONCERNING THE MODE OF ORIGIN OF THE MOON

In order to proceed, we must make some simplifying assumptions about the origin of the moon. Clearly the mode of origin of the moon needs to be considered, as we will use the moon as a reference planet against which to scale the other terrestrial planets. We consider five possible modes of origin for the moon (see Taylor, 1975 for general references):

1) Independent origin of the moon in a heliocentric orbit with a semimajor axis of 0.7 A.U. The moon was subsequently captured by the earth.

2) Independent origin of the moon in a heliocentric orbit with a semimajor axis of 1.0 A.U. The moon was subsequently captured by the earth. For our purposes, a geocentric origin ("double planet" hypothesis) is equivalent.

3) Independent origin of the moon in a heliocentric orbit with a semimajor axis of 1.6 A.U. The moon was subsequently captured by the earth.

4) Disintegrative capture of the moon from a large number of asteroidal sized differentiated bodies (Smith, 1974; Wood and Mitler, 1974). These differentiated bodies were stripped of their crusts by tidal interactions with the earth. The metal cores remained intact and passed through the earth system without capture. The disintegrated crustal material was then reassembled in earth orbit to form the moon.

5) Fission of the moon from the earth. Various means have been invoked to remove upper mantle material from the earth (e.g., Ringwood, 1970; Binder, 1978). The material was then reassembled in earth orbit to form the moon.

In order to use the moon as a scaling planet, our primary assumption must be that the moon formed in a similar manner to the other terrestrial planets. As it is not believed that the other terrestrial planets formed by disintegrative capture or by fission from other bodies, we discard modes 4 and 5 as unsuitable for direct scaling. For purposes of discussion we assume in the remainder of this paper that, regardless of how the moon was formed, it attained a geocentric orbit prior to global differentiation. The absence of chemical or isotopic evidence recording disturbance during capture more recently than 4.6 AE supports this assumption.

As we are comparing the terrestrial planets to the moon and to each other, we must also consider conditions in the early solar system as a function of time. An assumption is that the terrestrial planets and the moon all reached sizes comparable to their present sizes at the same time, so that the same heating mechanisms were operating on all of the bodies when their early crusts were formed.

THE METHOD OF RELATIVE SCALING

It is extremely difficult to model the absolute contributions of various heating mechanisms to each of the planets. Fortunately this exercise is not necessary. Because we assume that the moon formed at about the same time and by the same process that formed the other planets, and that the moon underwent a global melting event, we may use the moon as a standard against which to compare the other terrestrial planets. This approach allows us to compute relative rather than absolute effects.

In asking if the terrestrial planets underwent early global differentiation events, we are really asking about the amount of thermal energy deposited by various processes in the upper layer of each terrestrial planet. If the upper layer thickness dr is small compared to r (Fig. 1), this question is equivalent to asking about the amount of thermal energy deposited per unit surface area, assuming that possible

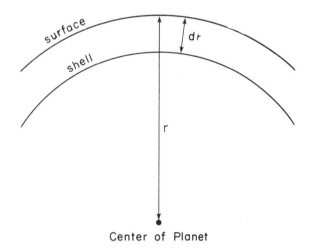

Center of Planet

Fig. 1. The thermal energy deposited in a thin shell in a planet per unit volume of that shell is equal to the thermal energy deposited per unit surface area of the planet if the shell thickness is small.

$$\text{Volume of shell} = V = \frac{4}{3}\pi r^3 - \frac{4}{3}\pi(r - dr)^3$$
$$\approx S dr.$$

S = surface area.

magma oceans on the terrestrial planets were of comparable thickness to the lunar magma ocean. Estimates of the depth of the lunar magma ocean vary from 300 km (Solomon and Chaiken, 1976), to total melting of the moon (Taylor, 1975). If the lunar magma ocean was deeper than comparable magma oceans on other terrestrial planets our approach yields lower bounds on the melting of the terrestrial planets. Our equations permit an estimation of the amount of thermal energy deposited on each planet per unit of planet surface area relative to the amount of thermal energy deposited on the moon per unit of lunar surface area.

Let λ_i equal the fraction of thermal energy input by mechanism i on the moon:

$$\lambda_i = H_i^m / \sum_j H_j^m \tag{1}$$

where H_i^m is the absolute amount of thermal energy input to the moon by mechanism i and the summation is over all possible mechanisms. Note that the sum of all λ_i must be equal to one. The amount of thermal energy deposited in a planet p relative to the moon from mechanism i is given by:

$$R_i^p = H_i^p / H_i^m. \tag{2}$$

The total amount of thermal energy input to planet p relative to the moon is a sum over the relative contributions of each of the mechanisms weighted by the fraction of thermal energy input to the moon for each mechanism:

$$F^p = \sum_i \lambda_i R_i^p. \tag{3}$$

We will usually be interested in contours of F in λ_i space. In our calculational scheme, the λ_i are unknowns, and the R's are calculated below. Note that if F = 1, the planet received as much thermal energy per unit surface area as did the moon and, therefore, should have undergone a global melting event as did the moon. It is also important to remember that, using our approach, the volume of the lunar and planetary magma oceans must be comparable for F = 1 (see Fig. 1).

HEATING MECHANISMS

We now consider the possible heating mechanisms. To be effective in melting an outer shell of a planet the heating mechanisms should have deposited their heat on or near the surface, have acted during or immediately after accretion of the terrestrial planets, and have generated a large amount of heat. Goles and Seymour (1980, in prep.) considered six mechanisms: enhanced solar luminosity, electromagnetic induction, shortlived radioactivities, adiabatic compression of nebular gases, accretion, and tidal dissipation. We have also considered thermal loss, autoreduction, and disproportionation, but find these mechanisms to be unimportant as major contributors to the generation of magma oceans. A magma ocean event appears to be required in order to trigger core formation: consequently gravitational potential energy liberated during core formation would augment

Table 1. Thermal mechanisms and their principal controlling parameters used for calculating thermal input per unit surface area relative to the moon.

Mechanism		Parameters	Comments
1) Solar luminosity	(Sl)	$\dfrac{1}{a^2}$	inverse square law for radiation
2) Al^{26}	(Al)	r	surface/volume law, assumed equal per unit volume
3) Accretion	(Ac$_1$)	r^2	r^4/unit area, for small planetesimals
	(Ac$_2$)	rρ	Grav potential/unit area, for large planetesimals
4) Adiabatic compression	(Ad)	rρ	mass/unit area
5) Tides	(Ti)		important only for earth and moon
6) Induction	(In)	$\dfrac{1}{a^2}\dfrac{1}{r}$	important only for Moon and Mercury

a = semimajor axis of heliocentric orbit
r = radius of body
ρ = density of body

planetary melting but is unlikely to operate as an independent mechanism capable of producing a magma ocean (e.g., Stevenson, 1980). A summary of the thermal mechanisms and their principal controlling parameters is given in Table 1, while Table 2 gives their numeric contributions relative to a moon in geocentric orbit at 1 A.U.

Enhanced solar luminosity

It has been suggested that stars of approximately solar mass undergo a brief period of superluminosity during the so-called Hayashi contraction (Hayashi, 1966). Sonett and Reynolds (1979) pointed out that more recent calculations of the pre-main sequence evolution do not display enchanced luminosity, and the

Table 2. Thermal input matrix (R$\tilde{\rho}$) for terrestrial planets relative to the moon for each of the thermal mechanisms.

Mechanism	Mercury	Venus	Earth	Moon*	Mars
Solar luminosity	6.68	1.91	1.0	1.0	0.43
Al^{26}	1.40	3.48	3.67	1.0	1.95
Accretion (1)	1.96	12.11	13.47	1.0	3.80
(2)	2.31	5.43	6.06	1.0	2.28
Adiabatic compression	2.31	5.43	6.06	1.0	2.28
Tides	0	0	1.0	1.0	0
Induction	4.77	0	0	1.0	0

*A hypothetical moon which attained geocentric orbit prior to global differentiation is the reference body for this calculation.

early stellar evolution sequence is still unknown. The time period during which the sun might have been superluminous was brief and was early in the sun's history. Ezer and Cameron (1962) and Hayashi (1966) demonstrated that a star of solar mass should exhibit luminosity greater than ten times the present solar luminosity no later than 10^5 years after the start of contraction.

Equation 2 is used to calculate the amount of thermal energy from solar luminosity incident on a planet relative to the moon. The solar flux and thus the thermal energy depends only on the semimajor axis of the planet and the moon:

$$R_{SL}^P = \frac{L}{4\pi a_p^2} \cdot \frac{4\pi a_m^2}{L} \tag{4}$$

or

$$R_{SL}^P = \left(\frac{a_m}{a_p}\right)^2 \tag{5}$$

where a_p and a_m are the semimajor axes of the planet and the moon, and L is the luminosity of the sun.

Electromagnetic induction

Electromagnetic induction, during a T Tauri-like period of the sun, requires a rapidly rotating sun with a strong magnetic field and greatly enhanced solar wind (Sonnett et al., 1970; Sonett and Reynolds, 1979). The time period in which the sun might have undergone a T Tauri phase was after the Hayashi contraction but before entry onto the main sequence, i.e., between 10^7 and 3×10^7 years (Ezer and Cameron, 1962).

We assume that induction is important only on small bodies lacking atmospheres and intrinsic magnetic fields, as an atmosphere or a highly conducting ionosphere would short out the induced currents. Thus electromagnetic induction was probably important only for the moon and Mercury. A crude scaling is given by Sonett et al. (1970):

$$R_{IN}^P = \left(\frac{a_m}{a_p}\right)^2 \left(\frac{r_m}{r_p}\right)^2 \tag{6}$$

where r_p and r_m are the radii of the planet and the moon respectively. This scaling assumes that the surface electrical conductivities of the terrestrial planets and the moon are similar.

Shortlived radioactivities

Shortlived radioactivities, produced in nuclear reactions in a time period close enough to the onset of accretion to permit preservation of isotopic anomalies, would have generated heat by radioactive decay. Of the possible lithophilic nuclides, only ^{26}Al appears to be a possible source for the early crustal heating. The time during which ^{26}Al could have been an important heat source is determined by its halflife, 7.2×10^5 years. Although this is not evidence that decay of ^{26}Al was an important contributor of thermal energy to the terrestrial planets or the moon, it is certain that ^{26}Al was "alive" at the time of its incorporation into Allende, and thus was a possible heat source (Lee *et al.*, 1977). We assume for simplicity that equal amounts of ^{26}Al per unit volume were present everywhere in the inner solar system. After normalizing to amount per surface area, i.e., correcting for surface/volume effects, the appropriate scaling is given by:

$$R_{AL}^P = \frac{r_p}{r_m}. \tag{7}$$

Adiabatic compression of nebular gases

As the planets grew, the nebular gases surrounding them could have generated thermal energy by adiabatic compression onto their surfaces. This process could have occurred at anytime during the accretional period, up to 10^8 years after the start of collapse of the solar nebula. Note that, if the sun underwent a T Tauri-like phase with electromagnetic induction operating, the attendant enhanced solar wind would have dispersed the solar nebula. Thus electromagnetic induction and adiabatic compression could not have operated simultaneously. Assuming similar background temperatures, Ostic (1965) showed that adiabatic compression depends on the final mass of the planet:

$$R_{AD}^P = \frac{r_p \rho_p}{r_m \rho_m} \tag{8}$$

where ρ_p and ρ_m are the densities of the planet and the moon. The factor of r is due to correction to unit surface area.

Accretion

The thermal contribution due to accretion depends strongly on growth conditions such as the size distribution of the accreting bodies and the relative velocities of the bodies. The relative velocities are due to differences in eccentricity and inclination, and are assumed to be equal on the average throughout the inner solar system. It appears to be unlikely that the maximum time period over which accretion could have operated exceeded 10^8 years (Weidenschilling, 1976).

Thus, the scaling of accretion depends upon the properties of the planetesimals. We have considered two possibilities. If the size/frequency curve in each feeding zone was steep, i.e., if there were many small bodies relative to large, and the second largest planetesimal in each feeding zone was small compared to the largest planetesimal, the accretion rate per unit surface area is of importance.

The accretion rate is assumed to be proportional to r^4. Thus the scaling for the first type of accretion is:

$$R^p_{AC1} = \left(\frac{r_p}{r_m}\right)^2.$$ (9)

We refer to this mode of accretion as "Mode 1".

If, on the other hand, the size/frequency distribution in each feeding zone was flat, and the second largest planetesimal in each feeding zone was large in comparison with the remainder of the planetesimals, the gravitational potential energy of the impacting planetesimals per unit surface area becomes important. The gravitational potential energy should depend upon the mass of the accreting planet, all other factors being equal on the average (e.g., average mass of planetesimal, average distance to planetesimal). After correction to unit surface area, the scaling for this second type of accretion is:

$$R^p_{AC2} = \left(\frac{r_p}{r_m}\right)\left(\frac{\rho_p}{\rho_m}\right).$$ (10)

We refer to this mode of accretion as "Mode 2".

Wetherill (1976) considered a particle in a box model for accretion. He suggested that once the earth reached a certain size relative velocities, and thus eccentricities, would be so great that planetesimals in the earth's feeding zone could begin to cross the orbit of Venus. The gravitational cross-section of the largest planetesimal (presumably the embryonic earth) would effectively be reduced, permitting the next largest planetesimal to grow to 0.07 earth masses, so that accretional heating would switch from Mode 1 to Mode 2. Greenberg (1979) pointed out that the Venus perturbation model is only one of a number of possible scenarios. He found that the uncertainties in the models are too great to derive with any confidence a size distribution curve near the end of accretion. We have therefore used both cases in calculating the results given below.

Tides

If we assume that the moon was captured subsequent to its formation by the earth, tidal dissipation in the earth and moon could have heated their outer shells. We assume that this mechanism operated only on the earth and the moon. As tidal heating is a complex process which depends on many unknowns, we assume, as did Goles and Seymour (1980, in prep.), that the effects were similar on the moon and the earth.

Loss of thermal energy

The terrestrial planets must have lost heat through thermal emission. The thermal output per unit surface area must have depended on the effective temperatures of the terrestrial planets. In lieu of adequate knowledge of this effect, we have assigned equal contributions for all of the terrestrial planets relative to the moon.

Furthermore, if thermal equilibrium is assumed and, if the planets achieved their present radii at approximately the same time, the ratio of thermal input to output is the same for each of the terrestrial planets, and we need not consider output further.

Other mechanisms

Other thermal mechanisms might have operated on the terrestrial planets. Such mechanisms as autoreduction (Kuskov and Khitarov, 1979) and disproportionation (Mao and Bell, 1977) require high pressure and temperature to operate and were demonstrably not responsible for the lunar magma ocean. The heat generated by these mechanisms is also deposited at depth. Core formation is unlikely to be the principal heat source responsible for melting the outer shell of a planet.

CHRONOLOGY OF HEATING MECHANISMS IN THE EARLY SOLAR SYSTEM

In order to combine the relative contributions from each heat source to yield the total amount of thermal energy input to a planet relative to the moon, using equation 3), it is necessary to know which combination of mechanisms should be included in the sum. We have therefore prepared a chronology of the early solar system to aid us in this choice. Figures 2 and 3 are graphical representations of

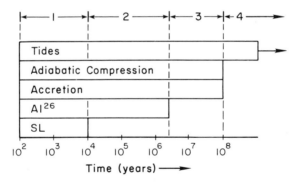

Fig. 2. Chronology of the Early Solar System—No Electromagnetic Induction. The length of the bar for each mechanism is dictated by the maximum time interval during which the given mechanism might have operated. Each time window represents a period of time during which the terrestrial planets might have attained their present radii, and possible magma oceans might have formed. The parenthesis around a mechanism indicates that the mechanism possibly provided thermal energy to the planets prior or subsequent to the time window, but not during the time window.

Window	Time	Mechanisms possibly operating
1	10^2–10^4	SL, AL, AC, AD, TI
2	10^4–2×10^6	—, AL, AC, AD, TI
3	2×10^6–10^8	—, —, AC, AD, TI
4	$10^8 \rightarrow$	—, —, (AC), (AD), TI

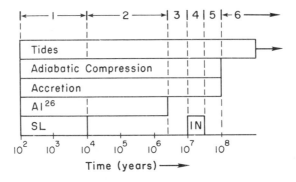

Fig. 3. Chronology of the Early Solar System—Electromagnetic Induction Possible. The length of the bar for each mechanism is dictated by the maximum time interval during which the given mechanism might have operated. Each time window represents a period of time during which the terrestrial planets might have attained their present radii, and possible magma oceans might have formed. The parenthesis around a mechanism indicates that the mechanism possibly provided thermal energy to the planets prior to or subsequent to the time window, but not during the time window.

Window	Time	Mechanisms possibly operating
1	$10^2 - 10^4$	SL, AL, AC, AD, TI, (IN)
2	$10^4 - 2 \times 10^6$	—, AL, AC, AD, TI, (IN)
3	$2 \times 10^6 - 10^7$	—, —, AC, AD, TI, (IN)
4	$10^7 - 3 \times 10^7$	—, —, AC, (AD), TI, IN
5	$3 \times 10^7 \rightarrow 10^{-8}$	—, —, AC, (AD), TI, (IN)
6		—, —, (AC), (AD), TI, (IN)

our chronology. The length of the bar for each mechanism is dictated by the maximum time interval during which the given mechanism might have operated. Note that the timescale on the horizontal axis is given in logarithmic units. The chronologies are divided into continuous time windows. Each time window represents a period of time during which the terrestrial planets might have attained their present radii, and possible magma oceans might have formed.

The two chronologies differ in their treatment of adiabatic compression of nebular gas and of electromagnetic induction. Because induction requires an enhanced solar wind, it seems probable that these two mechanisms cannot operate simultaneously as discussed earlier. Thus the chronology in Fig. 2 does not involve induction but permits the possibility of a contribution from adiabatic compression up to the time that accretion ends. The chronology in Fig. 3 permits adiabatic compression only up to the time window in which induction heating is possible.

RESULTS

If the terrestrial planets formed during time windows one or two in either of Figs. 2 or 3 (i.e., whether or not induction is possible) then the accretion time must have been shorter than approximately 2×10^6 years. Regardless of the possibility of enhanced solar luminosity, ^{26}Al, or tidal heating, the conclusion seems inescapable that accretion and adiabatic compression operating on such a short timescale should have easily produced magma oceans on all of the terrestrial planets. All terrestrial planets received more thermal energy from either mode of accretion and from adiabatic compression than did the moon.

For the other time windows we have prepared diagrams showing contours of F, the total amount of thermal energy input to the terrestrial planets relative to the moon (from equation 3), as functions of λ_i for each of the terrestrial planets. These diagrams are shown in Figs. 4a–4f. These ternary diagrams have two sets of lines for each contour value. The solid line corresponds to contour lines calculated using mode 2 accretion (related to the gravitational potential energy) and the dashed contour lines are calculated using mode 1 accretion (related to the accretional cross section).

The results in Figs. 4a and 4b correspond to the formation of the terrestrial planets in time windows three and four of Fig. 2. Accretion, adiabatic compression, and tides are the only mechanisms which contribute to heating the planets in these time windows. For Mercury and Mars the contour line for F = 1 is near $\lambda_{TI} = 0.6$ to 0.7. In other words, Mercury and Mars melted in this time window if tides contributed less than 60% to 70% of the total thermal energy to the moon. Venus would have melted if tides contributed less than 80% of the moon's thermal energy, and the earth would have melted regardless of the proportions of the three mechanisms.

The diagrams in Figs. 4c, 4d, 4e, and 4f correspond to the formation of the terrestrial planets in time windows three, four, five and six of Fig. 3. Because four mechanisms (accretion, adiabatic compression, tides, and induction) may contribute to heating the planets in these time windows, the results for each planet cannot be displayed on a single ternary diagram. We choose instead to present two ternary sections of this four-component space. Figs. 4c and 4d are a section along the hyperplane $\lambda_{AD} = 0$, i.e., no adiabatic compression. Figures 4e and 4f are a section along the hyperplane $\lambda_{TI} = 0$. (Figures 4a and 4b can be interpreted as a section along the hyperplane $\lambda_{IN} = 0$.) Note that the equations defining the contours are linear in λ_i space. Thus the contours must be planes, which aids in the visualization of the true three-dimensional figure portraying these time windows.

In Figs. 4c and 4d it can be seen that Mars melts if accretion (mode 2) supplied more than about 45% of the total thermal input to the moon. Earth would have melted if accretion (mode 2) supplied more than about 20% of the total thermal input to the moon. A similar condition holds true for Venus. Mercury cannot have escaped an early global melting event unless tidal heating contributed greater than 70% of the total thermal input to the moon.

In Figs. 4e and 4f it can be seen that Mars would have melted if induction contributed less than about 55% of the total thermal input to the moon. The earth and Venus would both develop magma oceans if induction contributed less than 80–85% of the total thermal input to the moon. Mercury melts far more extensively than the moon under all combinations of accretion, adiabatic compression, and induction.

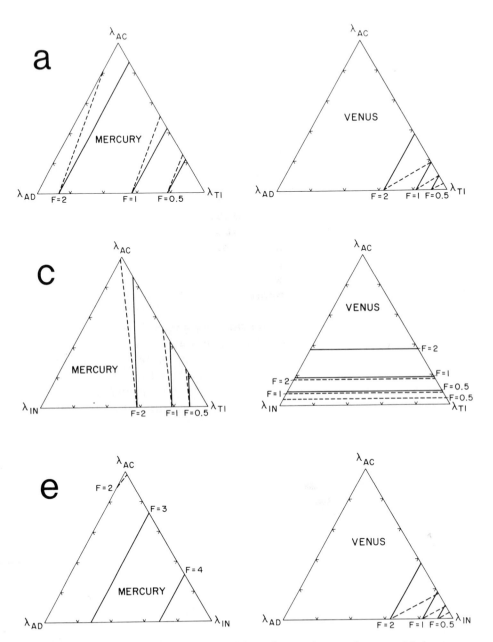

Fig. 4. Contours of F, the total amount of thermal energy input to the terrestrial planets relative to the moon as functions of λ_i for each of the terrestrial planets. The solid line for each contour value is calculated using mode 2 accretion and the dashed contour lines are calculated using mode 1 accretion.

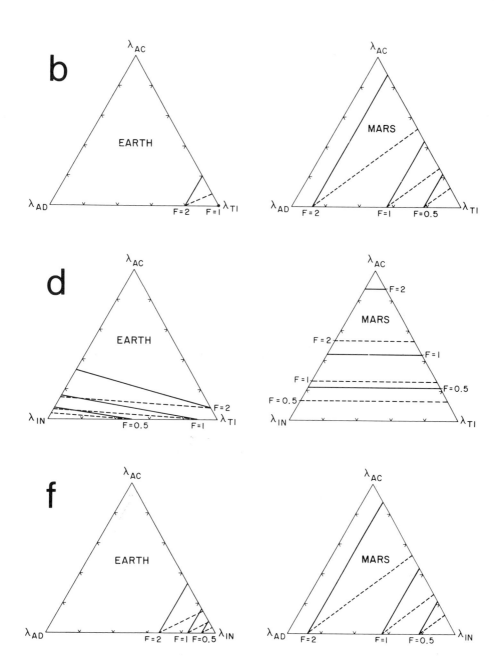

CONCLUSIONS

If the moon and planets formed within 2×10^6 years after the formation of the protosun, the conclusion seems inescapable that all of the terrestrial planets melted. Thermal energy from solar luminosity and ^{26}Al was probably available, but was not necessary in that accretional energy alone on so short a timescale is sufficient to melt large planets and satellites. A significant role for ^{26}Al as the principal heat source, if any (see Mittlefehldt, 1979) is restricted to small satellites and asteroids and could have operated only if these bodies formed on a shorter timescale than the terrestrial planets.

   If the moon and planets grew to their present sizes on a timescale in excess of 2×10^6 years, the outcome for each planet varies as summarized below.

1. Unless the moon received greater than 60–70% of its thermal input from tidal interactions with the earth, any combination of adiabatic compression, induction, and accretion sufficient to produce the lunar magma ocean would have melted Mercury. Note that although adiabatic compression and induction are precluded from operating simultaneously, they may operate sequentially. Thus the background temperature could be raised by adiabatic compression, while induction melted the planet subsequent to dissipation of the gas.
2. Unless the moon received greater than approximately 80% of its thermal input from tidal heating and/or induction, any combination of adiabatic compression and accretion sufficient to produce the lunar magma ocean would have melted Venus.
3. Unless the moon received greater than 80–90% of its thermal input from induction, any combination of tidal heating, accretion, and adiabatic compression sufficient to produce the lunar magma ocean would have melted the earth.
4. Unless the moon received greater than 50–60% of its thermal input from induction and/or tidal heating, any combination of accretion and adiabatic compression sufficient to produce the lunar magma ocean would have melted Mars.

   These results indicate that it is probable that all terrestrial planets underwent early global differentiation events such as is recorded in the returned lunar samples. If the age of the earth is interpreted as the time of core formation, and if segregation of metal requires extensive melting of the planet (e.g., Elsasser, 1963), then the earth must also have melted. There are, however, combinations of heating mechanisms (e.g., tides plus induction) which would result in the early global differentiation of the earth and the moon without requiring that all other terrestrial planets melted. Because of its relatively small mass and large semi-major axis compared to the other terrestrial planets, Mars is the best candidate for a terrestrial planet which did not experience an early global differentiation event. It is not impossible that the ancient southern cratered highlands of Mars may consist of primordial, albeit weathered, Mars material. The possibility exists that sample return missions to the terrestrial planets, in particular Mars, may provide the key to determining the time of formation of the planets and the heating mechanisms responsible for the early global melting of the moon and, presumably, the earth.

Acknowledgments—We have benefitted from discussions with G. Goles, R. Greenberg, F. Herbert, C. Sonett, and G. Wetherill. This work was supported by NASA grant NSG 7576.

REFERENCES

Binder A. B. (1978) On fission and devolatilization of a Moon of fission origin. *Earth Planet. Sci. Lett.* **41**, 381–385.

Elsasser W. M. (1963) Early history of the Earth. In *Earth Science and Meteorites* (J. Geiss and E. Goldberg, eds.), p. 1–30. North Holland, Amsterdam.

Ezer D. and Cameron A. G. W. (1962) A study of solar evolution. *Can. J. Phys.* **43**, 1497–1517.

Greenberg R. (1979) Growth of large, late-stage planetesimals. *Icarus* **39**, 141–150.

Hayashi C. (1966) Evolution of protostars. *Ann. Rev. Astron. Astrophys.* **4**, 171–192.

Kuskov O. L. and Khitarov N. I. (1979) Thermal effects of chemical reactions in the undifferentiated Earth. *Phys. Earth Planet. Inter.* **18**, 20–26.

Lee T., Papanastassiou D. A., and Wasserburg G. J. (1977) Aluminum-26 in the early solar system—fossil or fuel? *Astrophys. J.* **211**, L107–L110.

Mao H. K. and Bell P. M. (1977) Disproportionation equilibrium in iron-bearing systems at pressures above 100 kbar with applications to chemistry of the Earth's mantle. In *Energetics of Geological Processes* (S. K. Saxena and S. Bhattacharji, ed.), p. 236–249. Springer-Verlag, N.Y.

Mittlefehldt D. (1979) The nature of asteroidal differentiation processes: Implications for primordial heat sources. *Proc. Lunar. Planet. Sci. Conf. 10th*, p. 1975–1993.

Ostic R. G. (1965) Physical conditions in gaseous spheres. *Mon. Not. Roy. Astron. Soc.* **131**, 191–197.

Ringwood A. E. (1970) Origin of the Moon: The precipitation hypothesis. *Earth Planet. Sci. Lett.* **8**, 131–140.

Smith J. V. (1974) Origin of the moon by disintegrative capture with chemical differentiation followed by sequential accretion (abstract). In *Lunar Science V*, p. 718–720. The Lunar Science Institute, Houston.

Solomon S. C. and Chaiken J. (1976) Thermal expansion and thermal stress in the moon and terrestrial planets: Clues to early thermal history. *Proc. Lunar Sci. Conf. 7th*, p. 3229–3243.

Sonett C. P. and Reynolds R. T. (1979) Primordial heating of asteroidal parent bodies. In *Asteroids* (T. Gehrels, ed.), p. 822–848. Univ. Arizona Press, Tucson.

Sonett C. P., Colburn D. S., Schwartz K., and Keil K. (1970) The melting of asteroidal-sized bodies by unipolar dynamo induction from a primordial T-Tauri sun. *Astrophys. Space Sci.* **7**, 446–488.

Stevenson D. J. (1980) Core formation dynamics and primordial planetary dynamos (abstract). In *Lunar and Planetary Science XI*, p. 1088–1090. Lunar and Planetary Institute, Houston.

Taylor S. R. (1975) *Lunar science: A post-Apollo view.* Pergamon, N.Y. 372 pp.

Weidenschilling S. J. (1976) Accretion theory revisited. *Icarus* **27**, 161–170.

Wetherill G. W. (1976) The role of large bodies in the formation of the earth and moon. *Proc. Lunar Sci. Conf. 7th*, p. 3245–3257.

Wood J. A. and Mitler H. E. (1974) Origin of the moon by a modified capture mechanism, or, half a loaf is better than w whole one (abstract). In *Lunar Science V*, p. 851–853. The Lunar Science Institute, Houston.

Proc. Lunar Planet. Sci. Conf. 11th (1980), p. 1931–1939.
Printed in the United States of America

The first few hundred years of evolution of a moon of fission origin

Alan B. Binder

Institut für Mineralogie, Universität Münster, 4400 Münster, West Germany

Abstract—If the moon originated by fission, it probably had a temperature of 2000 to 4000°C, was deficient in iron and siderophile trace elements, was several times more massive than it currently is, and was orbiting the earth at about 3 earth radii immediately after it separated from the earth. Under these conditions, the moon should have undergone an explosive phase of retrograde boiling due to the large decrease in internal pressure the moon suffered as it detached from the earth. This process is proposed to have led to the loss of the majority of the original mass of the moon and the loss of 50 to 75% of the angular momentum of the early earth-moon system. The short, catastrophic mass loss phase was followed by a more quiescent period during which the superheated moon lost additional mass via evaporation and transfer to the earth through L_1. This second phase, which lasted a few hundred years, is proposed to have led to the loss of volatile elements from the moon via fractional evaporation. The second mass loss phase ended when the moon had reached the point in its cooling and initial crystallization sequence where the crust began to form.

INTRODUCTION

There is increasing evidence supporting the model that the moon originated by binary fission from the proto-earth. Wise (1963, 1969), O'Keefe (1968, 1969, 1970, 1972), and O'Keefe and Sullivan (1978) discussed dynamical models of the fission of a rapidly rotating proto-earth after core formation. Both Wise and O'Keefe pointed out *before Apollo 11* that the moon would be deficient, with respect to the earth, in *siderophile* and *volatile* elements if it formed by fission. Rubincam (1975) showed that the current inclination of the lunar orbit can be explained dynamically, even though the post-fission orbit was equatorial. The author and co-workers (Binder, 1974, 1975a, b, 1978, 1980; Binder and Lange, 1977, 1980) showed that the general petrology, thermal history, and physical properties of the moon can be explained in terms of a devolatilized and differentiated body of earth mantle composition.

Abt and Levy (1976) showed that fission is a common process in the formation of astronomical bodies. Prentice (1978) and Prentice and ter Haar (1979a, b) showed that the Laplace model for the origin of the solar system is dynamically sound. When the theoretical model of Prentice and ter Haar and the empirical data of Abt and Levy are taken together, it appears likely that a contracting nebula *must* throw off one or more rings of matter as it contracts and/or the central body *must undergo binary fission* at the end of contraction.

Further, Ringwood and Wänke and his co-workers (e.g., Rammensee and Wänke, 1977; Ringwood, 1978) indicated that the depletion of lunar siderophile elements can only be explained if these elements were removed from the lunar material in the presence of large amounts of iron, i.e., in the proto-earth as the iron core formed.

However, there are still questions related to the current deficiency of angular momentum in the earth-moon system if the moon formed by fission. The purpose of this paper is to integrate the models of O'Keefe (1969, 1970) and Binder (1978) in order to explain this angular momentum deficiency.

EARLY EARTH-MOON SYSTEM

Initial conditions and considerations

At the time when the moon had just fissioned from the earth, 1) both the earth and moon were very hot (2000–4000°C, Binder, 1978), 2) the earth-moon distance was about 3 earth radii (Wise, 1963), 3) the moon had no significant amount of iron which could form a core, 4) the moon's mass was up to 0.1 earth masses (O'Keefe, 1970), and 5) the earth-moon system had 2–4 times as much angular momentum as it currently does (Wise, 1963, 1969). O'Keefe (1969) and Binder (1978) pointed out that at that time the earth-moon system resembled a close binary star system. Such stellar systems are observed to exchange mass and lose mass and angular momentum through the first Lagrange point (L_1), Fig. 1.

O'Keefe (1969) suggested that mass loss through L_2 from a hot, evaporating moon might explain the angular momentum deficiency of the system and would lead to a depletion of volatile elements in the moon. Binder (1978) showed that

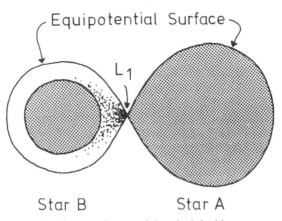

Fig. 1. Loss and transfer of matter in a semi-detached double star system. Matter is transferred from star A to star B and lost from the system as the atmosphere of A expands, fills the star's Roche lobe (defined by the equipotential surface drawn around the two stars), and pours through L_1 (first Lagrange point) to be captured by star B or to escape from the system.

the moon, but not the earth, could have lost matter through L_1 and that most of this volatile-rich material would have been captured by the earth, leaving a volatile depleted moon. In combining these models we note that 1) by (4) above the initial lunar mass could have been up to ~10 times its current value, 2) the thermal energy in a body is sufficient to totally evaporate it at temperatures greater than 2000°C (Binder, 1978), hence the moon could have lost most of its initial mass by the proposed mechanism, 3) the distance from the earth at which an orbiting body is lost from the earth, i.e., the radius of action (R_a), is 1.2×10^6 km or 3.2 current lunar orbit radii (Kuiper, 1961), and 4) material lost from the moon through L_1, but not captured by the earth would have escaped from the earth-moon system directly (as observed in binary stars) or due to tidal interactions and solar perturbations, thereby removing angular momentum from the system.

Angular momentum loss

In a normalized system where the unit of mass is 1 lunar mass (m), the unit distance is 1 earth radius (R), the unit of angular velocity is 1 per day (ω), and noting that the moment of inertia factor of the earth is very close to ⅓ and that the angular momentum of the moon due to its own rotation is negligible, the current angular momentum of the earth-moon system is

$$\frac{1}{3}(81m)R^2\omega + m(60R)^2 \frac{\omega}{27.3} = 159mR^2\omega. \qquad 1)$$

The first term in Eq. 1 is the angular momentum due to the rotation of the earth (mass = 81m) and the second is due to the orbital motion of the moon (radius = 60R and period = 27.3 days). According to Wise (1963, 1969), the angular momentum of the proto-earth prior to fission was 2 to 4 times the current value, or about 320 to $640mR^2\omega$. Hence, the system must have lost 160 to $480mR^2\omega$.

Any matter lost from the moon and ejected from the system would have gained angular momentum (from the earth and moon) until it reached the radius of action (R_a) at $3.2 \times 60R$ or 192R, after which it was lost from the system. From Kepler's third law one finds that the angular velocity at R_a is $\omega/156$. Thus, the angular momentum per unit mass loss at R_a is

$$m(192R)^2 \frac{\omega}{156} = 236mR^2\omega. \qquad 2)$$

Hence, in order to account for the loss of 160 to $480mR^2\omega$ of angular momentum, 0.7 to 2.0 lunar masses of material must have been ejected from the system.

Mass loss through L_1

As discussed by Binder earlier (1978), a hot, fissioned moon would have lost matter through L_1 since 1) the moon must have had a massive atmosphere of evaporated material (SiO, Mg, Fe, Al_2O, etc.) and 2) the scale height of the atmosphere was on the order of the lunar surface-L_1 distance. Hence the atmosphere would have filled the lunar Roche lobe and poured through L_1 into the earth's Roche lobe. The reverse did not happen because of the larger earth surface-L_1 distance and the smaller scale height of the terrestrial evaporated atmosphere.

This model of the mass loss mechanism was based on the assumption that the amount of mass lost was only a few tens of percent of the original lunar mass. However, since the initial lunar mass was up to ~10 times larger than its current value (O'Keefe, 1970), the model presented earlier (Binder, 1978) has been recalculated for initial lunar masses in the range of 1 to 10 m. The results are depicted in Fig. 2, from which it can be seen that the lunar surface-L_1 distance is always much smaller than the earth's surface-L_1 distance. Thus, the conclusion presented earlier that the moon, but not the earth, could have lost matter through L_1 is valid for the range of lunar masses of interest.

Also, as seen in Fig. 3, the distance from the lunar surface to L_1 is from one to a few scale heights of the atmosphere. Thus the pressure or density of the evaporated lunar atmosphere was large enough so that considerable amounts of

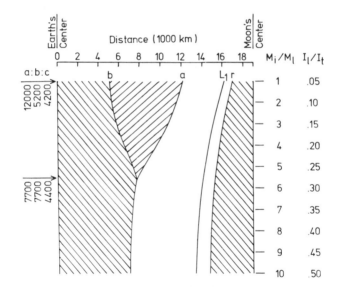

Fig. 2. Configuration of the earth-moon system just after fission as a function of the initial mass of the moon. The moon was about 19,000 km from the earth immediately after fission. The radii of the a and b equatorial axes of the triaxial earth as a function of the lunar mass are given by curves a and b, respectively. The earth was a triaxial, Jacobian ellipsoid for lunar masses between 1 and about 5.5. The lengths (in km) of the a, b, and c axes of the earth for the cases where the lunar mass was 1 and 5.5 are given in the left column. For lunar masses larger than 5.5, the earth was an oblate, Maclaurian spheroid. Curve L_1 gives the distance of the L_1 point from the centers of the earth and moon. Curve r gives the radius of the moon at the subearth point, i.e., under L_1. M_i/M_l is the initial mass of the post-fission moon (M_i) divided by the current mass of the moon (M_l). I_l/I_t is the ratio of the amount of angular momentum of the moon (I_l) to that of the total system (I_t).

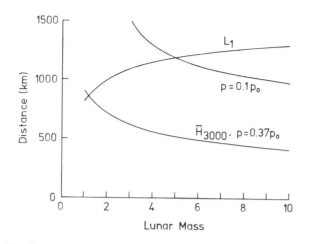

Fig. 3. The distance of L_1 from the surface of the moon as a function of the initial mass of the moon. Curve $\overline{H}_{3000}$ is the isothermal scale height of the lunar atmosphere for the case where the temperature was 3000°C. This curve also corresponds to the altitude above the lunar surface where the pressure (and density) in the atmosphere was 37% of its surface value. Curve $p=0.1p_0$ gives the altitude above the lunar surface where the pressure (and density) was 10% of its surface value.

evaporated material could have passed through L_1 into the earth's Roche lobe for all lunar masses considered. Note that the mechanism is much more efficient for small lunar masses (<5) than for larger ones.

Given these results one can evaluate O'Keefe's suggestion that lunar mass was lost through L_2. As shown in Fig. 4 the potential at L_2 is greater than at L_1. Thus, because evaporated lunar material poured through the lower potential L_1 point, the atmosphere could not have expanded to the L_2 point. This is exactly what is observed for semi-detached double stars (Kopal, 1978). Thus O'Keefe's model does not work exactly in the way he suggested and hence mass loss must be considered further.

Kopal (1978) has calculated a large number of orbits for particles escaping from an expanding star through L_1. Although the secondary to primary mass ratios used by Kopal are much larger than the range considered for the post-fission earth-moon system, the results are, qualitatively, applicable to the problem considered here. Kopal's calculations show that some of the matter transfered from the expanding star into the Roche lobe of the primary star will be lost from the system. However, Kopal's results suggest that only a small fraction of the lunar mass lost through L_1 would have been lost from the system. For the sake of discussion, if the amount of matter lost from the system were 10% of that lost from the moon, then, in order for the system to have lost 0.7 to 2.0 lunar masses of material calculated above, the initial mass of the moon was between 8 and 21 times its current value. While these values are perhaps acceptable, it does not seem likely that this mechanism is efficient enough to explain the observed angular momentum deficiency of the system.

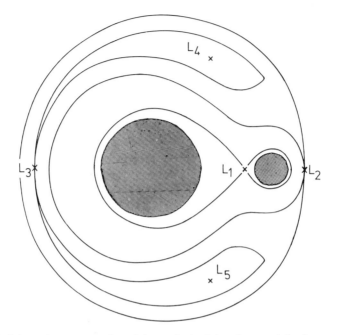

Fig. 4. Schematic representation of the equipotential surfaces and the Lagrange points in the earth and moon system for a representative lunar mass of 3 times its current value.

Catastrophic mass and momentum loss

Retrograde boiling of the moon during and immediately after fission could have led to a catastrophic loss of mass and momentum from the system. Extrapolation of the vapor pressure curves for SiO and Mg (the major components of the moon) given in Fig. 2 of Ringwood (1970) shows that the vapor pressures of these constituents at temperatures between 3000 to 4000°K is tens to hundreds of kilobars. Since the central pressure of the post-fission moon was only between 50 and 250 kb, a large fraction of the initial lunar mass would have changed to a vapor state during and immediately after fission. Since 1) the mean thermal velocity (1.4 km/s for a molecular weight of 40 at 3000°C) of the vaporized material is a large fraction of the escape velocity of the initial moon and 2) this rapidly expanding material could not have been quickly bled off through L_1, a considerable amount of the initial lunar mass must have explosively expanded beyond the limits of the lunar Roche lobe and thereby reaching L_2 or simply expanding beyond the equipotential surface tangent to L_2. This material could then have escaped the earth-moon system, carrying with it considerable amounts of angular momentum. Geometric considerations suggest that less than half of the material blown off the moon would have been captured by the earth. Given this rough estimate and the requirement that 0.7 to 2.0 lunar masses had to escape in order to account for the loss of angular momentum, the initial mass of the moon would have been only 2.5 to 5 lunar masses, a quite acceptable range according to the discussions given earlier.

EARLIEST EVOLUTION OF THE EARTH-MOON SYSTEM

According to the above discussions, the earliest evolution of the moon was as follows. During and for a short time (hours ?) after fission the moon underwent an explosive, retrograde boiling phase. During this phase the moon lost the majority of its initial mass, but did not change its basic composition. After the

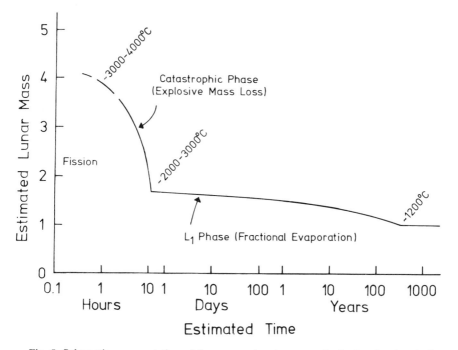

Fig. 5. Schematic representation of the proposed major events in the first few hundred years of lunar history. The fission process probably took place over at least one earth rotation (2.65 hrs). During fission and for a short period (several hours ?) after fission, the moon underwent a catastrophic phase in which the majority of the moon's mass was explosively lost from the moon. Most of this material was also probably lost from the earth-moon system, thereby accounting for most of the angular momentum deficiency of the current system. After the catastrophic phase, the moon continued to lose mass at a much slower rate through L_1. Though the majority of this matter was captured by the earth, some of it did escape the system—carrying additional angular momentum with it. Also, the material lost from the moon during this phase was enriched in volatile elements due to the fractional evaporation of the lunar materials. Hence, it was this phase which resulted in the depletion of lunar volatiles. This L_1 mass loss phase ended when the crust began to form a few hundred years after fission. The estimated temperatures of the moon at the beginning of each phase are given adjacent to the curve.

catastrophic phase, the cooler and less massive moon lost mass slowly through L_1 until the moon cooled to the liquidus and the evaporation mechanism shut off (Binder, 1978). This more quiescent L_1 phase lasted a few hundred years (Binder, 1978) and most of the mass lost during this phase was captured by the earth. It is proposed that the moon lost the majority of its volatile elements via fractional evaporation, subsequent loss through L_1, and capture by the earth during the L_1 mass loss phase (Binder, 1978). This sequence is schematically shown in Fig. 5.

Preliminary estimates suggest that the explosive phase alone could account for the current deficiency of angular momentum in the earth-moon system if the initial moon had a mass of 2.5 to 5 lunar masses. Since additional mass was lost from the system during the L_1 phase, not all of the angular momentum deficiency must be accounted for via the catastrophic phase. Thus the initial mass of the moon may have been in the range of 2 to 4 lunar masses. Note that it is this range where both mechanisms would have been very efficient. If this model is correct, then 0.7 to 2 lunar masses were ejected into heliocentric orbit and a similar amount was captured by the earth.

Finally, according to the models of Binder (1974, 1978), the time scale for the cooling of a superheated moon to the point in the lunar crystallization sequence where plagioclase saturation was reached and the crust began to form was only a few hundred years. As soon as the crust began to develop and the magma was capped, the evaporation-L_1 loss mechanism, already slowed by the ever lowering temperature of the lunar surface and atmosphere, shut off completely. However, by this time the earth-moon system had lost 50 to 75% of its initial angular momentum, the moon had lost most of its volatile elements (Binder, 1978), and about 80% of the moon had crystallized to form an olivine $\pm$ orthopyroxene lower mantle (Binder, 1974, 1976b). At that point there would have been a 150–200 km thick layer of residual magma which corresponds to the popular "magma ocean" from which the crust and mare basalt magma source region formed by one of a number of proposed models.

Acknowledgments—I thank Drs. M. Lange and J. O'Keefe for their critical reviews of the paper. This work was supported by the Deutsche Forschungsgemeinschaft.

REFERENCES

Abt H. A. and Levy S. C. (1976) Multiplicity amoung solar-type stars. *Astrophys. J. Suppl. Ser.* **30**, 273–306.

Binder A. B. (1974) On the origin of the moon by rotational fission. *The Moon* **11**, 53–76.

Binder A. B. (1975a) On the petrology and structure of a gravitationally differentiated moon of fission origin. *The Moon* **13**, 431–473.

Binder A. B. (1975b) On the heat flow of a gravitationally differentiated moon of fission origin. *The Moon* **14**, 237–245.

Binder A. B. (1976a) On the petrology and early development of the crust of a moon of fission origin. *The Moon* **15**, 275–314.

Binder A. B. (1976b) On the implications of an olivine dominated upper mantle on the development of a moon of fission origin. *The Moon* **16**, 159–173.

Binder A. B. (1978) On fission and the devolatilization of a moon of fission origin. *Earth Planet. Sci. Lett.* **41**, 381–385.

Binder A. B. (1980) On the internal structure of a moon of fission origin. *J. Geophys. Res.* In press.

Binder A. B. and Lange M. A. (1977) On the thermal history of a moon of fission origin. *The Moon* **17**, 29–45.

Binder A. B. and Lange M. A. (1980) On the thermal history, thermal state, and related tectionism of a moon of fission origin. *J. Geophys. Res.* In press.

Kopal U. (1978) *Dynamics of Close Binary Systems.* Reidel, Boston. 501 pp.

Kuiper B. P. (1961) Limits of completeness. In *Planets and Satellites* (B. P. Kuiper and B. M. Middlehurst, eds), p. 575–591. Univ. Chicago Press, Chicago.

O'Keefe J. A. (1968) Fission hypothesis for the origin of the moon. *Astron. J. Suppl.* **73**, 10.

O'Keefe J. A. (1969) Origin of the moon. *J. Geophys. Res.* **74**, 2758–2767.

O'Keefe J. A. (1970) The origin of the moon. *J. Geophys. Res.* **75**, 6565–6574.

O'Keefe J. A. (1972) The origin of the moon: Theories involving formation with the earth. *Astrophys. Space Sci.* **16**, 201–211.

O'Keefe J. A. and Sullivan E. C. (1978) Fission origin of the moon: Cause and timing. *Icarus* **35**, 272–283.

Prentice A. J. R. (1978) Origin of the solar system. I: Gravity contraction of turbulent protosun and the shedding of a concentric system of gaseous Laplacian rings. *The Moon* **19**, 341–398.

Prentice A. J. R. and ter Haar D. (1979a) Formation of the regular satellite systems and the rings of the major planets. *Moon and Planets* **21**, 43–61.

Prentice A. J. R. and ter Haar D. (1979b) Origin of the jovian ring and the Galiean satellites. *Nature* **280**, 300–302.

Rammansee W. and Wänke H. (1977) On the partition coefficient of tungsten between metal and silicate and its bearing on the origin of the moon. *Proc. Lunar Sci. Conf. 8th*, p. 399–409.

Ringwood A. E. (1970) Origin of the moon: The precipitation hypothesis. *Earth Planet. Sci. Lett.* **8**, 131–140.

Ringwood A. E. (1978) Origin of the moon (abstract). In *Lunar and Planetary Science IX*, p. 961–963. Lunar and Planetary Institute, Houston.

Rubincam D. P. (1975) Tidal friction and the early history of the moon's orbit. *J. Geophys. Res.* **80**, 1537–1548.

Wise D. U. (1963) An origin of the moon by rotational fission during core formation of the earth's core. *J. Geophys. Res.* **68**, 1547–1554.

Wise D. U. (1969) Origin of the moon from the earth: Some new mechanisms and comparisons. *J. Geophys. Res.* **74**, 6034–6045.

Proc. Lunar Planet. Sci. Conf. 11th (1980), p. 1941–1955.
Printed in the United States of America

The lunar magma ocean: A transient lunar phenomenon?

John W. Minear

Geology Branch, NASA Johnson Space Center, Houston, Texas 77058

Abstract—The time for a lunar magma ocean to solidify is controlled primarily by the thickness of the solid conduction boundary layer (crust) that develops at the surface of the ocean. However, little is known about the thickness of this boundary layer during crystallization of the magma ocean. Three models of crust formation and growth are considered in this paper: a chill margin, a plagioclase flotation layer, and a model based on crust destruction by meteorite impacts and regrowth by conduction cooling. Chill margin thickness is estimated by comparing the solid-liquid interface advance rate and the Stoke's settling velocity of crystals nucleating close to this interface. Rayleigh-Taylor instabilities may also limit the thickness of such a layer. The plagioclase flotation model assumes that the crust maintains a constant thickness during olivine crystallization and then grows as plagioclase begins to crystallize from the magma ocean. Solidification time is controlled by the crustal thickness during olivine crystallization; olivine crystallization may occur during either a very small or a large fraction of the total solidification time. The crust destruction-regrowth model assumes that planetesimals punch through the crust and destroy it at a rate depending on the impact flux. The crust then regrows at a rate dependent on conductive heat loss.

All of the models yield crustal thickness that are less than a kilometer for a substantial fraction of magma ocean solidification time. Based on these crustal growth models, the maximum solidification time for a 200 km thick magma ocean is 10^8 yrs. A more probable solidification time is about 6×10^7 yrs.

INTRODUCTION

The time for a lunar magma ocean to solidify is controlled by the heat loss at its surface. If the top of the ocean remains open (i.e., no thermal boundary crust develops and heat loss is by radiation), the ocean will solidify in a few decades (Herbert *et al.*, 1977). However, it does not seem likely that the top surface would remain a radiative boundary. Only a few millimeters of crust would effectively limit radiation. Some type of conduction boundary layer must exert the primary control over magma ocean solidification time.

The objective of this paper is to present estimates of conduction boundary (crust) thickness based on several models of the mechanism of crustal formation. Given the crustal thickness based on these models, heat flux, Q, through the crust, and consequently solidification time of the underlying magma ocean, can be estimated from the simple heat conduction relationship

$$Q = \frac{K(T_M - T_0)}{L}$$

where K is thermal conductivity, T_M and T_0 are the temperatures at the bottom (melt-solid interface) and top of the crust respectively, and L is the crust thickness. The latter parameter is assumed to be the only variable in the computations. During crust development L will vary by orders of magnitude. Thermal conductivity will vary between that for the early and late crystallizing phases; probably less than a factor of three. A value of $K = 2.5 \times 10^5$ ergs/°C·cm·sec was used in all of the computations. This corresponds to anorthite at about 1200°C and is less than the value for olivines or pyroxenes. The temperature $(T_M - T_0)$, decreases about one-third during solidification of the magma ocean.

A basic assumption in computing the heat loss is that a continuous, unbroken crust covers the magma ocean. In reality, any crust that forms must be broken and fractured by global tension (Solomon and Chaiken, 1976), meteorite impact, and dynamic forces exerted by the underlying magma ocean. Fractures will lead to more rapid cooling of the magma ocean because fresh magma is exposed or extruded on the lunar surface. This should last for a significant fraction of the time (if not for the entire time) for the ocean to solidify. By comparison, about 70% of the Earth's present heat loss is due to the effect of magma extrusion and upwelling at mid-ocean ridges rather than through old lithosphere. Consequently, solidification times computed from continuous crust models assumed in this paper can be regarded as maximum times given the particular crust formation mechanism. Even the model that is based on crust destruction by meteorite impact does not consider the effect of magma upwelling in fractures generated in newly formed crust.

CRUST FORMATION MODELS

Three different models for estimating conduction boundary layer thickness will be discussed. These are based on the formation of a chill margin, a plagioclase flotation layer, and a layer that is continually destroyed by meteorite impacts and regrows by conductive heat loss.

A. Chill-margin model

The simplest model of crust formation is the chill-margin model. Chemically, a chill margin is defined as an undifferentiated zone. Consequently two rates should be important in determining its thickness: the advance rate of the solid-melt interface, and the rate of crystal settling. Using these rates, the chill margin can be defined as the layer in which solid-melt interface advance exceeds the settling velocity of the fastest settling crystals.

Solid-melt interface advance rate was estimated from the cooling of a half space. Latent heat release on crystallization and convective heat transfer within the melt beneath the chill margin were ignored. Convective heat transfer retards the advance of a solid crust because heat brought from depth increases near surface temperatures. Thus, the actual solid-melt advance rate will be less than that estimated from the simple half-space model. Temperature in a half space with its surface maintained at 0°C as a function of depth and time is given by (Carslaw and Jaeger, 1959; p. 60)

$$T(x,t) = T_i \, erf \left\{ \frac{X}{2(kt)^{\frac{1}{2}}} \right\} \tag{1}$$

where k is the diffusivity and T_i is the initial temperature.

Initial and solidification temperatures for the half space were based on the solidus-liquidus relations for a Taylor-Jakeš composition magma ocean (Solomon and Longhi, 1977). Initial temperature was taken as 1550°C; the liquidus temperature. Solidification temperature was taken to correspond to 50% crystallization; the fraction of solid at which it was assumed crystals could no longer settle through the solid matrix. Based on the solidus-liquidus relations this temperature was taken as 1250°C.

Substituting 1550°C for T_0 and 1250°C for $T(x,t)$ and taking $k = .01 \, cm^2/s$ in (1) yields the depth and time for solidification

$$X = .18 \, t^{\frac{1}{2}}. \tag{2}$$

The solid-melt interface advance rate is then just

$$v_c = \dot{X} = .09 \, t^{-\frac{1}{2}}. \tag{3}$$

Crystal settling was assumed to be governed by Stoke's settling

$$v_s = \frac{2}{9} \cdot \frac{g\Delta\rho r^2}{\eta} \tag{4}$$

where g is acceleration of gravity, $\Delta\rho$ is solid-melt density contrast, r is crystal radius and η is viscosity. The greatest uncertainty in Eq. (4) is the crystal radius. Based on cumulate textures, crystal radius is of the order of millimeters to centimeters (Ryder and Norman, 1978). This is an upper limit giving the thinnest chill margin. A more realistic estimate of radius is one based on growth rate. Plagioclase growth rates estimated from Hawaiian lava lakes are about 10^{-9}–10^{-8} cm/s (Kirkpatrick, 1974). Experimentally determined growth rates for diopside are about 10^{-4} cm/s and for anorthite about 10^{-4}–10^{-2} cm/s (Kirkpatrick, 1975).

It is tacitly assumed that crystals nucleate away from the solid-liquid interface. Otherwise newly formed crystals would grow out from the interface and could not settle away from the interface. The width of the nucleation zone is unknown. We assume that it is 10% of the chill margin. This seems reasonable for chill margins a few centimeters to meters but probably overestimates the nucleation zone for thicker chill margins. Using the Hawaiian lava lake growth rate of 10^{-8} cm/s, a $\Delta\rho$ of .3 g/cm^3 and a melt viscosity of 10^2 poise, Stoke's settling velocities are plotted in Fig. 1. Also plotted in Fig. 1 is the solid-melt interface advance rate computed from Eq. (3). It can be seen that Stoke's settling velocities for the assumed growth rate exceed solid-melt interface advance velocities for chill margins thicker than about 0.2 km. Slow crystal growth rate and neglect of latent

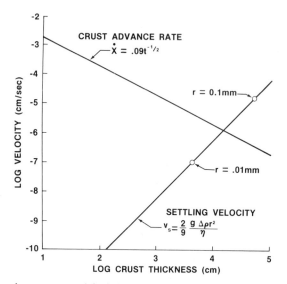

Fig. 1. Crust advance rate and Stoke's settling velocity vs. crust thickness. Crust advance rate is taken as the advance rate of the isotherm at which 50% crystallization has occurred (1250°C) in a cooling half space initially at 1550°C. For computing Stoke's settling velocity, crystals are assumed to nucleate within a zone equal to 10% of the crustal thickness and to grow at a rate of 10^{-8} cm/s. Stoke's settling velocities of crystals of constant radii of .01 and .1 mm are shown for comparison.

heat and magma convection have the effect of increasing the chill-margin thickness. The thicker the zone of nucleation, the thinner the chill-margin will be. The first two effects probably more than offset the effect of nucleation zone thickness uncertainy, particularly for chill-margin thicknesses greater than a few tens of meters. It is concluded that if a chill margin is assumed to be the conduction boundary layer at the top of a magma ocean then it remains less than a few tenths of a kilometer.

The fact that a chill margin is denser than the liquid from which it crystallized has been ignored in the previous discussion. A chill margin was assumed to remain on the surface of the magma ocean. In reality, the denser chill margin is gravitationally unstable and should sink. An estimate of the maximum thickness of such a crust can be made by assuming that the crust is a continuous unbroken layer. Pieces of a broken crust will sink immediately because they are not supported by lateral stresses.

A continuous crust will develop Rayleigh-Taylor instabilities at the melt-solid interface analogous to those that develop on buried salt layers (Daněs, 1964, Selig, 1965). The parameter of interest in this application is the time it takes for Rayleigh-Taylor instabilities to develop as a function of the thickness of the crust. Growth rate for the wavelength of maximum instability is given by (Selig, 1965)

$$w = 0.24 \, g \, (\rho_c - \rho_m) \, hc \cdot \left(\frac{\mu_m}{\mu_c}\right)^{1/3} \cdot \frac{1}{(\mu_m + \mu_c)} \tag{5}$$

where h_c is the crust thickness, μ is viscosity, ρ is density and the subscripts c and m refer to crust and melt respectively. The time for the instability to develop is

$$t = w^{-1}.$$

By Eq. (5) it can be seen that crust thickness and crust-melt viscosity contrast dominate instability growth time. The greatest uncertainty in computing t for a given h_c is the crust viscosity. μ_c is certainly much greater than μ_m so that w is approximately

$$w \approx 0.24 \text{ g } (\rho_c - \rho_m) \, h_c \cdot \frac{\mu_m^{1/3}}{\mu_c^{4/3}}. \tag{6}$$

A method of estimating μ_c is to use data from olivine creep experiments. For this one needs the temperature and stress within the crust. With the surface at 0°C and the crust-melt interface at 1550°C the average temperature of the crust is 775°C. The solidus for a typical magma-ocean composition is about 1000°C (Solomon and Longhi, 1977) yielding a homologous temperature for the crust of $T/T_m = 0.78$. Using the olivine deformation diagrams from Ashby and Verral (1977, Fig. 14) the maximum strain rate for $T/T_m \leq 0.8$ and shear stress less than 10^2 bars is 10^{-8}/s. This gives an effective viscosity of 10^{16} poise for a shear stress of 10^2 bars. Effective viscosity will be greater than this value unless $T/T_m > 0.8$ and or the shear stress is greater than 10^2 bars. Taking $(\rho_c - \rho_m)$ as 0.3 g/cm³, μ_m as 10^2 poise and μ_c as 10^{16} poise in Eq. (6) gives

$$w = 2.5 \times 10^{-20} \, h_c$$

which corresponds to a growth time of

$$t = \frac{0.4 \times 10^{20}}{h_c}.$$

If h_c is one kilometer then t is about 13 m.y. If μ_c is much greater than 10^{16} poise Rayleigh-Taylor instabilities will not develop in times (a few hundred million years) comparable to those reported to solidify magma oceans of 200–300 km thick (Minear and Fletcher, 1978; Solomon and Longhi, 1977). On the other hand, if μ_c is less than about 10^{12} poise, crusts as thin as one meter will develop instabilities in 6×10^4 yrs.

B. Plagioclase flotation crust

Plagioclase flotation has been and still remains the favorite mechanism for creating a lunar crust. A fact not generally appreciated is that a very substantial fraction of magma ocean thermal energy is lost by the time significant plagioclase begins to crystallize. This and the fact that lunar magma ocean compositions yield only about 25% plagioclase by volume indicate that plagioclase flotation may not be very important until late stages of ocean solidification. Relatively thick fragments of plagioclase crust such as the rockbergs suggested by Herbert *et al.* (1977) should affect ocean solidification time negligibly because of their small areal extent.

To quantify the effect of a plagioclase flotation crust a model for magma ocean fractional crystallization is needed. I use one similar to that described by Solomon and Longhi (1977). Volume fractions of potential olivine, pyroxene and plagioclase in the magma ocean are .50, .25 and .25 respectively. Liquidus temperatures are 1550°, 1150° and 1150°C for olivine, pyroxene and plagioclase respectively; solidus temperature is 1000°C. These temperatures correspond to a Taylor-Jakeš composition ocean. Olivine only is assumed to crystallize between 1550°C and 1150°C and to be completely crystallized at 1150°C. Pyroxene and plagioclase crystallize together and in equal proportions between 1150°C and 1000°C. Latent heats of crystallization of 845, 681 and 267 joules/gm for olivine, pyroxene and plagioclase respectively were used. The magma ocean beneath the crust conduction boundary layer is assumed to be in active convection, chemically homogeneous and of constant temperature.

Thermal energy that must be lost to crystallize a magma ocean is

$$Q_o = (C_p \rho_L (T_L - T_S) + \Delta L \rho_L) V \qquad (7)$$

where C_p is specific heat at constant pressure, ρ_L is liquid density, ΔL is the average latent heat of the ocean, T_L and T_S are liquidus and solidus temperatures and V is the ocean volume. Q_o is tabulated for different ocean thicknesses in Table 1.

Using the previously described fractional crystallization model, olivine crystallization releases about 70% of the total crystallization energy. This energy must be dissipated before significant plagioclase crystallization begins.

Growth of the plagioclase flotation layer was modeled in the following manner. An initial crust of thickness X_o was assumed. This crust corresponds to a thin chill margin. Crustal thickness remains constant at X_o until all of the olivine has crystallized. At this point the ocean temperature is 1150°C and 70% of the crystallization energy has been lost. From this point on, crustal thickness increases to simulate plagioclase flotation. An analytical method described in Carslaw and Jaeger (1959; p. 284–285) was used to approximate thickening of the crust. This method gives

Table 1. Energy of crystallization of magma areas.

Ocean thickness (km)	Volume (cm^3)	Q_o (ergs)
100	3.58×10^{24}	1.42×10^{35}
200	6.75×10^{24}	2.69×10^{35}
300	9.54×10^{24}	3.80×10^{35}
400	11.96×10^{24}	4.76×10^{35}

$\Delta L = 682$ joules/gm; $C_p = 1.2$ joules/gm°C; $\rho_L = 3.0$ gm/cm^3

$$X(t) = 2\lambda(k_s t)^{\frac{1}{2}} \tag{8}$$

where k_s is the diffusivity of the crust, $X(t)$ is crustal thickness with time, t. λ is given by

$$\lambda e^{\lambda^2} \text{erf}\lambda = \frac{C_p T_L}{L\pi^{\frac{1}{2}}} \tag{9}$$

where C_p is specific heat at constant pressure of the crust, T_L is liquidus temperature and L is the latent heat release that produces crustal growth by plagioclase flotation. For plagioclase flotation L is just the fraction of latent heat released by the crystallizing plagioclase. Using previously assumed values for C_p and T_L and $L = 132$ joules per gm of magma crystallized in Eq. (9), the graphical representation of Eq. (9) given by Carslaw and Jaeger yields $\lambda \approx 1.3$.

Equation (8) is based on the assumption that the top of the thickening crust is maintained at 0°C. The plagioclase flotation model assumes that the thickening crust described by Eq. (8) grows from the bottom of an initial crust of thickness X_o. The interface between the initial crust and the growth crust will be at some temperature greater than 0°C. Hence, the model based on Eq. (8) will overestimate the actual crustal growth rate and solidification time.

Heat flux through a plagioclase flotation crust that thickness according to Eq. (8) from an initial thickness of X_o is

$$\frac{dQ}{dt} = \frac{K(T_L - T_o)}{X_o + X(t)}. \tag{10}$$

Substituting Eq. (8) in Eq. (10) and integrating to yield the total heat flux in time t gives

$$\Delta Q = \frac{K(T_L - T_o)}{2\lambda^2 k_s} [2\lambda k_s^{\frac{1}{2}} t^{\frac{1}{2}} + X_o \log X_o - X_o \log(X_o + 2\lambda k_s^{\frac{1}{2}} t^{\frac{1}{2}})]. \tag{11}$$

Setting ΔQ in Eq. (11) equal to 30% of the total crystallization energy of the ocean, time, t, to lose this energy through a growing conduction boundary layer can be determined. This time is about 2×10^8 yrs and 4×10^8 yrs for 200 and 400 km thick magma oceans respectively. Initial crustal thickness, X_o, has negligible effect on the crystallization time. This is due to the fact that an initially thin crust will grow faster than an initially thick one, eventually overtaking the slower growing crust. After this point they will grow at the same rate.

Thickness of the plagioclase flotation crust as given by Eq. (8) reaches 50 km at about 10^7 yrs and 100 km at 4×10^7 yrs. These represent the upper bound on plagioclase crustal thickness for 200 and 400 km magma oceans (about ¼ plagioclase) respectively. Thus, a plagioclase flotation crust would reach its maximum thickness, represented by the total depletion of the plagioclase component at a few tens of millions of years after major olivine crystallization. It would not continue to thicken and reduce energy loss. Consequently, the solidification times for the plagioclase and pyroxene component of magma oceans given above represent maximum times.

To obtain the total time for solidification, the time to crystallize the olivine fraction with a constant crust thickness of X_o must be added to the pyroxene-plagioclase solidification time. Olivine crystallization time is

$$t_{OL} = \frac{E_{OL}X_o}{K(T_L - T_o)A} \tag{12}$$

where E_{OL} is the energy of olivine crystallization, $T_L = 1550°C$, $T_o = 0°C$, and A is the surface area of the moon. Total time of solidification for the plagioclase flotation model is shown in Fig. 2. Times represented by the solid portion of the curves are just t_{OL} from Eq. (12). An interesting feature illustrated by Fig. 2 is that olivine crystallization may occur either over a very small or a substantial fraction of the total time of ocean solidification. Its relative duration is determined by the crustal thickness before plagioclase becomes a significant crystallizing phase.

Dashed curves in Fig. 2 represent the plagioclase-pyroxene portion of solidi-fication. For thin initial crustal thicknesses the plagioclase flotation crust will thicken very rapidly and reduce the surface heat loss very rapidly. This is re-flected in the sharp slope changes of the curves for small X_o. For larger X_o, the effect of the increasing thickness of the plagioclase flotation crust is buffered by the initially thick crust.

As discussed before, solidification times given by the dashed curves represent maximum times because the plagioclase flotation crust was not limited in the model by the maximum potential plagioclase available. An estimate of the effect

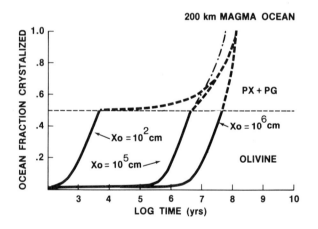

Fig. 2. Time for solidification of a 200 km thick magma ocean with a plagioclase flotation crust. The horizontal dashed line represents 50% crystallization. Below 50% crystalli-zation, olivine only crystallizes in the model; above 50% only pyroxene and plagioclase crystallize. Solid portions of the curves represent the time for all of the olivine to crystallize with a constant crustal thickness equal to the value given by X_o. Dashed portions of the curves represent the additional time to completely solidify the ocean with a continuously growing crust to represent plagioclase flotation. The dot-dash curve represents solidification time if the growing plagioclase flotation crust is limited to a thickness of 50 km which is attained at about 10^7 yrs.

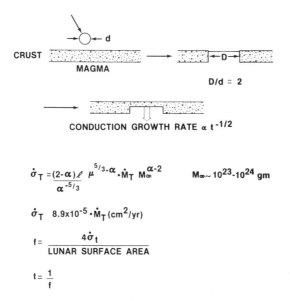

$$\dot{\sigma}_T = \frac{(2-\alpha)\ell}{\alpha^{-5/3}} \mu^{5/3-\alpha} \cdot \dot{M}_T \, M_\infty^{\alpha-2} \qquad M_\infty \sim 10^{23}\text{-}10^{24} \text{ gm}$$

$$\dot{\sigma}_T \;\; 8.9 \times 10^{-5} \cdot \dot{M}_T \, (\text{cm}^2/\text{yr})$$

$$f = \frac{4\dot{\sigma}_t}{\text{LUNAR SURFACE AREA}}$$

$$t = \frac{1}{f}$$

Fig. 3. Schematic of the crust destruction-regrowth model. Planetesimal impact into the magma ocean crust is assumed to destroy a circular region and expose fresh magma. A hole to planetesimal diameter ratio of 2 is shown. Crustal regrowth in the punctured areas is simulated by the cooling of a half space; cooling rate is proportional to $t^{-\frac{1}{2}}$. σ_T from Dohnanyi (1969) is the rate of change of the total cross sectional area of planetesimals hitting the moon's surface. The rate of lunar surface area destroyed by hole production is f and the time to punch through to entire lunar crust is f^{-1}.

of limiting the plagioclase flotation crust thickness is shown by the dot-dash curve for an X_0 of 10^5 cm. When the crust reaches a thickness of 50 km (at 10^7 yrs) it is held constant during the remainder of ocean solidification. This yields a total solidification time of about 60 m.y. for a 200 km thick ocean; solidification time for a 400 km thick ocean is about 120 m.y.

C. Crust destruction—regrowth model

This model is based on the destruction of the conduction boundary layer by meteorite impact and its regrowth by conduction cooling as depicted in Fig. 3. Destruction of material is not implied, only breaking up of a continuous solid surface-boundary layer overlying the magma ocean. The model is motivated by the fact that during a large fraction of the time to solidify a magma ocean, the crust will be thinner or of comparable thickness to impacting planetesimals. Goldreich and Ward (1973) estimate the diameter of post-condensation planetesimals as about 10 km. Such planetesimals should punch through a crust of comparable thickness leaving holes the diameter of or slightly larger than the planetesimals (Gehring, 1970). Fresh magma will be exposed in these holes and will cool by the growth of a conduction crust. In effect meteorite impacts continually thin the conduction crust; its net rate of growth depending on the relative rates of crustal destruction and regrowth. I have used Dohnanyi's formula (1969) for the cross sectional area of impacting bodies to estimate crustal destruction rate.

Consider a population of impacting planetesimals with a number density function given by

$$f(m) = Am^{-\alpha}. \tag{13}$$

The total cross-sectional areal flux of such a planetesimal population is for $\alpha > 5/3$ (Dohnanyi, 1969)

$$\dot{\sigma}_T = \frac{(2-\alpha)1\mu^{5/3-\alpha}}{\alpha - 5/3} \dot{M}_T \, M_\infty^{\alpha-2} \tag{14}$$

with

$$1 = (3\pi^{\frac{1}{2}}/4\rho)^{2/3}.$$

M_T, M_∞ and μ are the total, largest and smallest masses in kilograms, ρ is the density in kg/m^3 and $\dot{\sigma}_T$ is in m^2/sec. Reported values of α for the asteroids are less than 2 (Dohnanyi, 1969; Chapman, 1976). A value of 1.8 (Dohnanyi, 1971) was used in the following calculations. M_∞ is probably less than 10^{23}–10^{24}g (Wetherill, 1975). $\dot{\sigma}_T$ is not a strong function of M_∞, varying only about a factor of 2.5 for M_∞ ranging from 10^{22}–10^{24} g. Assuming a Goldreich-Ward (1973) type planetesimal (r = 5 km) with ρ = 2.0 g/cm^3, μ = 10^{18} g. Substituting these values into Eq. (14) gives

$$\dot{\sigma}_T = 8.9 \times 10^{-5} \dot{M}_T \tag{15}$$

If the planetesimals are assumed to all be of the same radius, r, then $\dot{\sigma}_T$ for a given $\dot{M}_T$ is given by

$$\dot{\sigma}_T = \frac{3\dot{M}_T}{\rho r}. \tag{16}$$

The rate of lunar surface destruction as a fraction of total lunar surface is

$$f = \frac{\dot{\sigma}_T}{A} \left(\frac{D}{d}\right)^2 \tag{17}$$

where A is lunar surface area, and D/d is the ratio of the diameter of the hole, D, produced in the crust by a planetesimal to the planetesimal diameter, d. Experiments (Gehring, 1970) indicate that D/d ranges from 2–4.5 as impact velocity ranges from 4–8 km/s which is less than the mean velocity reported by Wetherill (1975) for lunar impacts.

Time to destroy the lunar surface (f^{-1}) is shown as a function of impact rate in Fig. 4. Curves are shown for constant radius planetesimals and planetesimals with a range of radii described by Eq. (13). The stippled band represents the effect of different hole to planetesimal diameter ratios. Conduction crust thickness, computed from Eq. (2), corresponding to time along the x axis is shown along the top axis in Fig. 4. Impact rates during accretion are reported to range

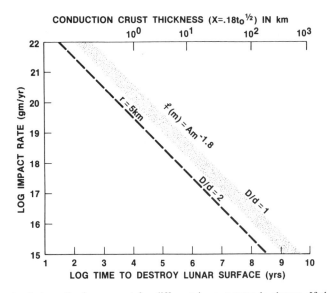

Fig. 4. Time to destroy the lunar crust for different impact rates is shown. If the impacting planetesimals have a number density function given by $f(m) = Am^{-1.8}$, crustal destruction times are represented by the stippled zone. The upper edge of the zone corresponds to the case in which the circular region of destroyed crust equals the planetesimal in diameter. The lower edge corresponds to a hole to planetesimal diameter ratio of 2. The dashed curve represents a population of impacting planetesimals of constant radius 5 km. The top axis shows the thickness of crust that will grow in the time given on the bottom axis.

from 10^{18}–10^{21} g/yr (Hartmann, 1979) corresponding to lunar accretion times of 100–.03 m.y. Impact rates greater than 10^{18} g/yr destroy the entire lunar crust in less than a million years. The maximum thickness of the crust that can form in this time is less than a few kilometers.

Thermal energy input into the magma ocean by the impacting bodies has been neglected. Some energy must certainly have been added to the ocean which would have slowed its solidification. Thermal energy input is difficult to estimate but certainly is less than the kinetic energy of impact. Computer simulation of planetary impacts (O'Keefe and Ahrens, 1977) suggest that only 15–30% of the impact energy goes into thermal energy for impacts into a solid half-space at impact velocities less than 10 km/sec. Impacts through a solid crust overlying a liquid magma ocean could partition more energy into crustal deformation and ejecta.

Impacts would have promoted cooling of the ocean by processes other than thinning the conduction boundary layer. These include cooling of magma ejected into the air and stirring of the magma which would promote upward convective heat transfer. It is not clear whether the heating or cooling effect is dominant, particularly in view of the orders of magnitude variation in impacting rate during early lunar formation. The following simple calculation suggests that cooling may dominate at low impact rates.

The objective of the calculation is to compare impact thermal energy input and energy lost through a conduction boundary layer of thickness X_0. The ratio of thermal energy input by impacts to that lost through a conduction crust is

$$R = \frac{FM\bar{v}^2 X_0}{2K(T_L - T_0)A \cdot t}$$

where M is the impact mass per year, F is the fraction of impact energy partitioned into heating taken to be 0.15, $\bar{v}$ is an average impact velocity of 4 km/sec, A is lunar surface area, t is 1 yr, and K, T_L and T_0 have been previously specified. X_0 is taken as the equilibrium crustal thickness for a given impact rate; e.g., 3 km for an impact rate of 10^{19} gm/yr (see Fig. 5). This is the maximum crustal thickness for a given impact rate and, thus, will underestimate the conductive heat loss. Values of R for different impact rates and the corresponding equilibrium crustal thicknesses are listed in Table 2. It can be seen that with the assumed average impact velocity meteorite impacts through a conduction boundary layer overlying a magma ocean result in net cooling of the ocean for impact rates of 10^{17} gm/yr or less.

For individual impacts, thermal energy input scales as r^3 and energy lost by conduction through the circular section of crust punched out by the impact scales as r^2. Therefore, R scales as r so that small impacts are more efficient coolers of the magma ocean than large ones.

DISCUSSION AND CONCLUSIONS

Results of the crust formation models are summarized in Fig. 5. If the chill-margin thickness is limited by the thickness at which the crystal settling velocity exceeds the solid-melt interface velocity the maximum thickness is a few tenths of a kilometer. If the chill-margin thickness is limited by the development of Rayleigh-Taylor instabilities, the thickness ranges from millimeters to hundreds of kilometers depending on the viscosity of the chill margin. A probable viscosity of 10^{16} poise, based on olivine creep data, yields a maximum crustal thickness of a kilometer. I conclude that if the boundary layer for the magma ocean was a chill-margin it was less than a kilometer thick. If covered by such a continuous

Table 2. Ratio, R, of thermal energy input due to meteorite impacts to energy lost through a conduction crust of thickness X_0.

Impact rate (gm/yr)	Crust destruction time (yrs.)	Equilibrium crust thickness, X_0 (km)	R
10^{19}	10^5	3	6
10^{18}	10^6	9	2
10^{17}	10^7	30	.6
10^{16}	10^8	90	.2

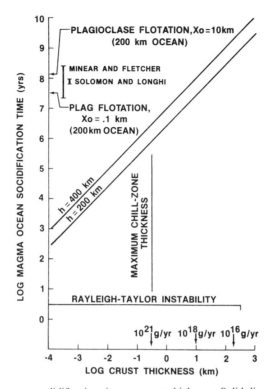

Fig. 5. Magma ocean solidification time vs. crust thickness. Solid diagonal curves represent the time for 200 and 400 km thick magma oceans to solidify if heat is lost by conduction through a crust of constant thickness X_o. Solidification time for the plagioclase flotation models are shown (on the y-axis) for initial crustal thickness of 0.1 and 10 km. For comparison, solidification times determined from crystallization models of magma oceans (Minear and Fletcher, 1978; Solomon and Longhi, 1977) are also shown. The vertical line represents maximal crustal thickness of about 0.5 km estimated from the chill-margin model. The horizontal line gives the range of crustal thickness predicted from Rayleigh-Taylor instability considerations of a dense chill margin crust; the range results from possible viscosity values of the solid crust. The greatest thickness corresponds to a crustal viscosity of 10^{20} poise. Maximum crustal thicknesses for the crust destruction-regrowth model are indicated by the arrows along the x-axis. Numbers by the arrows give planetesimal impact rates.

conduction boundary layer, a 400 km thick magma ocean would have solidified in less than 10^7 yrs.

A chill margin, chemically undifferentiated and more dense than the melt from which it crystallized, that forms a continuous conduction boundary layer on a magma ocean is at best an idealized concept. Breakup and sinking of pieces of such a layer would have continually exposed fresh magma. Therefore, solidification of the magma ocean could have taken considerably less than 10^7 yrs. If a conduction boundary layer is to be maintained at the surface of the magma ocean, this layer must be lighter than the magma. Such a layer could either be composed of relatively less dense mineral phases, i.e., a plagioclase flotation crust, or of a layer of less dense melt, i.e., a relatively silicic layer resulting from the depletion of mafic phases. Using a melt density of 2.83 gm/cm^3 (Minear and Fletcher, 1978) and densities of 3.3, 3.3 and 2.7 gm/cm^3 for olivine, pyroxene and plagioclase respectively, plagioclase must comprise about 80% of the crust if it is to be less dense than the melt. Thus, plagioclase must crystallize in substantial quantities to develop a thick flotation layer. At least 50% of the olivine should have crystallized and the temperature decreased by 400°C before significant plagioclase begins to crystallize (Solomon and Longhi, 1977). Before this point, assuming that the magma ocean is well stirred by convection, the ocean is above the plagioclase liquidus except possibly for a boundary zone beneath the solid crust. Temperature may decrease across this fluid boundary layer; in effect it may act thermally as part of the solid conduction crust. Thus, some plagioclase crystallization could occur in this layer at the top of the ocean as Longhi (1977) and Herbert *et al.* (1977) have suggested. Boundary layer flow theory and experiment suggest, however, that this layer should be quite thin (Eckert and Drake, 1959), on the order of tens of centimeters or less. It therefore seems that plagioclase crystallization in an upper melt boundary layer would be minor. Melt movement would tend to sweep crystals away from the solid-melt boundary and prevent the crust from growing.

With these considerations it seems probable that a solid conduction crust remained extremely thin until plagioclase became a significant crystallizing phase in the entire ocean (T $\leq$ 1150°C). The plagioclase flotation model of crustal formation yields solidification times of 60–120 m.y. for 200–400 km thick magma oceans if the crust remains less than a few tenths of a kilometer until significant olivine (75–100%) has crystallized. The time for a major fraction of olivine differentiation to occur is less than 10 m.y. (Fig. 2).

Impact rates and the amount of energy imparted to the magma ocean by impacts are the major source of uncertainty in solidification-time estimates based on the crust destruction-regrowth model. If energy input is ignored, reasonable impact rates yield crustal thicknesses less than a few kilometers, corresponding to a solidification times of about 3×10^7 and 8×10^7.

Based on the crustal growth models discussed, the maximum solidification time for a 200 km thick ocean is about 10^8 years. A more probable solidification time based on a plagioclase flotation crust model is about 6×10^7 years.

REFERENCES

Ashby M. F. and Verrall R. A. (1977) Micromechanisms of flow and fracture and their relevance to the rheology of the upper mantle. *Phil. Trans. Roy. Soc. London* **A288**, 59–95.

Carslaw H. S. and Jaeger J. C. (1959) Conduction of heat in solids. Oxford Univ. Press, Oxford. 510 pp.

Chapman C. R. (1976) Asteroids as meteorite parent bodies: The astronomical perspective. *Geochim. Cosmochim. Acta* **40**, 701–719.

Daněs F. (1964) Mathematical formulation of salt dome dynamics. *Geophys.* **19**, 414–424.

Dohnanyi J. S. (1969) Collisional model of asteroids and their debris. *J. Geophys. Res.* **74**, 2531–2554.

Dohnanyi J. S. (1971) Fragmentation and distribution of asteroids in physical studies of minor planets. NASA SP-267. 686pp.

Eckert E. R. and Drake M. (1959) *Heat and Mass Transfer*. McGraw-Hill. N.Y. 530 pp.

Gehring J. W. Jr. (1970) Theory of impact on thin targets and shields and correlation with experiments. In *High Velocity Impact Phenomena* (R. Kinslow, ed.), p. 105–155. Academic. N.Y.

Goldreich P. and Ward W. R. (1973) The formation of planetesimals. *Astrophys. J.* **183**, 1051–1061.

Hartmann W. K. (1979) Formative conditions controlling structure of planetary highlands (abstract). In *Papers Presented to the Conference on the Lunar Highlands Crust*, p. 42–44. Lunar and Planetary Institute, Houston.

Herbert F., Drake M. J., Sonett C. P., and Wiskerchen M. J. (1977) Some constraints on the thermal history of the lunar magma ocean. *Proc. Lunar Sci. Conf. 8th*, p. 573–582.

Kirkpatrick R. J. (1974) Nucleation and growth of plagioclase in Hawaiian lava lakes (abstract). In *Trans. Amer. Geophys. Union* **56**, 1198.

Kirkpatrick R. J., Robinson G. R. Jr., and Hays J. F. (1975) Kinetics of crystal growth from silicate melts: Diopside and anorthite (abstract). *Abstracts with Program,* Geol. Soc. Amer. **7**, 1147.

Longhi J. (1977) Magma oceanography: Chemical evolution and crustal formation. *Proc. Lunar Sci. Conf. 8th*, p. 601–621.

Minear J. W. and Fletcher C. R. (1978) Crystallization of a lunar magma ocean. *Proc. Lunar Planet. Sci. Conf. 9th*, p. 261–283.

O'Keefe J. D. and Ahrens T. J. (1977) Impact-induced energy partitioning, melting and vaporization on terrestrial planets. *Proc. Lunar Sci. Conf. 8th*, p. 3357–3374.

Ryder G. and Norman M. (1978) Catalog of pristine non-mare materials, Part 1. Non-anorthosites, JSC Publ. 14565. NASA Johnson Space Center, Houston.

Selig F. (1965) A theoretical prediction of salt dome patterns. *Geophys.* **30**, 633–643.

Solomon S. C. and Chaiken J. (1976) Thermal expansion and stress in the moon and terrestrial planets: Clues to early thermal history. *Proc. Lunar Sci. Conf. 7th*, p. 3229–3243.

Solomon S. C. and Longhi J. (1977) Magma oceanography: 1. Thermal evolution. *Proc. Lunar Sci. Conf. 8th*, p. 583–599.

Wetherill G. W. (1975) Slow accretion of the moon (abstract). In *Papers Presented to the Conference on Origins of Mare Basalts and Their Implications for Lunar Evolution*, p. 185–188. The Lunar Science Institute, Houston.

Proc. Lunar Planet. Sci. Conf. 11th (1980), p. 1957–1977.
Printed in the United States of America

Thermal evolution of Ganymede and Callisto: Effects of solid-state convection and constraints from Voyager imagery

Clifford H. Thurber, Albert T. Hsui*, and M. Nafi Toksöz

Department of Earth and Planetary Sciences, Massachusetts Institute of Technology, Cambridge, Massachusetts 02139

Abstract—New thermal models which include the effect of thermal convection are presented to demonstrate some of the important controlling factors for the evolution of Ganymede and Callisto. Convective heat transfer is incorporated into the models via a parameterized convection scheme. All the thermal models presented here are constructed under the assumption that the two satellites were initially composed of homogeneous mixtures of ice and silicates and with a relatively cold initial thermal state. First, models of two extreme cases of rheological behavior for an ice-silicate mixture are presented. These models demonstrate the large variations of evolutionary history due to different assumed rheological schemes. If deformation within the interior of these two satellites is dictated by the ice component, they will evolve very cold due to the efficient thermal convection. Additionally, the satellites will probably never reach a global differentiation stage. On the other hand, if interior deformation is dictated by the silicate component, the satellites will evolve very hot because convective heat transport will not be effective until the interiors reach a relatively warm thermal state. In this case, our model indicates that the satellites will be completely differentiated and an extensive layer of water exists within the satellites. Besides rheology, the effects of ice phase changes and crustal rigidity have also been investigated. From our model calculation, it is found that phase changes from ice II to ice V and ice II to ice VI have stabilizing effects on thermal convection. Consequently, thermal energy within the satellite interior will not escape as efficiently as in the case of no phase changes, resulting in a warmer evolutionary history. Rigidity of the crustal layer of these two satellites also has an important effect on their evolution. If the crust behaves as a viscous fluid on a long time scale, no water is able to exist underneath an icy crust. Rayleigh-Taylor instability requires all the water to migrate upward in a diapir manner. However, if the crustal layer is able to maintain sufficient rigidity on a long time scale, water may be able to exist beneath a primordial crust composed of an ice-silicate mixture.

INTRODUCTION

An abundance of information about the nature of the surfaces of the Galilean satellites has been obtained by the imaging experiments of the Voyager 1 and 2 fly-by missions (Smith *et al.*, 1979a,b). The surface characteristics of these bodies provide crucial new constraints for models of their thermal histories. Io and

---

* Present address: Department of Geology, University of Illinois at Urbana-Champaign, Urbana, Illinois 61801

Europa are both thought to be influenced by tidal interactions with Jupiter (Peale *et al.*, 1979, Cassen *et al.*, 1979). However tidal heating is probably not significant for either Ganymede or Callisto (Cassen *et al.*, 1980). For this reason we have chosen to model the thermal evolution of the latter two satellites only.

Recently Reynolds and Cassen (1979) have demonstrated the important effects which solid-state convection can have on the thermal evolution of icy satellites. Previous attempts at modeling the thermal histories of Ganymede and Callisto (for example Consolmagno and Lewis, 1976; Fanale *et al.*, 1975) have not taken these effects into account. We have included solid-state convective heat transfer in our models, using the approach of parameterized convection. Before presenting our models, we shall discuss the physical, chemical and geological data which are available as constraints on the thermal evolution of Ganymede and Callisto.

The radii of Ganymede and Callisto have been accurately determined by the Voyager missions (Smith *et al.*, 1979a). Both satellites are roughly the size of Mercury, with Ganymede being slightly larger than Callisto. Ganymede is also about 50 per cent greater in total mass, resulting in a slightly higher mean density (see Table 1). Cosmochemical arguments imply that the two satellites are composed primarily of water (either liquid or ice) and silicates (Consolmagno and Lewis, 1976). Assuming silicates with a density of 3.0 g/cm^3, we find that Ganymede is roughly 65 per cent silicate and 35 per cent water by mass, while Callisto is roughly 55 per cent silicate.

The Voyager missions have unveiled the general surface features of Ganymede and Callisto. Smith *et al.* (1979a,b) have made preliminary investigations of the photographs returned by the Voyager spacecraft. Their studies of crater frequencies and crater morphologies have yielded significant information about the early history of the crusts of these satellites. A brief discussion of their observations and interpretations follows.

Ganymede and Callisto both appear to possess surfaces composed of silicates and ice. However their surface features are distinctly different from each other. The surface of Callisto is uniformly very heavily cratered. Similar areas of dark, heavily cratered terrain are also found on Ganymede, but these regions are separated from each other by complex criss-crossing bright bands of what is termed grooved terrain. A general characteristic of both satellites is the absence of any significant topographic relief.

The heavily cratered (ancient) terrains of both satellites are interpreted as dating back to the end of the period of heavy bombardment, roughly four billion years ago. Crater densities on the grooved terrain of Ganymede are extremely variable, ranging from a density like that of the heavily cratered terrain to a

Table 1.

	Ganymede	Callisto
Mass (g)	1.48×10^{26}	1.07×10^{26}
Radius (km)	2640	2420
Mean density (g/cm^3)	1.93	1.79
% silicate (by mass)	65	55

density a factor of ten smaller. The implication is that formation of the grooved terrain began near the end of the heavy bombardment period, and continued for some period of time afterwards (probably a few hundred million years or less according to Smith *et al.*, 1979b). The grooved terrain is interpreted to have been formed as a result of crustal expansion, although Smith *et al.* (1979b) point out that it is difficult to establish firm proof for this hypothesis.

Finally, the observed crater morphologies provide a basis for estimating the rigidity (i.e., resistance to viscous relaxation) of the crusts of Ganymede and Callisto as a function of time. Larger craters (>150 km in diameter) are extremely subdued, lacking any significant topographic relief. Smaller craters (50 to 150 km in diameter) on the ancient terrains are found in a wide range of degradational states, from fairly fresh to featureless. Craters on Ganymede's grooved terrain are generally found to be much better preserved. The implication is that the crusts of both satellites were cooling and thickening with time as the heavy bombardment period ended. On Ganymede, the crust was clearly becoming more rigid during the period of formation of the grooved terrain. Successful thermal models must account for the existence of stable, thickening crusts on both Ganymede and Callisto about four billion years ago, as well as provide an explanation for the formation of the grooved terrain found exclusively on Ganymede.

MODEL FUNDAMENTALS

Many of the basic elements of our modeling procedure are taken from the work of Consolmagno (1975) and Consolmagno and Lewis (1976, 1977). These elements are briefly summarized here. We assume that the satellites are initially composed of a homogeneous mixture of water ice and silicates. The mean densities constrain the mass ratio of ice to silicate (see Table 1). It is possible that the heat generated during accretion may have been sufficient to result in some initial melting and differentiation of this ice-silicate mix if the accretionary process is a fast one (i.e., $<10^7$ years). For the purposes of this paper however, we assume that accretion is relatively slow and thus contributes insignificantly to the initial temperature profile.

Water ice is known to exist in a number of distinct stable phases. Figure 1 shows a phase diagram for ice (see Fletcher, 1970 or Hobbs, 1974) displaying the stability fields for seven solid phases and the liquid phase (i.e., water). Given a starting temperature profile, the satellite's initial internal structure (stable ice phase as a function of depth) can be determined.

Radiogenic nuclides provide the principal heat source within these satellites. Energy is available from the decay of potassium, thorium, and uranium (^{40}K, ^{232}Th, ^{235}U, and ^{238}U) in assumed chondritic abundances (Kaula, 1968). Due to the spatial and temporal variation of temperature, the phase of ice which is locally stable may change. Such phase changes have been modeled in detail—see Consolmagno (1975) for further discussion.

If a region within the satellite reaches the ice-liquid phase boundary, the silicates mixed with the ice are released as the ice melts. Given the assumed initial conditions, first melting generally is not at the center of a satellite but near to the

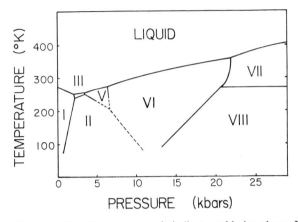

Fig. 1. Phase diagram of ice. Roman numerals indicate stable ice phase. Dashed lines are extrapolations. Data from Hobbs (1974).

surface. Thus the silicates will slowly segregate as melting proceeds towards the center, eventually forming a core. We assume that the radiogenic nuclides are retained by the silicates during the process of differentiation.

PARAMETERIZED CONVECTION

As discussed in the introduction, the thermal models of Consolmagno and Lewis (1976) fail to account for the possible effects of solid state convection. The work of Reynolds and Cassen (1979) suggests that solid state convection can have a profound effect on the thermal history of an icy satellite. One of the main purposes of this paper is to investigate the effect of convective heat transport on the thermal evolution of Ganymede and Callisto without worrying too much about the mechanical structure of convective patterns. Therefore we have chosen to adopt a parameterized convection approach to construct our models. This approach has been used to study the thermal evolution of various terrestrial planets (Sharpe and Peltier, 1978, 1979; Elsasser *et al.*, 1979; Cassen *et al.*, 1979; Schubert *et al.*, 1979). Recently Reynolds and Cassen (1979) utilized this method to analyze the stability of an ice crust overlying a liquid mantle for a Galilean-type icy satellite. In this paper we follow an approach similar to that described in Sharpe and Peltier, 1978. Even though the final equations are very similar, the derivations are not exactly identical.

Derivation of the heat transport equation

The general form of the equation of energy conservation, neglecting viscous dissipation, is given by (Landau and Lifshitz, 1959):

$$\rho C_p \frac{DT}{Dt} - \alpha T \frac{DP}{Dt} = \nabla \cdot (\kappa \nabla T) + H \tag{1}$$

where $\dfrac{D}{Dt} = \dfrac{\partial}{\partial t} + u \cdot \nabla$ is the Lagrangian derivative, u is the velocity vector, ρ is the density, C_p is the specific heat at constant pressure, T is temperature, t is time, α is the coefficient of thermal expansion, P is pressure, κ is thermal diffusivity and H is the uniform or radially varying heat production rate per unit volume. In spherical geometry (r, θ, ϕ) we can rewrite the following operators in component form:

$$\frac{D}{Dt} = \frac{\partial}{\partial t} + \frac{1}{r^2}(r^2 u) + \frac{1}{r \sin\theta}\left[\frac{\partial}{\partial\theta}(v \sin\theta) + \frac{\partial}{\partial\phi}(w)\right] \qquad (2)$$

where (u,v,w) are the components of the velocity vector, and

$$\nabla(\kappa\nabla T) = \frac{1}{r^2}\frac{\partial}{\partial r}\left(r^2\kappa\frac{\partial T}{\partial r}\right) + \frac{1}{r^2 \sin\theta}\left[\frac{\partial}{\partial\theta}\left(\kappa \sin\theta \frac{\partial T}{\partial\theta}\right) + \frac{\partial}{\partial\phi}\left(\kappa \frac{\partial T}{\partial\phi}\right)\right] \qquad (3)$$

Furthermore, since we are only interested in the radial variation of thermal energy, we define the following azimuthal averaging operator

$$\overline{A} = \frac{1}{4\pi}\int\int A \sin\theta \; d\theta \; d\phi \qquad (4)$$

Expanding Eq. (1) into component form via Eq. (2) and Eq. (3) and applying the averaging operator Eq. (4), we obtain

$$\frac{\partial\overline{T}}{\partial t} = \frac{1}{r^2}\frac{\partial}{\partial r}\left[r^2\left(\kappa\frac{\partial\overline{T}}{\partial r} - \overline{uT}\right)\right] - \frac{4\pi}{3}\rho\, G\frac{\alpha}{C_p} r\,\overline{uT} + \frac{H}{\rho C_p} \qquad (5)$$

where the variables with overbars represent the azimuthally averaged quantities. In deriving Eq. (5) we have used the relation

$$\rho = \frac{2\pi}{3}\rho^2\, G\,(r_0^2 - r^2) \qquad (6)$$

where r_0 is the outer radius of the satellite and G is the universal gravitational constant. Note that the second term on the right hand side of Eq. (5) is a heat production term. It represents the heat generated by work done on a particle as it moves along the pressure gradient.

The local heat flux can be expressed as $q_l = \kappa\dfrac{\partial\overline{T}}{\partial r} - \overline{uT}$. Consequently a local Nusselt number can be defined by

$$Nu_l = \left(\kappa\frac{\partial\overline{T}}{\partial r} - \overline{uT}\right)\Big/\left(\kappa\frac{\partial\overline{T}}{\partial r}\right) \qquad (7)$$

Hence, Eq. (5) can be rewritten as

$$\frac{\partial\overline{T}}{\partial t} = \frac{1}{r^2}\frac{\partial}{\partial r}\left[r^2 Nu_l\kappa\frac{\partial\overline{T}}{\partial r}\right] - \frac{4\pi}{3}\rho G\frac{\alpha}{C_p} r\overline{uT} + \frac{H}{\rho C_p} \qquad (8)$$

Following a quasi-steady state assumption which is intrinsic in the parameterized convection approach, and choosing sufficiently small time steps such that heat production becomes small, the local Nusselt number can be approximated by a global Nusselt number which is defined across the total thickness of a convecting region. The Nusselt number is calculated as

$$Nu = a(Ra)^b \tag{9}$$

where Ra is the Rayleigh number. The general expression for Ra is given by

$$Ra = \frac{g\alpha\Delta T(\Delta r)^3}{\kappa\nu} \tag{10}$$

where g is the acceleration of gravity, α is the coefficient of thermal expansion, ΔT is the temperature difference across the region, Δr is the radial extent of the region, κ is the thermal diffusivity, and ν is the viscosity.

We next proceed to non-dimensionalize Eq. (8) utilizing the following definitions:

$$T' = \frac{Tk}{H_r r_0^2} \qquad t' = t\frac{K_r}{r_0^2} \qquad r' = \frac{r}{r_0} \qquad \kappa' = \kappa/K_r \qquad u' = ur_0/K_r \tag{11}$$

where the primed quantities are the non-dimensional quantities. H_r and K_r are the reference heat production density and reference thermal diffusivity respectively; r_0 is the radius of the satellite as defined before. Substituting Eq. (11) into Eq. (8) and dropping the primes immediately we get

$$\frac{\partial\overline{T}}{\partial t} = \frac{1}{r^2}\frac{\partial}{\partial r}\left(r^2 Nu\kappa\frac{\partial\overline{T}}{\partial r}\right) - \frac{4\pi}{3}\frac{\alpha\rho Gr_0^2}{C_p}\overline{ruT} + H \tag{12}$$

The non-dimensional group $\frac{4\pi}{3}\alpha\rho Gr_0^2/C_p$ indicates the relative importance of compressional work. For Ganymede and Callisto this factor is estimated to be approximately 0.038 and 0.028 respectively. Therefore the second term on the right hand side of Eq. (12) can be neglected. Physically, dropping this term implies that heating due to compression is not important for these two satellites because of their relatively small size and low density. For large planets such as Venus or the Earth this term may not be small and hence not negligible. Eliminating the pressure work term, the final governing equation becomes

$$\frac{\partial\overline{T}}{\partial r} = \frac{1}{r^2}\frac{\partial}{\partial r}\left(r^2 Nu\kappa\frac{\partial\overline{T}}{\partial r}\right) + H \tag{13}$$

Note that Eq. (13) is identical to that used by Sharpe and Peltier (1978, 1979) even though it was derived somewhat differently.

Thermal profile conversion

It is immediately apparent that Eq. (13) is identical in form to a heat conduction equation. Consequently the solution will yield conductive-type thermal profiles. It is well recognized that thermal profiles within convective systems should be approximately adiabatic. Therefore the thermal profiles obtained using this approach will not be realistic, although the transient total heat budget within the satellite can be accurately approximated. In order to obtain realistic thermal profiles, the profiles which result from using Eq. (13) must be converted to possess an adiabatic gradient within a convecting region while keeping the total internal energy unchanged. A similar approach has been adopted by Cassen *et al.* (1979). This conversion is accomplished in the following way. Let a convective region be bounded by R_a and R_b as shown in Fig. 2. The conductive type profile for this region, found using Eq. (13), is the profile labeled f_c and bounded by temperatures Ta and Tb. Let the adiabatic profile we seek be f_a and the temperature at the top of the region, away from the thermal boundary layer, be T_c. Then f_a can be expressed as

$$f_a(r) = T_c \exp \left[\frac{2\pi}{3} \frac{\alpha\rho G}{C_p} (r_a{}^2 - r^2) \right] \tag{14}$$

where the physical parameters are defined as before. The total internal energy bounded by the two thermal profiles, f_c and f_a, can be respectively described as

$$E_c = 4\pi\rho C_p \int_{r_b}^{r_a} f_c \, r^2 \, dr$$

$$E_a = 4\pi\rho C_p \exp \left[\frac{2\pi}{3} \frac{\alpha\rho G}{C_p} r_a{}^2 \right] T_c \times \int_{r_b}^{r_a} r^2 \exp \left[-\frac{2\pi}{3} \frac{\alpha\rho G}{C_p} r^2 \right] dr \tag{15}$$

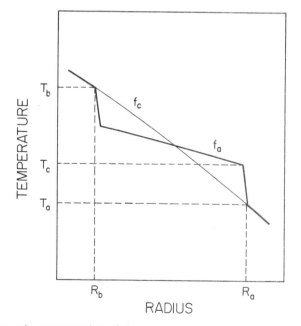

Fig. 2. Schematic representation of the temperature profile conversion method: the conductive-type profile f_c is converted to the adiabatic-type profile f_a which has the same total internal energy.

Now T_c may be calculated simply by equating E_c to E_a, since all other quantities are known. Once T_c is determined, the adiabatic profile that possesses the same internal energy as f_c can be computed using Eq. (14).

Examples of the profile conversion are illustrated in Fig. 3a and 3b. Figure 3a is for a layer of water, and Fig. 3b is for a silicate core. The solid lines show the conductive-type solution and the dashed lines are the converted (adiabatic) profiles. The extreme temperature change due to the conversion in the case of the silicate core is due to geometrical volume effects—less than a third of the core's actual mass is actually at a temperature above 1500°K in the original conductive profile.

It should be noted that the profile conversion is calculated assuming that the thermal boundary layer is small compared with the grid spacings used in the numerical computation. Our analysis indicates that this assumption is generally valid. Thus we find that the parameterized convection approach permits a proper treatment of the overall heat budget of a satellite, while the temperature profile conversion method allows us to reconstruct a more realistic profile of the internal temperature.

Rheology

One of the most important sources of uncertainty in our models concerns the viscosity relations which are needed in order to calculate the Rayleigh number

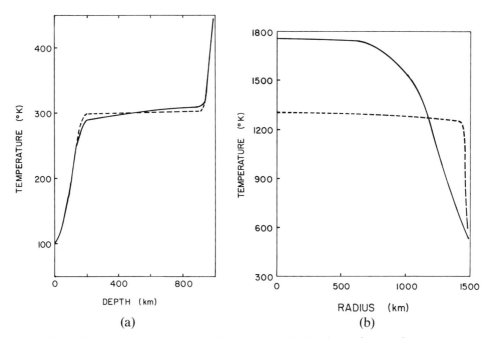

Fig. 3. Examples of temperature profile conversion. In a), a layer of convecting water, and in b), a convecting silicate core. Solid lines are the computed conductive-type profiles; dashed lines are the corresponding converted profiles.

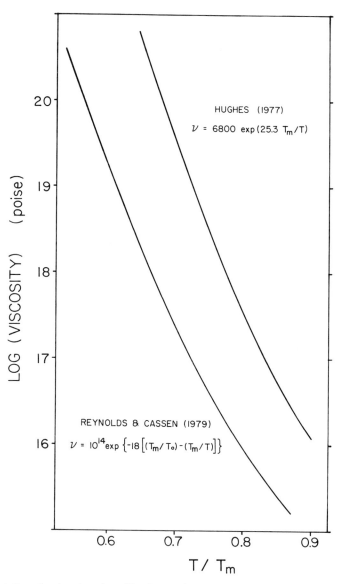

Fig. 4. Two relations for the viscosity of ice I—one from Reynolds and Cassen (1979) and the other derived from data of Hughes (1977). T_m is the ice melting temperature.

[Eq. (10)]. The rheology of ice I is poorly understood. The viscosity of the other higher-density phases of ice is completely unknown. We assume that all phases of ice follow the same viscosity relation. Since the crystalline structure and bonding are fundamentally similar for the different ice phases, this is a reasonable assumption for a first approximation.

We have chosen to test two different relations for the viscosity of ice as a function of temperature, as shown in Fig. 4. One is taken from the work of

Reynolds and Cassen (1979) which they used for calculating the stability of a layer of ice I to solid-state convection. The other is a linearized relation derived from data on strain rate as a function of stress for ice I given by Hughes (1977).

A related problem concerns the rheological behavior of mixtures of ice and silicates. We have assumed that the satellites were initially composed of a homogeneous mix of silicates and ice in some specified proportion. Thus the possibility of solid-state convection during the initial stages of evolution will be primarily controlled by the viscosity of the mixture. We would argue that the rheology should be dominated by the lower viscosity component of the mixture, that is ice, as long as the volume fraction of silicates is not too large. Both Ganymede and Callisto should in fact have silicate volume fractions substantially smaller than one-half. Rather than making this highly constraining assumption about the rheology, however, we have chosen to model two extreme cases for the rheology of an ice and silicate mix. We have constructed two sets of thermal models for Ganymede and Callisto, one in which ice dominates the rheology, and the other in which silicates dominate. These models are presented in the next section.

THERMAL EVOLUTION MODELS

A brief discussion of our assumptions about the initial conditions for Ganymede and Callisto is appropriate before presenting the types of models we have explored. The single most important assumption made about the initial state of these bodies is that they were essentially homogeneous following accretion. It could be argued that accretional heating might be sufficient to melt a substantial portion of the outer layers of these large satellites if they were accreted fast enough, as has been hypothesized in the case of the moon. In fact the relatively low melting temperature of ice would seem to make this even more likely to occur for the icy Galilean satellites. However, for simplicity, we assume a uniform initial internal temperature of 200°K, falling to the fixed surface value of 100°K (see Fig. 5 for an example).

Values for the density of the silicate phase and their radioactive content need to be assumed as well. Consolmagno and Lewis (1976) modeled two extreme cases for the silicate density, 2.5 g/cm^3 and 3.7 g/cm^3, and concluded that the value chosen made little difference on the resulting thermal histories. As a compromise, we assign the silicates a density of 3.0 g/cm^3. Like Consolmagno and Lewis (1976), we also assume chondritic abundances for uranium, potassium, and thorium, the key radiogenic nuclides considered.

Ice-dominant rheology

Given the initial conditions just described, and using the physical parameters of Consolmagno (1975) and the masses and radii listed in Table 1, thermal evolution models for Ganymede and Callisto were constructed under the assumption that for a mixture of ice and silicates with a silicate mass fraction of less than 0.75, the viscosity would be given by that of pure ice at that temperature. Both viscosity relations shown in Fig. 4 were tested. Even using the high range of ice viscosity values, neither satellite closely approaches the melting temperature of ice in their interiors at any time in their history. Figures 5a) and b) show the models obtained using the viscosity relation derived from Hughes (1977), giving the internal temperature as a function of depth initially (0.0 b.y.), after 1.0 billion years, and at the present (4.5 b.y.). Satellite-wide solid state convection begins early in their histories and continues up to the present.

Our prejudice is to reject these models as being in conflict with the constraints

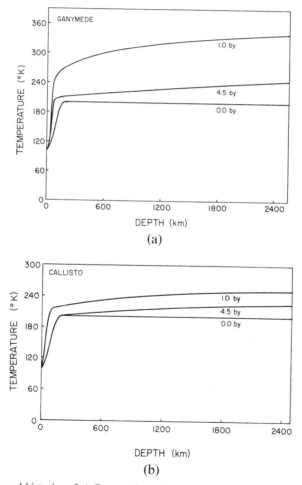

(a)

(b)

Fig. 5. Thermal histories of a) Ganymede and b) Callisto assuming ice-dominant rheology. Shown is the temperature as a function of depth in the satellites initially, after 1.0 billion years, and at the present (4.5 b.y.). No melting occurs.

derived from interpretation of the Voyager surface imagery. Based on these models, there is no significant difference between the evolution of Ganymede and Callisto. In particular, there is no hint of a period in which the crust of Ganymede alone should be subject to rifting and expansion. The models would also predict that the lithosphere of the satellites would be thinnest and weakest around 1.0 billion years, and only begin to thicken and strengthen after that. The crater morphology analysis of Smith *et al.* (1979b) implies a much shorter time scale for the evolution of the lithosphere of Ganymede and Callisto.

Silicate-dominant rheology

A second set of thermal evolution models was developed assuming the other extreme case for the rheology of an ice-silicate mixture. For a mixture of ice and silicates with a silicate mass fraction of more than 0.5, the viscosity is assumed to be given by that of pure silicate. A representative viscosity relation is taken from Oxburgh and Turcotte (1978).

Since both Ganymede and Callisto are modelled as initially homogeneous bodies with more than 50 per cent silicates by mass, this assumption prohibits any solid state convection from occurring in undifferentiated layers. That is, the melting temperature of ice will always be reached before convection becomes possible in such a mixture.

The resulting thermal evolution models are similar in nature to those presented by Consolmagno and Lewis (1976). Melting and differentiation occur early in the histories of the satellites. Both Ganymede and Callisto currently would possess moderately thin non-convecting crusts, large water mantles, and convecting silicate cores (see Fig. 6a and b).

We find these models to be unacceptable also. As in the first example, the evolutions of Ganymede and Callisto are basically indistinguishable. More importantly, this model predicts that both satellites would possess fairly thin solid crusts (lithospheres) for most of their histories. Such a crust would be extremely unstable, and subject to a variety of destructive tectonic processes, for example those discussed by Parmentier and Head (1979). These types of activity, for example diapirism, subduction, and volcanism, are simply not observed on the surfaces of Ganymede and Callisto. Furthermore, the variations in crater morphology could not be accounted for—an early period of lithosphere thickening and strengthening is absent from the models.

Phase changes and solid-state convection

Up to this point we have ignored any effects that the presence of different phases of ice might have upon solid-state convection. According to Schubert *et al.* (1975) there are two principal mechanisms through which a phase transition will interact with convection: thermal expansion due to latent heat, and phase boundary distortion due to both latent heat and advection. These effects can individually be

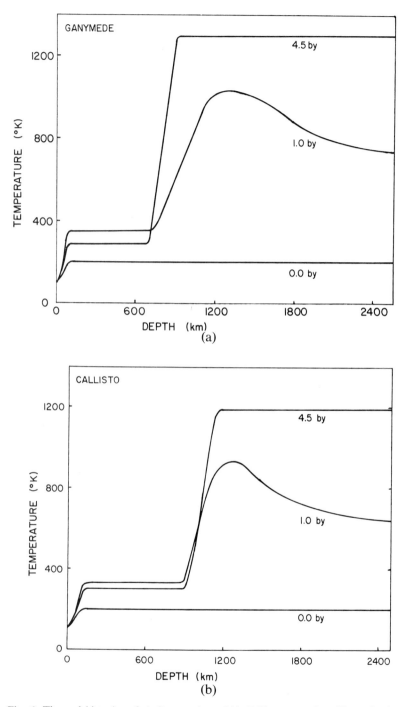

Fig. 6. Thermal histories of a) Ganymede and b) Callisto assuming silicate-dominant rheology (see Fig. 5). The present-day structure of both satellites would consist of thin crusts, large liquid water mantles, and convecting silicate cores.

either stabilizing or destabilizing. For the olivine-spinel phase transition in the earth, for example, the advection effect on the phase boundary is destabilizing, while the latent heat effects (phase boundary distortion *and* thermal expansion) are stabilizing (Schubert *et al.*, 1975).

In the case of the Galilean satellites, most of the relevant ice phase transitions should be unstable to solid state convection (P. Cassen, pers. comm.). Consolmagno and Lewis (1978) suggest that the ice VI–ice II phase transition would inhibit solid-state convection. In fact, a strong case can be made for the stability of the ice V–ice II and probably the ice VI–ice II phase transitions. The ice V–ice II phase boundary has a strongly negative Clapeyron slope (see Fig. 1). Since ice V is more dense than ice II, the transition from ice II to ice V must be endothermic. Thus the situation is comparable to the spinel-post spinel case discussed by Schubert *et al.* (1975).

The stability of the phase change can be investigated through the dimensionless parameters S and R_Q:

$$S = \frac{\Delta\rho/\rho}{\alpha d(\rho g/\gamma - \beta)} \qquad (16)$$

$$R_Q = \frac{\alpha g d^3 Q}{\kappa \nu C_p} \qquad (17)$$

along with the usual Rayleigh number Ra [see equation (10)]. The new parameters in Eq. (16) and (17) are: $\Delta\rho$, the density change between the two phases; γ, the Clapeyron slope; β, the local temperature gradient; and Q, the latent heat of the phase change. Using parameter values appropriate for Ganymede and Callisto for the ice V–ice II transition, S and R_Q can be estimated as roughly -5 and -100, respectively. Inspection of Fig. 4 of Schubert *et al.* (1975), which gives the critical Rayleigh number as a function of R_Q for a range of S from 0.0 to -1.5, suggests that this particular phase change would be stable against penetrative convection (i.e., convection through the phase transition boundary). The ice VI–ice II transition is probably stable as well. The style of convection which would be more likely to occur would be two-cell convection: separate convection cells in the ice II and ice V (or VI) regions.

Preliminary thermal models have been constructed for Ganymede and Callisto which incorporate a restriction of convection across these two phase transitions. As in the first set of models, we assume that ice dominates the rheology of ice-silicate mixtures. In addition, two possible assumptions about the long-term stability of the ice and silicate crust are examined. In one case, primordial crust overlying liquid water is presumed to be unstable due to the density difference between the crustal material and water. For the second case, the overlying crust is assumed to be stable.

Parmentier and Head (1980) discuss the Rayleigh-Taylor instability which would be expected if the crust above a liquid region behaves in a purely viscous fashion on a long time scale. This instability should in fact arise as soon as *any* significant amount of liquid water is present in these bodies, not only after a large

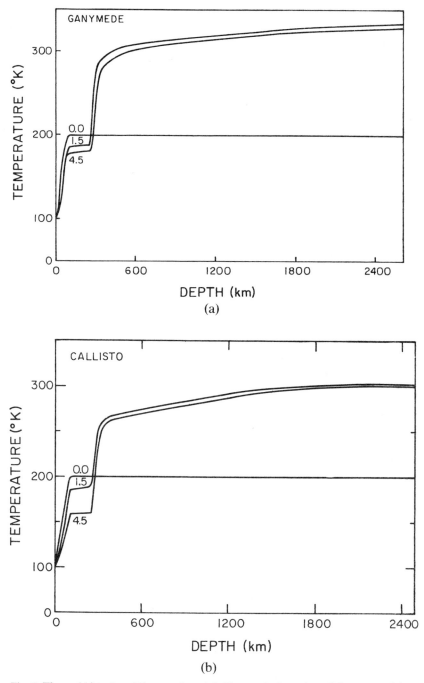

Fig. 7. Thermal histories of Ganymede and Callisto assuming primordial crust overlying liquid water behaves viscously on a short time scale. A steady state is reached in which ice melting is balanced by water diapirism.

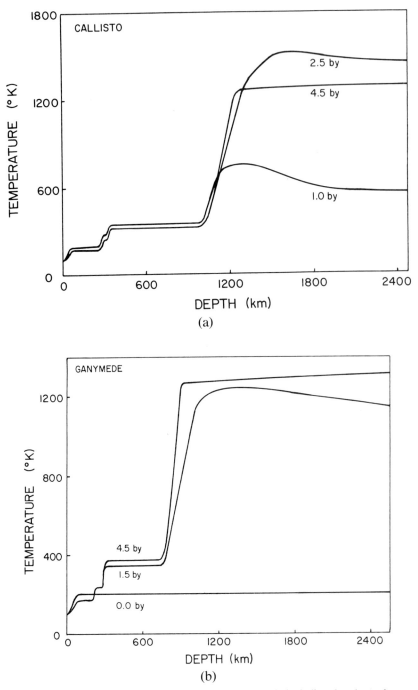

Fig. 8. Thermal histories of a) Callisto and b,c) Ganymede including the phase change barrier to convection (see text). Thick crusts, assumed to be stable, are present on both satellites, except for a period of crustal thinning on Ganymede in its history (shown in c).

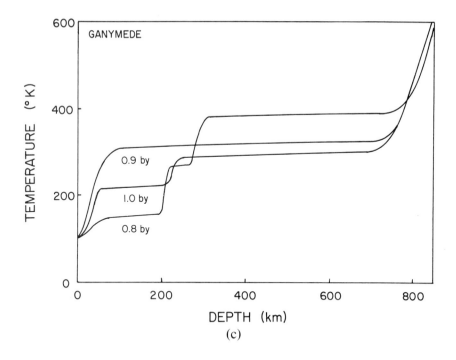

(c)

mantle of liquid water is formed. We have attempted to model this diapir type mechanism in the following manner: if the ice in a given layer in the interior of a satellite is partially molten, then this layer and the solid layer just above it are permitted to convect together—a Nusselt number appropriate to the solid layer is determined and used to compute the local rate of heat transport. In all likelihood this representation of the diapir mechanism will underestimate the enhancement of heat transport due to the Rayleigh-Taylor instability, given the rapid time scale on which the instability should develop (see Parmentier and Head, 1980, for discussion of the instability time scale).

Thermal models including this representation of the Rayleigh-Taylor instability were constructed for Ganymede and Callisto, yielding essentially identical thermal histories for the two bodies. Given the assumed initial conditions, melting of ice first occurs at a depth of about 250 km in both satellites. However, once melting begins the diapiric upwelling of the liquid water results in its promptly being refrozen. Thus a situation is reached whereby the freezing due to the diapir mechanism balances any additional melting, resulting in nearly steady-state temperature profiles for both Ganymede and Callisto (see Fig. 7a and b). The melting point of ice is not reached at any other place in these satellites.

As in the case of the ice-dominant-rheology models, it is difficult to reconcile these thermal models with the observed differences between the surfaces of Ganymede and Callisto. No endogenic cause for these differences is found in this set of models. Some primordial or external process would be required to account for the greater variety of surface features observed on Ganymede.

For a second set of models, it is assumed that the Rayleigh-Taylor instability will not occur (i.e., primordial ice and silicate crust overlying water is presumed to be stable on a long time scale). This is possible if the crustal layer is able to maintain sufficient rigidity on a large time scale. Even though this assumption is difficult to justify at present, it is premature to rule it out completely considering our current ignorance about the physical behavior of ice-silicate mixtures under temperature and pressure conditions similar to those near the surfaces of Ganymede and Callisto. Therefore, we are presenting the following models as an alternative if and only if the above assumption is valid. The resulting thermal histories are shown in Figs. 8a and b. In our model for Callisto (Fig. 8a) the outer 300 km of the satellite never reach the melting temperature of ice. Starting at the surface, we find a conducting layer, a layer of convecting ice and silicate, a non-convecting layer below the ice II–ice V transition, a liquid water mantle, and a silicate core which begins convecting after a few billion years. This model is consistent with the existence of a tectonically stable ancient crust on Callisto.

Superficially Ganymede appears to follow an identical evolutionary path (Fig. 8b). However our model predicts an episode of crustal thinning early in the satellite's history, as illustrated in Fig. 8c. At 800 million years we find a thick stable crust. At 900 million years, the crust has been thinned, and a net expansion of the satellite has occurred. By 1 billion years, the crust has thickened again.

A number of factors contribute to the cause of this crustal thinning period. Ganymede's higher silicate mass fraction implies a greater overall proportion of radiogenic nuclides and therefore generally more thermal energy available. The higher silicate fraction, added to Ganymede's greater total mass, implies a significantly larger release of potential energy during core formation (roughly a factor of two more than for Callisto). Finally, the higher pressures throughout Ganymede's interior result in the release upon melting of substantial quantities of very warm water—up to nearly 500°K. Such a crustal thinning episode could be responsible for the formation of Ganymede's grooved terrain.

DISCUSSION AND CONCLUSIONS

At the outset, our goal was to explore the effects of solid state convection on the thermal evolution of Ganymede and Callisto. An important factor that controls the efficiency of convective heat transport is the rheology of the interiors of the satellites. To demonstrate the sensitivity of thermal evolution models of a satellite to different rheological formulas, we have constructed two sets of models. The first set of models assumes that the rheological behavior within a satellite is dominated by the ice component. In such a case, material is allowed to creep at a relatively low temperature. Because of the highly efficient convective heat transport mechanism, the satellite remains relatively cool if it starts cold. Most of the satellite is found to remain well below the ice melting temperature. Consequently, global ice-silicate differentiation never occurs. On the other hand, if the rheological behavior of a satellite is dominated by the silicate component, a drastically different thermal evolution model will result. Because of the silicate-dominated rheology, the interior of a satellite will remain relatively rigid until the internal temperature reaches about half the melting temperature of silicate. By this point, however, the ice component would be well above its melting temperature. Thus ice-silicate separation would occur long before the satellite reaches this stage. Because of the relative inefficiency of conductive heat transport (the transport mechanism which will predominate in such a model for much of the history of the body) a satellite will evolve much hotter under such a rheological scheme. Our model shows that an extensive layer of water will exist in such a case. Based on these two models, it is apparent that rheology is a very important factor in determining the thermal evolution of an icy satellite. Since very little is known about the rheological behavior of ice-silicate mixtures, this property within the two satellites cannot be well constrained until more relevant laboratory data are available.

Besides rheological effects, we have also investigated the effects of phase changes of ice upon the thermal evolution of these two satellites. As pointed out previously, the phase changes from ice II to ice V and VI may have stabilizing effects on thermal convection. As a result, heat transport from the interior to the surface will not be as efficient as that without a phase boundary. This will yield a warmer satellite in general.

The stability of a crustal layer also has very important effects upon satellite thermal evolution. If the crustal layer behaves in a purely viscous fashion on a large time scale, it will not be able to sustain itself if water exists underneath. A complete overturn will take place as predicted by the Rayleigh-Taylor instability. On the other hand, if the crustal layer is able to maintain its rigidity on a long time basis, the onset of the Rayleigh-Taylor type of overturn may not take place.

If the crusts of the icy satellites indeed behave in this long-term rigid manner, then together with the effects of phase changes of ice, our study suggests that a primordial surface may exist throughout the history of Callisto. This may have some implications on the source of the large amount of silicate material observed on Callisto's surface. There are two possibilities for explaining this observation. First, the surface of Callisto may represent an undifferentiated primordial surface. Second, the silicate material may simply be the blanket of debris from impacting bodies. If it is the case that the crust of the satellite can maintain its rigidity, both explanations are equally acceptable. However, if the crust indeed behaves only viscously on long time scales, the first explanation will probably be invalid, as the primordial crust should be destroyed via the Rayleigh-Taylor instability.

Finally, the models presented in this paper are not meant to be definitive and well constrained. Instead, these models are presented in order to illustrate certain important effects upon the thermal evolution models of Ganymede and Callisto. Furthermore, the effects that have been studied here are by no means complete. Other important factors remain to be investigated. Foremost among these is the degree to which melting and differentiation occurred during the accretion stage. As in the case of the moon, for example, the outer few hundred kilometers or more of these satellites may have been melted during accretion. A second key uncertainty concerns the viscosity of ice I (poorly known) as well as that of the higher density phases of ice (essentially unknown). In the absence of new laboratory data, the treatment of the viscosity of ice will have to remain ambiguous. Most important however for more detailed modelling of icy satellites, additional data can be obtained from further analysis of the Voyager data and from future spacecraft observations.

Acknowledgments—We wish to thank Guy Consolmagno for his assistance with various aspects of this work. We also acknowledge Pat Cassen, Gerald Schubert, Stan Peale, Sean Solomon, and Hampton Watkins for helpful discussions. This research was supported by NASA Grant NSG-7081.

REFERENCES

Cassen P., Peale S. J., and Reynolds R. T. (1980) On the comparative evolution of Ganymede and Callisto. *Icarus* **41**, 232–239.

Cassen P. and Reynolds R. T. (1979) Is there liquid water on Europa? *Geophys. Res. Lett.* **6**, 731–734.

Cassen P., Reynolds R. T., Graziani F., McNellis J., Summers A. L., and Blalock L. (1979) Convection and lunar thermal history. *Phys. Earth Planet. Inter.* **19**, 183–196.

Consolmagno G. J. (1975) Thermal history models of icy satellites. Masters thesis, Massachusetts Institute of Technology. 202 pp.

Consolmagno G. J. and Lewis J. S. (1976) Structural and thermal models of icy Galilean Satellites. In *Jupiter* (T. Gehrels, ed.), p. 1035–1051. Univ. Arizona Press, Tucson.

Consolmagno G. J. and Lewis J. S. (1977) Preliminary thermal history models of icy satellites. In *Planetary Satellites* (J. Burns, ed.), p. 492–500. Univ. Arizona Press, Tucson.

Consolmagno G. J. and Lewis J. S. (1978) The evolution of icy satellite interiors and surfaces. *Icarus* **34**, 280–293.

Elsasser W. M., Olson P., and Marsh B. D. (1979) The depth of mantle convection. *J. Geophys. Res.* **84**, 147–155.

Fanale F., Johnston T., and Matson D. (1977) Io's surface and the histories of the Galilean satellites. In *Planetary Satellites*. (J. Burns, ed.), p. 379–405. Univ. Arizona Press, Tucson.

Fletcher N. J. (1970) *The Chemical Physics of Ice*. Cambridge Univ. Press. London. 271 pp.

Hobbs P. V. (1974) *Ice Physics*. Clarendon, Oxford. 837 pp.

Hughes T. (1977) West Antarctic ice streams. *Rev. Geophys. Space Phys.* **15**, 1–46.

Kaula W. M. (1968) An introduction to Planetary Physics. Wiley, N.Y. 490 pp.

Landau L. D. and Lifshitz E. M. (1959) *Fluid Mechanics*. Pergamon, London. 536 pp.

Oxburgh E. R. and Turcotte D. L. (1978) Mechanisms of continental drift. *Rep. Prog. Phys.* **41**, 1249–1312.

Parmentier E. M. and Head J. W. (1979) Internal processes affecting surfaces of low-density satellites: Ganymede and Callisto. *J. Geophys. Res.* **84**, 6263–6276.

Peale S. J., Cassen P., and Reynolds R. T. (1979) Melting of Io by tidal dissipation. *Science* **203**, 892–894.

Reynolds R. T. and Cassen P. M. (1979) On the internal structure of the major satellites of the outer planets. *Geophys. Res. Lett.* **6**, 121–124.

Schubert G., Cassen P., and Young R. E. (1979) Subsolidus convective cooling histories of terrestrial planets. *Icarus* **38**, 192–211.

Schubert G., Yuen D. A., and Turcotte D. L. (1975) Role of phase transitions in a dynamic mantle. *Geophys. J. Roy. Astron. Soc.* **42**, 705–735.

Sharpe H. N. and Peltier W. R. (1978) Parameterized convection and the earth's thermal history. *Geophys. Res. Lett.* **5**, 737–740.

Sharpe H. N. and Peltier W. R. (1979) A thermal history model for the earth with parameterized convection. *Geophys. J. Roy. Astron. Soc.* **59**, 171–203.

Smith B. A., Soderblom L. A., Johnson T. V., Ingersoll A. P., Collins S. A., Shoemaker E. M., Hunt G. E., Masursky H., Carr M. H., Davies M. E., Cook A. F. II, Boyce J., Danielson G. E., Owen T., Saqan C., Beebe R. F., Veverka J., Strom R. G., McCavley J. F., Morrison D., Briggs G. A., and Suomi V. E. (1979a) The Jupiter system through the eyes of Voyager I. *Science* **204**, 951–972.

Smith B. A., Soderblom L. H., Beebe R., Boyce J., Briggs G., Carr M., Collins S. A., Cook A. F. II, Danielson G. E., Davies M. E., Hunt G. E., Ingersoll A., Johnson T. V., Masursky H., McCavley J., Morrison D., Owen T., Sagan C., Shoemaker E. M., Strom R., Suomi V. E., and Veverka J. (1979b) The Galilean satellites and Jupiter Voyager II imaging science results. *Science* **206**, 927–950.

Proc. Lunar Planet. Sci. Conf. 11th (1980), p. 1979–1985.
Printed in the United States of America

Near surface magma movement

Otto H. Muller[1] and Michael R. Muller[2]

[1]Geology Department, Colgate University, Hamilton, New York 13346   [2]Department of
Mechanical, Industrial and Aerospace Engineering, Rutgers University, New Brunswick, New
Jersey 08903

Abstract—A widely held notion contends that magma comes to the surface through conduits which
remain open down to the source and within which nearly hydrostatic conditions exist. Recent work
on the form, growth and movement of pressurized cracks within elastic solids subjected to linear
stress gradients is reviewed which brings this notion into question. Although under certain conditions
such a continuous, nearly hydrostatic conduit is possible, most magma may reach the surface or near
surface environment within dike like bodies which have a tear drop shaped vertical cross section
and do not extend to the source area. The difficulties involved in creating a cylindrical pipe through
a solid which is below the melting point of the magma, and below the surface as well are discussed.
If magma moves through an elastically deformed crack these problems are not encountered. No
constraint on the elevation finally attained by the magma is imposed by this model.

INTRODUCTION

Volcanism occurs when magma, originating at depth, reaches the surface of a
planetary body. The classical model (Daly, 1933; Eaton and Murata, 1960; Yoder,
1976) contends that during eruption a continuous conduit extends from the source
of the magma to the point of egress (vent, fissure, etc.). The maximum elevation
of the top of the magma within this conduit is determined by the hydrostatic
("magmastatic"?) head needed to counteract the pressure in the source area
which is produced by the lithostatic load there. Because the density of crustal
and mantle rocks increases with depth, the deeper the source area, the higher the
magma column can extend, assuming the density of the magma increases more
slowly with depth than that of the country rock. Support for this model comes
from seismically determined source depths for Hawaiian volcanoes (Eaton and
Murata, 1960); volcano height vs. age relationships for oceanic volcanoes (Vogt,
1974); petrologically determined source depths for some lunar mare basalts (Kesson, 1975; Solomon, 1975); and hydrostatic considerations of lunar volcanism
during which the erupted magma was denser than the upper 20 km of lunar crust
(Solomon, 1975).

In this paper we question the general applicability of this model by showing
that (i) the creation of a cylindrical conduit requires the removal of material and
no satisfactory mechanism exists to initially accomplish this, and (ii) if the conduit
is crack shaped and the density of the magma is less than the country rock, a

brittle elastic model predicts that discrete "packets" of magma will depart from the magma source and move upwards individually. We also consider the case of a crack shaped reservoir containing magma which is as dense or denser than the country rock. We show that this crack will not move upwards, but may grow preferentially upwards if close enough to the surface.

CYLINDRICAL CONDUITS

The geometry of a cylindrical magma conduit would permit it to remain open even when the magma pressure within the pipe would not be sufficient to keep a crack open. Such a pipe is the simplest way to bring magma up from deep sources. Eroded volcanic piles often reveal circular vents, and the erosion of the country rock just beneath ancient volcanoes frequently exposes volcanic plugs and necks roughly cylindrical in form which served as magma conduits for the volcanoes (Williams, 1936; Williams and McBirney, 1979). Kimberlite mines have excavated many pipes which are conical or cylindrical in form. Therefore many people believe magma travels from its source area to the surface through cylindrical pipes. It is important to note that this evidence is derived from rock presently exposed at the surface which is thought to have been less than 5 km beneath the surface during eruption. If such cylindrical pipes extend to the source (which we presume to be in the mantle), their remains should exist in previously deep seated basement rock, such as that now exposed in the Canadian shield. Although this rock is riddled with dikes of all kinds, we are unaware of any well exposed cylindrical conduits there. Such indirect evidence is inconclusive, yet does suggest that crack shapes are most frequent, and may be the only shapes for magma conduits at depths greater than about 5 km.

To create a cylindrical magma conduit, the material which once occupied that volume must be displaced either downward or upward. If material is broken off from the rock above the top of the pipe and sinks into the magma (a process called stoping), the pipe will extend upward, but will fill up from below with the settling of the stoped blocks. Unless an open reservoir of fluid magma with sufficient volume to contain all of the material stoped from the pipe exists beneath the base of the pipe before its ascent begins, the settled blocks cannot be removed from the pipe. We believe such a reservoir is improbable. Without the reservoir this process may result in the upward migration of magma ("zone melting", Williams and McBirney, 1979), but it does not produce an open conduit from the source.

It is not possible to move material upward out of the conduit until the magma has reached or nearly reached the surface. This was first shown by G. K. Gilbert in 1877 in his classic study of laccoliths, and has recently been developed further by Pollard and others (Pollard and Johnson, 1973; Pollard and Holzhausen, 1979).

One possibility remains: the cylindrical conduits formed after magma breached the surface first as a fissure eruption. Delaney and Pollard (1980) have recently completed a detailed study of this process using field evidence. In their model the crack reaches the surface as a fissure eruption with greatest discharge near its center where it is widest. Erosion of the walls of the crack by the flowing magma is greatest here and this region widens further. Eventually most of the magma flow is channeled into this growing cylinder and the tips of the crack freeze shut. Depending on the ability of the magma to erode the country rock and the duration of volcanic activity, the diameter of the circular vent may grow to exceed the original length of the fissure, obliterating any evidence of it; or it may be only slightly greater than the original fissure width, producing a wide spot in the dike emplaced in the country rock and a cylindrical column in the volcanic pile which builds up around it. The depth to which this channel excavation proceeds is similarly a function of the erosive ability of the magma and the duration of volcanic activity. Kimberlites, characterized by highly gas charged magmas, appear to be particularly efficient at eroding conical channels which converge downward. This geometry suggests that erosion was more efficient and/or of longer duration near the surface, in agreement with this model. Is there any limit to the depth of cylindrical channels produced by this mechanism? Perhaps not. If not, it may permit the creation of cylindrical conduits from source to surface after the initial ascent has been accomplished by a pressurized crack. We leave this as a subject for future research and next consider how a pressurized crack makes it to the surface.

CRACK SHAPED CONDUITS

By elastically deforming the walls, magma may move into a crack without permanently removing material. The presence of magma under pressure within it may cause a crack to propagate in the same way that water under pressure is used to produce cracks in oil bearing strata. This process, called "magmafracturing" (Yoder, 1976), may be of fundamental importance in the rise of magma to the surface. As the crack extends, magma fills the newly created crack and causes it to extend further. Two geometries can be imagined: (i) A continuous conduit extending from the crack tip to the source region, kept filled with magma by the pressure in the source area, and (ii) A pocket of magma moving up through the crack, unconnected to the source region. We will see that the densities of the magma and country rock and the fracture toughness of the country rock will determine which geometry is selected.

We refer to the Linear Elastic Fracture Mechanics (LEFM) solutions of this problem (Weertman, 1971a,b, 1980; Pollard, 1976; Pollard and Muller, 1976; Secor and Pollard, 1975). The model consists of a vertical two dimensional crack (infinitely long in the third dimension) filled with a fluid of constant density, ρ_m, which exists within an ideal elastic solid of constant density, ρ_r, Poisson's ratio σ, and shear modulus, μ. Far from the crack the least principal compressive stress, S_x, is horizontal and increases linearly with depth, $-z$, so that $S_x = S_{xo}$

$- gz\rho_r$. The pressure within the fluid also increases linearly with depth: $P = P_o - gz\rho_m$. Solutions exist which describe the shape and behavior of a free crack and, using the method of images and a symmetrical stress gradient, a crack at the base of an elastic plate.

The free crack, with its center at the origin of a cartesian coordinate system (x,z) extends from $z = +l$ to $z = -l$. The crack at the base of the elastic plate has its base at the origin and extends from $z = 0$ to $z = l$. Shapes are defined by the equations for width:

$$W_{Free} = 2|\sqrt{1 - (z/l)^2}(C_1 + C_2(z/l))|$$
$$W_{Base} = 2\{\sqrt{1 - (z/l)^2}(C_1 + 2C_2/\pi) + (2C_2z^2/\pi l^2)\ln|(z/l)/1 - \sqrt{1 - (z/l)^2}|\}$$

where $C_1 = (1 - \sigma)(P_o - S_{xo})l/\mu$
$\qquad C_2 = (1 - \sigma)[g(\rho_r - \rho_m)]l^2/2\mu$
and $(P_o - S_{xo})$ is the difference between magma pressure and least compressive stress at $z = 0$.

For the cracks to remain open along their lengths these equations require that:

$$(P_o - S_{xo})_{Free} \geq [g(\rho_r - \rho_m)]l/2$$
$$(P_o - S_{xo})_{Base} \geq -[g(\rho_r - \rho_m)]l/\pi.$$

The LEFM solutions also yield equations for the mode I stress intensities, K_I, at the crack tips:

$$K_{I(Free)} = \sqrt{\pi l}\{(P_o - S_{xo}) \pm [g(\rho_r - \rho_m)]l/2\}$$
$$K_{I(Base)} = \sqrt{\pi l}\{(P_o - S_{xo}) \pm [g(\rho_r - \rho_m)]2l/\pi\}$$

where the $+$ sign holds for the upper crack tip, the $-$ sign holds for the lower crack tip, and a positive stress intensity indicates crack propagation.

Combining these results we see that the stress intensity at the upper tip of the free crack must be at least $\sqrt{\pi}g(\rho_r - \rho_m)/l^{1.5}$ and the stress intensity at the upper tip of a crack at the base of an elastic plate must be at least $\pi^{-\frac{1}{2}}g(\rho_r - \rho_m)/l^{1.5}$. Fracture occurs when the stress intensity exceeds the fracture toughness, K_{IC}, (also called the critical stress intensity) a material property discussed further below. Maximum stable lengths for these cracks are therefore given by:

$$L_{Max(Free)} = 2[K_{IC}/g(\rho_r - \rho_m)\sqrt{\pi}]^{2/3} = \text{(length from tip to tip)}$$
$$L_{Max(base)} = [K_{IC}\sqrt{\pi}/g(\rho_r - \rho_m)]^{2/3} = \text{(length from tip to plate base)}$$

The maximum stable length for the crack at the base of an elastic plate is $\pi^{2/3}/2$ or 1.07 times that for a free crack. When a crack growing up from such a base reaches its stable length, it will pinch off at its bottom and become a free crack.

What is meant by the stable length of a free crack? A crack longer than this length will have a fracture intensity at its upper end which is greater than the fracture toughness of the solid country rock there. The tenets of LEFM require that this rock must fracture, and doing so increases the length, and hence the stress intensity at the upper end, further. At the bottom end of the crack ($z = -1$ for the free crack; $z = 0$ for the base crack) the stress intensity becomes zero when the critical length is reached. A stress intensity of zero means that the magma pressure is just sufficient to hold open a preexisting crack. If the length of the crack were to increase, the magma pressure could not even do this, and the crack would close at its lower end. Therefore once the stable length has been reached LEFM suggests that its upper end will propagate as its lower end "zippers" shut. In this way the pressurized magma pocket contained within the swollen region of the crack will move upward through the gravitationally produced stress gradient.

If, as it moves upward, this tear drop shaped magma pocket encounters a host rock which is somewhat less dense but still denser than the magma, it will stop because the fracture toughness at its upper end is no longer exceeded. Its presence will perturb the stress field in its vicinity in such a way (Pollard, 1973) that other magma filled cracks moving up near it will be deflected towards it. As these latecomers arrive they will increase the volume and length of the crack until its new stable length is reached. The stress intensity at its upper end will again be equal to the fracture toughness of the rock and the stress of intensity at its lower end will once more be zero. The entire crack will move upward again, "zippering" itself shut as it moves. This process of pausing until additional magma comes up in new pockets from below could be repeated many times if the density of the rock through which the magma is moving progressively decreases.

We next consider what will occur if the magma pocket finds itself in a region of country rock having similar density. No finite stable crack length is predicted, and its net upward progress will cease. The crack will be elliptical in cross section, with equal stress intensities at upper and lower ends. We can imagine that additional magma pockets merge with it from below, causing it to grow equally at both ends while widening at its center to maintain its elliptical cross section as its volume increases.

We now assume that there is a free surface somewhere above our lengthening crack. As its length increases the effects of this free surface become more pronounced. Pollard and Holzhausen (1979) have shown this interaction to be a function of the ratio of the depth of the crack center to the crack length. Thus, when the upper crack tip is halfway between the crack center and the surface, the stress intensity there is 10% greater than it would be for that crack if the free surface did not exist. The stress intensity at the lower tip is also increased by the interaction, but to a lesser degree. Consequently, if the volume of the crack increases, its fracture toughness will be exceeded at the upper end first and it will grow upwards. The bottom crack tip will neither move nor pinch off. Can this crack grow through rock having a density less than that of the magma? The data presented by Pollard and Holzhausen (1979, their Fig. 11) suggest that it can. Even with a magma 3000 kg/m^3 denser than the country rock, they show that the surface effects are sufficient to produce a stress intensity at the upper end of the crack which is greater than that at its lower end, if the crack extends 87% of the way from its center to the surface. This may appear to contradict our previous result in which a similar crack grew into an elastic plate, pinched off, and migrated through it. The problem is resolved by noting that the length of the crack must be less than the maximum stable length before and after it interacts with the free surface for it to grow towards it.

We wish to emphasize that the ascent of a pocket of magma which is lighter than the country rock through which it moves takes place with no change in volume, whereas the growth of a magma filled crack in country rock of the same or lesser density requires additional magma for each increment of growth.

DISCUSSION

The elevation reached by magma during volcanism may not be related to the depth of the source region as is often assumed. In the model described in this paper a tendency for magma pockets to move upward or extend upward has come out of an LEFM analysis in which no hydrostatic arguments have been employed.

The model may not be directly applicable to volcanism on an early lunar crust floating on a magma "ocean", and so may not be useful in understanding the major mare basalt floods on the moon. The KREEP basalts, with densities less than that of the lunar crust (Solomon, 1975) might well have come to the lunar surface as described by this model if they were volcanic in origin.

The movement of magma within the brittle crust or lithosphere of Mars, Io, or other planetary bodies may be better described by the model reviewed in this paper than by the classical, open conduit model.

REFERENCES

Daly R. A. (1933) *Igneous Rocks and the Depths of the Earth,* McGraw-Hill, N.Y. 598 pp.

Delaney P. and Pollard D. D. (1980) Deformation of host rock and flow of magma during flow of minette dikes and breccia bearing intrusions near Shiprock, New Mexico. *U.S. Geol. Survey Prof. Paper 1182.* In press.

Eaton J. P. and Murata K. J. (1960) How volcanoes grow. *Science* **132,** 925–938.

Gilbert G. K. (1877) *Report on the geology of the Henry Mountains,* U.S. Geogr. Geol. Survey of the Rocky Mountains Region. 170 pp.

Kesson S. E. (1975) Melting experiments on synthetic mare basalts and their petrogenetic implications (abstract). In *Lunar Science VI,* p. 475–477. The Lunar Science Institute, Houston.

Pollard D. D. (1973) Derivation and evaluation of a mechanical model for sheet intrusions. *Tectonophys.* **19,** 233–269.

Pollard D. D. (1976) On the form and stability of open hydraulic fractures in the earth's crust. *Geophys. Res. Lett.* **3,** p. 513–516.

Pollard D. D. and Holzhausen G. (1979) On the mechanical interaction between a fluid-filled fracture and the earth's surface. *Tectonophys.* **53,** 27–57.

Pollard D. D. and Johnson A. M. (1973) Mechanics of growth of some laccolithic intrusions in the Henry Mountains, Utah, II. *Tectonophys.* **18,** 975–984.

Pollard D. D. and Muller O. H. (1976) The effect of gradients in regional stress and magma pressure on the form of sheet intrusions in cross-section. *J. Geophys. Res.* **81,** 975–984.

Secor D. T. and Pollard D. D. (1975) On the stability of open hydraulic fractures in the earth's crust. *Geophys. Res. Lett.* **2,** 510–513.

Solomon S. C. (1975) Mare volcanism and lunar crustal structure. *Proc. Lunar Sci. Conf. 6th,* p. 1021–1042.

Vogt P. R. (1974) Volcano height and plate thickness. *Earth Planet. Sci. Lett.* **23,** 337–348.

Weertman J. (1971a) Theory of water-filled crevasses in glaciers applied to vertical magma transport beneath oceanic ridges. *J. Geophys. Res.* **76,** 1171–1183.

Weertman J. (1971b) Velocity at which liquid-filled cracks move in the earth's crust or in glaciers. *J. Geophys. Res.* **76,** 8544–8553.

Weertman J. (1980) The stopping of a rising, liquid-filled crack in the earth's crust by a freely slipping horizontal joint. *J. Geophys. Res.* **82,** 967–976.

Williams H. (1936) Pliocene volcanoes of the Navajo-Hopi country. *Bull. Geol. Soc. Amer.* **47,** 111–172.

Williams H. and McBirney A. R. (1979) *Volcanology.* Freeman, Cooper, San Francisco, Calif. 397 pp.

Yoder H. S. (1976) *Generation of Basaltic Magma.* National Academy of Sciences, Washington, D.C. 265 pp.

Proc. Lunar Planet. Sci. Conf. 11th (1980), p. 1987–1998.
Printed in the United States of America

Fractures on Europa: Possible response of an ice crust to tidal deformation

Paul Helfenstein and E. M. Parmentier

Department of Geological Sciences, Brown University, Providence, Rhode Island 02912

Abstract—The surface of Europa contains a planetwide system of low albedo lineaments which have been interpreted as fractures in an icy crust. The pattern of fractures on the surface consists of radial and concentric fractures having the general appearance of tension cracks within a region near the antipode of the sub-Jupiter point. Outside this region, linear fractures intersect at angles near 60°, suggesting that they are conjugate shear fractures. The orientation of this pattern on the surface suggests that a principal axis of the deformation that produced the fractures was approximately radial to Jupiter. Fracturing may thus be consistent with an origin due to cyclical tidal deformation resulting from orbital eccentricity. Orbital eccentricity related to a relatively recent establishment of orbital resonance among the Galilean satellites may explain the presence of fractures in a relatively young, lightly cratered planetary surface.

INTRODUCTION

Recent Voyager images of Europa (Smith *et al.*, 1979) reveal a surface with low topographic relief and high albedo. The high albedo of the surface indicates the presence of icy materials. The most distinctive surface features are a planetary scale system of intersecting lineaments of lower albedo than the surface on which they occur. These lineaments have topographic relief of no more than a few tens of meters. The most pronounced topographic features so far described are planimetrically cuspate ridges that may be as high as a few hundred meters. These ridges have little albedo contrast with the surface on which they occur and can be clearly observed only under oblique viewing and lighting conditions of near-terminator images. The icy surface of Europa appears to be remarkably uncratered. At the resolution of Voyager images, only a few probable impact craters have been identified.

Europa is a planetary body with a diameter slightly greater than 3100 km, about the size of the moon. Assuming that the body is composed of a mixture of ice (density ≈1 gm/cm³) and silicates (density ≈2.5–3 gm/cm³) and that the body is differentiated into a silicate core and an ice crust, the known mass of the body is consistent with an ice crust ranging in thickness from 75 to 150 km (cf., Consolmagno and Lewis, 1978). There are no determinations of moment of inertia and therefore no constraints on the radial distribution of mass within the body. If Europa contained a metallic core, a thicker crust would be required.

Europa orbits Jupiter at a mean orbital radius of about 670×10^3 km with an orbital period of 3.55 days (Morrison *et al.,* 1977). The rotation of Europa is approximately synchronous. Three body resonance with Io and Ganymede results in a present-day orbital eccentricity of 0.01. The orbital eccentricity should have been much smaller prior to the establishment of this resonance which may have occurred within the last 500 m.y. (Yoder, 1979).

In this paper we explore the hypothesis that the global system of low albedo lineaments are fractures in an icy crust. The global nature of many of these lineaments suggests that they are fractures formed in response to a planetwide mode of deformation as might result from changes in the figure of the body due to tidal despinning, some form of tidal distortion, or a combination of these mechanisms.

The general pattern of fractures is consistent with a principal axis of deformation approximately radial to Jupiter. Within a region centered around approximately 20°S latitude and 170° longitude, fractures appear to be radial and concentric with a pattern that is suggestive of tension cracking. This region occurs approximately at the antipode of the present sub-Jupiter point. Linear intersecting fractures dominate well-resolved areas of the surface outside this region. Many of these fractures intersect with an angle of about 60–70°, suggesting that they are conjugate shear fractures. This regular pattern of fractures about an axis approximately radial to Jupiter suggests that the fractures may have formed in response to tidal deformation. This would have occurred after the stabilization of an almost crater-free surface and after the rotation of Europa had become nearly synchronous. Cyclical tidal distortions of an ice crust may be due to orbital eccentricity maintained by orbital resonances among the Galilean satellites as proposed as a mechanism of tidal heating in Io (Peale *et al.,* 1979). The geologically recent establishment of this resonance (Yoder, 1979) may explain the formation of fractures in a relatively young uncratered surface.

TECTONIC PATTERNS ON A DISTORTED LITHOSPHERE

The present study considers the distribution and orientation of fractures on the surface of Europa. We will assume that fracturing occurs in the outer portion of Europa's icy crust. Following the approach of earlier studies, surface tectonic patterns can be related to the distribution of stress within a thin, nearly spherical elastic shell.

Stresses in the rigid outer shell of a planetary body with a figure that deviates only slightly from a sphere can be determined by the superposition of stresses due to biaxial deformations of a spherical shell. For example, rotation causes flattening of a body along its axis of rotation. A decrease in the rotation rate reduces the flattening and corresponds to a biaxial stretching of the body along its rotation axis. If rotational flattening of the body is small, stresses due to a decrease in rotation rate are accurately represented by the stresses in a spherical shell biaxially stretched along the rotation axis. The collapse of a triaxial tidal distortion on a synchronously rotating satellite due to orbital recession may be treated in a similar manner by the superposition of two biaxial distortions.

Biaxial distortions of a planetary body may be caused by tidal despinning or a simple push-pull tide due to the disturbing potential of another planetary body. The stresses in a biaxially distorted, thin elastic shell were calculated by Vening-Meinesz (1947) and applied to fracturing of the earth's crust. Burns (1976) applied these results to examine the possible relationship of surface features on Mercury to tidal despinning. Melosh (1977) investigated this question in greater detail and included the effects of expansion or contraction of the planet on surface tectonic patterns. Melosh (1977) also pointed out that the theory of faulting in brittle materials (Anderson, 1951) could be applied to determine the types of tectonic features developed in response to stresses due to tidal despinning.

The distribution of stresses due to a biaxial distortion of a thin elastic shell enclosing constant volume is shown in Fig. 1a. Trajectories which show the principal stress directions are plotted on a Mercator projection with the symmetry axis of the deformation at 0° latitude and longitude. The directions of the principal stress occur along and perpendicular to these trajectories. The deformation axis is the axis of maximum stretching (shortening) with shortening (stretching) of half the magnitude along any two orthogonal axes to maintain constant volume within the shell.

Following a standard notation, deformations of a spherical shell are expressed in terms of the ratio of axial elongations along the three principal axes of the deformation (λ_1, λ_2, λ_3). With the convention that the λ_1 axis corresponds to the direction of maximum elongation magnitude and normalizing the other two elongations by this value, a biaxial deformation is expressed by the axial elongation ratios $(1, -1/2, -1/2)$. Due to the symmetry of the deformation, stress trajectories form small circles about the λ_1 axis. Within the shaded region both principal stresses are of the same sign while outside this region the principal stresses are of opposite sign. For stretching along the deformation axis tensional stresses occur within the shaded region with the maximum tensional stress perpendicular to the stress trajectories shown in Fig. 1a. Outside the shaded region, compressional stresses occur in the direction of the stress trajectories and tensional stresses perpendicular to them. For shortening along the symmetry axis tensional and compressional stresses are reversed.

The theory of faulting in brittle materials (Anderson, 1951) predicts the formation of thrust or normal faults if the two principal stresses are of the same sign and strike slip faults if the two principal stresses are of opposite sign. Normal and thrust faults should be oriented perpendicular to the largest tensional or compressional stress respectively. Strike slip faults should form by shear fracture at angles of 30–35° to the largest compressional stress. Therefore, for the stretching along the λ_1 axis, strike slip faults should form in the unshaded region of Fig. 1a with orientations of 30–35° to the stress trajectories. Normal faults parallel to stress trajectories or orthogonal sets of tension joints should form in the shaded

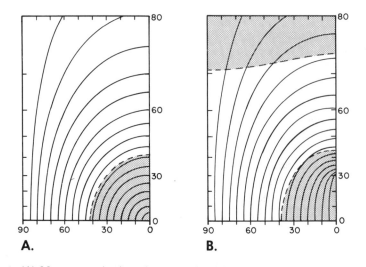

Fig. 1. (A) Mercator projection of stress trajectories for a biaxial deformation of an elastic spherical shell with a symmetry axis at 0° latitude and 0° longitude. Principal stresses are of the same sign in shaded region and of opposite sign in unshaded region. (B) Mercator projection of stress trajectories for tidal deformation of an elastic spherical shell. Axis radial to the planet is at 0° latitude and 0° longitude. Principal stresses are of the same sign in the shaded regions and of opposite sign in the unshaded regions.

region. For shortening along the λ_1 axis, thrust faults rather than normal faults would form in the shaded region, and strike slip faults in the unshaded region would have orientations of 55–60° relative to the stress trajectories.

These various provinces of faulting are discussed in more detail by Melosh (1977) who also considered the effect of planetary volume change on various faulting provinces. Volume change introduces isotropic global stresses proportional to $\theta^{1/3}$ where θ is the fractional volume change. Stresses due to biaxial distortion are proportional to the flattening of the distortion $\triangle f$, defined as the difference between the change in radius of the body perpendicular to and along the λ_1 or symmetry axis divided by the mean radius of the body. Negative $\triangle f$ thus corresponds to a stretching along the λ_1 axis and is equal to 3/2 of the elongation along this axis. The location of faulting provinces in terms of latitude relative to the λ_1 axis are shown in Fig. 2 as a function of $\triangle f/\theta^{1/3}$. This figure follows with only minor modifications of an earlier figure of Melosh (1977). The figure shows, for example, that stretching along the λ_1 axis results in normal faulting in a region surrounding this axis and strike slip faulting at lower latitudes. The effect of planetary expansion ($\theta > 0$) would extend the region of normal faulting to lower latitudes and would cause the shaded region of normal faulting in Fig. 1a to be larger.

The figure of a satellite orbiting a planet and rotating synchronously about an axis perpendicular to its orbital plane is triaxial and can be represented by the superposition of a biaxial tidal stretching along an axis radial to the planet and a biaxial rotational flattening along the rotation axis. A distortion of this form also describes the tidal deformation of a synchronously rotating satellite due to orbital recession or orbital eccentricity. The stresses due to such a deformation have been applied to consider tectonic features on the moon due to orbital recession (Gash, 1978). The ratio of axial elongations of such a triaxial deformation (Jeffreys, 1976) are $(1, -2/7, -5/7)$ where the λ_1 axis is radial to the planet and the λ_3 axis is along the satellite rotation axis. The stress trajectories for a triaxial deformation with these axial elongation ratios are shown in Fig. 1b and correspond to those in Fig. 1a for a biaxial deformation. The stress trajectories are plotted on a mercator projection with the λ_1 deformation axis at 0° latitude and longitude and the λ_3 axis at 0° longitude and 90° latitude. Within the shaded regions, the principal stresses are of the same sign. For stretching along the λ_1 axis, corresponding to an increase in the tidal distortion, stresses within the central shaded region are tensional and those within the shaded region at high latitude are compressional. For shortening along this axis, the stresses are of the opposite sign. The type and orientation of faults relative to the stress trajectories are the same as discussed for the biaxial distortion in Fig. 1a. The results of Fig. 2, while strictly applicable only to a biaxial deformation, show qualitatively how the sizes of various faulting provinces for triaxial deformation are modified by volume change of the planetary body.

PATTERNS OF FRACTURE ON THE SURFACE OF EUROPA

The pattern of fractures on the surface of Europa revealed by Voyager images is shown in the U.S.G.S. Preliminary Pictorial Map in Fig. 3a. Regional patterns

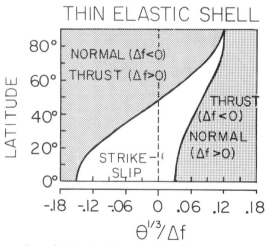

Fig. 2. Faulting provinces for biaxial deformation of a thin elastic shell due to flattening Δf and with fractional volume change θ. Latitude is measured with respect to a pole at the symmetry axis of the deformation. Modified from Melosh (1977).

of fracturing can be identified and characterized by the relative orientation of intersecting fractures. Within a region centered on 170° longitude and 20°S latitude, fractures form a crudely radial and concentric pattern. This region is roughly circular with an angular radius of about 30°. Outside this region, particularly in well resolved areas of the surface to the north and west, fractures are longer, more linear and intersect at acute angles.

These general features of the fracture pattern are expressed quantitatively by measurements of the intersection angles of fractures. Measurements of intersection angles made from the map of Fig. 3a are represented in the histograms of Fig. 4. The planetwide distribution of intersection angles shows a peak at angles of 60–70°. Excluding fractures within the region of radial and concentric fracturing significantly decreases the number of fracture intersections in the 75–90° range with a corresponding increase in the magnitude of the 60–70° peak. This regional tendency for fractures to intersect at angles of 60–70° suggests that they are conjugate shear fractures. Fractures within the region of radial and concentric fracturing with roughly 90° intersections are suggestive of tension cracking.

The orientation of fractures on the surface may provide constraints on the mechanism of deformation which caused them. As discussed earlier, the theory of faulting in brittle materials predicts that faults should form at particular angles relative to the directions of principal stress. To study the orientation of fractures on the surface of Europa, the stress trajectories in Fig. 1 have been superimposed on the pictorial map of Fig. 3a. Stress trajectories with the λ_1 deformation axis at the center of the region of radial and concentric fracturing are shown in Fig. 3b and Fig. 3c for a biaxial and triaxial deformation, respectively. It can be seen that the stress trajectories for these two cases are generally of similar form, the most significant differences occurring near the pole of the λ_1 axis.

In both cases, fractures outside the region of radial and concentric fracturing intersect the stress trajectories at acute angles. Preliminary measurements suggest that the majority of fracture intersections occur in the range of 30–70°. Few fracture intersection angles lie outside this range. Examples can be seen in which stress trajectories approximately bisect the angle between pairs of 60°-intersecting fractures. Within the region of radial and concentric fracturing, preliminary measurements as well as examination of the figures suggest that many fractures intersect the stress trajectories at high angles in the range of 70–90°. Few low albedo fractures are parallel to the stress trajectories. Intersections nearer 90° are more frequent for the triaxial stress trajectories.

For tidal despinning of Europa about its present axis of rotation, stress trajectories would form small circles about the north pole of the body and so would coincide with lines of geographic latitude. From Fig. 3a it can be seen that fractures do not intersect lines of latitude with any consistent range of orientations. This suggests that fractures were not produced solely by tidal despinning about the present axis of rotation.

It should be kept in mind that all fracture measurements were made from the U.S. Geological Survey Preliminary Pictorial map of Europa and that local map errors on the order of 10° (Smith *et al.,* 1979) may influence these results. Mea-

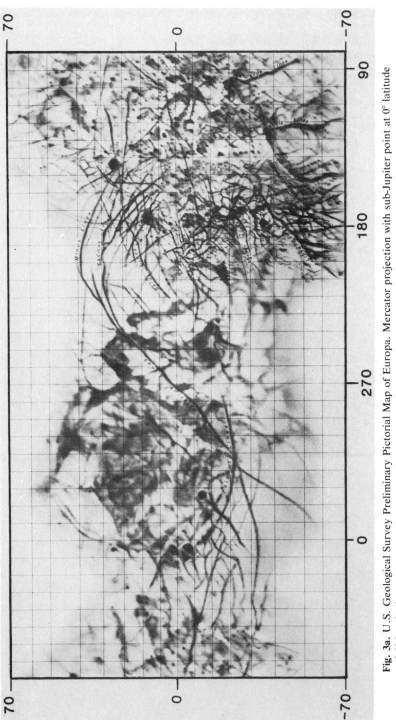

Fig. 3a. U.S. Geological Survey Preliminary Pictorial Map of Europa. Mercator projection with sub-Jupiter point at 0° latitude and 0° longitude.

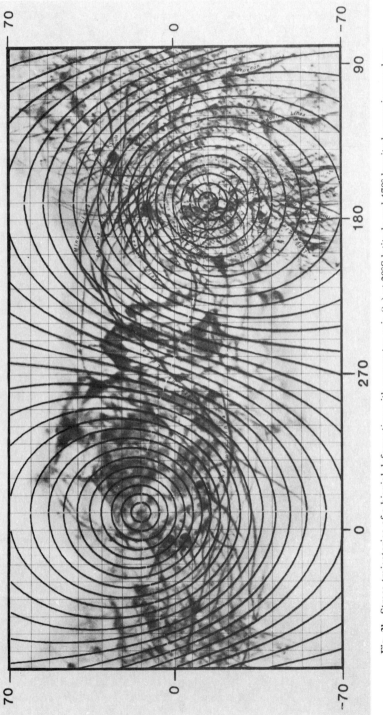

Fig. 3b. Stress trajectories of a biaxial deformation with symmetry axis (λ_1) at 20°S latitude and 170° longitude superimposed on the map of Fig. 3a.

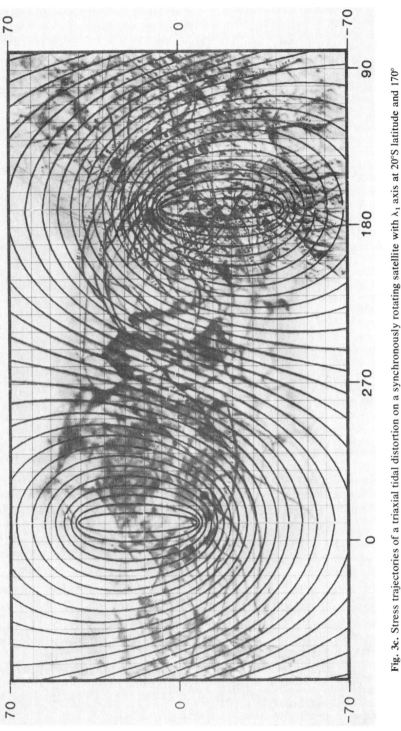

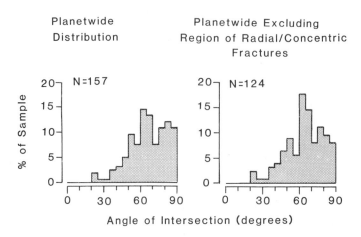

Fig. 4. Histograms of measured angles between intersecting fractures (see text for discussion).

surements from Voyager images of Europa are currently being made to further refine our analysis.

DISCUSSION

This pattern of fractures on the surface of Europa is broadly consistent with deformation of a tidal origin. This is strongly suggested by a principal axis of deformation approximately radial to Jupiter. Stretching along this axis is implied by our interpretation of tension cracking within the region of radial and concentric fracturing. Outside this region, shear fractures which often form 60°-intersecting sets and which have orientations in the range of 30–70° to stress trajectories are consistent with both stretching and shortening along this principal axis. Based on stresses in an unfractured, thin elastic shell, two distinct shear fracture orientations, 30–35° and 55–60° for brittle faulting, would be expected due to cyclical deformation along this axis. More realistically, shear fractures with a range of orientations about each of these ideal orientations might occur. The superposition of two such sets of fractures could produce a pattern of fractures with the observed 30–70° range of orientations relative to stress trajectories.

An orbital eccentricity of 0.01 would produce cyclical stresses with a magnitude of a few bars in a thin elastic shell overlying a fluid interior (Parmentier and Head, 1979). Thus a low longterm strength of surface materials under cyclical loading is required. Cyclical deformation would cause both tensional and compressional stresses in an area surrounding the λ_1 deformation axis. The absence of identifiable fractures parallel to stress trajectories, the expected orientation of thrust faults, within the region of radial and concentric fracturing appears to require that compressional stresses did not exceed the shear strength of surface materials in this region during shortening along the λ_1 axis. This may reflect the presence of a global isotropic tensional stress due to expansion of Europa's interior by the freezing of water at low pressure or by the warming of a silicate core.

Cyclical tidal deformation of a nearly synchronously rotating body due to orbital eccentricity is generally consistent with the pattern of fractures and also with the orbital evolution of the Galilean satellites discussed by Yoder (1979). The present eccentricity of Europa's orbit is maintained by a three body resonance between Io, Europa, and Ganymede. Prior to the establishment of this resonance by the orbital recession of Io and Europa, the eccentricity of Europa's orbit would have been less than 1/5 its present value. Yoder estimates from orbital mechanics considerations that the three body resonance could have been established as recently as 5×10^8 years ago. Tidal despinning and the damping of any initial free orbital eccentricity should have occurred shortly after the formation of Europa (Peale, 1977).

Yoder (1979) points out that a satellite in an eccentric orbit does not rotate synchronously and that the tidal distortion due to orbital eccentricity is not exactly radial. For synchronous rotation, the rotation rate of the satellite must be equal to its instantaneous orbital angular velocity. A radial tidal distortion does not provide the tidal torque required to vary the rotation rate of the satellite. The variation of rotation rate along the orbit also controls the form of the tidal distortion due to orbital eccentricity. If synchronous rotation were maintained with the instantaneous rotation rate of the satellite equal to its instantaneous orbital angular velocity, the tidal distortion would be of the triaxial form discussed earlier. For a slightly eccentric orbit, if the rotation rate were equal to the mean orbital angular velocity then the cyclical tidal distortion would be biaxial and would oscillate about the sub-planet point during the orbital motion. Yoder (1979) indicates that the amplitude of this oscillation is controlled by the libration of the satellite and by the delayed response of the variable tidal distortion due to dissipation within the satellite. The former would introduce an angular amplitude of the tidal distortion about the sub-planet point of 4 times the orbital eccentricity and the latter an amplitude of approximately $1/Q$ where $2\pi/Q$ is the specific dissipation per tidal cycle. Since the orbital eccentricity is small and for reasonable values of $Q \geq 10$, the amplitude of the oscillation of the λ_1 deformation axis would be less than 5°. This is consistent with the existence of a well defined region of radial and concentric fracturing on Europa.

Orbital recession of Europa due to tidal friction would result in the collapse of a tidal bulge and is another mechanism for inducing global systems of surface fractures. Strike-slip faulting outside the region of radial and concentric fracturing may be consistent with this mechanism. However, decreasing tidal distortion would cause compressional stresses within a region surrounding the λ_1-axis, and thrust faults parallel to the stress trajectories in Fig. 3c would be predicted. This does not appear to be consistent with the pattern of fractures within the region of radial and concentric fracturing.

Further constraints on the interpretation of fracture patterns will be provided by more detailed measurements of fracture orientations relative to stress trajectories. Geologic observations will also provide additional constraints, for example by detailed mapping of fracture offset relations with intersecting fractures. The low albedo and width of fractures also requires further study. Does the low albedo of many fractures represent opening due to planetary expansion and filling of fractures with more silicate rich material, or might the lower albedo of fractures be due to brecciation of surface material and changes in surface roughness? Finally, cuspate ridges must be considered as possible tectonic features. Considering the effect of viscous relaxation of topography on icy satellite surfaces, the high relief of these features may indicate that they are relatively young compared to other surface features. The orientation of cuspate ridges relative to stress trajectories may give further insight into their origin as well as providing additional constraints on global tectonic interpretations.

Although further study is required, we suggest that the pattern of fracturing on Europa is broadly consistent with an origin by tidal deformation due to orbital eccentricity. Orbital eccentricity due to the present orbital resonance could have developed recently relative to typical time scales of planetary surface evolution. This would be consistent with the formation of fractures on the lightly cratered surface of Europa.

Acknowledgments—This research was supported by National Aeronautics and Space Administration grant NSG-7605. The help of Nancy Christy and Sam Merrell in preparation of the manuscript is gratefully acknowledged. J. W. Head and H. J. Melosh provided helpful discussions. Robert Storm provided a critical and helpful review of the manuscript.

REFERENCES

Anderson E. M. (1951) *The Dynamics of Faulting.* Oliver and Boyd, Edinburgh. 206 pp.

Burns J. A. (1976) Consequences of the tidal slowing of Mercury. *Icarus* **28**, 453–458.

Consolmagno G. and Lewis J. S. (1978) The evolution of icy satellite interiors and surfaces. *Icarus* **34**, 280–293.

Gash P. J. S. (1978) Tidal stresses in the Moon's crust. *Mod. Geol.* **6**, 211–220.

Jeffreys H. (1976) *The Earth.* Cambridge Univ. Press, N.Y. 525 pp.

Melosh H. J. (1977) Global tectonics of a despun planet. *Icarus* **31**, 221–243.

Morrison D., Cruikshank D. P., and Burns J. A. (1977) Introducing the satellites. In *Planetary Satellites* (J. A. Burns, ed.), p. 3–17. Univ. Arizona Press, Tucson.

Parmentier E. M. and Head J. W. (1979) Internal processes affecting surfaces of low density satellites: Ganymede and Callisto. *J. Geophys. Res.* **84**, 6263–6276.

Peale S. J. (1977) Rotation histories of the natural satellites. In *Planetary Satellites* (J. A. Burns, ed.), p. 87–112. Univ. Arizona Press, Tucson.

Peale S. J., Cassen P., and Reynolds R. T. (1979) Melting of Io by tidal dissipation. *Science* **203**, 892–894.

Smith B. A., Soderblom L. H., Beebe R., Boyce J., Briggs G., Carr M., Collins S. A., Cook A. F. II, Danielson G. E., Davies M. E., Hunt G. E., Ingersoll A., Johnson T. V., Masurky H., McCauley J., Morrison D., Owen T., Sagan C., Shoemaker E. M., Strom R., Suomi V. E., and Veverka J. (1979) The Galilean satellites and Jupiter: Voyager 2 imaging science results. *Science* **206**, 927–950.

Vening-Meinesz F. A. (1947) Shear patterns of the Earth's crust. *EOS (Trans. Amer. Geophys. Union)* **28**, 1–61.

Yoder C. F. (1979) How tidal heating in Io drives the Galilean orbital resonance locks. *Nature* **279**, 767–770.

Proc. Lunar Planet. Sci. Conf. 11th (1980), p. 1999–2014.
Printed in the United States of America

Equations of state in planet interiors

Orson L. Anderson* and John R. Baumgardner

Department of Earth and Space Sciences, University of California, Los Angeles, California 90024

"There is more joy in heaven in a good approximation than in an exact solution."
—Nobelist Julian Schwinger, 1979

Abstract—A method for computing the density and temperature profiles of a planetary interior is proposed where laboratory data for the mechanical properties of the outer shell of the mantle are used as primary data. Also needed is the jump in density between the low-pressure phase (upper mantle) and the phase of the lower mantle.

The equation of state (EOS) used here is substantially different from that used by other workers. We use the isothermal EOS with 0 K values and put all thermal effects in the thermal pressure. Our thermal pressure has no reference to a Grüneisen parameter, being simply

$$P_{TH} = aT - b$$

We show that a is independent of T and V, and b, which is a small correction, is substantially independent of T and V.

However, the Grüneisen parameter is used to calculate the adiabatic contribution to the temperature rise across the lower mantle and core.

Using the simple EOS, we compute the density profile of the earth and show the result to be quite close to that found for the PEM (Dziewonski *et al.*, 1975). Temperature profiles for the earth and density and temperature profiles for Mars are presented.

INTRODUCTION

The purpose of this paper is to demonstrate that use of a particular equation of state (EOS) in the set of equations describing a planet interior enables one to specify directly the temperature dependence of the density distribution and to compute the thermal expansivity at depth in terms of the temperature distribution. To do this we select a member from the class of EOS for which the temperature function separates from the volume function. This class is known as the Hildebrand EOS (O. L. Anderson, 1979b; D. L. Anderson, 1967), and its members have the following form:

$$P(V,T) = f_1(V) + f_2(T) \qquad (1)$$

---

*also Institute of Geophysics and Planetary Physics, University of California, Los Angeles, California 90024 where $f_1(V)$ is the isothermal EOS, P_{ISO}, which arises by taking the volume derivative of the Helmholtz free energy at absolute zero, and $f_2(T)$ is the thermal pressure, P_{TH}.

D. L. Anderson (1967), in the introduction to his well-known seismic EOS, pointed out the virtues of analyzing the EOS in V,T space rather than in P,T space. He recommended the use of the Hildebrand EOS as an important conceptual device to separate pure volume and pure thermal effects in geophysical applications.

Three examples of P_{ISO} are the Birch-Murnaghan EOS (Birch, 1952), the Born-Mie (O. L. Anderson, 1970), and the method of potentials (Zharkov and Kalinin, 1975). There are other isothermal EOS but these three are most often found in the geophysical literature. Three constants (all measurable) define the $P_{ISO}(V)$ equation in the second degree,

$$P_{ISO}(V) = f(\rho/\rho_0, K_0, K_0'). \tag{2}$$

The constants ρ_0, K_0, and $K_0' = (dK/dP)_{P=0}$ represent respectively the values of the density, the isothermal bulk modulus, and dK/dP at zero pressure and at absolute zero.

For materials with Debye temperatures greater than room temperature by a factor of 3 or more, negligible error is made in using room-temperature values for ρ_0, K_0, and K_0' for the 0 K isothermal EOS Eq. (2) (O. L. Anderson, 1979a). For pressures at the earth's mantle the above-mentioned isothermal EOS yield essentially identical values for P_{ISO} and the results depend only slightly upon the choice of EOS for compressions found in the earth's core (O. L. Anderson, 1970). In this paper we shall use the Born-Mie EOS but the Birch-Murnaghan EOS or the method of potentials could have been used just as well.

For an EOS to belong to the class described by Eq. (1), the thermal pressure, P_{TH}, must be independent of V. P_{TH} is subject to several interpretations. The most common view is to assume that P_{TH} is a function of both V and T:

$$P_{TH}(V,T) = \frac{\gamma_e}{V} E_{TH} \tag{3}$$

where γ_e is the Grüneisen gamma (associated with the Mie-Grüneisen EOS) and E_{TH} is the thermal energy (Grüneisen, 1924; Barron, 1957). γ_e is a function of V and possibly T, and E_{TH} is independent of V at high T. In quasiharmonic lattice theory, it is found that γ_e is equal to the thermodynamic γ at high T (Barron, 1957) where

$$\gamma = \frac{\alpha K_T V}{C_V}, \tag{4}$$

and where C_V is specific heat, α is the coefficient of thermal expansivity, and K_T is K_0 corrected for T. For planet interiors where T greatly exceeds the Debye temperature, using Eq. (4) in Eq. (3) yields

$$P_{TH} = (\alpha K_T)\left(\frac{E_{TH}}{C_V}\right). \tag{5}$$

Since $E_{TH}/C_V = T$ at high T in the quasiharmonic approximation, we see that P_{TH} is roughly proportional to T, provided αK_T is independent of T and V. A more careful definition of P_{TH} is

$$P_{TH} = \int \left(\frac{\partial P}{\partial T}\right)_V dT = \int_0^T (\alpha K_T) dT$$

which integrates into the form (O. L. Anderson, 1979b)

$$P_{TH} = (\alpha K_T)^* T - b - CT^2 \qquad (6)$$

where $(\alpha K_T)^*$ is the high-temperature value, b is a small correction accounting for the behavior of αK_T at temperatures below θ, and CT^2 is the anharmonic correction to αK_T which is ordinarily quite small. O. L. Anderson (1979b) showed that $(\alpha K_T)^*$ is independent of volume for structurally dense minerals such as found in the earth's interior (but not for loose structures such as, for example, α-quartz). A plot of P_{TH} vs. T for the earth's mantle (O. L. Anderson and Sumino, 1980) is shown in Fig. 1, along with data for MgO and Al_2O_3 (O. L. Anderson, 1980) and NaCl and LiF (Boehler and Kennedy, 1980a, b). We see from Fig. 1 that at 3000 K, P_{TH} is of the order of 15 GPa, which is 11% of the pressure at the core-mantle boundary of the earth. For terrestrial planets, the anharmonic effects

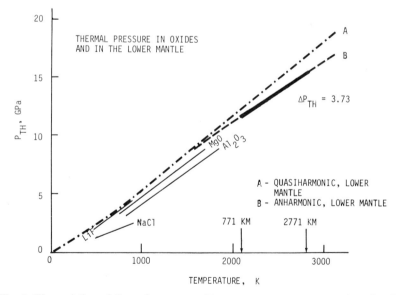

Fig. 1. The variation of thermal pressure with temperature at one atmosphere for the uncompressed lower mantle (O. L. Anderson and Sumino, 1980), MgO and Al_2O_3 (O. L. Anderson, 1980), and NaCl and LiF (Boehler and Kennedy, 1980 a,b). The curves are linear at intermediate temperature (800–1500 K) but do not extrapolate to zero, due to quantum corrections to α (this gives rise to the constant b). Above 2000 K there is evidence of anharmonicity, which is accounted for by the term CT^2 in Eq. (6). The various authors show P_{TH} to be independent of V over the range measured.

are sufficiently small that they may be ignored, so that to a good approximation (Fig. 1),

$$P_{TH} = aT - b \tag{7}$$

where a is independent of V, and b is sufficiently small that its pressure dependence can be ignored. The term a in Eq. (7) is

$$a = (\alpha K_T)^*. \tag{8}$$

In support of the approximation Eq. (7) of Eq. (6), Hardy (1980) has shown that high pressure greatly diminishes the anharmonicity observed at atmospheric pressure and high temperature. The quasiharmonic values of a and b found for the earth's mantle (O. L. Anderson and Sumino, 1980) are 0.0050 GPa/K and -1.6 GPa, and these will be used for $f_2(T) = P_{TH}$. Thus the EOS chosen for this study is not only a member of the Hildebrand class but is one in which P is proportional to T.

The function P_{ISO} selected for this analysis arises from the Born-Mie pair potential, ignoring all but nearest neighbors (O. L. Anderson, 1970). The isothermal pressure obtained is

$$P_{ISO} = \frac{3K_0}{3K_0 - 8} \left[(\rho/\rho_0)^{K_0'-4/3} - (\rho/\rho_0)^{4/3} \right]. \tag{9}$$

The change in pressure $\Delta P(\rho,T)$ in a planet from conditions at depth z to some depth (z + dz) will be calculated by using Eq. (9) along an isotherm, (0 K), to the density ρ, and then along an isochore (constant volume) going from P_{ISO} (ρ,0) to $P(\rho,T)$ using Eq. (7). This is the unique feature of the study, and this procedure differs considerably from the customary procedure in which the EOS parameters (ρ_0, K_0, K_0') are either taken to be temperature functions or taken to represent values at some (unknown) average high T. The customary approach amounts to the curve fitting of an EOS across isotherms. We emphasize that the parameters used in Eq. (9) will be those obtained from room-temperature experiments and will be uncorrected for high temperature. All temperature effects reside in Eq. (7).

Equations (7) and (9) together make up the EOS that is to be combined with the other equations defining a planet's interior. Equations appropriate to a radially symmetric planet follow:

$$\frac{dP}{dr} = -g(r)\rho(r) \tag{10}$$

$$\frac{dM}{dr} = 4\pi r^2 \rho(r) \tag{11}$$

$$\frac{dI}{dr} = \frac{8\pi}{3} r^4 \rho(r) \tag{12}$$

$$g(r) = \frac{GM(r)}{r^2} \tag{13}$$

where r, ρ, P, M, I, g, and G represent respectively radius, density, pressure, mass inside r, moment of inertia, gravitational acceleration, and the gravitational constant.

The relation between the heat flow Q(r) and r in a spherical body is

$$\frac{dQ}{dr} = H - \frac{2Q(r)}{r} \tag{14}$$

where H is the heat generated in a unit volume. In a crustal shell for which concentration of radioactive elements is assumed, H is given by $H = H_0 \exp[-(a - r)/D]$, where H_0 is the surface radioactive heat production and D is on the order of 10 km (Lachenbruch and Sass, 1978).

The temperature gradient is specified depending upon the thermal regime. For conducting shells,

$$\frac{dT}{dr} = -\frac{1}{k(r)} Q \tag{15}$$

where k is the thermal conductivity.

For convecting shells (in an exactly adiabatic compression)

$$\left(\frac{dT}{T}\right)_{ad} = \gamma d\ln\rho \tag{16}$$

where

$$\gamma/\gamma_0 = (\rho/\rho_0)^q \tag{17}$$

where γ_0 is the value at zero pressure, and q is constant independent of P, whose value is near unity. The superadiabatic component across a virtually adiabatic shell is (Brown and Shankland, 1980)

$$\alpha(\Delta T)_{sa} = \int_{\rho_1}^{\rho_2} \left(\frac{d\rho}{\rho} - \frac{dP}{K_S}\right) d\rho \tag{18}$$

where α is the coefficient of thermal expansivity appropriate to the average depth considered.

It can be shown that, within the accuracy required so as to replace (6) by (7), α at any depth z is given by

$$\alpha(\rho) = \frac{a}{K_{ISO}(\rho)} \tag{19}$$

where

$$K_{ISO}(\rho) = \rho\left(\frac{\partial P_{ISO}}{\partial\rho}\right)_T \tag{20}$$

This relates α to a in (7).

The appropriate equations above, (14) to (19), when applied to a shell of the planet, yield T(r), from which $P_{TH}(r)$ is determined by Eq. (7). Then ρ, P, M and

I are determined at r from equations (9) to (12). Integration of the equations begins at the surface, where the ρ_0, K_0, and K_0' are specified by the choice of the chemical composition. The model parameters of the inner shell [$\rho_i(0)$, $K_i(0)$] are chosen such that M and I vanish exactly at r = 0.

PHYSICAL PARAMETERS IN THE MANTLES OF PLANETS

Many authors have noted that specifying the chemical composition of a planet's mantle tightly constrains the properties of its core; see, for example, Solomon (1979) and Ringwood and Clark (1971). This also applies to the detailed structure of the mantle itself. In other words, once the chemical composition of the upper mantle is specified (so that ρ_0, K_0, and K_0' of the zero pressure phase are known) then, assuming chemical homogeneity, the physical properties of all its shells can in principle be specified. What is needed are the appropriate laboratory data from experiments that measure the pressure and temperature fields of the higher pressure phases for the chosen chemical composition.

From our point of view, this information can be considered as given because if the required experiments have not been done on the composition of interest, they could be with the present laboratory capability.

A vital laboratory datum is the change of density from the low-pressure phase to the higher phase identified with the lower mantle. The zero-pressure change in density, $\Delta\rho$, is defined as

$$\rho_2(0) = \rho_1(0) + \Delta\rho \tag{21}$$

where subscript 1 refers to the uncompressed initial phase (upper mantle) and subscript 2 refers to the uncompressed phase of the lower mantle. If $\rho_1(0)$, $K_1(0)$, and $\rho_2(0)$ are all known, then the other properties in the high-pressure phases can be estimated reasonably well from empirical relationships, as we shall discuss below. Thus, all the required uncompressed physical properties of the lower mantle are determinable from properties of the upper mantle plus knowledge of $\Delta\rho$.

We must still account for the transition zone which separates the upper and lower mantle. The physical properties of the transition zone can be estimated using data from experimental petrology where temperature measurements provide $\Delta T/\Delta P$ across the zone.

Since for the earth the lower mantle accounts for about 50% of the mass and 56% of the moment of inertia, the density distribution of the whole earth depends very much on the values of the parameters $\rho_2(0)$ and $K_2(0)$, so these should be estimated with care. A subsequent paper will deal with the sensitivity of the properties of the inner core to variations in the choice of $\rho_2(0)$.

THE MANTLE PARAMETERS OF THE EARTH

In all models of the earth discussed below a peridotite upper mantle is assumed. The mechanical properties of olivine are $\rho_0 = 3.31$, $K_0 = 128$ GPa, and $K_0' = 5.1$

(Kumazawa and O. L. Anderson, 1969). These properties of olivine must be mixed with those of pyroxene in roughly equal parts. Considering the values measured on orthopyroxene reported by Kumazawa (1969) of $\rho_0 = 3.38$ and $K_0 = 102.2$ GPa, with the estimate $K_0' = 5.3$ for orthopyroxene made by Chung (1973), the authors selected $\rho_0 = 3.33$, $K_0 = 119$ GPa, and $K_0' = 5.3$ for the mechanical properties of the peridotite upper mantle, or the low-pressure phase. $K_{0(2)}$ can be estimated from $K_{0(1)}$ and the density ratio, $\rho_{0(1)}/\rho_{0(2)}$, using the law of corresponding states (O. L. Anderson and Nafe, 1965; O. L. Anderson, 1966) diagrammed in Fig. 2. The corresponding formula is

$$ln(K_{0(2)}/K_{0(1)}) = x \; ln \, (\rho_{0(1)}/\rho_{0(2)}) \tag{22}$$

where $x \sim 3$ to 5. We see that the bulk modulus of the high-pressure phase is much higher than that of the initial phase.

For peridotite mantle models Liu (1977) found that $\Delta\rho = 0.88$ gm/cc across the whole transition zone. Using (21), $\rho_{0(2)} = 4.21$, and from (22), $K_{0(2)} = 264$ GPa where $x = 3.4$ (O. L. Anderson and Sumino, 1980). This solution corresponds to a phase transition but does not preclude a compositional difference between the upper and lower mantle.

However, K_0' of the second phase is lower than K_0' of the initial phase. This parameter can also be estimated from empirical relations between these two var-

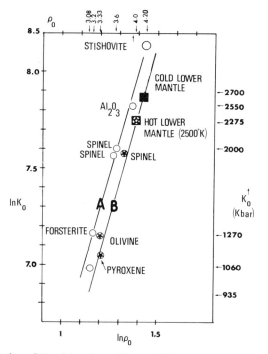

Fig. 2. The variation of K_0 with ρ_0 for oxides and silicates where $\bar{m}$ is near 21, showing that these materials obey the law of corresponding states. Data are referenced by Sumino and O. L. Anderson (in preparation).

iables as K_0' increases, K_0 decreases, as shown in Fig. 3. Using Fig. 3, an estimate of $K_{0(2)}'$ in the second phase is found once $K_{0(2)}$ of the second phase is selected. We find $K_{0(2)}' = 3.35$ agrees well with $K_{0(2)} = 264$ GPa.

With knowledge of $\rho_{0(2)}$, $K_{0(2)}$, and $K_{0(2)}'$, the isothermal trajectory (at room temperature) of the lower mantle, $P_{ISO(2)}$ $(\rho,0)$, can be found for the lower mantle.

The transition zone of the earth may likewise be specified in terms of the laboratory data. The isothermal trajectory of $P_{ISO(2)}$ through the transition zone begins at the phase boundary given by P* and T*, and this can be supplied by experimental results, either from actual measurements or from extrapolations of such measurements.

Experimental results for the transition in olivine have been obtained by Akaogi and Akimoto (1979). From their data it is found that the transition begins at T* = 1400 C and P* = 12.4 GPa for the composition $(Mg_{87}Fe_{13})_2SiO_4$. (This result corresponds to 11.5 GPa pressure for the olivine composition in their experiment.) For an olivine composition in which the iron content is greater, the pres-

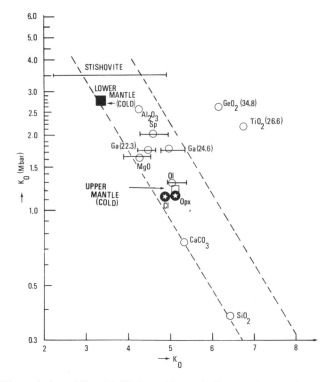

Fig. 3. The variation of K_0 with K_0' for oxides and silicates where $\overline{m}$ is near 21. $\overline{m}$ is given in parentheses. The bars indicate the range of reported K_0'. Data for garnet, GeO_2, and TiO_2 given for comparison. Sp = spinel; Ol = olivine; Opx = orthopyroxene; Ga = garnet; Cl = clinopyroxene. Data are referenced by Sumino and O. L. Anderson (in preparation).

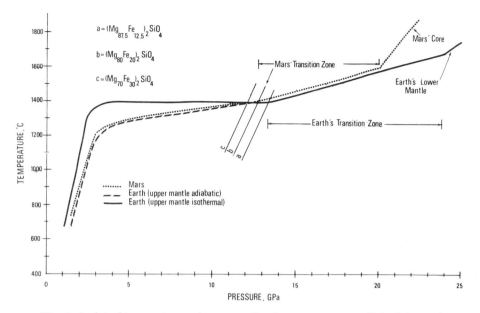

Fig. 4. A plot of temperature and pressure showing temperature profiles of the earth and Mars according to this model. The line separating out the lower phase (upper mantle) from the onset of the transition zone is shown for three compositions of olivine in a periodotite mantle: a, b, and c (Akaogi and Akimoto, 1979).

sure P* is somewhat lower. A P-T diagram showing the beginning of the transition zone, as a function of iron content, is shown in Fig. 4.

Akaogi and Akimoto (1979) found several transitions through the pressure range corresponding to the transition zone. For their peridotite model, the total temperature and pressure change throughout this zone is $\Delta T = 230°$ and $\Delta P = 11.3$ GPa. These values of ΔT and ΔP may then be applied directly to obtain the conditions at the top of the lower mantle. The transition zone is defined in the density distribution by the average slope $\Delta\rho/\Delta P = 0.88/11.3 = 0.08$ gm/cm^3/GPa.

COMPUTATION OF THE DENSITY PROFILE FOR THE EARTH

In the subsequent calculations of density and temperature distributions, little use will be made of existing seismic velocity models. The approach is to rely as much as possible on information obtained or obtainable from laboratory experiments for likely candidate minerals of the mantle. In the case of the earth, the resulting density distribution will be compared with density profiles associated with standard seismic models. This will demonstrate the accuracy of the method and the feasibility of applying this model to planets where seismic data are unavailable. The radius of the planet (a), the moment of inertia (I), and the planetary mass

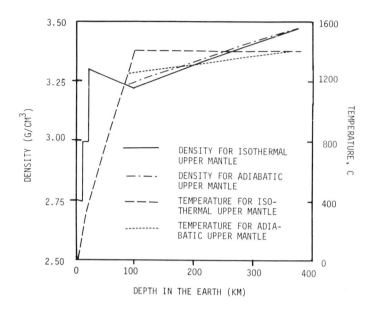

Fig. 5. Temperature and density profiles for the upper mantle according to the model in this paper. The density profile is affected only slightly by rather different temperature distributions in the upper mantle.

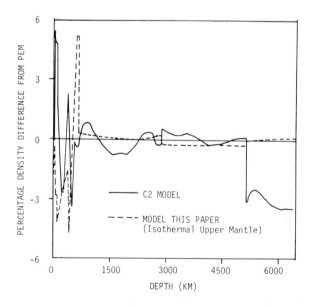

Fig. 6. A plot showing the percentage difference of the density from this model compared to the density of the PEM. Note that the density of this model tracks quite well the density of the PEM in the lower mantle and core, but differences are apparent in the upper mantle and transition zone. The difference between C2 (Hart *et al.*, 1977) and the PEM (Dziewonski *et al.*, 1975) is plotted to show comparison.

(M) will be assumed as known, and in the case of the earth the radius of the core will be assumed as known.

Above P* = 23.7 GPa in the lower mantle, an adiabatic increase of T is assumed following Eqs. (15) and (16). This calculation of T requires knowledge of γ and $\Delta\rho/\rho$. Guidance for selection of γ may be obtained by several theories or by evaluation of γ from seismic data for the earth's mantle (O. L. Anderson and Sumino, 1980). A superadiabatic $(\Delta T)_{sa}$ of 200 C across the lower mantle will be assumed (Brown and Shankland, 1980). We take $\gamma_0 = 1.5$ and $q = 1.35$ following Anderson (1979a). The lithosphere is assumed to act as a conductive layer, with an exponential decay of radioactive heating with depth in the crust, and a surface heat flow of 1.6 HFU. Thermal conductivities of 2.5 and 3.3 W/mC (.006 and .008 cal/sec cm C) are assumed for the crust and mantle, respectively. At the core-mantle boundary a jump of 500 C is assumed to account for convection boundary layers. While this is arbitrary, it is reasonable (Jeanloz and Richter, 1979). The temperature gradient in the outer core is computed using an adiabatic assumption along with a constant value of $\gamma = 1.4$ (Stevenson, 1980). The inner core's temperature gradient is taken as adiabatic with $\gamma = 1.4$ (Stevenson, 1980).

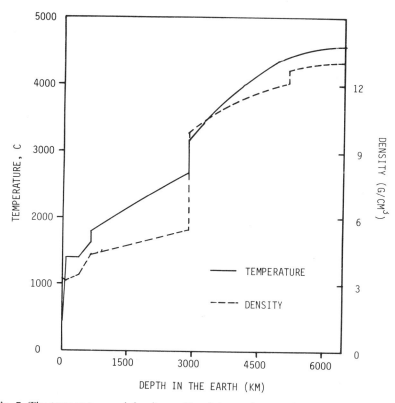

Fig. 7. The temperature and density profile of the earth according to this model. The isothermal assumption is used for the upper mantle.

The density distribution, and therefore I and M, of the mantle are now specified by the physical properties described above. The residual I and M of the planet are required to vanish at r = 0 by selecting the density distribution for the core. In our model this is done by adjusting the cold uncompressed density of the outer core.

In the first example calculation for the earth, the temperature distribution of the upper mantle is assumed to be isothermal, as suggested by Tozer (1967) and O. L. Anderson (1980). Thus between the lithosphere boundary and 380 km, T = 1400 C (Akaogi and Akimoto, 1979). In this case the lithosphere boundary lies at 98 km.

The second earth model treats the upper mantle as adiabatic. Here the heat flow and temperature distribution in the lithosphere are kept the same as in the first model but the lithosphere thickness is reduced some 12 km to give a T of 1250 C at the lithospheric boundary. Selecting $\gamma = 1.4$ yields T = 1400 C at 380 km, which satisfies our main boundary condition. The same value for the super-adiabatic components in the mantle and the core, and the same value of the jump at the core-mantle boundary are assumed. The comparison of the density distributions for the adiabatic and isothermal upper mantle is shown in Fig. 5. Note how insensitive the density distribution in the upper mantle is to the extreme cases of the temperature distribution. Both cases yield a minimum in the density curve in the upper mantle. Thus it is found that the difference between the adiabatic and isothermal upper mantles is trivial in the effect it has on the density distribution of the lower mantle and core, and on the mass and density of the earth. A plot of the percentage difference between this model and the PEM is given in Fig. 6. This is compared with the percentage difference between the C2 model (Hart *et al.*, 1978) and the PEM (Dizewonski *et al.*, 1975). It is seen that the present model is closer to the PEM than C2 is to the PEM. It is in the upper mantle where the difference between this model and the PEM is greatest, but it is here where the experimental information used in this model is best.

The temperature and density profiles of the earth computed by this method are shown in Fig. 7. We see that the central temperature is near 4500 C and at the top of the outer core it is near 3000 C. In the computation of the properties of the core, Eq. (7) was applied to the liquid part of the core as well as to the inner core. This assumption is considered reasonable, but it is unproven for a liquid metal. It will have to be verified by a basic theory of the liquid state.

The values of the physical parameters of the core which were adjusted to make the mass and moment of inertia vanish at r = 0, using the peridotite model, were: outer core (cold)—$\rho_0 = 7.0$, $K_0 = 150$ GPa; inner core (cold)—$\rho_0 = 7.38$, $K_0 = 150$ GPa. The cold value of $K_0' = 4.65$ was used for both inner and outer cores. These numbers are reasonable and are compatible with most models of an iron core diluted with a small amount of another element.

The value of α for the uncompressed upper mantle at 1400 C was taken to be 4×10^{-5}, the same as that of olivine (Suzuki, 1975). This leads to the value of α being approximately 2.7×10^{-5} for the upper mantle at a depth corresponding to 300 km. The corresponding a in (18) in the upper mantle is thus 0.005 GPa/degree. The parameter b was assumed to be 1.6 GPa (O. L. Anderson, 1980) following the results shown in Fig. 1. The values for a and b for the lower mantle are .0065 GPa/degree and 1.24 GPa, yielding $\alpha = 2.1 \times 10^{-5}$ and 1.1×10^{-5} for the top and bottom.

A MODEL FOR MARS

We use the mean density 3.933 gm/cm^3 (Bills and Ferrari, 1978) and 0.365 for the moment-of-inertia factor (Kaula, 1979). We use $\rho_0 = 3.44$ gm/cm^2 for the density of the uncompressed (cold) mantle (Goettel, 1980). This is to be compared with 3.33 to 3.41 (Okal and Anderson, 1978) and 3.47 to 3.58 (Johnston and Toksöz, 1977). Our model of the martian mantle is enriched in FeO over that of the earth.

We shall assume a peridotite upper mantle with equal parts of olivine and pyroxene, where the olivine component is $(Mg_{80}Fe_{20})_2SiO_4$, to obtain Goettel's value, $\rho_0 = 3.44$. With this value for ρ_0, we estimate $K_0 = 123$ GPa and we find $K_0' = 5.0$ from Fig. 3. A larger iron content means the transition pressure will be lowered relative to the earth as shown by curve b in Fig. 4. We take $P^* = 11.8$ GPa and $T^* = 1400$ C.

The assumed density structure includes a 50-km crust of density 3.00 and an upper mantle with the properties listed above. The beginning of the transition

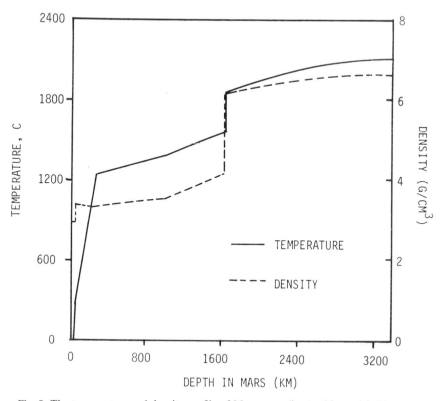

Fig. 8. The temperature and density profile of Mars according to this model. The transition zone of the mantle extends from 990 km to the core boundary at 1760 km.

zone corresponds to a depth of 990 km. A density structure through the transition zone similar to that for the earth is assumed except that higher values for ρ_0 and K_0 are used. The cold uncompressed densities for the core and its radius are chosen to satisfy the constraints of total mass and moment of inertia.

The assumed thermal structure resembles that of Toksöz *et al.* (1978). The lithosphere is 250 km thick with a temperature of 1250 C defining its lower boundary. The zone between the lithosphere and the top of the transition zone is taken as adiabatic with $\gamma = 1.5$. A temperature gradient similar to that used for the earth based on the experimental results of Akaogi and Akimoto (1979) is used for the transition zone. A value of $\Delta T = 300$ C is assumed at the core-mantle boundary to account for convection boundary layers. The core itself is taken as adiabatic with $\gamma = 1.4$ following the earth model. This approximation yields a temperature at the core's center of about 2100 C. The density and temperature distributions derived from this model are shown in Fig. 8.

Noteworthy is the fact that relatively few assumptions about the composition and properties of the bulk mantle lead directly to a solution for the radius and density of the martian core. The values obtained from this model are 1760 km for the core radius and 5.90 for the cold uncompressed core density. This value for the density is consistent with a Fe-FeS composition. As the calculated core temperature lies above the Fe-FeS liquidus, the core is predicted to be molten. Further, the core radius begins at a pressure within the transition zone of the iron-rich peridotite model (Fig. 4), so Mars is predicted not to possess a shell corresponding to the earth's lower mantle. The transition zone of Mars is calculated to contribute 16% of the planet's moment of inertia.

Acknowledgments—Support of the National Science Foundation (grant #EAR79-11212) is gratefully acknowledged. This is publication no. 2040, Institute of Geophysics and Planetary Physics, University of California, Los Angeles, CA 90024.

REFERENCES

Akaogi M. and Akimoto S. (1979) High pressure equilibria in a garnet lherzolite with special reference to Mg^{2+}–Fe^{2+} partitioning among constituent minerals, *Phys. Earth Planet. Inter.* **19**, 35–51.

Anderson D. L. (1967) A seismic equation of state. *Geophys. J. Roy. Astron. Soc.* **13**, 9–30.

Anderson D. L. (1972) Internal constitution of Mars. *J. Geophys. Res.* **77**, 789–795.

Anderson O. L. (1966) A proposed law of corresponding states for oxide compounds. *J. Geophys. Res.* **71**, 4963–4971.

Anderson O. L. (1970) Elastic constants of the central force model for three cubic structures: Pressure derivatives and equations of state. *J. Geophys. Res.* **25**, 2719–2240.

Anderson O. L. (1979a) The high-temperature acoustic Grüneisen parameter in the earth's interior. *Phys. Earth Planet. Inter.* **18**, 221–231.

Anderson O. L. (1979b) The Hildebrand equation of state applied to minerals relevant to geophysics. *Phys. Chem. Minerals* **5**, 33–51.

Anderson O. L. (1980) An experimental high-temperature thermal equation of state bypassing the Grüneisen parameter. *Phys. Earth Planet. Inter.* **22**, 173–183.

Anderson O. L. and Nafe J. E. (1965) A proposed law of corresponding states for oxide compounds, *J. Geophys. Res.* **70**, 3951–3962.

Anderson O. L. and Sumino Y. (1980) The thermodynamic properties of the lower mantle. *Phys. Earth Planet. Inter.* In press.

Barron T. H. K. (1957) On the thermal expansivity of solids at low temperatures. *Phil. Mag.* **46**, 720–734.

Bills B. G. and Ferrari A. J. (1978) Mars topography harmonics and geophysical implications. *J. Geophys. Res.* **83**, 3497–3508.

Birch F. (1952) Elasticity and constitution of the earth's interior, *J. Geophys. Res.* **57**, 277–286.

Boehler R. and Kennedy G. C. (1980a) Equation of state of sodium chloride up to 32 kbar and 500°C. *J. Phys. Chem. Solids,* **41**, 517–523.

Boehler R. and Kennedy G. C. (1980b) Thermal expansion of LiF at high pressures, *J. Phys. Chem. Solids.* In press.

Brown J. M. and Shankland T. J. (1980) Thermodynamic parameters in the earth as determined from seismic profiles. *Geophys. J.* In press.

Chung D. H. (1973) The equation of state of high-pressure solid phases, *Earth Planet. Sci. Lett.* **18**, 125–132.

Dziewonski A. M., Hales A. L., and Lapwood E. R. (1975) Parametrically simple earth models consistent with geophysical data. *Phys. Earth Planet. Inter.* **10**, 12–48.

Goettel K. A. (1980) Density of the mantle of Mars (abstract). In *Lunar and Planetary Science XI,* p. 333–335. Lunar and Planetary Institute, Houston.

Grüneisen E. (1924) The state of a solid body. *Handbuch der Physik, 10.* J. Spencer, Berlin (English translation, NASA RF 2-18-59W, February 1959).

Hardy R. J. (1980) Temperature and pressure dependence of intrinsic anharmonic and quantum corrections to the equation of state. *J. Geophys. Res.* In press.

Hart R. S., Anderson D. L., and Kanamori H. (1977) The effect of attenuation on gross earth models. *J. Geophys. Res.* **82**, 1647–1653.

Jeanloz R. and Richter F. M. (1979) Convection composition and the thermal state of the lower mantle. *J. Geophys. Res.* **84**, 5497–5504.

Johnston D. H. and Toksöz M. N. (1977) Internal structure and properties of Mars. *Icarus* **32**, 73–84.

Kaula W. M. (1979) The moment of inertia of Mars. *Geophys. Res. Lett.* **6**, 194–196.

Kumazawa M. (1969) The elastic constants of single-crystal orthopyroxene, *J. Geophys. Res.* **74**, 5973–5980.

Kumazawa M. and Anderson O. L. (1969) Elastic moduli pressure derivatives and temperature derivatives of single-crystal olivine and single-crystal forsterite. *J. Geophys. Res.* **74**, 5961–5972.

Lachenbruch A. H. and Sass J. H. (1978) Heat flow in the United States and the thermal regime of the crust. *Amer. Geophys. Union Monograph 20,* 626–675.

Liu L. (1977) Mineralogy and chemistry of the earth's mantle above 100 km, *Geophys. J. Roy. Astron. Soc.* **48**, 53–62.

Okal E. A. and Anderson D. L. (1978) Theoretical models for Mars and their seismic properties. *Icarus* **33**, 514–528.

Ringwood A. E. and Clark S. P. (1971) Internal constitution of Mars. *Nature* **234**, 89–92.

Solomon S. C. (1979) Formation, history and energetics of cores in the terrestrial planets. *Phys. Earth Planet. Inter.* **19**, 168–182.

Stevenson D. J. (1980) Applications of liquid state physics to the earth's core. *Phys. Earth Planet. Inter.* **22**, 42–52.

Suzuki I. (1975) Thermal expansion of periclase and olivine and their anharmonic properties. *J. Phys. Earth* **23**, 145–149.

Toksöz M. N., Hsui A. T., and Johnston D. H. (1978) Thermal evolutions of the terrestrial planets. *Moon and Planets* **18**, 281–370.

Tozer D. C. (1967) Towards a theory of thermal convection in the mantle. In *The Earth's Mantle,* (T. F. Gaskell, ed.), p. 328–353. Academic Press, London.

Zharkov V. N. and Kalinin V. A. (1971) *Equations of State for Solids at High Pressures and Temperatures.* Consultants Bureau, N.Y. 256 pp.

APPENDIX

Table A-1. Table of density, temperature, pressure, gravity, bulk modulus, and coefficient of thermal expansus versus depth for spherically symmetric earth, according to this model.

Radius km	Depth km	Density g/cm3	Temp deg C	Heat flux ncal/cm2-S	Gravity cm/sec2	Pressure Kbar	Bulk modulus	Alpha E-6 1/k
6371.	0.	1.030	5.	1613.0	983.0	0.0	1000.	0.0
6368.	3.	1.030	5.	1614.5	983.7	0.3	1001.	0.0
6368.	3.	2.750	5.	1614.5	983.7	0.3	700.	24.27
6360.	11.	2.746	196.	1220.1	984.3	2.5	695.	24.45
6360.	11.	2.996	196.	1220.1	984.3	2.5	695.	24.46
6350.	21.	2.994	386.	1058.3	984.9	5.4	694.	24.51
6350.	21.	3.297	386.	1058.3	984.9	5.4	1128.	44.31
6300.	71.	3.247	1045.	1050.0	986.7	21.5	1040.	48.10
6273.	98.	3.218	1399.	1045.5	987.9	30.2	990.	50.51
6273.	98.	3.217	1399.	1045.5	987.9	30.2	988.	50.60
6250.	121.	3.240	1399.	1041.7	988.9	37.5	1027.	48.71
6200.	171.	3.289	1399.	1033.3	991.1	53.7	1114.	44.89
6150.	221.	3.336	1399.	1025.0	993.3	70.1	1201.	41.62
6100.	271.	3.381	1399.	1016.7	995.5	86.8	1289.	38.78
6050.	321.	3.424	1399.	1008.3	997.6	103.7	1377.	36.30
6000.	371.	3.466	1399.	1000.0	999.7	121.0	1466.	34.12
5990.	381.	3.474	1399.	998.3	1000.1	124.4	1483.	33.71
5990.	381.	3.475	1399.	998.3	1000.1	124.4	1485.	33.67
5950.	421.	3.591	1431.	991.7	1001.7	138.6	1655.	30.20
5900.	471.	3.743	1472.	983.3	1003.2	156.9	1889.	26.46
5850.	521.	3.899	1512.	975.0	1004.3	176.1	2143.	23.33
5800.	571.	4.059	1552.	966.7	1004.9	196.1	2418.	20.68
5750.	621.	4.223	1591.	958.3	1004.9	216.9	2717.	18.40
5700.	671.	4.392	1630.	950.0	1004.4	238.5	3045.	16.42
5700.	671.	4.393	1780.	950.0	1004.4	238.5	3037.	21.40
5600.	771.	4.449	1824.	933.3	1002.8	282.9	3165.	20.54
5400.	971.	4.564	1915.	900.0	999.9	373.2	3433.	18.93
5200.	1171.	4.676	2003.	866.7	997.6	465.5	3705.	17.55
5000.	1371.	4.786	2088.	833.3	996.3	559.8	3979.	16.34
4800.	1571.	4.893	2171.	800.0	996.3	656.2	4256.	15.27
4600.	1771.	4.998	2251.	766.7	998.0	754.8	4537.	14.33
4400.	1971.	5.101	2329.	733.3	1002.1	855.8	4823.	13.48
4200.	2171.	5.203	2405.	700.0	1009.0	959.4	5113.	12.71
4000.	2371.	5.305	2480.	666.6	1019.6	1065.9	5409.	12.02
3800.	2571.	5.406	2554.	633.3	1034.9	1175.8	5713.	11.38
3600.	2771.	5.508	2627.	600.0	1056.1	1289.8	6026.	10.79
3485.	2886.	5.566	2668.	580.8	1071.6	1357.5	6211.	10.47
3485.	2886.	9.890	3169.	580.8	1071.6	1357.5	6180.	9.71
3400.	2971.	10.017	3234.	566.6	1053.0	1447.4	6490.	9.25
3200.	3171.	10.314	3390.	533.3	1006.8	1657.1	7250.	8.28
3000.	3371.	10.584	3536.	500.0	957.7	1862.7	7991.	7.51
2800.	3571.	10.830	3671.	466.6	906.1	2062.6	8708.	6.89
2600.	3771.	11.054	3796.	433.3	852.2	2255.3	9397.	6.39
2400.	3971.	11.258	3912.	400.0	796.2	2439.6	10053.	5.97
2200.	4171.	11.443	4018.	366.6	738.4	2614.1	10673.	5.62
2000.	4371.	11.609	4116.	333.3	679.2	2777.9	11253.	5.33
1800.	4571.	11.758	4206.	299.9	618.9	2929.9	11791.	5.09
1600.	4771.	11.890	4287.	266.6	557.9	3069.5	12283.	4.88
1400.	4971.	12.007	4360.	233.2	496.8	3195.8	12729.	4.71
1215.	5156.	12.102	4421.	202.3	441.0	3300.7	13098.	4.58
1215.	5156.	12.702	4421.	202.3	441.0	3300.7	13107.	4.58
1200.	5171.	12.707	4423.	199.8	435.8	3309.1	13127.	4.57
1000.	5371.	12.803	4473.	166.4	365.6	3411.8	13490.	4.45
800.	5571.	12.881	4514.	133.0	294.6	3497.0	13792.	4.35
600.	5771.	12.942	4545.	99.4	222.7	3564.3	14030.	4.28
400.	5971.	12.986	4569.	65.3	150.1	3613.1	14203.	4.22
200.	6171.	13.013	4583.	28.2	75.3	3643.0	14310.	4.19
0.	6371.	13.021	4587.	0.0	0.0	3651.3	14341.	4.18

Proc. Lunar Planet. Sci. Conf. 11th (1980), p. 2015–2030.
Printed in the United States of America

Time-dependent lunar density models

Floyd Herbert

Lunar and Planetary Laboratory, University of Arizona, Tucson, Arizona 85721

Abstract—Simple models of the geochemical and geophysical evolution of the moon were constructed, with the particular aim of investigating the mass density distribution within the moon as a function of time. The strongly inverted (i.e., densest material on top) density distribution resulting from fractional crystallization of the lunar magma ocean was found to rearrange itself into a highly stable stratification. Density stratification due to compositional variation dominated gravitational stability; thus, thermally driven convection was found to be suppressed in regions with relaxed (i.e., rearranged so as to minimize the gravitational potential energy) compositional density variation. The suppression of convection as a heat transfer process results in considerable modification of thermal evolution.

The models produced sufficient density segregation from magma oceans of a few hundred km thickness such that relaxation to the most stable rearrangement yielded a polar moment of inertia approximating that measured for the present-day moon. Thus, lunar models matching moment-of-inertia constraints but lacking a metallic core are motivated.

INTRODUCTION

The internal variation of density in the moon is a clue to past lunar evolution. Various external measurements constrain our knowledge of the density variation. The best-known such data are the mean lunar density, the moment of inertia coefficients, and the lateral variation of the lunar gravity field. Although these measurements tell us something about the present internal structure of the moon, they also represent the results of several billion years of change due to the thermal evolution of the moon.

Two processes involved in lunar thermal evolution strongly affect the density profile—chemical fractionation and subsolidus creep. The two processes strongly interact in the case of the moon because major fractionating minerals—olivine and pyroxene—are typically deposited in an inverted (highest density at the top) density distribution. This is because the least dense components crystallize out of the magma before the denser phases solidify. Subsequent convective rearrangement due to subsolidus creep tends to relax the inverted distribution to a gravitationally stable configuration, and may thereby carry particular mass elements to locations very different from their points of origin. This process has been discussed qualitatively by Kesson and Ringwood (1976) and Ringwood and Kesson (1976); the present work is an attempt to expand and quantify this analysis.

The most important fractionation undergone by lunar material was the lunar magma ocean event (Wood *et al.*, 1970). This is usually interpreted as the melting of the entire outer portion of the moon, though other views have been proposed (e.g., Wetherill, 1976). The typical scenario has the lunar magma ocean solidifying by fractional crystallization of dense mafic minerals such as olivines and pyroxenes, which settle to (or appear at) the ocean bottom, and by separation of less dense minerals such as plagioclase, which floats at the ocean surface. Because of the gravitational separation of the phases appearing at various times, fractionation of the mafic minerals results in the earliest-appearing of those rocks lying the deepest, with rock ages decreasing upwards.

In the case of olivine and pyroxene, which are likely to be the dominant mafic minerals crystallizing from the magma ocean, the phases appearing at the highest temperatures are the most iron-poor and magnesium-rich. Thus as the ocean cools and the process of solidification proceeds the crystallizing cumulates become progressively enriched in iron and are therefore progressively denser. The natural result of this process is an inverted density profile.

Because the cumulate minerals solidifying from the magma ocean are laid down at near-melting temperatures, they are likely to be fairly plastic. Since their density distribution is likely to be strongly inverted (as calculations described below show), relaxation of the density inversion by sub-solidus deformation should occur. The combination of relative high plasticity and severe density inversion leads to relaxation times (as estimated in the next section) that are short compared with the 4.6×10^9 year age of the moon.

The relaxation process, if operating only within the layer of mafic cumulate products of magma ocean solidification, does not lead to the lowest possible configuration of potential energy. The mean cumulate density will be at least as high as that of the underlying unmelted material, and thus the most dense cumulates will be significantly denser than the primordial lunar rock upon which they will come to rest. Further settling of these phases is therefore likely, with penetration all the way to the center of the moon quite possible. Because the unmelted material could well be considerably cooler than its (pressure-augmented) melting point, its viscosity could be much higher than that of the dense cumulates, however. Thus this second phase of relaxation would take a much longer time interval than the first. Indeed, it might be incomplete even in the present-day moon, so that there could still be denser phases sinking through lighter material with the gravitational potential not yet minimized.

Relaxation of an inverted density profile inevitably leads to a centrally concentrated mass distribution. In this way the lunar moment of inertia would be reduced below the value for a uniform sphere, as is observed. The low lunar moment of inertia has led to frequent speculation that the moon possesses an iron core; as will be seen below, quite reasonable models produce sufficiently low moments without any metal at all.

This analysis suggests that considerable transport of highly disparate materials could have occurred during lunar evolution. Because of the far-reaching consequences of this redistribution the following sections discuss some simple models of the processes just described.

MODEL DESCRIPTION

In order to investigate the combined action of the processes of chemical fractionation and sub-solidus convection some simplified models were constructed. The primordial moon was assumed to be composed of the minerals fayalite, forsterite and anorthite in proportions such that the mean density of the actual moon (3.34 g/cm³) was matched. For reasons discussed below, only these minerals were considered. Matching the mean density of the moon this way primarily adjusts the fayalite/forsterite ratio, since these are the principal components. An additional constraint was imposed that all of the anorthite in the lunar magma ocean accumulate into a crust of predetermined thickness, usually 50 km. This constraint mainly affects the anorthite abundance assumed. These two constraints, given magma ocean and anorthositic crustal thicknesses as model parameters, sufficed to determine the three-component composition. This procedure typically results in a highly iron-rich (20 to 25 mass percent FeO) composition. An iron abundance this high is primarily due to the fact that the measured mean lunar density (3.34 g/cm³) is significantly higher than the zero-pressure densities of either anorthite or forsterite (2.76 and 3.22 g/cm³, respectively). Thus matching the mean lunar density requires a large amount of fayalite, even though its assumed zero-pressure density is also quite high, at 4.39 g/cm³.

All densities of rock and magma were assumed to be given by the formula

$$\rho = \frac{\sum_i X_i m_i}{\sum_i X_i v_i} \tag{1}$$

where m_i is the mass per mole of the ith component, v_i is its molar volume (in either the liquid or solid state, as appropriate), and X_i is its concentration, expressed as a mole fraction.

The formation of the magma ocean was modelled by adjusting the temperature (assumed constant throughout the solid portion of the moon) until the magma was in equilibrium with the unmelted rock at the desired depth of the magma ocean base. The temperature of the unmelted interior was either set equal the initial magma temperature or, as in the two models discussed in detail later, was assumed to rise quadratically from 0°C at the center of the moon to the magma temperature at the solid-liquid interface. These cases are hereinafter referred to as "hot interior" and "cold interior" models, respectively. The composition of the initial magma was assumed to be the same as that of the unmelted interior. This is equivalent to neglecting the fraction of the lunar volume lying between the radii at which the (pressure-dependent) liquidus and solidus temperatures match the adopted temperature. The pressure dependence of the liquidus and solidus was assumed to be 10°C/kbar. The pressure dependence of the chemistry of the crystallizing olivine was assumed to follow by adjusting the equilibrium temperature using the same constant of proportionality:

$$T = T_{eq} + \frac{\partial T_s}{\partial P} P. \tag{2}$$

Here T is the actual temperature of the magma, T_{eq} is the temperature used in calculating the equilibrium constant and the distribution coefficients for the olivine composition, P is the pressure and $\frac{\partial T_s}{\partial P}$ is the variation with pressure of the solidus temperature.

The ocean was then allowed to cool, fractionally crystallizing olivine at the base of the magma ocean, with olivine composition computed from formulas given by Roeder and Emslie (1970):

$$\log \frac{X_{MgO}^{Ol}}{X_{MgO}^{Liq}} = \frac{3740}{T_{eq}} - 1.87$$

$$\log \frac{X_{FeO}^{Ol}}{X_{FeO}^{Liq}} = \frac{3911}{T_{eq}} - 2.50 \tag{3}$$

$$\log \frac{X_{FeO}^{Ol} X_{MgO}^{Liq}}{X_{FeO}^{Liq} X_{MgO}^{Ol}} = \frac{171}{T_{eq}} - 0.63.$$

Here X_i^{α} is the mole fraction of oxide i and phase α. Molar quantities are based on single-cation formulas.

For estimating the time duration of the magma ocean epoch, the heat flux out of the ocean was assumed to be limited by conduction through a growing plagioclase-rich crust assumed to float on top of the ocean. Following procedures discussed by Herbert et al. (1978), the crust was assumed to thicken at a rate proportional to the precipitation of olivine. As noted by those authors and others (e.g., Longhi, 1978; Warren and Wasson, 1979) the production of stable floating plagioclase may well be initially slower, subsequently accelerating, and thereby lower the solidification time of the magma ocean. For purposes other than estimating the lifetime of the lunar magma ocean, however, the constancy of the ratio of plagioclase to olivine crystallization is not important. In similar models in which the plagioclase crystallization was delayed until all the olivine had solidified, the principal difference was a much shorter solidification time and a higher concentration of radionuclides in the crust. Not all the radionuclides were frozen out there, however, as about half the original quantity wound up in the last-freezing iron-rich olivine layers which ultimately sank into the lunar interior.

An approximation of the calculation is that the coprecipitation of pyroxene was ignored. Although pyroxene crystallization is very important to the chemistry of the magma and cumulates (Drake, 1976; Warren and Wasson, 1979), particularly at great depth, it is ignored here for simplicity. The order of crystallization of magnesium-rich phases before iron-rich phases is the same for pyroxene as it is for olivine, and thus the effect of pyroxene crystallization on the density profile of cumulates is the same as that of olivine. Since the primary focus of the present work is upon density profiles, the neglect of pyroxene chemistry is assumed to be justified by the great simplification thereby produced.

Sub-solidus mass motion is modelled in a highly simplified description. The distribution of matter is assumed to be spherically symmetric, varying only as a function of radius. Mass motion is assumed to occur in discrete shells and is simulated by switching the contents of adjacent shells. While this approximation produces an important simplification, however, it should be kept in mind that for mass flow in the real moon the interchange of mass will inevitably be incomplete. Even if the amount of material left behind by the bulk of a mass shell is small compared to the total, it may be chemically important.

A very rough estimate is made of the time required for shell overturn by using a simple viscous flow model. Shell viscosity μ is calculated from the formula

$$\mu = \mu_o \exp\left(\frac{T_m(P)}{T_m(0)} \cdot \frac{a}{T}\right), \tag{4}$$

where $T_m(P)$ is the pressure-adjusted (as in Eq. 2) melting point of the rock. The quantity μ_o (462 N s/m²) and the constant a (62928°K) are taken from Turcotte *et al.* (1979). The mean mass motion velocity $\overline{v}$ is derived from the laminar flow velocity profile in a pipe of length L and radius R, (Landau and Lifshitz, 1959), yielding

$$\overline{v} = \frac{\Delta P R^2}{8\mu L}. \tag{5}$$

The pressure differential ΔP due to bouyancy is taken as $gL\Delta\rho$, where g is the local acceleration of gravity and $\Delta\rho$ is the density differential. The values of L and R to be used in the formula correspond roughly to the size expected for a convection cell.

The size of a convection cell is rather difficult to estimate, but is likely to be approximately equal to the distance over which the viscosity, distance from the center of the moon or some other fundamental quantity changes significantly. In these models the convection cell size was taken as one tenth of the local radius. Figure 1 shows a range of overturn times resulting from Eq. 5.

The use of a pipe flow formula is little more accurate than dimensional analysis, but a very rough estimate of the overturn time is probably adequate for the purposes of these models. The overturn time is approximated by the ratio $L/\overline{v}$.

The simultaneous overturn of material at various points in the moon is handled by keeping track of the time a given pair of shells, the densest of which is at the top, have been together. Once this time has exceeded the estimated overturn time for the pair, they reverse positions and their accumulated time parameter is set to zero. If the topmost of a pair of shells is the least dense, their joint accumulated time parameter is reset to zero at each evolutionary time step. Sub-solidus flow driven by both thermal and compositional density differences is modelled by this technique.

Conductive heat transfer and radionuclear heat generation were handled conventionally. At each time step, the temperature change due to the input of heat from decay of radionuclides and conduction from adjacent shells was computed before mass transport was assumed to occur. The algorithm for calculating heat conduction was the Crank-Nicholson implicit finite difference technique (Acton, 1970) adapted for spherical symmetry and variable thermal conductivity and radius step size. The thermal model parameters are given in Table 1.

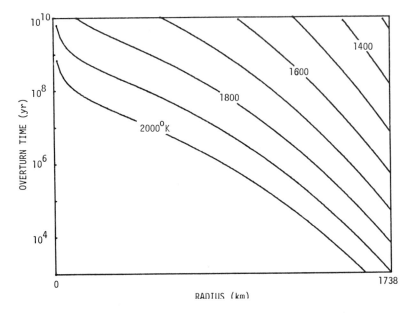

Fig. 1. Representative overturn times as calculated from Eq. (5). Calculations assumed $\Delta\rho$ = 0.1 g/cm³, and L = R = 0.1 R_{moon}. Curves are labelled by the local temperature assumed. The assumed zero-pressure melting temperature $T_m(O)$ was 1500°K.

The radionuclide concentration was assumed to be ''chondritic'' (Kaula, 1968). Redistribution of radionuclides from a melt was assumed to be governed by

$$X_i^s = DX_i^l, \tag{6}$$

where X_i^s and X_i^l are the mole fractions of radionuclide i in solid and liquid, respectively, and D is a constant distribution coefficient. Values of D < 1 concentrate the radionuclide in the residual liquid and consequently into the last-freezing solid.

For reasons of simplicity, lunar evolution was divided into two eras: the lunar magma ocean era and the sub-solidus convection era. In actuality, since the lunar magma ocean may have lasted for as long as 2 × 10⁸ years, and since some of the hotter regions of the model moon can overturn in less time than that, the distinction is somewhat artificial. Nevertheless, creep modelling in the calculations of this work only begins after magma ocean solidification.

The result of one such calculation, hereafter called Model 1, is shown in Figs. 2 and 3, which are plots of density with radius. Model 1 assumes a 300 km magma ocean to have been created above a cold interior. The abundances of the three assumed components were adjusted as previously outlined in order to yield a 50 km anorthite crust and a mean lunar density of about 3.34 g/cm³. The weight percentages of the five corresponding oxides are given in Table 2. Figure 2 shows the lunar density profile immediately following the final solidification of the magma ocean. The inverted density distribution resulting from a 300 km depth magma ocean is clearly visible. The noticeable decrease in density outward from the center is due to the combination of falling pressure and rising temperature in the unmelted interior.

The very high values of the density just below the crust are due to the high degree of enrichment of the residual magma in iron. Iron is enriched in the residual magma because the much higher melting temperature of forsterite causes it to preferentially crystallize out before most of the fayalite. These high density layers will also contain the bulk of the elements incompatible with silicate crystal lattices, in particular the major radionuclides ^{235}U, ^{238}U, ^{232}Th and ^{40}K.

Figure 3 shows the density profile at a time corresponding to the present epoch. The last-freezing iron-rich layers have moved from the top of the mafic cumulate

Table 1. Thermal model parameters.

Property	Material	Value
Density[1]	Anorthite	2760 kg/m³
	Fayalite	4393
	Forsterite	3213
Thermal conductivity	Anorthite	2 W/m/°K
	Olivine	4
Specific heat	(all)	1200 J/kg/°C
Enthalpy of crystallization[1,2]	Anorthite	2.7×10^5 J/kg
	Fayalite	4.52×10^5
	Forsterite	9.5×10^5
Relative volumetric thermal	Anorthite	1.45×10^{-5} °C^{-1}
expansion coefficient[3]	Fayalite	2.6×10^{-5}
	Forsterite	3.9×10^{-5}
Relative volumetric	Anorthite	1.0×10^{-3} kbar^{-1}
compressibility[3]	Fayalite	9.1×10^{-4}
	Forsterite	7.9×10^{-4}
Radionuclide concentration[4]	^{238}U	1.19×10^{-8} kg/kg
(now)	^{235}U	8.64×10^{-11}
	^{232}Th	4×10^{-8}
	^{40}K	9.7×10^{-8}
Half-life[5]	^{238}U	4.51×10^9 yr
	^{235}U	7.1×10^8
	^{232}Th	1.41×10^{10}
	^{40}K	1.26×10^9
Heat generation rate[4]	^{238}U	2.98×10^3 J/kg/yr
(per kg of *radionuclide*)	^{235}U	1.8×10^4
	^{232}Th	8.4×10^2
	^{40}K	8.8×10^2
Radionuclide partition coef.	(all)	0.03

[1] Robie *et al*, (1978)
[2] McConnell *et al.* (1967)
[3] Clark (1966)
[4] Kaula (1968)
[5] Lederer *et al.* (1967)

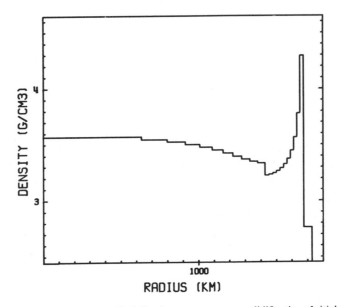

Fig. 2. Model lunar density profile following magma ocean solidification. Initial magma ocean depth was 300 km, overlying an interior of the same temperature as the magma.

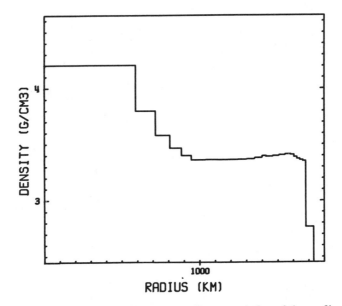

Fig. 3. Model lunar density profile corresponding to evolution of the profile shown in Fig. 2 up to the present time.

Table 2. Model compositions—oxide percentages by weight.

Oxide	Model 1	Model 2
CaO	3.3	2.0
Al_2O_3	5.9	3.6
SiO_2	38.1	38.3
FeO	24.8	23.8
MgO	27.9	32.4

layer to the lunar interior. The less dense layers have largely rearranged themselves into a gravitationally more stable stratification, although the cooling due to conductive loss to the surface has slowed the process to the point where the distribution has not completely relaxed. The relative completeness of relaxation is highly dependent on the assumptions regarding creep viscosity and flow geometry. This strong dependence is due to the competition between the cooling of the outer layers and the deformation due to compositional density differences, and is probably the most model-dependent result of this work. Fluid effects were not considered, though the initial upward flow was sufficiently rapid to allow some pressure-release melting to occur.

However, although the radionuclides are largely carried into the core in the model, a more realistic model would require that the transfer be incomplete. Mass flow driven by the higher density of the final cumulates would only continue until the bulk of the material had sunk. Some would remain, to comprise such materials as KREEP.

Figure 4 shows a contour plot of temperature as a function of time and position within the moon. The contours (which are hand-smoothed approximations of a computer-generated plot) are labelled by temperature in °C. The cooling of the mantle and the rising core temperature (due to radionuclides swept downwards by the iron-rich layers) are visible.

The current surface heat flow according to this model should be about 10 mW/m^2. The actual global average value is estimated by Keihm and Langseth (1977) to be 18 mW/m^2, though measurements vary from 14 to 21mW/m^2. Sonett and Duba (1975) have estimated a heat flux of about 25 to 35 mW/m^2, based on sub-crustal temperature estimates.

The final model moment of inertia coefficient (I/MR^2, where I is the polar moment of inertia, M is the lunar mass and R the radius) was 0.391. Current estimates of the lunar moment of inertia coefficient lie near 0.3905 ± 0.0023 (Ferrari *et al.*, 1980). Total radius change up to the present was an expansion of about 2 km between about 3.5 and 2×10^9 years ago followed by a 3 km contraction.

Model 2, which differs from Model 1 in assuming a 600 km deep magma ocean, also assumes an initially cold interior and is illustrated in Figs. 5 and 6. Figure 5 shows the density profile immediately following complete magma ocean solidification. The assumed composition required to match a 50 km anorthite crust

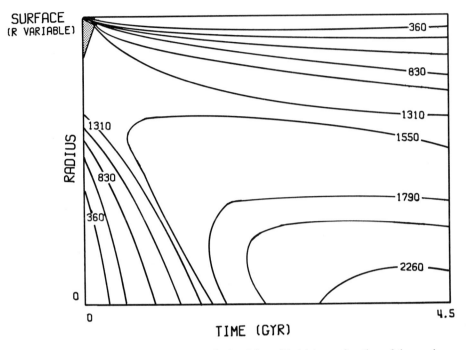

Fig. 4. Contour plot of temperature calculated from Model 1 as a function of time and position within the moon. Contours are hand-smoothed approximations and are labelled in °C. The stippled region represents the magma ocean.

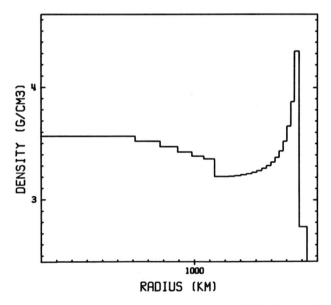

Fig. 5. Model lunar density profile following magma ocean solidification. Initial magma ocean depth was 600 km, overlying a cool interior.

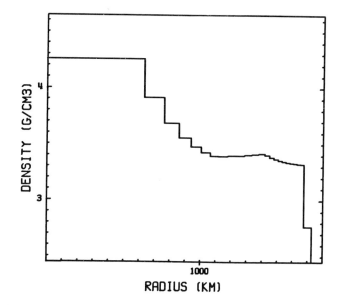

Fig. 6. Model lunar density profile corresponding to evolution of the profile shown in Fig. 5 up to the present time.

and the measured current lunar mean density is given in Table 2. Figure 6 shows the relaxation of the density profile of Fig. 5, after 4.6×10^9 years of model evolution.

Figure 7 shows a contour plot of temperature calculated from Model 2 as a function of time and position within the moon. Contours are smoothed and labelled with temperature in °C. Compared with the results of Model 1, mantle temperatures are somewhat cooler.

This model yielded a final polar moment of inertia coefficient of 0.388. Total radius change from 3.5×10^9 years ago to the present was a 5 km contraction.

Models similar to Models 1 and 2 except for having initially hot interiors were also constructed. In the main they were similar to the cold interior models except that they typically underwent an overall contraction of approximately 10 km.

DISCUSSION

Remelting of cumulates via pressure release in a rising sub-solidus flow to form the mare basalt source regions has been discussed by Kesson and Ringwood (1976) and Ringwood and Kesson (1976). In both models discussed in the previous section (but particularly in Model 2) the temperatures of some shells did indeed rise above the solidus early in their thermal histories, verifying Ringwood and Kesson's expectations within the accuracy limits of the present calculation.

Because of the great concentration of radionuclides into the last-freezing cumulates, which descend into the center of the model moon, later radionuclear

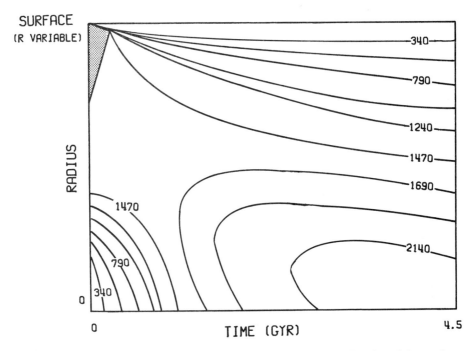

Fig. 7. Contour plot of temperature calculated from Model 2 as a function of time and position within the moon. Contours are hand-smoothed approximations and are labelled in °C. The stippled region represents the magma ocean.

heating causes a considerable rise in central temperatures. In fact all models constructed so far, including particularly Model 1, attained core temperatures well above the solidus. The reason for this is that the stratification of compositional density variation resulting from subsolidus flow is stable against thermal convection, at least until fairly complete melting occurs. Thus thermal convection cannot transport heat out of the core as fast as it is generated by the radionuclides presumably concentrated there. This situation is contrary to that more customarily assumed, in which subsolidus thermal convection controlled by the highly temperature-dependent viscosity maintains the lunar interior at a constant fraction of the local solidus temperature.

The results of these models are in qualitative agreement with the decrease in seismic velocities at depths of 300 to 500 km (e.g., see Goins *et al.*, 1979). A decrease in both compressional and shear wave velocities with depth is consistent with density increasing with depth. That the shear wave velocity decreases faster with depth than the compressional is consistent with a weakening of the rigidity of the lunar interior with increasing depth, as one would expect if the temperature were rising with depth faster than the solidus. The models of the present work lead to just this situation, which is contrary to the outcome of viscosity-controlled thermal convection as discussed in the preceding paragraph.

Partial melting of the lunar interior by radionuclides held localized by compositional density stratification would account for the zone of high seismic attenuation found by Nakamura *et al.* (1973, 1979). A partially molten lunar interior would also be consistent with the observations of Sonett and Wiskerchen (1977), who found a low frequency electromagnetic lunar transfer function that was consistent with a high (electrical) conductivity core of 400 km radius. Even partial melting of silicate rocks is usually associated with a sharp electrical conductivity increase (Waff and Weill, 1975), and thus a partially molten lunar interior could electromagnetically mimic a somewhat smaller metallic core.

On the basis of earlier calculations, Herbert (1980) speculated that the constraint on lunar thermal evolution developed by Solomon and Chaiken (1976) and Solomon and Longhi (1977), which was based on tectonic evidence against significant (>1 km) lunar expansion or contraction in the last 3.5×10^9 years, might be weakened by the possibility of concentration of radionuclides in the lunar interior. The reasoning behind this statement was that the radionuclides of the magma ocean would be largely concentrated into the last-forming rocks of the magma ocean, because of the general compatibility of the major heat-producing radionuclides (^{40}K, ^{235}U, ^{238}U and ^{232}Th) with silicate crystal matrices. As has been already discussed, it is just these late-crystallizing rocks that ultimately flow down into the lunar core. The early-crystallizing mafic rocks, which would be significantly depleted in radionuclides, ultimately form the outermost layers of the moon. Thus, a net downward transport of heat sources would occur. Deeper burial of radionuclides increases the quasi-equilibrium internal temperatures by increasing the mean thermal gradient. This effect would tend to offset the net cooling trend, which in the models of the aforementioned authors leads to the observationally disallowed global lunar shrinkage.

However, the lunar thermal models discussed in the previous section all resulted in expansion or contraction greater than 1 km, especially the hot interior models. In addition, very deep magma oceans produce initial lunar models that have an insignificant amount of initial cool material to be expanded by later heating. Thus, model results so far in this study tend to confirm the overall conclusion of Solomon and Chaiken (1976) and Solomon and Longhi (1977). As those authors pointed out, however, the balancing of expansional and contractional forces is rather delicate, so that further exploration of such parameters as the rock (thermal) conductivity and distribution coefficients for radionuclides crystallizing within rocks is necessary to the drawing of firm conclusions. Moreover, partial melting of the lunar interior produces expansion which has not yet been incorporated into the models described here.

The calculations done so far have yielded upper mantles at late times with too high a sub-solidus viscosity to permit complete relaxation of the density distribution. This result occurs because conduction of heat through the crust cools the outer mantle more quickly than it does the interior. Thus is formed a rheologically (that is, with a high viscosity resulting from low temperatures, rather than by means of compositional differences) defined lithosphere at fairly early times (depending on the model, perhaps roughly 4×10^9 years ago). The earliness of the lithospheric stiffening is important to the (not isostatically supported) mascons, as was pointed out by Kesson and Ringwood (1976).

The stiffening of the lithosphere while the density profile is still relaxing is likely to lead to considerable lateral inhomogeneity in the density and the resulting gravitational field. While the models of this paper do not treat non-spherically-symmetric effects and instead model mass flow by interchange of shells, real planetary mass flow occurs via Rayleigh-Taylor instabilities. In such cases underdense rising columns will be present at some locations and overdense descending plumes will occur elsewhere. Such large-scale inhomogeneities will not only produce local gravitational anomalies, but, given the relatively large (of the order of 1 g/cm^3) density variations under discussion here, could in concept be responsible for the observed offset between the lunar mass and figure centroids.

The fayalite abundance assumed by the model calculations was higher than values normally assumed because the measured mean lunar density is well above that of the other assumed components (anorthite and forsterite). If the fayalite/forsterite ratio is lowered, the measured mean lunar density is no longer well matched. For example, approximating with these components the model lunar composition proposed by Ganapathy and Anders (1974), results in a model lunar density of about 3.1 g/cm^3 and a moment of inertia coefficient of about the same as for Models 1 and 2 for corresponding magma ocean depths. Clearly such a model is underdense (though its moment of inertia coefficient is still reasonable); a more plausible composition is likely to include more iron. Whether the oxidation state of the iron would be such that it is all present as olivine (as in this work) or as metal is not resolvable in the present analysis. Alternatively, the low density of the approximation to the Ganapathy and Anders composition might be due to the oversimplified (three component) composition assumed. That is, other minerals than olivine might be denser for the same iron/magnesium ratio.

The generally good match between the moment of inertia of the moon and most of the models constructed so far is interesting. However, it should be noted that the first and second density moments can be well matched by a tremendous variety of density distributions. But the fact that density profile relaxation models lead to moments at or below the measured value suggests that any iron core existing in the moon must be very small, if the evolution of the lunar magma ocean was at all as is usually thought.

There is also the question of whether there actually ever was a lunar magma ocean. Wetherill (1976) views this occurrence as merely the stochastic superposition of very numerous impact melts, highly separated in time. Hartmann (1980) has argued that the modification of the evolution of any magma ocean by the terminal accretionary bombardment of the moon must have been extreme. In particular, the terminal bombardment would have continually reexcavated the lunar surface during the magma ocean era, mixing magma and floating crust ejecta. This stirring would have greatly promoted heat transfer and thus greatly shortened the duration of the magma ocean (Minear, 1980). The mixing of crust and quickly quenched ejected magma might also lower the mean density of the floating crust, causing it to founder and sink.

Over the bombardment history one can define a mean depth of excavation as a function of time, defined as the mean depth below which material has been undisturbed since that time. If the mean excavation depth decreased to less than

10 km or so within the time the magma ocean was still largely molten, probably the models of this work (and most other magma ocean modelling exercises) are still largely valid. If, however, the mean excavation depth was of the order of 100 km for several hundred million years, as Hartmann (1980) suggests, the magma ocean picture (and the models in this paper) are in doubt.

Regardless of the outcome of that particular debate, it is clear from simple chemical fractionation models that density variation due to melt crystallization is sizeable. The models constructed in the present work demonstrate that these density variations significantly modify subsolidus flow driven by thermal bouyancy. Thus, compositional variation must be incorporated into any realistic lunar thermal models.

Acknowledgments—Conversations with M.J. Drake have been highly informative and stimulating. Reviews of this manuscript by C. R. Fletcher and S. C. Solomon have improved it considerably. This work was funded by NASA Grant NGR 03-002-370. Computations used equipment funded by grants from NSF and NASA.

REFERENCES

Acton F. S. (1970) *Numerical Methods that Work*. Harper and Row, N.Y. 541 pp.

Clark S. P. (ed.) (1966) *Handbook of Physical Constants* (revised edition). Geol. Soc. Amer. Memoir 97. N.Y.

Drake M. J. (1976) Evolution of major mineral compositions and trace element abundances during fractional crystallization of a model lunar composition. *Geochim. Cosmochim. Acta.* **40**, 401.

Ferrari A. J., Sinclair W. S., Sjogren W. L., Williams J. G., and Yoder C. F. (1980) Geophysical parameters of the Earth-Moon system. *J. Geophys. Res.* In press.

Ganapathy R. and Anders E. (1974) Bulk compositions of the Moon and Earth, estimated from meteorites. *Proc. Lunar Sci. Conf. 5th*, p. 1181–1206.

Goins N. R., Toksöz M. N., and Dainty A. M. (1979) The lunar interior: A summary report. *Proc. Lunar Planet. Sci. Conf. 10th*, p. 2421–2439.

Hartmann W. K. (1980) Dropping stones in magma oceans: Effects of early lunar cratering. In *Proc. Conf. Lunar Highlands Crust* (J. J. Papike and R. B. Merrill, eds.), p. 155–171. Pergamon, N.Y.

Herbert F. (1980) Time-dependent lunar density models (abstract). In *Lunar and Planetary Science XI*, p. 432–434. Lunar and Planetary Institute, Houston.

Herbert F., Drake M. J., and Sonett C. P. (1978) Geophysical and geochemical evolution of the lunar magma ocean. *Proc. Lunar Planet. Sci. Conf. 9th*, p. 249–262.

Kaula W. M. (1968) *An Introduction to Planetary Physics; The Terrestrial Planets*. Wiley, N.Y. 490 pp.

Keihm S. J. and Langseth M. G. (1977) Lunar thermal regime to 300 km. *Proc. Lunar Sci. Conf. 8th*, p. 499–514.

Kesson S. E. and Ringwood A. E. (1976) Mare basalt petrogenesis in a dynamic moon. *Earth Planet. Sci. Lett.* **30**, 155–163.

Landau L. and Lifshitz E. M. (1959) *Fluid Mechanics*. Addison-Wesley, Reading, Massachusetts. 536 pp.

Lederer C. M., Hollander J. M., and Perlman I. (1967) *Table of Isotopes*. Wiley, N.Y. 594 pp.

Longhi J. (1978) Pyroxene stability and the composition of the lunar magma ocean. *Proc. Lunar Planet. Sci. Conf. 9th*, p. 285–306.

McConnell R. K. Jr., McClaine L. A., Lee D. W., Aronson J. R., and Allen R. V. (1967) A model for planetary igneous differentiation. *Rev. Geophys.* **5**, p. 121–172.

Minear J. W. (1980) The lunar magma ocean: A transient lunar phenomenon? *Proc. Lunar Planet. Sci. Conf. 11th.* This volume.

Nakamura Y., Lammlein D., Latham G., Ewing M., Dormar J., Press I., and Toksöz N. (1973) New seismic data on the state of the deep lunar interior. *Science* **181**, 49–51.

Nakamura Y., Latham G. V., Dorman H. J., Ibrahim A. B. K., Koyama J., and Horvath P. (1979) Shallow moonquakes: Depth, distribution and implications as to the present state of the lunar interior. *Proc. Lunar Planet. Sci. Conf. 10th,* p. 2299–2309.

Ringwood A. E. and Kesson S. E. (1976) A dynamic model for mare basalt petrogenesis. *Proc. Lunar Sci. Conf. 7th,* p. 1697–1722.

Robie R. A., Hemingway B. S., and Fisher J. R. (1978) Thermodynamic properties of minerals and related substances at 298.15 K and 1 bar (10^5 Pascals) pressure and at higher temperatures. *U.S. Geol. Survey Bull.* **1452,** 456 pp.

Roeder P. L. and Emslie R. F. (1970) Olivine-liquid equilibrium. *Contrib. Mineral. Petrol.* **29,** 275–289.

Solomon S. C. and Chaiken J. (1976) Thermal expansion and thermal stress in the moon and terrestrial planets: Clues to early thermal history. *Proc. Lunar Sci. Conf. 7th,* p. 3229–3243.

Solomon S. C. and Longhi J. (1977) Magma oceanography: 1. Thermal evolution. *Proc. Lunar Sci. Conf. 8th,* p. 583–599.

Sonett C. P. and Duba A. (1975) Lunar temperature and global heat flux from laboratory electrical conductivity and lunar magnetometer data. *Nature* **258**, p. 118–121.

Sonett C. P. and Wiskerchen M. J. (1977) A lunar metal core? *Proc. Lunar Sci. Conf. 8th,* p. 515–535.

Turcotte D. L., Cooke F. A., and Willeman R. J. (1979) Parameterized convection within the moon and the terrestrial planets. *Proc. Lunar Planet. Sci. Conf. 10th,* p. 2375.

Waff H. S. and Weill D. F. (1975) Electrical conductivity of magmatic liquids: Effects of temperature, oxygen fugacity and composition. *Earth Planet. Sci. Lett.* **28**, 254–260.

Warren P. H. and Wasson J. T. (1979) Effects of pressure on the crystallization of a "chondritic" magma ocean and implications for the bulk composition of the moon. *Proc. Lunar Planet. Sci. Conf. 10th,* p. 2051–2083.

Wetherill G. W. (1976) The role of large bodies in the formation of the earth and moon. *Proc. Lunar Sci. Conf. 7th,* p. 3245–3257.

Wood J. A., Dickey J. S. Jr., Marvin U. B., and Powell B. N. (1970) Lunar anorthosites and a geophysical model of the moon. *Proc. Apollo 11 Lunar Sci. Conf.,* p. 965–988.

Proc. Lunar Planet. Sci. Conf. 11th (1980), p. 2031–2041.
Printed in the United States of America

On constraining lunar mantle temperatures from gravity data

Susan Pullan and Kurt Lambeck

Research School of Earth Sciences, Australian National University, Canberra 2600 Australia

Abstract—An estimate of the temperature for the upper lunar mantle is obtained from an inversion of gravity data for density anomalies and the associated stress-state of the moon's interior and the comparison of this stress-state with flow laws and an estimate of likely strain rates. The resulting temperatures are upper limits and at depths of about 300 km, they must be at least 200–300°C less than those proposed by Duba *et al.* (1976) in order to be consistent with the maintenance of lunar gravity anomalies and topography over the past 3–4 billion years.

INTRODUCTION

Estimates of the present-day temperature of the lunar interior are usually based on one or more of the following arguments. (1) An inversion of magnetic data for electrical conductivity profiles (e.g., Dyal *et al.*, 1976) followed by a second inversion for temperature (e.g., Duba *et al.*, 1976; Huebner *et al.*, 1979). Both inversions are non-unique while the latter requires quite crucial assumptions regarding the lunar mantle composition. (2) An interpretation of the two Apollo surface heat flow observations, usually based on a steady state thermal model for the upper mantle (e.g., Keihm and Langseth, 1977). (3) Thermal evolution calculations (e.g., Toksöz *et al.*, 1978), yielding results that are also highly model dependent and that are usually constrained by the observed heat flow data.

Another approach to estimating the lunar temperature is from the present-day stress state of the moon deduced from the lunar gravity and topography observations (Lambeck and Pullan, 1980). The procedure is outlined in Fig. 1 where the principal assumptions associated with each step are also indicated. The observed lunar gravity field indicates the existence of lateral density variations and of a non-hydrostatic stress state. The density structure cannot be determined uniquely from the gravity data alone but the characteristics of the gravitational potential power spectrum point to these anomalies being near the surface, an interpretation that is also in accordance with the correlations seen between gravity and certain topographic features. The second step in the inversion is from density anomalies to stress. An approximate method has been adopted in which

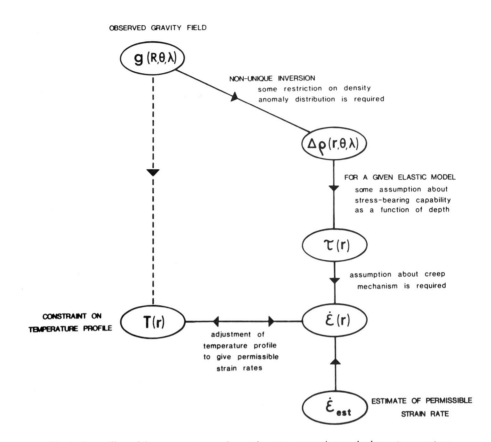

Fig. 1. An outline of the arguments used to arrive at a constraint on the lunar temperature regime from gravity data, with the major assumptions indicated at the appropriate steps.

the moon is modelled as a spherically layered body, initially in a hydrostatic state, and subjected to a surface load deduced from the gravity or from the topography. Seismic data indicate that the outer few hundred km of the moon is fairly homogeneous and characterized by a very high Q. Based on these results and on the evidence that the present day moon does not differ substantially from the moon of 3 AE ago, the stress state is calculated for a number of simple two-layer models with elastic "lithospheres" of varying thicknesses overlying an inner region that is too weak to support any significant stress-differences. The resulting stress profiles are then inverted for strain-rates using flow laws for dry olivine. These flow laws and consequently the strain-rates are strongly temperature dependent and, together with an estimate of the total strain that the moon has experienced, provide a limit to the present day lunar temperature profile. In this paper the same approach is used with the emphasis being on the assumptions that have been made and the consequence of these assumptions on the deduced selenotherm. The main result is that the temperatures in the upper regions of the moon may be considerably lower, by some 200–300°C, than given by the conductivity models of Duba *et al*. (1976), and at or below the lower limits given by Keihm and Langseth (1977).

THE OBSERVATIONAL EVIDENCE

Phillips and Lambeck (1980) discuss the observational evidence for the lunar gravity field. The anomalous gravitational potential at selenocentric distance r, latitude ϕ and longitude λ is expressed as a series of spherical harmonics

$$\Delta U(r,\phi,\lambda) = \frac{GM}{r} \sum_{l=2}^{\infty} \left(\frac{R}{r}\right)^l \sum_{m=0}^{l} (\overline{C}_{lm} \cos m\lambda + \overline{S}_{lm} \sin m\lambda)\overline{P}_{lm}(\sin\phi) \qquad (1)$$

where $\overline{P}_{lm}(\sin\phi)$ are fully normalized Legendre polynomials, $\overline{C}_{lm}$ and $\overline{S}_{lm}$ are the corresponding Stokes co-efficients, R is the lunar radius, M is the mass of the moon and G is the gravitational constant. The dimensionless power spectrum of the gravitational potential is defined by

$$V_l^2(\Delta U) = \sum_m (\overline{C}_{lm}^2 + \overline{S}_{lm}^2). \qquad (2)$$

The relatively slow decay of this power spectrum with increasing degree l is consistent with a model of near surface density anomalies of short wavelength. This does not preclude the existence of lateral density variations at greater depth but these must be relatively small for otherwise the decay of the spectrum would be more rapid.

The anomalous near-surface density layer $\sigma_g(\phi,\lambda)$ can also be expanded in spherical harmonics and its power spectrum is related to the potential spectrum according to

$$V_l^2(\sigma_g) = \left(\frac{2l+1}{3}\right)^2 (\overline{\rho})^2 \, V_l^2(\Delta U) \qquad (3)$$

where $\overline{\rho}$ is the mean density of the moon. The average crustal density is denoted by ρ_c. The surface load due to the observed lunar topography is $\sigma_r(\phi,\lambda) = \rho_c h(\phi,\lambda)$ and has a power spectrum of

$$V_l^2(\sigma_t) = (\rho_c)^2 V_l^2(h) \qquad (4)$$

where $V_l^2(h)$ is the power spectrum of the observed topography. According to Bills (1978), a smoothed estimate of this spectrum is

$$V_l^2(h) = 1.5 \times 10^{-6}/l(l+1). \qquad (5)$$

Comparing the two estimates σ_g and σ_t of the near-surface load leads to $\sigma_t > \sigma_g$, indicating that the topography is at least partly isostatically compensated. In the following discussion the two surface density layers σ_g and σ_t are used as measure of the load stressing the moon.

An approximate method has been adopted for calculating the stress-differences set up in an elastic sphere due to a harmonic surface load σ_g or σ_t. In keeping with the statistical approach, the maximum stress-difference $\tau_{lm}(r)$ is computed for each harmonic and is averaged over a spherical surface of radius r. The average maximum stress-difference due to a number of harmonics is given by

$$\bar{\tau}(r) = \left\{ \sum_l \sum_m \tau_{lm}^2(r) \right\}^{\frac{1}{2}}. \tag{6}$$

These stress-differences reflect the conditions in the body below the load, while the stress-differences near the surface due to the topography will be of the order of $\rho g h$. At any point on the shell of radius r the maximum stress-difference may exceed the value given by Eq. (6) and a reasonable upper limit is obtained by

$$\tau_{max} \approx \sum_l \sum_m \tau_{lm}(r) \tag{7}$$

which may exceed the estimate Eq. (6) by a factor of about four. In Lambeck and Pullan (1980) the stress values that are given are inadvertently referred to as stress-differences whereas they actually are shear-stresses. Hence all stress-difference estimates given in that paper should be increased by a factor of 2.

The stresses associated with each harmonic in the surface load have been estimated from the average distortional strain energy E_s (see, for example, the formulation by Kovach and Anderson, 1967) according to $\tau \approx (8/3)(\mu E_s)^{\frac{1}{2}}$ where μ is the rigidity. By using the methods developed by Kaula (1963) and applied to the moon by Arkani-Hamed (1973) it is possible to carry out a more rigorous inversion of gravity for both density anomalies and deviatoric stresses without going through the surface layer representation but, in view of the evidence that the density anomalies are mainly surficial, this is not warranted for the present purpose.

The magnitude and distribution of the stress-differences beneath the surface load depend on the elastic properties of the body. In constructing the models used for these calculations we have been guided largely by the seismic evidence. Published seismic models all indicate an upper mantle of a few hundred km that is characterized by an extremely high shear Q. Below ~300 km (Nakamura et al., 1976), or perhaps 500 km (Goins et al., 1978), and extending to depths of ~1000 km the Q may be reduced but it is still much higher than for the earth's upper lithosphere. These high Q values suggest that the lunar mantle is volatile-poor and that partial melting is not widespread. The occurrence of moonquakes at depths of ~800 km is further supporting evidence for a relatively cold and thick lunar lithosphere. On this basis, the average stress-differences have been computed for a series of models having elastic lithospheres of 1000 km, 500 km, 400 km and 300 km thicknesses (see Fig. 2). Thus the thickness of the tectonic lithosphere is estimated to be from 30 to 100% of that of the seismic lithosphere. If an analogy with the earth is valid, then 30–50% of the seismic thickness appears as a reasonable choice. The distribution of the stress-differences with depth varies according to the degree of the load—the low degree loads tend to stress the

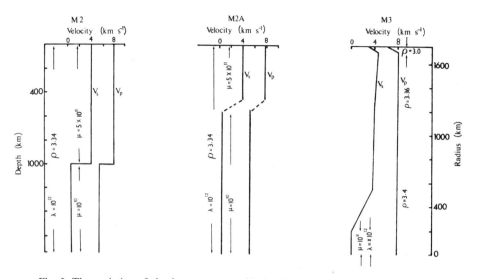

Fig. 2. The variation of elastic parameters with depth for the models considered in this work (units for λ, μ, ρ in cgs). Models M2B and M2C are equivalent to M2A, except that in these models the transition layer is centred at a depth of 400 km and 300 km respectively.

interior while higher degrees of the load stress mainly the layers close to the surface. If the deeper regions of the body are incapable of supporting such stresses, the low degree components of the stress-differences become concentrated in the outer elastic "lithosphere" just above the rigidity interface.

Figure 3 illustrates $\bar{\tau}(r)$ for five different spherical models (Fig. 2) loaded at their surfaces by the density layer σ_g. Only harmonics up to degree 12 have been considered as these loads are the dominant contributors to the stress-differences in the lower regions of the elastic lithosphere. The results in Fig. 3 demonstrate two points: first, that the stress-difference profiles become highly peaked as the stress-bearing lithosphere is thinned, and second, that the magnitude of the stress-differences is only of the order of a few tens of bars with the estimate of an upper limit [τ_{max} as defined in Eq. (7)] approaching perhaps 200 bars. These values are compatible with those obtained by Arkani-Hamed (1973) and also with the lack of evidence of any significant tectonic activity on the lunar surface. For the surface load σ_t, the stress-differences in the lunar interior are approximately twice those illustrated here if there is no isostatic compensation. To obtain an estimate of the contributions of the higher degree harmonics, the topographic spectrum can be used to estimate $\tau \approx \rho gh$

$$\tau_{(l>12)} \approx \rho_c g R \sqrt{\sum_{l=13}^{\infty} V_l^2(h)}.$$

With Eq. (5) this leads to an estimate of $\tau_{(l>12)} \approx 30$ bars, and the higher degree terms ($l > 12$) of the load can be justifiably ignored.

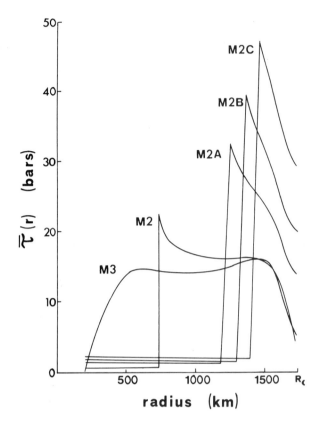

Fig. 3. Stress-differences as a function of depth evaluated according to Eq. 6, for the lunar models described in Fig. 2.

Strain rates in the mantle will in general be stress, temperature and pressure dependent and the following power law is adopted,

$$\dot{\epsilon} = A\tau^n \exp(-(E^* + PV^*)/RT)$$

where $\dot{\epsilon}$ is the strain rate, E^* and V^* are the activation energy and volume, R is the gas constant, P is the pressure and T the temperature. The constants A and n and the parameters E^* and V^* depend on the chemical and mineralogical composition of the material as well as on the mechanisms that dominate the creep process. For a given distribution of temperature with depth this flow law permits the calculated stress-differences given in Fig. 3 to be converted to strain rates in the lunar mantle. For the moon, $PV^* \ll E^*$ and the choice of the activation volume is not critical. However, the choice of E^* and n is critical for, with the stress-differences, temperatures and pressures considered here, a change in E^* of only 10% results in a change in the strain rate of three orders of magnitude. The strain rate profiles have been calculated using experimentally derived flow laws for dry olivine and the validity of the results depends critically on two assumptions: that dry olivine is an appropriate model of the lunar interior, and that the flow laws can be extrapolated from laboratory to geological conditions. Certainly the lunar mantle is not composed exclusively of olivine, but probably also contains pyroxenes, garnet, etc. Goetze (1978) did not consider that such a mixture would significantly effect the creep strength of the material but experimental data on the creep of pyroxenes is sparse. The extrapolation of laboratory determined rheologies to geological problems is unavoidable in many areas of the geological and geophysical sciences if any progress is to be made at all, yet it is fraught with problems. In laboratory experiments loading cycles are short and the measured strain rates are very fast compared to those acting on geologic time scales. Hence any extrapolation to geological problems remains extremely uncertain. However, it can be expected that on geological time scales, creep will become significant at much lower temperatures than in the laboratory (Carter, 1976), and that secondary creep mechanisms not observed on a short time scale may become operative (Paterson, 1976). Thus strain rates calculated from the experimentally derived flow laws could be considered as lower limits. Several flow laws for dry olivine have been published (e.g., Kirby and Raleigh, 1973; Durham and Goetze, 1977; and Post, 1977) and, when applied to the stress-difference profiles given in Fig. 3, give strain rates that vary over three or four orders of magnitude. Only Post's (1977) rheology, which results in the lowest calculated strain rates, will be considered here. The other rheologies will lead to higher strain rates and hence require lower temperatures if the stresses are to be maintained. For the selenotherm based on the inversion of electrical conductivity profiles by Duba *et al.* (1976) (see Fig. 4), Post's rheology and the stress-difference profiles (Fig. 3) for the models given in Fig. 2 lead to strain rates in the elastic lithosphere that are of the order of 10^{-16} sec^{-1} (Fig. 5).

Present strain rates on the moon are not observable, but the general absence of major tectonic activity over the last 3 AE would imply low values. It appears that by late Pre-Imbrian time the crust and upper mantle temperatures were probably low enough for subsequent isostatic readjustments to be small. The following period, Imbrian time, was characterized by the basaltic volcanism which led to the formation of the mass concentrations in the near side circular maria. These mascons do not appear to have reached an isostatic balance although the concentric graben structures and compressional wrinkle ridges point to a mild form of tectonic failure at the surface (see Phillips and Lambeck, 1980 for a further discussion).

A measure of total strain since the time of the last major bombardment of the lunar surface may be deduced from the present lunar topography spectrum and Lambeck and Pullan (1980) estimated an average strain rate of 10^{-19}–10^{-21} sec^{-1}. Solomon and Chaiken (1976) argue that the absence of surface features that can be attributed to planetary changes in volume, leads to the constraint that the lunar radius has changed by no more than 1 km since the emplacement of the oldest mare surfaces 3.8 AE ago. This constraint also points to an average strain rate over the last 3–4 AE of the order of 10^{-20} sec^{-1}. These values would be upper limits to present day strain rates since any major relaxation would have occurred early in this time period.

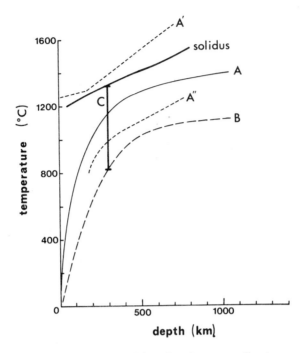

Fig. 4. Estimates of the selenotherm according to
A) Duba *et al.* (1976)
B) this work
C) Keihm and Langseth (1977)

IMPLICATIONS

The strain-rates calculated using Duba *et al.*'s (1976) selenotherm clearly exceed the estimated upper limits of the present day values for all models considered (see Fig. 5) but before drawing any conclusions it may be useful to reiterate some of the consequences of the assumptions made in obtaining these values. First, density anomalies may, and undoubtedly do, occur below the upper mantle and crustal layer to which they have been restricted in the inversion. Hence the stress and strain-rates may be underestimated. Second, the manner in which stresses due to individual spherical harmonics in the load are summed leads to average values for the maximum stress-difference but maximum values may exceed these by a factor of 4 (compare the expressions 6 and 7). Third, if an uncompensated topography is taken to be the load the stress differences may be further underestimated by a factor of about 2. Since strain-rate is proportional to (stress-difference)^3 these three points indicate that the results in Fig. 5 could be underestimated by as much as three orders of magnitude. Fourth, the extrapolation of the flow laws to geological conditions may lead to an underestimation of the calculated strain-rates while the "observed" strain-rates represent upper limits.

Taken together, the discrepancy between the computed and "observed" strain-rates may be even greater than indicated in Fig. 5 and this implies that the stress-differences should have relaxed and that much of the topography and gravity anomalies should have vanished during the last 4 billion years. One way to avoid this is to lower the selenotherm. No attempt has been made here to carry out a formal inversion for temperature, but we have introduced a rather *ad hoc* upper limit for lunar mantle temperatures (Fig. 4) on the basis that it leads to "acceptable" strain rates. The essential point is that the temperature near 300 and 400 km depth must be at least 200–300°C less than that proposed by Duba *et al.* (1976)—that is, at or below the lower limit suggested by Keihm and Langseth (1977) and similar to the present day temperatures proposed by Toksöz *et*

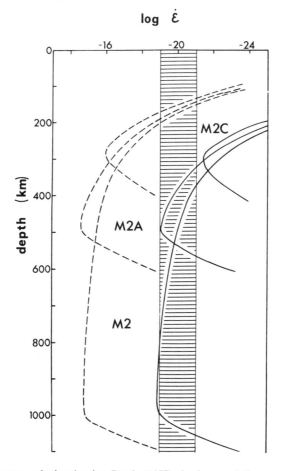

Fig. 5. Strain rates calculated using Post's (1977) rheology and the stress-difference profiles of Fig. 3—for the nominal temperature profile B (solid lines) and Duba *et al.*'s (1976) selenotherm (dashed lines).

al. (1978). This conclusion appears unavoidable unless the extrapolation of the laboratory-determined flow laws to lunar conditions is entirely inappropriate.

The above arguments for a relatively cool moon are for the present-day, although it is implied that the moon has not undergone any significant thermal evolution for the last 3 AE. This is contradictory to the thermal evolution model proposed by Toksöz et al. (1978) in which the lithosphere was less than 100 km thick 4 AE ago and would have increased to only a little more than 100 km by 3.2 AE ago. The high selenotherms proposed by Toksöz et al. for the period of mare formation require that the relatively thin and warm lithosphere can support considerably greater stress-differences than those estimated to exist in the moon now, for not only will there have been some relaxation of stress, but the stress-differences would have been concentrated in a much thinner layer. The present calculations are not very helpful in the case of a thin lithosphere since deformation will be controlled by the brittle strength of the crustal material rather than by the creep strength. Solomon and Head (1979) used a model of a thin spherical elastic shell overlying a fluid interior to compute the stresses in the lithosphere due to mascon loading, and obtained stress-differences of the order of 0.5–1 kbar. These are, however, lower limits for two reasons. They consider the present load, not that at the time of mascon formation, and they have not considered the low degree harmonics in the load, though it is these that most stress the shell (see Lambeck and Pullan, 1980). Kuckes (1977) has considered he overall non-equilibrium figure of the moon (i.e., the second degree harmonics), and estimated that these alone would produce stresses of the order of 400 bars in a 100 km thick elastic shell.

Arkani-Hamed (1974) concluded that the thickness of the lunar lithosphere had to be about 400 km at the time of the formation of the Serenitatis mascon to support the associated stress-differences, which he estimates are presently around 70 bars; a conclusion that is in accord with the present study. More recently Delano et al. (1980) have argued that the mare basalts originated at depths of 400 to 500 km, a further indication that in the interval 3.9–3.2 AE ago the lithosphere may have been thicker by a factor of about 4 than assumed by Toksöz et al., and that the subsequent thermal evolution of the upper part of the lunar mantle has been less important than suggested by their model.

REFERENCES

Arkani-Hamed J. (1973) Density and stress distribution in the moon. The Moon 7, 84–126.

Arkani-Hamed J. (1974) Stress constraint on the thermal evolution of the moon. Proc. Lunar Sci. Conf. 5th, p. 3127–3134.

Bills B. G. (1978) A harmonic and statical analysis of the topography of the Earth, Moon and Mars. Thesis, California Institute of Technology, Pasadena. 263 pp.

Carter N. L. (1976) Steady state flow of rocks. Rev. Geophys. Space Phys. 14, 301–360.

Delano J. W., Taylor S. R., and Ringwood A. E. (1980) Composition and structure of the deep lunar interior (abstract). In Lunar and Planetary Science X, p. 225–227. Lunar and Planetary Institute, Houston.

Duba A., Heard H. C., and Schock R. N. (1976) Electrical conductivity of orthopyroxene to 1400°C and the resulting selenotherm. *Proc. Lunar Sci. Conf. 7th*, p. 3173–3181.

Durham W. B. and Goetze C. (1977) Plastic flow of oriented single crystals of olivine. I. Mechanical data. *J. Geophys. Res.* **82**, 5737–5753.

Dyal P., Parkin C. W., and Daily W. D. (1976) Structure of the lunar interior from magnetic field measurements. *Proc. Lunar Sci. Conf. 7th*, p. 3077–3095.

Goetze C. (1978) The mechanisms of creep in olivine. *Phil. Trans. Roy. Soc. London* **A288**, 99–119.

Goins N. R., Toksöz M. N., and Dainty A. M. (1978) Seismic structure of the lunar mantle: An overview. *Proc. Lunar Planet. Sci. Conf. 9th*, p. 3575–3588.

Huebner J. S., Duba A., and Wiggins L. B. (1979) Electrical conductivity of pyroxene which contains trivalent cations: Laboratory measurements and the lunar temperature profile. *J. Geophys. Res.* **84**, 4652–4656.

Kaula W. M. (1973) Elastic models of the mantle corresponding to variations in the external gravity field. *J. Geophys. Res.* **68**, 4967–4978.

Keihm S. I. and Langseth M. G. (1977) Lunar thermal regime to 300 km. *Proc. Lunar Sci. Conf. 8th*, p. 499–514.

Kirby S. H. and Raleigh C. B. (1973) Mechanisms of high-temperature, solid state flow in minerals and ceramics and their bearing on the creep behaviour of the mantle. *Tectonophys.* **19**, 165–194.

Kovach R. L. and Anderson D. L. (1967) Study of the energy of the free oscillation of the earth. *J. Geophys. Res.* **72**, 2155–2168.

Kuckes A. F. (1977) Strength and rigidity of the elastic lunar lithosphere and implications for present-day mantle convection in the moon. *Phys. Earth Planet. Inter.* **14**, 1–12.

Lambeck K. and Pullan S. (1980) Inferences on the lunar temperature from gravity, stress state and flow laws. *Phys. Earth Planet. Inter.* **22**, 12–28.

Nakamura Y., Latham G. V., Dorman H. J., and Duennebier F. K. (1976) Seismic structure of the moon: A summary of current status. *Proc. Lunar Sci. Conf. 7th*, p. 3113–3121.

Paterson M. S. (1976) Some current aspects of experimental rock deformation. *Phil. Trans. Roy. Soc. London* **A283**, 163–172.

Phillips R. J. and Lambeck K. (1980) Gravity fields of the terrestrial planets: Long wavelength anomalies and tectonics. *Rev. Geophys. Space Phys.* **18**, 27–77.

Post R. L. (1977) High-temperature creep of Mt. Burnet dunite. *Tectonophys.* **42**, 75–110.

Solomon S. C. and Chaiken J. (1976) Thermal expansion and thermal stress in the moon and terrestrial planets: Clues to early thermal history. *Proc. Lunar Sci. Conf. 7th*, p. 3229–3243.

Solomon S. C. and Head J. W. (1979) Vertical movement in mare basins: Relation to mare implacement, basin tectonics, and lunar thermal history. *J. Geophys. Res.* **84**, 1667–1682.

Toksöz M. N., Hsui A. T., and Johnston D. H. (1978) Thermal evolutions of the terrestrial planets. *Moon and Planets* **18**, 281–320.

Proc. Lunar Planet. Sci. Conf. 11th (1980), p. 2043–2058.
Printed in the United States of America

The bulk composition of the moon based on geophysical constraints

W. Roger Buck and M. Nafi Toksöz

Department of Earth and Planetary Sciences, Massachusetts Institute of Technology, Cambridge,
Massachusetts 02139

Abstract—The bulk composition of the moon is estimated using geophysical constraints within a framework of evolution consistent with petrology, geochemistry and geophysics. The constraints used are: 1) the upper and lower mantle seismic velocities; 2) nearside crustal thickness and seismic velocities; 3) center-of-figure to center-of-mass offset; 4) mean density; 5) moment-of-inertia. The crust is taken to have been derived from only the outer four hundred kilometers of a homogeneously accreted moon. The deep interior is taken to have never melted completely. Assumed oxide compositions are converted to normative mineral assemblages via pressure dependent normative schemes. Densities are extrapolated to pressures and assumed temperatures for the interior using the Birch-Murnaghan equation of state. Seismic velocities are extrapolated using temperature and pressure derivatives taken from work on single crystals and aggregates. Compositions which match all constraints lead us to the following conclusions: 1) the Mg/Si ratio must be much lower than for the upper mantle of the earth or for the CI chondritic meteorites; 2) Al content and Mg/Si ratio are linked but the best fits are for compositions with less than five weight percent Al_2O_3; 3) cores are required ranging from one to two weight percent of the bulk moon; 4) the lower mantle seismic velocity drop is consistent with several weight percent Fe-FeS in this region.

INTRODUCTION

It has long been known that the bulk composition of the moon is different from that of the earth simply on the basis of the mean density of the body. This can be accounted for by a lower iron content for the moon and the moon having either no metallic core or a very small one. Seismic results from the Apollo program indicated that the crust of the moon is much thicker than that of the earth and this seems to imply either a compositional difference or a different path of evolution for the bodies. Although we do not have as much geophysical information on the moon as for the earth, we have one great advantage in using that data to infer composition and that is the small size of the moon. The maximum central pressure in the moon is about 40 kilobars while the pressures in the lower mantle of the earth exceed 1600 kilobars. Therefore, laboratory data on the physical properties of minerals at high pressure can be applied with confidence to test assumed compositions against the geophysical constraints.

Many models for the bulk composition of the moon have been advanced on the basis of petrological, geophysical and cosmochemical constraints (Taylor and

Jakeš, 1974; Ganapathy and Anders, 1974; Taylor and Bence, 1975; Ringwood and Kesson, 1977; Wänke et al., 1977; Morgan et al., 1978). These models have varied widely in abundances assigned to several important constituents. For example, the Mg/Si weight ratios have varied slightly from 0.86 for Morgan et al. (1978) to 0.96 for Ringwood and Kesson (1977), but the Al_2O_3 abundances have varied much more widely from 3.7 weight percent for Ringwood and Kesson (1977) to 11.5 for Ganapathy and Anders (1974). Revision of estimates of geophysical parameters such as moment-of-inertia and seismic velocity structure, in recent years, constrain acceptable models to narrower bounds than were possible before. To test a broad range of models against the geophysical constraints, we converted the chemical abundances suggested for the bulk moon into mineralogical abundances for a layered moon and then found the resulting seismic and density profiles. The conversion to mineral abundances with depth rests on assumptions of (1) the composition and thickness of the crust, (2) the differentiation history of the moon, and (3) a normative scheme for converting oxide abundances to normative mineral abundances dependent on the temperature and pressure range. We will discuss the knowledge we have of the crust and mantle which justify the assumptions we use, then we will outline the modelling procedure used. The results are given in terms of ranges of acceptable compositions and a model which produces the best fit to the observables is presented.

THE CRUST

Petrologic and geochemical information about the highlands combined with seismic information about the crust can put narrow bounds on the average composition of the crust. The average composition of the lunar surface material has been determined from study of returned samples and spectroscopic measurements done from orbit (Taylor and Bence, 1975). The composition determined for the highlands is anorthositic gabbro which contains about 26 wt. % Al_2O_3. The mare basalts, though covering a significant part of the surface, comprise an estimated 1% of the volume of the crust (Head, 1979). Therefore the highlands are considered to be representative of the outer portion of the crust to an uncertain depth. The seismic velocity profile for the crust matches the profile resulting from a crust completely made from anorthositic gabbro, but could also be satisfied by a composition lower in Al_2O_3 (Toksöz et al., 1972; Liebermann and Ringwood, 1976). There is a velocity discontinuity at about 20 km depth which marks either a compositional change (Wood, 1975) or the depth of crack closure (Todd et al., 1973). The former view is now the prevalent one (Simmons et al., 1980) and we will consider the implications of different average crustal compositions.

Determinations of mean crustal thickness depend on the seismic estimates of nearside thickness and on either measurements of gravity and topographic variation from orbit or the offset of center-of-figure to center-of-mass. The thickness at the nearside mare sites of Apollo 12 and 14 have been estimated to be ~55 km by Toksöz et al. (1972). There has been discussion of the possibility of this determination being too high by about 10 km (Koyama and Nakamura, 1979), but Goins (1978) points out that this estimate is extremely well determined. He also notes that the highland site of Apollo 17 has a crustal thickness of 75 km though this determination is less well constrained.

One method for estimating the mean thickness is to use the detailed gravitational and topographic data and the seismically determined thicknesses with assumptions of the crustal and mantle density to model variations in crustal thickness. This has been done by Bills and Ferrari (1977) who show that a mean crustal thickness of 70 km is consistent with the data and an assumed composition of anorthosite. A cruder way to estimate the mean thickness is to use the center-of-figure to center-of-mass offset for the moon, which is 2.55 km (Kaula *et al.*, 1974), and the nearside thickness estimates. This is useful to show the effect of changes in the assumed crustal density on the thickness calculation. We relate the assumed crustal density $\rho 1$ to the difference between far and nearside thickness ΔT through the center-of-mass to center-of-figure offset ΔC by:

$$\Delta T = \frac{2\Delta C \ [R_2^3(\rho_2-\rho_1) + R_1^3\rho_1]}{R_2^3(\rho_2-\rho_1)}$$

where ρ_2 is the average density beneath the crust and where R_1 is the radius of the moon. R_2 is the radius of the subcrustal sphere which $= R_1-(Tn+T/2)$, where Tn is the thickness of the nearside crust. We do this calculation with densities of different materials which satisfy the seismic velocity profile. Table 1 shows the mean crustal thicknesses (T) thus determined with a 55 km assumed nearside thickness. The variations of Al_2O_3 abundance in these cases is also calculated. The crust will have a greater average thickness for the higher density and lower Al_2O_3 content cases. Therefore, the total Al_2O_3 content of the crust does not vary much over the seismically acceptable range of compositions.

THE MANTLE

Our knowledge of the mantle of the moon is derived from seismic data, petrologic work on the returned samples and modelling using other geophysical constraints such as is done in the present work. The seismically determined structure of the interior is shown in Fig. 1. We will use it to discuss the different regions of the interior and discuss its derivation from seismic data along with other implications of that data. Then the inferences from the petrologic and geochemical work on the returned samples will be briefly reviewed. The results from modelling will be covered in a later section of this paper.

The lunar seismic data indicate one velocity discontinuity in the mantle (Nakamura *et al.*, 1976; Dainty *et al.*, 1976; Goins *et al.*, 1978). The different workers found different depths for the discontinuity and slightly different velocity profiles. We use the most recently published profile which is that of Goins *et al.* (1978) shown in Fig. 2. Also shown on the figure are extrapolations of seismic velocities for olivines and pyroxenes of differing Mg/Mg+Fe mole ratios using the data

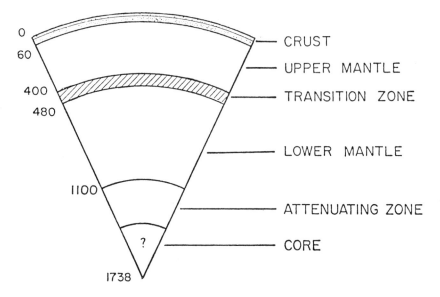

Fig. 1. The seismically determined structure of the moon. From Goins *et al.*, 1978.

shown in Table 1. We see that the velocity drops between 400 and 500 km imply several possibilities. Suggestions which have been made are a compositional change, a phase change or the onset of a small degree of partial melting (Goins, 1978). There seems to be too small a change in shear wave attenuation between the upper and lower mantles to have partial melting produce the velocity drop. In the modelling for this paper we have addressed the question of phase changes and we would expect an increase, not a decrease, in velocity at this level independent of the assumptions on differentiation of the interior (see Fig. 3). The major suggestion for a compositional change is that the iron content of the silicates increases at this level, i.e., the Mg/Mg+Fe ratio goes down (Dainty *et al.*, 1976). We suggest that another way to accomplish this is for there to be an increase in the content of free reduced iron and iron sulfide (Fe−FeS).

As noted before the crust is covered by a volumetrically small amount of basalts. These rocks are very different in age and composition from the highland material. Radiometric dating has shown that the highland material crystallized soon after the presumed date of origin of the moon and that the basalts are younger and crystallized over a wider range in time (Papanastassiou and Wasserburg, 1971). Another important difference between these two categories is their patterns of rare earth element abundances (REE). Not only do different lunar rock types show different overall abundances of REE, but mare basalts are depleted to differing degrees in europium (Eu) relative to other REE, while highland rocks are enriched in this element (Taylor, 1976). Several hypotheses have been advanced to account for this, but the fact that Eu in the reduced (divalent) state is readily accommodated into the feldspar lattice has led to the prevalent view that early large scale differentiation of the crust depleted part of the interior in Eu (Philpotts and Schnetzler, 1970). The negative Eu anomaly in some basalts indicates association with this depleted area.

Work on basalt petrogenesis indicates that the upper mantle is a source region

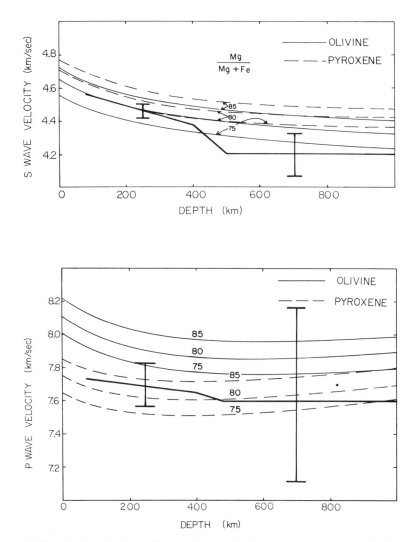

Fig. 2. The seismic velocity profiles for the mantle of the moon are shown as the bold lines and are from Goins *et al.*, 1978. The brackets indicate the maximum possible variations in velocity for a given depth and are not error bars. The average velocity values are well determined. The curved lines are extrapolations of STP velocity values for olivines and pyroxenes of the Mg/Mg + Fe molar ratio shown. The data in Table 2 were used to make the extrapolations.

Table 1. Crustal thicknesses, densities and Al_2O_3 contents consistent with the lunar density and center-of-mass to center-of-figure offset.

Model crustal composition	1	2	3	4[a]
Assumed crustal density	2.90	2.95	3.00	3.05
Mean subcrustal density	3.40	3.39	3.38	3.37
Mean crustal thickness (km)	74	76	79	84
Al_2O_3 content of model crusts in weight percent	29.0	26.5	22.7	20.0
Al_2O_3 content of crust expressed as weight percent of moon[b]	2.84	2.76	2.50	2.38

[a] 1 = material more aluminous than anorthositic gabbro
2 = anorthositic gabbro
3 = troctolite
4 = material less aluminous than troctolite

[b] Defined as $\dfrac{(\text{mass of } Al_2O_3 \text{ in crust})}{\text{mass of moon}} \times 100$

of mare basalts (Kesson, 1975; Walker *et al.*, 1975). Minimum depths for the source regions extend down to 400 km (Ringwood, 1977) and it has been suggested that the depth range from 400 to 500 km may also be a major source area (Delano *et al.*, 1980). The Mg/Mg+Fe ratio of the source regions indicates a range from 0.75 to 0.80 (Ringwood and Essene, 1970; Morgan *et al.*,(1978) which matches the seismically determined range. The basalt types indicate origin at different depths and that there are areas of different composition. Resolving the composition and depth of the source regions is complicated since the basalts show evidence of not being derived directly from partial melting of the source region but also show effects of fractional crystallization and contamination en route to the surface (Ringwood, 1977). The data do suggest that there may be an increase in olivine content with depth in the upper mantle and that incompatible elements are concentrated in the top part of the upper mantle (Taylor, 1978).

The hypothesis that the moon was molten to a depth of 400 to 500 km near the time of its origin and that the crust differentiated from this material can explain many of the observed seismic and petrologic features (Taylor, 1978; Longhi, 1978). This region has been termed the "magma ocean" after the work of Wood *et al.* (1970) who realized that a plagioclase rich crust would have to have been differentiated from a large volume of molten material. Much work has been done to model the mineralogy with depth resulting from crystallization of magma oceans of different depths and compositions (Wood, 1975; Minear and Fletcher, 1978; Longhi, 1978). Common features of the models are that Mg-rich olivine would be the first phase to crystallize, settling to the bottom of the magma ocean, and that incompatible elements would be concentrated in the last region to crystallize which was near the top of the upper mantle. The Eu anomaly pattern between the highlands and the mare basalts is also qualitatively reproduced by these models (Taylor, 1978). It has been suggested that the decrease in seismic velocities at the 400 to 500 km depth is what would be expected on going from a Mg enriched olivine layer to primitive undifferentiated mantle material below (Taylor, 1978). We will consider this possibility.

MODELLING SCHEME

Now we turn to the assumptions made in testing compositions against the constraints. We will present the assumptions used in doing the modelling and the

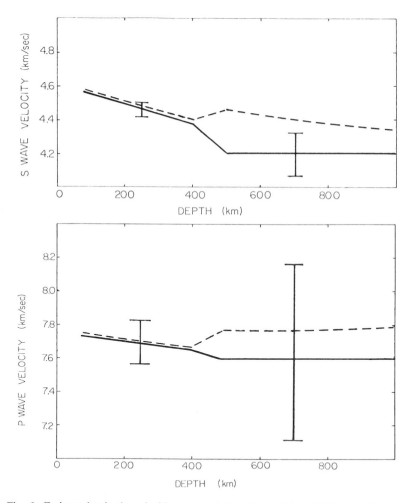

Fig. 3. Estimated seismic velocities at depth for the model of Table 3 without any increased iron content in the lower mantle. Two weight percent Fe-FeS in the lower mantle will produce a match to observed velocities, including a larger drop in S than P velocities.

estimated uncertainties of the geophysical data. In the next section we will discuss the dependence of the results on these assumptions. We are only modelling the major elements, specifically the oxides SiO_2, Al_2O_3, MgO and FeO. The more minor oxides (TiO_2, Cr_2O_3, Na_2O) have little effect on the output of the modelling and are taken from the model of Morgan *et al.* (1978). The CaO abundance is assumed to be linked to the abundance of Al_2O_3 as described below.

The first step is to assume the moon accreted homogeneously. This is justified on the basis of the difficulties heterogeneous accretion has in explaining chemical features of the returned samples (Taylor, 1973) and on the grounds that it is the only way to consistently compare different compositions. We then must remove a crust from some portion of the moon. The most important feature about the crust we remove is the total Al_2O_3 content of it and we have shown (Table 1) that this is relatively insensitive to the composition assumed. We picked troctolite as our average composition, and so took it to be 75 km thick.

The portion of the interior we take to have been involved in the differentiation to form the crust, the magma ocean depth, is difficult to estimate. Thermal modelling considering thermal stress has yielded estimates of the initial depth of moonwide melting between 200 and 400 km (Solomon, 1977). The area immediately below a solidified magma ocean of depth 400 km has been suggested as a region of melting of undifferentiated material to explain features of some mare basalts (Delano *et al.*, 1980). This has also been shown to be qualitatively consistent with thermal modelling (Ahern and Turcotte, 1980). Finally we choose the 400 km depth since it marks the only mantle seismic discontinuity.

Once the oxide abundances of the crust have been removed from the upper mantle we convert the upper and lower mantle compositions into normative mineral assemblages. This is done using the CAMS projection scheme of O'Hara (1968) applied to the different pressure regions (see Stolper, 1980) to predict stable phases. The norm calculation also takes into account substitution of Al_2O_3 into the pyroxene structure via data from Akella (1976), but this has neglible effect on the densities of the assemblages.

The seismic velocities at depth for the mineral assemblages are determined using simple averaging of velocity values shown in Table 2 and extrapolated to the pressures and assumed temperatures of the interior using the derivatives shown in the same table. The temperatures are taken from the work of Toksöz *et al.* (1978), but different temperature distributions have a minor effect on the extrapolated values. Figure 3 shows the result of such calculations for a composition which matches all constraints except seismic. The velocities in the lower mantle are higher not lower than those of the upper mantle because of the onset of the garnet stability field (see Green and Ringwood, 1967). This result is qualitatively independent of the magma ocean depth. Increasing the Al_2O_3 content of a model composition will cause this increase to be greater. We are forced to invoke one of the methods previously mentioned to lower the velocities. We choose to add Fe-FeS until we match the lower mantle velocities. This is consistent with data that show that some mare basalts are saturated with both iron and sulfur (Gibson *et al.*, 1976) and because sulfur is a cosmically abundant

Table 2. Seismic velocities and temperature derivatives of velocity used to calculate seismic velocities of normative minerals at depth in the moon.

	Seismic Velocities (km/sec)			
Mg/Mg+Fe	Olivine [e]		Pyroxene [j]	
mole %	V_p	V_s	V_p	V_s
90	8.30	4.80	7.96	4.82
85	8.20	4.72	7.86	4.76
80	8.10	4.64	7.77	4.70
75	7.99	4.55	7.67	4.65
	Garnet		*Spinel*	
60	8.90[i]	5.0[i]		
75			9.26[h]	5.04[h]

	Velocity Derivatives			
	$\partial V_p/\partial P$[a]	$\partial V_p/\partial T$[b]	$\partial V_s/\partial P$[a]	$\partial V_s/\partial T$[b]
Olivine	10.8[c]	4.84[d]	3.60[e]	3.40[d]
Pyroxene	13.0[c]	4.15[f]	4.33[f]	3.39[g]
Garnet	7.84[i]	3.93[i]	2.17[i]	2.15[i]
Spinel	−0.47[h]	5.4[f]	5.18[h]	3.11[f]

[a] $\times 10^{-3}$ km/sec kilobar
[b] $\times 10^{-4}$ km/sec °C
[c] Christensen (1974)
[d] Simmons and Wang (1971)
[e] Kumazawa and Anderson (1969)
[f] Birch (1969)
[g] Birch (1943)
[h] Wang and Simmons (1972)
[i] Wang and Simmons (1974)
[j] Kumazawa (1969)

element (Cameron, 1978). The model shown in Fig. 3 requires two weight percent to accomplish this. It is worth noting that the larger drop in shear velocity relative to P-wave velocity results since the P-velocity increases more on going into the garnet pressure region.

To test compositions against the constraints of mean density and moment-of-inertia, we estimate the densities at depth. Producing density profiles from the mineralogical models first requires the estimation of the effect of temperature. The thermal expansion coefficient and its first and second derivatives with temperature are calculated for each mineral assemblage by simple averaging of the coefficients for the individual minerals. The density decrease caused by the internal tempratures is calculated using these parameters and the temperature profile. The isothermal bulk modulus of mineral assemblages is calculated in the same way and adjusted to the temperature of the interior. The error in doing these extrapolations is very small because experimental data exist for the pressure and temperature range we are considering. The major source of error is due to some uncertainty in the temperature profile we used. This could produce a maximum error of 0.5 percent in the density at depth.

The pressure effect on density is calculated using the second order Birch-Murnaghan equation of state (t = temperature in °C).

$$\rho = \frac{3\ K_0\ (t)}{2} \left[\left(\frac{\rho}{\rho_0(t)} \right)^{7/3} - \left(\frac{\rho}{\rho_0(t)} \right)^{5/3} \right]$$

where ρ = density at a given pressure ρ for material with a temperature corrected zero pressure density $\rho_0(t)$ and bulk modulus, $K(t)$. Pressure is calculated by numerically integrating the equation of hydrostatic equilibrium:

$$\rho(z) = \int_{R-Z}^{R} \rho(r)\ g(r)\ dr \qquad \begin{array}{l} R = radius \\ Z = depth \end{array}$$

The density distribution thus calculated was used to calculate the mean density and moment of inertia of the model moons.

The geophysical constraints are known with differing levels of confidence. The best known is the mean density which is 3.344 ± 0.003 (Bills and Ferrari, 1977). We require all models to match this value exactly. The seismic velocities, especially for the upper mantle, are well known. Our major uncertainty in using the seismic data is the uncertainty in mineral seismic velocity values from laboratory measurements. We do not have data on the minor mineral phases such as the aluminous pyroxenes, and this combined with the uncertainties in the values for the major phases and the temperature profile could lead to errors of up to 0.05 km/sec for velocity at depth. The least well determined constraint is the moment-of-inertia factor (I/MR^2). The accepted value for the factor has changed markedly in the last ten years, since more data have been analyzed, from over 0.400 (Kaula, 1970) to 0.395 (Kaula et al., 1974) to the presently accepted value of 0.391 ± 0.002 (Blackshear and Gapcynski, 1977). We therefore put more weight on matching the seismic velocity values than I/MR^2. Models are required to be within the uncertainty range for the moment-of-inertia. Cores are added to accomplish this. The core material is taken to be Fe-FeS of STP density ~6.0 g/cm^3. The origin of the core is taken to be from the molten material which would fall to the bottom of the magma ocean. This would have existed as a molten layer above the undifferentiated interior until the lower mantle heated sufficiently for the commencement of solid state convection at which time the dense molten material would work its way in to form the core.

What we have outlined is an inverse problem. Our constraints, in order of weights, are: 1) mean density 2) upper mantle seismic velocities 3) lower mantle velocities and 4) moment-of-inertia factor. The model features which are important to matching these constraints are: 1) Al_2O_3 content 2) Mg/Si ratio 3) Mg/Mg+Fe ratio 4) Fe-FeS amount in lower mantle and 5) core size. Our system is underdetermined so we take out Al_2O_3 content and fix it at different values and vary the other parameters. Al_2O_3 contents below 4.0 weight percent are not considered since there is insufficient material then available to make the crust. The calcium content is fixed along with the Al_2O_3 content using the Ca/Al weight ratio of 1.10 which is the average value for all chondrite classes and eucrites and howardites determined by Ahrens and von Michaelis (1969), who have pointed out the remarkable constancy of this ratio for these meteorite classes.

RESULTS

The first and most important of our results is that the Mg/Si ratio for the bulk moon must be less than that of the C1 chondrites and the values assumed in many compositional models (Fig. 4). This result is completely independent of the assumed path of differentiation for the interior. The result is also independent to large measure of the weights assigned to the geophysical parameters. Models which did not incorporate the seismic velocities but only matched the mean density and moment-of-inertia have lowerMg/Si ratios even if we take the upper limit of I/MR^2 (.393). The result that cores are required is likewise found for the whole possible range for I/MR^2. Finally, using only the seismic and mean density constraints also requires Mg/Si to be less than 0.91.

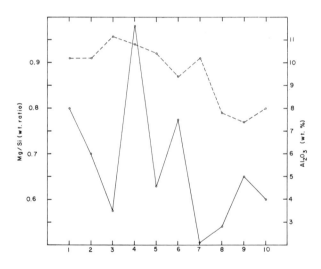

Fig. 4. Mg/Si weight ratios and Al_2O_3 contents for the silicate portions of various models of the moon and for two classes of meteorites. The numbers on the abscissa are:
1) Taylor and Jakeš (1974)
2) Taylor and Bence (1975)
3) Ringwood and Kesson (1977)
4) Ganapathy and Anders (1974)
5) Wänke et al. (1977)
6) Morgan et al. (1978)
7) Cl chondrites with iron and sulfur removed; data from Mason (1971)
8) Hyperstene chondrites, taken to be representative of ordinary chondrites (Mason, 1971)
9) Model matching geophysical data with 5.0 weight percent Al_2O_3 (see text)
10) Model matching geophysical data with 4.0 weight percent Al_2O_3 (see text)
Note: Dashed line is Mg/Si ratio and solid line is Al_2O_3 content.

In using all the constraints as described in the last section we find that the best fits to the seismic velocities can only be achieved for Al_2O_3 contents near the lower limit of 4.0 weight percent. The higher Al_2O_3 contents also necessitate extremely low Mg/Si ratios and the seismic velocities cannot be as well satisfied with such ratios even though the Mg/Mg + Fe is adjusted to make for the best fit possible. This effect can be understood simply from the fact that aluminum will cause more high density garnet and spinel in the regions below 180 km (>8 kilobars pressure). These minerals have higher seismic velocities than olivines and pyroxenes and so more Fe-FeS will be added to the lower mantle to match the seismic velocities. The only way we have to make large adjustments in the density is to lower the Mg/Si ratio. Al_2O_3 contents of from 4.0 to 5.0 weight percent give the best fit to all constraints. We choose 5.0 as our preferred value since this allows for some inefficiency in the crustal differentiation process. The Mg/Si ratio for this case is 0.77 and the Mg/Mg+Fe molar ratio is 0.79. A core of radius of 300 km is also required to bring the moment-of-inertia factor into the range of its uncertainty. For this model we take the upper limit of the range and $I/MR^2 = 0.393$. To match the central value of $I/MR^2 = 0.391$ would require a larger core and lower Mg/Si ratio. The oxide abundances of our preferred model are given in Table 3 and the mineralogy resulting from our normative calculation is also given there. The density profile for this model is shown in Fig. 5. Those who have more confidence in the determination of I/MR^2 will prefer the model with 4.0 weight percent Al_2O_3 which matches seismic velocities with $I/MR^2 = 0.392$ and Mg/Si = 0.80 and Mg/Mg+Fe = 0.78. Our alternative method for

Table 3. Oxide abundances and normative mineral assemblages for a model which matches all geophysical constraints (Weight per cents).†

			a.	b.	c.
SiO_2	48.37	Jadeite	0.00	0.85	0.98
TiO_2	0.40	Plagioclase	1.31	0.00	0.00
Al_2O_3	5.00	Ca-Tschermak	0.00	0.00	1.06
Cr_2O_3	0.30	Diopside	5.15	5.30	12.49
MgO	29.02	Mg-Tschermak	0.00	0.00	3.51
FeO	12.90	Hypersthene	54.53	56.74	44.43
CaO	3.83	Olivine	37.80	35.85	25.32
Na_2O	0.15	Pyrope-Almandine	0.00	0.00	9.24
Total	99.97	Grossular	0.00	0.00	1.10
		Ilmenite	0.76	0.76	0.76
		Chromite	0.52	0.52	0.44
		Total	100.07	100.02	99.34
		STP Density (g/cm³)	3.38	3.39	3.42

†. To this we must add 2 weight percent Fe-FeS of STP density 6.0 g/cm³

a. Normative assemblage for the plagioclase stability field of the upper mantle.

b. Normative assemblage for the upper mantle below the stability field of plagioclase. See text for details of the removal of the crust from the upper mantle.

c. Normative assemblage for the lower mantle.

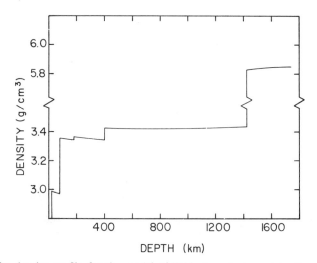

Fig. 5. The density profile for the model of Table 3 made to match all geophysical constraints.

producing the seismic velocity drop for the lower mantle (reduction to liberate iron in the magma ocean) gives virtually the same results except that the FeO content is increased and the free Fe-FeS content is reduced. These models have slightly smaller and denser cores and denser lower mantles.

CONCLUSION

The Mg/Si ratio for models which best fit the geophysical data are much closer to the values of ordinary chondrites than for the Cl chondrites. Al_2O_3 contents are linked to the Mg/Si ratio but only the lower range of aluminum contents are acceptable. Cores are required but amount to only 1–2 weight percent of the moon. The simplest model of lower mantle velocity drop involves accretion of about 2 weight percent Fe-FeS with the silicates.

The bounds we put on the bulk composition of the moon bear on the question of the origin of the moon but certainly do not resolve it. Kaula (1977) gives a good review of the possible modes of origin. Our bounds on the Mg/Mg + Fe ratio can be used to estimate the condensation temperature of that material. Our bounds on the bulk composition of the moon are outside those determined for the upper mantle of the earth (Ringwood, 1975). We wish to stress that the bounds on the bulk composition of the entire mantle of the earth (Anderson *et al.,* 1971; Ahrens, 1980) are not outside the bounds we have found for the moon.

Acknowledgments—We would like to thank Sean Solomon for many helpful discussions and David Johnston for assistance with the computer programs and procedures used. This research was supported by NASA Grant NSG-7081.

REFERENCES

Ahern J. L. and Turcotte D. L. (1980) The effects of magma migration on the evolution of the moon (abstract). In *Lunar and Planetary Science XI*, p. 4–5. Lunar and Planetary Institute, Houston.

Ahrens L. H. and Von Michaelis H. (1969) The composition of stony meteorites III. Some inter-element relationships. *Earth Planet. Sci. Lett.* **5**, 395–400.

Ahrens T. J. (1980) Dynamic compression of earth materials. *Science* **207**, 1035–1041.

Akella J. (1976) Garnet pyroxene equilibria in the system $CaSiO_3$-$MgSiO_3$-Al_2O_3 and in a natural mineral mixture. *Amer. Mineral.* **61**, 589–598.

Anderson D. L., Sammis C. G., and Jordan T. H. (1971) Composition and evolution of the mantle and core. *Science* **171**, 1103–1112.

Bills B. G. and Ferrari A. J. (1977) A lunar density model consistent with gravitational, librational, and seismic data. *J. Geophys. Res.* **82**, 1306–1314.

Birch F. (1943) Elasticity of igneous rocks at high temperatures and pressures. *Bull. Geol. Soc. Amer.* **54**, 263–286.

Birch F. (1969) Density and composition of the upper mantle; first approximation as an olivine layer. *Amer. Geophys. Union Monograph* **13** p. 18–36.

Blackshear W. T. and Gapcynski J. P. (1977) An improved value of the lunar moment of inertia. *J. Geophys. Res.* **82**, 1699–1701.

Cameron A. G. W. (1978) The primitive solar accretion disk and the formation of the planets. In *The Origin of the Solar System* (S. F. Dermott, ed.), p. 49–73. Wiley, New York.

Christensen N. I. (1974) Compressional wave velocities in possible mantle rocks to 30 kilobars. *J. Geophys. Res.* **79**, 407–412.

Dainty A. M., Toksöz M. N., and Stein S. (1976) Seismic investigation of the lunar interior. *Proc. Lunar Sci. Conf. 7th*, p. 3057–3075.

Delano J. W., Taylor S. R., and Ringwood A. E. (1980) Composition and structure of the deep lunar interior (abstract). In *Lunar and Planetary Science XI*, p. 225–227. Lunar and Planetary Institute, Houston.

Ganapathy R. and Anders E. (1974) Bulk compositions of the moon and earth estimated from meteorites. *Proc. Lunar Sci. Conf. 5th*, p. 1181–1206.

Gibson E. K. Jr., Usselman T. M., and Morris R. V. (1976) Sulfur in the Apollo 17 basalts and their source regions. *Proc. Lunar Sci. Conf. 7th*, p. 1491–1505.

Goins N. R. (1978) Lunar seismology: the internal structure of the moon. Ph.D. Thesis, M.I.T., Boston. 666 pp.

Goins N. R., Toksöz, M. N., and Dainty A. M. (1978) Seismic structure of the lunar mantle: an overview. *Proc. Lunar Planet. Sci. Conf. 9th*, p. 3575–3588.

Green D. H. and Ringwood A. E. (1967) The stability field of aluminous pyroxene peridotite and garnet peridotite and their relevance in upper mantle structure. *Earth Planet. Sci. Lett.*, **3**, 151–160.

Head J. W. (1979) Lava flooding of early planetary crusts: Geometry, thickness, and volumes of flooded impact basins (abstract). In *Lunar and Planetary Science X*, p. 516–518. Lunar and Planetary Institute, Houston.

Kaula W. M. (1970) The gravitational field of the Moon. *Science* **166**, p. 1581–1588.

Kaula W. M. (1977) On the origin of the moon, with emphasis on bulk composition. *Proc. Lunar Sci. Conf. 8th*, p. 321–333.

Kaula W. M., Schubert G., Lingenfelter R. E., Sjogren W. L., and Wollenhaupt W. R. (1974) Apollo laser altimetry and inferences as to lunar structure. *Proc. Lunar Sci. Conf. 5th*, p. 3049–3058.

Kesson S. E. (1975) Mare basalts: melting experiments and petrogenetic interpretations. *Proc. Lunar Sci. Conf. 6th*, p. 921–944.

Koyama J. and Nakamura Y. (1979) Re-examination of the lunar seismic velocity structure based on the complete data set (abstract). In *Lunar and Planetary Science X*, p. 685–687. Lunar and Planetary Institute, Houston.

Kumazawa M. and Anderson O. L. (1969) The elastic moduli pressure derivatives and temperature derivatives of single crystal olivine and single crystal forsterite. *J. Geophys. Res.* **74**, 4317–4328.

Kumazawa M. (1969) The elastic constants of single-crystal orthopyroxene. *J. Geophys. Res.* **74,** 5973–5980.

Lieberman R. C. and Ringwood A. E. (1976) Elastic properties of anorthosite and the nature of the lunar crust. *Earth Planet. Sci. Lett.* **31,** 69–74.

Longhi J. (1978) Pyroxene stability and the composition of the lunar magma ocean. *Proc. Lunar Planet. Sci. Conf. 9th,* p. 285–306.

Mason B. (1971) *Handbook of Elemental Abundance in Meteorites.* Gordon and Breach, N.Y. 555 pp.

Minear J. W. and Fletcher C. R. (1978) Crystallization of a lunar magma ocean. *Proc. Lunar Planet. Sci. Conf. 9th,* p. 263–283.

Morgan J. W., Hertogen J., and Anders E. (1978) The Moon: Composition determined by nebular processes. *Moon and Planets* **18,** 465–478.

Nakamura Y., Latham G. V., Dorman H. J., and Duennebier F. K. (1976) Seismic structure of the moon; a summary of current status. *Proc. Lunar Sci. Conf. 7th,* p. 3113–3121.

O'Hara M. J. (1968) The bearing of phase equilibria on synthetic and natural systems on the origin and evolution of basic and ultrabasic rocks. *Earth Sci. Rev.* **4,** 69–133.

Papanastassiou D. A. and Wasserburg G. J. (1971) Lunar chronology and evolution from Rb-Sr studies of Apollo 11 and 12 samples. *Earth Planet. Sci. Lett.* **11,** 37–52.

Philpotts J. A. and Schnetzler C. C. (1970) Apollo 11 lunar samples: K, Rb, Sr, Ba and rare earth concentrations in some rocks and separated phases. *Proc. Apollo 11 Lunar Sci. Conf.,* p. 1471–1486.

Ringwood A. E. (1975) *Composition and Petrology of the Earth's Mantle.* McGraw-Hill, New York. 618 pp.

Ringwood A. E. (1977) Basaltic magmatism and the composition of the moon, I. Major and heat-producing elements. *The Moon* **16,** 389–423.

Ringwood A. E. and Essene E. (1970) Petrogenesis of Apollo 11 basalts, internal constitution and origin of the moon. *Proc. Apollo 11 Lunar Sci. Conf.,* p. 769–799.

Ringwood A. E. and Kesson S. (1977) Composition and origin of the moon. *Proc. Lunar Sci. Conf. 8th,* p. 371–389.

Simmons G., Batzle M. L., and Harlow A. L. (1980) Thermal modification of microcracks in lunar rocks and revised estimates for the elastic properties of the shallow moon (abstract). In *Lunar and Planetary Science XI,* p. 1030–1032. Lunar and Planetary Institute, Houston.

Simmons G. and Wang H. (1971) *Single Crystal Elastic Constants and Calculated Aggregate Properties: A Handbook,* 2nd edition. M.I.T. Press, Boston. 370 pp.

Solomon S. C. (1977) The relationship between crustal tectonics and internal evolution in the moon and Mercury. *Phys. Earth Planet. Inter.* **15,** 135–145.

Stolper E. (1980) Mineral assemblages in planetary interiors: predictions based on estimates of planetary composition (abstract). In *Lunar and Planetary Science XI,* p. 1100–1103. Lunar and Planetary Institute, Houston.

Taylor S. R. (1973) Geochemistry of the lunar highlands. *The Moon* **7,** 181–195.

Taylor S. R. (1976) Geochemical constraints on the composition of the moon. *Proc. Lunar Sci. Conf. 7th,* p. 3461–3477.

Taylor S. R. (1978) Geochemical constraints on melting and differentiation of the moon. *Proc. Lunar Planet. Sci. Conf. 9th,* p. 15–23.

Taylor S. R. and Bence A. E. (1975) Evaluation of the lunar highland crust. *Proc. Lunar Sci. Conf. 6th,* p. 1121–1141.

Taylor S. R. and Jakeš P. (1974) The geochemical evolution of the moon. *Proc. Lunar Sci. Conf. 5th,* p. 1287–1305.

Todd T., Richter D. A., Simmons G., and Wang H. (1973) Unique characterization of lunar samples by physical properties. *Proc. Lunar Sci. Conf. 4th,* p. 2639–2662.

Toksöz M. N., Press F., Anderson K., Latham G., Ewing M., Dorman J., Lammlein D., Nakamura Y., Sutton G., and Duennebier F. (1972) Velocity structure and properties of the lunar crust. *The Moon* **4,** 490–504.

Toksöz M. N., Hsui A. T., and Johnston D. H. (1978) Thermal evolution of the terrestrial planets. *Moon and Planets* **18**, 281–320.

Walker D., Longhi J., and Hays J. F. (1975) Differentiation of a very thick magma body and implications for the source regions of mare basalts. *Proc. Lunar Sci. Conf. 6th*, p. 1103–1120.

Wang H. and Simmons G. (1972) Elasticity of some mantle crystal structures, 1. Pleonaste and hercynite spinel. *J. Geophys. Res.* **77**, 4379–4392.

Wang H. and Simmons G. (1974) Elasticity of some mantle crystal structures: 3. spessartite-almandine garnet. *J. Geophys. Res.* **79**, 2607–2613.

Wänke H., Baddenhausen H., Blum K., Cendales M., Dreibus G., Hofmeister H., Kruse H., Jagoutz E., Palme C., Spettel B., Thacker R., and Vilcsek E. (1977) On the chemistry of lunar samples and achondrites. Primary matter in the lunar highlands. A re-evaluation. *Proc. Lunar Sci. Conf. 8th*, p. 2191–2213.

Wood J. A. (1975) Petrogenesis in a well-stirred magma ocean. *Proc. Lunar Sci. Conf. 6th*, p. 1087–1102.

Wood J. A., Dickey J. S., Marvin U. B., and Powell B. W. (1970) Lunar anorthosite and a geophysical model of the moon. *Proc. Apollo 11 Lunar Sci. Conf.*, p. 965–988.

Proc. Lunar Planet. Sci. Conf. 11th (1980), p. 2059–2074.
Printed in the United States of America

Dynamic properties of mare basalts: Relation of equations of state to petrology

Thomas J. Ahrens and J. Peter Watt

Seismological Laboratory, California Institute of Technology, Pasadena, California 91125

Abstract—New shock compression and release adiabat data to 160 GPa are presented for the low-titanium mare basalt 12063 and compared to previous results for the high-titanium mare basalt 70215. While the two rocks have very similar zero-pressure densities, above 100 GPa 12063 is considerably less compressible (less dense by up to 0.2 Mg/m³). Equation of state modelling of the high-pressure phase region using addition of mineral Hugoniots is limited by the very sparse data for calcium-bearing pyroxenes. An estimate of the high-pressure (perovskite) phase Hugoniot of wollastonite (ρ_{hpp} = 4.28 Mg/m³, $K_o \approx$ 270 GPa) predicts smaller compressibility than for enstatite. Addition of component oxide Hugoniots adequately reproduces the 70215 data, but yields model Hugoniots denser than the 12063 data by up to 0.30 Mg/m³ above 100 GPa. Thus, rocks with complex mineralogies can have significantly different dynamic behaviors even though their oxide components are very similar. A likely explanation for the sharply-differing dynamic properties of the two rocks is the 10% higher ilmenite content of 70215. The present 12063 high-pressure data lie essentially midway between the 70215 data and those obtained previously for anorthosite, the dominant constituent of the lunar high-lands. Thus, the behavior of lunar surface material under dynamic compression is terrane dependent; this dependence is found to extend to different regions of the maria, and may affect the determination of relative ages of different rocks units from areal crater densities.

INTRODUCTION

We report results of a study of the dynamic properties of the low-titanium mare basalt 12063 to complement previously published work on the high-titanium mare basalt 70215 (Ahrens *et al.*, 1977).

Characterization of the response of planetary surface materials to shock impact is of importance in relating areal crater densities to the relative ages of various flow units on planetary surfaces; for example, on the unsampled regions of the lunar maria (Neukum, 1977), on the mercurian intercrater plains (Trask and Guest, 1975), and on the various martian volcanic terranes (Neukum and Hiller, 1980). The effect of known, or inferred, regional differences in chemistry on the response of a rock unit to a given flux of impactors, and hence on the areal density, or apparent age, has not yet been investigated because of a lack of supporting data on rock properties under shock, and detailed cratering calculations have not yet been carried out although Neukum and Wise (1976, Fig. 3) consider in general terms the possible effects of differences in target properties.

Previous impact cratering research (for example, O'Keefe and Ahrens, 1975,

1976, 1977; Ahrens and O'Keefe, 1977) has demonstrated that the dynamic behavior of a lunar highland anorthositic terrane is dominated by a series of shock-induced phase changes at stresses below 100 GPa, and that these transformations affect the degree of meteorite penetration, the extent of shock-induced melting, the final crater volume, and the near-field ejecta distribution. In general, it is expected that for a series of rock types, the lower shock impedance rocks will exhibit older apparent cratering ages.

Jeanloz and Ahrens (1978, 1980a) have investigated the shock behavior of a calcic anorthosite, 60025, thought to be representative of a significant portion of the lunar highland crust. Ahrens *et al.* (1977) studied the dynamic properties of the high-titanium (12.4 wt% TiO_2) mare basalt 70215, believed to be typical of the oldest exposed mare rocks, with an age of 3.8 G.y. (Kirsten and Horn, 1974), and found the mare material considerably more compressible than the highland anorthosite.

In this paper, we investigate the dynamic behavior of the low-titanium (5 wt.% TiO_2) mare basalt 12063. Papanastassiou and Wasserburg (1971) report an age of 3.3 G.y. for this rock. The chief differences between 12063 and 70215 are higher pyroxene content (64% vs. 58%) and lower ilmenite content (8% vs. 18%) for 12063. Zero-pressure bulk density (3.21 Mg/m^3) and crystal density (3.36 Mg/m^3—average of seven samples) yield an average porosity of 4.5% for 12063, compared with a bulk density of 3.33 Mg/m^3 (ten samples), crystal density of 3.40 Mg/m^3 (seven samples), and an average porosity of 2.4% for 70215. Both rocks are about 6% denser than the densest terrestrial pyroxene-plagioclase rock which has been studied by shock techniques (McQueen *et al.*, 1967). Shock compression data are obtained for 12063 to shock pressures of 160 GPa, and additional data for 70215 in the mixed-phase region from 30 to 100 GPa. The data are analyzed in terms of the mineralogies of the two rocks.

EXPERIMENTAL

The experimental techniques used here have been described elsewhere (Ahrens *et al.*, 1977; Jeanloz and Ahrens, 1977; Jackson and Ahrens, 1979) and will be only briefly discussed.

Samples were cut from 12063.74 and 70215.14, .217 in the form of rectangles (3–6 mm thickness, 8–17 mm lateral dimensions). The surfaces normal to the shock propagation direction were ground and lapped to flatness and parallelism generally within ±5 μm. "Intrinsic" (crystal) and bulk densities were determined (generally to within ±0.001 Mg/m^3) by the Archimedean method using toluene and from the dimensions and mass of the samples, respectively. The samples were mounted on 2024 aluminum, tungsten, or tantalum driver plates. Lexan or fused quartz arrival and buffer mirrors were mounted on the sample assemblies which were then impacted with lexan projectiles bearing flyer plates of the same material as the driver plates. The projectiles were launched from either a 40 mm propellant gun (for projectile velocities less than 2.5 km/sec) or from a two-stage light gas gun (for projectile velocities up to 6.5 km/sec). Data in the form of streak records from an image converter streak camera were analyzed both visually and with a scanning microphotometer in the manner described by Jeanloz and Ahrens (1978, 1980b) and Jackson and Ahrens (1979) with corrections applied for projectile tilt and bowing and non-uniform streak camera writing rate, where appropriate. The projectile velocity and the shock velocities in the sample and in the buffer mirror were used to calculate the Hugoniot and release states in the sample using the impedance match technique (Walsh and Christian, 1955) and the Riemann integral equation (Rice *et al.*, 1958; Lyzenga and Ahrens,

1978). The equations of state of McQueen *et al.* (1970) for 2024 aluminum, tantalum, and tungsten, from Wackerle (1962) and Jackson and Ahrens (1979) for fused quartz, and from Carter and Marsh (1980) for lexan were used in the impedance match solution.

RESULTS AND DISCUSSION

Table 1 presents the experimental data and the calculated Hugoniot and release states for the seven experiments on 12063 and the three additional experiments on 70215.

In Fig. 1, we plot sample shock velocity, u_s, versus sample particle velocity, u_p, for both 12063 and 70215, including the earlier data of Ahrens *et al.* (1977). Both rocks exhibit a distinct change of u_s-u_p slope in the 2.25–3.0 km/sec u_p range, indicating a transformation to a high-pressure phase assemblage. The three additional 70215 data are consistent with the linear trends evidenced by the original 70215 data. Both legs of the u_s–u_p curves are well-described by linear relationships:

$$70215 \ (u_p < 3.0 \ \text{km/sec}): u_s = 5.401 + 0.701 \ u_p \quad (r^2 = 0.985) \quad (1)$$
$$70215 \ (u_p > 3.0 \ \text{km/sec}): u_s = 1.349 + 2.065 \ u_p \quad (r^2 = 0.968) \quad (2)$$
$$12063 \ (u_p < 2.25 \ \text{km/sec}): u_s = 4.843 + 0.905 \ u_p \quad (r^2 = 0.986) \quad (3)$$
$$12063 \ (u_p > 2.25 \ \text{km/sec}): u_s = 3.434 + 1.544 \ u_p \quad (r^2 = 0.996) \quad (4)$$

where r^2 is the correlation coefficient.

We also plot at $u_p = 0$ in Fig. 1 bulk velocities, V_Φ, calculated from ultrasonic measurements. Mizutani and Osako (1974) measured compressional and shear velocities in 70215 (density = 3.37 Mg/m^3) to 9 kbar (0.9 GPa) and observed the rapid increase of velocity with pressure at low pressures typically caused by the initial closing of cracks and compression of pores. We use their 9 kbar data to approximate the bulk velocity of the dense 70215. The only ultrasonic data available for 12063 are the ambient pressure measurements of Warren *et al.* (1971) which yield a range of V_Φ values between 2.0 and 2.6 km/sec for propagation in three orthogonal directions on a sample of density 3.10 Mg/m^3, considerably more porous than the samples used here (Table 1). Thus, these results are not useful for comparison with the data in Fig. 1. Accordingly, we plot V_Φ calculated from the 10 kbar velocities of Kanamori *et al.* (1971) for 12052 and 12065 (density 3.27 Mg/m^3), two pigeonite-containing basalts with ~5% more pyroxene than 12063 (Papike *et al.* 1976).

Figure 2 presents pressure versus density results from 12063 and 70215, including the earlier data of Ahrens *et al.* (1977). Also shown are the high-pressure phase Hugoniots for Centreville diabase (McQueen *et al.*, 1967) and for a terrestrial anorthite (An$_{95}$) (Jeanloz and Ahrens, 1978, 1980a). Up to pressures of about 40 GPa, the 12063 and 70215 data are fairly similar, with 70215 only slightly more compressible. This difference in compressibility becomes much more pronounced in the region above 90 GPa, such that the initial crystal density difference of 0.03 Mg/m^3 at STP increases to more than 0.2 Mg/m^3 above 90 GPa. Correcting for the additional heating in 12063 at high pressures caused by the

Table 1. Experimental data and calculated Hugoniot and release states for 12063 and 70215.

Shot	Driver material	Sample thickness (mm)	Bulk density (Mg/m³)	Crystal density (Mg/m³)	Projectile velocity (km/sec)	Hugoniot state				Release state			
						Shock velocity (km/sec)	Particle velocity (km/sec)	Pressure (GPa)*	Density (Mg/m³)	Buffer shock velocity (km/sec)	Buffer particle velocity (km/sec)	Pressure (GPa)*	Density (Mg/m³)
12063													
434	Al	3.371 (±0.015)	3.184 (±0.001)	3.352 (±0.001)	0.777 (±0.010)	5.208 (±0.137)	0.385 (±0.007)	6.4 (±0.1)	3.438 (±0.011)	2.921 (±0.164)	0.376 (±0.104)	1.3 (±0.4)	3.438 (±0.012)
433	Al	3.632 (±0.010)	3.260 (±0.001)	3.368 (±0.001)	1.476 (±0.050)	5.518 (±0.085)	0.730 (±0.027)	13.1 (±0.5)	3.758 (±0.025)	3.489 (±0.157)	0.738 (±0.100)	3.1 (±0.6)	3.758 (±0.025)
478	W	3.678 (±0.003)	3.196 (±0.001)	3.376 (±0.002)	1.932 (±0.010)	6.156 (±0.082)	1.572 (±0.005)	30.9 (±0.3)	4.291 (±0.026)	5.615 (±0.030)	2.092 (±0.019)	14.0 (±0.2)	4.015 (±0.030)
479	W	3.678 (±0.003)	3.224 (±0.001)	3.358 (±0.001)	2.441 (±0.010)	6.700 (±0.100)	1.963 (±0.010)	42.4 (±0.6)	4.561 (±0.034)	6.147 (±0.366)	2.431 (±0.233)	17.9 (±2.8)	4.382 (±0.193)
67	Ta	3.035 (±0.003)	3.169 (±0.001)	3.360 (±0.003)	4.734 (±0.010)	8.824 (±0.086)	3.520 (±0.008)	98.4 (±0.8)	5.272 (±0.040)	9.004 (±0.700)	4.251 (±0.446)	45.8 (±8.4)	5.005 (±0.352)
86	Ta	3.274 (±0.005)	3.210 (±0.002)	3.355 (±0.003)	5.354 (±0.010)	9.575 (±0.032)	3.935 (±0.006)	120.9 (±0.4)	5.449 (±0.040)	8.753 (±0.059)	4.819 (±0.037)	93.0 (±1.3)	4.729 (±0.084)
68	Ta	3.038 (±0.003)	3.207 (±0.001)	3.380 (±0.003)	6.462 (±0.010)	10.678 (±0.241)	4.708 (±0.023)	161.2 (±2.9)	5.736 (±0.124)	11.031 (±0.399)	5.542 (±0.254)	73.1 (±6.0)	5.487 (±0.200)
70215													
430	W	5.956 (±0.003)	3.318 (±0.001)	3.430 (±0.001)	2.397 (±0.010)	6.782 (±0.064)	1.913 (±0.009)	43.0 (±0.4)	4.621 (±0.022)	6.221 (±0.228)	2.478 (±0.145)	18.4 (±1.8)	4.360 (±0.146)
432	W	5.885 (±0.005)	3.307 (±0.002)	3.432 (±0.001)	1.873 (±0.040)	6.378 (±0.065)	1.505 (±0.033)	31.7 (±0.7)	4.329 (±0.030)	5.354 (±0.210)	1.926 (±0.134)	12.3 (±1.3)	4.164 (±0.125)
45	Ta	3.924 (±0.003)	3.291 (±0.002)	3.451 (±0.003)	4.647 (±0.010)	8.492 (±0.118)	3.452 (±0.10)	96.5 (±4.1)	5.545 (±0.250)	7.523 (±0.173)	4.043 (±0.109)	67.0 (±3.4)	5.202 (±0.268)

* 1 GPa = 10 kbar

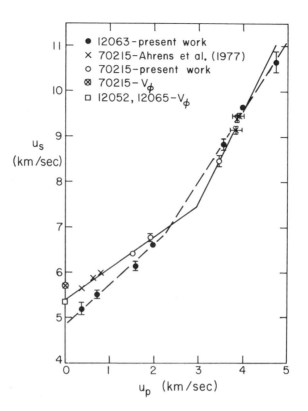

Fig. 1. Sample shock velocity (u_s) versus sample particle velocity (u_p) for mare basalts 12063 and 70215. The straight lines are fits to the two regions for each rock. Shown at $u_p = 0$ is the bulk velocity, V_ϕ, for 70215 and for two Apollo 12 basalts, 12052 and 12065.

additional 2.1% porosity compared to 70215 accounts for about one-third of the observed difference in the two high-pressure Hugoniots. Both mare rocks are considerably more compressible than the Centreville diabase, which is in turn, more compressible than the non-porous terrestrial anorthite (An$_{95}$) or the porous lunar highland anorthosite described by Jeanloz and Ahrens (1978, 1980a).

The release adiabat data at low pressures show that from Hugoniot states near 7 GPa, 70215 releases to less than its initial density, while 12063 appears to undergo irreversible compression, presumably to a series of shock-induced phases. Thus the low-pressure phase (lpp) untransformed regime appears to extend to only about 7 GPa in 12063, somewhat lower than the 14 GPa estimated from the 70215 low-pressure data (Ahrens *et al.*, 1977). The release data at 13, 31, and 43 GPa demonstrate a remarkable similarity for the two rocks.

The steep release adiabats for the 12063 data at 161 and 98 GPa are similar to those found for the three 70215 Hugoniot points around 120 GPa, while the 70215 datum at 96 GPa has a much shallower release path, similar to those in the mixed-phase (mp) region, although the density undertainty is large (Table 1). Thus, the

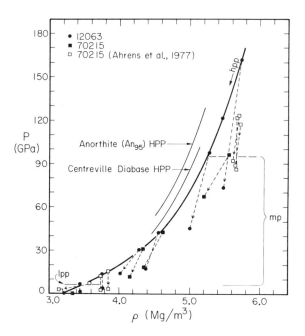

Fig. 2. Hugoniot and release adiabat data for 12063 and 70215. The previous 70215 data of Ahrens *et al.* (1977) are shown as open squares. The inferred low pressure phase (lpp), mixed-phase (mp), and high pressure phase (hpp) regimes are indicated. Also shown are Hugoniots for Centreville diabase (McQueen *et al.*, 1967) and anorthite (Jeanloz and Ahrens, 1978, 1980a).

mixed-phase region extends from about 7 to 95 GPa. The release path for the 12063 datum at 121 GPa (Table 1) appears to be anomalous. Several explanations for the often observed phenomenon of release paths steeper than the Hugoniot in silicates are summarized by Jeanloz and Ahrens (1978) and references therein; however, it does not appear possible to assess the roles of the various mechanisms such as material strength effects or entropy production during release.

The new 70215 datum at 96 GPa occurs at a lower density than would be inferred from the earlier estimated Hugoniot (Ahrens *et al.*, 1977, Fig. 3), implying less compression in the mixed-phase region. The new 70215 data at 31 and 42 GPa support this conclusion.

We attempt to model the high-pressure equation of state in the same manner as was done for 70215, by adding specific volumes calculated from the high-pressure Hugoniots of the constituent minerals at a given pressure:

$$\frac{1}{\rho(P)} = \sum_i \frac{m_i}{\rho_i(P)} \tag{5}$$

where $\rho(P)$ is the whole-rock density at pressure P, $\rho_i(P)$ is the density of the i'th mineral component at P, and m_i is the mass fraction of the i'th component. This differs from the usual mixture method of addition of constituent oxide Hugoniots, which has been demonstrated to approximately account for the dynamic behavior of many simple silicates up to pressures of 150 GPa (McQueen, 1968; Al'tshuler and Sharipdzhanov, 1971). Addition of mineral Huogniots was shown to adequately predict the high-pressure densities of well-characterized terrestrial rocks, such as Westerly granite and Centreville diabase, for which detailed chemical, mineralogical, and Hugoniot data exist (Ahrens *et al.*, 1977).

The high-pressure behavior of minerals needed for such modelling is described by the zero-pressure density, ρ_0, and by the parameters c and s in the linear u_s–u_p fit ($u_s = c + su_p$), such as were given in Eqs. (1)–(4) for 70215 and 12063. Table 3 of Ahrens *et al.* (1977) presents parameters for a wide range of relevant minerals and indicates that a linear u_s–u_p relation is a good fit to the available data ($r^2 > 0.96$ in all cases). The pressure-density relation is then readily derived from the Rankine-Hugoniot equations:

$$P = \rho c^2(\rho/\rho_0 - 1)/[s + (1 - s)\rho/\rho_0]^2. \tag{6}$$

The densities calculated from Eq. (6) for the various constituent minerals are then summed using (5) to obtain the whole-rock density.

Table 2 presents major element (Willis *et al.*, 1971) and modal (Papike *et al.*, 1976) data for 12063. We assign plagioclase, olivine, and ilmenite compositions based on the 12063 microprobe results of Papike *et al.* (1976); opaques were assumed to be entirely ilmenite. The 12063 modal analysis of Dungan and Brown (1977) is similar to that used here, while the analysis of James and Wright (1972) has about 10% more olivine at the expense of pyroxene, most likely because the olivine composition was fixed at $Fa_{30}Fo_{70}$, whereas Table 2 gives the olivine composition as $Fa_{91}Fo_9$.

Also shown in Table 2 are zero-pressure densities for the components of 12063 calculated from the densities of the end-members of the solid solutions. In addition, we require an estimate of the dynamic behavior of fayalite (Fa). Using the data of McQueen and Marsh (1966) and Marsh (1980) for Rockport fayalite

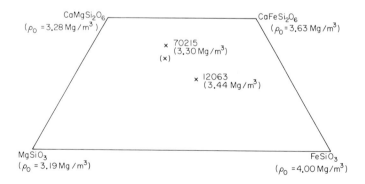

Fig. 3. Compositions of the quadrilateral pyroxenes of 12063 and 70215. The densities inferred from the addition of mineral Hugoniots are included for comparison with the densities of the four quadrilateral end-members. The bracketed point is obtained from the pyroxene composition calculations of Dymek *et al.* (1975).

Table 2. Major element and modal data and zero-pressure densities for 12063.

Oxide	wt %[a]	Mode[b]		ρ_0 (Mg/cm³)
SiO$_2$	43.48	pyroxene	63.7 see text	(3.38)
TiO$_2$	5.00	plagioclase	22.2 An$_{88}$Ab$_{12}$[c]	2.76
Al$_2$O$_3$	9.27	olivine	2.8 Fa$_{91}$Fo$_9$[d]	4.26
FeO	21.26	opaques	8.1 Il$_{100}$[e]	4.75
MnO	0.28	quartz	1.6	2.65
MgO	9.56	mesostasis	1.6	—
			100.0	3.37
CaO	10.49			
Na$_2$O	0.31			
K$_2$O	0.06			
P$_2$O$_5$	0.14			
Cr$_2$O$_3$	0.44			
S	0.09			
	100.38			

[a] Willis *et al.* (1971), Table 1
[b] Papike *et al.* (1976), Table 9
[c] Papike *et al.* (1976), Table 13—average of 12063.35, .13, .18
[d] Papike *et al.* (1976), Table 14
[e] Papike *et al.* (1976), Table 16

($u_p > 2.0$ km/sec), c and s were found to be 3.784 and 1.418, respectively ($r^2 = 0.997$). The zero-pressure density of 4.26 reported for Rockport fayalite corresponds to a composition of Fa$_{89}$Fo$_{11}$, assuming no porosity. These parameters are taken to be those of the Fa$_{91}$Fo$_9$ in 12063.

For both 70215 and 12063, the pyroxene chemistry is complicated by the presence of Ti and Al, which are not accommodated by the pyroxene quadrilateral (CaMgSi$_2$O$_6$—CaFeSi$_2$O$_6$—MgSiO$_3$—FeSiO$_3$) (Papike *et al.*, 1974; Papike and Cameron, 1976). Additionally, existing high-pressure phase data for the end-member quadrilateral pyroxenes are still few and rather poorly defined—only enstatites (MgSiO$_3$) and bronzites ((Mg$_{\sim0.85}$Fe$_{\sim0.15}$)SiO$_3$) of varying quality (Jeanloz and Ahrens, 1977 and references therein) and three datum points for hedenbergite (CaFeSi$_2$O$_6$) (Simakov *et al.*, 1974) have been reported. Accordingly, the procedure described above was modified to *estimate* the density of the pyroxene components of 12063 and 70215 needed for compatibility with the observed whole-rock density at a given pressure. For 70215, Ahrens *et al.* (1977) estimated pyroxene densities of 3.30 Mg/m³ and 5.53 Mg/m³ at ambient and 120 GPa pressures, respectively, and concluded that these values were consistent with the limited ambient and high-pressure data for the few pyroxenes that have been measured by shock techniques.

A similar calculation for 12063 yields an inferred zero-pressure density of 3.44 for the pyroxene component (average of values of 3.41 Mg/m³ using the modal analysis of Papike *et al.* (1976) and 3.47 Mg/m³ using the modal analysis of

Dungan and Brown (1977). The modal analysis of James and Wright (1972) yields a value of 3.36 Mg/m^3.

In Fig. 3 we plot the 70215 and 12063 inferred zero-pressure pyroxene densities in the pyroxene quadrilateral. Pyroxene compositions were calculated from the oxide analyses of Papike *et al.* (1976, Table 12e) for 12063 and Dymek *et al.* (1975, Table 1) for 70215. The method was that of Papike *et al.* (1974), Papike and Cameron (1976), and Papike and White (1979) which uses a change balance equation based on cation substitutional pairs to distribute the oxides between the quadrilateral components of the pyroxene and Al-, Cr-, and Na-containing py-roxenes, termed "others." The average of the results from the oxide analyses for 12063 yields a quadrilateral pyroxene composition of $Wo_{27}En_{32}Fs_{41}$, while the weighted average of the medium-calcium, high-calcium, and iron pyroxenes for 70215 yields a quadrilateral composition of $Wo_{39}En_{36}Fs_{25}$, somewhat different from that calculated from a weighted average of the quadrilateral components as determined by Dymek *et al.* (1975)—$Wo_{34}En_{38}Fs_{28}$—the bracketed composition in Fig. 3. This difference is likely due to different methods of assigning compo-sitions by the two sets of authors. Accordingly, we adopt the 70215 pyroxene composition calculated here for consistent comparison with 12063.

The computed pyroxene compositions are plotted on Fig. 3, along with the inferred zero-pressure densities given above. These densities are seen to be con-sistent with the densities of the four end-member pyroxenes. It should be re-membered, however, that the inferred densities include the non-quadrilateral Ti- and Al-containing pyroxene components. For 12063, the calculation divides the pyroxenes into 0.91 quadrilateral—0.09 "others." For 70215, the division is 0.75 quadrilateral—0.25 "others" (0.85 quadrilateral—0.15 "others" using the cal-culations of Dymek *et al.* (1975). The higher non-quadrilateral component for 70215 could account for the apparent low density of 70215 relative to enstatite ($MgSiO_3$) and diopside ($CaMgSi_2O_6$) on the quadrilateral.

The non-quadrilateral pyroxene components for the various 12063 and 70215 analyses are tightly clustered. Using the ternary classification of Papike *et al.* (1974) for non-quadrilateral pyroxenes—Ti in the M1 site, Al in the tetrahedral site, and Na in the M2 site, we find the following "others" compositions:

$$12063: \quad {}^{VI}Ti^{4+}_{33\pm5} \, {}^{IV}Al^{3+}_{65\pm4} \, {}^{VIII}Na^+_{1\pm1}$$
$$70215: \quad {}^{VI}Ti^{4+}_{33\pm4} \, {}^{IV}Al^{3+}_{64\pm2} \, {}^{VIII}Na^+_{2\pm1}$$

This demonstrates that by far the dominant substitutional couple in the non-quadrilateral pyroxene component is Ti-Al, and this component can be written as $R^{2+}Ti_{0.5}R^{2+}_{0.5}(SiAl)O_6$, where R^{2+} = Mg, Ca, or Fe^{2+}. This is consistent with the observations of Papike and Cameron (1979) for the lunar basalts. They also note that the proportion of "others" pyroxenes in mare basalts is higher than in other basalt groups, reflecting rapid cooling histories.

The densities inferred for the pyroxene component of 12063 at high pressures using Eqs. (5) and (6), the modal analyses in Table 2, and the linear u_s–u_p fits from Table 3 of Ahrens *et al.* (1977) are 5.10 Mg/m^3 at 100 GPa and 5.58 Mg/m^3 at 160 GPa. The modal analyses of Papike *et al.* (1976) and Dungan and Brown

(1977) both yield the same results, while the analysis of James and Wright (1972) yields pyroxene density estimates lower by about 0.15 Mg/m³. In Fig. 4, we plot these pyroxene density estimates as well as the estimated high pressure Hugoniot for the pyroxene component of 70215. For comparison, we also plot the high pressure phase data of Simakov *et al.* (1974) for hedenbergite and the high pres-

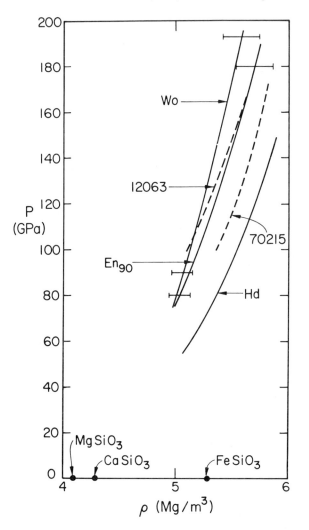

Fig. 4. High-pressure phase Hugoniots estimated for the pyroxene components of 12063 and 70215 using addition of mineral Hugoniots. The curve for hedenbergite (Hd) is from Simakov *et al.* (1974). The wollastonite (Wo) and enstatite (En₉₀) Hugoniots are calculated from third-order finite strain theory using measured or estimated parameters for $MgSiO_3$ and $CaSiO_3$ in the perovskite phase. The horizontal bars represent uncertainties in the calculated Hugoniots due to the uncertainties in the parameters. The zero pressure densities are either experimental or estimated values for the perovskite forms of $MgSiO_3$, $CaSiO_3$, and $FeSiO_3$.

sure (perovskite) Hugoniot for enstatite (En_{90}) as estimated by Jeanloz and Ahrens (1977). We have also calculated a theoretical Hugoniot for the high pressure phase of wollastonite. Wollastonite has been shown to transform to a perovskite phase at pressures above 16.0 GPa (Liu and Ringwood, 1975). The theoretical Hugoniot was calculated using 3rd-order finite strain theory (see, for example, Jackson and Ahrens, 1979, for application to Hugoniot calculations) with the following parameters—ρ_0 (perovskite) = 4.28 Mg/m^3 (Liu, 1979), $K_0 \approx 270$ GPa (estimate from bulk modulus-molar volume and bulk sound speed-mean atomic weight systematics by Liebermann (1976), $K_0' = 3.5 \pm 1.0$, $\gamma_0 = 1.5 \pm 0.5$, $\gamma = V_0(\rho_0/\rho)$, $E_{tr} \approx 1.0$ kJ/g. The unreferenced values are similar to those assumed by Jeanloz and Ahrens (1977) for $MgSiO_3$ (perovskite). The horizontal error bars on the Wo and En_{90} theoretical Hugoniots represent the uncertainty arising from the combined uncertainties in the assumed parameters. The zero-pressure densities for the perovskite phases of enstatite, wollastonite, and ferrosilite are taken from the experimental data of Liu (1979) for $MgSiO_3$, Liu and Ringwood (1975) for $CaSiO_3$ and from the estimated molar volume of $FeSiO_3$ (Yagi *et al.*, 1978). The inferred high-pressure densities of the calcium-bearing pyroxenes in both 12063 and 70215 are in agreement with the limited existing shock compression data, and thus the dynamic properties of both rocks appear to be explainable by addition of component mineral Hugoniots.

The high-pressure Hugoniots of 12063, 70215, and Centreville diabase were further modelled by using the addition of oxide Hugoniots (McQueen *et al.*, 1967; Al'tshuler and Sharipdzhanov, 1971) and the oxide analyses of Willis *et al.* (1971), Dymek *et al.* (1975), and Birch (1961). All oxides with less than 1 wt% concentration were ignored and the remaining oxides re-normalized to 100%. High-pressure Hugoniot data were thus required for SiO_2, TiO_2, Al_2O_3, CaO, MgO, FeO, and (for Centreville diabase) Fe_2O_3. All calculations were centered at ambient pressure oxide densities. Linear u_s–u_p fits for SiO_2 and TiO_2 were taken from Ahrens *et al.* (1977). Linear fits were carried out for Al_2O_3 using the single-crystal data of McQueen and Marsh (1966) and the dense polycrystal data of Ahrens *et al.* (1968), for MgO using the single-crystal data of McQueen and Marsh (1966) and Ahrens (1966), and for Fe_2O_3 using the data of McQueen and Marsh (1966). The high-pressure regions of CaO and FeO were fit using the recent data of Jeanloz and Ahrens (1980b) which demonstrate the existence of previously unknown phase transformations, established to be a B1-B2 transformation for CaO (Jeanloz *et al.*, 1979). In all cases, a linear u_s–u_p relation is a good fit to the data ($r^2 > 0.98$).

The high-pressure mixed-oxide Hugoniots, labelled "MO," are shown in Fig. 5. The zero-pressure densities of the high-pressure phases were calculated from the measured or inferred densities of the high-pressure phases of the oxides. The 70215 high-pressure data are in good agreement with the mixed-oxide Hugoniot, while the 12063 and Centreville data are in poor agreement, being displaced to lower densities by 0.15–0.30 Mg/m^3. This discrepancy is larger than would have been found before the new CaO and FeO data were obtained. One explanation could be lack of complete transformation to the high-pressure phases. This could

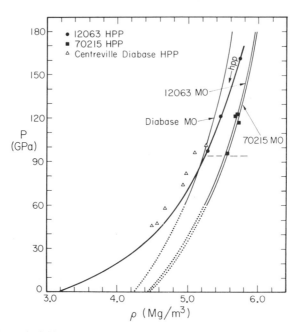

Fig. 5. Theoretical high-pressure phase Hugoniots for 12063, 70215 and Centreville diabase using addition of constituent oxide Hugoniots. The mixed-oxide (MO) calculation reproduces only the 70215 data satisfactorily.

be plausible for the lower range of the Centreville data, but appears very unlikely for the 160 GPa 12063 datum. Thus, the 12063 results are anomalous compared with the 70215 data, a point not evident from the addition of *mineral* Hugoniot results presented above.

Examination of the oxide analyses of 12063 and 70215 shows that the two rocks are fairly similar (12063 is somewhat higher in SiO_2, Al_2O_3, and FeO, and lower in TiO_2 and CaO than 70215), so that the similarity of the mixed-oxide Hugoniots in Fig. 2 is not surprising. The differences between the two measured Hugoniots strongly suggest that the dynamic properties are influenced not so much by oxide composition as by the distribution of the oxides among the various phases.

Previous mixed-oxide modelling has focussed on single minerals, two-component physical mixtures or alloys, or rocks with relatively simple mineralogies such as dunites (McQueen *et al.*, 1967; Al'tshuler and Sharipdzhanov, 1971) with the result that the oxide assemblages treated successfully are relatively simple cases.

Ahrens *et al.* (1973) modelled the dynamic behavior of a gabbroic anorthosite, 15418, and Frederick diabase (Birch, 1960) in terms of the dominant minerals (plagioclase and pyroxene) and found good agreement between the theoretical and experimental Hugoniots for the diabase [which has composition and experimental Hugoniot very similar to Centreville diabase (McQueen *et al.*, 1967, Table 2 and Fig. 32)]. The high-pressure region of the lunar sample, 15418, was modelled as 0.74 anorthite (An_{93})—0.26 pyroxene (En_{64}). Since the experimental data for 15418 extended to only 28 GPa, it is not clear how adequate a representation of the high-pressure region the calculated Hugoniot is. Additionally, the simplified composition used in the modelling ignored the 21 wt% CaO in the pyroxene component. The utility of equation of state modelling using addition of mixed-oxide Hugoniots applied to complex materials such as 12063 and 15418 cannot be further tested until additional shock compression data are obtained for calcium-containing silicates, particularly pyroxenes.

CONCLUSION

Shock compression and release adiabat data have been obtained to 160 GPa for the low-titanium mare basalt 12063 and compared with previous data for the high-titanium mare basalt 70215. Although the two rocks have zero-pressure densities differing by only 0.03 Mg/m^3, 12063 is considerably less compressible at high pressures than 70215—up to 0.2 Mg/m^3 less dense above 90 GPa. Both rocks achieve higher shock densities than the densest terrestrial plagioclase-pyroxene rocks. The mixed-phase regime is inferred to extend from about 7 GPa to 95 GPa in 12063. The release paths of the two rocks are very similar—steep release adiabats in the high-pressure phase region and shallower adiabats in the mixed-phase regime.

Additional data obtained for 70215 in the mixed-phase region and at the lower end of the high-pressure phase regime show that 70215 is less compressible at these pressures than was inferred from the original high-pressure Hugoniot and release adiabat data.

Equation of state modelling of the high-pressure phase region of 12063 using the method of addition of mineral Hugoniots previously applied to 70215 is complicated by the complexity of the pyroxene component. The limited shock compression data for pyroxenes, especially those containing calcium, and significant amounts of non-quadrilateral components (Ti- and Al-containing), make direct addition of mineral Hugoniots unfeasible. Instead, we have estimated the density the pyroxene component must have at a given pressure to produce the measured whole-rock shock density. The relative densities of the pyroxene constituents of 12063 and 70215 are consistent with their position on the standard pyroxene quadrilateral and with the limited high-pressure shock compression data for pyroxenes.

The non-quadrilateral components of the pyroxenes have compositions tightly clustered, in accord with the extensive basaltic pyroxene observations of Papike and White (1979), and in agreement with their conclusion that Ti-Al pyroxenes are by far the dominant non-quadrilateral component of pyroxene in lunar basalts.

Modelling of the high-pressure regime of 12063 and 70215 using addition of component oxides produces two theoretical Hugoniots only slightly separated, because of the rather similar oxide compositions of the two rocks. The 70215 experimental data are in good agreement with a mixed-oxide model, while the 12063 data are up to 0.30 Mg/m^3 less dense than is predicted from the oxides. Thus, while mixed-oxides may satisfactorily account for the high-pressure dynamic behavior of many minerals and rocks, rocks with complex mineralogies such as 12063 appear to require more sophisticated modelling, incorporating as yet unavailable information about the dynamic properties of chemically complex pyroxenes, especially those containing calcium. Acquisition of such data could lead to a sound framework for predicting the high-pressure equations of state of lunar and other planetary surface rocks displaying diverse petrologies. Additionally, high-pressure properties of Ca- and Al-bearing oxides and silicates are of interest because these materials are early high-temperature condensates from the solar nebula (Grossman, 1972).

Construction of a Hugoniot for the high-pressure (perovskite) phase of wollastonite using a measured density (4.28 Mg/m^3) and bulk modulus (270 GPa) estimated from elasticity systematics suggests that wollastonite is somewhat less compressible than enstatite. Existing Hugoniot data for wollastonite (Marsh, 1980) do not extend to sufficiently high pressures to confirm the static experimental observation of a transformation to a perovskite structure and to enable the predicted Hugoniot to be tested; however, experiments in progress should provide the necessary data.

The most likely explanation for the sharply-differing dynamic properties of the high-pressure phase assemblages of 12063 and 70215 is the difference in titanium and ilmenite contents of the two rocks. The higher ilmenite (18% vs. 8%) and titanium (12% vs. 5%) contents of 70215 would be expected to produce a higher shock density at a given pressure, as observed.

Finally, the observed dynamic behavior of the low-titanium basalt is approximately midway between that of anorthite and that of the oldest exposed mare basalts, in accord with the speculation of Jeanloz and Ahrens (1978) that the dynamic properties of these two rocks likely bracket the properties of all major lunar rock types. The present results demonstrate that even within the lunar maria, there exist significant differences in material behavior under impact. The effect of these differences on determination of relative crater ages deserves further investigation.

Acknowledgments—We appreciate the careful construction and execution of these experiments by E. Gelle, J. R. Long, and M. Long. J. J. Papike kindly made available the pyroxene composition calculation program and L. Levien provided helpful insights into pyroxene crystal chemistry. Supported by NASA grant NSG-9019. Contribution #3443, Division of Geological and Planetary Sciences, California Institute of Technology, Pasadena, California 91125.

REFERENCES

Ahrens T. J. (1966) High-pressure electrical behavior and equation of state of magnesium oxide from shock wave measurements. *J. Appl. Phys.* **37**, 2532–2541.

Ahrens T. J., and O'Keefe J. D. (1977) Equations of state and impact-induced shock wave attenuation on the moon. In *Impact and Explosion Cratering* (D. J. Roddy, R. O. Pepin, and R. B. Merrill, eds.), p. 639–656. Pergamon, N.Y.

Ahrens T. J., Gust W.H., and Royce E. B. (1968) Material strength effect in the shock compression of aluminum. *J. Appl. Phys.* **39**, 4610–4616.

Ahrens T. J., Jackson I., and Jeanloz R. (1977) Shock compression and adiabatic release of a titaniferous basalt. *Proc. Lunar Sci. Conf. 8th*, p. 3437–3455.

Ahrens T. J., O'Keefe J. D., and Gibbons R. V. (1973) Shock compression of a recrystallized anorthosite rock from Apollo 15. *Proc. Lunar Sci. Conf. 4th*, p. 2575–2590.

Al'tshuler L. V. and Sharipdzhanov I. I. (1971) Additive equations of state of silicates at high pressures. *Izv. Earth Phys.* **3**, 11–28.

Birch F. (1960) The velocity of compressional waves in rocks to 10 kilobars, Part 1. *J. Geophys. Res.* **65**, 1083–1102.

Birch F. (1961) The velocity of compressional waves in rocks to 10 kilobars, Part 2. *J. Geophys. Res.* **66**, 2199–2224.

Carter W. J. and Marsh S. P. (1980) Hugoniot equations of state of polymers. *J. Chem. Phys.* In press.

Dungan M. A. and Brown R. W. (1977) The petrology of the Apollo 12 ilmenite basalt suite. *Proc. Lunar Sci. Conf. 8th*, p. 1339–1381.

Dymek R. F., Albee A. L., and Chodos A. A. (1975) Comparative mineralogy of Apollo 17 mare basalts: Samples 70215, 71055, 74255, and 75055. *Proc. Lunar Sci. Conf. 6th*, p. 49–77.

Grossman L. (1972) Condensation in the primitive solar nebula. *Geochim. Cosmochim. Acta* **36**, 597–619.

Jackson I. and Ahrens T. J. (1979) Shock wave compression of single-crystal forsterite. *J. Geophys. Res.* **84**, 3039–3048.

James O. D. and Wright T. L. (1972) Apollo 11 and 12 mare basalts and gabbros: Classification, compositional variations, and possible petrographic relations. *Bull. Geol. Soc. Amer.* **83**, 2357–2382.

Jeanloz R. and Ahrens T. J. (1977) Pyroxenes and olivines: Structural implications of shock-wave data for high pressure phases. In *High-Pressure Research: Applications to Geophysics* (M. H. Manghnani and S. Akimoto, eds.), p. 439–462. Academic, N. Y.

Jeanloz R. and Ahrens T. J. (1978) The equation of state of a lunar anorthosite: 60025. *Proc. Lunar Planet. Sci. Conf. 9th*, p. 2789–2803.

Jeanloz R. and Ahrens T. J. (1980a) Anorthite: Thermal equation of state to high pressures. *Geophys. J. Roy. Astron. Soc.* In press.

Jeanloz R. and Ahrens T. J. (1980b) FeO and CaO: Hugoniot equations of state of two oxides. *Geophys. J. Roy. Astron. Soc.* In press.

Jeanloz R., Ahrens T. J., Mao H. K., and Bell P. M. (1979) B1-B2 transition in calcium oxide from shock-wave and diamond-cell experiments. *Science* **206**, 829–830.

Kanamori H., Mizutani H., and Hamano Y. (1971) Elastic wave velocities of Apollo 12 rocks at high pressures. *Proc. Lunar Sci. Conf. 2nd*, p. 2323–2326.

Kirsten T. and Horn P. (1974) Chronology of the Taurus-Littrow region III: Ages of mare basalts and highland breccias and some remarks about the interpretation of lunar highland rock ages. *Proc. Lunar Sci. Conf. 5th*, p. 1451–1475.

Liebermann R. C. (1976) Elasticity of the ilmenite-perovskite phase transformation in $CdTio_3$. *Earth Planet. Sci. Lett.* **29**, 326–332.

Liu L.-G. (1979) High pressure phase transformations in the joins Mg_2SiO_4–Ca_2SiO_4 and MgO–$CaSiO_3$. *Contrib. Mineral. Petrol.* **69**, 245–247.

Liu L.-G. and Ringwood A. E. (1975) Synthesis of a perovskite-type polymorph of $CaSiO_3$. *Earth Planet. Sci. Lett.* **28**, 209–211.

Lyzenga G. and Ahrens T. J. (1978) The relation between the shock-induced free-surface velocity and postshock specific volume of solids. *J. Appl. Phys.* **49**, 201–204.

Marsh S. P. (1980) Los Alamos Scientific Laboratory Shock Hugoniot Data. Univ. California Press, Berkeley. 658 pp.

McQueen R. G. (1968) Shock-wave data and equations of state. In *Seismic Coupling*, p. 53–106. Proc. Advanced Research Projects Meeting, Stanford Research Institute. National Technical Information Service, Springfield, Virginia.

McQueen R. G. and Marsh S. P. (1966) In *Handbook of Physical Constants* (S.P. Clark, ed.), p. 156. Geol. Soc. Amer., N.Y.

McQueen R. G., Marsh S. P., and Fritz J. N. (1967) Hugoniot equation of state of twelve rocks. *J. Geophys. Res.* **72**, 4999-5036.

McQueen R. G., Marsh S. P., Taylor J. W., Fritz J. N., and Carter W. J. (1970) The equation of state from shock wave studies. In *High Velocity Impact Pheonomena* (R. Kinslow, ed.), p. 294–419 and Appendices. Academic, N. Y.

Mizutani H. and Osako M. (1974) Elastic-wave velocities and thermal diffusivities of Apollo 17 rocks and their geophysical implication. *Proc. Lunar Sci. Conf. 5th*, p. 2891–2901.

Neukum G. (1977) Lunar cratering. *Phil. Trans. Roy. Soc. London* **A285**, 265–272.

Neukum G. and Hiller K. (1980) Martian ages. *J. Geophys. Res.* In press.

Neukum G. and Wise D. U. (1976) Mars: A standard crater curve and possible new time scale. *Science* **194**, 1381–1386.

O'Keefe J. D. and Ahrens T. J. (1975) Shock effects from a large impact on the moon. *Proc. Lunar Sci. Conf. 6th*, p. 2831–2844.

O'Keefe J. D. and Ahrens T. J. (1976) Impact ejecta on the moon. *Proc. Lunar Sci. Conf. 7th*, p. 3007–3026.

O'Keefe J. D. and Ahrens T. J. (1977) Impact induced energy partitioning, melting and vaporization on terrestrial planets. *Proc. Lunar Sci. Conf. 8th*, p. 3357–3374.

Papanastassiou D. A. and Wasserburg G. J. (1971) Lunar chronology and evolution from Rb-Sr studies of Apollo 11 and 12 samples. *Earth Planet. Sci. Lett.* **11**, 37–62.

Papike J. J. and Cameron M. (1976) Crystal chemistry of silicate minerals of geophysical interest. *Rev. Geophys. Space Phys.* **14**, 37–80.

Papike J. J. and White C. (1979) Pyroxenes from planetary basalts: Characterization of "other" than quadrilateral components. *Geophys. Res. Lett.* **6**, 913–916.

Papike J. J., Cameron K. L., and Baldwin K. (1974) Amphiboles and pyroxenes: Characterization of *other* than quadrilateral components and estimates of ferric iron from microprobe data. *Geol. Soc. Amer. Abstracts with Programs* **6**, 1053.

Papike J. J., Hodges F. N., Bence A. E., Cameron M., and Rhodes J. M. (1976) Mare basalts: Crystal chemistry, mineralogy and petrology. *Rev. Geophys. Space Phys.* **14**, 475–540.

Rice M. H., McQueen R. G., and Walsh J. M. (1958) Compressibility of solids by strong shock waves. *Solid State Physics* **6**, 1–63.

Simakov G. V., Pavlovskiy M. N., Kalashnikov N. G., and Trunin R. F. (1974) Shock compressibility of twelve minerals. *Izv. Earth Phys.* **8**, 11–17.

Trask N. J. and Guest J. E. (1975) Preliminary geologic terrrain map of Mercury. *J. Geophys. Res.* **80**, 2461–2477.

Wackerle J. (1962) Shock-wave compression of quartz. *J. Appl. Phys.* **33**, 922–937.

Walsh J. M. and Christian R. H. (1955) Equation of state of metals from shock-wave measurements. *Phys. Rev.* **97**, 1544–1556.

Warren N., Schreiber E., Scholtz C., Morrison J. A., Norton, P. R., Kumazawa M., and Anderson O. L. (1971) Elastic and thermal properties of Apollo 11 and Apollo 12 rocks. *Proc. Lunar Sci. Conf. 2nd*, p. 2345–2360.

Willis J. P., Ahrens L. H., Danchin R. V., Erlank A. J., Gurney J. J., Hofmeyer P. K., McCarthy T. S., and Orren M. J. (1971) Some interelement relationships between lunar rocks and fines, and stony meteorites. *Proc. Lunar Sci. Conf. 2nd*, p. 1123–1138.

Yagi T., Mao H. K., and Bell P. M. (1978) Effect of iron on the stability and unit-cell parameters of ferromagnesian silicate perovskites. *Carnegie Inst. Wash., Yearb.* **77**, 837–841.

Proc. Lunar Planet. Sci. Conf. 11th (1980), p. 2075–2097.
Printed in the United States of America

Impact cratering in viscous targets: Laboratory experiments

R. Greeley[1]*, J. Fink[1], D. E. Gault[2], D. B. Snyder[1], J. E. Guest[3], and P. H. Schultz[4]

[1]Center for Meteorite Studies and Department of Geology, Arizona State University, Tempe, Arizona 85281; [2]Murphys Center for Planetology, Murphys, California 95247; [3]University of London Observatory, Mill Hill Park, London NW7 2QS, England; [4]Lunar and Planetary Institute, Houston, Texas 77058

Abstract—Martian multilobed craters (''splosh'' craters, rampart craters, etc.) may involve fluidization of ejecta as a result of entrained water, melted/vaporized ice, and/or aerodynamically decelerated ejecta. To determine the effects of target viscosity and yield strength on the formation and morphology of impact craters, 75 experiments were carried out in which target properties and impact energies were varied. The following sequence was observed in high speed motion pictures of the experiments: 1) initial impact of projectile; 2) excavation of crater and rise of ejecta plume; 3) formation of a transient central mound which collapses generating a surge of material which can partly override the plume deposit; 4) oscillation of the central mound (in high energy impacts and fluid targets) with progressively smaller surges of material leaving the crater. The oscillating mound may ''freeze'' as either a peak or a depression. Dimensional analysis of the experimental results indicates: 1) dimensions of the central mound are proportional to the energy of the impacting projectile, and to the inverse of both the yield strength and viscosity of the target material; 2) for most of the impact experiments, effects of target viscosity appear to have been relatively minor and the target muds behaved essentially as inviscid fluids; the cratering process thus represented a balance between projectile kinetic energy and the potential energy required to excavate the initial transient cavity (i.e., gravity scaling); 3) extrapolation of these experimental results to large martian craters requires that the effective viscosity of the surface layer(s) must be less than 10^{10} poise, which is compatible with estimates for terrestrial debris flows. Multilayer targets influence the ejecta morphology: a thin ''dry'' layer (i.e., ''regolith'' or icy crust) on top of the viscous mixture retarded the emplacement of ejecta, whereas a thin fluid (water) layer on top of the mixture lubricated the ejecta and enhanced flow. These results may be applicable to interpretation of martian craters and to impacts into outer planet satellites composed of ice-silicate mixtures such as Callisto and Ganymede.

1.0 INTRODUCTION

One of the more striking discoveries of the Viking mission is the presence of unusual forms of impact crater ejecta, facies, or deposits. Although differences between lunar and martian craters were first indicated by Mariner 9 images, the high quality Viking Orbiter images revealed an unexpectedly wide range of ejecta

---

* also Space Sciences Division, NASA Ames Research Center, Moffett Field, California 94035

morphologies, from typical lunar and mercurian forms to types suggestive of ejecta emplacement as a fluidized mass (Carr *et al.*, 1977). These crater forms, variously termed "splosh" craters, ejecta-flow craters, and rampart craters, all termed *multilobed craters* herein, do not occur on the "dry" atmosphere-free planets Mercury and Moon, and it has been suggested that water and/or the atmosphere of Mars may somehow be responsible for the flow-like nature of the ejecta. Several photogeologic studies of martian craters have attempted to determine geometric crater relationships (diameter versus ejecta lobe size, etc.) and establish their occurrence as functions of parameters such as latitude, elevation, terrain type and geologic age (Mouginis-Mark, 1979; Mutch and Woronow, 1980; Johansen, 1978; and others). In addition to the photogeologic studies, some investigators have studied the effects of the thin martian atmosphere on ejecta emplacement (Schultz and Gault, 1979) and considered possible terrestrial analogs (Roddy *et al.*, 1979).

Our approach to the problem has been two-fold: 1) laboratory experiments using the NASA Ames Vertical Ballistic Gun Facility, and 2) photogeologic studies to provide a basis of comparison for the laboratory results. Exploratory experiments involving cratering in viscous clay have been reported previously (Gault and Greeley, 1978), as has consideration of viscous deformation of craters (Scott, 1967); results of our photogeologic studies will be reported later. Here we discuss impacts in clay targets under controlled conditions, and discuss the implications for the formation of multilobed ejecta deposits, central peaks and central pits. Not only may these experiments be applicable to impact craters on Mars and Earth, but they may have implications for cratering on the volatile-rich outer planet satellites and general cratering processes as well.

2.0 LABORATORY EXPERIMENTS

The primary objective of the experiments was to determine the effect that target rheological properties have on crater and crater ejecta morphology. The Ames Facility is ideally suited for such experiments because target and impact conditions can be controlled, and because high speed motion pictures obtained during impact permit assessment of the processes of crater formation and ejecta emplacement. The facility consists of a target chamber 2.5 m in diameter by 2.5 m high that can be evacuated to martian atmospheric pressures. In these experiments, chamber pressure ranged from about 15 to 50 mb. These pressures were utilized in order to prevent freezing of the water contained in the target material. Impact velocities ranged from 5×10^2 m s^{-1} to 5.5×10^3 m s^{-1}; impact projectiles included 3.175×10^{-3} m, 4.77×10^{-3} m, and 6.35×10^{-3} m glass, aluminum and steel spheres, giving a range of impact energies from 6.9 joules to 2.2×10^3 joules. The target material was contained in one of two cylindrical buckets (small = 0.60 m in diameter by .205 m deep; large = 0.91 m in diameter by 0.46 m deep) fitted in a flat floor 2.5 m in diameter; target material was placed in the bucket so that it formed a nearly continuous surface with the floor. Ideally, the target should be of very large or of infinite dimension, so as to reduce or eliminate boundary effects. Such conditions cannot be achieved in the laboratory, and the target buckets used in these experiments are the same as in previous investigations (e.g., Gault and Wedekind, 1977) to allow comparisons with other target materials. Two types of target materials were used: potters clay mixed with either silicon oil or with water. Silicon oil was used as a fluid because it does not freeze and it is relatively stable to atmospheric pressures of about 15 mb. Clay/water targets were used for more complex models involving mud, and multilayered ice/viscous mud targets.

A series of 75 successful impacts was completed in which three parameters were varied: target viscosity, impact energy and angle of impacting projectile with respect to target (Table 1). The first series was divided into subsets A through J. In each subset, target viscosity was held nearly constant while impact energies and angles were varied. In the last shot of each subset, a 1 cm layer of dry clay was emplaced on the mud to produce a multilayered target. After the shot, the dry clay was mixed with the underlying mud to increase the viscosity for the next subset, and the procedure was repeated. The rheological properties (viscosity and yield strength) of the target material were measured for each subset of experiments, using a sample collected from the target bucket. For the homogeneous, non-layered, targets (#1–8, 11–14, 17–27, 31, and 33–36) the apparent viscosity was measured directly with a Brookfield viscometer Model HBT, *in situ,* both immediately before and after impact using a disk-type spindle of known surface area at a rotational speed of 10 RPM. Plastic viscosities and yield strengths were calculated for batches of slurry extracted every 2 to 7 runs, whereas apparent viscosities were measured during every run. The rheological behavior of the target materials is discussed in detail in Appendix I.

During impact, the cratering events were filmed with high speed motion pictures (400 frames s^{-1}). After each shot, color and black and white photographs were taken and various measurements were made of the crater and ejecta deposits. Subsequent analyses of the motion pictures enabled measurement of various transient features such as ejecta plume angle.

3.0 EXPERIMENTAL RESULTS

In this section we present a general model of crater formation and ejecta emplacement observed for impacts in viscous targets and discuss possible correlations.

3.1 General laboratory model

Analyses of high speed motion pictures lead to a general model of cratering in viscous media (Figs. 1,2). Following the terminology of Gault *et al.* (1968) for a three-stage cratering process, during the initial *compression stage* in which jetting occurs, there is no apparent difference between impacts in dry targets and viscous targets. During the *excavation stage,* however, a slight bulge on the surface of the target material is pushed ahead of the expanding ejecta plume; in

Table 1. Impact crater experiment data. Pre-impact data: target density (ρ = kg·m⁻³), apparent viscosities at various rotation rates [η_A(RPM) = kg·m⁻¹·s⁻¹], calculated plastic viscosities (η_P) and yield strengths (τ = kg·m⁻¹·s⁻²); type of slurry fluid (oil = O; water = W); target configuration (R = uniform slurry; P = powder surface layer; W = water surface layer; S = sand surface layer; N = newspaper surface layer; I = ice surface layer; D = doublet: two shots into same target; T = two clay layers separated by dry powder or newspaper); projectile mass (m = kg), velocity (v = m·s⁻¹), angle of impact (A_s = degrees); chamber pressure at time of impact (P = mm Hg). Post impact data: central peak height (h_m = m), apex angle (A_m = m) and basal diameter (D_m); ejecta plume angle (A_p) and plume deposit diameter (D_p); surge deposit diameter (D_s); and crater diameter (D_c). Not all data were obtainable for each run.

		Target properties							Impact conditions				Crater features							
Subset	Run #	ρ kg·m⁻³	η_P kg·m⁻¹·s⁻¹	η_A kg·m⁻¹·s⁻¹	RPM	τ kg·m⁻¹·s⁻²	Clay base	Target config.	m ×10⁻⁵ kg	v ×10² m·s⁻¹	A_s deg.	P mm Hg	h_m m	D_m m	A_m deg.	D_p m	A_p deg.	D_s m	D_c m	Shot number
Series I																				
A	1*	1540	2.26	10.6	10.0	4.74	O	R	4.56	0.55	90	19.0	0.107	0.178	58	0.44	73	—	—	790634
	2*	1540	2.26	10.6	10.0	4.74	O	R	37.35	1.80	90	19.5	0.425	—	58	—	75	1.34	—	790635
	3*	1550	2.26	10.4	10.0	4.74	O	R	37.35	1.84	90	18.0	0.383	0.672	64	—	77	1.29	—	790636
	4	1550	2.26	11.5	10.0	4.74	O	P	37.40	2.06	90	16.0	0.325	0.463	55	0.71	60	—	—	790701
B	5*	1560	2.77	13.8	10.0	7.72	O	R	4.60	1.15	90	30.0	0.132	0.229	70	0.46	73	—	—	790702
	6*	1560	2.77	14.7	10.0	7.72	O	R	37.23	1.57	90	15.0	0.355	0.621	55	1.45	74	1.25	—	790703
	7	1560	2.77	14.4	10.0	7.72	O	R	37.23	1.71	15	12.0	0.215	0.386	56	1.14	67	1.00	—	790704
	8	1560	2.77	15.5	10.0	7.72	O	R	37.36	1.36	15	20.0	0.241	0.471	52	1.32	68	1.00	—	790705
	9	1580	—	—	—	—	O	P	37.36	2.09	90	25.0	—	—	—	—	—	—	—	790706
	10	1580	—	—	—	—	O	P	161.50	1.49	90	16.0	0.696	0.401	—	—	58	0.55	—	790708
C	11*	1580	2.29	72.3	5.0	6.43	O	R	4.56	1.09	90	18.0	0.089	0.216	80	0.57	73	—	0.290	790709
	12*	1580	2.29	65.6	5.0	6.43	O	R	37.36	1.87	90	18.0	0.301	0.617	61	1.60	74	0.98	—	790710
	13*	1580	2.29	38.4	5.0	6.43	O	R	15.93	1.55	90	10.0	0.190	0.345	59	0.80	74	—	—	790711
	14*	1580	2.29	32.0	5.0	6.43	O	R	10.55	1.72	90	19.0	—	—	—	1.50	—	—	—	790713
	15	1580	2.29	—	—	6.43	O	R	29.97	2.06	15	15.0	0.181	0.305	68	0.50	68	—	—	790716
	16	1580	2.29	—	—	6.43	O	R	29.04	0.81	15	12.0	0.057	0.127	99	—	66	—	0.170	790719
	17	1580	2.14	36.8	10.0	4.60	O	P	37.29	3.78	90	9.0	0.200	0.397	66	0.57	47	—	—	790721
D	18*	1640	9.20	44.8	10.0	33.97	O	R	15.88	1.07	90	14.0	0.090	0.228	100	0.51	73	—	0.360	790722
	19*	1640	9.20	61.6	10.0	33.97	O	R	10.56	1.72	90	16.0	0.156	0.412	96	—	70	—	0.600	790723
	20	1640	9.20	—	—	33.97	O	R	30.00	2.03	15	7.0	0.051	0.156	110	—	67	—	0.290	790725
	21*	1640	9.20	48.0	10.0	33.97	O	R	15.90	5.24	90	4.5	0.248	0.436	79	0.80	67	—	0.560	790726
	22	1640	9.20	25.6	5.0	33.97	O	R	30.00	4.33	15	6.0	0.026	0.091	130	—	60	—	0.250	790728
	23	1640	9.20	96.0	5.0	33.97	O	W	37.27	2.19	90	4.0	0.278	0.366	58	0.630	70	—	—	790731
	24	1640	9.20	96.0	5.0	33.97	O	W	106.20	1.71	90	19.0	0.379	0.521	48	1.320	75	1.04	—	790732
	25	1640	9.20	155.0	5.0	33.97	O	P	37.36	5.35	90	10.0	0.304	0.385	67	—	43	—	—	790733

Group	No.																			
E	26*	1720	10.87	160.0	10.0	39.81	O	R	15.82	1.62	90	21.0	—	—	—	—	71	—	0.343	790734
	27*	1720	10.87	176.0	10.0	39.81	O	R	37.29	1.57	90	19.0	0.050	0.237	131	0.570	76	—	0.520	790735
	28	1720	10.87	—	—	39.81	O	R	105.50	1.48	90	22.0	0.160	0.319	81	0.708	77	—	0.776	790736
	29	1720	10.87	—	5.0	39.81	O	R	29.90	1.95	15	21.0	—	—	—	0.855	66	—	0.315	790737
	30	—	—	—	—	—	O	P	106.30	1.51	90	28.0	—	—	—	1.850	—	—	—	790739
F	31	1760	—	122.0	10.0	—	O	R	15.94	1.54	90	20.0	—	—	—	—	—	—	—	790740
	32	1760	—	—	—	—	O	R	15.91	1.80	90	21.0	—	—	—	—	59	—	0.273	790741
	33	1760	—	275.0	5.0	—	O	R	37.65	1.79	90	20.0	—	—	—	—	—	—	—	790742
	34	1760	—	256.0	5.0	—	O	R	29.91	2.03	90	20.0	—	—	—	—	—	—	—	790744
	35	1760	—	243.0	5.0	—	O	D	37.62	2.11	90	20.0	—	—	—	—	62	—	0.456	790745
	36	1760	—	237.0	5.0	—	O	D	15.94	1.62	90	19.0	—	—	—	—	—	—	0.210	790746
	37	1760	—	—	—	—	O	R	12.59	1.47	15	20.0	—	—	—	—	65	—	0.143	790747
G	38	1800	13.75	256.0	5.0	0.11	W	R	4.69	1.01	90	30.0	—	—	—	—	64	—	0.266	790749
	39	1800	13.75	179.0	5.0	0.11	W	R	37.57	0.55	90	30.0	—	—	—	—	68	—	0.473	790750
	40	1800	13.75	198.0	5.0	0.11	W	W	12.52	1.02	90	100.0	—	—	—	—	68	1.23	0.325	790751
H	41	1770	1.93	57.6	5.0	6.92	W	R	4.67	1.64	90	30.0	0.018	0.068	—	—	65	—	0.227	790752
	42	1770	—	—	2.5	210.00	W	R	15.93	2.22	15	32.0	0.050	0.275	120	—	62	—	0.271	790754
	43	1770	—	115.0	2.5	210.00	W	R	37.59	2.07	90	31.0	0.077	0.382	144	—	63	—	0.511	790755
	44	1770	—	102.0	2.5	210.00	W	R	105.49	1.53	90	30.0	—	—	—	0.726	68	—	—	790756
	45	1770	—	141.0	2.5	210.00	W	R	37.29	2.29	90	26.0	—	—	—	0.717	70	—	0.900	790757
I	46	1730	1.56	54.4	5.0	12.42	W	T	15.96	0.92	15	26.0	—	—	—	—	68	—	0.224	790758
	47	1730	1.56	57.6	5.0	12.42	W	R	4.61	1.61	90	27.0	0.032	0.200	150	—	69	—	0.269	790759
	48	1730	1.56	70.4	5.0	12.42	W	R	15.68	1.85	90	26.0	0.054	0.343	144	—	70	—	0.466	790760
	49	1730	1.52	60.8	5.0	12.42	W	R	37.28	1.92	90	14.0	—	—	—	—	69	—	0.583	790762
	50	1730	1.52	80.0	5.0	12.42	W	R	4.67	5.49	90	10.0	—	—	—	—	68	—	0.432	790763
J	51	1800	—	173.0	5.0	—	W	T	37.28	2.17	90	16.0	—	—	—	0.470	65	—	0.191	790764
	52	?	—	80.0	5.0	—	W	T	37.26	2.05	90	12.0	—	—	—	—	69	—	0.469	790765
	53	1550	1.61	39.2	10.0	22.96	W	R	3.69	1.03	90	50.0	0.081	0.191	93	0.313	72	—	—	790766
	54	1550	1.61	45.6	10.0	22.96	W	R	3.78	2.12	90	40.0	0.120	0.236	85	0.448	75	—	0.236	790767
	55	1550	—	45.6	5.0	—	W	T	37.29	2.17	90	16.0	0.216	0.417	63	0.681	68	—	0.477	790768

Table 1. (*Continued*)

		Target properties							Impact conditions				Crater features							
Subset	Run #	ρ kg·m⁻³	η_P kg·m⁻¹s⁻¹	η_A kg·m⁻¹s⁻¹	RPM	τ kg·m⁻¹s⁻²	Clay base	Target config.	m ×10⁻⁵kg	v ×10²m·s⁻¹	A_s deg.	P mm Hg	h_m m	D_m m	A_m deg.	D_p m	A_p deg.	D_s m	D_c m	Shot number
Series II																				
	56	1604	—	84.0	10.0	—	O	T	37.63	1.40	90	120.0	—	—	—	0.762	—	—	0.585	791207
	57	1604	—	110.0	10.0	—	W	T	4.59	1.56	90	110.0	—	—	—	—	—	—	0.356	791208
	58	1614	—	—	—	—	W	R	4.59	1.50	90	60.0	—	—	—	—	—	—	0.380	791209
	59	1614	—	—	—	—	W	N	4.63	1.60	90	120.0	—	—	—	—	—	—	0.330	791210
	60	1614	—	230.0	10.0	—	O	T	37.66	2.06	90	54.0	—	—	—	—	—	—	0.380	791211
	61	1747 / 1611	—	310.0 / —	10.0 / —	—	W	T	37.66	2.00	90	90.0	—	—	—	—	—	—	0.410	791212
	62	1761 / 1699	—	880.0 / —	10.0 / —	—	W	T	37.66	2.20	90	120.0	—	—	—	0.430	—	—	0.290	791213
	63	1761 / 1666	—	760.0 / —	10.0 / —	—	W	T	37.68	2.18	90	80.0	—	—	—	0.650	—	—	0.540	791214
	64	1755 / 1690	—	1040.0 / 280.0	10.0 / —	—	W	T	37.61	2.06	90	40.0	—	—	—	—	—	—	0.380	791216
	65	1840 / 1740	—	1400.0 / 445.0	10.0 / —	—	W	T	37.64	2.13	90	80.0	—	—	—	0.470	—	—	0.305	791217
	66	1745	—	480.0	10.0	—	W	T	37.60	2.07	90	—	—	—	—	—	—	—	0.219	791218
	67	2020	—	1150.0	10.0	—	W	R	4.43	1.90	90	40.0	—	—	—	—	—	—	0.203	791219
	68	1720 / 1770	—	620.0 / 560.0	10.0	—	W	I	37.65	2.20	90	2.00	—	—	—	—	—	—	0.140	791220
	69	1720	—	620.0	10.0	—	W	I	37.71	2.21	90	6.00	—	—	—	0.380	—	—	0.330	791221
	70	1670	—	570.0	10.0	—	W	I	4.72	2.09	90	10.00	—	—	—	—	—	—	0.210	791222
	71	1610	—	225.0	10.0	—	W	I	4.45	1.70	90	45.00	—	—	—	—	—	—	0.241	791224
	72	1638	—	220.0	10.0	—	W	I	4.45	1.80	90	25.00	—	—	—	—	—	—	0.162	791225
	73	1638 / 1638	—	240.0 / 240.0	10.0	—	W	I	37.69	2.18	90	6.00	—	—	—	—	—	—	0.318	791226
	74	1620	—	280.0	10.0	—	W	R	4.41	2.10	90	10.00	—	—	—	0.430	—	—	0.203	791227
	75	1610	—	155.0	10.0	—	W	S	4.56	2.01	90	80.00	—	—	—	0.254	—	—	0.203	791228

* denotes shots used in Figs. 5–7.

some cases, this bulge is preserved as a remnant of the initial crater rim. The angle the expanding ejecta plume makes with the target surface appears to be inversely proportional to target strength or cohesion. Thus, the plume angle is quite steep in water, decreases with thicker clay slurries, and is lowest in dry sand (Fig. 3). The initial ejecta plume appears to be composed of discrete parcels of ejecta which quickly merge to form a relatively continuous sheet (Fig. 2). The expanding plume then tears into discrete segments when tensional stresses exceed material cohesion. This, too, appears to be qualitatively related to viscosity, ranging from a nearly continuous ejecta plume for water, to large "plates" or clots for very fluid clay, to small pieces for more viscous clay, and finally to individual grains for dry sand targets. The effect on ejecta morphology is shown in Fig. 4, in which the more fluid targets display well defined continuous outer boundaries progressing to the more familiar lunar-type "dry" ejects deposits consisting of various facies (continuous ejecta, discontinuous ejecta, bright rays).

The greatest and perhaps most significant differences between impacts in dry and viscous targets occur during the *post-impact modification* stage (Fig. 2). When the transient cavity reaches its maximum depth, isostatic adjustment begins immediately; the rapidity and magnitude of the adjustment being greater with decreasing viscosity. In relatively fluid targets, the recovery of the floor is so rapid that a rebounding central mound develops, its maximum height varying directly with impact energy and inversely with target viscosity and strength. Gravitational collapse of the central mound generates a surge of material that in some cases is sufficient to surmount the crater rim, sending a blanket of material over earlier emplaced ejecta plume deposits, although the radial extent of this surge is less than that of the plume deposit. The extent of the surge deposit is a function of central mound height, crater rim height, and fluidity of the target material.

Depending upon impact energy and target viscosity, the central mound may oscillate, forming a series of central mounds of decaying heights. Each of these mounds may produce a surge deposit of lesser and lesser extent, with the later units not having sufficient energy to surmount the crater rim. Oscillation was greatest in the very fluid targets, and in most cases the crater rim was not preserved as a topographic form after oscillation ceased. Oscillation of the central mound ceases when the energetic forces from the impact are dissipated and shear stresses are less than the yield strength of the target. The mound may "freeze" in a negative position (forming a central depression), depending also on the target strength.

Immediately after some shots, the target material in and near the impact zone actively degassed; evidently the impact released volatiles from the clay-water and clay-oil mixtures, fluidizing the target material. The effect was two-fold: in 74 percent of the impacts there was a decrease in target viscosity after impact (measured *in situ* before and after impact), and the post-impact fluidized material smoothed out irregularities in the floors of some craters, giving them a flooded appearance.

Thus, in this general model for impacts into viscous targets at laboratory scales

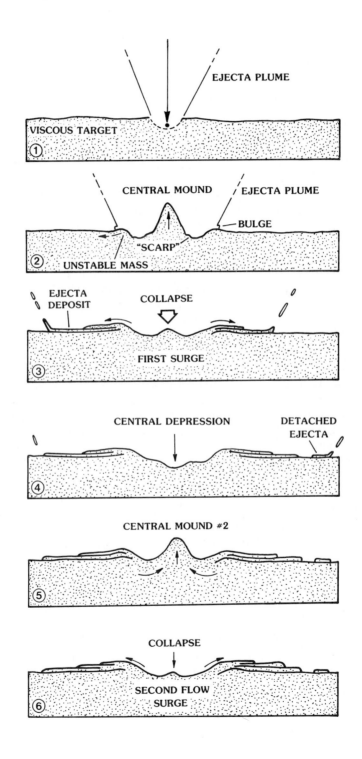

it is possible to produce multiple tiers of ejecta deposits, central peaks, flat floors, and central pits, all as functions of impact energies and target viscosity and strength. We point out, however, that after the generation of the first central mound in the experiments, subsequent oscillations may have been influenced by reflected waves from the walls and floor of the target bucket.

Impact at angles as low as 15° above the surface had no apparent effect on either the ejecta pattern or the crater morphology. In some experiments involving both normal and oblique impact angles, the collapsing central mound was asymmetric (perhaps because of slight inhomogenieties in the target) which resulted in asymmetric ejecta flow deposits.

The oscillation of the central mound may well be influenced by boundary effects of the target bucket, particularly in the high energy rounds. To determine these effects we recently ran a series of experiments ranging from very low energy shots in which the size of the crater was less than 1/20 of the bucket diameter, to high energy shots which clearly showed influence of the bucket walls. In all cases, oscillating central mounds could be produced if the ratio of impact energy to viscosity was high enough. (These results will be presented in more detail in a later paper.)

3.2 Dimensional analyses and functional relationships

In order to provide a basis for scaling to full size impacts, parametric relationships were sought in dimensionless form. Because our early experiments aimed to determine qualitative relationships, we varied many parameters and used slightly different recording techniques during different subsets of experiments. Of the total of 75 shots, 14 share sufficient conditions to be directly comparable. In the following discussion we consider only those experiments involving vertical impacts into homogeneous clay-oil targets. No multilayer targets are included.

Impact experiments into water (Gault and Wedekind, 1978), dry sand and soils, and rock (Gault, 1973; Gault and Wedekind, 1977) show that crater dimensions depend on impact energy, gravitational acceleration, and target strength. Crater diameter, for example, scales with the ¼ power of the impact energy for inviscid targets such as water and approximately with the ⅓ power of energy for targets

---

Fig. 1. Sequence of impact cratering in viscous targets derived from analysis of high speed motion pictures (compare to Fig. 2): 1) formation of ejecta plume; 2) rebound of floor produces a central mound; ejecta plume expands outward, pushing a small "bulge" of target material across the surface; ejecta plume begins to tear into discrete segments; 3) ejecta plume forms a deposit on the surface; gravitational collapse of central mound sends a surge of material up and over the transient crater rim to form a deposit overlying the ejecta plume deposits; 4) central mound oscillates to form a central depression, which in some cases "freezes" to leave a central pit; late-stage ejecta plume consists of discrete segments deposited as detached ejecta lobes; 5,6) oscillating central mound sends successive lobes of material over the rim to form terraced, superposed ejecta deposits. (Numbers refer to sequence, not time).

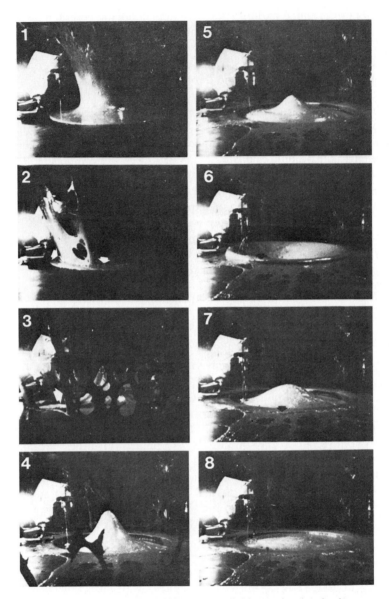

Fig. 2. Sequential frames from shot #6 (790703) showing impact in target having apparent viscosity of 150 poises (15 kg-m^{-1}-s^{-1}): 1) ejecta plume; 2) white arrows identify "bulge" of material being pushed ahead of expanding ejecta plume; ejecta plume begins to tear; 3) rising central mound is visible through "windows" in the expanding ejecta plume; 4) outline of target bucket visible; 5) central mound collapses, sending a surge of material (arrow in Frame 6) out of the bucket to be superimposed on the ejecta plume deposit (arrow in Frame 7); 8) due to low strength of target material, no crater topography was preserved. Device on the left of each frame is the viscometer.

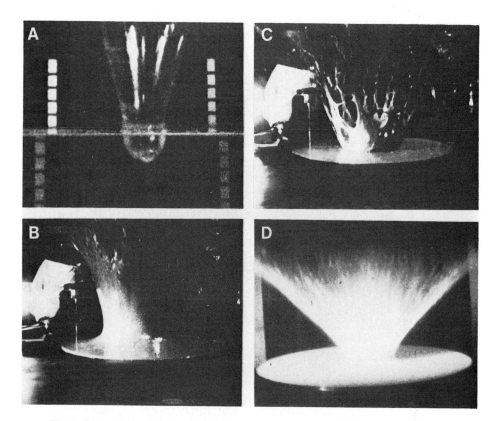

Fig. 3. Comparison of ejecta plume angles (A_p) for impacts in different target materials showing that increase in viscosity (η) leads to decrease in plume angle. A) Water ($\eta \sim 10^{-2}$P; $A_p = 75°$); B) "wet" clay slurry ($\eta \sim 10^2$P; $A_p = 70°$); C) stiff clay slurry ($\eta \sim 10^3$P; $A_p = 66°$); D) dry sand ($A_p = 58°$).

with high strength such as solid granite or basalt (Gault, 1973). Our current experiments employ target materials which are viscous and have a finite yield strength. Earlier exploratory impacts into viscous targets indicated that crater dimensions were proportional to the impact energy and to the inverse of the target viscosity (Gault and Greeley, 1978). To determine the relative influences of impact energy, target viscosity and yield strength on crater dimensions, we used dimensional analyses and combined the experimental variables into dimensionless groups. We considered six variables to be most important in the cratering process: impact energy (E); target viscosity (η), yield strength (τ), and density (ρ); gravitational acceleration (g); and some crater dimension (*l*) such as height of the central mound (h_m), average diameter of the ejecta plume deposit (D_p) or diameter of the final crater (D_c).

Of the many possible dimensionless groups, we first considered one which relates a crater dimension such as h_m to the energy of impact and the target weight

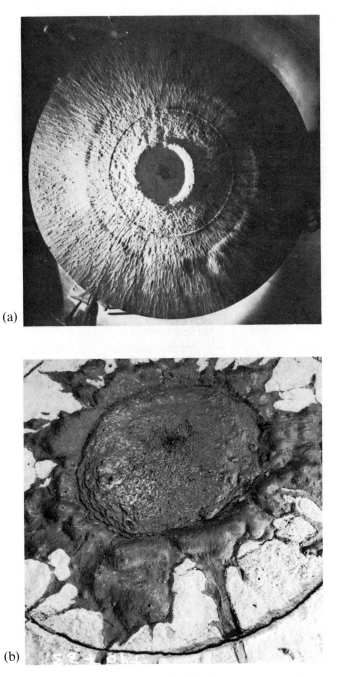

(a)

(b)

Fig. 4. Comparison of ejecta morphology for "dry" sand target (a) showing ragged "lunar" type deposits with that of an impact in viscous target (b) showing well defined ejecta boundary. (a) involved impact of pyrex sphere into dry pumice powder, (b) involved impact into clay-oil mixture.

(ρg). The group $\dfrac{E}{\rho g h_m^4}$ implies that the fourth power of the peak height should be proportional to the energy and the inverse of target weight. Thus, where this "gravity scaling" applies, large increases in impact energy, decreases in target density or decreases in gravitational acceleration all lead to relatively small increases in central peak height. This prediction was experimentally corroborated for the transient crater diameter by impacts into water (Gault and Wedekind, 1978).

A second scaling factor includes the influence of target strength, $\dfrac{E}{\tau h_m^3}$, which implies that for strength-dominated impacts, crater dimensions should scale as the ⅓ power of the energy divided by strength. This functional relationship was observed experimentally for impacts into targets of solid basalt and granite (Gault, 1973).

A third dimensionless group relates energy to viscosity, $\dfrac{\rho E}{\eta^2 h_m}$, and indicates that crater dimensions scale as the inverse of the square of viscosity. This second power dependence suggests that when viscosity effects operate, they have stronger influence on crater morphology than energy, gravity, target strength or density.

In our experiments we sought to determine the conditions under which gravity, target viscosity, or strength would dominate. We can determine the relative influences of yield strength and viscosity by comparing a viscous stress, $\eta^2/\rho h_m^2$, to yield strength, τ. For a typical experiment (all units m-k-s), $\tau \simeq 10$, $\eta \simeq 20$, $\rho \simeq 1500$, and $h_m \simeq 0.2$ (Table 1), so that $\eta^2/\rho h_m^2 = \dfrac{400}{(1500)(0.04)} \simeq 6 \simeq \tau$. Thus, strength and viscous effects should be comparable. Now considering gravitational stress, $\rho g h_m$, for the same conditions where $g \simeq 10$, we find $\rho g h_m \simeq 3000 \gg \tau \simeq \eta^2/\rho h_m^2$. Gravity effects should thus predominate in our experiments so that central mound height should scale as the ¼ power of the impact energy. The one exception is run #790735 (Figs. 5–7), where the calculated viscous stress was much larger than the gravity stress, due to the high target viscosity and low height of the central mound. In any such cases where strength or viscous effects apply, peak height should scale as a greater power of the impact energy, closer to ⅓ or 1.

In Figs. 5–7 we have plotted central peak height vs. different energy terms E/ρg, E/τ and ρE/η^2. Power law curves of the form $h = k(E^*)^n$, where E^* is the energy term and k and n are constants, were then fitted to the data. Figure 5 compares the effects of gravity and impact energy on mound height:

$$h_m = 0.7 \left(\frac{E}{\rho g} \right)^{0.28}, \text{r (correlation coefficient)} = 0.82.$$

Mound height exhibits only a slightly greater dependence on energy (0.28 vs. 0.25) than the fourth root scaling expected if gravity alone were modifying the

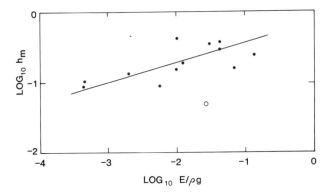

Fig. 5. Plot of maximum central mound height (h_m) as a function of gravity (g), target density (ρ) and energy (E); correlation coefficient = 0.82. Open circle = shot #790735 (see text).

crater morphology, and this difference is probably not statistically significant here.

Figure 6 shows the effect of the yield strength and indicates a closer correlation with the theoretically predicted ⅓ root scaling for mound height:

$$h_m = 0.8(E/\tau)^{0.32}, r = 0.94.$$

The higher exponent value and the better correlation both suggest that yield strength exerts a significant influence on the cratering process.

Figure 7 shows central mound height scaled against energy and viscosity. The correlation of r = 0.95 is better than that for the gravity scaling in Fig. 5, although

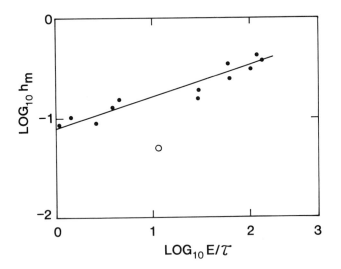

Fig. 6. Plot of maximum central mound height (h_m) as a function of target yield strength (τ) and energy (E); correlation coefficient = 0.94.

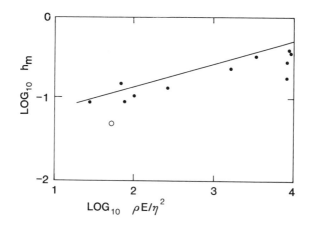

Fig. 7. Plot of maximum central mound height (h_m) as a function of target density (ρ), energy (E) and apparent viscosity (η); correlation coefficient = 0.95.

the power law exponents, n, are closer to the 0.25 characteristic of gravity scaling than to the 1.0 for viscous scaling:

$$h_m = (.04)\left(\frac{\rho E}{\eta^2}\right)^{0.28}, r = 0.95.$$

Thus it appears that in these experiments the target behaved more as an inviscid fluid than as a viscous one. In those runs where viscous forces dominated, the mound height was greatly reduced or nonexistent. We infer that in order to form "ejecta flow" craters through the agent of an oscillating central mound (see next section) gravitational effects must greatly outweigh viscous ones. Choosing a typical martian multilobed crater, ~18 km diameter Yuty crater, we can roughly compute gravitational and viscous stresses as follows. Based on our experiments (Table 1) central mound height is approximately equal to ⅓ the crater diameter so that Yuty might have had a 6 km high transient central mound. Assuming the bulk density of martian surface material to be at least 1500 kg-m^{-3} (measured value for martian soil, Moore *et al.*, 1977) and g = 3.8 m-s^{-2}, we calculate the conditions under which $\rho gh \gg \eta^2/\rho h_m^2$. Substitution yields a maximum apparent viscosity of 10^9 kg-m^{-1}-s^{-1} or 10^{10} poises. Johnson (1970, p. 513) calculated a plastic viscosity (see Appendix) of about 10^3 poises and an apparent viscosity of about 10^4 poises for a debris flow in California. Thus if the rheological behavior of the impacted martian crust were comparable to that of terrestrial debris flows, we could expect gravitational effects to predominate over viscous ones.

4.0 DISCUSSION AND PLANETARY IMPLICATIONS

The general objective of the experiments was to determine the morphology of impact craters formed in viscous targets. The results described above apply to the laboratory models. Extrapolation to planetary scales requires a host of dimensionless parameters, few of which can be satisfied simultaneously. Nonetheless, the experiments provide qualitative insight into the general processes involved and, through the isolation and study of individual parameters, some understanding can be gained of the processes involved in the formation of martian multilobed craters and of the effects of target viscosity on morphology (Boyce and Roddy, 1978). Photogeologic studies show that there is more than one type of martian multilobed crater (Carr *et al.,* 1977; Mouginis-Mark, 1979, Mouginis-Mark and Head, 1979; and others) and therefore it may not be appropriate to attempt to derive a single model of formation. Rather, once the morphologic types are fully classified, individual models may be required to explain each class. Until such a classification is developed, we prefer to discuss elements of multilobed craters and to suggest the implications of our experiments for those elements.

The primary characteristic of multilobed craters is the flow-like ejecta. Martian ejecta deposits consist of several facies, not all of which are necessarily present around all craters: 1) bright ray patterns; 2) deposits from secondary cratering; 3) isolated, detached flow lobes; and 4) flow lobes of continuous ejecta (which can occur in multiple tiers). Radial striations can occur on some or all of the flow lobes.

Bright ray patterns (Fig. 8) are similar to those observed around lunar and mercurian craters. Although rare on Mars, where rays do occur, they commonly are overlain by ejecta flow lobes or other ejecta deposits. Rays may be present on craters of equal size and located on apparently similar geologic material in the same area, suggesting that changes in target properties influence their development. We suggest that the bright rays are formed by ejection of an upper "dry" regolith or surface layer during the initial high velocity phase of ejecta excavation, comparable to impact cratering on the moon, and consist of both ejecta and mixing of local materials. The initial ejecta would be thrown farthest (ejected at highest velocities), followed by excavation of deeper, viscous material to form the flow-type deposits. Thus, the presence and degree of development of bright rays could be a function not only of stage of preservation (age), but the presence and thickness of a "dry" upper layer.

Detached ejecta deposits isolated from the main part of the ejecta field occur around many martian multilobed craters. These patches appear to be *primary* features rather than remnants of an eroded continuous ejecta blanket (Arvidson *et al.,* 1976). In the experiments the ejecta plume for fluid targets tears into descrete clots (Gault and Greeley, 1978), which impact the surface as separate units (Fig. 2). At the scale of martian craters, such clots may not remain as a single coherent large mass but disperse into a system of smaller clots (Schultz and Gault, 1979). Nevertheless, dependent on the degree of dispersion, upon impact they could coalesce and produce the detached ejecta clumps observed around certain martian craters.

The main ejecta deposit (Fig. 1) can be derived from two sources: 1) the initial ejecta plume, and 2) post-ejecta plume deposits. The initial ejecta plume deposits are emplaced ballistically on the surface. Once on the surface, however, they

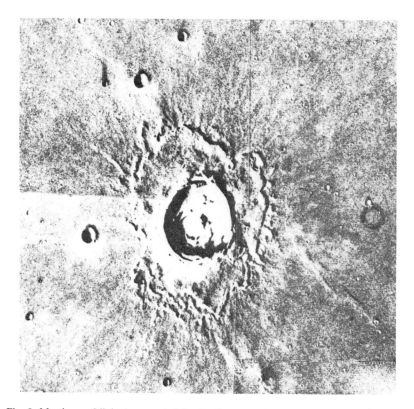

Fig. 8. Martian multilobed crater (~8 km in diameter) showing bright ray pattern interpreted to result from initial ejecta being derived from an upper "dry" regolith layer; deeper excavation penetrates volatile rich material that is emplaced as a fluidized mass which overrides part of the bright ray material. (VO 10A 21,22,23).

may behave similarly to landslides assuming the following conditions: 1) the ejecta mass consists of fragmental debris, possibly containing entrained volatiles [gases, water, melt-water (Carr *et al.*, 1977)], 2) the region around the crater is still seismically active (Schultz and Gault, 1975) from the impact as the ejecta plume material is being deposited, and 3) the mass is oversteepened on the rim of the crater; all three conditions favor mass wasting processes (Sharpe, 1938). In addition, possible early arrival of certain ejecta fractions in layered lithologies may form a lubricating layer of material (Schultz and Gault, 1979) that would enhance mass wasting. The third factor, oversteepening, results from: 1) the higher gravity on Mars causing deposition closer to the crater rim than in the lunar case, 2) the presence of an atmosphere which may cause deceleration of the finer size ejecta, including dispersed water droplets and vapor (Schultz and Gault, 1979) and its subsequent deposition closer to the crater, and 3) as suggested by the experiments, the ejecta plume angle varies inversely with the viscosity; thus the ballistic trajectory of low viscosity plume material would be at a high angle with deposition near the rim. Given these conditions, we assume that ejecta plume deposits can

be treated essentially as landslides, thus accounting for the flow-like characteristics previously described (Carr *et al.*, 1977). Although we envision these deposits to be emplaced early in the sequence, the unstable deposits could fail and
flow conceivably any time in the history of the crater, and more than one failure
could occur in the same mass to produce overlapping flow lobes.

The second source for the main mass of ejecta could be the collapsing central
mound; the size of the rebounding central mound and the extent of the surge
deposit resulting from its collapse vary inversely with the viscosity of the target.
Although this deposit would be superposed on the ejecta plume deposit, its lateral
extent would be less. It is conceivable, however, that the surge deposit could
extend beyond the continuous plume deposits if it were derived from a deep,
highly fluidized substratum. Although some estimates of ejecta blanket areas
versus crater areas and volumes (Mutch and Woronow, 1980; Mouginis-Mark and
Carey, 1980) have been made, a rigorous analysis of the volumes has not been
performed. However, first order estimates (Fig. 9) of ejecta volume for several
martian multilobed craters exceed by several times the volume of the crater bowl
(even taking into account the transient crater volume and "bulking" of the ejecta),
suggesting that flow of subsurface material during the excavation stage has occurred.

An oscillating central mound, similar to that observed in the experiments, could

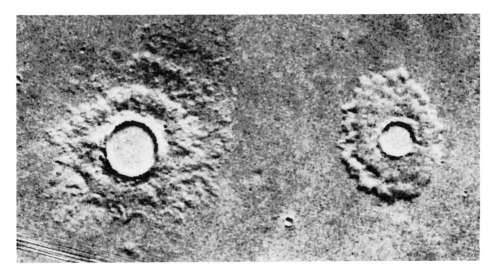

Fig. 9. Two martian craters for which estimated ejecta volumes exceed maximum calculated volume of crater bowl, even taking into account "bulking" of ejecta and uncertainties inherent in determining vertical lengths from shadow measurements. Crater
at left is 4.8 km in diameter by 0.2 km deep, giving a maximum crater volume (assuming
a hemispherical bowl shape) of 28.9 km^3, ejecta volume is estimated to be 38.8 km^3; 3.8
km diameter crater at right has a minimum volume of 1.4 km^3, a maximum volume of
7.1 km^3, and an ejecta volume of 38 km^3. Thus, some mechanism is required other than
"normal" excavation to account for excess volume. We interpret excess to result from
subsurface flow of material during uplift of the central mound.

Fig. 10. A 35 km-diameter crater southeast of Amazonis Planitia showing well-developed interior terraces and slumping of the wall in large competent blocks, and lunar-type secondary craters; martian craters of this type appear to lack well developed multiple flow lobes (although an outer scarp of ejecta is present), and may reflect impact into materials having less volatiles than those in which multilobed craters form. (VO Frame 635A82).

account for the formation of the multiple tiers of ejecta lobes. We point out that in the experiments vigorous oscillations involved very fluid targets in which the crater rim was rarely preserved. Although some of the oscillations in the experiments were probably due to waves reflected from the walls and floor of the target bucket, in some respects this could be analogous to reflected energy from a competent rock stratum underlying unconsolidated volatile-rich material. We note that a model involving an oscillating central peak has been proposed to explain certain multi-ringed basins (Murray, 1980).

The emplacement mechanisms for the main ejecta mass described here involve flow of debris comparable to mass wasting. Striations observed on some ejecta lobes of martian craters may be shear lines similar to those that occur in landslide masses. However, striations on certain ejecta lobes appear to cross continuously both the inner lobes and outer lobes. This continuity may reflect shear lines propagated upward through overlying lobes, or as suggested by Mouginis-Mark (1980) and Singer and Schultz (1980), they may reflect scour marks by late-stage ejecta emplacement.

Although a planet-wide analysis has not been carried out, some martian craters show extensive wall slumping and interior terraces, tend to have less well developed multiple ejecta flow lobes, and tend to display secondary crater fields more typical of lunar and mercurian impact craters (Fig. 10). We interpret these craters to have formed in more competent materials than that of the typical multilobed crater. Where such craters exist near multilobed craters on the same geological unit, it suggests that the target properties may have changed with time such as loss or gain of subsurface volatiles, with or without a change in atmospheric density. Conversely, the walls of many multilobed craters lack multiple terraces and have shallow, flat floors that appear to be flooded, perhaps as a result of local target degassing from the impact and its fluidization, comparable to that observed in the experiments.

5.0 CONCLUSIONS

The impact experiments in viscous targets produced results that at least qualitatively suggest mechanisms of formation for some of the martian multilobed craters and their associated features. Furthermore, some of the results may have a bearing on problems of large impact craters (i.e., basins) and the formation of multiple rings and impacts into "magma oceans" or accumulations of semimolten lava. The experiments almost certainly have implications for impacts into icy and ice-silicate bodies such as Ganymede and Callisto. The difficulty in applying the experimental results lies in scaling time, size and material properties to planetary conditions. Such problems, however, occur in all modeling, and can be overcome partly through dimensional analysis. We are continuing these analyses, and concurrently conducting impacts into targets including layers of differing viscosities to simulate multilayered ice/silicate mixtures, and other more complex planetary crusts.

Acknowledgments—Many people contributed to the successful completion of the experiments described here. We thank first the staff of the NASA-Ames Vertical Ballistic Gun Facility for their ready willingness to accomodate our needs; C. Wilbur and P. Spudis aided in the fabrication of special target buckets; R. Leach and M. Plummer assisted with the measurements of the rheological properties of the various mixtures; T. Timmcke assisted in some of the measurements taken from the motion pictures; V. Sisson, Planetary Geology Summer Intern, assisted with the first series of experiments. Finally, we thank all who got their hands in the mud and who helped clean up after the experiments. This work was supported by NASA grants NSG-7429, NSG-7415, and NAGW-56.

The Lunar and Planetary Institute is operated by the Universities Space Research Association under Contract No. NSR 09-051-001 with the National Aeronautics and Space Administration. This paper constitutes the Lunar and Planetary Institute Contribution No. 417.

REFERENCES

Arvidson R. E., Coradini M., Carusi A., Caradini A., Fulchignoni M., Federico C., Funiciello R., and Salomone M. (1976) Latitudinal variation of wind erosion of crater ejecta deposits on Mars. *Icarus* **27**, 503–516.

Boyce J. M. and Roddy D. J. (1978) Martian rampart craters: crater processes that may affect diameter-frequency distributions (abstract). Reports of Planetary Geology Program 1977-1978, p. 162–165. NASA TM-79729.

Carr M. H., Crumpler L. S., Cutts J. A., Greeley R., Guest J. E., and Masurky H. (1977) Martian impact craters and emplacement of ejecta by surface flow. *J. Geophys. Res.* **82**, 4055–4065.

Gault D. E. (1973) Displaced mass, depth, diameter, and effects of oblique trajectories for impact craters formed in dense crystalline rocks. *The Moon* **6**, 32–44.

Gault D. E. and Greeley R. (1978) Exploratory experiments of impact craters formed in viscous-liquid targets: Analogs for martian rampart craters. *Icarus* **34**, 486–495.

Gault D. W., Quaide W. L., and Oberbeck V. R. (1968) Impact cratering mechanics and structures. In *Shock Metamorphism of Natural Materials* (B. M. French and N. M. Short, eds.), p. 87–99. Mono, Baltimore.

Gault D. E. and Wedekind J. A. (1977) Experimental hypervelocity impact into quartz sand, II: Effects of gravitational acceleration. In *Impact and Explosion Cratering* (R. O. Pepin and R. B. Merril, eds.), p. 1231–1244. Pergamon, N.Y.

Gault D. E., and Wedekind J. A. (1978) Experimental impact "craters" formed in water: Gravity scaling realized. *EOS* (*Trans. Amer. Geophys. Union*) **59**, 1121.

Grimshaw R. W. (1971) *The Chemistry and Physics of Clays*. Wiley, N.Y. 1024 pp.

Hulme G. (1974) Interpretation of lava flow morphology. *Geophys. J. Roy. Astron. Soc.* **39**, 361–383.

Johansen L. A. (1978) Martian splosh cratering and its relation to water. *Proc. Second Colloqium on Planetary Water and Polar Processes*, p. 109–110. Hanover, N.Y.

Johnson A. M. (1970) *Physical Processes in Geology*. Freeman, San Francisco. 577 pp.

Moore F. (1965) *Rheology of Ceramic Systems*. MacLaren, London. 78 pp.

Moore H. J., Hutton R. E., Scott R. F., Spitzer C. R., and Shorthill R. W. (1977) Surface materials of the Viking Landing Sites. *J. Geophys. Res.* **82**, 4497–4523.

Mouginis-Mark P. (1979) Martian fluidized crater morphology: variations with crater size, latitude, altitude and target material. *J. Geophys. Res.* **84**, 8011–8022.

Mouginis-Mark P. (1980) An emplacement sequence for martian fluidized ejecta craters (abstract). In *Lunar and Planetary Science XI*, 753–755. Lunar and Planetary Institute, Houston.

Mouginis-Mark P. J. and Carey D. L. (1980) Volume estimates of fluidized ejecta deposits in the northern plains of Mars (abstract). In *Lunar and Planetary Science XI*, 759–761. Lunar and Planetary Institute, Houston.

Mouginis-Mark P. J. and Head J. W. (1979) Emplacement of martian rampart crater ejecta blankets: A morphological analysis (abstract). In *Lunar and Planetary Science X*, 870–872. Lunar and Planetary Institute, Houston.

Murray J. B. (1980) Oscillating peak model of basin and crater formation. *Moon and Planets* **22**, 269–291.

Mutch P., and Woronow A. (1980) Martian rampart and pedestal craters ejecta-emplacement: Coprates Quadrangle. *Icarus* **41**, 259–268.

van Olphen H. (1977) *An Introduction to Clay Colloid Chemistry*. Wiley, N.Y. 318 pp.

Roddy D. J., Arthur D. W. G., Boyce J. M., Pike R. J., and Soderblom L. A. (1979) Martian impact cratering: Preliminary report (abstract). Reports of Planetary Geology Program 1978–1979, p. 187–189. NASA TM-80339.

Schultz P. H. and Gault D. E. (1975) Seismic effects from major basin formations on the Moon and Mercury. *The Moon* **12**, 159–177.

Schultz P. H. and Gault D. E. (1979) Atmospheric effects on martian ejecta emplacement. *J. Geophys. Res.* **84**, 7669–7687.

Scott R. (1967) Viscous flow of craters. *Icarus* **7**, 139–148.

Sharpe C. F. S. (1938) *Landslides and Related Phenomena*. Columbia Univ. Press. N.Y. 137 pp.

Shaw H. R., Wright T., Peck D., and Okamura R. (1968) The viscosity of basaltic magma: An analysis of field measurements in Makaopuhi Lava Lake, Hawaii. *Amer. J. Sci.* **266**, 225–264.

Singer J. and Schultz P. H. (1980) Secondary impact craters around lunar, mercurian and martian craters (abstract). In *Lunar and Planetary Science XI*, 1042–1043. Lunar and Planetary Institute, Houston.

van Wazer J. R., Lyons J. W., Kim K. Y., and Caldwell R. E. (1963) *Viscosity and Flow Measurements*. Wiley, N.Y. 406 pp.

Appendix
RHEOLOGY OF EXPERIMENTAL TARGET MATERIALS

An underlying assumption of our experiments was that the formation of multilobe craters requires a target which can behave both as a fluid to produce the observed flow patterns around obstacles and as a solid to produce the steep flow lobe fronts. Without specifying the composition of such a material, we can characterize its rheology by the Bingham model. A Bingham material deforms either as an elastic solid or a linearly viscous fluid, depending upon whether the applied shear stress is greater or less than a critical value called the yield strength, τ_y. For stress greater than τ_y, strain rate is proportional to applied stress and the proportionality constant is the plastic viscosity, η_p; below τ_y, no permanent deformation occurs. Bingham materials are thus characterized by these two material constants, τ_y and η_p. This plastic viscosity has the same dimensions as Newtonian viscosity and is a measure of a Bingham body's resistance to flow. Clay suspensions were selected as target materials because they commonly exhibit Bingham behavior.

Four groups of clay suspensions were made by combining one of two commercially available clay bodies with either water or silicon oil. The suspensions were mixed at the test facility using a power drill with a stirring attachment. The consistencies of the slurries were varied by adding more dry clay or more fluid (oil or water) to the containers and restirring.

In general, clay slurries exhibit a continuous range of flow properties from viscous to plastic, depending primarily on the concentration of suspended particles and their degree of interaction (Grimshaw, 1971). At volume concentrations of less than about 2 percent solids, clay slurries have a Newtonian viscosity which varies linearly with the concentration (Moore, 1965). At increasing clay concentrations the particles begin to inferfere with one another, leading to an exponential variation of viscosity with concentration. At about 30 percent solids many clay suspensions begin to develop an internal structure and an associated yield strength. This yield value increases roughly as the third power of the concentration (Moore, 1965). At high concentrations some clay suspensions may show strongly nonlinear behavior, in which the strain rate is proportional to the applied stress raised to some power ($\epsilon = |\sigma|^n$; ϵ = strain rate; σ = stress), the sign of n depending on the mineralogy of the clay particles. Time dependent effects are sometimes observed, caused by changes in yield strength accompanying increased duration of shearing stress.

These relationships between rheologic properties and other factors were determined under ideal laboratory conditions. Measurements made during our experiments were much less well controlled and several of the required experimental procedures could lead to inadvertant rheologic changes. It is impossible to monitor continuously all of the effects, and rheologic measurements must be taken for every batch of slurry used in the runs. Here we list some of the changes in viscosity and yield strength that might be expected in the preparation of our target materials.

During our experiments we frequently altered the concentration of the slurries, thus changing the rheologic properties. After energetic shots, much clay was expelled from the target bucket and had to be replaced. Homogenization of the slurry was then accomplished by stirring at very high speeds which could cause both the viscosity and yield strength to decrease in proportion to the amount of time of stirring, perhaps because interparticle forces are reduced. As the mixed clay is allowed to rest, its strength and viscosity may rise by an amount dependent upon the time of standing. In addition, large amounts of air are inadvertantly mixed into the slurry during stirring which also decreases both rheologic parameters. This effect is partly reversed by exsolution of the gas during periods when the clay is not being worked. Depressurization of the chamber before impact also promotes degassing which often leads to stiffening of the clay. In those experiments where a stiff surface layer of sand or clay was added, mixing in some of the stiffer component inevitably occurred after impact, raising the bulk viscosity of the slurry. Similarly, targets with icy surface layers also led to a higher viscosity and strength after impact.

Very high shear stresses are generated locally at the site of impact, where nonlinear effects might cause the actual viscosity to be different from that recorded either immediately before or after the shot. However, our rheologic measurements indicated that these power law effects were minimal in the slurries we used. The temperature of the slurries ranged over about 8°C during the course of the experiments. Within this range, the viscosities would not be greatly influenced (van Olphen, 1977).

Rheologic measurements

Rheologic data were collected using a Brookfield Model HBT Synchrolectric Viscometer with a Helipath attachment which allows measurement in highly viscous fluids. The viscometer rotates a spindle in the fluid and measures the torque necessary to overcome the viscous resistence to the induced movement. The spindles consist of a cylinder with either a disk or t-bar at the base. Neither of these geometries allows simple computations of the viscosity (van Wazer *et al.*, 1963). The torque readings indicate the "instrument viscosity" (Shaw *et al.*, 1968) which can be converted to the absolute viscosity for Newtonian fluids and the apparent viscosity for non-Newtonian fluids. The apparent viscosity is the ratio of total shear stress to shear rate and is not a material constant for non-Newtonian fluids. At a stated shear rate, however, it provides a measure of the fluid's resistance to flow that allows intercomparisons with other similar fluids. For Bingham materials, plastic viscosity and yield strength must be determined by indirect methods.

We attempted to calculate the Bingham properties of the clay slurries by following a procedure described by Hulme (1974). First the instrument readings were converted to absolute force values through calibration with a known mass and the force of gravity. Then, for the disk-type spindles, the area was measured and stress was computed by dividing the force by this area. Strain rate was equated with the measured rate of rotation of the shaft, and a plot of shear strain rate versus stress was constructed. These plots conformed to those expected for a Bingham material, and values of yield strength and plastic viscosity could be measured directly from the plots.

Although this procedure indicated Bingham-like properties, it has three inherent problems. First, the computations were only made for batch samples collected every 2 to 8 shots during the first series of experiments (runs 1–55) and thus represent average values which might not adequately reflect rheologic conditions specific to each run. Second, surface areas could only be measured for the disk-type spindles which could not be used for the more viscous target materials. Third, the actual strain rate varies widely across the radius of the cup holding the test fluid, and the calculated values were valid only at the shaft. Despite these limitations, we were able to characterize the relative rheologic properties of the target materials for the first series of experiments.

In subsequent experiments, including those currently underway, we are attempting to eliminate some of these difficulties. As an alternative to the calculated plastic viscosity values, we are using instrument viscosity data collected for every target material. Relative yield strengths are being measured directly by using the cone indentation method (Moore, 1965). We do not yet have a means of computing the effective areas of the t-bar spindles, but if the same spindle is used for all measurements, then the *relative* viscosities should produce internally consistent results. Presumably, the functional relationships indicated by these data will also apply to absolute measurements, differing by some as yet unknown constant. At this stage we are investigating these functional relationships. In our ongoing experiments we are attempting to determine the absolute values of the variables so that we can better scale our results to martian conditions.

Experimental data

Table 1 lists the experimental data. Impact conditions and target properties were set prior to each shot; crater morphology was measured after impact, and the geometry of the transient structures such as central peaks and surge waves was observed in high speed motion pictures. Each successful run (no misfires) was assigned a number. Runs 1–55 (series 1) were carried out in June and July, 1979. Rounds 56–75 (series 2) were conducted in December 1979. In the first set of experiments, plastic viscosities and yield strengths were calculated for average batches of target materials, as described earlier. Apparent viscosities at a single RPM value were measured for each target. During the second set, complete determinations were made for each shot. In multilayer configurations the properties of both slurries were determined.

Proc. Lunar Planet. Sci. Conf. 11th (1980), p. 2099–2128.
Printed in the United States of America

Meteor Crater: Energy of formation—implications of centrifuge scaling

R. M. Schmidt

Shock Physics and Applied Math Organization, Boeing Aerospace Company, P.O. Box 3999,
Seattle, Washington 98109

Abstract—Scaling results derived from centrifuge impact experiments indicate that the energy of formation for Meteor Crater, Arizona, is in the range of 22 to 61 MT with corresponding impact velocities of 7 to 25 km/sec. Scaling relationships for both crater volume and crater radius are given for the centrifuge experiments as well as for various other experimental and computational data sets. These results indicate that the density and the angle of internal friction of the target are the dominant material properties controlling crater size. Cohesion, tensile strength and details of complex fracture mechanisms are shown to be significant only for values of total energy less than the order of 1 KT. At large scaled-energy, gravity controls final crater size since rupture zone dimensions probably scale with something greater than the cube-root of yield, on the order of 1/2.8. Crater dimensions on the other hand scale with something less than the cube-root of yield, on the order of 1/3.6. The general analytical forms of the derived scaling relationships provide information on the role of specific dependences of projectile mass, energy, velocity (specific energy) and momentum. These dependences are different for each of the different material types considered: dry sand, "hydrodynamic" limestone and water. The dry sand results are used to estimate the energy of Meteor Crater. The "hydrodynamic" limestone response and especially the water data are shown to approach quarter-root scaling, which was used as a limiting behavior to calculate a lower bound on the energy of formation for velocities in excess of 7 km/sec.

1. INTRODUCTION

Initial conditions leading to the formation of very large impact craters are subject to considerable speculation. This is brought about for various reasons. A primary obstacle has been the impossibility of performing appropriate experiments to investigate the role of large size on cratering mechanics. Recent work on explosive cratering (Schmidt, 1977, 1978; Schmidt and Holsapple 1978a, 1979, 1980a; Holsapple and Schmidt 1979) has demonstrated the utility of performing subscale experiments on a geotechnic centrifuge to develop scaling rules for very large energy events.

An extension of this technique to impact cratering is presented here. Experiments have been performed using a projectile gun mounted directly on the centrifuge rotor to launch projectiles into a suitable soil container undergoing centripetal accelerations in excess of 500 G. The pump tube of a two-stage light-gas gun was used to attain impact velocities of approximately 2 km/sec. (Higher velocities will be obtainable with the completion of the light-gas gun installation currently under development.) To complement the centrifuge investigation, a set of 1-G baseline experiments with impact velocities up to 7 km/sec were performed using the Vertical Gun Ballistic Range (VGBR) at the NASA Ames Research Center (Schmidt, *et al.*, 1979). From the combined results of both the high-G and the 1-G experiments, a scaling relationship for impact cratering was developed. Dimensionless quantities were used allowing generalization of the observed behavior. The resulting functional form is shown to suitably explain both experimental data for various soil media and the finite-difference continuum code results of Bryan *et al.* (1980). Furthermore, these computational results support the validity of using the same functional form to correlate the role of impact velocity at constant mass and constant gravity over the range of 2–25 km/ sec.

Arguments are presented to support using the derived scaling relationship to estimate a range of possible initial conditions for Meteor Crater, Arizona. This application is intended to complement current work of Bryan *et al.* (1978, 1980), Orphal *et al.* (1980), Roddy (1978), Roddy *et al.* (1980) and others. Meteor Crater is a well-preserved, relatively young terrestrial impact crater that has been extensively studied and its morphology and geologic setting are well documented by Roddy (1978), Roddy *et al.* (1975), Shoemaker and Kieffer (1974), Shoemaker (1963) and others.

The following analysis indicates that the kinetic energy leading to the formation of Meteor Crater was considerably greater than the presently accepted estimate of 3.85–5.0 MT (Roddy, 1978; Shoemaker and Kieffer, 1974). These earlier estimates were based upon scaling arguments using the TEAPOT ESS and the SEDAN nuclear craters, respectively, as explosive analogs. Holsapple (1980) shows that the scaled equivalent depth of burst for an impact event is considerably less than the actual depth of burst for either SEDAN or TEAPOT ESS. Hence, these approximations would underestimate the energy required for the impact event.

Another factor contributing to this larger estimate is the role of soil angle of internal friction. For very large explosive events, field data indicate that cohesion and tensile strength play an insignificant role in crater formation (Schmidt and Holsapple, 1980b). In contrast, soil angle of internal friction leads to a significant induced strength effect that governs final crater size (Holsapple and Schmidt, 1979). These same mechanisms are assumed to control impact cratering and are shown to be consistent with the results presented below. The calculated results of Bryan *et al.* (1978), using a kinetic energy of 4.5 MT, support the earlier estimates and are based upon a material model with zero angle of internal friction.

Bryan *et al.* (1978) has also summarized fifteen different previous estimates of meteorite mass, impact velocity, and kinetic energy reported in the literature over the past thirty years. These range from as little as 0.07 MT, with eleven estimates in the range of 1.1 to 8.1 MT and three others much larger. Cook (1964) estimated 20 MT. Earlier, Opik (1958, 1961) calculated energies of 70 and 67 MT using independent theoretical models for penetration, crater volume, and fragment size. His analysis also addressed the dependence of mechanical efficiency on impact velocity in the vaporization regime.

2. DIMENSIONLESS PARAMETERS

Following the derivations of previous work (Schmidt and Holsapple, 1978b, 1980a; Holsapple and Schmidt, 1979), a set of dimensionless parameters based upon an assumed set of governing variables are defined. In the study of impact cratering, dependent variables of interest are the volume V, the radius r and the depth h of the apparent crater. Included among the independent variables is the projectile description: mass W, density δ, and velocity U. A projectile equivalent radius can be defined as

$$a = \left[\frac{3W}{4\pi\delta}\right]^{1/3} \tag{1}$$

and can be used as an alternative to the mass W as an independent variable. In some cases, it is useful to replace the impact velocity with the specific energy,

$$Q_e = \frac{U^2}{2}. \tag{2}$$

This provides an independent variable that has a direct counterpart for explosive cratering.

The target medium is described as a continuum with density ρ, cohesion c, and angle of internal friction ϕ. Environmental variables include the atmospheric pressure P and gravity g.

These variables involve independent units of mass, length and time; thus any dependent variable expressed as a dimensionless parameter π_i can be shown to be a function of just five dimensionless quantities (Buckingham, 1914; Bridgeman, 1949, and others)

$$\pi_i = F(\pi_2, \pi_3, \pi_4, \pi_5, \pi_6) \tag{3}$$

Typical dependent parameters π_i of interest are

$$\pi_v = \frac{V\rho}{W} \qquad \text{(cratering efficiency)} \tag{4a}$$

$$\pi_r = r\left(\frac{\rho}{W}\right)^{1/3} \qquad \text{(scaled radius)} \tag{4b}$$

$$\pi_h = h\left(\frac{\rho}{W}\right)^{1/3} \qquad \text{(scaled depth).} \tag{4c}$$

The independent parameters are

$$\pi_2 = \frac{3.22\,ga}{U^2} = \frac{g}{Q_e}\left(\frac{W}{\delta}\right)^{1/3} \qquad \text{(gravity-scaled size)} \qquad (5a)$$

$$\pi_3 = \rho/\delta \qquad \text{(density ratio)} \qquad (5b)$$

$$\pi_4 = \tan\phi \qquad \text{(angle of internal friction)} \qquad (5c)$$

$$\pi_5 = \frac{2c}{\delta U^2} \qquad \text{(cohesion/energy density)} \qquad (5d)$$

$$\pi_6 = \frac{P}{\rho ga} \qquad \text{(pressure ratio).} \qquad (5e)$$

Any two impact experiments are called similar when each of the independent dimensionless parameters (Eqs. 5a–5e) has the same value for both experiments. For certain size regimes in a given target material, crater size is expected to be independent of one or more of the parameters given by Eqs. (5a–5e). Without some form of experimental evidence, however, no reduction in the list of variables can be made.

3. EXPERIMENTAL TECHNIQUE

Two series of impact experiments were conducted. The first set, a baseline control study consisting of shots 39-X through 48-X shown in Table 1, was performed at fixed 1-G conditions on the Ames Vertical Gun Ballistic Range (VGBR). These experiments were designed to investigate the role of projectile properties using a single target material consisting of dry dense Ottawa Flintshot sand. This soil type was chosen to provide impact results that could be compared directly with explosively formed craters obtained previously by Piekutowski (1974, 1975, 1980) at 1-G conditions and by Schmidt and Holsapple (1978a, 1979, 1980a, 1980b) using a centrifuge. Except for shot 48-X, the Flintshot sand targets were fabricated using a pluviation technique to obtain maximum density of approximately 1.80 gm/cc. The sand density for shot 48-X approximately matched the lower density (1.65–1.70 gm/cc) of the #24 quartz sand as used in both the experiments by Gault and Wedekind (1977) and those reported by Oberbeck (1977).

Impact velocities ranged from 2.10 to 7.25 km/sec using both the 30-caliber powder gun and the hydrogen gas gun. Projectiles, except for one aluminum sphere (shot 47-0), were either lexan cylindrical slugs or nylon spheres.

Three shots were fired into an air-filled target chamber held at one atmosphere pressure. A one-mil diaphragm isolated the launch tube, which was evacuated to approximately 1 mm Hg. These control shots provided data on the effect of atmospheric pressure on 1-G laboratory-scale impact crater formation (Holsapple, 1979).

The second set of experiments was performed using the Boeing 600-G geotechnic centrifuge (Schmidt, 1978). A powder gun was mounted on the rotor hub as shown in Fig. 1a. Cylindrical polyethylene slugs were launched by firing 50 grains (3.3 gm) of Bullseye pistol powder contained in a .375 H&H magnum brass shell. The firing pin was activated remotely using an E-106 explosive

Fig. 1. (a) Top photo shows powder gun mounted on rotor hub. Explosively driven firing pin is contained in housing on the left. (b) Bottom photo shows solid-state 4-MHz counter circuit and mount for break-wire grids, used to determine projectile impact velocity, in place on centrifuge rotor.

(a)

(b)

Table 1. Summary of experimental impact conditions and data for the size and shape of the apparent crater.

Shot number	g Gravity (G)	P_e Pressure (mm Hg)	Soil type	ρ Soil density (gm/cc)	Projectile material	L/D (cm/cm)	W Projectile mass (gm)	E Kinetic energy (erg)	Q Specific energy (erg/gm)	δ Projectile density (gm/cc)
19-1	1	760	Permaplast clay	1.53	Polyethylene	.810/1.22	0.888	3.68 E10	4.15 E10	0.94
19-2	309	760	Permaplast clay	1.53	Polyethylene	.830/1.21	0.900	1.39 E10	1.55 E10	0.94
19-3	1	760	Iron grit	4.16	Polyethylene	1.22/1.22	1.335	1.93 E10	1.44 E10	0.94
35-O	111	760	Flintshot sand	1.794	Polyethylene	1.22/1.22	1.335	2.24 E10	1.67 E10	0.94
35-X	464	760	Flintshot sand	1.791	Polyethylene	1.22/1.22	1.341	2.32 E10	1.73 E10	0.94
39-X	1	1	Flintshot sand	1.799	Lexan	.798/.787	0.454	9.32 E10	2.05 E11	1.17
40-O	1	760	Flintshot sand	1.810	Lexan	.775/.782	0.442	5.53 E10	1.25 E11	1.19
40-X	1	1	Flintshot sand	1.809	Lexan	.775/.782	0.439	8.76 E10	2.00 E11	1.18
43-O	1	1	Flintshot sand	1.809	Nylon	Sphere	0.019	5.02 E 9	2.63 E11	1.16
43-X	1	1	Flintshot sand	1.804	Lexan	.800/.784	0.448	8.20 E10	1.83 E11	1.16
44-O	1	1	Flintshot sand	1.806	Nylon	Sphere	0.019	5.10 E 8	2.64 E10	1.16
45-O	1	1	Flintshot sand	1.806	Lexan	.775/.772	0.429	1.41 E10	3.28 E10	1.18
45-X	1	1	Flintshot sand	1.805	Nylon	Sphere	0.019	5.59 E 8	2.88 E10	1.16
46-O	1	1	Flintshot sand	1.810	Lexan	.775/.775	0.464	3.45 E10	7.45 E10	1.17
46-X	1	760	Flintshot sand	1.801	Lexan	.775/.775	0.422	9.30 E 9	2.20 E10	1.15
47-O	1	1	Flintshot sand	1.804	Aluminum	Sphere	0.373	3.61 E10	9.68 E10	2.77
47-X	1	760	Flintshot sand	1.807	Lexan	.790/.787	0.460	2.21 E10	4.80 E10	1.20
48-X	1	1	Flintshot sand	1.708	Lexan	.787/.772	0.439	1.52 E10	3.46 E10	1.19
68-O	464	760	Flintshot sand	1.804	Polyethylene	1.22/1.22	1.339	2.10 E10	1.57 E10	0.95
69-O	464	760	Flintshot sand	1.804	Polyethylene	1.21/1.22	1.333	2.18 E10	1.64 E10	0.94
70-O	464	760	Banding sand	1.681	Polyethylene	1.21/1.22	1.337	2.12 E10	1.58 E10	0.95
71-O	523	760	KAFB Alluvium	1.596	Polyethylene	1.22/1.22	1.336	2.14 E10	1.60 E10	0.94

Table 1. (*Continued*)

Shot number	a Equivalent radius (cm)	U Projectile velocity (cm/sec)	V Crater volume (cc)	r Crater radius (cm)	h Crater depth (cm)	α Aspect ratio	π_2 $\dfrac{3.22\,ga}{U^2}$	π_V $\dfrac{V\rho}{W}$	π_r $r\left(\dfrac{\rho}{W}\right)^{1/3}$	π_h $h\left(\dfrac{\rho}{W}\right)^{1/3}$	π_6 $\dfrac{P}{\rho ga}$
19-1	.609	1.80 E5	59.7	3.20	3.06	1.05	5.94 E−8	103	3.84	3.67	1090
19-2	.611	1.76 E5	61.7	3.52	2.83	1.24	1.93 E−5	105	4.20	3.38	3.53
19-3	.696	1.75 E5	299	10.7	1.60	6.69	7.18 E−8	932	15.6	2.34	352
35-O	.698	1.83 E5	74.6	5.80	1.17	4.96	7.31 E−6	100	6.40	1.29	7.33
35-X	.698	1.86 E5	33.7	4.42	1.03	4.29	2.96 E−5	45.0	4.87	1.13	1.76
39-X	.452	6.41 E5	1140	14.1	4.09	3.45	3.47 E−9	4520	22.3	6.47	1.65
40-O	.446	5.00 E5	638	11.9	2.48	4.80	5.64 E−9	2610	19.0	3.97	1260
40-X	.446	6.32 E5	924	12.9	4.06	3.18	3.53 E−9	3810	20.7	6.51	1.66
43-O	.159	7.25 E5	107	6.31	1.55	4.07	9.56 E−10	10200	28.8	7.08	4.66
43-X	.452	6.05 E5	1020	13.6	3.78	3.60	3.90 E−9	4110	21.6	6.01	1.64
44-O	.159	2.30 E5	31.4	4.25	1.06	4.01	9.49 E−9	2980	19.4	4.84	4.67
45-O	.442	2.56 E5	435	9.97	2.62	3.81	2.13 E−8	1830	16.1	4.23	1.68
45-X	.159	2.40 E5	32.7	4.27	1.09	3.92	8.72 E−9	3110	19.5	4.97	4.67
46-O	.455	3.86 E5	700	11.7	3.28	3.57	9.65 E−9	2730	18.4	5.16	1.63
46-X	.444	2.10 E5	305	9.00	2.21	4.07	3.18 E−8	1300	14.6	3.58	1275
47-O	.318	4.40 E5	693	11.8	3.18	3.71	5.19 E−9	3350	20.0	5.38	2.34
47-X	.451	3.10 E5	451	10.2	2.48	4.11	1.48 E−8	1770	16.1	3.91	1250
48-X	.445	2.63 E5	606	10.4	4.46	2.33	2.03 E−8	2360	16.4	7.01	1.76
68-O	.697	1.77 E5	32.2	4.39	0.83	5.29	3.26 E−5	43.4	4.85	0.92	1.75
69-O	.696	1.81 E5	33.9	4.37	1.03	4.24	3.11 E−5	45.9	4.83	1.14	1.75
70-O	.695	1.78 E5	34.3	4.44	1.07	4.15	3.22 E−5	43.1	4.79	1.15	1.88
71-O	.697	1.79 E5	32.4	4.04	1.83	2.21	3.59 E−5	38.7	4.29	1.94	1.75

detonator initiated electrically through slip rings. Projectile velocity was measured for each shot using a set of three one-mil copper break-wire grids. A 4-MHz solid-state counter circuit (74C926) with an intermediate latch output gave transit times along paths of 7.82 cm and 11.73 cm. The unit is battery powered and mounts on the centrifuge rotor arm, as shown in Fig. 1b, approximately 60 cm from the muzzle exit and 40 cm ahead of the target impact surface. The stored data are read from an LED display after the centrifuge is brought to rest. Velocity determination with the specific installation geometry and counter frequency is within ±2%. The velocity variation between successive shots for common loading conditions is within this inherent indeterminacy.

There was no effect of muzzle blast on the crater or surrounding soil with the powder gun in this firing location. The air pressure within the centrifuge housing is a nominal one atmosphere. Projectile deceleration between the velocity measurement station and the target surface is of the order of 1–2% (Holsapple, 1979).

Various soil types were used for the high-G experiments with dry dense Ottawa Flintshot being the standard. Shot 70-0 was made in a fine grain Ottawa Banding sand for a particle-size comparison. One shot was fired into KAFB desert alluvium (Schmidt, 1978; Schmidt and Holsapple 1979; Holsapple and Schmidt, 1979), which exhibits slight cohesion. Iron grit with a nominal particle size of 700 microns was used to provide information on increased target density (4.16 gm/cc). Two shots into Permaplast oil-base clay samples completed the brief soil survey.

The soil target container used for the dry Ottawa sand in both sets of experiments is 46 cm in diameter and 14 cm deep. For the crater sizes produced in these impact experiments, the container was sufficiently large to preclude any adverse boundary effects on crater dimensions. All the centrifuge tests produced craters with diameters and depths that were less than 25% of the respective container dimensions. The 1-G shots at the Ames VGBR produced craters with diameters that were less than 60% of the container diameter and crater depths that were less than 30% of container depth. Based on target container-size studies performed during the development of the range, Wedekind (pers. comm., 1980) confirmed that these latter ratios were well within the maximum allowable for dry sand. In those earlier studies, a 56-inch diameter container with a 24-inch depth was used to reproduce craters formed in the smaller VGBR standard container, which has a 24-inch (61-cm) diameter and an 8-inch (20-cm) depth.

An implicit confirmation of the adequacy of the container size is the constant π_2 test discussed below in Section 4. Two shots (44-0 and 46-0) produced craters whose volumes differed by a factor of 22 in a test of similarity based upon velocity scaling. This test proved successful, confirming that the larger crater (shot 46-0) as well as the smaller crater did not experience any adverse boundary effects, even though the craters and the container boundaries were not geometrically similar.

Previous work on explosive cratering supports these conclusions. Control shots performed by Piekutowski (1980) in dry Ottawa sand at 1-G in a 60-inch (152-cm) diameter tank, 36 inches (91-cm) deep were in good agreement with centrifuge shots (Schmidt and Holsapple, 1980a) using explosives in the same soil container described above for the impact shots.

The two clay shots (19-1 and 19-2) were performed in a stainless steel bowl approximately hemispherical in shape (19 cm radius). The actual depth is 15 cm and the bottom is a 10-cm radius flat surface. These clay containers are thought to be adequate in size based upon previous explosive work (Holsapple and Schmidt, 1979). Preliminary code calculations presently underway by Thompson (pers. comm., 1980) have, however, raised questions regarding the possibility of adverse boundary effects in regard to clay targets exhibiting negligible porosity and low "strength." For the study described here, the two clay shots were merely exploratory in nature and are not used in the quantitative analysis to follow.

4. EXPERIMENTAL RESULTS

Table 1 summarizes initial conditions and crater size for the twenty-two impact experiments performed in the two series described above. Figure 2 is a composite plot comparing crater volume data (shown as cratering efficiency) from these

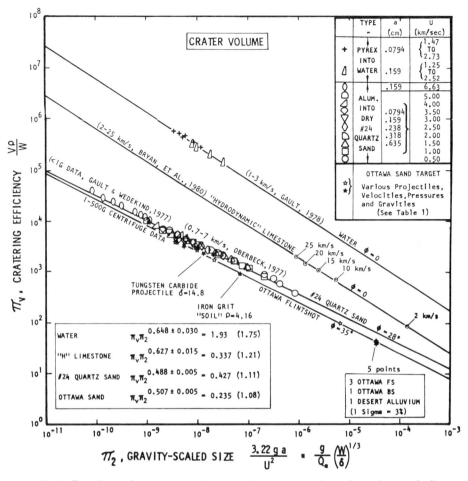

Fig. 2. Experimental and computational results for crater volume due to hypervelocity impact. The power-law correlation for each of four different target materials demonstrates the suitability of using a single dimensionless parameter (π_2) to explain the dependence upon projectile size and density, impact velocity, and gravity. (The number in parenthesis is the variance factor for one standard error of estimate (68%) as is the uncertainty shown for the exponent on π_2.)

experiments to published results for other materials. A simple power law of the form

$$\pi_v\pi_2{}^\alpha = A = \text{constant} \qquad (6)$$

is seen to adequately describe each of the four different data sets.

Holsapple (1979) argues that, if the overlying air is important for large terrestrial impact simulation, centrifuge experiments need to be performed with ambient air at one-atmosphere pressure. The Ames VGBR 1-G shots showed distinct differences between the evacuated shots ($\pi_6 \simeq 1$) and the air shots ($\pi_6 \simeq 1000$). The role of ambient air pressure for large scaled size is not completely understood and may depend upon target characteristics. However, for a fixed soil type, it is desirable to have a curve of constant π_6; hence, the high-G shots in air ($\pi_6 \simeq 1$) are plotted in conjunction with the 1-G evacuated shots.

All of the materials shown have negligible cohesion, hence, $\pi_5 = 0$. Each material set is characterized by a friction angle π_4 as shown on each curve. The value of 35° for dry dense Ottawa sand was obtained from low- and high-pressure triaxial tests (Schmidt and Holsapple, 1979). The value of 28° shown for the #24 quartz sand impact experiments (Gault and Wedekind, 1977; Oberbeck, 1977) were inferred from an initial void ratio based upon an initial density of 1.70 gm/cc. Direct-shear data for Ottawa sand of various initial void ratios (Taylor, 1948, p. 347–351.) was then used to approximate the friction angle for the smaller initial density.

The curve labeled ' "hydrodynamic" limestone results from computer code calculations performed by Bryan et al. (1980) and summarized in Table 2. For these calculations, the friction angle can be considered to be effectively zero due to its omission from the hydrodynamic equation-of-state description. His additional assumption of no shear strength was supported out to ballistic extrapolation times of 0.5 seconds by a control calculation incorporating a shear modulus of 350 kbar and a von-Mises yield strength of 0.2 kbar. As will be discussed below, this lack of shear strength probably models the post-shock degraded state quite well. His omission of the angle of internal friction, however, does not seem to be supported by either the explosive or the impact results obtained from either the centrifuge or the VGBR.

The water data (Gault, 1978) shown to the top of the plot is over an impact velocity range of 1.0–3.0 km/sec at 1-G for two different pyrex projectile sizes, 0.079 and 0.159 cm radius. It also is characterized by both zero friction angle and zero cohesion. With the exception of the Ottawa-sand data, each of the curves for a fixed target material is also for a constant ratio of soil density to projectile density π_3. For the water data, $\pi_3 = 0.44$; for the "hydrodynamic" limestone data, $\pi_3 = 0.34$; and for the #24 quartz sand, $\pi_3 = 0.60$.

Table 2. A parametric set of code calculations for impact crater dimensions in a "hydrodynamic" limestone. The projectile mass was held constant at 1.67×10^{11} gram and only the impact velocity was varied. (Bryan et al., 1980).

U Projectile velocity (km/sec)	V Crater volume (m³)	r Crater radius (m)	h Crater depth (m)	π_2 $\dfrac{3.22\,ga}{U^2}$	π_V $\dfrac{V\rho}{W}$	π_r $r\left(\dfrac{\rho}{W}\right)^{1/3}$	π_h $h\left(\dfrac{\rho}{W}\right)^{1/3}$
2	5.39×10^6	217	75.6	1.36×10^{-4}	87.1	5.49	1.91
10	4.57×10^7	444	154	5.43×10^{-6}	740	11.2	3.89
15	6.85×10^7	505	179	2.41×10^{-6}	1110	12.8	4.53
20	9.58×10^7	567	198	1.36×10^{-6}	1550	14.3	5.01
25	1.29×10^8	631	214	8.68×10^{-7}	2090	16.0	5.41

In the case of the Ottawa sand, limited data for different projectile densities indicate that for a fixed cohesionless target material, cratering efficiency is independent of π_3 (Schmidt *et al.*, 1979). Previously, Dienes and Walsh (1970) concluded from a computational study of late-stage equivalence that "changing projectile density at constant projectile mass leaves unchanged the overall effects on the target." Their result, based upon calculations of small-scale hypervelocity impact into metals, does not include the additional constraint observed here for gravity dominated response—that π_2 be constant also. This latter restriction provides a useful scaling rule for impact velocity dependence. For fixed gravity, similarity can be achieved by holding the quantity a/U^2 constant as can be seen by comparing shots 44-0 and 46-0. Both were performed at 1-G conditions, and the velocity ratio for the two shots was equal to the square root of the projectile radius ratio giving a common value for π_2 and hence π_v. The actual crater volumes varied by a factor of approximately 22 which, within experimental uncertainty, corresponds to the ratio of projectile masses. This velocity scaling role verified by the constant π_2 test allows scaling size with velocity as well as with increased gravity. Its validity has been demonstrated over the range of 0.7–7 km/sec for crater volume as shown in Fig. 2. Extension to impact velocities up to 25 km/sec will now be addressed.

This experimental behavior is also supported by the computed results of Bryan *et al.* (1980). For his "hydrodynamic" limestone, only impact velocity was varied for conditions of fixed target, fixed impactor and fixed terrestrial gravity. As can be seen in Fig. 2, a straight line provides a good fit to the data over a velocity range which encompasses no-melt conditions (U = 2 km/sec) through the onset of vaporization (U = 25 km/sec). The different slope and larger cratering efficiencies compared to those for Ottawa sand are attributed to the material response in the absence of strength effects, shear modulus, cohesion and especially friction angle. It must be pointed out, however, that these velocity results are for constant projectile mass and constant gravity. The dimensionally invariant form can only be confirmed by additional code calculations. These should include varying mass at fixed gravity and fixed velocity for different velocities. Preliminary results based upon variation of gravity in the ballistic phase (fixed mass and fixed velocity) are not conclusive (Bryan, pers. comm., 1979).

The difference between the limestone curve and the water data may be due to differences in the idealized mathematical models used for the calculations as opposed to the experimentally observed behavior of water. In terms of the controlling parameters given by Eqs. (5a–5e), only π_3 differs for the two materials, 0.34 versus 0.44. Both $\pi_4 = \pi_5 = 0$ and π_6 are approximately zero for both sets of results.

As suggested above, no significant dependence of cratering efficiency upon π_3 is expected. Two pathological shots, shown in Fig. 2 emphasize this conclusion. One, shot 19-3, is a polyethylene projectile fired into iron grit, $\pi_3 = 4.43$. The second, labeled "tungsten-carbide projectile" was fired into #24 quartz sand, $\pi_3 = 0.12$ (Wedekind, pers. comm., 1979). This variation of approximately a factor of 40 in π_3 shows no significant deviation from the curves shown in Fig. 2 for polyethylene into Ottawa sand, $\pi_3 = 1.91$ and for aluminum into #24 Quartz sand, $\pi_3 = 0.63$. Aside from a possible dependence on π_3, it is suggested that the ballistic link-up time of 0.5 sec used in the limestone calculations may not be sufficient to develop full crater size. A code calculation of the 100-ton TNT tangent-above MIDDLE GUST III event, currently in progress, required the finite-difference phase to be carried out to times greater than 200 msec before the size of the final crater stabilized (Schuster, pers. comm., 1980). Scaling this value to a time corresponding to 4.5 MT indicates a ballistic link-up time on the order of 3–7 sec. In general, less material strength also requires greater link-up time for fixed energy (Schuster, pers. comm., 1980). Premature link-up times could cause the resulting crater to be smaller in size due to the flow field not being fully developed prior to ballistic extrapolation.

Figures 3 and 4 show the scaled crater radius as a function of π_2. As shown above for crater volume, the scaled crater radius can also be described in terms

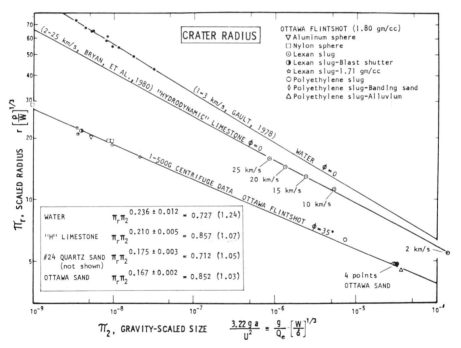

Fig. 3. Experimental and computational results for crater radius due to hypervelocity impact. Again for radius as shown above for crater volume in Fig. 2, a power-law correlation provides a good fit to the individual data sets confirming the role of the gravity-scaled-size dimensionless parameter (π_2) to explain the dependence upon projectile size and density, impact velocity, and gravity. A comparison plot of crater radius for Ottawa sand and for #24 quartz sand is shown separately in Fig. 4 on an expanded scale for clarity. (The number in parenthesis is the variance factor for one standard error of estimate (68%) as is the uncertainty shown for the exponent on π_2.)

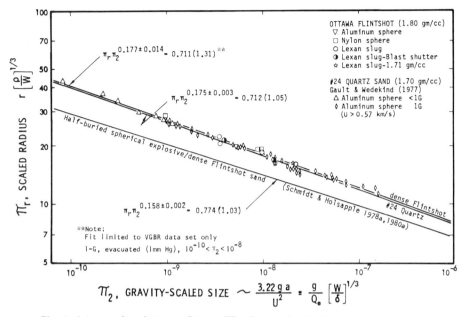

Fig. 4. A comparison between Ottawa Flintshot sand and #24 quartz sand shows that the crater radius due to hypervelocity impact is identical within the bounds of experimental uncertainty. (The number in parenthesis is the variance factor for one standard error of estimate (68%) as is the uncertainty shown for the exponent on π_2.)

of a single dimensionless parameter which adequately explains the dependence upon projectile size and density, impact velocity, and gravity,

$$\pi_r \pi_2{}^{\beta} = B = \text{constant.} \tag{7}$$

The distinction between #24 quartz sand and the Ottawa sand is shown on an expanded scale in Fig. 4 where the range of π_2 values correspond to just the Ames VGBR data. Note the nearly identical response of the less-than-1-G and the 1-G #24 quartz-sand data of Gault and Wedekind (1977) with the 1-G Ottawa-sand data obtained on the Ames VGBR. Half-buried spherical explosive results are also shown for comparison.

A blown-up section of Fig. 2 for crater volume is shown in Fig. 5. This again allows a direct comparison between discrete data points for the less-than-1-G and the 1-G #24 quartz sand with those for the 1-G Ottawa sand. The final comparison between these two sands is given in Fig. 6. Here crater depth shows a significant difference which is attributed to the small difference in friction angle producing greater shear strength at depth for the Ottawa sand. This is supported by shot 48-X, plotted as an open star on the curves. For this case, the Ottawa Flintshot was placed by pouring, resulting in a lower density (hence lower friction angle) on the same order as that of the #24 quartz sand. The high-G centrifuge data was not included in the comparisons for radius, volume and depth shown in

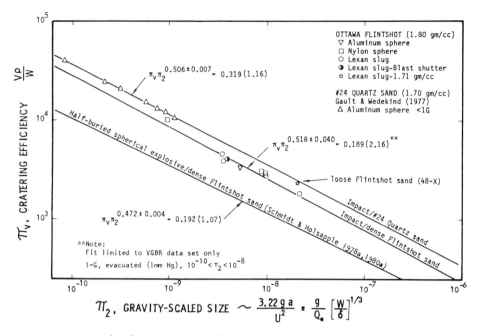

Fig. 5. A comparison between Ottawa Flintshot sand and #24 quartz sand shows greater cratering efficiency for the latter. This is attributed to the smaller friction angle associated with the lower density of the #24 quartz sand. The volume difference is primarily due to differences in crater depth as seen in Fig. 6. When Ottawa Flintshot sand was placed at the comparable density (shot 48-X), it showed good agreement with the #24 quartz sand curve. Experimental uncertainties shown are for one standard error of estimate.

Figs. 4 through 6. Therefore, the coefficients and exponents in the scaling relationships differ slightly from those shown in Figs. 2 and 3. This was intentional to emphasize a direct comparison between two sand types under common experimental conditions using the Ames VGBR.

5. STRENGTH MODEL

Moore *et al.*, (1965) and Gault (1973) report cratering results for small-scale hypervelocity impact into basalt, granite and limestone. In all cases they observed a significant reduction in effective target strength with increasing crater size. Comparable results for very large impact experiments into rock don't exist; however, field data from various large high-explosive shots in rock indicate that cohesion, tensile strength and fracture mechanisms have negligible influence on final crater size for yields above 1 KT. Figure 7 is a composite plot showing envelopes for cratering efficiency versus scaled size. Hypothetically, the upper region is the envelope for all half-buried, high-explosive, spherical charge configuration. The curves shown for alluvium, clay, and Ottawa sand result from an analytical strength model for apparent crater volume based upon small-scale, high-G centrifuge experiments (Holsapple and Schmidt, 1979). The lower bound shown by the dotted line is a locus of field data for cratering tests in competent rock. The MTCE series (Carlson and Jones, 1965) consist of 1000, 4000 and 16,000 pounds of TNT in a vesiculated basalt. The FLAT TOP I event was 40,000 pounds of TNT half-buried in limestone (Rooke and Davis, 1964). The cratering efficiency for the rock is seen to increase with yield approaching the curve for dry dense Ottawa sand at a yield of 100–500 tons TNT.

The same trend is seen for tangent-above explosive spheres shown in the lower part of the plot. The broken line through the rock field data appears to fair into the centrifuge Ottawa sand line at a yield of 0.5–1 KT of TNT (Schmidt and Holsapple, 1980b). MINERAL ROCK and MINE ORE both were 100 tons of TNT fired in Cedar City granite (Davis, 1970a,b). MIXED COMPANY III was 500 tons of TNT in a weathered sandstone (Roddy, 1973). The consistency of these rock data with respect to the dry-sand behavior supports the speculation that the cratering efficiency of competent rock will not exceed that of dry sand and at large yield will equal that of dry sand. This is reasonable from two points of view. One, as event size increases, more fault planes and fracture sites are encompassed by the crater. These tend to reduce the "average" cohesion and tensile strength of the rock. Second, as event size increases, the volume of frac-

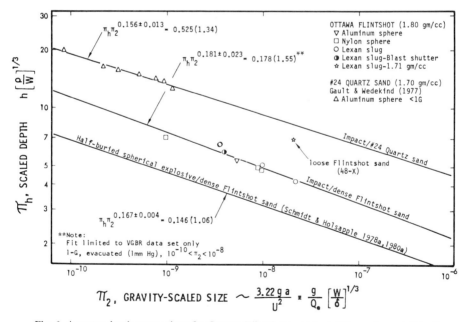

Fig. 6. A crater depth comparison for Ottawa Flintshot sand and #24 quartz sand (Gault & Wedekind, 1977). The significant difference in the two curves for depth are attributed to a difference in the angle of internal friction for the two materials, 35° for the Ottawa sand and approximately 28° for the #24 quartz sand. Experimental uncertainties shown are for one standard error of estimate.

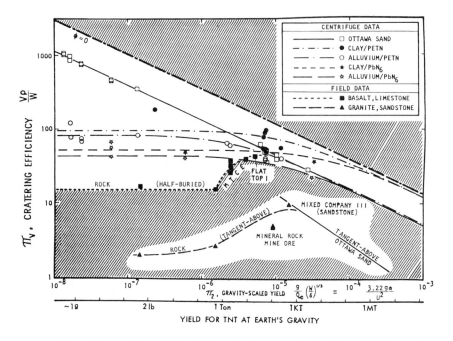

Fig. 7. Material-strength model for high-explosive cratering in all geological media. Upper region is for half-buried sphere (Holsapple and Schmidt, 1979). Lower region is for tangent-above sphere where the solid line for Ottawa sand is also based upon recent centrifuge experiments (Schmidt and Holsapple, 1980a). Large-yield field data for various events in rock geology is represented by the bold dotted lines.

tured rock increases until some critical transition size above which, ejection becomes dominated by gravity and not by the amount fractured due to the shockwave passage. Code calculations underway for MIDDLE GUST III (Wagner, pers. comm., 1980) as well as one for a hypothetical 0.5 MT nuclear burst in competent saturated tuff (Bergstresser, pers. comm., 1980) both indicate rupture and fracture zones larger in size than the final crater volume.

Therefore, in the absence of any fluidization effect such as suggested by Melosh (1979) or grain-size effects, it seems plausible that as event size increases, cratering efficiency in competent rock becomes asymptotic to that for dry granular materials with comparable angles of internal friction.

Figure 8 depicts analogous regimes for impact cratering. To the left of the transition region, the cratering efficiency is dominated by density, cohesion, tensile strength and fracture mechanisms. These lesser values of π_2 correspond to experiments in the small-scale 1-G laboratory size regime, such as those conducted on the Ames VGBR. The expected form of the dependence is indicated on the plot based upon an analogy to the explosive cratering behavior shown in Fig. 7. The data of Moore and Gault (1963) in conjunction with centrifuge results and other data for various competent soils and rock support the cohesive-tensile strength model shown schematically to the left of the transition region ($\pi_2 \simeq 10^{-7}$) in Fig. 8.

Based upon the trends seen in Fig. 2, the expected behavior in the gravity-dominated regime is shown to the right of the transition. In this size regime, the response is determined only by π_2 and by the angle of internal friction π_4. For a cohesionless material, no transition exists and this simple straight line behavior extends to values of π_2 as small as 10^{-10} as shown by the less-than-1-G data of Gault and Wedekind (1977).

6. SCALING OF IMPACT CRATERING

The results of the previous section show that at large scaled size the cratering efficiency of competent soils can be assumed to be independent of density, cohesion, tensile strength, and fracture mechanisms. In which case, the response for an arbitrary soil can be represented by the experimentally obtained relationships given by Eqs. (6) and (7). The expression for crater volume, Eq. (6), can be

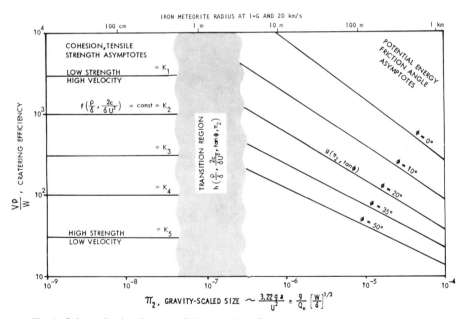

Fig. 8. Schematic showing two distinct regimes for impact cratering behavior. To the left is a cohesion/tensile strength dominated regime characteristic of small-scale impact experiments into competent materials. As scaled size is increased, these strength parameters cease to be important and are dominated by the angle of internal friction in the gravity regime to the right. The transition region is not well understood, but its upper bound does not extend beyond kinetic energies of 1 KT based upon the high-explosive results shown in Fig. 7. The important point here is that for large scaled size π_2, behavior is solely dependent upon angle of internal friction as shown.

expanded using the π-group definitions, Eqs. (4a) and (5a), in conjunction with the equivalent projectile radius given by Eq. (1) as follows:

$$V\left(\frac{\rho}{W}\right)\left[\frac{2g}{U^2}\left(\frac{W}{\delta}\right)^{1/3}\right]^\alpha = A. \tag{8}$$

Solving for crater volume gives

$$V = A\,2^{-\alpha}\rho^{-1}\delta^{\alpha/3}g^{-\alpha}W^{1-\alpha/3}U^{2\alpha}. \tag{9}$$

for any dimensionally consistent set of units. From this, the explicit volume dependence on each independent variable can be obtained in terms of the slope of the particular curve of interest, such as those given in Fig. 2. Using the coefficients for Ottawa sand, A = 0.235 and α = 0.507, as an example, Eq. (9) becomes

$$V = 0.165\rho^{-1}\delta^{0.17}g^{-0.51}W^{0.83}U^{1.01}. \tag{10}$$

Therefore at fixed values of ρ, δ, g and W, the volume is seen to be proportional to the 1.01 power of impact velocity. Alternately, at fixed impact velocity with all other variables held fixed, the volume is proportional to the 0.83 power of projectile mass.

The volume dependence upon energy E using the expression for total kinetic energy of the projectile at fixed velocity

$$E = \frac{1}{2}WU^2 \tag{11}$$

to eliminate the mass W from Eq. (9), is

$$V = A\,2^{1-4\alpha/3}\rho^{-1}\delta^{\alpha/3}g^{-\alpha}E^{1-\alpha/3}U^{8\alpha/3-2}. \tag{12}$$

Again using the Ottawa sand data for illustration, Eq. (12) becomes

$$V = 0.294\rho^{-1}\delta^{0.17}g^{-0.51}E^{0.83}U^{-0.65}. \tag{13}$$

Here the energy dependence is identical to the mass dependence at fixed values of the other variables. However, the velocity dependence at fixed energy is seen to be significantly different than that for fixed mass. In the case of Eq. (10) above, increasing the velocity at fixed mass increases the total energy, and as expected, the crater volume would increase also. In contrast, increasing velocity at fixed energy will result in a smaller total crater volume. For this case of holding the energy fixed and increasing the impact velocity, stresses increase and the process becomes less efficient. There is increased heat loss resulting from irreversible compression and other dissipation mechanisms, thereby causing a larger partition of the total energy into internal energy of the surroundings. An analogous result for explosive cratering showing the same trend for specific energy dependence has been observed by Schmidt and Holsapple (1978a, 1980a) using lead azide and PETN explosives.

Alternately, the energy dependence at fixed mass can be obtained by eliminating the velocity from Eq. (9) using Eq. (11),

$$V = A\rho^{-1}\delta^{\alpha/3}g^{-\alpha}E^\alpha W^{1-4\alpha/2}. \tag{14}$$

For the Ottawa-sand data this becomes

$$V = 0.235\rho^{-1}\delta^{0.17}g^{-0.51}E^{0.51}W^{0.32} \tag{15}$$

demonstrating that the energy dependence at fixed mass is quite different from that at fixed velocity. This is an important distinction which if not specified can lead to an observed energy exponent anywhere in the range of 0.51 to 0.83 depending on how the kinetic energy is varied (i.e., mass vs. velocity).

A fourth representation of common interest for impact cratering is to determine the energy dependence at fixed momentum and vice versa. Using the definition of total projectile momentum

$$H = WU = 2 E/U \tag{16}$$

to eliminate the velocity U from Eq. (12) gives the following expression for crater volume

$$V = A \, 2^{4\alpha/3-1}\rho^{-1}\delta^{\alpha/3}g^{-\alpha}E^{7\alpha/3-1}H^{2-8\alpha/3}. \tag{17}$$

The Ottawa-sand results reduce this expression to

$$V = 0.188\rho^{-1}\delta^{0.17}g^{-0.51}E^{0.18}H^{0.65}. \tag{18}$$

Hence for the case of Ottawa sand, as well as for the #24 quartz sand, the crater volume is seen to depend primarily upon projectile momentum. (For $\alpha = 3/7$, the energy dependence at fixed momentum would be nil.) For explosive cratering, Holsapple and Schmidt (1980a) have derived theoretical limits on the value of the exponent α using dimensional analysis in conjunction with assumptions regarding the sign of the volume change as each independent variable increases or decreases in value. The same theoretical approach, currently being applied to impact cratering (Holsapple and Schmidt, in prep.), indicates that the limits derived for explosive cratering also apply for the impact problem.

In particular, their results for materials exhibiting effects due to cohesion and tensile strength, but limited to rate-independence and no inherent size properties, show the exponent α is bounded,

$$0 \leq \alpha \leq \tfrac{3}{4}. \tag{19a}$$

As can be readily seen from Eq. (12), $\alpha = \frac{3}{4}$ corresponds to quarter-root scaling, whereas $\alpha = 0$ corresponds to cube-root scaling. It is also interesting to note that for the case of quarter-root scaling, $\alpha = 3/4$, Eq. (12) shows that crater volume must be independent of impact velocity. This has never been observed and is not expected because loss mechanisms associated with shock compression, generally increase with increased impact velocity. Gault's (1978) volume data for water approaches but does not result in quarter-root scaling. The dimensional invariant correlation for his data shown in Fig. 2 has a value of $\alpha = 0.648 \pm 0.030$, which, using Eq. (12) gives an exponent on energy of 0.78 ∓ 0.01 and more importantly a velocity dependence of -0.27 ± 0.08 at fixed energy.

The water results for radius are closer to the quarter-root scaling as can be seen using Eq. (22) below. However, the usual energy balance (K.E. = P.E.) leading to quarter-root scaling applies only to volume. Both radius and depth can be and usually are different functions of π_2. Furthermore they do not have any *a priori* restriction based upon energy balance.

For the case of very large scaled size, where material cohesion and tensile strength do not influence crater dimensions, it seems reasonable that the constraints on the exponent α might be more on the order of

$$\frac{1}{2} \le \alpha \le \frac{3}{4}. \tag{19b}$$

This reduction on the range of α is based on the observed behavior of the various materials represented in Fig. 2. For materials with angle of internal friction small or equal to zero, α approaches the value 0.75 and the scaling approaches quarter root. As the angle of internal friction increases, the exponent α tends towards the value 0.5 as seen by the two different sand data sets.

The same type of analysis can be applied to Eq. (7), the relationship for crater radius as a function of projectile size and density, impact velocity, target density, and gravity. The result for radius in terms of projectile mass and impact velocity is

$$r = B \, 2^{-\beta} \rho^{-1/3} \delta^{\beta/3} g^{-\beta} W^{(1-\beta)/3} U^{2\beta}. \tag{20}$$

Using the values $B = 0.852$ and $\beta = 0.167$ for the Ottawa sand shown in Fig. 3, Eq. (20) becomes

$$r = 0.759 \, \rho^{-0.33} \delta^{0.06} g^{-0.17} W^{0.28} U^{0.33}. \tag{21}$$

The energy-velocity dependence is obtained by using Eq. (11) to eliminate the mass W giving

$$r = B \, 2^{(1-4\beta)/3} \rho^{-1/3} \delta^{\beta/3} g^{-\beta} E^{(1-\beta)/3} U^{(8\beta-2)/3}. \tag{22}$$

Again using the coefficients for Ottawa sand, the energy-velocity dependence is

$$r = 0.920 \, \rho^{-0.33} \delta^{0.06} g^{-0.17} E^{0.28} U^{-0.22}. \tag{23}$$

The energy-mass dependence is obtained as above by using Eq. (11) to eliminate the velocity U from Eq. (20) as follows:

$$r = B \, \rho^{-1/3} \delta^{\beta/3} g^{-\beta} E^{\beta} W^{(1-4\beta)/3}. \tag{24}$$

For the Ottawa-sand data this becomes

$$r = 0.852 \, \rho^{-0.33} \delta^{0.06} g^{-0.17} E^{0.17} W^{0.11} \tag{25}$$

producing an analogous result to that for crater volume given by Eq. (15) above.

Last, the energy-momentum dependence for crater radius can be obtained by using Eq. (16) to replace the velocity U above to give

$$r = B \, 2^{(4\beta-1)/3} \rho^{-1/3} \delta^{\beta/3} g^{-\beta} E^{(7\beta-1)/3} H^{(2-8\beta)/3}. \tag{26}$$

This expression then becomes

$$r = 0.789 \, \rho^{-0.33} \delta^{0.06} g^{-0.17} E^{0.06} H^{0.22} \tag{27}$$

for the Ottawa sand behavior. An estimate of the bounds on the exponent β for crater radius can be inferred by analogy to Eq. (19b) as

$$\frac{1}{6} \le \beta \le \frac{1}{4}. \tag{28}$$

Equations (9), (12), and (17), give the mass-velocity, the energy-velocity, the energy-mass and the energy-momentum dependences for crater volume. Likewise Eqs. (20), (22), (24) and (26) give these same respective dependences for crater radius. The role of the remaining variables: target density, projectile density and gravity is, of course, also included in these expressions. As shown by Holsapple and Schmidt (1980), a specific dependence upon any given variable such as energy is ambiguous without stating which other independent variables are being held fixed. These expressions are general insofar as they apply to any material whose cratering behavior can be expressed as a simple power law such as those given by Eqs. (6) and (7) and shown in Figs. 2 and 3. For the case of impact cratering the use of a power law of this form is justified by its suitability in correlating the data.

7. METEOR CRATER ANALYSIS

An estimate for the energy of formation of Meteor Crater can be made using the energy-velocity scaling relationship of the previous section. Equation (12) can be inverted and solved for the energy E in terms of apparent crater volume V, impact velocity U, gravity g and densities ρ and δ

$$E_V = V^{3/(3-\alpha)} A^{3/(\alpha-3)} 2^{(3-4\alpha)/(\alpha-3)} \rho^{3/(3-\alpha)} \delta^{\alpha/(\alpha-3)} g^{3\alpha/(3-\alpha)} U^{(8\alpha-6)/(\alpha-3)} \tag{29}$$

The coefficients A and α are experimentally determined constants that depend primarily upon the value of the soil angle of internal friction. For a friction angle of $35°$, which is appropriate for the dry sandstone, the coefficients for the Ottawa sand curve in Fig. 2, A = 0.235 and α = 0.507, will be used. Hence Eq. (29) can be written as

$$E_v = 21.5 \, V^{1.20} \rho^{1.20} \delta^{-0.20} g^{0.61} U^{0.78}. \tag{30}$$

The values for the pre-erosion dimensions of Meteor Crater are taken from Roddy (1978). The apparent crater volume is 7.5×10^{13} cm^3 and the apparent crater radius is 5.11×10^4 cm. The average density is taken to be 2.3 gm/cc, the meteorite density to be 7.86 gm/cc and gravity to be 981 cm/sec^2. Equation (30) then reduces to

$$E_v \text{ (ergs)} = 2.59 \times 10^{19} \, U^{0.78} \tag{31}$$

giving a scaling relationship based upon crater volume for energy as a function of assumed impact velocity in cm/sec.

Likewise, Eq. (22) can be inverted and solved for the energy E in terms of crater radius r, impact velocity U, gravity g and densities ρ and δ

$$E_r = r^{3/(1-\beta)} B^{3/(\beta-1)} 2^{(1-4\beta)/(\beta-1)} \rho^{1/(1-\beta)} \delta^{\beta/(\beta-1)} g^{3\beta/(1-\beta)} U^{(8\beta-2)/(\beta-1)} \tag{32}$$

Setting B = 0.852 and β = 0.167 from the Ottawa sand VGBR and centrifuge data shown in Fig. 3, Eq. 31 becomes

$$E_r = 1.35 \ r^{3.60} \rho^{1.20} \delta^{-0.20} g^{0.60} U^{0.80}. \tag{33}$$

Using Roddy's (1978) value of 5.11 × 10⁴ cm for the radius, the same values for the densities and for gravity as used immediately above for the volume calculation,

$$E_r \ (\text{ergs}) = 1.33 \times 10^{19} \ U^{0.80}, \tag{34}$$

for impact velocity in cm/sec. This crater-radius scaling result is in reasonable agreement with the crater-volume relationship given by Eq. (31). The exponent on velocity differs slightly as does value of the constant.

The discrepancy between the two relationships is due to slight shape differences between Meteor Crater and the high-G centrifuge craters. The shape of Meteor Crater (Shoemaker, 1963) can be compared with that of 464-G centrifuge shot 69-0 in Fig. 9. Also shown is a profile of the 1.2 KT TEAPOT ESS nuclear crater, which has been used as an explosive analog for Meteor Crater by Shoemaker (1963), Roddy *et al.* (1975), Roddy (1978) and others.

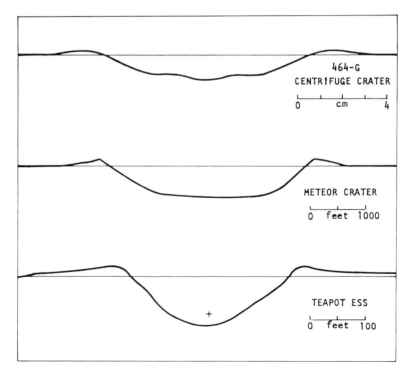

Fig. 9. Comparison of Meteor Crater cross-sectional profile (Shoemaker, 1963) with typical cross section of a high-G centrifuge impact crater. The latter profile can be seen to compare more favorably with the cross section of Meteor Crater than does the TEAPOT ESS 1.2 KT nuclear crater cross section.

Table 3. Energy of formation for Meteor Crater for various impact velocities.

Impact velocity (km/sec)	Scaling model for angle-of-internal friction $\phi = 35°$		Lower bound based on quarter-root scaling at constant mass	
	Meteorite mass (10^{12} gm)	Energy (MT)	Meteorite mass (10^{12} gm)	Energy (MT)
7	3.8	22	3.8	22
10	2.5	30	2.0	24
15	1.5	41	0.97	26
20	1.1	51	0.60	28
25	0.81	61	0.39	29

Roddy (1978) reports even greater discrepancy between radius (diameter) scaling and volume scaling due to what he refers to as the "shape factor problem." Previous work on explosive cratering by Schmidt and Holsapple (1979) has resulted in better correlation of scaled energy with crater volume than with crater radius. It seems reasonable that because of stability arguments, crater radius and especially crater depth might depend more on cohesion and other soil properties than would crater volume. Therefore preference will be given to crater volume for scaling.

Using Eqs. (31) and (11), energy of formation and meteorite mass for various velocities are calculated and shown in Table 3. Also shown is a tabulation of an estimated lower bound for energy of formation. Both sets of values are plotted in Fig. 10. The range of 1.3–7 km/sec corresponds to the actual velocity range tested and is subject only to the assumption that the Ottawa sand provides a limiting behavior for sandstone (or any other competent soil with comparable friction angle). For velocities greater than 7 km/sec, the upper curve is a continuation of the Ottawa sand relationship, Eq. (31), subject to the additional assumption that the same velocity dependence holds. Since this is beyond the range of velocities tested, a theoretical lower bound for energy of formation was devised.

If strength effects due to soil angle of internal friction and/or velocity dependence at fixed energy become less significant with increased velocity, quarter-root scaling can be considered a limiting behavior. Thus, the energy for impact velocity in excess of 7 km/sec was calculated using quarter-root scaling at constant mass from that for 7 km/sec. As mentioned above, this limit has never been observed nor is it expected because of increased irreversible shock heating and increased melt and vaporization with increased velocity. Nevertheless it does provide a lower bound for the energy of formation.

Holsapple (1980) shows that the equivalent depth of burst for very large energy impact craters is on the order of 1.5 charge radii. He demonstrates a volume equivalence based upon, among other variables, the specific energies of both the impact event and the equivalent explosive event. His analysis indicates that the various desert buried-nuclear events are not applicable to use in scaling approx-

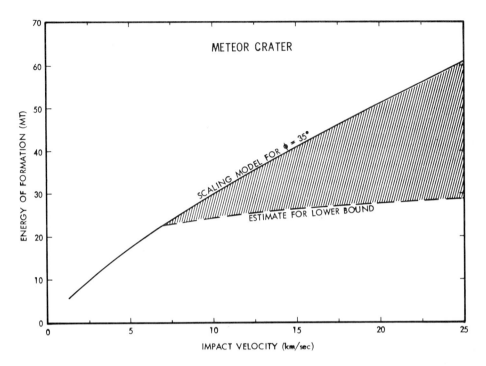

Fig. 10. Energy of formation for Meteor Crater as a function of meteorite impact velocity. Upper curve is based upon Ottawa sand data with $E \propto U^{0.78}$. Curve for lower bound is based upon the lesser velocity dependence seen for quarter-root scaling at constant mass, $E \propto U^{0.19}$. The divergence with increasing velocity represents the uncertainty in the mechanical efficiency due to effects of melting, vaporization, irreversible shock heating, etc.

imations for Meteor Crater, since they were not performed at the appropriate depth of burst.

Comparison of crater volume and crater radius for both impact and explosive phenomena is shown in Figs. 11 and 12. These data support Holsapple's (1980) contention that the equivalent depth of burst for an explosive event is on the order of 1.5 charge radii (assumed to be the vaporization radii in the case of nuclear).

8. DISCUSSION

Additional centrifuge experiments are needed to explore layer effects at large scaled size as well as to determine the effect of a water table in the Coconino sandstone (Roddy, 1978). The results described here used a homogeneous granular continuum to approximate the complex stratigraphy that comprised the pre-impact target conditions of Meteor Crater.

The use of the centrifuge to simulate large-scale explosive cratering behavior has been demonstrated in various applications. A subscale simulation of the 0.5 KT shallow-buried nuclear event, JOHNIE BOY, has been reported by Schmidt (1978). Simulations of both the tangent-below JANGLE HE2 20-ton TNT event and the deeper-buried STAGECOACH III 20-ton TNT cratering event have been successfully performed by Schmidt and Holsapple (1980b). There appears to be no significant theoretical or practical limitation on scaling, when the small-size high-G centrifuge results are compared with explosive field events as large as 1 KT. No grain-size effects or significant rate effects have been observed. General scaling results for explosive cratering based on centrifuge experiments are in good agreement with available field data (Schmidt, 1977, 1978; Schmidt and Holsapple, 1978, 1980b; Holsapple and Schmidt, 1979).

All of these explosive cratering experiments were performed using a soil as nearly identical to the prototype soil as possible. In principle, exact simulation can only be achieved using the centrifuge when identical soil and energy source type are used as discussed by Schmidt and Holsapple (1980a). However in practice as shown by Holsapple and Schmidt (1979) the requirement on identical soil can be relaxed for certain size regimes. In particular their results for half-buried spheres indicate that for prototype explosive charge size

$$W_{explosive} > \delta \left[\frac{c}{\rho g (\tan \phi + 0.1)} \right]^3 \qquad (35)$$

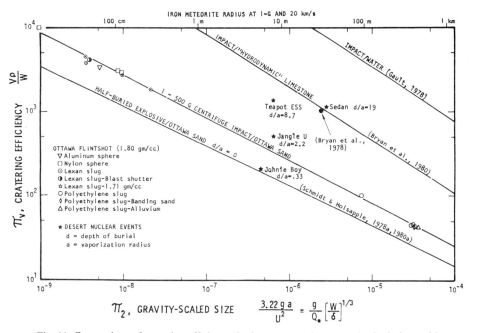

Fig. 11. Comparison of cratering efficiency for impact experiments and calculations with explosive results for various depths of burst. The data is seen to support Holsapple's (1980) contention that scaled equivalent depth of burst is on the order of d/a = 1.5.

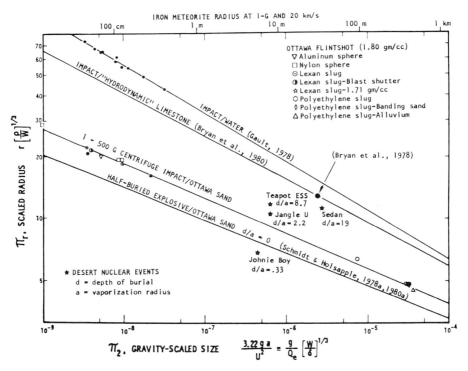

Fig. 12. Comparison of scaled crater radius for impact with explosive results for various depths of burst. The data for radius is also seen to support Holsapple's (1980) contention that scaled equivalent depth of burst corresponding to impact is on the order of d/a = 1.5.

crater volume does not depend upon cohesion c. This same conclusion can be drawn for tangent-above spheres and as shown in Fig. 7 includes competent rock as well as cohesive soils.

A completely analogous behavior is tentatively postulated for impact cratering as shown in Fig. 8. Namely, at sufficiently large scaled size, cratering-efficiency is not dominated by density, material cohesion, tensile strength or fracture behavior. This is particularly important in understanding the role of small-scale 1-G laboratory experiments (Moore and Gault, 1963; Croft *et al.*, 1979 and others) where very complex behavior is reported. For these experiments, density, cohesion, tensile strength, fracture behavior, fault plane orientation, etc., do dominate the impact cratering behavior; whereas at large scaled size these variables have negligible influence on the response. Furthermore extrapolation of results for π_2 less than approximately 10^{-7} does not appear to be justified since the functional dependence upon π_2 changes due to the reduction in the number of governing variables.

This behavior gives rise to an extension of the scaling which encompasses the concept of an equivalent material. That is, above some size transition regime, only the angle of internal friction dominates. This is supported by shots 19-3, 47-

0 and 71-0, as well as by the tungsten-carbide projectile onto sand, where specific variation in material compressibility does not seem to affect the impact cratering phenomenon at these lower velocities.

One of the important assumptions in this analysis is the implicit velocity scaling inferred by the form of the independent π_2 variable. This dimensionless parameter for scaled size has a velocity squared term in the denominator which suitably correlates velocity dependence as discussed above in Section 4 and as demonstrated explicitly by the constant π_2 test performed at fixed 1-G conditions. Velocity scaling for crater volume using the π_2 parameter as the sole velocity dependence has been experimentally verified over the range of 0.7–7.0 km/sec by the experiments described here in conjunction with those of Gault and Wedekind (1977) and Oberbeck (1977) as shown in Fig. 2. Likewise, Fig. 3 shows the suitability of velocity scaling for crater radius which has been verified over the velocity range of 1.5–7.0 km/sec by the same sets of experiments.

The minimum velocity for this hypervelocity scaling regime is determined by projectile and/or target material properties. Gault and Wedekind (1977) show that for an aluminum projectile into sand, this minimum velocity is on the order of 1.5 km/sec. For velocities less than this, the functional form of the energy dependence is different. This transition can be seen in Fig. 4 by the 12 points to the far right which lie above the straight line fit to the data. These points correspond to impact velocities of 0.6 to 1.2 km/sec. For plastic projectiles into sand targets, it is expected that this lower bound on velocity would be less than that for aluminum projectiles. Likewise, for more competent targets, the lower bound on velocity might also be less for aluminum projectiles. For the experiments performed here to obtain the Ottawa sand cratering behavior, the lowest velocity was 1.8 km/sec and the corresponding projectiles were polyethylene. This is well above the lower bound for hypervelocity impact conditions for either crater volume or crater radius.

The validity of velocity scaling for velocities greater than 7.0 km/sec is less well established. The extension into the melt regime and into the vaporization regime (7.0–25 km/sec) has only been tenuously supported by the limited set of calculations of Bryan *et al.* (1980). More code calculations would be desirable in the range of 1.8–7.0 km/sec to allow comparison to recent code calculations in the higher velocity regimes (Bryan *et al.* 1978, 1980; Orphal *et al.*, 1980; Roddy *et al.*, 1980; and others).

The results of the experiments described here indicate that the energy of formation of any large impact crater depends upon the impact velocity. This dependence, shown in Fig. 10 for the case of Meteor Crater, is consistent with analogous results for the specific energy dependence of explosives and is expected to persist to impact velocities in excess of 25 km/sec. The good agreement for specific energy Q_e dependence as accounted for by the π_2 parameter gives additional confidence in the use of this gravity-scaled size parameter to scale effects due to varying impact velocity. In addition, different dependences for mass-velocity, energy-velocity, energy-mass, and energy-momentum are clearly seen for the various experimental results.

Acknowledgments—The author would like to acknowledge many stimulating and useful discussions with K. A. Holsapple of the University of Washington who also participated in most of the experiments described above. In addition, mention must be made of the efforts of H. E. Watson, B. Ausen and C. R. Wauchope in performing the experiments. Most of the data reduction and all of the artwork was done by C. R. Wauchope. This work was performed for the National Aeronautics and Space Administration, contract NASW-3291, under the direction of W. L. Quaide. The author would also like to thank the reviewers, M. Austin, S. K. Croft, and D. E. Gault whose questions and comments contributed to the discussion.

REFERENCES

Bridgeman P. W. (1949) *Dimensional Analysis,* Yale Univ. Press, New Haven, Conn. 113 pp.

Bryan J. B., Burton D. E., Cunningham M. E., and Lettis L. A. Jr. (1978) A two-dimensional computer simulation of hypervelocity impact cratering: Some preliminary results for Meteor Crater, Arizona. *Proc. Lunar Planet. Sci. Conf. 9th,* p. 3931–3964.

Bryan J. B., Burton D. E., Lettis L. A. Jr., Morris L. K., and Johnson W. E. (1980) Calculations of impact crater size versus meteorite velocity (abstract). In *Lunar and Planetary Science XI,* p. 112–114. Lunar and Planetary Institute, Houston.

Buckingham E. (1914) On physical similar systems. Illustrations of the use of dimensional equations. *Phys. Rev. IV,* **4,** 345–376.

Carlson R. H. and Jones G. D. (1965) High explosive cratering studies in hard rock; Project MTCE. Boeing document D2-90704-1, Boeing Aerospace Co., Seattle.

Cook C. S. (1964) The mass of the Canyon Diablo Meteorite (letter). *Nature* **204,** 867.

Croft S. K., Kieffer S. W., and Ahrens T. J. (1979) Low-velocity impact craters in ice and ice-saturated sand with implications for Martian crater count ages. *J. Geophys. Res.* **84,** 8023–8032.

Davis L. K. (1970a) MINE SHAFT series, subtask N123 calibration cratering series. T. R. N–70–4, U.S. Army Waterways Experiment Station, Vicksburg, Miss.

Davis L. K. (1970b) MINE SHAFT series events MINE UNDER and MINE ORE subtask N121, crater investigations. T. R. N–70–8, U.S. Army Waterways Experiment Stations, Vicksburg, Miss.

Dienes J. K. and Walsh J. M. (1970) Theory of impact: Some general principles and the method of Eulerian codes. In *High-Velocity Impact Phenomena* (R. Kinslow, ed.), p. 46–104. Academic, N.Y.

Gault D. E. (1973) Displaced mass, depth, diameter and effects of oblique trajectories for impact craters formed in dense crystalline rocks. *The Moon* **6,** 32–44.

Gault D. E. (1978) Experimental impact "craters" formed in water: Gravity scaling realized (abstract). *EOS (Trans. Amer. Geophys. Union)* **59,** 1121.

Gault D. E. and Wedekind J. A. (1977) Experimental hypervelocity impact into quartz sand—II, effects of gravitational acceleration. In *Impact and Explosion Cratering,* (D. J. Roddy, R. O. Pepin, and R. B, Merrill, eds.), p. 1231–1260. Pergamon, N.Y.

Holsapple K. A. (1979) Impact experiments with ambient atmospheric pressure (abstract). EOS *(Trans. Am. Geophys. Union)* **60,** 871.

Holsapple K. A. (1980) The equivalent depth of burst for impact cratering (abstract). In *Lunar and Planetary Science XI,* p. 456–458. Lunar and Planetary Institute, Houston.

Holsapple K. A. and Schmidt R. M. (1979) A material-strength model for apparent crater volume. *Proc. Lunar Planet. Sci. Conf. 10th,* p. 2757–2777.

Holsapple K. A. and Schmidt R. M. (1980) On the scaling of crater dimensions I: Explosive processes. *J. Geophys. Res.* In press.

Melosh H. J. (1979) Acoustic fluidization: A new geologic process? *J. Geophys. Res.* **84,** 7513–7520.

Moore H. J. and Gault D. E. (1963) Relations between dimensions of impact craters and properties of rock targets and projectiles. Astrogeologic Studies Annual Progress Report, Aug. 1961—Aug. 1962, part B, Lunar and Planetary Investigations, p. 38–79. U.S. Geol. Survey.

Moore H. J., Gault D. E., and Heitowit E. D. (1965) Change of effective target strength with increasing size of hypervelocity impact craters. *Proc. 7th Symposium on Hypervelocity Impact,* p. 35–45.

Oberbeck V. R. (1977) Application of high explosion cratering data to planetary problems. In *Impact and Explosion Cratering* (D. J. Roddy, R. O. Pepin, and R. B. Merrill, eds.), p. 45–65. Pergamon, N.Y.

Opik E. J. (1958) Meteor impact on solid surface. *Irish Astron. J.* **5,** 14; Armagh. No. 24.

Opik E. J. (1961) Notes on the theory of impact craters. In *Proceedings of the Geophysical Laboratory—Lawrence Radiation Laboratory Cratering Symposium* (M. D. Nordyke, ed.), p. 51–528. Lawrence Livermore Laboratory, Livermore, Calif.

Orphal D. L., Borden W. F., and Larson S. A. (1980) Calculations of impact melt generation and transport (abstract). In *Lunar and Planetary Science XI,* p. 833–835. Lunar and Planetary Institute, Houston.

Piekutowski A. J. (1974) Laboratory-scale high-explosive cratering and ejecta phenomenology studies. Air Force Weapons Laboratory report AFWL–TR–72-155, Albuquerque, New Mexico. 328 pp.

Piekutowski A. J. (1975) A comparison of crater effects for lead azide and PETN explosive charges. Air Force Weapons Laboratory Report AFWL–TR–74–182, Albuquerque, New Mexico. 140 pp.

Piekutowski A. J. (1980) A compendium of data from laboratory-scale high-explosive cratering and ejecta studies. Univ. Dayton Research Institute Report UDR–TR–80–10, Dayton, Ohio. 660 pp.

Roddy D. J. (1973) Geologic studies of the Middle Gust and Mixed Company Craters. *Proceedings of the Mixed Company Middle Gust Results Meeting,* 13–15 March 1973, Vol. II, p. 79–124. Defense Nuclear Agency DNA 3151P2.

Roddy D. J. (1978) Pre-impact geologic conditions, physical properties, energy calculations, meteorite and initial crater dimensions and orientations of joints, faults and walls at Meteor Crater, Arizona. *Proc. Lunar Planet. Sci. Conf. 9th,* p. 3931–3964.

Roddy D. J., Boyce J. M., Colton G. W., and Dial A. L. Jr. (1975) Meteor Crater, Arizona, rim drilling with thickness, structural uplift, diameter, depth, volume, and mass-balance calculations. *Proc. Lunar Sci. Conf. 6th,* p. 2621–2644.

Roddy D. J., Schuster S. H., Kreyenhagen K. N., and Orphal D. L. (1980) Computer code simulations of the formation of Meteor Crater, Arizona: Calculations MC–1 and MC–2. *Proc. Lunar Planet. Sci. Conf. 11th.* This volume.

Rooke A. D. Jr. and Davis L. K. (1964) Crater Measurements; Project 1.9, Ferris Wheel Series, Flat Top Event, POR–3008 (WT–3008), August 1966; Defense Atomic Support Agency, Washington, D.C.

Schmidt R. M. (1977) A centrifuge cratering experiment: Development of a gravity-scaled yield parameter. In *Impact and Explosion Cratering* (D. J. Roddy, R. O. Pepin, and R. B. Merrill, eds.), p. 1261–1278. Pergamon, N.Y.

Schmidt R. M. (1978) Centrifuge simulation of the JOHNIE BOY 500 ton cratering event. *Proc. Lunar Planet. Sci. Conf. 9th,* p. 3877–3889.

Schmidt R. M. and Holsapple K. A. (1978a) Centrifuge crater scaling experiments I: Dry granular soils. Defense Nuclear Agency Report DNA 4568F, Washington, D.C.

Schmidt R. M. and Holsapple K. A. (1978b) A gravity-scaled energy parameter relating impact and explosive crater size (abstract). *EOS (Trans. Amer. Geophys. Union)* **59,** 1121.

Schmidt R. M. and Holsapple K. A. (1979) Centrifuge crater scaling experiments II: Material strength effects. Defense Nuclear Agency Report DNA 4999Z. Washington, D.C.

Schmidt R. M. and Holsapple K. A. (1980a) Theory and experiments on centrifuge cratering. *J. Geophys. Res.* **85,** 235–252.

Schmidt R. M. and Holsapple K. A. (1980b) Centrifuge crater scaling experiments III: HOB/DOB effects. Defense Nuclear Agency Report, Contract DNA 001–78–C–0149, Washington, D.C.

Schmidt R. M., Watson H. E., and Wauchope C. R. (1979) Projectile density/target density correlation for impact cratering (abstract). *EOS (Trans. Amer. Geophys. Union)* **60,** 871.

Shoemaker E. M. (1963) Impact mechanics at Meteor Crater, Arizona. in *The Moon, Meteorites and Comets* (B. M. Middlehurst and G. P. Kuiper, eds.), p. 301–306. Univ. Chicago Press, Chicago.

Shoemaker E. M. and Kieffer S. W. (1974) Synopsis of the geology of Meteor Crater. In *Guidebook to the Geology of Meteor Crater, Arizona*. Prepared for the 37th Ann. Mtg. Met. Soc., Aug. 1974. 66 pp.

Taylor D. W. (1948) *Fundamentals of Soil Mechanics*. Wiley, N.Y. 700 pp.

Proc. Lunar Planet. Sci. Conf. 11th (1980), p. 2129–2144.
Printed in the United States of America

Formation of bowl-shaped craters

Andrew J. Piekutowski

University of Dayton Research Institute, Dayton, Ohio 45469

Abstract—Study of impact craters formed on planetary and lunar surfaces has shown that bowl-shaped craters form the largest number and simplest class of impact structure observed. The study described in this paper uses high-explosive charges to form laboratory-scale bowl-shaped craters in several types of granular media. Although the crater formation process initiated by the impact of a hypervelocity body is not identical with that initiated by the detonation of a high explosive, the processes are related in terms of crater morphologies and structural deformations. High-speed films of the experiments are used to obtain crater growth rate and particle displacement data. Results of work performed to date are presented and discussed. Included in these results are quantitative stress, strain, displacement, and velocity data. These data and data currently being obtained will be compared with particle displacement and velocity data from large explosion experiments which produced bowl-shaped craters. Results of these comparisons and other comparisons of the morphological features and structural deformations of large explosion and impact craters will be used to develop a time-sequence description of the formation of a large bowl-shaped impact crater.

1. INTRODUCTION

Continuing studies of impact features formed on planetary and lunar surfaces have shown that craters formed on these bodies can be classified according to their size and shape (Oberbeck and Quaide, 1967; Gault *et al.*, 1975; Cintala *et al.*, 1976; Pike, 1976). Bowl-shaped craters form the largest number and simplest class of impact structures observed in these studies. The complexities of the impact cratering process, however, continue to limit our understanding of processes associated with crater formation, particularly for those craters exhibiting the more complex structural features.

Gault *et al.* (1968) and others have developed models used to describe the mechanics of crater formation. Oberbeck (1975) developed a crater growth model for use in modeling ejecta emplacement. Maxwell (1977) has formulated a description of the crater formation process using the so-called Z model. Since these models, and others not mentioned, were developed after careful study of high-speed films of numerous impact and explosion cratering experiments and thorough examination of the craters produced by the experiments, it is no surprise that these models agree in their overall description of the crater formation process. However, these experiments do not provide sufficient detailed dynamic data to permit various aspects of crater formation to be quantified. The study described in this paper uses a quarter-space test container to expose the crater and the region surrounding the crater to view while the crater is forming. As a result, a variety of dynamic particle displacement, velocity, stress, and strain data can be obtained.

One wall of the quarter-space container is constructed of a thick piece of clear plastic and permits crater formation to be photographed "in section" with high-speed cameras. A large number of dyed sand grain tracer particles are placed in the cratering medium at the medium-plastic wall interface to provide targets whose motions are followed. Analysis of the collective motions of these tracer grains permits the effects of many crater formation related processes to be quantified. In this paper, results of a partial analysis of the motions of these tracers are presented for several experiments; results of a fairly complete analysis of the motions produced during one experiment employing a shallow depth of burst are also presented. These results and the results of additional experiments will be used to develop a quantitative description of the formation of large bowl-shaped craters.

2. EXPERIMENTAL PROCEDURES

Lead azide and pentaerythritol tetranitrate (PETN) high-explosive charges were used to produce bowl-shaped craters in various granular media. Two different charge placements were used: (1) half-buried and (2) completely buried with the top of the charge tangent to the preshot surface. Spherical and hemispherical lead azide charges weighing 1.7 and 0.85 gm, respectively, were used. Two weights of spherical PETN charges—0.400 g and 1.265 g—and two weights of hemispherical PETN charges—0.200 gm and 0.632 g—were also used. The spherical 1.7 g lead azide and 0.400 g PETN charges each released 2.25×10^{10} ergs when detonated (Piekutowski, 1975a). Bomb calorimeter measurements were not made to determine the amount of energy released by the other weights and shapes of charges. It was assumed that the energy released by those charges was proportional, on a weight basis, to the previously mentioned charges.

Dry Ottawa Flint Shot (average grain diameter ~0.55 mm), dry Ottawa Banding Sand (average grain diameter ~0.17 mm), and Kirtland Air Force Base desert alluvium (average moisture content ~4.5%) were used as cratering media. All media were placed in the test containers using procedures which produced the maximum *in situ* density of the test bed (Piekutowski, 1974; Piekutowski, 1975b). Densely packed Ottawa Flint Shot was selected as the base line medium. Similar quartz sands have been used for numerous laboratory-scale impact cratering experiments performed at the NASA Ames Vertical Ballistic Gun Range (Gault *et al.*, 1968; Braslau, 1970; Oberbeck, 1971; Stöffler *et al.*, 1975; Gault and Wedekind, 1977; Holsapple, 1980; and Schmidt, 1980) and by other investigators for various explosion cratering experiments (Johnson *et al.*, 1969; Piekutowski, 1974, 1975a, 1975b, 1975c, 1977, 1980a; Andrews, 1975, 1977; and Schmidt and Holsapple, 1978, 1980). Use of Ottawa Flint Shot as the base line medium in this study facilitates correlation of the results of this study with the results of the large number and variety of experiments just cited.

Spherical charges were detonated in test beds prepared in a large round (1.52 m diameter, 0.61 m deep) and a small round (0.76 m diameter, 0.23 m deep) test tank to produce baseline morphological and crater dimension data. Considerable use was also made of data obtained from earlier experiments performed in the round containers using the three test media and charge types and weights (Piekutowski, 1980b). Hemispherical charges were detonated in the quarter-space tank to provide two types of data: (1) documentary data to describe crater growth rate and final crater dimensions and (2) detailed particle displacement data for use in describing the behavior of the cratering medium during the formation of bowl-shaped craters. Data from the quarter-space experiments were routinely compared with data from similar experiments in the round tanks to insure that results of experiments performed in the quarter-space tank were not aberrant.

The quarter-space container shown in Fig. 1a is 0.76 m long, 0.38 m wide, and 0.23 m deep. The container is constructed of 1.25 cm thick aluminum plate. A portion of the front wall has been removed to permit the installation of a 5 cm thick clear Plexiglas window. The window is bolted to the front wall and extends approximately 6 cm above the preshot surface (see Figs. 1b and 1c). A filler plate is installed in place of the window during preparation of the test beds. When installed, the filler plate is flush with the inside and top surface of the front wall of the tank. After slightly overfilling the quarter-space tank, excess cratering medium was carefully scraped away and a cover plate bolted to the top of the tank. The filled and closed container was supported in bearings on short shafts attached to the end plates of the quarter-space tank and rotated until the front wall was horizontal. Further work with the upturned test bed continued after the filler plate was removed.

Preparation of the container was complete when the Plexiglas window was bolted to the tank wall (a rabbet machined around the edge of the Plexiglas permitted the interior surface of the window to be flush with the interior surface of the test container), the test container rotated to its initial position, and the cover plate removed. The flat faces of the hemispherical charges were cemented to the window before the window was installed on the test container. A small cavity formed in the cratering medium provided space for the charge. Fiducial lines, shown in Fig. 1c, were engraved on the inside surface of the Plexiglas window for use as references during subsequent measurements.

When dyed sand grain tracers were required for an experiment, they were installed before the window was fixed in place. Positioning of the grains was facilitated by the use of a special fixture placed in the opening formerly occupied by the filler plate. The fixture had provisions for precisely locating a drilled plate approximately 0.25 mm above the test bed surface. This plate contained 260 funnel-shaped holes drilled at the intersections of a grid system 8 mm square and formed the heart

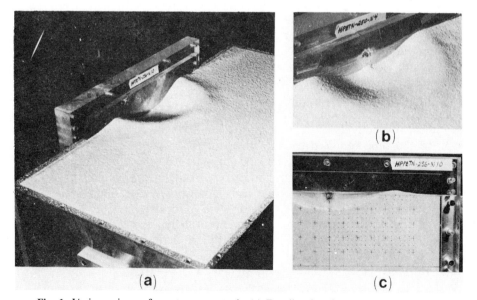

Fig. 1. Various views of quarter-space tank. (a) Details of tank construction showing tapped holes used to attach cover plate. (b) View of crater formed by a half-buried charge. (c) View of Plexiglas window showing method of attachment to tank wall. Also shown are fiducials and final positions of a number of dyed sand grain tracers. Results of analysis of motions of these grains are presented in later figures.

of the fixture. Four colors of sand grains were placed in 13 rows of 20 grains each by dropping the appropriate colored grain through a hole in the fixture. Use of specific colors at specific locations in the grid system simplified subsequent identification of grains which experienced large and/or irregular displacements.

A high-speed camera was used to record the crater formation process. When documenting crater development (no tracers) the normal field of view extended approximately 15 cm to either side of the charge, with the test bed occupying two-thirds the height of the field of view. In the particle displacement experiments, the field of view just extended beyond the area which had been seeded with tracer particles (see Fig. 1c). Framing rates of approximately 6400 pictures per second and color film were used for both types of experiments.

Oblique angle and stereo-pair photography were used to document the crater. Crater dimension data for all experiments were obtained with special profilometers. A series of equally-spaced profile measurements were made at 45° intervals around the crater. A series of transient crater profile data were also obtained from high-speed films of the documentary and particle displacement experiments. Crater dimension data (volume, depth, and radius) were determined from the profile measurements using a moment-area method of computing these data. Coordinates of dyed sand grains in selected frames of the particle displacement films were recorded on computer cards with use of a Benson and Lehner semiautomatic film reader. Motions of single grains and groups of grains were further analyzed with use of several computer programs.

3. RESULTS

Data presented in this paper concentrate on the results of experiments performed in Ottawa Flint Shot. Although some experiments were performed in the other media described in the preceeding section, work with these media was in progress at the time this paper was written.

High-speed films of seven quarter-space experiments in Ottawa Flint Shot were analayzed to determine the rate of growth of the crater as a function of explosive composition, weight, and depth of burst. Series of transient crater profiles were obtained and analyzed for each film to provide crater growth-rate data. Data from five of these experiments are presented in Fig. 2. In this figure, the ratio of the instantaneous crater volume, depth, and radius to the final crater volume, depth, and radius, respectively, is plotted as a function of the ratio of time after detonation to the time required to form the crater. Crater formation time, shown schematically in Fig. 2, was arbitrarily defined to be the time after detonation when a slight inflection of the interior surface of the ejecta plume was observed in the region of the developing crater rim. Hemispherical lead azide and PETN charges, each releasing 1.12×10^{10} ergs, were used to produce the craters analyzed for Fig. 2. Films of two experiments utilizing the larger hemispherical PETN charges did not provide useful data for this plot since the craters grew beyond the field of view of the camera and it was not possible to determine crater formation time.

Vertical scatter of plotted values at the earliest stages of crater formation result from the uncertainty of the exact time of detonation with respect to the interframe time of the camera. The effect of this uncertainty on the value of T/T_{max} is greatest for the first few frames after detonation. Some dimensional characteristics of the craters used to develop Fig. 2 are presented in Table 1. As shown in Table 1, crater volumes and formation times vary considerably. Yet, the crater

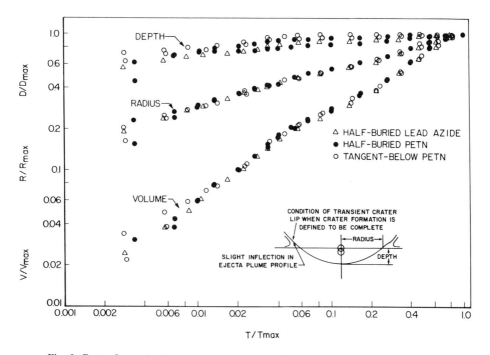

Fig. 2. Rate of growth of volume, depth, and radius of bowl-shaped craters produced by the detonation of small high-explosive charges in densely packed Ottawa Flint Shot.

volume and radius ratios plot very tightly. Crater depth ratios are strongly influenced (in this figure) by the fact that the explosive charge initially occupies a significant portion of the distance determined to be the maximum crater depth. Crater formation times for the two experiments with the larger PETN charges were in excess of the maximum times shown in Table 1.

Displacement histories of a large number of dyed sand grain tracers used in one of the experiments with a shallowly buried PETN charge, N10, are shown in Fig. 3. The shallow depth of burial of the explosive charge should produce a crater that closely resembles a crater which would be produced by a hypervelocity impact in the same medium (Oberbeck, 1971). Inaccuracies in the determination of the positions of the tracers resulting from film reading and magnification errors, etc., are typically on the order of 0.1 mm although occasional errors of 0.25 mm or more may occur.

Initial motions of grains nearest the charge were radially away from the charge. Alteration of this initial trajectory occurred as crater formation progressed. Grains directly below the charge came to rest and then exhibited a small amount of rebound. Grains near the surface were ejected from the crater. Grains at intermediate locations experienced a variety of motions. Most were subjected to gradually developing forces that resulted in the grains exhibiting varying amounts of displacement toward the surface and comparatively large displacements away from the charge. Extensions of lines drawn through the trajectories of the various

Table 1. Characteristics of craters formed in quarter-space tank.

Event	Explosive/ Configuration	Formation Time* (msec)		Volume (cm³)	Depth (mm)	Radius (mm)
N1	Lead Azide Half-Buried	58	76	425	31	93
N4	PETN Half-Buried	47	75	232	22	75
N7	PETN Half-Buried	47	89	268	26	77
N10	PETN Tangent-Below	59	96	388	28	90
N11	PETN Tangent-Below	58	96	394	30	91

  * First value shown is formation time as defined in Fig. 2. Second value is time when ejecta veil has passed over crater rim. This time is equivalent to the formation time used by Gault *et al.* (1968).

tracers indicate, rather clearly, that the effective center of disturbance or motion-producing source was located in a very small region at the base of the charge.

Displacement data for the surface grain closest to the charge can be used to estimate the probable time of detonation of the charge with respect to the inter-frame time of the high-speed camera. In Fig. 4a, the displacement of this grain, in two directions, is plotted for the time interval it was in view of the camera. Time of detonation, for this experiment, was approximately 85 μsec after the time when frame zero was exposed.

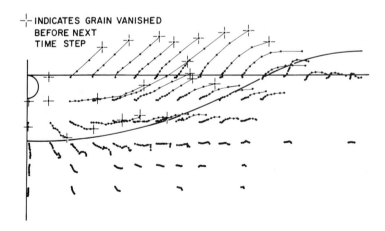

Fig. 3. Trajectories of selected dyed sand grain tracers during the formation of a bowl-shaped crater. The crater was produced by the subsurface detonation of a 0.200 g hemispherical PETN charge in 1.80 g/cm³ Ottawa Flint Shot. Final crater profile shown as solid line. Motion of all but surface grains has ceased for the last time shown.

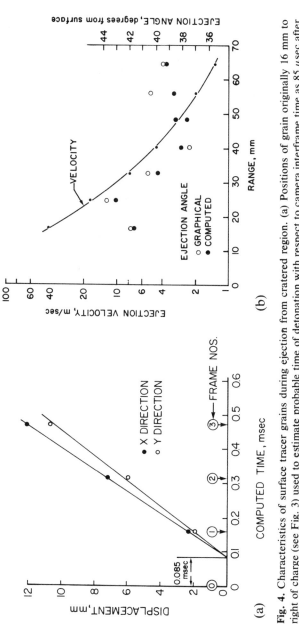

Fig. 4. Characteristics of surface tracer grains during ejection from cratered region. (a) Positions of grain originally 16 mm to right of charge (see Fig. 3) used to estimate probable time of detonation with respect to camera interframe time as 85 μsec after frame zero (last frame which shows charge intact). (b) Velocity and ejection angle of seven surface grains as a function of original distance from charge.

Average ejection velocities of seven surface grains were determined using average upward and outward incremental velocity components to determine a resultant velocity magnitude. Resultant ejection velocities and ejection angles for these seven grains are shown in Fig. 4b. Two values for the ejection angle of each grain are presented. Computed values were determined using the components of the average incremental velocities of the individual grains. Graphical values were obtained by measuring the angle which the average trajectory of each grain made with respect to the original surface. Assumed grain trajectories are shown by the thin lines in Fig. 3. Results of a least-squares fit to the velocity data presented in Fig. 4b indicate that the ejection velocity of surface material decreased with increasing distance from the charge as a function of distance (range) to the -2.51 power. Significant acceleration or deceleration of the motions of the grains whose velocities are plotted in Fig. 4b did not appear to occur after these grains were launched into flight. However, the final velocity and direction of the grains furthest from the charge was affected by the collision of these grains with material in free flight above them. These collisions appeared to occur very shortly after the grains were displaced.

Displacement and velocity histories of a number of grains in the fourth row below the surface are presented in Fig. 5. Values shown are for the horizontal component of motion of the grains since this was the primary direction of motion of these grains. One and possibly two measured positions were incorrect for the grain originally at a range of 121 mm. Estimated values of displacement and corresponding values of velocity for this grain are shown with an "X" on these curves.

The abscissa of each plot in Fig. 5 is placed at appropriately scaled vertical distance from an upper reference line corresponding to the axis of the crater ($X = 0$). Consequently, average velocity of the disturbance propagating in the horizontal direction can be determined directly using the line drawn diagonally through the various displacement curves. Using the estimated time of detonation determined in Fig. 4a, the solid diagonal line corresponds to a velocity of 167 m/sec for the disturbing wave. This value compares with a value of 204 m/sec (broken line) published by Wetzel and Vey (1970) for the propagation velocity of a compressive wave traveling through a dense Ottawa sand. Values of peak grain velocity were estimated by extrapolating the velocity curve back to the time when the grain was estimated to have experienced passage of the compressive wave.

Peak velocity data shown in Fig. 5 have been plotted in Fig. 6 as a function of range from the axis of the crater. If the disturbing wave is assumed to present a planar front during passage through the region around the grain, then peak stress experienced by material in the region around the grain can be calculated using the relationship and values presented in the upper right of Fig. 6. A least-squares fit to the data presented in Fig. 6 showed that peak stress decayed exponentially with range as e to the $-0.029x$ power, where x is the range in mm.

Analysis of the motions of a large number of the tracer grains was made with use of a computer program. In addition to performing simple calculations to determine the coordinates of each grain (and corresponding velocity compo-

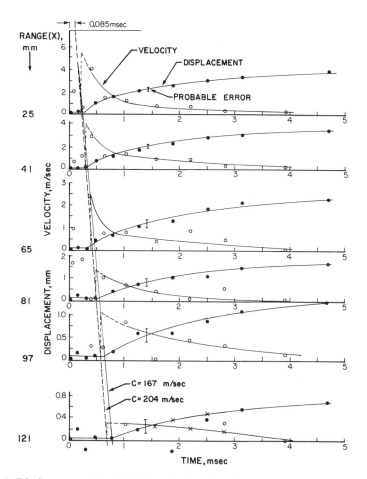

Fig. 5. Displacement and velocity histories for selected tracer grains in fourth row from surface (see Fig. 3). Four different vertical scales are used. Dual units for each scale as well as original ranges to the tracer grains are identified to the left of the curves. Probable error in determining the position of the grains is shown with an error bar on each displacement curve. Further discussion of this figure is provided in the text.

nents), computations were made for the purpose of determining the behavior of material within a region bounded by each set of four grains initially located at the corners of the squares in the grid system. Initial positions of the grains were determined using the last frame of film before detonation (frame zero). Subsequent positions of the grains were determined for approximately 15 additional frames of film. Consecutive frames of film were read for very early times with the time interval between frames gradually increasing as the crater grew and the displacements and level of activity decreased.

Volumetric strain is a property useful in describing the general behavior of the cratered medium during crater formation. If the area bounded by lines drawn

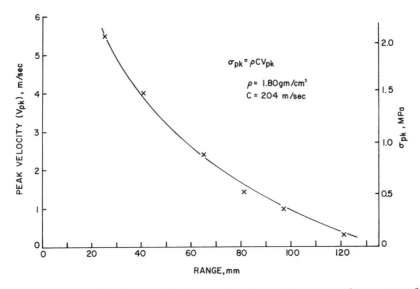

Fig. 6. Estimate of peak stress levels experienced at various ranges from center of crater. Peak velocities were obtained from curves presented in Fig. 5.

through pairs of grains forming the corners of a square is the cross-sectional area of a ring of material parallel to the surface, the volume of the ring can be calculated by multiplying this area by $2\pi R_{CG}$, where R_{CG} is the radius to the center of gravity of the cross-sectional area. Changes in the volume of a ring can be determined by comparing the instantaneous volume of the ring with the initial volume of the ring. A Monte Carlo computer technique and typical values of film reading error ($\sim$0.1 to 0.25 mm) were used to determine that volume of the rings could be calculated with about 3% error.

A series of transient crater profiles and levels of volumetric strain for a portion of the cratered region are presented in Fig. 7. The compressive wave front shown in Fig. 7a has a stress level of approximately 0.5 MPa during passage of the wave through this region. Wave front shape, as reflected in levels of volumetric strain, appears to be more or less vertical during passage through the four rows of cross-sectional areas at the top of Fig. 7b. Spall of surface material is also evident in Fig. 7b. For the experiment shown in Fig. 7, spall is believed to result from ground-induced motions since the cratering medium around the charge should shield the test bed from airblast-induced motions. Extent of the regions which

Fig. 7. Volumetric strain of uncratered media at various times after detonation. Percentage of formation time and final crater volume shown at left of each figure. Transient and final crater profiles are shown as solid and dashed lines, respectively. Approximate position of stress wave front (see Figs. 5 and 6) is shown in (a) at a range of about 80 mm.

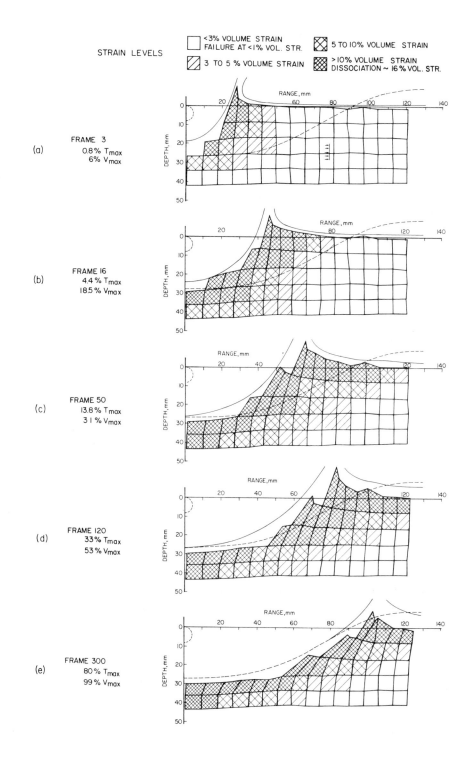

STRAIN LEVELS

☐ <3% VOLUME STRAIN
FAILURE AT <1% VOL. STR.

▨ 3 TO 5 % VOLUME STRAIN

▨ 5 TO 10% VOLUME STRAIN

▨ >10% VOLUME STRAIN
DISSOCIATION ~ 16% VOL. STR.

(a) FRAME 3
0.8 % T_{max}
6 % V_{max}

(b) FRAME 16
4.4 % T_{max}
18.5 % V_{max}

(c) FRAME 50
13.8 % T_{max}
31 % V_{max}

(d) FRAME 120
33 % T_{max}
53 % V_{max}

(e) FRAME 300
80 % T_{max}
99 % V_{max}

will experience greater than 3% volumetric strain have been more or less identified in Fig. 7c (less than 14% of the time required to form the crater). Continued
spallation of surface material provides relief for the small lithostatic and inertial
forces which prevent further strain of material in the region just below the preshot
surface. The extent of the spalled region can be rather well defined by observing
the positions of the surface tracer grains in Fig. 7a and subsequent portions of
Fig. 7 with respect to the preshot surface (solid line) in each of the figures. Spall
has become a significant near-crater phenomenon in Fig. 7d. Crater formation
(crater volume) is 99% complete in Fig. 7e. A small contribution to the total
crater volume occurred when the material in the region below the crater and a
range of approximately 40 mm, exhibited a small amount of collapse. The general
behavior of material in this collapsed region is shown by the motions of the
appropriate tracer grains in Fig. 3.

4. DISCUSSION

Comparison of results of experiments performed in the normal half-space environment (air over cratering medium) and in the quarter-space environment indicate that use of the Plexiglas-wall in the quarter-space does not significantly affect
the crater formation process. Data compared included: (1) crater dimensions, (2)
morphological features of craters, and (3) subsurface deformations. Craters produced in the quarter-space container were typically smaller (10 to 20%) than their
counterparts in the half-space container. This is not surprising since the Plexiglas
wall accepts a portion of the energy which would be available for crater formation
if the normal half-space environment could be split in two. The sectioned crater
shown in Fig. 8a was produced by a surface detonation of a lead azide charge
in densely packed Ottawa Flint Shot. Dark sand has been poured over the surface
of the crater to enhance definition of the crater boundaries. Sectioning occurred
after formation of the crater in the large round tank. Permanent structural deformations of the layers and columns exhibited in this section view compare in
magnitude and extent with those exhibited by the zone boundaries in Fig. 7e.

Views of the crater formation process obtained with use of the quarter-space
tank show the cratering flow field, as evidenced by tracer grain motion, is relatively simple. Once initiated, the flow field remains relatively unchanged during
further development of the crater. The quarter-space films indicate that deposition
and partitioning of energy into various processes associated with crater formation
certainly occurs in less than 0.1% of the time required to form the crater. Actual
time required for these processes to occur is probably on the order of 0.02% (10
to 20 μsec) of the total crater formation time as defined in Fig. 2.

Crater growth-rate data presented in Fig. 2 indicate, for a limited range of near-
surface charge configurations, that processes which govern the geometry of the
evolving crater are almost independent of the properties of the source which
motivates formation of the crater. Properties of the source and partitioning of
available source energy only govern the size of the crater. Therefore, it is con-

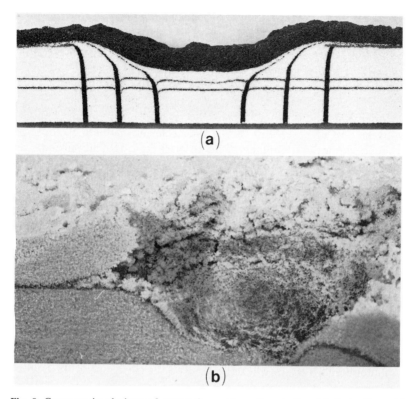

Fig. 8. Cross-sectional views of craters formed in earlier studies. (a) Crater formed in dense, dry Ottawa Flint Shot. Deformations of columns and layers similar to structural deformations exhibited in Fig. 7e (localized deformations of layers at intersection with columns occurred during installation of columns). (b) Crater formed in a loosely-packed, wet Ottawa Flint Shot. This crater clearly exhibits separation, but not the down dropping indicated by the motions of tracer grains in Fig. 3 and Fig. 7e.

cluded that any bowl-shaped crater produced in Ottawa Flint Shot by any type or size of slightly to shallowly buried explosive would evolve according to the relationships presented in Fig. 2. Previous work by Schmidt and Holsapple (1978) demonstrated that the shapes and morphological features of explosion craters produced in Ottawa Flint Shot at elevated gravity were approximately the same as those for explosion craters produced at terrestrial gravity (Piekutowski, 1974, 1975b). Crater growth-rate data obtained from experiments in Ottawa Banding Sand and the desert alluvium will (1) provide new relationships which indicate that crater growth rate is medium sensitive or (2) produce curves similar to those presented in Fig. 2, indicating that the geometries of evolving bowl-shaped craters are the same, regardless of cratering medium properties.

Flow fields defined by the tracer grain trajectories during crater formation were similar for all experiments. Craters produced by events N1 and N10 (see Table 1) had approximately the same dimensions although N1 was produced by half-buried lead azide charge and N10 was produced by a shallowly buried PETN charge. Trajectories of the tracer grains used in both events (results of N10 presented in Fig. 3) were identical except for a slightly shallower ejection angle of several surface grains nearest the charge of event N1.

Data presented in Fig. 7 indicate that bulking of uncratered medium occurred in a significant portion of the region surrounding the crater. In many of the zones near the crater floor and preshot surface, the sand grains dissociated and became discrete missiles of ejecta. As a result, the cratering medium in these regions behaved much like a fluid. Bulking of uncratered material and subsequent readjustment of bulked material produced a significant portion of the upthrust region surrounding the crater. Ejecta deposited on the surface, in this region, contributed to less than half the total lip height. Volume gained through repositioning of bulked material accounted for more than half the final crater volume. Experiments performed by Andrews (1975) in dry Ottawa Flint Shot showed that 55% of the material originally in the cratered region remains in a region that does not extend beyond 2 crater radii.

Motion of grains in the cratered region is initiated by passage of a stress wave. Previous work with media having some cohesive strength, e.g., wet Ottawa Flint Shot, has repeatedly shown that the production of ejecta ceases rather abruptly when crater growth has reached a certain stage of development. Flight of ejecta in these experiments is not characterized by the smooth, continuous flows characteristically observed for experiments in dry sand. Termination of the production of ejecta undoubtedly occurs when the level of the diverging stress wave drops below a stress level which the wet sand is capable of sustaining without fragmenting. Energy which is deposited in the wet sand after passage of the stress wave appears to dissipate through lifting of entire regions of the near-crater environment, rather than through the continued motions of individual grains. As a result, the significant amounts of repositioning of uncratered medium do not occur. In fact, Andrews (1977) has shown, for some experiments in wet sand, that the mass of material ejected from the crater was equivalent to the mass of material which would be required to fill the crater. Craters produced in wet, homogeneous Ottawa Flint Shot tend to exhibit rather steep crater walls and upthrust regions around the crater. Both these characteristics are shown in the crater shown in Fig. 8b. Downward motion of the upthrust region was prevented by the insertion of a wedge of material from some other part of the crater.

Acknowledgment—I wish to express my sincere appreciation to Dr. W. L. Quaide for his assistance and understanding during the early phases of this work. I am also indebted to Dr. D. Roddy for his continued interest in this work and for his helpful discussions of experimental results. Special appreciation is due Mr. S. Hanchak for the care and assistance he provided during the experiments and to Mr. R. Watt and Mrs. N. Brantley for their efforts and assistance in making the film readings possible. Finally, I would like to thank Drs. R. M. Schmidt and K. A. Holsapple for their careful reviews of this paper. This work was supported by NASA under Grant NSG 7444.

REFERENCES

Andrews R. J. (1975) Origin and distribution of ejecta from near-surface laboratory-scale cratering experiments. AFWL-TR-74-314, Air Force Weapons Laboratory, Albuquerque, New Mexico.

Andrews R. J. (1977) Characteristics of debris from small-scale cratering experiments. In *Impact and Explosion Cratering* (D. J. Roddy, R. O. Pepin, and R. B. Merrill, eds.), p. 1089–1100. Pergamon, N. Y.

Braslau D. (1970) Partitioning of energy in hypervelocity impact against loose sand targets. *J. Geophys. Res.* **75**, 3987–3999.

Cintala M. J., Head J. W., and Mutch T. A. (1976) Martian crater depth/diameter relationships: Comparisons with the moon and Mercury. *Proc. Lunar Sci. Conf. 7th*, p. 3575–3587.

Gault D. E., Guest J. E., Murray J. B., Dzurisin D., and Malin M. C. (1975). Some comparisons of impact craters on Mercury and the moon. *J. Geophys. Res.* **80**, 2444–2460.

Gault D. E., Quaide W. L., and Oberbeck V. R. (1968) Impact cratering mechanics and structures. In *Shock Metamorphism of Natural Materials* (B. M. French and N. M. Short, eds.), p. 87–99. Mono, Baltimore.

Gault D. E. and Wedekind J. A. (1977) Experimental hypervelocity impact into quartz sand-II, effects of gravitational acceleration. In *Impact and Explosion Cratering* (D. J. Roddy, R. O. Pepin, and R. B. Merrill, eds.), p. 1231–1244. Pergamon, N. Y.

Holsapple K. A. (1980) The equivalent depth of burial for impact cratering. *Proc. Lunar Planet. Sci. Conf. 11th*. This volume.

Johnson S. W., Smith M. A., Franklin E. G., Moraski L. K., and Teal D. J. (1969) Gravity and atmospheric pressure effects on crater formation in sand. *J. Geophys. Res.* **74**, 4838–4850.

Maxwell D. E. (1977) Simple Z model of cratering, ejection, and the overturned flap. In *Impact and Explosion Cratering* (D. J. Roddy, R. O. Pepin, and R. B. Merrill, eds.), p. 1003–1008. Pergamon, N. Y.

Oberbeck V. R. (1971) Laboratory simulation of impact cratering with high explosives. *J. Geophys. Res.* **76**, 5732–5749.

Oberbeck V. R. (1975) The role of ballistic erosion and sedimentation in lunar stratigraphy. *Rev. Geophys. Space Phys.* **13**, 337–362.

Oberbeck V. R. and Quaide W. L. (1967) Estimated thickness of a fragmental surface layer of Oceanus Procellarum. *J. Geophys. Res.* **72**, 4697–4704.

Piekutowski A. J. (1974) Laboratory-scale high-explosive cratering and ejecta phenomenology studies. AFWL-TR-72-155, Air Force Weapons Laboratory, Albuquerque, New Mexico.

Piekutowski A. J. (1975a) A comparison of cratering effects for lead azide and PETN explosive charges. AFWL-TR-74-182, Air Force Weapons Laboratory, Albuquerque, New Mexico.

Piekutowski A. J. (1975b) The effects of variations in test media density on crater dimensions and ejecta distributions. AFWL-TR-74-326, Air Force Weapons Laboratory, Albuquerque, New Mexico.

Piekutowski A. J. (1975c) The effect of a layered medium on apparent crater dimensions and ejecta distribution in laboratory-scale cratering experiments. AFWL-TR-75-212, Air Force Weapons Laboratory, Albuquerque, New Mexico.

Piekutowski A. J. (1977) Cratering mechanisms observed in laboratory-scale high-explosive experiments. In *Impact and Explosion Cratering* (D. J. Roddy, R. O. Pepin, and R. B. Merrill, eds.), p. 67–102. Pergamon, N. Y.

Piekutowski A. J. (1980a) Cratering experiments in wet and dry layered media. UDR-TR-80-49, University of Dayton Research Institute, Dayton, Ohio.

Piekutowski A. J. (1980b) A compendium of data from laboratory-scale high-explosive cratering and ejecta studies. UDR-TR-80-10, University of Dayton Research Institute, Dayton, Ohio.

Pike R. J. (1976) Simple to complex impact craters: the transition on the moon (abstract). In *Lunar Science VII*, p. 700–702. The Lunar Science Institute, Houston.

Schmidt R. M. (1980) Meteor Crater: Energy of formation—implications of centrifuge scaling. *Proc Lunar Planet. Sci. Conf. 11th*. This volume.

Schmidt R. M. and Holsapple K. A. (1978) Centrifuge cratering experiments I: Dry granular soils. Report DNA 4568F, Defense Nuclear Agency, Washington, D. C.

Schmidt R. M. and Holsapple K. A. (1980) Theory and experiments on centrifuge cratering. *J. Geophys. Res.* **85,** 235–252.

Stöffler D., Gault D. E., Wedekind J., and Polkowski G. (1975) Experimental hypervelocity impact into quartz sand: distribution and shock metamorphism of ejecta. *J. Geophys. Res.* **80,** 4062–4077.

Wetzel R. A. and Vey E. (1970) Axisymmetric stress wave propagation in sand. *ASCE J. Soil. Mech. Found. Div.* **SM5,** 1763–1786.

Proc. Lunar Planet. Sci. Conf. 11th (1980), p. 2145–2158.
Printed in the United States of America

Impact-induced water loss from serpentine, nontronite and kernite

Mark B. Boslough, Ray J. Weldon, and Thomas J. Ahrens

Seismological Laboratory, California Institute of Technology, Pasadena, California 91125

Abstract—Preliminary experiments have been conducted to study shock-release of volatiles from minerals. Impact-induced loss of bound water from hydrous minerals has been observed, using infrared absorption and X-ray powder diffractometer techniques. Serpentine ($Mg_3Si_2O_5(OH)_4$) and nontronite (.5 $Ca_{0.7}Fe_4[(Si_{7.3}Al_{0.7})O_{20}](OH)_4 \cdot nH_2O$) were shocked and recovered from pressures of up to 38 GPa, using one-dimensional shock reverberation techniques. Kernite ($Na_2B_4O_7 \cdot 4H_2O$) was impacted by a spherical pyrex projectile traveling at 4.89 km sec^{-1}, which produced a peak pressure of ~33 GPa. The infrared absorption spectra indicate that some of the bound water from these three minerals was released as a result of shock compression and subsequent rarefaction. This evidence is supported by the recovery of small amounts of vapor from the serpentine shocked to 23.5 GPa and the nontronite shocked to 18 GPa. The recovered vapor is inferred to be water from the shocked minerals. X-ray diffraction spectra indicate no major changes in the unit cell dimensions of the two silicates, except for a decrease in the lattice constant in the c-direction of the nontronite, consistent with the loss of interlayer water.

INTRODUCTION

Central to any model of the formation of the terrestrial planets is the degassing history. Extensive evidence supports the hypothesis of Fanale (1971) that the loss of volatiles from the solid earth was early and rapid. Arrhenius *et al.* (1974) and Benlow and Meadows (1977) demonstrated that a primitive atmosphere will form as a result of strong shock compression and heating associated with the accretion of large planetesimals. Shimizu (1979), Jakosky and Ahrens (1979), and others have used such an atmosphere as a starting point for recent models of atmospheric evolution. The concept of an impact-created early atmosphere is not inconsistent with the idea of a later slow migration of volatiles from the mantle to the surface (Walker, 1977). In fact, energy derived from accretional impacts may have mobilized volatiles into energetically favorable states for later removal from the solid mantle.

Lange and Ahrens (1979) and Lange (1980) have calculated the shock pressure necessary to release water from the hydrous minerals brucite and serpentine with varying initial porosity. These calculations were based on the entropy-gain method of Ahrens and O'Keefe (1972) and made use of thermodynamic and shock wave data for brucite and serpentine. The Grüneisen parameters used in the entropy calculations had to be estimated because they have not been determined experimentally for these minerals. Thus the calculated pressures required for vapor release are dependent on the Grüneisen parameters used and are subject to the uncertainties in them. These uncertainties are difficult, if not impossible, to estimate without determining the Grüneisen parameters experimentally.

We chose a more direct method of determining the efficiency of impact-induced devolatilization of hydrous minerals, which bypasses theoretical calculations altogether. Our approach was to carry out exploratory shock-vaporization experiments which allowed shock-induced volatile release to be observed in the laboratory. The information obtained should help to constrain theories of atmosphere formation and evolution.

EXPERIMENTAL TECHNIQUES AND SAMPLE PREPARATION

Six successful shock-recovery experiments were conducted on three hydrous minerals. The critical data are summarized in Table 1. Three experiments were performed on serpentine ($Mg_3Si_2O_5(OH)_4$), and two on nontronite (0.5 $Ca_{0.7}Fe_4$ $[(Si_{7.3}Al_{0.7})O_{20}](OH)_4 \cdot nH_2O$) using methods similar to those described by Gibbons et al. (1975). In each of these experiments the target was shock-loaded by a flat-faced metal projectile accelerated to high velocity by a propellant gun. Three of the experiments—two serpentine and one nontronite—were carried out on sample disks sandwiched between two polished metal plates and pressed into metal target cylinders in an arrangement similar to that described in detail by Gibbons (1974). A 20-mm caliber gun was used to launch the projectiles in these three experiments.

One serpentine and one nontronite experiment employed a modified target assembly designed to isolate the solid sample and to capture any vapor liberated (Fig. 1). The presence of captured vapor is sufficient to show that volatile release has occurred. When vapor escapes from the sample it expands into an evacuated chamber and can be condensed through a valve into a glass ampoule and weighed. A 20-mm caliber gun was used to accelerate the projectile in the serpentine experiment, and a 40-mm caliber gun was used in the nontronite experiment.

In all the serpentine and nontronite experiments, the flyer plates and target cylinders were stainless steel-304 (68% Fe, 21% Cr, 9% Ni). Peak shock pressures in all cases were determined by applying graphical impedence matching techniques to the Hugoniot data for the flyer plate and target material (McQueen et al., 1970).

One impact experiment was carried out on a kernite ($Na_2B_4O_7 \cdot 4H_2O$) crystal. The projectile, a 3.18-mm diameter pyrex sphere, was fired from a light gas gun into the single crystal block of kernite (approximately 15cm on edge). This configuration produces a two-dimensional, axially symmetric impact much like a cratering experiment. Since the kernite crystal was much larger than the projectile, the shock wave generated was similar to a natural event. The ejecta fragments from the impact were captured by an aluminum witness plate and

Table 1. Impact vaporization recovery experiments.

Shot #	20–544*†	20–551*	20–552*	40–492***†	20–550*	AVG-790909***
Sample	Serpentine	Serpentine	Serpentine	Nontronite	Nontronite	Kernite
Sample Mass	.3989 g	.0146 g	.0142 g	.3000 g	.0161 g	—
Heat Treatment	100°C, 24 hr	100°C, 24 hr	100°C, 24 hr	none	none	none
Impactor	SS-304	SS-304	SS-304	SS-304	SS-304	Pyrex sphere 3.18 mm dia.
Velocity	1.10 km sec^{-1}	1.37 km sec^{-1}	1.64 km sec^{-1}	0.875 km sec^{-1}	1.36 km sec^{-1}	4.89 km sec^{-1}
Peak Pressure	23.5 GPa	30.3 GPa	37.8 GPa	18.1 GPa	30.0 GPa	28–38 GPa

* Performed at Caltech, 20 mm gun

** Performed at Caltech, 40 mm gun

*** Performed at Ames Vertical Gun Range

† Performed in target modified to capture vapor

recovered for analysis. Because no Hugoniot data for kernite exist, an exact calculation of the peak pressure induced could not be made. To estimate the peak pressure we used the measured Hugoniot of gypsum, which is similar in density and hardness to kernite, and applied the graphical impedence matching method.

The principal mineral chosen for study was serpentine, which is believed to have been one of the most abundant hydrous silicates present in the accreting planetesimals because it is a major hydrous constituent of carbonaceous chondrites (Kaula, 1968). Serpentine contains 15% H_2O by weight. Lange (1980) estimated that serpentine, with an initial porosity of less than 10%, loses water at shock pressures between 40 and 60 GPa.

The source of the serpentine was Warren County, New York. Electron microprobe data showed the composition of the samples to be homogeneous and close to the magnesium-rich end-member. The serpentine is mainly composed of the polymorph antigorite, as determined by X-ray diffraction and infrared absorption. The mean measured density of the specimens to be shocked was 2.53 g cm^{-3}. Because the single crystal density of antigorite is 2.6 g cm^{-3} (Deer *et al.*, 1966), we conclude that the upper bound for the porosity of our samples is 3%. Some of the density difference is probably due to impurities—the actual pore space is considered to be very small.

The surfaces of the serpentine sample disks were polished to provide good contact with the steel driver plates. Samples for all three serpentine experiments were dried in an evacuated oven at 100°C for 24 hours before they were shocked, to remove as much unbound and adsorbed H_2O as possible.

The second mineral we studied was a smectite, nontronite. Due to its chemistry (Baird *et al.*, 1976; Toulmin, 1977) and overall spectral characteristics (Singer *et al.*, 1979; Weldon *et al.*, 1980), nontronite is believed to be a good candidate for the major constituent of the martian fines. The nontronite was collected in Riverside, California and was the same material investigated by the Viking Inorganic Chemistry Team. The sample was a fine powder and was pressed into the steel target cylinders to a density of about 2.7 g cm^{-3}. The nontronite samples were not subjected to a heat treatment, so interlayer and adsorbed water was not removed before the samples were shocked.

We chose to use kernite in the two-dimensional impact experiment because it is a hydrous mineral that could be obtained in a sufficiently large homogeneous block. An appropriately-sized single crystal of kernite was collected from the U.S. Borax Mine near Boron, California. The crystal was encased in concrete with a smooth face exposed. No heat treatment was conducted prior to the experiment.

ANALYSIS AND RESULTS

The solid material recovered from the shock experiments was analysed using infrared absorption. Particular attention was paid to the absorption bands known to be associated with the vibration of molecular water and hydroxyl ions. X-ray diffraction of the shocked minerals was also employed, so that structural changes could be characterized.

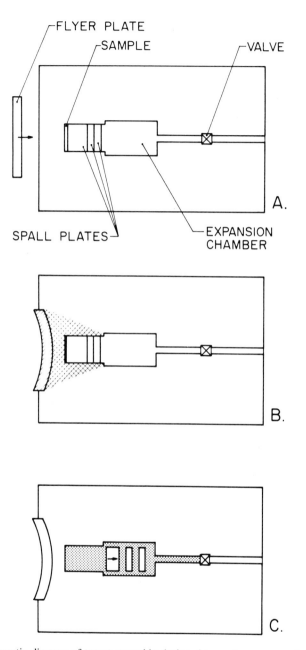

Fig. 1. Schematic diagram of target assembly designed to capture vapor released from minerals shocked to high pressure. (A) Before impact. Flyer plate approaches from left at high velocity. Sample disk is in place in cylindrical well with spall plates pressed into the well behind it to provide a high impedence rear surface. (B) At impact. Shock wave propagates into steel cylinder, reverberating the sample disk up to the Hugoniot pressure of steel. (C) After impact. Sample is released to low pressure, and any vapor liberated fills the expansion chamber. Spall plates and shocked sample material are carried into the expansion chamber by momentum gained from the shock wave.

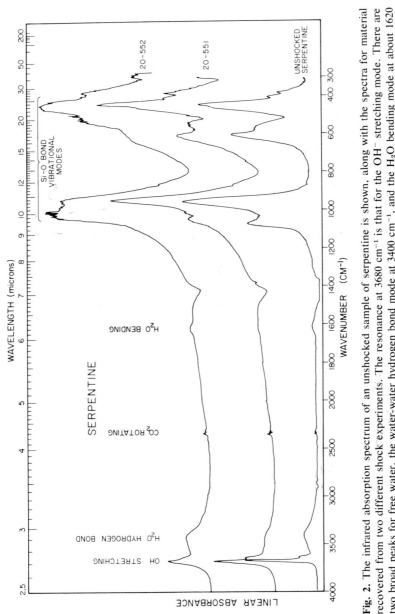

Fig. 2. The infrared absorption spectrum of an unshocked sample of serpentine is shown, along with the spectra for material recovered from two different shock experiments. The resonance at 3680 cm^{-1} is that for the OH$^-$ stretching mode. There are two broad peaks for free water, the water-water hydrogen bond mode at 3400 cm^{-1}, and the H$_2$O bending mode at about 1620 cm^{-1}. The large peaks at the low wavenumber end of the spectra are associated with the Si-O bond. Note that the relative peak height of the OH$^-$ stretching resonance decreases with increasing shock pressure, indicating a loss of structural water. The small peak near 2400 cm^{-1} is due to CO$_2$ in the sample chamber.

Samples to be used for infrared analysis were ground into a fine powder. The powdered samples were mixed with KBr, reground, and pressed into pellets. In each case, 1 mg of sample was used with 100 mg of KBr. Care was taken to standardize the sample preparation.

Infrared spectra for material recovered from shots 20–551 (serpentine, 30 GPa), and 20–552 (serpentine, 38 GPa) are shown in Fig. 2, with the spectrum of unshocked serpentine. Because the spectrum of the material recovered from shot 20–544 (serpentine, 24 GPa) was identical to the spectrum for unshocked serpentine, it was not included. Shocked samples show a decrease in the intensity of the absorption peak at 3680 cm^{-1} relative to the Si-O stretching and bending peaks at the low-wavenumber end of the spectrum. The decrease is more marked for the serpentine shocked to the higher pressure. There is also a broadening of absorption peaks, which is indicative of non-uniform local potentials. This could be brought about by increased internal disorder. The broad peaks at 3400 cm^{-1} and 1620 cm^{-1} are due to absorption by molecular water (Farmer, 1974), and do not seem to be dependent on whether the sample was shocked. This suggests that they are due to water adsorbed on the surface of the material, the amount of which is subject to change during post-experiment sample treatment, such as the length of time the shocked material was left in the target cylinder before it was recovered. When infrared spectra were measured after vacuum heating the sample pellets at 100°C, the broad molecular water peaks were all diminished; however, the other absorption peaks were unaffected. Upon further investigation we found that the only peaks sensitive to sample preparation and handling were the molecular water absorptions at 3400 and 1620 cm^{-1}.

X-ray powder diffraction spectra for the shocked and unshocked serpentine were obtained using a General Electric model 700 diffractometer with a copper source, operated at a voltage of 45 kV and a current of 17 mA. Portions of these spectra corresponding to the 7.25 interlayer lattice constant are shown in Fig. 3. Any structural changes due to shock loading the serpentine do not appear to be within the resolution of this instrument.

The infrared spectra of nontronite have the qualitative characteristics of serpentine (Fig. 4). The free water peak at 3400 cm^{-1}, however, includes a contribution from interlayer water, so changes may be more significant in this case. Both this peak and the OH^{-} peak, which corresponds to structural water, show marked decreases in the specimen from shot 20–550 (30 GPa). Heating the nontronite KBr pellets decreased the 3400 cm^{-1} peaks to about the same low level, which was only slightly below that peak of unheated shocked nontronite. As for the serpentine, the OH^{-} peaks were unaffected by heat treatment.

According to X-ray powder diffraction data for the nontronite from shot 20–550, the shock-loading caused a significant collapse of the basal layer. The other d-spacings were relatively unchanged, suggesting that the unit cell parameters in the silicate sheets were mostly undisrupted and the major shock-induced structural change in the nontronite was the decrease in the distance between the layers.

The infrared spectrum of ejecta fragments recovered from the kernite shot (AVG-790909), along with the spectrum of unshocked kernite, is shown in Fig.

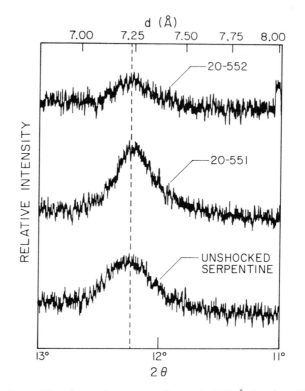

Fig. 3. The X-ray diffraction peak corresponding to the 7.25 Å (interlayer) lattice constant in serpentine. This peak is shown for the unshocked serpentine and for serpentine shocked to 30 and 38 GPa. No change in the position of this peak is apparent, indicating that there is no change in the distance between the silicate layers.

5. The recovered fragments were imbedded in the aluminum witness plate above the impact site, and probably came from the near field of impact where material was most heavily shocked. The kernite spectra look similar to the nontronite spectra in the region of the OH$^-$ and H_2O absorption peaks near 3400 cm^{-1}. The peak due to OH$^-$ stretching at 3550 cm^{-1} does not show as marked a decrease for the shocked kernite as for the shocked nontronite. The free water peak of the kernite spectrum was broadened somewhat.

Volatile release has been corroborated by direct measurement in shots 20–544 and 40–492. In 20–544, 2.62 mg of gas was recovered, amounting to 0.66% of the sample weight. Serpentine contains 15% H_2O by weight, so 4.4% of the available water in the sample was released, assuming this is the only source of the vapor recovered. Similar measurements were made for nontronite from 40–492 showing a recovery of 9.17 mg of gas, amounting to 3.06% of the sample weight.

DISCUSSION

The differences in the infrared spectra of shocked and unshocked serpentine can be explained by the loss of structural water. Greater water loss would be expected for higher shock pressures. Our data support this over a limited pressure range (Fig. 2). The loss of hydroxyl is accompanied by a small amount of structural modification of the silicate sheets. This is manifested in the broadening and overall shape-change of the peaks associated with the Si-O bond (those below 1200 cm^{-1}).

Even though released vapor was directly detected, the infrared spectrum of material recovered from shot 20–544 showed no evidence of water loss. This apparent inconsistency can be explained by the fact that the sample was too large. The extra-large sample was used in this experiment to increase the total volume of vapor released so it could easily be detected in our gas collector. The sample was so large, however, that only about 10% of its total mass was shocked to the peak pressure of 23.5 GPa. In most of the sample, the shock wave was attenuated by dilatation waves from the sides of the target cylinder, so the material never reached the calculated peak pressure. Because the violent motions associated with the subsequent rarefaction effectively pulverized and homogenized the solid material, it was impossible to recover only the highly shocked portion for infrared analysis. It is unlikely that the vapor detected in shot 20–544 was adsorbed water because that had already been removed by heating of the sample just prior to the shot.

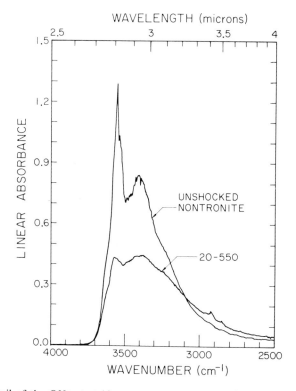

Fig. 4. Detail of the OH$^-$ stretching resonance at 3550 cm^{-1} and the H$_2$O hydrogen bond resonance at 3400 cm^{-1} in the infrared spectra of unshocked nontronite and nontronite shocked to 30 GPa. Note that both peaks are smaller for the shocked nontronite.

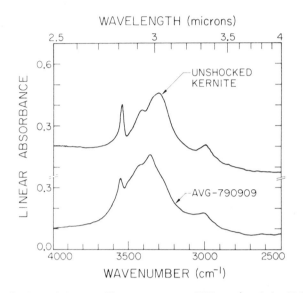

Fig. 5. Detail of the OH⁻ stretching resonance at 3550 cm⁻¹ and the H_2O hydrogen bond resonance between 3300 and 3400 cm⁻¹ in the infrared spectra of unshocked and shocked kernite. Note the decrease in the peak height of the OH⁻ stretching mode and the change in shape of the H_2O peak for the shocked kernite.

In all three shock wave experiments on serpentine we detected the release of water either directly or by obtaining infrared spectra of the recovered solid. The peak pressures achieved in these experiments were well below the 60 GPa estimated by Lange (1980) to be necessary for incipient vaporization of water from serpentine. In his calculations the shock heating was assumed to be uniform behind the shock front. If, however, the dissipation of energy in the shock wave is heterogeneous, there will be small regions of shocked material where the temperature and internal energy is much greater than average (Grady, 1977). The specific entropy in these localized regions might then surpass that required to release water upon rarefaction, even though the average specific entropy behind the shock front is below this value.

Shock loading also caused a major loss of structural water from the nontronite as indicated by the reduction in the OH⁻ peak height in the spectrum of sample 20–550 (Fig. 4). The shock should also have driven off much of the less tightly bound interlayer water. Because the interlayer water is easier to remove, a decrease in the molecular water peak relative to the OH⁻ peak might be expected. However, in shot 20–550 it appears that more of the hydroxyl water was lost. This can be explained if we consider that after the impact the shocked material is in intimate contact with the released water vapor. Some of the released water will be reabsorbed as the system begins to return to equilibrium. At post-shock temperatures, water is prevented from returning to the structural positions by energetic obstacles to diffusion processes. However, water is readily diffused and reabsorbed *between* the layers in nontronite. Collapse of the basal layer of the nontronite from 14.9 to 11.7 Å, shown by the X-ray diffraction data, suggests that a loss of about 2/3 of the interlayer water occurred upon impact. X-ray and infrared data indicate that both interlayer and structural water were released. It appears that some vapor was reabsorbed as interlayer water but not enough to return the interlayer water to the original quantity. The vapor recovered from 40–492 is a significant fraction of the total mass of the sample. Most of this vapor probably came from interlayer and adsorbed water, as the nontronite was not desiccated by a heat treatment before it was shocked.

It is important to remember that the serpentine and nontronite samples were not brought to their peak pressure in the same way they would be for an impact in nature. In reality, material would be loaded to high pressure by a single shock wave produced by the impact of two bodies. In these experiments mineral samples were reverberated to high pressure by consecutive reflected shock waves over a finite time interval (Jeanloz, 1980). Material loaded by this process remains closer to the isentrope than material loaded to the same pressure by a single shock. Thus multiply-shocked material does not generally reach as high a temperature or specific entropy as naturally shocked material and will tend to retain a greater percentage of its volatiles. For this reason we believe our experiments provide a minimum estimate of shock-induced volatile release.

Another difference between experimental and natural shock-loading is the sample container. In nature the shocked mineral would not be surrounded by stainless steel. Thus we should also consider the post-shock temperature of the stainless steel container and its effect on the amount of volatilization. Post-shock temperatures of stainless steel have been measured for shock pressures in the range considered here (Raikes and Ahrens, 1979). The post-shock temperatures corresponding to shock pressures of 16.0, 23.0, and 43.0 GPa, were measured to be 145, 195, and 325°C, respectively. Experimental work in the kinetics of dehydration of serpentine (Brindley and Hayomi, 1963) have shown that the effects of these temperatures on serpentine for the times required for the targets to cool (~10 minutes) are very small. In shots 20–544 and 40–492, most of the shocked sample did not remain in contact with the hot steel; immediately after the shot, the sample was scattered into the expansion chamber along with the spall plates (Fig. 1). In the nontronite shots a color change accompanied the release of volatiles (Weldon *et al.*, 1980). The part of the sample that remained in more intimate contact with the shock-heated portion of the stainless steel container showed no more color change than the portion that exploded into the expansion chamber. This suggests that the heating of the sample by the stainless steel was not significant.

Comparison of the infrared spectra of the shocked and unshocked kernite (Fig. 5) reveals differences similar to those between the two nontronite spectra in Fig. 4. The intensity of the OH^- stretching vibration diminished slightly, relative to the molecular H_2O vibration in the shocked sample. The molecular H_2O absorption peak broadened and changed shape, indicating a change in the local environment of the unbound water. A reasonable interpretation is that part of both the structural and nonstructural water was released by the impact and later reabsorbed by the shocked kernite.

Because of the nature of the experiment, it is impossible to give an accurate account of the shock history of the kernite ejecta fragments we studied. Material from the near field of an impact is ejected at very high velocity, often higher than the projectile velocity (Gault *et al.*, 1962). If the kernite ejecta we recovered came from near the point of impact it was effectively shocked twice—once during the primary impact and once when it imbedded in the aluminum witness plate. Some of the ejecta fragments, however, may have originated at points away from the impact. The important result of the kernite experiment is that it gave some evidence that water was rearranged in a hydrous mineral by a shock experiment that closely models a natural impact.

CONCLUSION

Five experiments presented here show strong evidence of shock-induced water loss from hydrous silicates. Released water was recovered in the two lowest pressure experiments, which were designed to capture vapor. The target design in the three higher pressure experiments did not allow collection of released vapor; however, infrared and X-ray spectra of the recovered solid indicate that water was indeed released. The quantity of water released increases with shock pressure over the range studied. These data also contain information on the type of water (hydroxyl or molecular) that was removed from the silicates by shock-loading.

The minimum shock pressure required for incipient vaporization of water from serpentine is less than 23.5 GPa, which is well below the 60 GPa estimated by Lange (1980). While more detailed and extensive experimentation must be carried out before more quantitative results can be obtained, we believe the difference is significant. Interlayer water is lost from nontronite at pressures as low as 18 GPa, and a significant quantity of water is lost from the structure at 30 GPa.

We have observed that structural, interlayer and absorbed water in hydrous minerals can be released by shock pressures generated by two types of impacts, one which resembles the type which would occur in nature, a fact which has important consequences regarding the formation of planetary hydrospheres and atmospheres. We believe that impacts played a major role in the release of volatiles from materials in the proto-planetary objects during their accretional stages. There is a net mobilization of water in hydrous minerals by impact release and subsequent reabsorption into less tightly bound configurations. In this way water can be made more available for reactions and more easily mobilized for later episodes of volatization by accretional and endogenic processes.

Acknowledgments—We appreciate the experimental assistance of J. Long, E. Gelle and M. Long at Caltech and J. Astralfa, E. Brooks, R. Krause and O. Koontz at NASA Ames Research Center. We are indebted to G. Rossman, S. Hill and J. Vizgirda for assistance in infrared and X-ray sample analysis. We thank J. Bauer, W. Harrison, F. Hörz, and R. Schaal for their helpful comments. Supported under NGLO5-002-105. Contribution #3446, Division of Geological and Planetary Sciences, California Institute of Technology, Pasadena, California 91125.

REFERENCES

Ahrens T. J. and O'Keefe J. D. (1972) Shock melting and vaporization of lunar rocks and minerals. *The Moon* **4**, 214.-249.

Arrhenius G., De B. R., and Alfvén H. (1974) Origin of the ocean. In *The Sea*, vol. 5 (E. D. Goldberg, ed.), p. 839–861. Wiley-Interscience, N.Y.

Baird A. K., Toulmin P., Clark B. C., Rose H. J., Keil K., Christian R. P., and Gooding J. L. (1976) Mineralogic and petrologic implications of Viking geochemical results from Mars: Interim Report. *Science* **194**, 1288–1293.

Benlow A. and Meadows A. J. (1977) The formation of the atmospheres of the terrestrial planets by impact. *Astrophys. Space Sci.* **46**, 293–300.

Brindley G. W. and Hayami R. (1964) Kinetics and mechanisms of dehydration and recrystallization of serpentine-1. *Clays and Clay Minerals* **12**, 35–47.

Deer W. A., Howie R. A., and Zussman J. (1966) *An Introduction to Rock Forming Minerals*. Langman, London. 528 pp.

Fanale F. P. (1971) A case for catastrophic early degassing of the Earth. *Chem. Geol.* **8**, 79–105.

Farmer V. C. (1974) *The Infrared Spectra of Minerals*. Mineralogical Society, London. 539 pp.

Gault D. E., Shoemaker E. M., and Moore H. J. (1962) Spray ejected from the lunar surface by meteroid impact. NASA TN D-1767. U.S. Geol. Survey, Menlo Park, Calif. 39 pp.

Gibbons R. V. (1974) Experimental effects of high shock pressure on materials of geological and geophysical interest. Ph.D. thesis, Calif. Inst. Technol., Pasadena. 216 pp.

Gibbons R. V., Morris R. V., Hörz F., and Thompson T. D. (1975) Petrographic and ferromagnetic resonance studies of experimentally shocked regolith analogs. *Proc. Lunar Sci. Conf. 6th*, p. 3143–3171.

Grady D. E. (1977) Processes occurring in shock wave compression of rock and minerals. In *High Pressure Research Applications in Geophysics* (M. H. Manghnani and S. Akimoto, eds.), p. 389–438. Academic, N.Y.

Jakosky B. M. and Ahrens T. J. (1979) The history of an atmosphere of impact origin. *Proc. Lunar Planet. Sci. Conf. 10th*, p. 2727–2739.

Jeanloz R. (1980) Shock effects in olivine and implications for Hugoniot data. *J. Geophys. Res.* In press.

Kaula W. M. (1968) *An Introduction to Planetary Physics: The Terrestrial Planets*. Wiley, N.Y. 490 pp.

Lange M. A. (1980) The evolution of an impact generated atmosphere (abstract). In *Lunar and Planetary Science XI*, p. 596–598. Lunar and Planetary Institute, Houston.

Lange M. A. and Ahrens T. J. (1979) Impact vaporization of water in hydrous minerals (abstract). *EOS (Trans. Amer. Geophys. Union)* **60**, 308.

McQueen R. G., Marsh S. P., Taylor J. W., Fritz J. N., and Carter W. J. (1970) The equation of state of solids from shock wave studies. In *High Velocity Impact Phenomena* (R. Kinslow, ed.), p. 293–417. Academic, N.Y.

Raikes S. A. and Ahrens T. J. (1979) Measurement of post-shock temperatures in aluminum and stainless steel. *High Pressure Science and Technology*, Vol. II (K. D. Timmerhaus and M. S. Barber, eds.), p. 889–894. Plenum, N.Y.

Shimizu M. (1979) An evolutional model of the terrestrial atmosphere from a comparative planetological view. *Precambrian Res.* **9**, 311–324.

Simikov G. V., Pavlovskiy M. N., Kalashnikov N. G., and Trunin R. F. (1974) Shock compressibility of twelve minerals. *Phys. Solid Earth* (Academy of Sciences, USSR, Izvestiya), English edition #8, p. 488–492. Amer. Geophys. Union, Washington, D.C.

Singer R. B., McCord T. B., and Clark R. N. (1979) Mars surface composition from reflectance spectroscopy: A summary. *J. Geophys. Res.* **84**, 8415–8426.

Toulmin P., Baird A. K., Clark B. C., Keil K., Rose H. J., Evans P. H., and Kelliher W. C. (1977) Geochemical and mineralogic interpretation of the Viking inorganic chemical results. *J. Geophys. Res.* **82**, 4625–4634.

Walker J. C. G. (1977) Origin of the atmosphere: history of the release of volatiles from the solid
   earth. In *Chemical Evolution of the Early Precambrian* (C. Ponnamperuma, ed.), p. 1–11. Academic, N.Y.
Weldon R. J., Boslough M. B., and Ahrens T. J. (1980) Shock-induced color changes in nontronite:
   a possible Martian surface process (abstract). In *Lunar and Planetary Science XI*, p. 1234–1235.
   Lunar and Planetary Institute, Houston.

Proc. Lunar Planet. Sci. Conf. 11th (1980), p. 2159–2189.
Printed in the United States of America

Control of crater morphology by gravity and target type: Mars, Earth, Moon

Richard J. Pike

U.S. Geological Survey, Menlo Park, California 94025

Abstract—Both gravitational acceleration and target characteristics have influenced the shape of impact craters across the solar system. Contrasts in crater shape that are attributable to lithologic contrasts in their targets are evident for fresh craters on at least three planets, in decreasing order of certainty: Earth, the moon, and Mars. Observations available at this time suggest that the responsible differences in target characteristics involve layering rather than rock strength. Variations in target stratification may account for deviation of craters on Mercury, Mars, Callisto, and Ganymede from an inverse statistical relation between gravitational acceleration, g, and D_t, the mean diameter marking the transition from simple to complex craters. Callisto, Ganymede, and Mars are inferred to be weak or layered targets, whereas Mercury is inferred to be a strong or homogeneous target. Size-dependence of crater morphology on Mars, as determined from a new sample of 230 fresh craters from Viking Orbiter pictures, closely resembles that for lunar craters, except that D_t is about 6 km rather than about 19 km. Complex craters on Mars, which also have an unusually steep depth-to-diameter (d/D) trend, average 13 percent deeper on rough terrain than on plains. The simple-to-complex transition on Mars is paralleled by a change in the texture of ejecta from ballistic to fluid flow at D $\approx$ 4 km, but the ejecta of large martian craters (>80 km) also has a primarily ballistic character. A revised analysis of data for the moon reveals that the simple-to-complex transition occurs in smaller mare craters than it does in upland craters: complex craters on the maria average 12 percent shallower than complex upland craters. Preliminary topographic data from Venus Pioneer radar suggest that possible impact craters on Venus could have a d/D ratio much like that of ringed complex meteorite craters on Earth.

INTRODUCTION

Impact craters are the outstanding surface features on solid planets (and satellites) in the solar system. Although the formational process—hypervelocity collision of a planet with much smaller bodies (asteroids, planetesimals, comet nuclei, meteoroids, and their fragments)—is thought to be essentially the same everywhere and over a wide range of impact energy, the resulting landforms are not all alike (e.g., Schultz, 1976). Impact craters can differ systematically in morphology by planet (Hartmann, 1972), location on a planet (Dence, 1972), size (Gilbert, 1893), and relative age (Baldwin, 1949). The contrasts in crater shape, some subtle and gradational but others blatant and abrupt, probably reflect one or more external influences on the impact process. These might include density, velocity, and angle of impact of the projectile; strength, structure, and physical state of the target material; the planet's gravitational acceleration, atmospheric density, and thermal history; and the efficacy of postimpact processes such as erosion, sedimentation, isostasy, and magmatism (Quaide *et al.*, 1965).

The many proposed influences on hypervelocity impact can be narrowed down to the most likely alternatives. One approach is measuring and otherwise documenting morphologic characteristics of freshly-formed impact craters and making interplanetary and intraplanetary comparisons of the findings. The work of Gault *et al.* (1975), Schultz (1976), Head (1976), Malin and Dzurisin (1977), Pike (1977), Smith and Hartnell (1978), Wood *et al.* (1978), and Mouginis-Mark (1979a) exemplify recent progress toward the goal. This paper elaborates on issues in crater morphology that I have previously raised only in abstracts and orally (Pike and Arthur, 1979; Pike *et al.*, 1980). It presents new results, primarily for Mars but including recent data for Earth and the moon as well, that deal with the possible influence of gravitational acceleration and target properties on the shape of impact craters. Older morphologic data for mercurian craters and recent (and preliminary) data for craters on Ganymede, Callisto, and (probably) Venus provide further context for the problems addressed here. The paper incorporates an emphasis that is absent in most current writing about impact craters: comparing terrestrial meteorite craters, the only readily available ground-truth, with impact craters on the planets (Dence, 1964; Von Englehardt *et al.*, 1967; Wilshire *et al.*, 1972; Milton *et al.*, 1972; Milton and Roddy, 1972; Roddy, 1979; Grieve and Robertson, 1979; Kieffer and Simonds, 1980).

PREVIOUS WORK

The interior morphology of freshly formed impact craters on a planet is more complicated with increasing crater size (Gilbert, 1893). This size/morphologic continuum is not perfectly gradational, but rather contains marked discontinuities or contrasts that can be used to evaluate some of the alternative influences on crater shape (Mackin, 1969). One of the two most evident contrasts is that between large craters and small basins. The other, that between simple and complex craters, is the aspect of crater shape described here. These observations apply primarily to fresh-appearing craters, because a surface morphology that is unaffected by postcratering degradation best reflects the formational processes that are of interest. Exactly what constitutes a "fresh" impact crater varies from study to study. However, the working definition current among planetologists, regardless of the precise criteria of selection, is a crater whose characteristics have changed little or none since formation. This definition is implemented here by criteria specified in the section on new martian data, below. I have assumed that the spirit of the definition has been adhered to by all workers whose morphologic data I have compared with mine.

Simple and complex craters

Dence (1964) first recognized the fundamental and qualitative difference between simple and complex meteorite craters on Earth. The distinction was immediately applied to fresh-appearing impact craters on the moon (Quaide *et al.*, 1965). Although observations supporting the difference in crater shape are partly lithologic and structural for Earth but entirely topographic for the moon, they quite evidently describe the same phenomenon on both planets (e.g., Roddy, 1977). Simple lunar craters encompass Wood and Andersson's (1978) morphologic types ALC and BIO; complex craters correspond to their types TRI and TYC. The two-part classification, which was extended to include multiringed structures and basins on both Earth and the Moon (Dence, 1972; Hartmann, 1972) applies to impact craters on all six (observed) impacted bodies larger than Alamthea: Earth, the moon, Mars, Mercury, Ganymede, and Callisto.

The morphologic contrast is illustrated here for Mars. A simple crater (Fig. 1A) usually is a bowl-shaped depression that has a smooth and highly circular rim crest (C = 0.86 here; cf. Pike, 1977), an interior devoid of major topographic features, and a depth/diameter (d/D) or aspect ratio, of roughly 1/5 (I adopted the d/D notation of Wood and Andersson [1978] here rather than the $R_i D_r$ notation of earlier publications [e.g., Pike, 1974], because of the need for the subscripted term D_t and because d/D is simpler). The upper wall of a simple crater commonly reveals stratification and evidence of mass-wasting. The crater may have a very small flat floor (although not in this example), presumably of mass-wasted debris (Pike, 1977) but possibly (on Mars) containing eolian sediment. A complex crater (Fig. 1B) has one or more of the following interior features: irregular or scalloped rim-crest outline, although still more or less radially symmetric; a broad floor that is flat and level overall, but that often is partly occupied or obscured by other morphologic features; a centrally disposed hill, peak, or pit; single or multiple slices or blocks of slumped material that may appear to have originally occupied scalloped segments in the upper wall of the crater; continuous terraces on the crater wall that represent wholesale failure of the rim *en bloc*; and a d/D that varies widely with diameter, from about 1/5 for small fresh complex craters to about 1/150 for large complex craters, depending upon the planet.

The simple-to-complex transition

Simple craters are small and complex craters are large. Morphologic studies have demonstrated that this generalization now applies to six planets and satellites (for example, Earth: Grieve and Robertson, 1979; the moon: Wood and Andersson, 1978; Mercury: Malin and Dzurisin, 1977; Mars: Carr *et al.*, 1977; and Ganymede and Callisto: Pike *et al.*, 1980). The transition from simple to complex morphologies may be expressed quantitatively. One of the more readily measured contrasts diagnostic of the transition is the change in d/D from about 1/5 to a lower and more variable fraction with increasing crater size.

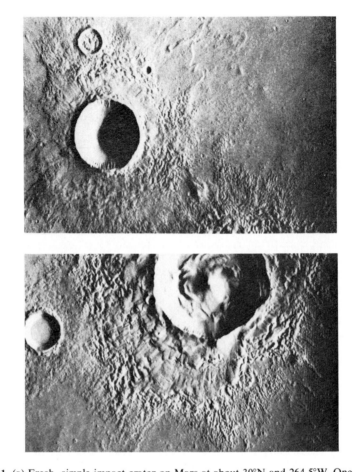

(a)

(b)

Fig. 1. (a) Fresh, simple impact crater on Mars at about 30°N and 264.5°W. One of the large (8.2 km across) and deep (1.9 km) simple craters in western Elysium Planitia discussed in text, it has the smooth, rounded interior typical of smaller simple craters and ejecta that shows characteristics of emplacement by fluid flow. Viking Orbiter frame No. 645A09. (b) Fresh, complex impact crater on Mars at about 31°N and 267.3°W, close to the simple crater pictured above. About 14.6 km across, it has a prominent central peak, scalloped rim crest, slump blocks, a small terrace, a comparatively modest depth (1.3 km), and a flat floor. Ejecta has well-developed flow characteristics: lobes and striations. Viking Orbiter frame No. 645A01.

The d/D results for fresh craters on four planets are summarized in Fig. 2; supporting data are given in Tables 1 and 2. A classification, into simple or complex, of all craters from which Fig. 2 is derived has been made for Mars (n = 230; Table 1), Earth (n = 30, Table 2), and the moon (n = 339, Table 2). The martian data, which are new, are discussed at length below. The d/D results of Pike (1974) for the moon are updated and improved here by adding more craters (largely from Pike, 1976) and by dividing the craters into simple and complex types *before* fitting the regression lines. These changes have been made to maintain a high-quality lunar data set that can be relied upon as a basis

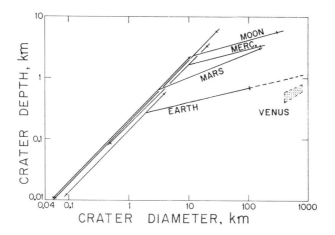

Fig. 2. Depth (d) vs. diameter (D) for simple (steep slope) and complex (gentle slope) impact craters on five planets. Lines are least-squares fits (Table 1, 2). Martian curves, undifferentiated by terrain type, are new from Viking Orbiter shadow-lengths (this paper); lunar curves are revised from those in Pike (1974) using data in Pike (1976); mercurian curves are from Malin and Dzurisin (1977); terrestrial curves mostly from data in Pike (1976); preliminary data for Venus from Pioneer Venus radar, reported by Kerr (1980).

Table 1. Depth/diameter for fresh impact craters on Mars: A summary of least-squares results.[1]

Terrain type	Crater type	Correl. coeff.	Sample size	Intercept (km)	Standard error (km)	Slope	± Std. error	Source
Mixed	Simple	0.96	21	0.187	+0.039, −0.032	0.962	0.064	2
Mixed	Complex	0.63	41	0.456	+0.098, −0.081	0.348	0.068	2
Mixed	Simple	0.92	49	0.162	+0.044, −0.035	1.003	0.063	3
Mixed	Complex	0.83	8	0.410	+0.047, −0.043	0.330	0.090	3
Mixed[5]	Simple	0.99	125	0.204	+0.025, −0.022	1.019	0.008	4
Plains	Simple	0.99	102	0.206	+0.024, −0.021	1.017	0.009	4
Cratered Terrain	Simple	0.99	23	0.195	+0.028, −0.024	0.999	0.024	4
Mixed[5]	Complex	0.93	105	0.415	+0.063, −0.055	0.395	0.016	4
Plains	Complex	0.90	51	0.467	+0.065, −0.057	0.334	0.023	4
Cratered Terrain	Complex	0.95	54	0.396	+0.058, −0.050	0.423	0.020	4

[1] Equations in form: log depth = log intercept + slope × log diameter
[2] Pike and Arthur (1979), photometric shadow lengths (data only)
[3] Pike *et al.* (1980), photogrammetry
[4] This paper, shadow lengths on Viking Orbiter prints
[5] Subdivided below by terrain type

Table 2. Depth/diameter for fresh impact craters on four planets:
A summary of least-squares results.[1]

Planet and terrain	Crater type	Correl. coeff.	Sample size	Intercept (km)	Standard error (km)	Slope	± Std. error	Source
EARTH	Simple	0.99	18	0.137	+0.029, −0.024	1.02	0.035	2
	Complex	0.68	12	0.23	+0.07, −0.06	0.24	0.08	
MERCURY	Simple	—	ca 80	0.176	— —	0.98	—	3
	Complex	—	ca 99	0.910	— —	0.26	—	
MOON								
Mare	Simple	0.99	179	0.195	+0.030, −0.026	1.013	0.008	2
Upland	Simple	0.99	89	0.192	+0.027, −0.024	1.022	0.009	2
Mare	Complex	0.94	24	0.841	+0.070, −0.065	0.332	0.025	2
Upland	Complex	0.92	47	1.008	+0.092, −0.084	0.313	0.020	2
PHOBOS	Simple	0.98	25	0.179	+0.060, −0.045	0.938	0.043	4

[1] Equations in form: log depth = log intercept + slope × log diameter
[2] This paper, data mostly from Pike (1976)
[3] Malin and Dzurisin (1977)
[4] Data from Thomas (1978, Fig. 20)

of comparison for craters on other planets. The craters used by Malin and Dzurisin (1977) in their d/D analysis for Mercury have not yet been formally classified into simple and complex types. Malin and Dzurisin's results are used in Fig. 2, however, because their selection criteria (1977, p. 377–378) seem to have been essentially those adopted here. Preliminary d/D measurements have been made for what appear to be a few large craters from Pioneer Venus radar data (reported in Kerr, 1980). The craters on Venus may not be as fresh as their presumably complex counterparts on other planets (Fig. 2). However, they evidently are fresh enough to have the strong relief that is needed for them to be detected in the orbiter's radar returns (only ~25 km horizontal resolution and 200 m vertical accuracy). Highly degraded craters on Venus probably would not show up on the results obtained from currently operational radar systems. These tentative data are included here because they are the first d/D estimates ever made for probable Venusian craters. Depth/diameter data for terrestrial meteorite craters (see Pike, 1976 and 1977) were compiled from many sources, most of which are cited in Grieve and Robertson (1979). All of the craters on Earth are at least slightly degraded, but this problem has been largely overcome, particularly for the small craters, by supplementing topographic data with information gathered from drill cores to get usable estimates of initial depth.

Figure 2 shows that simple craters on four planets are rather similar with respect to d/D, but that complex craters differ substantially. Fresh complex impact craters are deepest on the moon, followed by those of Mercury, Mars, and Earth, in that order. The diameter of intersection for each pair of curves can be used as one of several quantitative estimates for the location of the simple-to-complex transition of craters on that planet.

The transition from simple to complex morphologies is not abrupt, but rather is gradational over a modest crater-size range on each planet. Central peaks, flat floors, evidence for wall failure, and diminished d/D ratios do not all appear in complex craters at the same diameter. The transition may be expressed statistically as the geometric mean, D_t, of several diameter values that characterize specific morphologic contrasts (Pike *et al.,* 1980). Depth/diameter is but one of these. The other types of contrasts (Table 3) are exemplified by presence of swirl-textured floors in craters on the moon (modal diameter: Smith and Hartnell, 1978), observation of wall terracing in craters on Mercury (diameter at 50 percent frequency: Cintala *et al.,* 1977), craters displaying ballistic versus fluid-flow ejecta (median diameter of overlap: Boyce, 1979; this paper, Fig. 8), and occurrence of central uplifts in meteorite craters (median diameter of overlap: Grieve and Robertson, 1979; this paper, Fig. 4).

The salient result emerging thus far from the many studies of crater morphology is that mean transition diameter, D_t, is not the same on all planets. The difference is shown graphically in Fig. 3, a diagram that summarizes statistically the findings of various investigators (Table 3). Modified only slightly from the first version published in Pike *et al.* (1980), Fig. 3 furnishes a context for the new observations reported here. It also presents an interpretation of past morphologic results in terms of gravitational acceleration and target characteristics, insofar as any such general picture can be proposed at this time. Fig. 3 is fully explained in the caption.

The mean transition diameter, D_t, for impact craters on the four planets plotted as open circles in Fig. 3 varies roughly with the inverse of gravitational acceleration at the surface, g (both variables expressed logarithmically). The relation is approximate, for neither Mercury nor Mars falls on the regression line fitted to the four means. However, D_t values for Earth (3.1 km) and the moon (18.7 km) differ by a factor of six, exactly the ratio of their values of g. This correlation is too close to be accidental. The factor-of-six difference has been evident only since the detailed compilations of astroblemes by Dence (1972) and by Grieve and Robertson (1979) unequivocally established D_t for terrestrial craters as lying between 3 and 4 km rim diameter, and since Pike (1977) calculated D_t for the moon (revised here) as being at about 17.5 km.

Crater depth exemplifies individual variables that may scale inversely with g. From Fig. 2 and the equations in Table 2, it is evident that complex craters average six times deeper on the moon than they do on Earth. This already has been pointed out by Pike (1977), Cooper (1977), and by Grieve and Robertson (1979), and is discussed further by Pike (1980). I interpret the factor-of-six difference as indicating that crater depth could well scale inversely with g on all planets—allowing duly for perturbing factors to explain anomalies such as those of Mars and Mercury in Fig. 2. Venus may be a case in point. The recent d/D estimates for probable impact craters on Venus plot below the equation (which has a large dispersion) for terrestrial complex craters in Fig. 2. Thus, even if the data reported by Kerr (1980) represent somewhat degraded craters, these shallow features on Venus still could be roughly as deep as would be expected for a planet having a g similar to that of Earth, nearly 1000 cm/sec². Additional data are needed to test this possible d/D similarity between craters on Earth and their counterparts on Venus before any general dependence of impact crater depth on g can be affirmed.

There are at least two reasons for emphasizing the Earth-Moon relations in

Table 3. Summary of morphologic data by which values of D_t were calculated for four planets.

Morphologic aspect	Statistic	Source	Sample size	Diameter (km)
EARTH (8 morphologic aspects: D_t = 3.1 km)				
Overall transition	Median D overlap	Grieve and Roberston, 1979	80	3.5
Depth/diameter (d/D)	Intersection D of curves	This paper	30	1.9
Occurrence of simple craters:				
Crystalline rock	Largest D	Grieve and Robertson, 1979	31	4.5
Sedimentary rock	Largest D	Grieve and Robertson, 1979	35	2.4
Occurrence of complex craters:				
Crystalline rock	Smallest D	Grieve and Robertson, 1979	31	5.0
Sedimentary rock	Smallest D	Grieve and Robertson, 1979	35	2.5
Gravity anomaly/diameter	Intersection D of curves	Gordin et al., 1979	29	4.0
Rim height/diameter	Intersection D of curves	Pike, 1976, and unpublished data	22	2.3
MOON (23 morphologic aspects: D_t = 19 km)				
Overall transition	Misc. D (see p. 118 of source)	Mutch et al., 1976	N.A.	20
Terraces/occurrence (maria)	50% frequency D	Smith and Hartnell, 1978	51	20
Terraces/occurrence (uplands)	50% frequency D	Smith and Hartnell, 1978	139	31
Types TRI + TYC/ALC+ BIO+TRI+TYC	50% frequency D	Wood and Andersson, 1978	2598	20
Swirl-textured floors/D	Modal D	Smith and Hartnell, 1978	221	25
Simple crater/occurrence	Largest D	Wood and Andersson, 1978	2598	27
Rimwall slope/diameter	Intersection D of curves	Pike, 1977	106	17
Rim-crest circularity/diameter	Inflection D of distribution	Pike, 1977	200	12
Rim-crest evenness/diameter	Inflection D of distribution	Pike, 1977	21	14
Cumulative morphology Index/D	Inflection D of distribution	Pike, 1975	152	9
Impact melt, exterior/D	Onset D of abundance	Hawke and Head, 1977	100	15
Apparent D/rim D	Intersection D of curves	R. J. Pike, unpublished data	144	15
Scallops/occurrence	Modal D	Wood and Andersson, 1978	2441	28
Central peaks/occurrence (maria)	50% frequency D	{ Wood and Andersson, 1978 / Smith and Hartnell, 1978 }	N.A.	19
Central peaks/occurrence (uplands)	50% frequency D	{ Wood and Andersson, 1978 / Smith and Hartnell, 1978 }	N.A.	27

Aspect	Measurement	Reference	D	
Scallops + terraces (maria)	50% frequency D	Wood and Andersson, 1978	N.A.	19
Scallops + terraces (uplands)	50% frequency D	Wood and Andersson, 1978	N.A.	21
Overall transition	Onset D of terracing	Mackin, 1969	—	20
Depth/diameter	Intersection D of curves	Wood and Andersson, 1978	1534	12
Bowl shape/flat floors	50% frequency D	{Wood and Andersson, 1978 / Pike, 1975}	2598 / 152	10
Rim height/diameter	Intersection D of curves	Pike, 1977	162	21
Rim width/diameter	Intersection D of curves	Pike, 1977	162	30
Floor diameter/D	Intersection D of curves	Wood and Andersson, 1978	1075	22

MERCURY (18 morphologic aspects: $D_t = 16$ km)

Aspect	Measurement	Reference	D	
Rim height/diameter	Intersection D of curves	Cintala, 1979	30	16
Depth/diameter	Intersection D of curves	Gault et al., 1975	130	9
Depth/diameter	Intersection D of curves	Malin and Dzurisin, 1977	178	10
Central peaks/occurrence	50% frequency D	Malin and Dzurisin, 1978	178	16
Central peaks/occurrence	50% frequency D	Gault et al., 1975	130	11
Central peaks/occurrence	50% frequency D	Smith and Hartnell, 1978	152	12
Central peaks/occurrence (cratered terrain only)	50% frequency D	Cintala et al., 1977	657	17
Central peaks/occurrence (plains only)	50% frequency D	Cintala et al., 1977	336	20
Terraces/occurrence	50% frequency D	Gault et al., 1975	130	12
Terraces/occurrence	50% frequency D	Smith and Hartnell, 1978	130	23
Terraces/occurrence (cratered terrains only)	50% frequency D	Cintala et al., 1977	657	29
Terraces/occurrence (plains only)	50% frequency D	Cintala et al., 1977	336	31
Overall transition	—	Malin and Dzurisin, 1977	178	10
Scallops/occurrence (cratered terrain only)	Modal D	Cintala et al., 1977	657	25
Scallops/occurrence (plains only)	Modal D	Cintala et al., 1977	336	25
Bowl shape (cratered terrain only)	50% frequency D	Cintala et al., 1977	657	17
Bowl shape (plains only)	50% frequency D	Cintala et al., 1977	336	19
Wall-failure	50% frequency D	Malin and Dzurisin, 1978	145	10

Table 3. (*Continued*)

Morphologic aspect	Statistic	Source	Sample size	Diameter (km)
MARS (10 morphologic aspects: D_t = 5.8 km)				
Depth/diameter	Intersection D of curves	This paper	230	3.1
Occurrence of simple craters	Largest D	This paper	125	10.2
Occurrence of complex craters	Smallest D	This paper	105	3.4
Ballistic-to-fluid-flow ejecta	Median D of overlap	This paper	230	4.1
Bowl shape to flat floor	Median D of overlap	This paper	230	6
Bowl shape to central peak	Median D of overlap	This paper	230	6
Bowl shape to slump blocks	Median D of overlap	This paper	230	6.5
Bowl shape to scalloped rim	Median D of overlap	This paper	230	6.5
Bowl shape to terraces	Median D of overlap	This paper	230	8
Bowl shape to peak pits	Median D of overlap	This paper	230	7.5

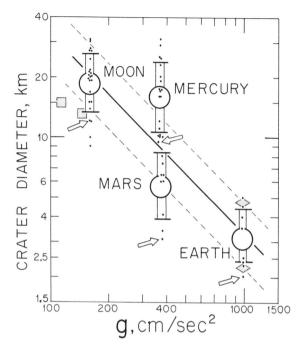

Fig. 3. The simple-to-complex transition for fresh impact craters on six planets and satellites, interpreted in terms of gravitational acceleration, g, and target characteristics. Each of the 59 small dots is a statistically derived crater diameter that expresses a different aspect of the transition, such as inflection of the depth/diameter distribution (arrows), frequency or onset of central peaks, and median overlap of simple and complex distributions (see text and Table 3). For clarity, the dots for Mars are displaced slightly to the right of those for Mercury. There are 23 morphologic aspects for the moon, 18 for Mercury, 10 for Mars (revised for this paper from Pike *et al.*, 1980), 8 for Earth, and one each (approximate onset of central peaks; see text and Pike *et al.*, 1980) for Callisto (upper square) and Ganymede (lower square). Large circles are centered on geometric means, D_t, for the 8, 10, 18, and 23 diameters of morphologic aspects for the four planets indicated. Bars show one standard deviation. Solid line, least-squares fit to four geometric means: $\log D_t = \log 3300 - 1.00 \log g$, interpreted as indicating targets of average strength or degree of stratification (see text). Slope of -1.00 is consistent with dimensionalities of plotted variables D_t and g. D_t value for Earth changes dramatically when craters are subdivided by type of target. Upper diamond: D_t for meteorite craters in crystalline rock only (see shaded rectangles in Fig. 4); lower diamond: D_t in sedimentary rock only (see Fig. 4). Dashed lines are extrapolations, down to lower values of g, of disparate terrestrial D_t values for crystalline and sedimentary rock, at a slope of -1.00: similar to slope of least-squares fit to the four whole-planet values of D_t. Upper dashed line interpreted to represent planets with resistant or homogeneous target characteristics. Lower dashed line interpreted to represent planets with weak or strongly layered targets. Martian data and estimates of D_t for impact craters on the two Galilean satellites are consistent with weak and layered, volatile-rich, strata (see discussion in text).

Figs. 2 and 3 over the deviant D_t values for Mars and Mercury: (1) Lunar rocks (largely basalts, anorthosites, and their fragmental derivatives) have known terrestrial equivalents, whereas samples of rocks on Mars and Mercury have not yet been obtained. Indeed, the peculiar characteristics of crater ejecta on Mars suggest that the physical properties of rocks at the martian surface—whatever their composition—differ significantly from those on Earth, the moon, or Mercury. (2) Differences in crater shape that might be ascribed to characteristics of the space environment are less likely to occur on the earth and the moon than between these two bodies and Mars and Mercury. For example, the earth and its moon are so close to one another that they probably are not subject to significant differences in the velocity or mass-density distribution of impacting bodies, whereas these parameters conceivably could vary for more distant planets (Hartmann, 1977). For these reasons, I contend further that the deviation of Mars ($D_t \simeq 6$ km) and Mercury ($D_t \simeq 16$ km) from the solid line in Fig. 3 reflects influences on crater shape *in addition to* g, and thus does not necessarily disprove the dependence of D_t on g. One of these influences is geology (e.g., physical properties) of the target (Cooper, 1977; Cintala *et al.*, 1977; Wood *et al.*, 1978; Mouginis-Mark, 1979a; Kieffer and Simonds, 1980).

Target properties

The effects of lithology on diameter of the simple-to-complex transition are much better established for terrestrial meteorite craters than for craters on any other planet (Fig. 4). For other rock-correlated differences, see the recent review by Kieffer and Simonds (1980). From the compilations of data by Dence (1972) and Grieve and Robertson (1979), it is clear that the transition falls at a much smaller median diameter (about 2.25 km) for craters in sedimentary rock than for craters

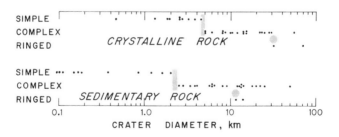

Fig. 4. Dependence of the simple-to-complex transition (essentially presence or absence of a central uplift) in terrestrial impact craters (shaded bars) on target characteristics. Transition in sedimentary rock occurs at D = 2.25 km, whereas it occurs at 4.75 km in crystalline rock. Transition from complex craters to ringed complex craters (shaded circles) exhibits similar trend. Comparison with lunar data in Fig. 9 suggests that the observed contrast in morphology reflects layering of the sedimentary targets rather than difference in overall rock strength. Arrays of "proven" and "probable" meteorite craters compiled from data of Grieve and Robertson (1979). N = 80.

in crystalline rock (about 4.75 km). Although data on ringed impact structures that have been excavated wholly in either one type of target or the other are few (typical targets for these large craters being a veneer of sediments on a crystalline basement complex), the transition from complex craters to ringed structures also seems to depend upon rock type (Fig. 4).

Analogous contrasts in morphology have been sought in craters on other planets. The usual procedure is to examine craters on different topographic units, typically "smooth terrain" (maria and plains) and "rough terrain" (upland and cratered terrains) (Quaide *et al.*, 1965; Cintala *et al.*, 1977; Malin and Dzurisin, 1977; Smith and Hartnell, 1978; Wood *et al.*, 1978; Mouginis-Mark, 1979a) and to equate physiographic contrasts of the targets with differences in physical properties or structure. Results thus far indicate that target characteristics do indeed affect crater shape. Although intraplanetary differences can be pronounced, as they seem to be in the case of Mars (e.g., Mouginis-Mark, 1979a), the terrain-dependent results often are subtle, ambiguous, or even contradictory.

An example from the moon illustrates some of the problems with appraisals of terrain effects on the planets. According to equations 3–6 of Wood and Andersson (1978), fresh simple craters (their types ALC and BIO) on the lunar maria are 9 to 14 percent deeper (over the diameter range 2 to 20 km) than similar craters on the highlands, whereas my equations (Table 2) put the difference at less than half a percent. Both results cannot be correct. I contend that those of Wood and Andersson (1978) reflect a problem common to large samples of craters on planets with surfaces of greatly varying age. A sample of uniformly fresh craters can be obtained readily from surfaces that, as in the case of the lunar maria, are young. The task is much more difficult for rough, older terrains. Slightly older craters can easily be incorporated preferentially into a sample of fresh upland craters, which probably is what befell Wood and Andersson's sample of Class 1 (their freshest) craters for the lunar highlands. Older craters inadvertently find their way into highland samples because (1) the time-stratigraphic horizon that limits mare craters to exclusively young ones is absent from the older highlands surfaces, and (2) morphologic differences between Class 1 craters and the only slightly more degraded Class 2 craters can be subtle and difficult to judge on rough terrains. The statistical dispersion of crater depth calculated by Wood and Andersson (1978) bears out my thesis. Standard deviations, which are 56 percent (for Type ALC craters) and 78 percent (for Type BIO craters) greater on the uplands than on the maria, almost certainly reflect the inclusion of somewhat older and shallower craters in the upland sample. Thus it is clear that any terrain-related difference that might exist in the shape of fresh craters from different terrain categories can be masked by morphologic contrasts that reflect only relative age. Moreover, fresh impact craters can differ morphologically by virtue of having formed on uneven surfaces rather than on smooth surfaces, a contrast that is not necessarily related to physical properties of the target materials (Pike, 1977, Fig. 7; Wood and Andersson, 1978, p. 3676). A final caveat follows from this example. Namely, a large sample of craters emphatically does not guarantee either objectivity of the data or statistical validity of the results.

Just how lithology has brought about the factor-of-two difference in D_t on Earth is not known with certainty, but at least three (interrelated) aspects may be important: rock strength, stratification, and volatile content (see Kieffer and Simonds, 1980, and their extensive reference list). One possible explanation involves mainly rock strength and volatiles. On the basis of cratering experiments demonstrating that an impact of a given energy will generally excavate more efficiently in soft rock than hard rock (e.g., Cooper, 1977), Pike and Arthur (1979) suggested that the deviation of Mars and Mercury from the linear D_t:g relation (Fig. 3) might reflect crustal materials of significantly different strengths. They proposed that Mercury presented an unusually resistant target to an impacting body (upper dashed line in Fig. 3), compared to the moon and Earth—a conclusion foreshadowed, although not drawn explicitly, by the findings of Cintala *et al.*, 1977), whereas Mars was a less resistant target (lower dashed line in Fig. 3). Wood *et al.* (1978) ascribed some *interior* morphologic features of complex martian craters to the presence of subsurface volatiles. Indeed, the results of cratering experiments (Cooper, 1977) demonstrate that water-saturated materials are much more easily excavated than dry materials. Thus the likelihood that the martian crust contains abundant volatiles (Carr *et al.*, 1977; Gault and Greeley, 1978; Cintala and Mouginis-Mark, 1980) is consistent with the weak-target interpretation for many of the data plotted in Fig. 3.

Preliminary estimates of the simple-to-complex transition for two Galilean satellites, both of which are believed to have largely water-ice crusts (see reviews by Morrison and Cruikshank, 1974; and Johnson, 1978), also are consistent with a weak-target interpretation of Fig. 3 (Pike *et al.*, 1980). The D_t values for Callisto and Ganymede in Fig. 3 were estimated solely from the prevalence of central peaks in small impact craters. I examined ten high-resolution (~3 km/line-pair) Voyager I images for each satellite, and on each picture measured the rim-crest diameter, D, of the three smallest craters for which a central peak could be seen with a 10X magnifier. This method probably estimates a characteristic crater size that is somewhat larger than the onset diameter for peaks because I could still detect peaks virtually at the limit of resolution (in craters as small as 8 km on Ganymede and 10 km on Callisto). The resulting median crater diameters, 15 km on Callisto and 13 km on Ganymede, fall only slightly below the weak-target model extrapolated to low values of g from terrestrial data in Fig. 3. Results from the 60 craters measured suggest that surface materials of the two Galilean satellites are even weaker than those of Mars. This is quite consistent with the subdued crater topography so clearly evident in the Voyager pictures (Smith *et al.*, 1979a and 1979b).

An alternative explanation of the data in Fig. 3, advanced here, is that stratification—or rather the degree of stratification—of the target and not its strength alone modifies the relation between D_t and g. The alternative follows from the inference that, for a given energy of impact, a complex crater will form more readily in a highly stratified target than in a homogeneous target (Roddy, 1977; Cooper, 1977). The overall concept is not new. A layered target, albeit *specifically* consisting of unconsolidated material overlying bedrock, has been invoked to explain flat floors and central peaks in impact craters for some time (Wegener, 1921; Sabaneyev, 1962; Oberbeck and Quaide, 1967; Head, 1976). I propose here (1) that stratification can merely *determine* (and as a factor distinctly subordinate to g) the crater *size* at which the complex features appear (Fig. 4), *not initiate* the features themselves (Pike, 1980), as Wegener (1921) and those who adopted his model maintained. Moreover, (2) just what constitutes a layered target needs to be clarified and broadened: a stratified target need not consist exclusively of a weak surface layer and a strong substrate, as currently seems to be the working definition in planetary studies. Stratified targets on Earth often comprise alternating beds of (weak) sedimentary rock, with varying degrees of strength. The sedimentary sequence often is thick enough so that the underlying basement complex of (strong) crystalline rock does not impinge upon formation of the impact crater. Well-known complex craters excavated in targets of this type include Flynn Creek, Tennessee (Roddy, 1979), Steinheim Basin, Bavaria (Von Englehardt *et al.*, 1967), and Sierra Madera, Texas (Wilshire *et al.*, 1972). Some extraterrestrial layered targets may be somewhat similar. For example, sequences of lava flows alternating with eolian, fluvial, or glacial sediments could well have accumulated on Mars, and layers of ice, interspersed with deposits of impact ejecta, may comprise much of the upper crusts of Callisto and Ganymede.

The diversity of crater morphology revealed by Viking Orbiter pictures (Carr *et al.*, 1977) suggests that detailed analysis of martian craters on contrasting terrains might assist in choosing between rock strength and rock layering, or otherwise contribute toward eventual solution of the simple-to-complex problem. Indeed, some general terrain-related effects on crater morphology on Mars already have been identified (Wood *et al.*, 1978; Johansen, 1979; Mouginis-Mark, 1979a). I have examined the morphology of fresh-appearing martian craters for which rim-to-floor depth (d) and rim diameter (D) could be measured from Viking Orbiter data. The first results were published for 73 craters where d/D data were extracted by a detailed photometric technique (Pike and Arthur, 1979). Related findings for 57 craters where d/D measurements came from photogrammetry have appeared more recently (Pike *et al.*, 1980). This paper presents the results of the most recent survey of martian craters, those whose depths were measured from shadow lengths on Viking Orbiter prints.

THE NEW DATA

Criteria

The 230-crater sample analyzed here (Fig. 5) resulted from a systematic perusal of all enlarged (six-inch) Viking Orbiter prints (shade-corrected versions only) in the U.S. Geological Survey Menlo Park file in February 1979. Except for some of the earlier spacecraft revolutions, the collection was relatively complete through frames A900 and B705. In order to obtain consistent and reliable data on crater depth/diameter as well as on photogeologic characteristics, stringent standards for both image quality and crater suitability were maintained. Criteria of selection included image clarity and resolution, sun-elevation angle and sun azimuth, shadow density and length, and freshness of the crater form. Comparatively few of the thousands of available pictures, especially those covering high latitudes, proved suitable for both measurement of crater depths and identification of diagnostic morphologic features.

Only fresh craters were accepted so that the analytical results would be comparable to those for craters on the moon (Pike, 1974; Wood and Andersson, 1978) and the other planets (e.g., Malin and Dzurisin, 1977; Smith and Hartnell, 1978). Selection of the sample was intentionally biased toward fresh craters, although I did not erect a formal relative-age classification such as used by Baldwin (1949) or by Wood and Andersson (1978). In the spirit of the discussion earlier in the paper, I defined freshness of crater form as overall crispness of the rim and other interior morphologic features, minimal superposition of much smaller impact craters on both the interior and on the ejecta blanket, and where resolvable, presence of fresh-appearing ejecta. Otherwise fresh-appearing craters that

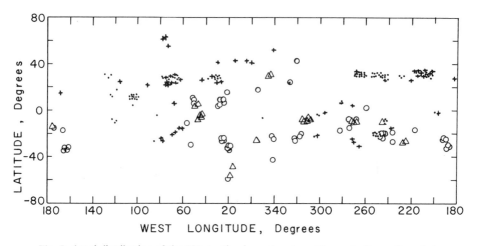

Fig. 5. Areal distribution of the 230 martian impact craters. Concentrations of symbols indicate coincidence of high-quality images with occurrence of fresh craters in this sample. There are 102 simple craters on plains (dots), 23 simple craters on cratered terrain (triangles), 51 complex craters on plains (crosses), and 54 complex craters on cratered terrains (circles). For distribution by photogeologic units see Table 4; for distribution by size (rim diameter, D) see Fig. 6.

seemed to contain significant fill (for example, an unduly broad flat floor that embayed and partially buried the central peak or pit) were omitted. Age of the photogeologic unit (Scott and Carr, 1978) on which the crater was located was not a criterion.

Resolution and image quality had to be sufficient for an unambiguous classification of a crater as either simple or complex and for clear evidence of either ballistic or fluid-flow textures (for this distinction, see Carr *et al.*, 1977) in the ejecta. For example, craters that were obviously large enough to contain peaks, terraces, and other complex features (see Fig. 7, below), but both lacked these features *and* also appeared only on low-quality images, were rejected. In order to obtain the best possible d/D data, and following lunar practice, I rejected simple craters for which the sun-elevation angle fell below 15° or exceeded 25°. Somewhat lower sun angles were often accepted for large complex craters to insure that the shadow tip extended onto the flat floor. In no case were sun angles higher than 30° or lower than 10° permitted. Considerable experimentation showed that crater depths derived from outside these limits often were unreasonably low or high (e.g., depths for simple craters that yielded a d/D of 0.4; twice that evident for fresh lunar craters). Because anomalously high or low depths also resulted from some pictures where the sun azimuth departed significantly from 90° or 270°, deviations of more than two or three degrees from due east or west were cause for rejection of a picture. I do not yet understand this effect, but mention it to indicate the care taken to exclude all possible sources of error from the data. The shadows cast in crater interiors had to be dense and sufficiently crisp to permit unambiguous measurement of their lengths. Moreover, any substantial difference between shadow configuration in orthographic and rectilinear versions of the prints, which could indicate false "shadows" resulting from manipulation of the grey scale during image processing, prompted rejection of the crater. Finally, shadow margins in simple craters had to lie at, or approximately at, the center of the crater (18 exceptions discussed below) and had to be smoothly rounded, not squared off. An important exception to the latter rule applies to those obviously otherwise simple craters that had very small flat floors (e.g., Pike, 1977; Wood and Andersson, 1978). Edges of the broad, squared-off shadows typically observed in complex craters had to be cast from the rim crest onto the flat floor, and not from a high rimwall terrace or onto a terrace located part-way up the inner wall. The above criteria excluded many otherwise suitable martian craters from the sample.

The data

The size of the new sample for Mars is comparable to that used to establish the d/D relation of fresh lunar craters from Apollo data (Pike, 1974). Morphologic data on most of the 230 martian craters that survived all the tests for exclusion should gain in both precision and accuracy what they lack in sheer volume. There are some exceptions. Data on 39 simple craters less than 0.9 km across—roughly the limit of resolution for identifying morphologic features in craters on the bulk of the Viking Orbiter images—are far less precise and complete than data on larger craters. These small craters were included to extend the d/D relation down to as small a crater size as possible. Virtually no morphologic information is available on the under-0.9 km-diameter craters, although all of them probably have the simple morphology. A pocket optical comparator sufficed to measure diameters and shadow lengths for such craters, the smallest of which were only about two or three pixels across. The craters for which depth and diameter were determined range from D = 53 m and d = 10 m to D = 155 km and d = 3.23 km. Diameters were calculated from the SEDR data sheets and edge data rather than from the printed bar scales, which cross-checking revealed were occasionally incorrect (M. H. Carr and D. W. G. Arthur, pers. comm., 1979–1980). As for complex craters on the moon (Pike, 1976), depths of complex craters on Mars are average values, not maximum figures from the longest shadows.

Depths for 18 especially well-depicted simple craters on high-resolution images in which the shadow tip did not strike the exact crater center, usually (but not always) the lowest point, were adjusted upward empirically. This was done because eight of the craters were of the large simple type discussed below, and thus of special interest. The correction, which relates percentage of crater depth to percentage of crater radius covered by shadow, was derived from the averaged interior geometry of four fresh simple craters on the moon (Macrobius A and B, Cauchy CA, and a small unnamed crater on map 40A4S1) that are particularly well contoured on Lunar Topographic Orthophotomaps. The average increase in crater depth was about 18 percent of the value measured from Viking Orbiter pictures. Because the interior form of the fresh *simple* craters differs little from planet to planet (Fig. 2), the lunar-based correction introduced little or no subjectivity into the martian sample. Validity of the procedure is confirmed indirectly by the fact that omission of the 18 depth-corrected craters does not change the d/D equations for simple martian craters (Table 1).

In addition to depth and diameter, I ascertained the presence of the following nine morphologic features described earlier (cf. Fig. 1A and 1B): bowl shape, flat floor, central peak (or central depression), "slump" block, scalloped rim crest, rimwall terrace, cratered central peak (peak pit), predominantly ballistic ejecta, and predominantly fluid-flow ejecta. In some few cases, identification of one or two of these features is uncertain, owing to marginally acceptable resolution, poor development or uncertain expression of the feature, or obscuration of a possible feature by interior shadow.

Crater classification

Before analyzing the data, the 230 craters (Fig. 5) were grouped as to photogeologic background on Mars, either plains (n = 153) or rough (cratered) terrain (n = 77, and overall morphologic type, either simple (n = 125) or complex (n = 105). Craters in the plains subset fall on nine of the geologic units mapped by Scott and Carr (1978), all but one of which (unit Npm) are of Hesperian age or younger (Table 4). Most (79) of the plains craters are on one of the youngest terrains, unit Apc. Craters in the cratered-terrain subset fall on four of Scott and Carr's Noachian-age map units (Table 4). All but two of these craters are located on unit Nhc and Nplc, old heavily-cratered terrain in the uplands. The definition of a simple crater is fairly conservative, and usually requires a crisp, rounded shadow tip on a bowl-shaped and clearly featureless bottom. Most departures from this elementary form identify a crater as complex. The presence of a small flat floor, which because of its often multigenetic character is not a diagnostic criterion of morphologic complexity, can be an exception to this rule of crater taxonomy (see Pike, 1977 and Wood and Andersson, 1978). In practice, grouping the fresh impact craters in this sample as either simple or complex was not difficult. Other classifications of complex martian craters are, of course, possible (e.g., Johansen, 1979; Mouginis-Mark, 1979a).

Table 4. Photogeologic terrain type of 230 fresh craters on Mars.[1]

n	Symbol	Possible or probable geology
		PLAINS MATERIALS
		(Mostly Amazonian and Hesperian in age)
6	Aps	Eolian and volcanic materials
1	Apt	Young lava plains
15	Avy	Young basaltic lava flows
79	Apc	Eolian-veneered lava plains
23	Hpr	Lava plains
1	ANch	Fluvial and eolian deposits
17	Hprg	Volcanic flows
10	Hvo	Shield volcanoes and flow units
2	Npm	Cratered lava plains
		ROUGH (CRATERED) TERRAIN MATERIALS
		(Mostly Noachian in age)
1	HNk	Diverse origins
39	Nplc	Ancient lava flows and impact ejecta
1	HNbr	Ejecta and ancient terrain
35	Nhc	Oldest exposed surface on Mars

[1] After Scott and Carr (1978).

GENERAL RESULTS

An analysis of the new martian data is presented in Figs. 2, 3, 6, 7, and 8, and in Table 1. The findings are consistent with the overall picture of size-dependence in crater morphology that was outlined previously (Pike and Arthur, 1979; Pike *et al.*, 1980). The diameter of the simple-to-complex transition, D_t, for fresh craters on all terrain types has been recalculated (Fig. 3). The D_t for Mars now is 5.75 km (n = 10 variables), down from the 6.3 km value of Pike *et al.* (1980) and the nearly 7 km value of Pike and Arthur (1979), a trend that reflects increasing attention to detail and better sampling of craters on Mars. I do not expect this figure to decrease much further. The new D_t value does not disrupt the inverse relation between crater morphology and gravity-cum-target strength (or layering) (Fig. 3). In fact, the data point for Mars now is a better fit to the line extrapolated back from terrestrial impact craters in sedimentary rock. The other, more specific new results are addressed below.

INTERIOR MORPHOLOGY

Depth/diameter results

Linear depth/diameter fits (Table 1) to all 230 craters are roughly equivalent to those derived previously for smaller samples (Pike and Arthur, 1979; Pike *et al.*, 1980). The new expressions differ from the most recent results of Cintala and

Mouginis-Mark (1980) in that their trend for large (equivalent to my complex) craters is much steeper than mine and their equation for small (my simple) craters is somewhat lower and less steep than mine. All in all, fresh simple craters on Mars have about the same interior geometry (d/D) as fresh simple craters on other solid planets (Fig. 2). It is also clear from the equations plotted in Fig. 2 that complex martian craters, which fall midway between those on the moon and Mercury and those on Earth (and perhaps Venus), have a much steeper d/D relation (slope = 0.395) than is observed elsewhere in the solar system (0.24–0.332). Cintala and Mouginis-Mark (1980) offered one possible explanation for this, probably real, steepness in terms of substrate volatiles.

Terrain-correlated differences in martian craters are ambiguous with respect to depth/diameter. The new results for Mars are broken down by terrain and crater type in Figure 6 and in Table 1. The two sets of log-linear distributions have much in common. In both cases the transition from simple to complex craters occurs at about 3.5 km diameter. However, the equations for craters on plains units cross at a crater diameter of 2.5 km, whereas they intersect at 3.3 km for craters on the rough terrains. The difference may be significant, but conceivably could reflect too small a sample. Simple craters on the plains also are about five percent deeper than those elsewhere on the planet, a small difference that may

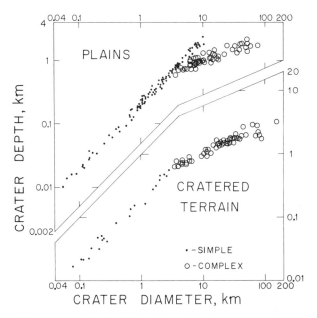

Fig. 6. Depth/diameter results for 230 fresh martian impact craters illustrate the morphologic transition from simple to complex. Simple craters on cratered terrains (n = 23) are no larger (3.4 km across) than the smallest complex craters (n = 53), whereas simple craters on plains (n = 102) reach diameters (maximum = 10.3 km) well beyond that (3.6 km) of smallest complex craters (n = 52). Depth/diameter equations (Table 1) differ considerably more for complex craters than for simple craters. Depths measured from shadow lengths on Viking Orbiter pictures.

reflect nothing more than the previously cited tendency for slightly older and more degraded craters to be incorporated into crater samples from rough, old terrains (cf. Table 2, Fig. 9; Pike, 1974; and Wood and Andersson, 1978, p. 3676). The d/D distribution for complex craters in rough terrains has a steeper slope (0.423) than that for the plains (0.334), although nowhere near the 0.557 observed by Cintala and Mouginis-Mark (1980) from their d/D data. Complex craters on plains also are shallower than those elsewhere on Mars. The difference between the two equations (Table 1) varies from zero at D = 5 km to about 27 percent at D = 100 km. Moreover, the considerable scatter for both subsets of complex craters suggests caution in making geologic conclusions based on these contrasts. The most convincing difference between fresh impact craters on the two principal types of martian terrain may lie not in the d/D equations themselves but rather in the fact that simple craters on the plains extend well past the intersection diameter, to about 10 km D, whereas such large simple craters are absent from cratered terrains.

Other interior features

Size-morphology arrays for nine geomorphic aspects of crater interiors, undifferentiated by terrain type (Fig. 7), confirm results obtained previously for a similar, but smaller, sample of craters (Pike and Arthur, 1979). Morphologic features appear in fresh complex craters on Mars at different crater sizes. Flat floors, central peaks, and low d/D mark the onset of complexity in craters about 3 km to 3.5 km across. Rim-crest scalloping, accompanied by slump blocks, appears in slightly larger craters (at about 4 km diameter). The final features observed with increasing size of fresh complex craters are cratered central peaks (or central pits) and terraced rimwalls, at about 6 km to 8 km diameter. The morphologic data of Wood *et al.* (1978) for 173 fresh martian craters, when arrayed as in Fig. 7 and undifferentiated by terrain type, also are generally consistent with this ranking. The much more voluminous data of Mouginis-Mark (1979a) (compare his Fig. 7 and 8) show that central peaks occur consistently in smaller craters than do scallops and terraces (i.e., slump blocks, see Cintala *et al.*, 1977 and Wood *et al.*, 1978) up to a crater diameter of about 9 or 10 km, at which size the frequency of occurrence reverses. However, Mouginis-Mark's higher frequency of wall failure (terraces plus scallops) than central peaks above this crater diameter does not indicate a true reversal of the peaks-first trend. Rather his Fig. 8, the graph for wall failure, reflects augmenting of scallops by terraces, which my Fig. 7 shows as appearing in craters around 7 km to 8 km across.

An aside on graphic presentation is in order here. Onset diameters, which mark the first occurrence of a feature in a crater, are statistically unsound parameters of the simple-to-complex transition because (1) they only indicate the lower end of the transition range and (2) they are unique cases subject to wide variance with minor increase in the size of the sample. Accordingly, diameter/percentage-frequency curves give the most reliable transition diameters. Such diagrams are

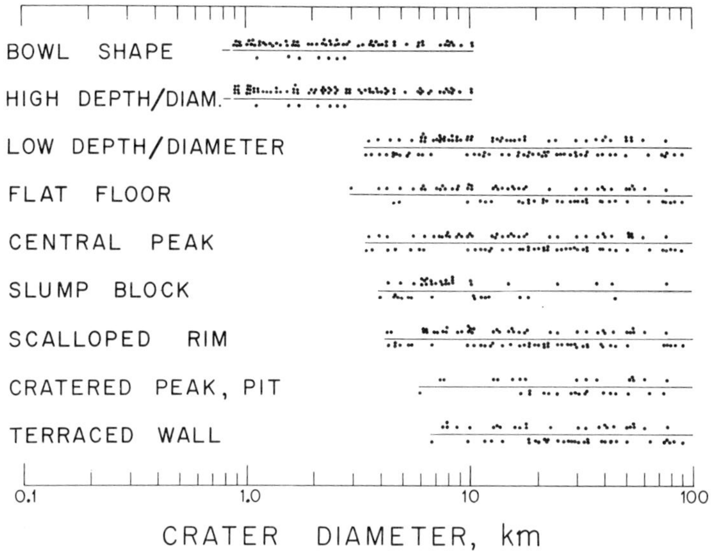

Fig. 7. Diameter/morphology arrays for nine interior characteristics of fresh martian craters. Size-dependence of crater morphology—the simple-to-complex transition—is pronounced, but terrain dependence of crater shape is more subtle (see text). Dots show unambiguous identifications of morphologic characteristics on Viking Orbiter images for most of the same 230 craters plotted in Fig. 6. Data above lines: craters on plains units (see Table 4); data below lines: craters on cratered terrain.

not given here for the data in Fig. 7, however, because the absence of large (~4–10 km across) simple craters in rough terrains would yield misleadingly different frequency curves for the occurrence of morphologic features in complex craters on the two major terrain types of Mars. This is a subtle, but crucial point, because frequency curves customarily are interpreted in terms of the appearance of *complex*-crater features. Fig. 7 shows that the occurrence of complex-crater features is the *same* on *both* terrain types; rather it is the occurrence of *simple* craters that is so different. Regardless of what technique is chosen to display and abstract the data, however, the relative order in which the features are observed in complex craters with increasing crater size remains as described in the last paragraph.

Perhaps the most important result evident in Fig. 7 is that central peaks appear in complex martian craters at smaller diameters, by a factor of two, than do terraced walls. The same relative order is clearly evident for craters on the moon (Smith and Sanchez, 1973; Wood and Andersson, 1978) and on Mercury (Cintala et al., 1977; Smith and Hartnell, 1978). From the meager terrestrial data (Roddy, 1977), the sequence also may characterize meteorite craters on Earth and their experimental analogs. Together these data suggest that the first manifestations of complexity in impact craters are a flat floor, a central peak, and a low d/D ratio. Early appearance of these features is inconsistent with formation of complex craters by collapse of the transient-cavity rim into wall terraces and consequent deep centripetal sliding. Rather, terrace formation appears to be a later expression of complexity, not the initiating mechanism (Head, 1976; Pike, 1980).

Subdivision of the size/morphology arrays in Fig. 7 by terrain type reveals nothing that might explain the absence of large simple craters on the rough terrains or their presence on the plains, but it does suggest more possible terrain-related differences in the onset and prevalence of features within complex craters. For example, both flat floors and central-peak pits persist down to smaller crater sizes on the plains than they do in cratered terrains, an observation that agrees with the findings of Wood *et al.* (1978) and, in the case of the peak pits, with preliminary data of C. A. Hodges (pers. comm., 1980). Although central peaks and other features in Fig. 7 seem to appear at roughly the same crater sizes, Wood *et al.* (1978) found wall failure to occur more often in smaller craters on the plains. Moreover, according to the voluminous data of Mouginis-Mark (1979a, Figs. 7 and 8), central peaks and wall failure occur systematically in smaller craters in the plains than on the cratered terrains. If the latter two terrain-related contrasts are real, and do not reflect problems in obtaining a crater sample of uniform relative age from terrains of greatly varying age and roughness (e.g., Pike, 1974; Wood and Andersson, 1978), then all nine morphologic aspects of complex impact craters in Fig. 7 could be said to exhibit at least some systematic terrain-correlation (wall failure includes three aspects: slump blocks, scalloped rim crest, and terraces): namely, complex craters on Mars appear at smaller sizes on plains units than on cratered terrains.

EXTERNAL MORPHOLOGY: EJECTA

The simple-to-complex transition for internal morphology coincides with a change in the texture of ejecta surrounding fresh martian impact craters. The mechanism of ejecta emplacement seems to depend strongly on crater size (Fig. 8). Craters less than about 4 km across tend to have ejecta that displays the usual ballistic characteristics typical of craters on the moon and Mercury, whereas larger craters

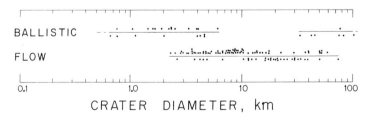

Fig. 8. Characteristic morphology of ejecta surrounding fresh impact craters on Mars. Ballistic ejecta is observed for craters 30 km ≤ D ≤ 6 km, whereas ejecta dominated by features usually ascribed to fluid flow is observed for craters 2 km ≤ D ≤ 75 km across. Overlap of the distributions about a median diameter of 4 km coincides approximately with the transition from simple to complex interior morphology. The largest craters, D ≥ 80 km, have mainly ballistic ejecta. For remarks on data, see Fig. 7.

have ejecta textures that indicate a nonballistic mode of emplacement—probably by surface flow (Carr *et al.,* 1977). This 4 km-diameter threshold is close to the 5 km-diameter result obtained by Boyce (1979) for craters at the Viking Lander I site. There is similar overlap in ejecta type for both terrain types distinguished in Fig. 8, generally between crater diameters of 2.5 km and 6 km, around the median value (of overlapping craters) of 4 km, again, close to the results of Boyce (1979). The ballistic-versus-flow dichotomy in ejecta texture is not identical with the simple-to-complex contrast, however, even though the two types of changes are observed at similar crater sizes: most martian craters over 2.5 km across on plains units show evidence of flow *regardless* of whether they are simple *or* complex. Thus whatever controls formation of simple craters over 4 km across, presumably gravitational acceleration and target strength, does not necessarily control texture of the ejecta. Fluidized ejecta is thought to result from the incorporation of abundant subsurface volatiles in the cratering process (Carr *et al.,* 1977; Gault and Greeley, 1978). Detailed work by Mouginis-Mark (1979a) and by Singer and Schultz (1980) has identified other variables that may contribute to observed differences within the population of fluidized-ejecta craters on Mars.

The observations in Fig. 8 also suggest that there may be an upper size limit to fresh martian flow-ejecta craters, at a diameter of about 80 km. Identification of both ballistic and fluid-flow textures in the ejecta surrounding the fresh 55 km-diameter crater Bamberg by Mouginis-Mark (1979b) further indicates that the fluid-flow regime dominating smaller rampart craters on Mars may pass into a ballistic regime in larger craters. If these preliminary conclusions are correct, I suggest that the 4 km to 80 km size-range reflects a balance between volatile content of the target rocks, as well as depth to the volatile-rich layer, and heat released by the impact. In craters less than about 4 km across, volatile-rich strata may be too deep to be tapped by the impact (see also Boyce, 1979; Cintala and Mouginis-Mark, 1980), and in craters over 80 km across volatiles in this layer may be entirely vaporized by the great amounts of energy available. In both cases, ballistic ejecta dominate; in the latter instance, the volume of ballistic ejecta would be so much greater than that of any flow ejecta that flow features would be obscured or overwhelmed by ballistic features. Thus in neither very small nor very large craters does fluid flow dominate, a conclusion also reached by Cintala and Mouginis-Mark (1980). In the diameter range 4 km to 80 km, however, the balance between mass and depth to concentration of target volatiles and the energy level of impact may be just that required to generate the geomorphic characteristics that have been attributed to fluid flow.

With respect to texture of the ejecta, the impact craters examined in this study do not have a strong affinity for one major terrain type or the other. Craters surrounded by fluid-flow ejecta exhibit a 2:1 preference for plains units over cratered terrain (Fig. 8) but if the large simple craters, all of which fall on the plains, are ignored then the ratio drops to about 4:3. This probably is not a significant difference, particularly since flow patterns are much more difficult to see on rough terrain. Results for the two most numerous categories in Mouginis-Mark's (1979a) 1558-crater study are equally ambiguous on this score. His Type

1 craters (single ejecta facies) are found more on cratered ("ancient") terrain than they are on cratered and smooth plains (60%:47%), but his Type 2 craters (double ejecta facies) occur in just the opposite (17%:31%) relative frequency. For the most common varieties of fluidized-ejecta craters on Mars, terrain dependence remains uncertain. More detailed work, perhaps using revised geologic units from the new geologic map of Mars (Scott, 1980), is needed.

LARGE SIMPLE CRATERS

The problem of large simple craters on Mars probably is one of explaining their absence from cratered terrains (assuming this absence is not a sampling fluke; Fig. 5) rather than their presence on the plains. This perspective follows from Fig. 2, which shows that simple craters larger than the d/D intersection are common on all planets for which both simple and complex craters have been identified. Mars is probably no exception. More detailed data for craters on the moon (Fig. 9) suggest that differences in the target, although affecting some aspects of crater shape, do not wholly preclude large simple craters from forming on one type of terrain or another. The problem for Mars may be approached by

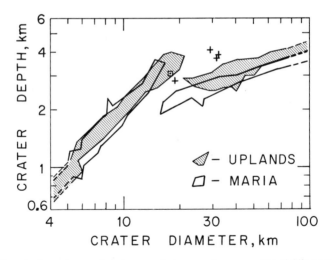

Fig. 9. Terrain-dependence of crater morphology on the moon. Depth/diameter results for 203 mare craters vs. 136 upland craters. Simple craters do not differ much in d/D (0.5%), but upland complex craters are on the average 12% deeper than mare complex craters. The simple-to-complex transition occurs on maria in craters about 16 km across, whereas it occurs in upland craters about 21 km in diameter. Square (Diophantus) and crosses (with increasing size: Dionysius, Proclus, Kant, Auzout) are craters with a morphology transitional from simple to complex configurations. To show relations in detail, plotted fields outline only craters 4.2 km ≤ D ≤ 95 km; d/D equations, revised from those in Pike (1974) using data in Pike (1976), are given in Table 2.

examining the spatial distribution of similar-sized fresh craters on both terrain types considered here.

The 22 large simple craters (4.3 km $\leq$ D $\leq$ 10.3 km) all share a common geologic environment: young mid-latitude (16.5°N to 33.5°N) volcanic plains. Seventeen of them are concentrated in the belt between 25°N and 31°N and between longitude 228°W and 265°W, in western Elysium Planitia (Fig. 5), amid what appears to be a vast field of scattered small volcanic cones like those described by Hodges (1979). Three more craters are clustered within 200 km of each other in northeastern Chryse Planitia. One other lies 100 km west of Hecates Tholus, and another is 100 km south of Jovis Tholus. Fifteen of the craters have been excavated in unit Apc (Scott and Carr, 1978), fairly recent lava plains partly and thinly veneered with eolian sediments. Proximity of fresh volcanic constructs suggests that the lavas in which the craters formed are very young indeed. These target conditions, however, are not unique to large simple craters. Sixteen complex craters within the same size range share the same geologic environment: unit Apc. These craters lie between 5°N and 29°N, 12 of them between 23°N and 28°N. Half of them cluster to the north of Kasei Vallis, three more in Chryse Planitia, and the remaining five are disposed singly: one 300 km north of Jovis Tholus, two in Amazonis Planitia, one in Isidis Planitia, and one south of Hephaestus Fossae.

The two disparate groups of similar-sized craters do not seem to reflect apparent or inferred differences in geology, even though the complex craters all lie outside the belt of large simple craters in western Elysium Planitia (the closest being about 650 km to the south). If the Elysium concentration did not arise as the result of special target characteristics, then other alternatives, such as mass-density of impacting bodies, may be invoked. It has been proposed that simple craters formed primarily by dense meteoroids and complex craters by light meteoroids (Quaide *et al.*, 1965; Milton and Roddy, 1972; Wetherill, 1977). Could the concentration of large, deep simple craters in Elysium indicate the strewn-field of a swarm of unusually large iron-rich bolides, or is the problem one of sampling? This question cannot be answered until the locations of more large simple craters on Mars are studied, and the possibility of bias through inadequate photo coverage has been evaluated in more detail.

COMPARISON WITH EARTH AND MOON

Terrain-related contrasts in the topography of craters on the moon, which until recently have proved elusive, provide a key piece of data. Neither Quaide *et al.* (1965) nor Pike (1974) identified such a difference, but Wood and Andersson (1978) found that upland craters in their class TRI (typically 15 km $\leq$ D $\leq$ 50 km) were systematically deeper (by an average of about 18%) than those on mare surfaces. Figure 9 is a revision of my 1974 results using photogrammetric data (Pike, 1976), with some recent (unpublished) amplifications. Although simple craters on the maria do not differ markedly (less than half a percent) from simple upland craters in d/D (Table 2), complex upland craters 15 to 100 km across average 12% deeper than similar-sized craters on the maria (Table 2). What is even more interesting, however, is that the simple-to-complex transition on the moon occurs at a crater diameter of about 16 km on the maria and at about 21 km on the uplands (Fig. 9). This difference may correspond to the terrestrial contrast documented in Fig. 4.

The terrain-related differences in crater sizes for the transition on Earth and Moon may indicate presence or absence of significant layering in the target materials. On both planets, the smaller onset diameter occurs in demonstrably layered rocks: sandstones, shales, and limestones on Earth and mare-basalt flows on the moon. The larger onset diameter, on the other hand, is observed on both planets in what are known or inferred to be less sharply stratified materials: gneisses, schists, metavolcanics, and granites on Earth and highland breccias on the moon. These intraplanetary contrasts in D_t thus do not seem to reflect simply hard versus soft targets. Were that the case, then the transition on the moon should occur in smaller craters on the softer highland breccias than in the hard mare basalts. This is not what is observed, however, in Fig. 9.

The subtle terrain-correlated differences in size-morphology summarized here for craters on Mars suggest, tentatively, that plains materials could differ stratigraphically from those of cratered terrains. Certainly all martian terrains may be layered to some extent in that they include volatile-rich strata at depths that have affected the morphology of most impact craters over 4 km across (Boyce, 1979; Cintala and Mouginis-Mark, 1980). The difference perhaps is one of degree rather than of outright presence or absence of stratification. If the correlation between morphology and crater size that has been observed on the moon and Earth (Figs. 4 and 9) also applies to Mars, then the plains, where complex craters may occur at slightly smaller sizes than they do on cratered terrains, are inferred to be the more strongly layered of the two types of targets. The source of the enhanced stratification of plains materials could have arisen from their being successions of lava flows (perhaps intercalated with sediments), being young (and hence little reworked by impact), and being preferentially veneered by eolian and perhaps by other recent unconsolidated sediments. Further work with Viking pictures may enable these and perhaps other alternatives to be sorted out and tested.

Conrol of the simple-to-complex threshold-diameter in craters by layering of the target on Earth, Moon, and now perhaps Mars is consistent with an alternative interpretation of the D_t:g relation (Fig. 3). Rather than reflecting simply hard versus soft targets (Pike and Arthur, 1979; Pike *et al.*, 1980), deviations of the mean threshold diameter D_t from the log-log linear relation with g could indicate significant broad-scale layering in planetary crusts, as suggested by authors from Wegener (1921) to Head (1976). However, the layering itself does not necessarily trigger the formation of complex craters (Pike, 1980), and as suggested above, stratification does not always mean presence of a weak layer over a strong substrate. According to morphologic data available for craters on six solid planets and satellites, Mercury may be reinterpreted as essentially non-layered (cf. Cintala *et al.*, 1977)—perhaps the result of a planet-wide melting episode or some other type of thermal event, possibly brought about through long proximity to the sun; the earth and its moon each have both layered and non-layered—or rather strongly layered and weakly layered—portions of their crusts; and Callisto, Ganymede, and Mars are interpreted as strongly layered. Again, some of the variance in the D_t:g data in Fig. 3 may well reflect other influences not considered in detail, such as projectile characteristics, but the arguments advanced here suggest that deviation from the mean D_t:g trend could have resulted largely from the degree of target stratification.

CONCLUSIONS

The results presented in this paper, together with findings reached by other investigators, suggest that *both* gravitational acceleration *and* target characteristics have affected the morphology of impact craters across the solar system. Neither of these important influences alone can explain all of the many observations on craters. The emerging picture of morphology and process, with morphology expressed by the characteristic crater size (D_t) where the transition from simple to complex types occurs, is unclear in that the relative importance of gravity and target type differs from planet to planet. For example, intraplanetary contrasts in crater shape that reflect differences in target materials seem to be large on Earth, significant but minor on Mars and the moon, and minimal on Mercury. Whole-planet differences in crater shape between Earth and the moon, on the other hand, seem entirely dependent on g. However, the whole-planet values of D_t observed on Mercury and Mars vary from values predicted by gravity-scaling alone, an effect that may indicate strong contrasts in the surface materials of the two planets. Specific findings on crater morphology reached in this study, several of which are consistent with the results of other workers, include the following:

(1) D_t, the characteristic size that divides most fresh-appearing impact craters with a simple form from those with a complex form is determined for four planets. D_t is defined as the geometric mean of several (statistically-derived) crater diameters that express individual aspects of the simple-to-complex transition, such as changes in depth/diameter and the occurrence of central peaks.

(2) D_t is not the same for craters on each planet: ~3 km on Earth, ~6 km on Mars from a new sample of 230 craters on Viking Orbiter images, ~16 km on Mercury, and ~19 km on the moon. Preliminary estimates of D_t for Ganymede and Callisto are, respectively, ~13 km and ~15 km.

(3) Fresh impact craters vary in morphology according to the terrain on which they form on at least three planets, in decreasing order of certainty: Earth, Moon, Mars. At this time, stratification of target rocks, including the effects of volatiles, seems to explain the observations at least as well as strength *per se* of the materials.

(4) The appearance of terraces, central peaks, and other morphologic features within martian craters 3 to 8 km across occurs in a size-dependent sequence

resembling that observed for craters on the moon. Only the crater size at which the simple-to-complex transition occurs, D_t, is different (cf. Wood *et al.*, 1978). The sequence, in which central peaks are observed in smaller craters than are rimwall terraces, is not consistent with formation of complex craters by centripetal slump and deep sliding of rim terraces (cf. Head, 1976; Pike, 1980).

(5) Occurrence of the transition on both martian plains and cratered terrain in smaller craters ($D_t \approx 6$ km) than would be expected from the inverse statistical relation between D_t and g alone ($D_t \approx 9$ km) shows that some other influence applies. Comparison of similar morphologic data for craters on different terrains of Earth and the moon suggests that all martian terrains may be significantly stratified.

(6) The morphologic transition may occur at a slightly smaller crater size on martian plains than on the cratered terrains, a difference I ascribe to plains materials being more highly stratified (cf., Wood *et al.*, 1978).

(7) Complex craters on both the moon and Mars exhibit marked terrain-related differences in d/D, whereas simple craters do not. Complex martian craters in rough terrains average 13 percent deeper than plains craters. Revision of the lunar d/D relation reveals a similar terrain-correlated difference: the morphologic transition occurs in slightly smaller craters on the maria than it does in upland craters, a result of upland craters being significantly (an average of 12 percent) deeper than those on the maria (cf., Wood and Andersson, 1978).

(8) Cratered terrains on Mars may be deficient in large simple impact craters (3.5 km $\leq D \leq 10$ km). Other than acknowledging a possible sampling deficiency, at this time I have no firm explanation for the observation that 17 of the 22 large simple craters examined here are localized in western Elysium Planitia.

(9) A change from ballistic to fluid-flow ejecta occurs in martian craters on both plains and cratered terrains at about 4 km diameter (cf., Boyce, 1979). The difference is not necessarily related to the simple-to-complex transition in interior morphology. A second change in ejecta texture, back to a mainly ballistic regime for craters over 80 km across, suggests that the quantity of crustal volatiles that can become involved in an impact may have a finite limit (cf., Cintala and Mouginis-Mark, 1980).

(10) Preliminary determinations of the simple-to-complex transition for craters on Callisto and Ganymede are consistent with either low target strength or significant layering in the rocks of these bodies. Crustal volatiles are important in either case. Conversely, Mercury is inferred to be either a very strong target, stronger than the moon, or else to have nonlayered (and dry) rocks at the surface (cf., Cintala *et al.*, 1977).

(11) The first d/D results for what appear to be large impact craters on Venus (reported in Kerr, 1980) lie just below the d/D trend for ringed complex craters on Earth. More data are needed before interpreting this possible coincidence in terms of the overall D_t:g model proposed here.

The most fruitful area for further research in crater morphology now lies in ascertaining the relative importance of gravity and target type on each planet, rather than in attempting to ascribe all morphologic observations to either one competing explanation or the other. Another related problem is whether or not variations in velocity and/or mass-density of the impacting projectiles might better explain some of the morphologic data. Finally, detailed observations of craters on Venus, Ganymede, Callisto, and perhaps the larger satellites of Saturn will test some of the ideas discussed or proposed here.

Acknowledgments—Supported by NASA Geology Programs office. Mike Carr and Dai Arthur brought to my attention the many intricacies of Viking Orbiter images and their processing. I am indebted to Karl Blasius, Mike Carr, Mark Cintala, and Hank Moore for reviews that substantially improved the presentation. This paper is dedicated to the memory of Alfred Wegener, insightful investigator of crater morphology, on the occasion of the 100th anniversary of his birth.

REFERENCES

Baldwin R. B. (1949) *The Face of the Moon*. Univ. Chicago Press, Chicago. 239 pp.

Boyce J. M. (1979) A method for measuring heat flow in the martian crust using impact crater morphology. Repts. Planetary Geol. Program 1978–1979, NASA TM 80339, p. 114–118.

Carr M. H., Crumpler L. S., Cutts J. A., Greeley R., Guest J. E., and Masursky H. (1977) Martian impact craters and emplacement of ejecta by surface flow. *J. Geophys. Res.* **82**, 4055–4065.

Cintala M. J. (1979) Mercurian crater rim heights and some interplanetary comparisons. *Proc. Lunar Planet. Sci. Conf. 10th*, p. 2635–2650.

Cintala M. J. and Mouginis-Mark P. J. (1980) Martian fresh carter depths: More evidence for substrate volatiles? (abstract). In *Lunar and Planetary Science XI*, p. 143–145. Lunar and Planetary Institute, Houston.

Cintala M. J., Wood C. A., and Head J. W. (1977) The effects of target characteristics on fresh crater morphology: Preliminary results for the moon and Mercury. *Proc. Lunar Sci. Conf. 8th*, p. 3409–3425.

Cooper H. F. (1977) A summary of explosion cratering phenomena relevant to meteor impact events. In *Impact and Explosion Cratering* (D. J. Roddy, R. O. Pepin, and R. B. Merrill, eds.), p. 11–44. Pergamon, N. Y.

Dence M. R. (1964) A comparative structural and petrographic study of probable Canadian meteorite craters. *Meteoritics* **2**, 249–270.

Dence M. R. (1972) The nature and significance of terrestrial impact structures. *International Geological Congress, XXIV Session, Section 15*, p. 77–89.

Gault D. E. and Greeley R. (1978) Exploratory experiments of impact craters formed in viscous-liquid targets: Analogs for martian rampart craters? *Icarus* **34**, 486–495.

Gault D. E., Guest J. E., Murray J. B., Dzurisin D., and Malin M. C. (1975) Some comparisons of impact craters on Mercury and the moon. *J. Geophys. Res.* **80**, 2444–2460.

Gilbert G. K. (1893) The Moon's face; a study of the origin of its features. *Bull. Phil. Soc. Wash.* **12**, 241–292.

Gordin V. M., Dabizha A. I., Krass M. S., Mikhailov V. O., and Myasnikov V. P. (1979) Geophysical and geomechanical aspects of the study of meteorite structures. *Phys. Earth Planet. Inter.* **20**, 1–11.

Grieve R. A. F. and Robertson P. B. (1979) The terrestrial cratering record I. Current Status of observations. *Icarus* **38**, 212–229.

Hartmann W. K. (1972) Interplanet variations in scale of crater morphology—Earth, Mars, Moon. *Icarus* **17**, 707–713.

Hartmann W. K. (1977) Relative crater production rates on planets. *Icarus* **31**, 260–276.

Hawke B. R. and Head J. W. (1977) Impact melt on lunar crater rims. In *Impact and Explosion Cratering* (D. J. Roddy, R. O. Pepin, and R. B. Merrill, eds.), p. 815–841. Pergamon, N. Y.

Head J. W. (1976) The significance of substrate characteristics in determining morphology and morphometry of lunar craters. *Proc. Lunar Sci. Conf. 7th*, p. 2913–2929.

Hodges C. A. (1979) Some lesser volcanic provinces on Mars. Repts. Planetary Geol. Program 1978–1979, NASA TM 80339, p. 247–249.

Johansen L. A. (1979) The latitude dependence of martian splosh cratering and its relationship to water. Repts. Planetary Geol. Program 1978–1979, NASA TM 80339, p. 123–125.

Johnson T. V. (1978) The galilean satellites of Jupiter: Four worlds. *Ann. Rev. Earth Planet. Sci.* **6**, 93–125.

Kerr R. A. (1980) Venus: Not simple or familiar, but interesting. *Science* **207**, 289–293.

Kieffer S. W. and Simonds C. H. (1980) The role of volatiles and lithology in the impact cratering process. *Rev. Geophys. Space Phys.* **18**, 143–181.

Mackin J. H. (1969) Origin of lunar maria. *Bull. Geol. Soc. Amer.* **80**, 735–748.

Malin M. C. and Dzurisin D. (1977) Landform degradation on Mercury, the moon, and Mars: Evidence from crater depth/diameter relationships. *J. Geophys. Res.* **82**, 376–388.

Malin M. C. and Dzurisin D. (1978) Modification of fresh crater landforms: evidence from the moon and Mercury. *J. Geophys. Res.* **83**, 233–243.

Milton D. J., Barlow B. C., Brett R., Brown A. R., Glikson A. Y., Manwaring F. A., Moss F. J., Sedmik E. C., Van Son J., and Young G. A. (1972) Gosses Bluff impact structure, Australia. *Science* **175**, 1199–1207.

Milton D. J. and Roddy D. J. (1972) Displacements within impact craters. *International Geological Congress, XXIV Session, Section 15*, p. 119–124.

Morrison D. and Cruikshank D. P. (1974) Physical properties of the natural satellites. *Space Sci. Rev.* **15**, 641–739.

Mouginis-Mark P. J. (1979a) Martian fluidized crater morphology: Variations with crater size, latitude, altitude, and target material. *J. Geophys. Res.* **84**, 8011–8022.

Mouginis-Mark P. J. (1979b) Ejecta emplacement of the martian impact crater Bamburg. *Proc. Lunar Planet. Sci. Conf. 10th*, p. 2651–2668.

Mutch T. A., Arvidson R. E., Head J. W., Jones K. L., and Saunders R. S. (1976) *The Geology of Mars*. Princeton Univ. Press, N. J. 400 pp.

Oberbeck V. R. and Quaide W. L. (1967) Estimated thickness of a fragmental layer of Oceanus Procellarum. *J. Geophys. Res.* **72**, 4697–4704.

Pike R. J. (1974) Depth/diameter relations of fresh lunar craters: revision from spacecraft data. *Geophys. Res. Lett.* **1**, 291–294.

Pike R. J. (1975) Size-morphology relations of lunar craters: Discussion. *Mod. Geol.* **5**, 169–173.

Pike R. J. (1976) Crater dimensions from Apollo data and supplemental sources. *The Moon* **12**, 463–477.

Pike R. J. (1977) Size-dependence in the shape of fresh impact craters on the moon. In *Impact and Explosion Cratering* (D. J. Roddy, R. O. Pepin, and R. B. Merrill, eds.), p. 489–509. Pergamon, N. Y.

Pike R. J. (1980) Formation of complex impact craters: Evidence from Mars and other planets. *Icarus*. In press.

Pike R. J. and Arthur D. W. G. (1979) Simple to complex impact craters: The transition on Mars. Repts. Planetary Geol. Program 1978–1979, NASA TM-80339, p. 132–134.

Pike R. J., Roddy D. J., and Arthur D. W. G. (1980) Gravity and target strength: Controls on the morphologic transition from simple to complex impact craters. Repts. Planetary Geol. Program 1979–1980, NASA TM 81776, p. 108–110.

Quaide W. L., Gault D. E., and Schmidt R. A. (1965) Gravitative effects on lunar impact structures. *Ann. N. Y. Acad. Sci.* **123**, p. 563–572.

Roddy D. J. (1977) Large-scale impact and explosion craters: Comparisons of morphological and structural analogs. In *Impact and Explosion Cratering* (D. J. Roddy, R. O. Pepin, and R. B. Merrill, eds.), p. 185–246. Pergamon Press, N. Y.

Roddy D. J. (1979) Structural deformation at the Flynn Creek impact crater, Tennessee: A preliminary report on deep drilling. *Proc. Lunar Planet. Sci. Conf. 10th*, p. 2519–2534.

Sabaneyev P. F. (1962) Some results deduced from simulations of lunar craters. In *The Moon* (Z. Kopal and Z. K. Mikhailov, eds.), p. 419–431. Academic, N. Y.

Schultz P. H. (1976) *Moon Morphology*. Univ. Texas Press, Austin. 626 pp.

Scott D. H. (1980) Mars geologic map, 1:15 million scale. Repts. Planetary Geol. Program, NASA TM 81776, p. 372.

Scott D. H. and Carr M. H. (1978) Geologic map of Mars. U. S. Geol. Survey Misc. Geol. Inv. Map I–1083.

Singer J. and Schultz P. H. (1980) Secondary impact craters around lunar, mercurian, and martian craters (abstract). In *Lunar and Planetary Science XI*, p. 1042–1044. The Lunar and Planetary Institute, Houston.

Smith E. I. and Hartnell J. A. (1978) Crater size-shape profiles for the moon and Mercury: Terrain effects and interplanetary comparisons. *The Moon and Planets* **19**, 479–511.

Smith E. I. and Sanchez A. G. (1973) Fresh lunar craters: Morphology as a function of diameter, a possible criterion for crater origin. *Mod. Geol.* **4**, 51–59.

Smith B. A., Soderblom L. A., Johnson T. V., Ingersoll A. P., Collins S. A., Shoemaker E. M., Hunt G. E., Masursky H., Carr M. H., Davies M. E., Cook A. F. II, Boyce J., Danielson G. E., Owen T., Sagan C., Beebe R. F., Veverka J., Strom R. G., McCauley J. F., Morrison D., Briggs G. A., and Suomi V. E. (1979a) The Jupiter system through the eyes of Voyager 1. *Science* **204**, 951–972.

Smith B. A., Soderblom L. H., Beebe R., Boyce J., Briggs G., Carr M., Collins S. A., Cook A. F. II, Danielson G. E., Davies M. E., Hunt G. E., Ingersoll A., Johnson T. V., Masursky H., McCauley J., Morrison D., Owen T., Sagan C., Shoemaker E. M., Strom R., Suomi V. E., and Veverka J. (1979b) The Galilean satellites and Jupiter: Voyager 2 imaging science results. *Science* **206**, 927–950.

Thomas P. C. (1978) *The Morphology of Phobos and Deimos*. Ph.D. thesis, Cornell Univ., Ithaca, N. Y. 272 pp.

von Englehardt W., Bertsch W., Stöffler D., Groschopf P., and Reiff W. (1967) Anzeichen für meteoritischen Ursprung des Beckens von Steinheim. *Naturwissenschaften* **54**, 198–199.

Wegener A. (1921) Die Entstehung der Mondkrater. Braunschweig, Sammlung Vieweg, Tayesfragen. *Natur. u. Tech.* **55**, 48 pp.

Wetherill G. W. (1977) The nature of the present interplanetary crater-forming projectiles. In *Impact and Explosion Cratering* (D. J. Roddy, R. O. Pepin, and R. B. Merrill, eds.), p. 613–615. Pergamon, N. Y.

Wilshire H. G., Offield T. W., Howard K. A., and Cummings D. (1972) Geology of the Sierra Madera cryptoexplosion structure, Pecos County, Texas. *U. S. Geol. Survey Prof. Paper 599–H*, 42 pp.

Wood C. A. and Andersson L. (1978) New morphometric data for fresh lunar craters. *Proc Lunar Planet. Sci. Conf. 9th*, p. 3669–3689.

Wood C. A., Head J. W., and Cintala M. J. (1978) Interior morphology of fresh martian craters: The effects of target characteristics. *Proc. Lunar Planet. Sci. Conf. 9th*, p. 3691–3709.

Proc. Lunar Planet. Sci. Conf. 11th (1980), p. 2191–2205.
Printed in the United States of America

Central peaks in mercurian craters: Comparisons to the moon

W. Hale and J. W. Head

Department of Geological Sciences, Brown University, Providence, Rhode Island 02912

Abstract—Central peak craters on Mercury were examined to determine what relationships exist between central peak morphology and morphometry and the rim diameter and substrate of the parent crater. These are then compared to analogous relationships in lunar central peak craters. The mercurian data set consists of 140 fresh central peak craters from 15–175 km in diameter, developed on cratered terrains and smooth plains from all latitudes.

A linear relationship is demonstrated between central peak diameter (D_{cp}) and rim crest diameter (D_{rc}), $D_{cp} = 0.17\,D_{rc} + 1.97$. This is statistically indistinguishable from a similar relationship defined earlier for the moon. This observation suggests that gravity is not the controlling factor in central peak formation.

A scheme previously used to characterize the morphologies of central peaks in lunar craters has been extended to Mercury. Under this system, peaks are classified by complexity (simple or complex) and geometry (linear, symmetric or arcuate). On Mercury, complexity appears unresponsive to substrate controls, while geometry is somewhat responsive, in that arcuate central peaks occur more frequently on smooth plains units. In addition, central peaks show no strong tendency to become more complex or more symmetric at larger crater diameters. This trend is also observed on the moon, and suggests that the transition from central peak craters to peak ring basins is not describable as a simple expansion of the central peak to form a ring.

INTRODUCTION

Analysis of orbital images from the moon, Mercury and Mars have demonstrated the significance of impact processes on these planets. The pervasiveness of craters and basins as surface features makes an understanding of planetary response to hypervelocity impact phenomena a prerequisite to the overall understanding of the evolution of planetary surfaces.

Earlier studies of lunar craters have revealed a pronounced change in morphology with increasing crater diameter from simple bowl-shaped features to more complex forms with flat floors, terraced walls and central structures (Baldwin, 1963; Dence, 1965; Hartmann and Wood, 1971; Smith and Sanchez, 1973; Pike, 1975; Cintala *et al.*, 1977; Wood and Andersson, 1978). At still larger diameters (140–200 km), peak rings replace peaks as the central features, defining the impact structure as a basin rather than a complex crater (Hartmann and Wood, 1971; Wood and Head, 1976; Head, 1977, 1978; Hodges and Wilhelms, 1978). This progression in morphology with increasing crater diameter is also observed on Mercury (Gault *et al.*, 1975; Cintala *et al.*, 1977) and Mars (Hartmann, 1972; Cintala *et al.*, 1976; Smith, 1976; Wood *et al.*, 1978). A key question in cratering studies continues to be the mode of formation of central structures. Studies of central peaks, the smallest and simplest central features found in impact structures, could provide clues to the formational mechanisms of the entire morphologic sequence of central features.

Although virtually all workers now associate central peak origins directly with crater formation, the constructional process remains the subject of debate. Central peaks may be formed by the interaction of impact-generated shock and rarefaction waves concentrated at the sub-impact point during the early stages of the cratering event, resulting in a dynamic rebound of target material (Baldwin, 1963; Milton and Roddy, 1972; Pike, 1977; Head, 1978). Alternatively, central peak formation has been associated with gravity-induced failure of the transient cavity during the late stages of the cratering event (Dence, 1968; Mackin, 1969; Gault *et al.*, 1968, 1975).

Lunar craters with central peaks have been characterized by Baldwin (1963), Wood (1973), Smith and Sanchez (1973), Cintala *et al.*, (1977), Smith and Hartnell (1978), Wood and Andersson (1978), Hale and Head (1979a,b). Studies of mercurian craters (Cintala *et al.*, 1977) suggest that Mercury may have target material properties similar in some ways to the lunar maria. However, Mercury has a gravitational field strength (3.68 m/sec^2) twice as great as the moon's and a modal impact velocity perhaps 1.5–2 times greater (Gault *et al.*, 1975; Hartmann, 1977). Thus comparative studies of the morphology and morphometry of central peaks on Mercury and the moon provide an opportunity to evaluate the relative contributions to central peak formation by gravity and modal impact velocity.

Several important relationships have been established between various characteristics of mercurian central peaks and their parent craters, including an increasing frequency of occurrence in craters from 15–35 km in diameter, reaching 100% occurrence at the latter value (Cintala *et al.*, 1977). The transition from craters to basins, characterized by the disappearance of the central peak and appearance of a peak ring, has been suggested to occur at diameters from 90–130 km (Wood and Head, 1976). However, more detailed information on mercurian central peaks is required to directly compare them to their lunar counterparts. The purpose of the present study is to examine systematically central peaks in fresh mercurian craters in order 1) to determine if a linear relationship exists between central peak diameter and parent crater diameter, similar to one derived earlier for the moon (Hale and Head, 1979a); 2) to characterize mercurian central peak morphology as a function of rim diameter; 3) to compare mercurian and lunar central peak morphometry and morphology in order to evaluate the role of gravity and modal impact velocity in central peak formation; and 4) to provide information on the nature of the crater to basin transition on both planets.

DATA

Mercury statistics

Craters with central peaks resolvable on the 0.15–4 km resolution (Davies and Batson, 1975) vidicon images acquired by the Mariner 10 spacecraft were examined for this study. To minimize the effects of erosion and degradation, only the morphologically freshest craters with central peaks were utilized. Fresh mercurian craters correspond to class 1 and 2 lunar craters (Arthur *et al.*, 1963; Wood *et al.*, 1977) and are characterized by sharply defined rims. The data set consists of 140 such craters ranging from 15–175 km in diameter and spanning all terrain types in the equatorial, mid-latitude and polar regions imaged by Mariner 10.

Previous workers have shown the importance of crater diameter (Pike, 1977), substrate type (Cintala *et al.*, 1977) and crater degradation state in characterizing crater groups and establishing intergroup relationships. In addition to these parameters, crater location, central peak diameter and peak morphology were determined for each case. Since central peaks commonly occur as irregular clusters, central peak diameter may be more difficult to measure than rim crest diameter. An earlier study of lunar central peaks (Hale and Head, 1979a) utilized a best fit circle enclosing all of the exposed central peak mass (Fig. 1) and this convention is extended to Mercury. This circle is fit to the base of a major peak and is not a measure of primary floor roughness which frequently covers large sections of the crater floor. Studies of the floor units of lunar craters report the presence of material which embays large blocks of debris as well as the base of the crater walls and central peaks. This material has been interpreted to be solidified impact melt (Howard and Wilshire, 1975; Hawke and Head, 1977). Since only the exposed central peak is measured in this study, the possibility of drowning of significant portions of the central peak mass by impact melt must be examined. Central peaks have steep slopes ranging from 18° to 38°, and assuming a slope angle of 20°, even 200 m of additional melt equivalent to the total melt thickness on the floor of Copernicus (Howard and Wilshire, 1975) would reduce the peak diameter by only 1 km. This would amount to a reduction in excess of 15% of the total measured diameter only in craters with central peaks less than 6 km across. Craters with such small peaks comprise <15% of the total data set. Thus no strong effect on the overall relationships between central peak diameter and crater diameter is expected as a result of impact melt drowning of the central peak base. Exceptions to this should occur only at small peak diameters.

Lunar statistics

Lunar central peak morphology and morphometry as characterized by Hale and Head (1979a) Provides the basis of comparisons to Mercury. This lunar data set consists of 175 fresh craters from nearside, polar and farside regions. It was derived from measurements made on Lunar Orbiter and Apollo images and NASA/DMA Lunar Topographic Orthophotomaps.

In order to compare the lunar and mercurian data sets, the effects of differences in resolution of Lunar Orbiter images and Mariner 10 images must be evaluated. Mariner 10 imaged the moon at a resolution of 1.3 km/pixel, a scale approximately that of most of the Mercury coverage. Schultz (1977) compared Mariner 10 and Lunar Orbiter images of the same area on the moon and concluded that significant differences in the ability to identify surface features occur only for features less

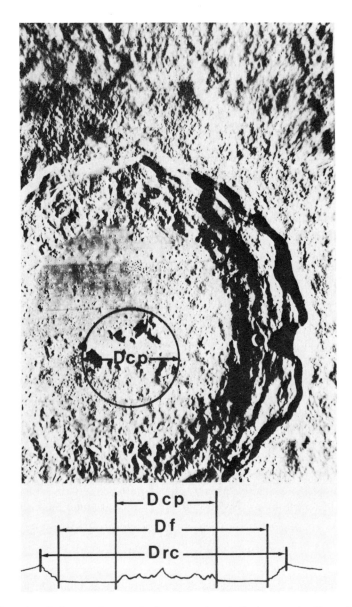

Fig. 1. Central peaks frequently occur in clusters, as here in the lunar crater Copernicus (D_{rc} = 96 km. LOIV 121H2). Central peak diameter (D_{cp}) is taken to be the diameter of a best fit circle enclosing all of the exposed central peak mass.

than 4–5 pixels (5–6 km) in diameter. Since only 15% of the craters in this study have central peaks with diameters ≤5–6 km, the effects of resolution differences should not significantly affect the comparisons of central peaks in mercurian and lunar craters.

Morphometry of mercurian central peaks

A linear relationship between central peak diameter (D_{cp}) and rim diameter (D_{rc}) in craters from 17–175 km in diameter has been demonstrated for the moon (Hale and Head, 1979a). In order to determine if a similar relationship exists for mercurian central peak craters, central peak diameters obtained in this study were plotted against the rim diameters of their parent craters (Fig. 2). Despite some degree of scatter, a linear fit to these data yields a statistically significant (correlation coeff. = 0.88) relationship between the rim diameter (D_{rc}) and peak diameter (D_{cp}) defined as $D_{cp} = 0.17\ D_{rc} + 1.97$ for the fresh crater population. This relationship includes craters developed on all terrain types and from all latitudes.

Morphology of mercurian central peaks

Central peaks in mercurian craters display a range of morphologies similar to those found in lunar craters. Single massive mountain peaks rising out of a flat floor are common, but more complicated forms such as elongate or arcuate ridges are also observed. Further, central peaks commonly occur as clusters of small peaks which may be concentrated near the crater center or elongated across the

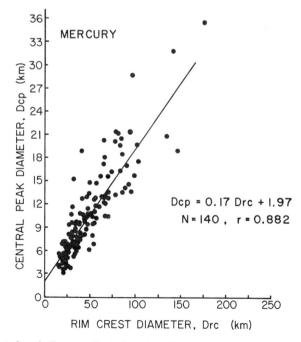

Fig. 2. Central peak diameter (D_{cp}) plotted against rim crest diameter (D_{rc}) for 140 fresh mercurian craters. The line represents a linear fit to this data defined by the equation $D_{cp} = 0.17\ D_{rc} + 197$, with a correlation coefficient, $r = 0.882$.

floor. These morphologic variations may be characterized under a system first developed to describe similar variations in lunar central peak craters (Hale and Head, 1979a). Under this scheme central peaks are classified by morphological complexity (simple or complex) and geometry (symmetric, linear or arcuate). *Simple peaks* are defined as single peaks or single coherent ridges while *complex peaks* consist of peak clusters of all geometries (Fig. 3). The same population may be divided by geometry. *Symmetric peaks* are defined as a single mountain peak or symmetrically oriented cluster, *linear peaks* consist of elongated ridges or clusters and *arcuate peaks* consist of arcuate ridges or clusters (Fig. 4).

The distribution of these peak types as a function of crater diameter is shown in Fig. 5A. A comparison of the distribution of simple versus complex craters shows that there is no strong tendency towards complex over simple peaks with increasing crater diameter. Further, a comparison of the diameter distribution of different geometric types reveals that there is no strong tendency for peaks to become more symmetric at larger diameters. Central peak morphology distributions for different substrate types can be seen in Table 1. Cratered terrain consists of several subunits including heavily cratered terrain, hilly and lineated terrain and intercrater plains (Trask and Guest, 1975) while smooth plains consist of more sparsely cratered, low relief regions cut by ridges or scarps (Trask and Guest, 1975).

Mercury: Comparisons to the moon

In order to facilitate comparisons of central peak morphometry on Mercury and the moon, central peak diameter is plotted against parent crater rim diameter for each planet's data set (Fig. 6). Linear fits to each planet's data set are also shown in Fig. 6. The data set for Mercury completely overlaps the lunar one, and the fits appear to be similar. An F-test was performed on both data sets, and the results indicate that these two relationships are in fact statistically *indistinguishable* within the real scatter of the data. Thus mercurian craters appear to have the same overall rim diameter/central peak diameter relationship as lunar central peak craters. The fall-off of central-peak diameters in both data sets at small crater diameters may reflect in part a relative reduction of exposed peak diameter due to partial drowning by impact melt.

The morphologic trends for central peak complexity and geometry with increasing crater diameter are similar on the moon and Mercury (Figs. 5A and B). Simple peaks are approximately as abundant as complex peaks on Mercury and the moon, and neither shows any strong tendency towards increasing peak complexity with increasing rim diameter. Further, there is no strong tendency towards increasingly symmetric peaks at larger rim diameters on either planet. Central peak complexity appears relatively insensitive to substrate controls on Mercury (Table 1), as on the moon. Peak geometry on Mercury also appears relatively unresponsive to substrate control, with the possible exception of arcuate peaks, which appear to be preferentially developed on smooth plains (Table 1). This is

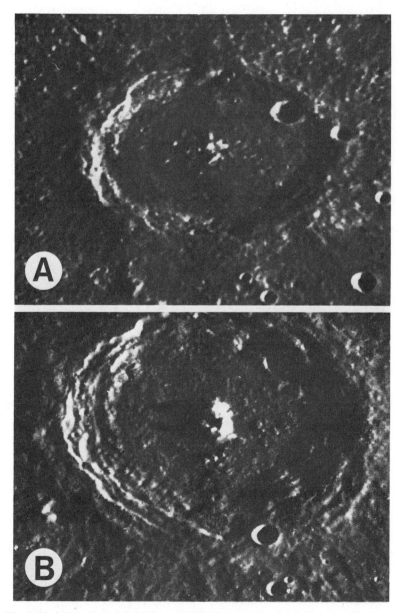

Fig. 3. Variations in morphologic complexity in mercurian central peaks. (A) *Complex* peak—Verdi, D_{rc} = 144 km (Mariner 10 FDS 0000166). (B) *Simple* peak—Brahms, D_{rc} = 100 km (Mariner 10 FDS 000080).

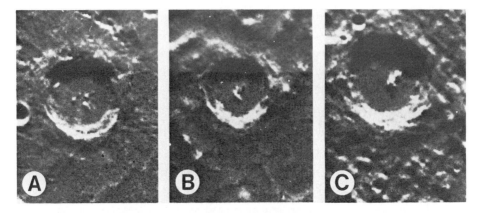

Fig. 4. Variations in mercurian central peak geometry. A) *Symmetric* peak—unnamed crater at +26°, 186°, D_{rc} = 59 km (Mariner 10 FDS 0000195). B) *Arcuate* peak—unnamed crater at +36.5°, 178.5°, D_{rc} = 50 km (Mariner 10 FDS 0000194). C) *Linear* peak—Nervo, D_{rc} = 63 km (Mariner 10 FDS 0000193).

in contrast to the lunar case, where linear central peaks are preferentially developed in highland terrains (Hale and Head, 1979; Hale, 1980).

DISCUSSION

Central peaks in fresh mercurian craters display a linear relationship between central peak diameter and rim crest diameter. This relationship is indistinguishable, within the scatter of the data, from an analogous one derived earlier for central peaks in lunar craters. These observations have important implications for the applicability of two presently proposed models for central peak formation.

One model for central peak formation involves *gravity-induced centripetal collapse* of the crater rim, leading to deep seated slumping of crater walls and the formation of terraces which converge at the crater center to uplift a central peak (Shoemaker, 1963; Quaide *et al.,* 1965; Dence, 1968; Gault *et al.,* 1975). Previous workers have attempted to use interplanetary variations in the diameter of onset and the diameter above which all fresh craters display peaks to evaluate the effects of changing gravity on the cratering process. However, Cintala *et al.* (1977) have shown that both these diameters vary with terrain type over a single planetary surface. Thus it is inappropriate to use them on a planetwide basis to directly evaluate the effects of gravity on central peak formation.

To better evaluate the gravity-driven model for peak formation, the tentative scaling relationships so far derived between kinetic energy, gravity and crater diameter must be discussed. Gault *et al.* (1975) have analytically derived exponential relationships between the kinetic energy of impact (KE), surface gravity (g), and diameter of the excavation cavity (D) for excavation diameters >1 km.

A. MERCURY

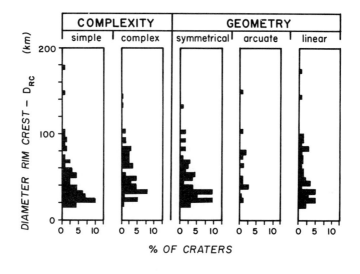

B. MOON

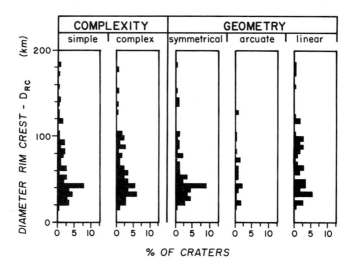

Fig. 5. Diameter distributions of morphologic complexity and geometry for central peaks in craters on A) Mercury (N = 140), B) Moon (175). For each planet, the number of complex or simple peaks in each 5 km bin was normalized to the total number of craters in that data set (N). This process was then repeated for number of peaks of each geometric type.

Table 1. Substrate variations.

	Cratered Terrain		Smooth Plains	
Complexity				
Simple	(63)	52%	(11)	57.8%
Complex	(57)	47%	(7)	36.8%
IND	(1)	.8%	(1)	5.4%
Geometry				
Symmetrical	(63)	52%	(9)	47.5%
Linear	(44)	36.5%	(6)	31.5%
Arcuate	(12)	9.9%	(4)	21%
IND	(2)	1.6%	(0)	0%

Gault and Wedekind (1977) have derived similar relationships experimentally for aluminum spheres impacting into quartz sand targets. These are, respectively,

$$D \propto g^{-1/4} F (KE)^{1/4} \tag{1}$$

and

$$D \propto \left(\frac{KE}{g} \right)^{0.181 \pm 0.009} \tag{2}$$

where F is a constant related to the partitioning of energy during impact.

Equation (2) implies that the real exponential dependence between these three factors is always less than the ¼ root scaling expected analytically (Eq. 1). This difference cannot be accounted for by the simple addition of strength effects (Gault and Wedekind, 1977).

In order to compare the differences in D due to increased g and increased modal impact velocity on Mercury relative to the moon, Eq. (1) and (2) may be converted into ratio relationships (for constant projectile mass):

$$D\left(\frac{Merc}{moon} \right) = \left(\frac{g_{moon}}{g_{Merc}} \right)^{1/4} \left(\frac{V^2 Merc}{V^2 moon} \right)^{1/4} \tag{3}$$

and

$$D\left(\frac{Merc}{moon} \right) = \left(\frac{g_{moon}}{g_{Merc}} \right)^{1/4} \left(\frac{V^2 Merc}{V^2 moon} \right)^{0.181 \pm 0.009} \tag{4}$$

respectively, where V = modal impact velocity. Utilizing values for V at Mercury of 1.5 and 2 times lunar velocities (Hartmann, 1977), Eq. (3) yields an excavation diameter ratio for Mercury relative to the moon of 1-1.15. Utilizing these same velocity values for the experimentally determined values for the exponent, Eq. (4) yields diameters for mercurian craters 1.11 times larger than those formed on the moon. These ratios suggest that the higher mercurian gravitational field, which

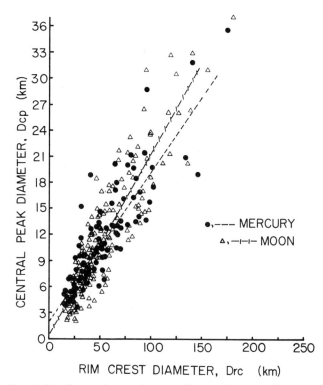

Fig. 6. Combined plot of central peak diameter (D_{cp}) to rim crest diameter (D_{rc}) for mercurian (circles) and lunar (triangles) craters. Linear fits to each planet's data are also shown: Mercury (dashed line), Moon (slashed line).

would act to restrain cavity growth, can be more than compensated for by the addition of energy to the system in the form of increased impact velocity. If modal impact velocities differ by 1.5-2 times from Mercury to the moon, such compensation will yield (for a constant projectile mass) craters with roughly similar initial diameters on both planets.

Gault *et al.* (1975) have argued that if central peaks form by gravity-driven centripetal slumping of the crater rim, central peaks on Mercury should be approximately twice as high as lunar peaks for a given crater diameter. This would be expected if central peaks were affected only by gravity ($g_{Merc} \approx 2\, g_{moon}$) while rim diameter scales according to Eq. (3) or (4). Such greater elevations, if present, would imply greater volumes of exposed uplifted material, which should be mirrored in increased central peak diameters on Mercury for the same crater diameter. Malin and Dzurisin (1976), however, report that central peak heights relative to rim diameters are similar on the moon and Mercury. Further, this study indicates that the central peak diameter/rim diameter relation on Mercury is indistinguishable from the lunar case. These observations suggest that the gravitational field strength of the planet is not the dominant force controlling central peak formation.

Important relationships have been suggested between energy and peak height (Pike, 1977), and central peak or peak ring formation (Wood, 1973; Pike, 1977; Head, 1978; Hodges and Wilhelms, 1978), as well as excavation cavity diameter. These relationships, as well as mapping of terrestrial impact and explosion craters with central uplifts (Milton and Brett, 1968; Milton *et al.*, 1972; Wilshire *et al.*, 1972; Milton and Roddy, 1972; Roddy, 1977; Dence *et al.*, 1977) have led to a model for central peak formation which involves *dynamic rebound* of the target material initiated by the interaction of shock and rarefaction waves concentrated at the sub-impact point early in the cratering event (Milton and Roddy, 1972; Pike, 1977; Ullrich *et al.*, 1977; Head, 1978; Hale and Head, 1979). In this model, peak formation processes scale to kinetic energy of impact in a fashion similar to crater diameter [see Eqs. (1) and (2)]. Such a model implies that for a given projectile mass impacting on both the moon and Mercury, roughly similar central peak and rim diameters would be expected on both planets. Any variations in projectile energy (mass or velocity) which would vary crater rim diameter would also (and proportionally) vary central peak diameter in this model. The resulting mercurian crater would be morphometrically indistinguishable from a lunar crater formed by a lower energy event. Further, the overall central peak/rim diameter relation would be expected to remain the same on Mercury and the moon, despite differences in surface gravity. The present study reports just such observations (Fig. 6). The interplanetary constancy (over the moon and Mercury) of a peak height to rim diameter relation (Malin and Dzurisin, 1976, 1978), central peak distributions (Cintala *et al.*, 1977) and a peak ring to rim diameter relation (Head, 1978) further support this model. Thus, of the presently proposed processes for central peak formation, dynamic rebound best fits current observations.

Central peak morphologies follow similar trends on both the moon and Mercury (Fig. 5A and B). In both cases, no strong trend towards increasing central peak complexity or symmetry is observed with increasing crater diameter. This is in contrast to a proposed model for the transition from central peaks to peak rings, which suggests a break-up of a symmetric central peak mass with peaks moving outward to form a ring (Hartmann and Wood, 1971; Head, 1978; Hodges and Wilhelms, 1978). In terms of the classification scheme presented here, this would predict an increasing abundance of complex, symmetric peaks at larger diameters. Such a trend is not observed on either Mercury or the moon. In addition, the slope of the central peak diameter/rim diameter relation determined here for Mercury (equivalent to that for the moon) is markedly different from the slope of a peak ring diameter/rim crest diameter relation derived earlier for both planets (Head, 1978). The two relations do not intersect in the diameter range at which transitional central peak basins occur (Fig. 7 and Hale and Head, 1979b). Thus, the transition from central peak to peak rings, and hence the transition from craters to basins may not be the result of a simple, gradual transition, but may instead reflect a more abrupt shift in formational mechanisms.

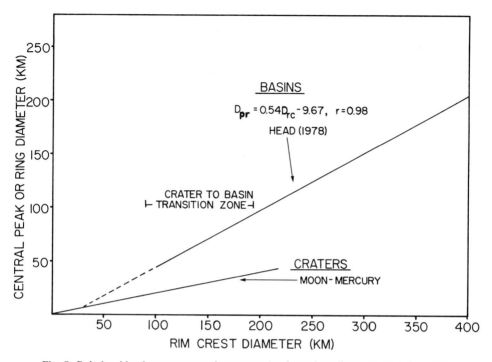

Fig. 7. Relationships between central structure (peak or ring) diameters and rim crest diameters for basins on Mercury, Mars, and the moon (Head, 1978), and for central peak craters on Mercury and the moon. The transitional crater to basin zone, from 90–200 km is also shown. Note the difference in slope of the two relationships. A projection of the basin line to small diameters yields an intersection with the crater line at a rim diameter at 35 km, far below the transitional zone.

Acknowledgments—This work was carried out under NASA Grant NGR-40-002-116 from the Office of Space Science Lunar and Planetary Programs, which is gratefully acknowledged. Thanks are extended to M. Cintala and P. Mouginis-Mark for helpful discussions and R. J. Pike, E. Smith and M. Malin for thoughtful reviews. Additional thanks go to N. Christy for aid in preparation of the manuscript.

REFERENCES

Arthur D. W. G., Agnieray A. P., Horvath R. A., Wood C. A., and Chapman C. R. (1963) The system of lunar craters, Quadrant I. *Commun. Lunar and Planetary Lab.*, Univ. Arizona **2**, 71–78.

Baldwin R. B. (1963) *The Measure of the Moon*. Univ. Chicago Press, Chicago. 448 pp.

Cintala M. J., Head J. W., and Mutch T. A. (1976) Characteristics of fresh martian craters as a function of diameter: Comparison with the moon and Mercury. *Geophys. Res. Lett.* **3**, 117–120.

Cintala M. J., Wood C. A., and Head J. W. (1977) The effects of target characteristics on fresh crater morphology: Preliminary results for the moon and Mercury. *Proc. Lunar Sci. Conf. 8th*, p. 3409–3425.

Davies M. W. and Batson R. M. (1975) Surface coordinates and cartography of Mercury. *J. Geophys. Res.* **80**, 2417–2430.

Dence M. R. (1965) The extraterrestrial origin of Canadian craters. *Ann. N.Y. Acad. Sci.* **123**, 941–969.

Dence M. R. (1968) Shock zoning at Canadian craters: Petrography and structural implications. In *Shock Metamorphism of Natural Materials* (B. M French and N. M. Short, eds.), p. 169–184. Mono, Baltimore.

Dence M. R., Grieve R. A. F., and Robertson P. B. (1977) Terrestrial impact structures principal characteristics and energy considerations. In *Impact and Explosion Cratering* (D. J. Roddy, R. O. Pepin, and R B. Merrill, eds.), p. 247–275. Pergamon, N.Y.

Gault D. E., Quaide W. L., and Oberbeck V. R. (1968) Impact mechanics and structures. In *Shock Metamorphism of Natural Materials* (B. M. French and N. M. Short, eds.), p. 87–99. Mono, Baltimore.

Gault D. E., Guest J. E., Murray J. B., Dzurisin D., and Malin M. C. (1975) Some comparisons of impact craters on Mercury and the moon. *J. Geophys. Res.* **80**, 2444–2460.

Gault D. E. and Wedekind J. A. (1977) Experimental hypervelocity impact into quartz sand-II, Effects of gravitational acceleration. In *Impact and Explosion Cratering* (D. J. Roddy, R. O. Pepin, and R. B. Merrill, eds.), p. 1234–1244. Pergamon, N.Y.

Hale W. S. and Head J. W. (1979a) Central peaks in lunar craters: Morphology and morphometry. *Proc. Lunar Planet. Sci. Conf. 10th*, p. 2623–2633.

Hale W. S. and Head J. W. (1979b) Lunar central peak basins: Morphology and morphometry in the crater to basin transition zone. NASA TM 80339, p. 160–163.

Hale W. S. (1980) Orientations of linear central peaks in lunar craters: Implications for regional structural trends. In *Proc. Conf. Lunar Highlands Crust.* (H. J. Papike and R. B. Merrill, eds.), p. 197–209. Pergamon, N.Y.

Hartmann W. K. and Wood C. A. (1971) Moon: Origin and evolution of multi-ringed basins. *The Moon* **3**, 2–78.

Hartmann W. K. (1972) Interplanet variations in crater morphology. *Icarus* **18**, 708–713.

Hartmann W. K. (1977) Relative crater production rates on planets. *Icarus* **31**, 260–276.

Hawke B. R. and Head J. W. (1977) Impact melt on lunar crater rims. In *Impact and Explosion Cratering* (D. J. Roddy, R. O. Pepin, and R. B. Merrill, eds.), p. 815–841. Pergamon, N.Y.

Head J. W. (1977) Origin of outer rings in lunar multi-ringed basins: Evidence from morphology and ring spacing. In *Impact and Explosion Cratering* (D. J. Roddy, R. O. Pepin, and R. B. Merrill, eds.), p. 583–573. Pergamon, N.Y.

Head J. W. (1978) Origin of central peaks and peak rings (abstract). In *Lunar and Planetary Science X*, p. 485–487. Lunar and Planetary Institute, Houston.

Hodges C A. and Wilhelms D. (1978) Formation of lunar basin rings. *Icarus* **34**, 294–323.

Howard K. A. and Wilshire H. G. (1975) Flows of impact melt at lunar craters. *J. Res. U. S. Geol. Surv.* **3**, 237–251.

Malin M. C. and Dzurisin D. (1976) Modification of fresh crater landforms: Evidence from Mercury and the moon (abstract). In *Papers Presented to the Conference on Comparisons of Mercury and the Moon*, p. 21. The Lunar Science Institute, Houston.

Malin M C. and Dzurisin D. (1978) Modification of fresh crater landforms: Evidence from the Moon and Mercury. *J. Geophys. Res.* **83**, 233–243.

Mackin J. H. (1969) Origin of lunar maria. *Bull. Geol. Soc. Amer.* **80**, 734–748.

Milton D. J. and Brett R. (1968) Gosses Bluff astrobleme, Australia—the central uplift region (abstract). *Geol. Soc. Amer. 64th Ann. Meeting*, p. 82. Tucson.

Milton D. J., Barlow B. C., Brett R., Brown A. R., Glikson A. Y., Manwaring F. A., Moss F. J., Sedmik E. C. E., Van Son J., and Young G. A. (1972) Gosses Bluff impact structure, Australia. *Science* **175**, 1199–1207.

Milton D. J. and Roddy D. J. (1972) Displacements within impact craters. *International Geological Congress, XXIV Session*, Planetology, p. 119–124.

Pike R. J. (1975) Size-morphology relations of lunar craters: Discussion. *Mod. Geol.* **5**, 169–173.

Pike R. J. (1977) Size-dependence in the shape of fresh impact craters on the Moon. In *Impact and Explosion Cratering* (D. J. Roddy, R. O. Pepin, and R. B. Merrill, eds.), p. 489–509. Pergamon, N.Y.

Quaide W. L., Gault D. E., and Schmidt R. A. (1965) Gravitative effects on lunar impact structures. *Ann. N.Y. Acad. Sci.* **123**, 563–572.

Roddy D. J. (1977) Large-scale impact and explosion craters: Comparisons of morphological and structural analogs. In *Impact and Explosion Cratering* (D. J. Roddy, R. O. Pepin, and R. B. Merrill, eds.), p. 185–246. Pergamon, N.Y.

Schultz P. H. (1977) Endogenic modification of impact craters on Mercury. *Phys. Earth Planet. Inter.* **15**, 202–219.

Shoemaker E. M. (1963) Preliminary analysis of the fine structure of the lunar surface. NASA Tech. Rep. 32–700, p. 75–134.

Smith E. I. and Hartnell J. A. (1978) Crater size-shape profiles for the Moon and Mercury: Terrain effects and interplanetary comparisons. *Moon and Planets* **19**, 479–511.

Smith E. I. and Sanchez A. G. (1973) Fresh lunar craters: Morphology as a function of diameter, a possible criterion for crater origin. *Mod. Geol.* **4**, 51–59.

Smith E. I. (1976) Comparison of crater morphology-size relationship for Mars, Moon, and Mercury. *Icarus* **28**, 543–550.

Trask N. J. and Guest J. E. (1975) Preliminary geologic terrain map of Mercury. *J. Geophys. Res.* **80**, 2461–2477.

Ullrich G. W., Roddy D. J., and Simmons G. (1977) Numerical simulations of a 20-ton TNT detonation on the earth's surface and implications concerning the mechanics of central uplift formation. In *Impact and Explosion Cratering* (D. J. Roddy, R. O. Pepin, and R. B. Merrill, eds.), p. 959–982. Pergamon, N.Y.

Wilshire H. G., Offield T. W., Howard K. A., and Cummings D. (1972) Geology of the Sierra Madera cryptoexplosion structure, Pecos County, Texas. *U. S. Geol. Survey Prof. Paper 599–H*. 42 pp.

Wood C. A. (1973) Central peak heights and crater origins. *Icarus* **20**, 503–506.

Wood C. A. and Andersson L. (1978) New morphometric data for fresh lunar craters. *Proc. Lunar Planet. Sci. Conf. 9th*, p. 3669–2689.

Wood C. A. and Head J. W. (1976) Comparison of impact basins on Mercury, Mars and the moon. *Proc. Lunar Sci. Conf. 7th*, p. 3629–3651.

Wood C. A., Head J. W., and Cintala M. J. (1978) Interior morphology of fresh martian craters: The effects of target characteristics. *Proc. Lunar Planet. Sci. Conf. 9th*, p. 3691–3709.

Proc. Lunar Planet. Sci. Conf. 11th (1980), p. 2207–2219.
Printed in the United States of America

Variations in interior morphology of 15-20 km lunar craters: Implications for a major subsurface discontinuity

René A. De Hon

Department of Geosciences, Northeast Louisiana University, Monroe, Louisiana 71209

Abstract—Craters vary in morphology as a function of crater diameter, age, and mode of origin. This study concentrates on the morphology of young lunar impact craters within a limited size range. Elimination of morphologic variations generally attributed to crater size or age leaves a small population which should nearly reflect the varying properties of the lunar substrate. The sample consists of 17 craters 15-20 km in diameter with both simple and complex morphologies.

While depth/diameter ratios do not obviously differ between mare and highland subsets, apparent depth, rim height, and profile data do differ distinctly. Highland craters tend to be deep, simple, and bowl-shaped. Mare craters tend to be shallow and flat-floored. Rim heights of complex mare craters are typically greater than those of simple craters.

Differences of highland and mare crater morphologies are attributed to variations in the thickness of the lunar megaregolith. Highland craters in this size range do not penetrate the megaregolith. The depth and morphology of complex craters are controlled by the discontinuity at the transition from highly brecciated megaregolith to more coherent crystalline material of the upper crust.

INTRODUCTION

Crater morphology provides important insights into the processes that shape planetary surfaces, the geologic history of planets, and the nature of subsurface materials. Crater morphology often may be viewed as a function of one of the following: crater diameter, crater age, crater origin, property of the impacting projectile, or property of the target material. Crater morphometry is generally described as a function of crater diameter. A statistical treatment, which is required by the large number of craters on lunar and planetary surfaces, is valuable in establishing the trends in crater morphology and morphometry.

The transition from simple, bowl-shaped craters to complex, flat-floored craters as crater diameter increases is well documented on both the earth (Dence, 1972) and on the moon (Baldwin, 1963; Pike, 1967, 1976, 1977a; Hörz and Ronca, 1971; Smith and Sanchez, 1973; Wood and Andersson, 1978). Studies of this type tend to focus on youthful-appearing craters over a wide range of crater diameters. Youthful in this sense is defined by either observed superposition relationships or by morphologic criteria (Pohn and Offield, 1970). The transition from simple to complex lunar craters (Quaide *et al.*, 1965) at approximately 15 km diameter is attributed to one or more of the following mechanisms:

1. Modification of a bowl-shaped transient cavity during terminal stages of the cratering event or by isostatic compensation (Gault *et al.,* 1968; Pike, 1977a; Settle and Head, 1979).
2. Changes in style of excavation due to physical characteristics of the target (Quaide and Oberbeck, 1968; Head, 1976; Cintala *et al.,* 1977; Smith and Hartnell, 1978).
3. Changes in style of excavation due to gravity-related factors which vary from planet to planet (Gault *et al.,* 1975; Pike *et al.,* 1980).
4. Changes in style of excavation due to physical characteristics of the projectile (Quaide *et al.,* 1965; Roddy, 1980).

Recognition of degradational trends in crater morphology is usually the result of comparing craters within restricted size ranges but of widely varying ages (Baldwin, 1963; Pohn and Offield, 1970; Head 1975; Wood *et al.,* 1977; Wood, 1979). Changes in morphology are observed as a function of increasing duration of exposure to processes of degradation. In regions where convenient stratigraphic horizons are scarce or where low resolution does not permit observation of superposition, degradational trends are employed as a basis for crater age assignments.

Further variations in crater morphology may be attributed to modification by secondary processes (such as volcanism) or to formation of crater-form structures by non-impact processes (such as volcanism or collapse). Studies of alternate modes of origin tend to focus on a small number of craters with one or more distinctive features that deviate from normal crater morphologic trends (De Hon, 1971; Smith and Sanchez, 1973; Pike, 1974; Schultz, 1976; Wood, 1978; Head and Wilson, 1979).

The transition from simple to complex craters marks a major variation in crater morphology that is incompletely understood. For this reason a restricted study was designed to eliminate, as much as possible, variations in morphology that are a function of age, diameter, or alternate modes of origin. Even with these factors reduced, craters within the same size and age class exhibit considerable variations in morphology (Fig. 1). The diameter range of craters used in this study is chosen to coincide with the transition from simple to complex craters.

METHOD

This study concentrates on the morphology and morphometry of craters within a limited size range (diameters of 15–20 km) and a limited age range (Copernican and Eratosthenian). Young craters (Table 1) are used in order to avoid the effects of crater degradation. Superposition relationships demonstrate that most of the craters in the study are Copernican or Eratosthenian in age. Critera for freshness based on morphometric properties only were not used in crater selection. Craters on rugged highland terrain are not easily assigned ages; therefore, a few highland pre-Eratosthenian craters may be included in the data set. Inclusion of a few craters that may be pre-Eratosthenian in age is considered to be less detrimental to the results than the exclusion of post-Imbrian craters that may not fit idealized models of fresh crater morphology. The restricted diameter range limits gross size

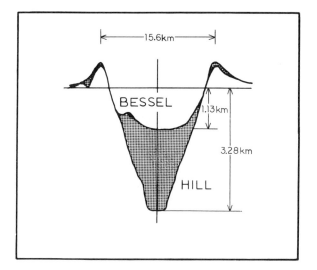

Fig. 1. Superimposed cross-sections of lunar craters Bessel and Hill. Bessel is a complex mare crater in this size range. Hill is a simple highland crater. Note the small difference in rim heights and the large difference in depth. Vertical exaggeration is 5 times normal. Bessel is enlarged slightly more (<1.5%) to match the apparent diameter of Hill.

effects and accentuates those properties associated with the simple to complex crater transition. Finally, only those craters within the current LTO map coverage are used to assure consistency of quantitative data. Two or three profiles were constructed for all craters selected, basic morphometric data were collected, and local surface topography was determined. The initial data were searched for correlations with topography and geologic province.

The crater parameters measured in this study are essentially the same as those defined by Pike (1976). The diameter (D_r) is the diameter of the best-fit circle matched to the rim crest. Minor rim irregularities due to rim slumping do not affect the measurement. Apparent diameter (D_a) is determined from the best-fit circle matched to the internal contour that approximates the average exterior surface elevation. Depth (R_i) is the rim-crest elevation less the floor elevation. Rim height (R_e) is the elevation of the rim-crest less the average exterior surface elevation. Apparent depth (R_a) is the rim to floor depth (R_i) minus the rim height (R_e). Elevations of the exterior surface, rim-crest, and floor are averages of four to eight determinations.

MORPHOLOGIC AND MORPHOMETRIC COMPARISONS

Seventeen craters were identified within the age and location criteria of the investigation (Fig. 2). Basic morphometric data are collected in Table 1. Although the diameter range (D_r) is quite restrictive, the depth (R_i) varies by a factor greater than two; apparent depth (R_a) and rim height (R_e) vary by almost three. Five craters occur on mare surfaces, and 12 craters occur on highland terrains. The highland craters are further subdivided into those formed on relatively level surfaces and those located on rugged slopes.

Table 1. Individual crater parameters.

Crater	Rim-crest diam. (km)	Depth (km)	Apparent diam. (km)	Apparent depth (km)	Rim height (km)
Maria					
Bessel	15.4	1.88	12.6	1.13	0.75
Peirce	18.8	2.56	15.7	1.94	0.62
Dawes	16.6	2.41	15.0	1.71	0.70
Diophantus	17.8	3.05	15.4	2.55	0.50
Carmichael	20.4	3.75	17.4	3.10	0.65
Highlands					
Respighi*	17.8	2.29	13.4	1.99	0.30
Conon*	20.9	2.88	18.9	2.50	0.38
Unnamed†	16.2	2.35	13.9	2.19	0.16
Daly*	15.7	2.41	12.8	2.01	0.40
Geissler	17.8	2.92	14.8	2.42	0.50
Black*	19.4	3.21	17.7	3.04	0.18
Liouville	16.6	2.83	14.4	2.41	0.43
Unnamed‡	15.4	2.67	13.4	2.33	0.34
Glaisher	16.0	2.87	12.5	2.22	0.65
Dario*	18.2	3.46	15.5	3.06	0.41
Isidorus D	15.3	3.18	13.2	2.77	0.41
Hill	15.6	3.93	13.2	3.28	0.65

* Crater occurs on irregular terrain.
† 83C3
‡ 65D3

Highland craters

Highland craters generally have simple, bowl-shaped morphology with concave floors or narrow, flat-floors (Fig. 3). Shallow craters with broad, flat-floors are not common. The extremes of crater morphologies are typified by craters Respighi and Hill (Fig. 4). Respighi is of uncertain age and occurs on extremely rugged terrain between the Crisium and Smythii basins (Fig. 2). Respighi exemplifies the influence of relatively steep slope and rugged terrain on the form of the crater. Typically, the profiles of these craters are asymmetrical and have widely varying rim elevations relative to the local surface. It is difficult to assign an exact rim diameter to such craters as the increased rim height on the uphill side of the crater displaces the rim crest and increases the radius of the crater. In like manner, the rim height cannot be measured from a simple level datum plane. Rim height (R_e) values in Table 1 are derived from elevation differences between the rim crest and the trend of the pre-crater surface. The floor of Respighi is flat and is concordant with nearby patches of plains; hence, the crater might be partially filled with later plains-forming materials. Hill is located east of Mare Serenitatis (Fig. 2) o the ejecta blanket of the basin. Hill exhibits characteristics typical of

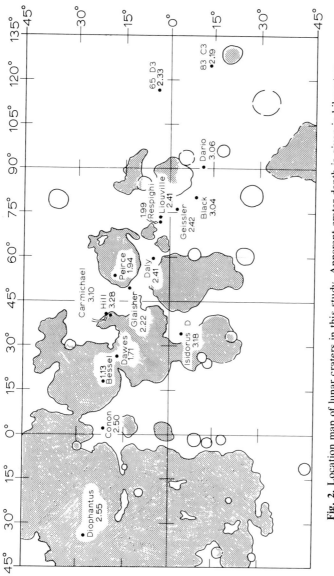

Fig. 2. Location map of lunar craters in this study. Apparent crater depth is given in kilometers.

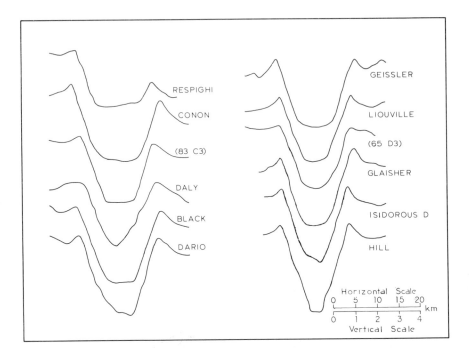

Fig. 3. Profiles of highland craters. Craters Respighi–Dario occur on steeply sloping highland surfaces, Geissler–Hill on level terrain. Vertical exaggeration is 5 times normal.

highland craters in this size range formed on relatively level terrain. Such craters are symmetrical in profile (Fig. 4), and rim elevations are nearly constant along the rim crest. Rim slumping is absent or inconspicuous. Slumps, if present in highland craters of this size, are usually thin veneers of debris that do not form well-defined terraces. The floors are poorly-defined, concave, or small and flat. As a group, highland craters exhibit significantly lower rim heights than mare craters (Table 1). The deepest crater, Hill, has a funnel-shaped profile (Figs. 1, 3, and 4).

Mare craters

Craters formed on mare surfaces (Fig. 5) exhibit the greatest range of basic morphology. Primarily as a result of the relatively flat surface on which they form, mare craters have uniform rim elevations. All mare craters treated in this study are flat-floored, but the floors may contain small hummocks or central mounds (Fig. 4). Bessel is shallow, has a broad flat-floor, and displays a well-developed wall slump. Carmichael is relatively deep, has no slumps, and displays irregular floor development. Intermediate craters Dawes and Pierce (Fig. 5) have prominent wall slumps which create terraces. Pierce and Diophantus (Fig. 5) have small central peaks or mounds. In contrast to highland craters, mare craters are characterized by consistently high rim-height values (Table 1).

Comparisons and contrast

A comparison of the depth/diameter ratios (R_i/D_r) fails to display any significant differences between mare and highland craters (Fig. 6 and Table 2). Both crater groups are superimposed and exhibit a wide range of values which overlap the trends of both simple and complex craters. When apparent depth/diameter ratios (R_a/D_r) are compared, the two groups begin to form separate clusters. Two deep mare craters, Carmichael and Diophantus, overlap the cluster of highland craters. The most distinctive separation of crater groups is seen in plots of rim height and diameter. The mare craters exhibit consistently greater rim heights which cluster on the trend defined by Pike (1977). On the other hand, highland craters display a wide range of rim heights and all craters but one plot well below normal trends (Fig. 6). Depth-to-diameter ratios of highland and mare craters overlap because, for this data set, there is an inverse relationship between rim height and apparent depth. As apparent depth decreases, rim height increases.

Apparent crater diameter/rim-crest diameter ratios (D_a/D_r) remain constant regardless of the terrain (Table 2). Both mare and highland craters have D_a/D_r values near 0.85 (Pike, 1977b). This suggests that significant enlargement of mare craters by slumping is not important. If complex mare craters formed by modi-

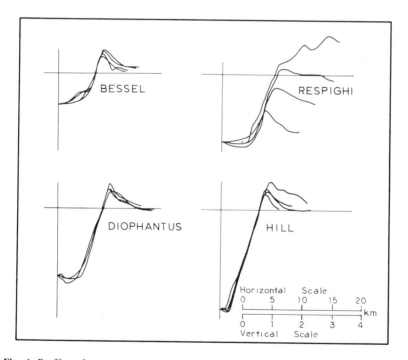

Fig. 4. Profiles of representative craters. Several profiles for each crater are superimposed for comparison of crater symmetry. Respighi, age uncertain, exemplifies the influence of steep slope and rugged terrain on the form of the crater. Vertical exaggeration is 5 times normal.

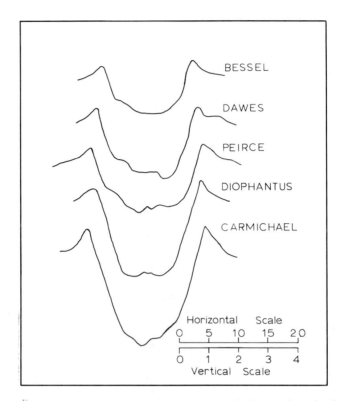

Fig. 5. Profiles of mare craters. Wall slumping or floor relief is prominent in all craters. Deepest crater, Carmichael, occurs on thin mare basalt overlying rim ejecta of the Serenitatis basin. Vertical exaggeration is 5 times normal.

fication of originally simple craters, then the D_a/D_r ratio would change by rim-wall slumping (Pike, 1977b). While wall terracing is common in mare craters, it is sufficiently restricted to limited portions of the crater wall so as not to affect the diameter measurements. The diameters are close to their original (pre-slump) dimensions.

DISCUSSION AND CONCLUSIONS

On the basis of both morphology and morphometry it is clear that there are significant differences between 15–20 km diameter lunar craters formed on mare and highland surfaces. Highland craters tend to be simple, deep, and low-rimmed. Mare craters exhibit a wide range of depths, but as a group, all are flat-floored, high-rimmed, and wall slumps and central mounds are common. While some previous data (Cintala *et al.* 1977) and new data (Pike, 1980) indicate that differences exist between highland and mare craters, the differences are not obvious in other data sets. Several factors are responsible for past failures to appreciate these differences fully. Simple depth-to-diameter ratios in early crater studies, the most common function derived, do not portray these differences (Pike, 1974). Mare craters are easier to measure, since there is a level datum for reference, and thus pre-Apollo data sets tend to be weighted with more mare craters. Finally, because the maria have a low crater density, complex craters on the mare are rare, but smaller simple craters are not.

A dependence of crater morphology on terrain is an inescapable conclusion. The variations between craters are not time dependent, as the craters in this study are not degraded. The variations are not controlled by properties of the projectile, as crater morphologic types in this size range are not randomly distributed. The variations in crater morphology are not the result of isostatic adjustments, as positive gravity anomalies are detected only with craters much

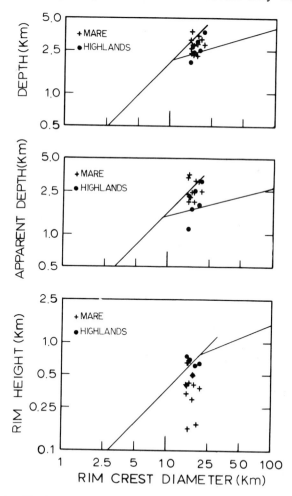

Fig. 6. Relationships between crater diameter and several vertical relief measurements. Regression lines after Pike (1974, 1977a).

Table 2. Statistical comparison of crater parameters by terrain type.

	Maria N=5			Highlands N=12		
	Mean	Range	Std. error	Mean	Range	Std. error
Rim diam. (D_r)	17.8	5.0	0.86	17.1	5.6	0.51
Apparent diam. (D_a)	15.2	4.8	0.77	14.5	6.4	0.58
Crater depth (R_i)	2.73	1.87n	0.51	2.92	1.64	0.14
Apparent depth (R_a)	2.08	1.97	0.34	2.55	1.29	0.13
Rim height (R_e)	0.64	0.25	0.42	0.40	0.49	0.04
R_i/D_r	0.15	0.06	0.01	0.17	0.12	0.01
R_a/D_r	0.12	0.08	0.02	0.15	0.10	0.01
R_e/D_r	0.04	0.01	0.00	0.02	0.03	0.00
D_a/D_r	0.86	0.08	0.02	0.85	0.16	0.01

larger than those employed in this study (Phillips *et al.*, 1976). Complex craters in this study are not formed by modification of simple bowl-shaped cavities simply by rim slumping, as the rim heights of complex craters would be significantly lowered by slumping (Cintala, 1979). Instead, for this size range, flat-floored craters have rim heights significantly greater than simple, bowl-shaped craters. The dichotomy of crater types within this size range suggests that target properties are the major control of crater morphology.

Early in lunar history the surface was completely blanketed by intercalated ejecta from basins and craters (Howard, 1974). The resulting fragmental layer of ejecta and autochthonous breccias now comprises a megaregolith (Hartmann, 1973) which overlies more coherent materials of the lunar crust. Estimates of average thickness of the megaregolith based on theoretical considerations range from 1 to 3 km (Short and Forman, 1972; Hartmann, 1973; Hörz *et al.*, 1976; Aggarwal and Oberbeck, 1979; Duplechin and DeHon, 1980). Maximum thicknesses near basin rims may exceed 5–8 km. Head (1976) arrived at a value of 2–3 km for the highlands based on the transition of simple craters to complex craters. Thompson *et al.* (1979) estimate a 2 km minimum thickness for the lunar highland breccias based on the absence of large blocks in the rims of highland craters. Golombek (1979) concluded that the widths of lunar grabens are controlled by a mechanical discontinuity between the megaregolith and basement materials at depths of 1 to 3.9 km. The thickness of intrabasin megaregolith depends on the amount of fall-back material, proximity to inner basin rings, and the length of time the basin floor is exposed to cratering. Young basins will have appreciably less magaregolith than older basins. A shallower depth of origin for rilles within basins is consistent with a thinner local megaregolith (Golombek, 1979).

Mare craters, by nature of their location, excavate a surface layer of basalt. The basalt layer is responsible for blocky rim materials as identified in infrared and radar studies (Thompson *et al.*, 1979). These basalts are 250–500 m thick at most craters used in this study (De Hon and Waskom, 1976; De Hon, 1979). The basalt thickness is inconsequential in the total depth of the mare craters; hence, it is unimportant in controlling the morphology of these craters. Deep craters on mare basalt, such as Carmichael and Diophantus which are transitional in character, are formed in localities characterized by thick megaregolith. Carmichael is located on thin basalts mantling Serenitatis basin rim materials. Diophantus is located on basalt that veneers thick ejecta deposits of the Imbrium basin inner ring.

The transition from highly brecciated megaregolith to more coherent crystalline materials of the upper crust comprises a major physical discontinuity which appears to control the morphology of craters (Head, 1976; Cintala *et al.,* 1977; Hodges and Wilhelms, 1978; De Hon, 1978; Aggarwal and Oberbeck, 1979). Highland craters 15–20 km in diameter excavate material from depths of at least as deep as the crater floors which are 2 to 3.2 km below the surface (Fig. 2 and Table 1). A simple, bowl-shaped morphology suggests that the target material is of uniform properties with depth. The megaregolith is thicker in the highlands than the depth of excavation of the 15–20 km diameter craters. Mare craters are 1.1 to 3.1 km deep (Fig. 2 and Table 1). The flat floor suggests that the growth of the crater cavity during formation is limited by the megaregolith-crystalline rock discontinuity which is not reached in highland craters of the same diameter. The greater rim height of complex craters might be caused by steeper ejection angles as excavation reaches the discontinuity.

Crater size does not necessarily determine cratering mechanics, although diameter may be convenient morphometric property to measure. The transition from simple craters to complex craters near 15 km diameter is only a statistical mean from a large data set, and both types of craters co-exist in the 15–20 km range. Crater morphology is primarily controlled by target characteristics at depth, specifically the depth to a major discontinuity. Inasmuch as the thickness of the megaregolith is variable, the transition from simple to complex craters is triggered at variable depths, and thus appears in a range of crater sizes.

Acknowledgments—This research is supported by NASA Grant NGR 7534 at Northeast Louisiana University. The valuable assistance of Dale Olson, Stephen Cowgill, and James Sargent ·is acknowledged. Thanks are due Jesse Carter for help in drafting some of the illustrations. This work profited greatly from the comments and criticisms of M. J. Cintala, C. A. Wood, and R. J. Pike. Special thanks to M. Denise Sharp for preparation of the manuscript.

REFERENCES

Aggarwal H. R. and Oberbeck V. R. (1979) Monte Carlo simulation of lunar megaregolith and implications. *Proc. Lunar Planet. Sci. Conf. 10th,* p. 2689–2705.
Baldwin R. B. (1963) *The measure of the moon.* Univ. Chicago Press, Chicago. 488 pp.
Cintala M. J. (1979) Mercury crater rim heights and some interplanetary comparisons. *Proc. Lunar Planet. Sci. Conf. 10th,* p. 2635–2650.
Cintala M. J., Wood C. A., and Head J. W. (1977) The effects of target characteristics on fresh crater

morphology: Preliminary results for the moon and Mercury. *Proc. Lunar Sci. Conf. 8th*, p. 3409–3425.

De Hon R. A. (1971) Cauldron subsidence in lunar craters Ritter and Sabine. *J. Geophys. Res.* **76**, 5712–5718.

De Hon R. A. (1978) Stratigraphy and crater morphology (abstract). In *Lunar and Planetary Science IX*, p. 232–234. Lunar and Planetary Institute, Houston.

De Hon R. A. (1979) Thickness of the western mare basalts. *Proc. Lunar Planet. Sci. Conf. 10th*, p. 2935–2955.

De Hon R. A. and Waskom J. D. (1976) Geologic structure of the eastern mare basins. *Proc. Lunar Sci. Conf. 7th*, p. 2729–2746.

Dence M. R. (1972) Nature and significance of terrestrial impact structures. *Proc. International Geological Congress 24th*, p. 77–89. Harpell, Quebec.

Duplechin M. C. and De Hon R. A. (1980) Thickness of lunar farside basin ejecta (abstract). In *Lunar and Planetary Science XI*, p. 247–249. Lunar and Planetary Institute, Houston.

Gault D. E., Guest J. E., Murray J. B., Dzurisin D., and Malin M. (1975) Some comparisons of impact craters on Mercury and the moon. *J. Geophys. Res.* **80**, 2444–2460.

Gault D. E., Quaide W. L., and Oberbeck V. R. (1968) Impact cratering mechanics and structures. In *Shock Metamorphism of Natural Materials* (B. M. French and N. M. Short, eds.), p. 87–99. Mono, Baltimore.

Golombek M. P. (1979) Structural analysis of lunar grabens and the shallow crustal structure of the Moon. *J. Geophys. Res.* **84**, 4657–4666.

Hartmann W. K. (1973) Ancient lunar mega-regolith and subsurface structure. *Icarus* **18**, 634–636.

Head J. W. (1975) Processes of lunar crater degradation: changes in style with geologic time. *The Moon* **12**, 299–329.

Head J. W. (1976) The significance of substrate characteristics determining morphology and morphometry of lunar craters. *Proc. Lunar Sci. Conf. 7th*, p. 2913–2929.

Head J. W. and Wilson L. (1979) Alphonsus-type dark-halo craters: Morphology, morphometry, and eruption conditions. *Proc. Lunar Planet. Sci. Conf. 10th*, p. 2861–2898.

Hodges C. A. and Wilhelms D. E. (1978) Formation of lunar basin rings. *Icarus* **34**, 294–323.

Hörz F., Gibbons R. V., Hill R. E., and Gault D. E. (1976) Large scale cratering of the lunar highlands: Some Monte Carlo model considerations. *Proc. Lunar Sci. Conf. 7th*, p. 2931–2945.

Hörz F. and Ronca L. B. (1971) A classification of impact craters. *Mod. Geol.* **2**, 65–69.

Howard K. A. (1974) Fresh lunar impact craters: Review of variations with size. *Proc. Lunar Sci. Conf. 5th*, p. 61–69.

Phillips R. J., Abbot E. A., Conel J. E., Johnson T. V., and Saunders R. S. (1976) Isostatic adjustment in lunar history (abstract). In *Lunar Science VII*, p. 688–690. The Lunar Science Institute, Houston.

Pike R. J. (1967) Schroeter's Rule and the modification of lunar impact morphology. *J. Geophys. Res.* **72**, 2099–2106.

Pike R. J. (1974) Depth/diameter relations of fresh lunar craters: Revision from spacecraft data. *Geophys. Res. Lett.* **1**, 265–271.

Pike R. J. (1976) Crater dimensions from Apollo and supplemental sources. *The Moon* **15**, 463–477.

Pike R. J. (1977a) Size-dependence in the shape of fresh impact craters on the moon. In *Impact and Explosion Cratering* (D. J. Roddy, R. O. Pepin, and R. B. Merrill, eds.), p. 489–509. Pergamon, N.Y.

Pike R. J. (1977b) Apparent depth/apparent diameter relation for lunar craters. *Proc. Lunar Sci. Conf. 8th*, p. 3427–3436.

Pike R. J. (1980) Control of crater morphology by gravity and target type: Mars, Earth, Moon. *Proc. Lunar Planet. Sci. Conf. 11th*. This volume.

Pike R. J., Roddy D. J., and Arthur D. W. G. (1980) Gravity and target strength: Controls on the morphologic transition from simple to complex craters. *Reports of the Planetary Geology Program 1979–1980*, p. 108–110. NASA TM 81776.

Pohn H. A. and Offield T. W. (1970) Lunar crater morphology and relative age determination of lunar geologic units, Part I, classification. *U.S. Geol. Survey Prof. Paper 700–C*, p. C–153–162.

Quaide W. L., Gault D. E., and Smith R. A. (1965) Gravitative effects on lunar impact structures. *Ann. N.Y. Acad. Sci.* **123,** 563–572.

Quaide W. L. and Oberbeck V. R. (1968) Thickness determinations of the lunar surface layer from lunar impact craters. *J. Geophys. Res.* **73,** 5247–5270.

Roddy D. J. (1980) Theoretical and observational support for formation of flat-floored central uplift craters by low-density impacting bodies. In *Lunar and Planetary Science XI,* p. 943–945. Lunar and Planetary Institute, Houston.

Schultz P. H. (1976) *Moon Morphology.* Univ. Texas Press, Austin. 626 pp.

Settle M. and Head J. W. (1979) The role of rim slumping in the modification of lunar impact craters. *J. Geophys. Res.* **84,** 3081–3096.

Short N. M. and Forman M. L. (1972) Thickness of impact crater ejecta on the lunar surface. *Mod. Geol.* **3,** 69–91.

Smith E. I. and Hartnell J. A. (1978) Crater size–shape profiles for the moon and Mercury: Terrain effects and interplanetary comparisons. *The Moon and Planets* **19,** 479–511.

Smith E. I. and Sanchez A. G. (1973) Fresh lunar craters: Morphology as a function of diameter, a possible criterion for crater origin. *Mod. Geol.* **4,** 51–159.

Thompson T. W., Roberts W. J., Hartmann W. K., Shorthill R. W., and Zisk S. H. (1979) Blocky craters: Implications about the lunar megaregolith. *The Moon and Planets* **21,** 319–342.

Wood C. A. (1978) Lunar concentric craters (abstract). In *Lunar and Planetary Science IX,* p. 1264–1266. Lunar and Planetary Institute, Houston.

Wood C. A. (1979) Crater degradation through lunar history. In *Lunar and Planetary Science X,* p. 1373–1375. Lunar and Planetary Institute, Houston.

Wood C. A. and Andersson L. (1978) New morphometric data for fresh lunar craters. *Proc. Lunar Planet. Sci. Conf. 9th,* p. 3669–3689.

Wood C. A., Head J. W., and Cintala M. J. (1977) Crater degradation of Mercury and the moon: Clues to surface evolution. *Proc. Lunar Sci. Conf. 8th,* p. 3503–3520.

Proc. Lunar Planet. Sci. Conf. 11th (1980), p. 2221–2241.
Printed in the United States of America

Martian double ring basins: New observations

Charles A. Wood*

Center for Earth and Planetary Studies, Smithsonian Institution, Washington, D.C. 20560, and
Geophysics Branch, Goddard Space Flight Center, Greenbelt, Maryland 20771

Abstract—Eighteen previously unknown martian basins have been detected on Viking photographs. The smallest basins have diameters 50 to 100 km less than the smallest known basins on Mercury or the moon. On the latter two planets and the earth, basin morphology varies with increasing diameter: central peak (CP) basins have both central peaks and fragmentary peak rings; peak ring (PR) basins have only concentric rings of peaks; and multi-ring (MR) basins have two or more concentric rings. The new Viking results show that on Mars the morphology sequence is PR-CP-PR-MR, and two CP basins occur within the larger PR basin diameter interval. The reasons for the existence of the anomalous small PR basins and the two large CP basins are uncertain, but may be related to unique crustal properties on parts of Mars, or alternatively to unique properties of some impacting bodies.

The diameter distribution of martian basins exhibits three distinct slope segments, with inflections at the same diameters that changes in basin morphology occur. Embayed and ghost craters, rilles, and mare ridges occur on basin floors, suggesting that basins have been the sites of igneous extrusions as on the moon.

Peak rings within martian and lunar basins are almost always composed of isolated peaks or short mountainous arcs, but on Mercury the rings are commonly complete. This remarkable difference may reflect lateral homogeneous physical properties in Mercury's crust or increased efficiency in peak ring production due to high modal velocities for impacting bodies.

INTRODUCTION

Nearly 20 years ago Hartmann and Kuiper (1962) recognized lunar basins—large circular depressions with distinctive concentric rings—as a major class of planetary landform. Since then similar impact basins have been identified on the other terrestrial planets and on the jovian satellites Callisto and Ganymede. Despite intensive study of basins there is still considerable controversy on the origins of rings, depths of excavation, and transient cavity diameters. One observation that was first documented for lunar basins (Stuart-Alexander and Howard, 1970; Hartmann and Wood, 1971) and later extended to basins on Mars and Mercury (Wood and Head, 1976) is that basin morphology apparently varies systematically with basin diameter. *Central peak (CP) basins,* having both a central peak and a discontinuous ring of peaks, appear to be transitional between normal craters and basins. At larger diameters the central peak disappears leaving a *peak ring (PR)*

---

* Senior Resident Research Associate, National Research Council

basin whose inner ring is generally more massive and continuous than in CP basins. Development of a third (and/or additional) ring is diagnostic of *multi-ring (MR) basins*. Wood and Head (1976) found that for the moon, Mars and Mercury CP basins have diameters (D) of 100–175 km, PR basins range from D = 120–600 km, and MR basins are larger than 350 km.

I now describe 18 martian basins, newly discovered on Viking photographs. Because this paper is primarily concerned with comparisons of basin morphology, photographs of 21 basins are included as documentation of the trends discussed. The new basins range in diameter from 45 to 205 km and include five with CP morphology (previously only 1 martian CP basin was known). Additionally, two previously known PR basins, Lyot and Herschel, are shown to have central peaks and thus are now reclassified as CP basins. Most of these new basins are smaller than basins detected on Mars using Mariner 9 data (Wilhelms, 1973; Wood and Head, 1976), and many are considerably smaller than any basin previously known on Mercury or the moon. The diameter-morphology sequence for Mars, incorporating all known basins, does not conform to the sequence proposed from Mariner photography, cautioning that basin morphology may not be purely diameter dependent, but also may be strongly influenced by other factors.

NEW OBSERVATIONS OF MARTIAN DOUBLE RING BASINS

A search of Viking mosaics available at the National Space Science Data Center, Goddard Space Flight Center, yielded 18 new basins and improved views of most previously known martian basins (Table 1). This listing is not complete because various areas of the planet are not shown on the available mosaics. Multi-ring basins have not been re-examined, this report deals only with double ring (CP and PR) basins. The distribution of presently known martian basins is shown in Fig. 1.

Small double ring basins (D ≤ 100 km)

Ten of the newly-found basins are significantly smaller than previously known martian basins (all with D > 135 km), and all but one have PR morphology. These small basins, four of which are illustrated in Fig. 2, generally have large, but broken inner rings, and smooth floors. One of the largest members of this group, 5-Kd (Fig. 2d), is an excellent example of a PR basin. The massive and continuous (except where completely missing) inner ring is similar to rings in basins twice its diameter. The floor is smooth except for some parts of the bench between rim and ring. Basin 11-Tu (Fig. 2c) is a strange object with a complex interior composed of an elevated circular rough zone surrounding a central pit. Because of the pit—a central peak substitute (Wood *et al.*, 1978)—11-Tu is considered a CP basin. The inner rings of the other small basins are commonly composed of rounded hills or mountains, (e.g., 16-Qg, Fig. 2b), although 12-Xq's ring (Fig. 2a) includes linear elements. Most of these small basins are relatively

Table 1. Martian basins

No. in Fig. 1	Designation	MC	Lat.	Long.	Terrain Unit	D_{pr}	D	Basin Type	Photo No.
1*	26-Sx	26	−34.2°	41.8°	Nplc	20	45	PR	—
2*	23-Fa	23	−29.7	190.9	Nhc	22	50	PR?	23SE
3*	12-Xq	12	+18.2	355.1	Nhc	30	52	PR	212A28,30
4*	16-Qg	16	−22.9	163.3	Nhc	25	57	PR	635A92
5*	24-Du	24	−37.8	129.5	Nplc	25	58	PR	56A33
6*	11-Tu	11	+23.5	33.8	Nplc	20	60	CP?	211-5034
7*	16-Wc	16	−26.7	173.0	Nhc	33	62	PR	—
8*	24-Dt	24	−39.6	127.9	Nplc	36	73	PR	56A33
9*	5-Kd	5	+35.2	324.6	Nhc	40	95	PR	211-5741
10*	23-Rs	23	−9.2	209.5	Nhc	45	100	PR	23NW
11*	Mie	7	+48.4	220.5	Aps	33	105	CP	211-5510
12*	Arrhenius	29	−40.3	237.0	Nplc,HNk	60	115	PR	211-5673
13	Liu Hsin	24	−54.5	171.5	Nhc	55	135	CP	526A42
14*	Moreux	5	+42.0	315.5	HNk	40	140	CP	211-5664
15	16-Re	16	−25.2	164.5	Nhc	67	145	PR	635A92
16*	Bakhuysen	20	−22.2	344.3	Nplc	65	150	CP	211-5803
17	Gale	23	−5.3	222.0	Y	85?	150	CP	7650473
18*	Holden	19	−26.3	33.9	Nplc	65	150	CP	211-5755
19*	29-Eb	29	−63.6	192.0	Aps	90	160	PR	429B50,52
20	Ptolemaeus	24	−46.2	157.5	Nplc	73	165	PR	526A27-29
21	Phillips	30	−66.5	44.9	Nhc	95	175	PR	527B16,36
22	Molesworth	23	−27.7	210.7	Nhc	87	180	PR	631A17
23	Lowell	25	−52.3	81.3	Z	90	190	PR	211-5736
24	Kaiser	27	−46.4	340.5	Nhc	95	200	PR	94A40,42
25*	Schmidt	30	−72.3	77.5	Nhc	85	200	PR	211-5674
26*	Secchi	28	−58.2	258.0	Nhc,Nm	120	205	PR	6318163
27	Lyot	5	+50.5	331.0	Apc	125	215	CP	211-5819
28	Kepler	29	−47.0	218.5	Nplc	115	210	PR	97A97,99
29	Galle	26	−51.0	30.8	HNbr	100	220	PR	P17022
30	Herschel	22	−14.6	230.2	Nhc	150	285	CP	101A49
31	Antoniadi	13	+22.0	299.0	Nhc	195	390	PR	7003743
32	Schiaparelli	20	−3.2	343.6	Nhc,Hprg	230	460	PR	669K32
33	Huygens	21	−14.0	304.2	Nhc	260	470	PR	623A72-75
34	South Polar	30	−82.9	266.4	Nhc	670	850	MR?	211-5627
35	Argyre	26	−49.5	42.7	Nplc,Nhc	640	1375	MR	211-5428
36	Isidis	13,14	+13.1	272.5	Nhc	1170	3000	MR	—
37	Hellas	27,28	−42.1	292.2	Nhc,Nplc	1285	3675	MR	—

NOTES. *No. in Fig. 1:* * = newly recognized basin. *Designation:* Number-letter designations are from Batson *et al.* (1979): 26-Sx is crater Sx on Mars Chart (MC) 26. *Terrain Unit:* Geologic unit in which basin formed. See Scott and Carr (1978) for description and interpretation of each unit. Y = Nplc, HNpd, Hpr; Z = Nplc, Nhc, Aps. D_{pr}: Peak ring diameter in km. *D:* Basin diameter in km. *Basin Type:* CP = central peak and peak ring; PR = peak ring only; MR = multi-rings. Double ring basins (CP and PR) are further classed as small (No. 1–10), medium (11–18), large (19–29) and very large (30–33). *Photo No.:* Viking mosaic numbers are preceded by "211" or "P". Rev. and frame numbers are for individual Viking frames: 56A33 is frame 33 taken during orbit 56 of Viking Orbiter 1 (A). No photos for — .

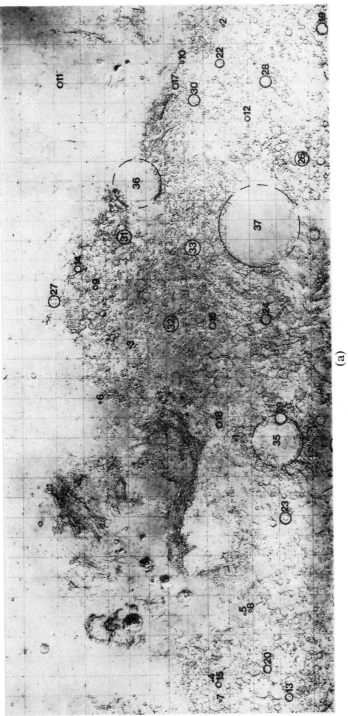

(a)

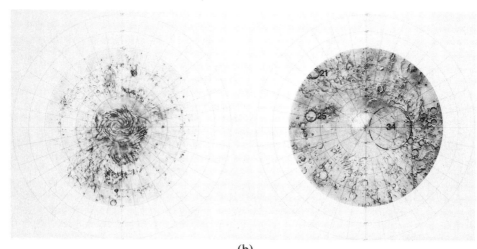

(b)

Fig. 1. Distribution of martian basins; numbered as in Table 1. Base map is USGS 1:25 m shaded relief map of Mars.

fresh, displaying remnants of ejecta, and none appear to have been modified by unusual erosional processes (cf. Schultz and Glicken, 1979).

Medium size double ring basins (100 < D < 150 km)

There are eight basins in this diameter range and six are CP types. At least three different varieties of inner ring morphology are present in this group:

1. Liu Hsin (Fig. 3b) is the martian type-example of a PR basin (Wood and Head, 1976), and Mie (Fig. 3a) and Moreux (Fig. 3c) are similar. Mie lacks a prominent central peak, however, and Moreux's peak ring is defined only by a clump of hills, a low scarp and a suggestive annular albedo feature.
2. The three largest basins in this group, with diameters of 150 km, all have central peak structures and weakly developed peak rings. Gale (not illustrated) has a small but prominent central peak and well defined arcs of a central ring that is highlighted by a low albedo annulus. Holden (Fig. 4b) and Bakhuysen (Fig. 4a) both have eccentrically located peaks that partially define incomplete central pits. The inner ring of Bakhuysen is largely defined by a few small hills and the edge of an interior zone of floor roughness that appears to be somewhat elevated above the much smoother bench zone. A nearly radial crater chain, 125 km long, is located 90 km to the SW of the crater's rim. Only a few hills on opposite sides of the central pit and a low ridge hint at a peak ring within Holden. The bench area is smooth with a narrow, lunar-like rille along at least 90° of its southern sector. A large

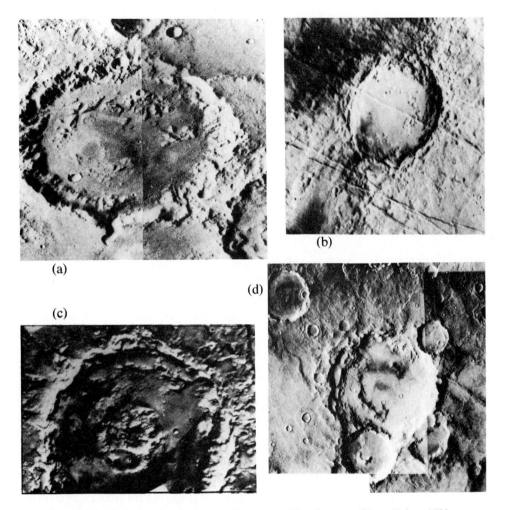

Fig. 2. Small double ring basins of Mars: (a) 12-Xq, diameter (D) = 52 km, Viking frames 212 A28,30. (b) 16-Qg, D = 57 km, Viking frame 635 A92 (c) 11-Tu, D = 60 km, from Viking mosaic 211-5034. (d) 5-Kd, D = 95 km, Viking mosaic 211-5741.

subradial valley (Erythraea Fossa) extends 220 km beyond the SE rim of Holden, the much wider and more nearly radial Uzbai Vallis extends to the SW, and a smaller crater chain trends to the north.

3. Included in this group of largely CP basins are two of PR morphology. The inner ring of Arrhenius (not shown) is represented by only a few small hills, but basin 16-Re (Fig. 3d) has a very well defined inner ring composed of arcuate mountain ranges. The entire floor of 16-Re is covered by a smooth mare-like unit with abundant north-south striking mare ridges. There is no evidence for central peaks in either of these basins.

Large double ring basins (150 < D < 250 km)

Peak rings of basins with diameters of 150 to 250 km are generally massive, but short, mountainous arcs. Lowell (Fig. 5d) has a nearly complete (360°) mountainous ring, but most large basins have continuous arcs that comprise less than 90°, with the remainder of the rings being traced by low hills and a sense of

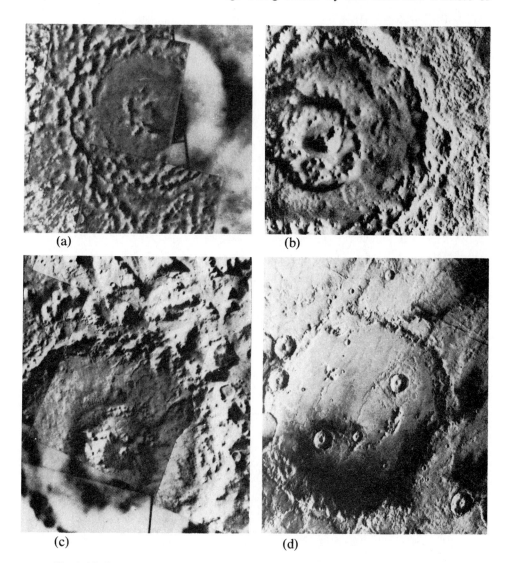

(a)   (b)

(c)   (d)

Fig. 3. Medium size double ring basins: (a) Mie, D = 105 km, Viking mosaic 211-5510. (b) Liu Hsin, D = 135 km, Viking frame 526 A42. (c) Moreux, D = 140 km, Viking mosaic 211-5664. (d) 16-Re, D = 145 km, Viking frame 635 A92.

(a)

(b)

Fig. 4. (a) Bakhuysen (D = 150 km, Viking mosaic 211-5803) and (b) Holden (D = 150 km, Viking mosaic 211-5755) are medium size double ring basins, like those in Fig. 3, but they are unusual in having fragmentary central pits instead of massive central peaks.

circular symmetry. Most of the large basins have smooth, lunar mare-like materials (occasionally showing mare ridges, e.g., Ptolemaeus, Fig. 5b) within their peak rings and in parts of their bench zones, but the benches also often contain hills and mountain masses suggestive of higher and older surfaces. Archimedian style craters (craters embayed or filled by smooth, mare-like materials) occur on the floors of Ptolemaeus, 29-Eb, Molesworth (Fig. 5c) and Kaiser (Fig. 5e), implying emplacement of floor materials significantly after basin formation.

Lyot (Fig. 6a), with a diameter of 215 km, is morphologically unlike any other basin in the large or very large basin groups. Lyot has a continuous and mountainous peak ring, similar to that of Lowell (Fig. 5d), and a large, irregular central peak. The overall morphology of Lyot is reminiscent of the much smaller basin Liu Hsin (Fig. 3b). Although it has small patches of smooth material, most of the floor of Lyot is blocky and rough, again similar to Liu Hsin.

Very large double ring basins (250 < D < 500 km)

The very largest martian double ring basins are characterized by inner rings that are much less conspicuous than those in smaller basins. Currently, these largest basins are only depicted in low resolution Viking photographs, which, combined with their apparently old and degraded state, hinders interpretation. The inner rings of these basins lack the mountainous character common in smaller basins, and instead appear to be ridge-like (Antoniadi and Herschel). Schiaparelli (Fig. 7b) does not have a single ring, but rather, like Caloris on Mercury, has a broad zone of anastomosing wrinkle ridges. Huygens' (Fig. 7c) inner ring is defined by a few ridges and scarps but appears to be best marked as the boundary of a lower, relatively smooth region. At least one flat-floored, concentric rille (reminiscent of those around the lunar Humorum basin) occurs in the bench zone. Herschel (Fig. 7a) is the most remarkable of these very large basins for it possesses an indisputable central peak complex. In fact, Herschel is the largest authenticated crater in the solar system with a central peak.

Possible other basins

The basins described above and listed in Table 1 are ones that I have examined on Viking or good Mariner 9 photographs. Other basins may well exist; in particular, South, Korolev and Milankovic appear to have basin structure in the shaded relief drawings of Batson *et al.* (1979), and Croft (1979) lists five additional basins that I have not been able to check on Viking photographs. Much more ancient and highly speculative basins have been proposed by Schultz and Glicken (1979), but even if these features once existed they now provide little significant morphological information. Malin (1976) also listed some features not included in Table 1, but his criterion for the use of the term "basin" was simply large diameter and many of his entries lack double ring structures.

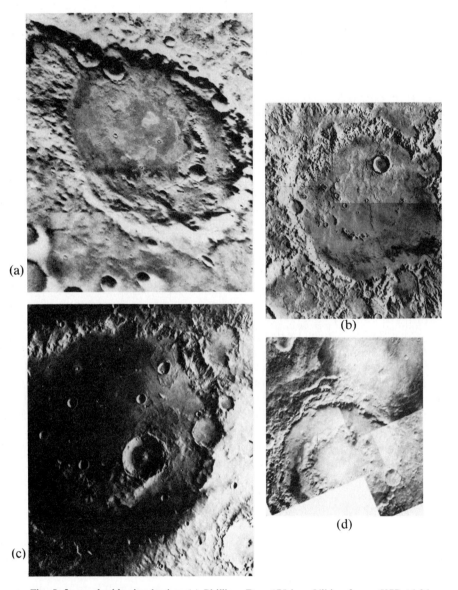

Fig. 5. Large double ring basins: (a) Phillips, D = 175 km, Viking frame 527B 16,36. (b) Ptolemaeus, D = 165 km, Viking frame 526 A27-29; note concentric structure within crater breaking NE rim (upper left) of Ptolemaeus. (c) Molesworth, D = 180 km, Viking frame 631 A17. (d) Lowell, D = 190 km, Viking mosaic 211-5736; see mosaic 211-5141 for a lower resolution but more complete view of Lowell. Note that a floor ridge in a crater on Lowell's eastern (left) rim lies along the trace of the basin rim.

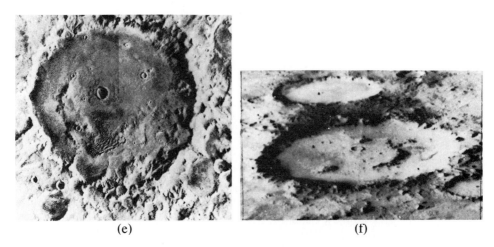

(e)   (f)

(e) Kaiser, D = 200 km, Viking frames 94 A40,42. (f) Galle, D = 220 km, Viking mosaic P17022; see also 211-5428.

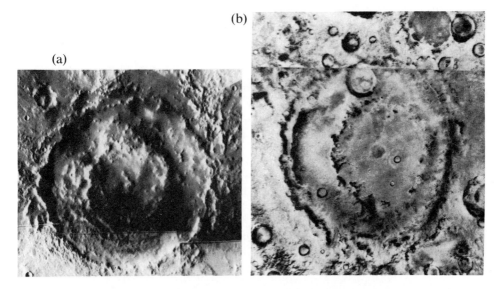

(b)

(a)

Fig. 6. Large double ring basins: (a) Lyot, D = 215 km, Viking mosaic 211-5819. Both Lyot and Herschel (Fig. 7a) have central peaks, unlike other basins in their diameter groups. (b) Kepler, D = 210 km, Viking frames 97 A97,99 (rectified). The center of Kepler's inner ring is offset 15 km from the center of the basin rim.

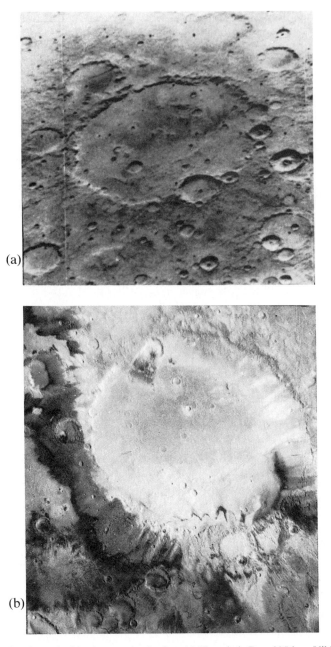

Fig. 7. Very large double ring martian basins: (a) Herschel, D = 285 km. Viking frame 101 A49. (b) Schiaparelli, D = 460 km, Viking frame 669 K32.

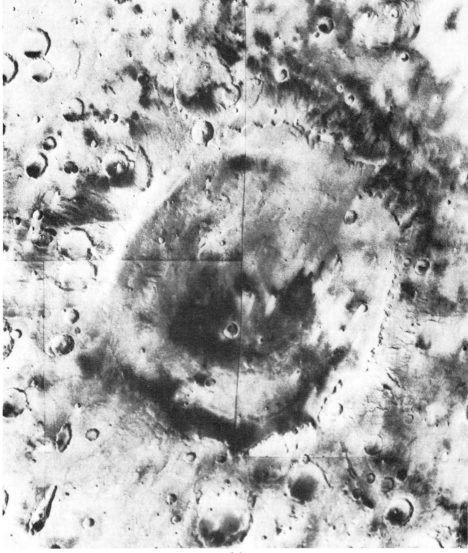

(c)

(c) Huygens, D = 470 km, Viking frames 623 A72-75.

DISCUSSION OF BASIN OBSERVATIONS

These new observations reveal that diameter dependent differences in the morphology of martian basins are more complex than previously proposed (Wood and Head, 1976). The simple progression with increasing diameter from central peak craters to CP basins to PR basins to MR basins observed for the moon and

Mercury (Wood and Head, 1976) does not *appear* to be valid for Mars. Table 2 shows, in contrast to the lunar and martian cases, that CP basins are not restricted to a single diameter interval, and that PR basins occur at smaller diameters than CP basins. Nonetheless, martian CP basins are strongly concentrated in the 100 to 150 km diameter interval, and 13 out of 15 (87%) of the larger basins are PR types. Thus, for basins greater than 100 km diameter, the diameter dependent sequence in basin morphology seen on Mercury and the moon does *generally* hold on Mars. Accepting this, the obvious questions are: What is the origin of the PR basins with diameters less than 100 km? Why are there two CP basins (Lyot and Herschel) at large diameters where only PR structure is seen on Mercury and the moon?

Small PR basins

The smallest basins on the moon and Mercury have diameters of 140 km and 90 km, respectively (Wood and Head, 1976). The considerably smaller diameters and PR morphology of the small martian basins (Fig. 2) raises the question of whether these basins may have been produced by some different process that also forms concentric rings. As an example, two separate types of inner rings are found in lunar craters: nested rims within craters a few hundred meters in diameter (Quaide and Oberbeck, 1968), and concentric craters, typically 6–7 km wide (Wood, 1978). Because of the small sizes of craters with these features, neither type is likely to be confused with basins. A potentially more confusing analog exists on Mars, where many craters up to about 200 km in diameter (Hodges *et al.*, 1980) have central pits (Wood *et al.*, 1978). These pits often (especially in large craters) lack raised rims or have only low hills, whereas basin inner rings are usually defined by rather large hills and arcuate mountain ranges.

It is difficult to argue that the small PR basins were formed by processes different from the larger basins because: (1) The morphology of some of the small basins is nearly identical to the larger ones; compare 5-Kd (95 km, Fig. 2d) and Phillips (175 km, Fig. 5a). (2) The distribution of the small PR basins (nos. 1 to

Table 2. Characteristics of martian double ring basins.

| Basin Class | Diameter Range | Basin Type | | D_{pr}/D | Peak Ring Continuity | Peak Ring Morphology |
		PR	CP			
small	45–100	9	1	0.45±.07	180°–360°	large hills
medium	100–150	2	6	0.43±.10	90°	small hills
large	150–250	10	1	0.50±.05	90°–180°	mt. ranges
very large	250–500	3	1	0.51±.02	180°–360°	ridges

Diameters in km; Basin Type: PR = peak ring; CP = central peak. D_{pr} = diameter (km) of peak ring; D = diameter (km) of basin.

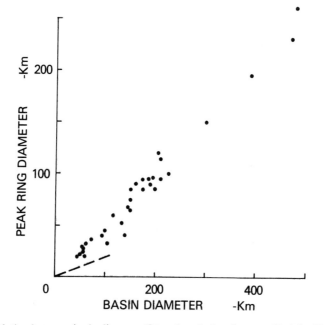

Fig. 8. Relation between basin diameter (D) and peak ring diameter (D_{pr}) for 33 martian double ring basins:

$$D_{pr} = 0.54D - 8.19 \text{ km}.$$

Dashed line is central pit diameter (D_p) versus crater diameter (D) for 15 central pit craters (from Wood *et al.*, 1978):

$$D_p = 0.17D - 0.13 \text{ km}.$$

10 in Fig. 1) appears no different from the distribution of the larger basins, arguing against control by a unique terrain type. Additionally, various of the small **PR** basins have neighboring craters of comparable diameter and degradational state that lack basin morphology. (3) The relation between basin diameter (D) and peak ring diameter (D_{pr}) is continuous (Fig. 8) from the smallest (45 km) to the largest (470 km) double ring basin, implying that all belong to the same population,

$$D_{pr} = 0.54D - 8.19 \text{ km} \qquad (r = 0.98, n = 33).$$

I conclude that small double ring basins on Mars are most likely to be formed by the same general processes as the larger basins.

The relationship between peak ring and basin diameters argues against the proposal that central pits evolve with increasing crater diameter into peak rings (Hodges, 1978), because the central pit diameter to crater diameter relation (Wood *et al.*, 1978) is considerably different (dashed line in Fig. 8). Further evidence against that proposal is (1) the occurrence of central pits within three basins: 11-Tu, Holden and Bakhuysen, and (2) the observation that peak rings occur on all planets, but central pits are unique to Mars, Ganymede and Callisto.

POSSIBLE ORIGINS OF THE ANOMALOUS BASINS

There is little doubt that basins are simply end members of the impact crater diameter spectrum, and that their morphological peculiarity (concentric and radial structure) is but the final stage in the diameter dependent evolution of crater morphology. The origins of basin rings, however, remain a controversial topic. Each of the various models of basin ring formation (summarized by McKinnon and Melosh, 1980) assumes that differences in morphology from central peak craters to CP, PR and ultimately MR basins is a diameter dependent sequence, reflecting either increasing depth of excavation and sequential intersection of critical layers (Hodges and Wilhelms, 1978), or little understood relations between morphology and energy (e.g., Dence and Grieve, 1979). Thus, the anomalous small PR basins and the two large CP basins require a special origin. Explanations may center around the character of the impacted target or of the projectile.

On Earth, target strength (sedimentary versus crystalline rock) strongly influences the diameter of transition from simple to complex crater morphology (Grieve and Robertson, 1979) and similar target-morphology effects have been documented for craters on Mars (Wood *et al.*, 1978), the moon, and Mercury (Cintala *et al.*, 1978). It is unlikely, however, that target characteristics can account for the small PR basins because of the close association of similar size craters and basins (as mentioned above). Additionally, the most unusual target characteristic of Mars—the imprisoned subsurface volatile layer—produces characteristic structures (fluidized craters, central pit craters) that appear to be globally distributed on Mars (Mouginis-Mark, 1979; Hodges *et al.*, 1980), whereas most craters of small basin diameter (50 to 100 km) lack basin morphology. In other words, fluidized and central pit craters are ubiquitous on Mars but small basins are rare. If small basins are common on Ganymede and Callisto—the only other known bodies with important subsurface volatile deposits—it will be difficult to refute the claim that substrate characteristics are important in the formation of small basins.

It may be that anomalous characteristics of the projectiles, not of the target, produced the anomalous basins. Theoretical (Roddy *et al.*, 1980) and experimental (O'Keefe and Ahrens, 1980) studies suggest that low density impacting bodies can form shallow craters with complex interior morphology (central peaks and possibly peak rings). Thus, if the main sequence of impact craters CP, PR and MR basins were formed by impacts of relatively dense projectiles (e.g., Apollo objects or asteroids), perhaps the anomalous basins resulted from impacts of less dense bodies (e.g., comets). Following up on this speculation, we can estimate the relative frequency of comet to asteroid impacts on Mars for two different diameter intervals. Between 50 and 100 km diameter there are 10 anomalous small basins and 1244 normal martian craters (R. Arvidson and E. Guinness, pers. comm.), suggesting that low density impacting objects were rare. For the diameter range (200 to 300 km) of the two anomalously large CP basins (Lyot and Herschel) there are approximately 18 other craters and PR basins (counted from Batson *et al.*, 1979), yielding a comet impact frequency of 10%. The two estimates of low to high density impactors could be used to suggest that comet impacts become increasingly important on Mars at large crater diameters. This speculation is weakened, however, by the observation that anomalous basins are not observed on the other terrestrial planets, which should have been cratered by even more comet impacts than would have Mars (Hartmann, 1977).

BASIN MORPHOLOGY ON MARS AND OTHER PLANETS

Rings

The morphology of martian peak rings varies with basin diameter (Table 2). In small double ring basins the rings are comprised of large conspicuous hills and mountains that span 180° to 360° of the ring. Medium diameter basins are predominantly CP types and their peak rings are often alignments of inconspicuous and isolated hills. Dark annular albedo features often accentuate small scale roughness associated with rings. Except for Liu Hsin, actual segments of the inner rings of medium size basins usually have a cumulative arc length of less than 90°. Inner rings within large basins are no longer circles defined by isolated peaks but tend to be arcuate mountain rings. Except for Lowell and Lyot, which have complete inner rings, large basin rings usually extend only 90° to 180°. The rings of very large martian double ring basins are defined by ridges and occasional hills, are nearly complete (180° to 360°), but are low and inconspicuous.

Roughly similar diameter dependent differences in ring morphology are seen in basins on Mercury and the moon. One intriguing interplanetary difference in basin morphology is that peak rings are often complete circles on Mercury. Lowell (Fig. 5d) and Schrödinger are commonly given as examples of martian and lunar PR basins, respectively, but in fact the completeness of their rings is nearly unique on those planets. In contrast, ten basins on Mercury have virtually unbroken peak rings and most other mercurian inner rings have continuous arcs of 270° to 360°. If peak ring continuity is related to substrate homogeneity (as is outer ring development; Head and Solomon, 1980), Mercury must have a remarkably homogeneous crust, whereas the moon, and especially Mars, must have sharp lateral variations in crustal properties. Alternatively, it may be that the higher modal velocities for impacts on Mercury (Hartmann, 1973) translates into a greater efficiency in peak ring formation.

Peaks

The Viking observations of Martian basins may help explain an apparently anomalous observation on Venus. Radar images of Venus reveal numerous circular features that are often interpreted as impact craters, and many have radar bright centers that have been compared with central peaks (Campbell *et al.*, 1979). These central spot craters are as wide as 280 km, much larger than any previously known central peak crater on Mars, Moon or Mercury. The large central spot craters on Venus may actually be basins similar to the martian CP basin Herschel, with a central peak that returns a radar bright spot, and a peak ring too low to produce a radar return.

Plains

The floor of nearly every martian basin is smooth, and lunar-like mare ridges and ghost craters occur. These observations suggest that (1) the smooth material was emplaced after impacts had occurred on the basin floor, and (2) the smooth material forms ridges and buries craters similarly to lunar mare basalts. This interpretation was previously reached for the larger martian basins (Scott and Carr, 1978; Wood and Head, 1976) and also appears true for the smaller basins.

DIAMETER FREQUENCY DISTRIBUTION

A cumulative frequency-diameter plot for the martian basins listed in Table 1 exhibits three different slope segments (Fig. 9). It is unlikely that the relatively abrupt changes in slope are due to selection or degradation effects because the inflections occur at diameters that subdivide basins into different morphological classes. Figure 9 illustrates that there is an excess of small, medium and large double ring basins (D < 250 km) compared to the number of very large double ring basins and multi-ring basins. Additionally, there is a relative deficiency of

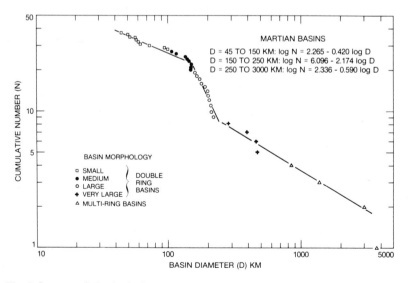

Fig. 9. Log cumulative basin frequency versus log basin diameter for all martian basins in Table 1.

small and medium size double ring basins (D < 150 km). The similar distributions of the latter two morphological types are further evidence that the anomalously small PR basins are part of the overall basin continuum. Correlation of basin morphology and relative abundance for Mars, Mercury and the moon (Wood, in prep.) adds a new and as yet little understood element to basin investigations.

CONCLUSIONS

Examination of Viking photomosaics has resulted in the doubling of the number of known basins on Mars; more undoubtedly await detection. Basins have been classified according to ring morphology and a diameter dependent sequence apparently exists:

- a) small basins (45 < D < 100 km) are almost exclusively (9 out of 10) PR types;
- b) medium size basins (100 < D < 150 km) dominantly (6 of 8) have CP morphology;
- c) large basins (150 < D < 250 km) are PR types with one exception (10 of 11);
- d) very large double ring basins (250 < D < 500 km) have PR morphology (1 CP type out of 4) with ridge rings rather than peak or massif rings;
- e) multi-ring basins (only 4) occur at D > 500 km.

This morphology sequence does not conform to the CP-PR-MR progression on the moon, Mercury and Earth. The small martian basins, however, are 50 to 100 km smaller than any basin on Mercury and the moon, supporting the view that they are unique to Mars. On Earth there are significant terrain influences on crater and basin morphology that encourage the speculation that localized unique characteristics of the martian crust led to basin formation at diameters where craters would normally be formed. Alternatively, following theoretical and experimental evidence, the anomalous PR basins (and perhaps the two large CP basins) may have resulted from unique characteristics of the impacting body (e.g., low density comet). Both interpretations have obvious weaknesses.

Abrupt changes in the slope of the diameter distribution of martian basins (Fig. 9) are correlated with changes in basin morphology.

Martian basins commonly have smooth floors, ridge and rille structures, and embayed and ghost craters, all reminiscent of the lunar mare. Although there are undoubtedly many processes that form smooth surfaces on Mars this evidence plus the obvious abundance of volcanic landforms (Scott and Carr, 1978) implies that small to very large basins on Mars have localized igneous extrusions.

Peak ring morphology and continuity are generally similar for basins on Mars and the moon, but on Mercury peak rings are much more complete. This dramatic difference in peak ring structure may reflect a homogeneous crust or higher modal impact velocities for Mercury.

Acknowledgments—I thank Steve Croft for a copy of his thesis, Ed Strickland for the computer enhanced photograph of Schiaparelli, Herb Frey for basin discussions, and Carlos Madera and Barbara Lueders for rapid and precise typing. René DeHon, Henry Moore and Ted Maxwell provided useful reviews, which were richly supplemented by comments by Dick Pike and his colleagues at Menlo Park. NSSDC provided Viking images.

REFERENCES

Batson R. M., Bridges P. M., and Inge J. L. (1979) *Atlas of Mars*. NASA ST-438, 146 pp.

Campbell D. B., Burns B. A., and Boriakoff V. (1979) Venus: Further evidence of impact cratering and tectonic activity from radar observations. *Science* **204,** 1424–1427.

Cintala M. J., Wood C. A., and Head J. W. (1977) The effects of target characteristics on fresh crater morphology: Preliminary results for the moon and Mercury. *Proc. Lunar Sci. Conf. 8th,* p. 3409–3425.

Croft S. K. (1979) Impact craters from centimeters to megameters. Ph.D. thesis, Univ. Calif., Los Angeles. 264 pp.

Dence M. R. and Grieve R. A. F. (1979) The formation of complex impact structures (abstract). In *Lunar and Planetary Science X,* p. 292–294. Lunar and Planetary Institute, Houston.

Grieve R. A. F. and Robertson P. B. (1979) The terrestrial cratering record: I. Current status of observations. *Icarus* **38,** 212–229.

Hartmann W. K. (1973) Martian cratering, 4, Mariner 9 initial analysis of cratering chronology. *J. Geophys. Res.* **87,** 4096–4116.

Hartmann W. K. (1977) Relative crater production rates on planets. *Icarus* **31,** 260–276.

Hartmann W. K. and Kuiper G. P. (1962) Concentric structures surrounding lunar basins. *Commun. Lunar and Planetary Lab. Univ. Arizona* **1,** 51–66.

Hartmann W. K. and Wood C. A. (1971) Moon: Origin and evolution of multi-ring basins. *The Moon* **3,** 3–78.

Head J. W. and Solomon S. C. (1980) Lunar basin structure: Possible influence of variations in lithospheric thickness (abstract). In *Lunar and Planetary Science XI,* p. 421–423. Lunar and Planetary Institute, Houston.

Hodges C. A. (1978) Central pit craters on Mars (abstract). In *Lunar and Planetary Science IX,* p. 521–522. Lunar and Planetary Institute, Houston.

Hodges C. A., Shew N. B., and Clow G. (1980) Distribution of central pit craters on Mars (abstract). In *Lunar and Planetary Science XI,* p. 450–452. Lunar and Planetary Institute, Houston.

Hodges C. A. and Wilhelms D. E. (1978) Formation of lunar basin rings. *Icarus* **34,** 294–323.

Malin M. C. (1976) Comparison of large crater and multiringed basin populations on Mars, Mercury, and the Moon. *Proc. Lunar Sci. Conf. 7th,* p. 3589–3602.

McKinnon W. B. and Melosh H. J. (1980) Multi-ringed basins in the solar system: A "new" paradigm (abstract). In *Lunar and Planetary Science XI,* p. 708–710. Lunar and Planetary Institute, Houston.

Mouginis-Mark P. (1979) Martian fluidized crater morphology: Variations with crater size, latitude, altitude and target material. *J. Geophys. Res.* **84,** 8011–8022.

O'Keefe J. D. and Ahrens T. J. (1980) Cometary impact calculations: Flat floors, multi-rings and central peaks (abstract). In *Lunar and Planetary Science XI,* p. 830–832. Lunar and Planetary Institute, Houston.

Quaide W. L. and Oberbeck V. R. (1968) Thickness determinations of the lunar surface layer from lunar impact craters. *J. Geophys. Res.* **73,** 5247–5270.

Roddy D. J., Kreyenhagen K., Schuster S., and Orphal D. (1980) Theoretical and observational support for formation of flat-floored central uplift craters by low-density impacting bodies (abstract). In *Lunar and Planetary Science XI,* p. 943–945. Lunar and Planetary Institute, Houston.

Schultz P. H. and Glicken H. (1979) Impact crater and basin control of igneous processes on Mars. *J. Geophys. Res.* **84,** 8033–8047.

Stuart-Alexander D. E. and Howard K. A. (1970) Lunar mare and circular basins—a review. *Icarus* **12,** 440–456.

Scott D. H. and Carr M. H. (1978) Geologic Map of Mars. U.S. Geol. Survey Misc. Geol. Inv. Map I-1083.

Wilhelms D. E. (1973) Comparison of martian and lunar multi-ringed circular basins. *J. Geophys. Res.* **78,** 4084–4096.

Wood C. A. (1978) Lunar concentric craters (abstract). In *Lunar and Planetary Science IX,* p. 1264–1266. Lunar and Planetary Institute, Houston.

Wood C. A. and Head J. W. (1976) Comparison of impact basins on Mercury, Mars and the moon. *Proc. Lunar Sci. Conf. 7th,* p. 3629–3651.

Wood C. A., Head J. W., and Cintala M. J. (1978) Interior morphology of fresh martian craters: The effects of target characteristics. *Proc. Lunar Planet. Sci. Conf. 9th,* p. 3691–3709.

Proc. Lunar Planet. Sci. Conf. 11th (1980), p. 2243–2259.
Printed in the United States of America

A comparison of secondary craters on the Moon, Mercury, and Mars

Peter H. Schultz[1] and Jill Singer[2]

[1]Lunar and Planetary Institute, Houston, Texas 77058   [2]Department of Geological Sciences,
Rice University, Houston, Texas 77001

Abstract—Comparison of secondary crater populations around selected martian, lunar and mercurian craters reveals significant differences that are interpreted as the effects of substrate on the cratering flow field and ejecta size. An unnamed 34 km-diameter crater in Chryse Planitia exhibits relative number densities of secondaries comparable to mercurian and lunar craters. Similarly, the maximum extent of the ejecta flow lobes nearly matches the maximum extent (2 crater radii from the rim) of the continuous ejecta facies of similar-size mercurian and lunar craters (adjusted for gravity). In contrast, the crater Arandas (24 km in diameter) exhibits very few secondaries larger than 2% of the primary diameter. The inner ejecta lobe also approximates the continuous ejecta facies of lunar/mercurian craters but the outer lobe extends 4 crater radii from the rim. Arandas secondaries larger than 0.5 km tend to be circular in contrast with secondaries around the Chrysie crater and lunar/mercurian craters. The more circular Arandas secondaries suggest higher ejection angles. Although high ejection angles can reduce the ballistic range and alter the over-all secondary crater distribution, such effects appear not to be sufficient to account for the paucity of Arandas secondaries. The unusual target lithology resulting in high ejection angles also may produce greater fractions of fine-size ejecta. Sufficiently small ejecta will be decelerated by aerodynamic drag and incorporated in near-rim ejecta deposits, rather than forming secondaries. This proposal is consistent with the observed multi-phased sequence of ejecta emplacement for certain martian craters where the outer ejecta lobes have overridden the inner ejecta lobes.

INTRODUCTION

Craters on the Moon, Mars, and Mercury display contrasting styles of ejecta facies that presumably reflect different cratering environments (i.e., gravity, atmosphere, target properties). Two fundamental and distinct problems must be addressed when considering ejecta morphology: ballistic history and ejecta emplacement. On the Moon and Mercury, distinct secondary craters clearly record the effects of the ballistic history. On Mars, ejecta flows reveal the importance of post-ballistic ejecta emplacement, and most discussions have concentrated on this process. Secondary craters also surround martian impact craters, but the Viking images reveal highly variable populations (Carr, *et al.,* 1977; Schultz and

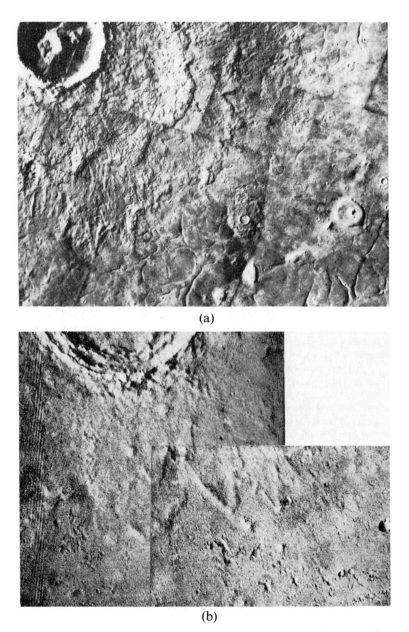

Fig. 1. Comparison of ejecta facies of selected martian craters. (a) The 24 km-diameter martian crater Arandas situated on the northern fractured plains near latitude +42.8°, longitude 345.1°W. Viking Mosaic 211-5031. Ejecta facies of an unnamed crater (MC-11, Wx; labeled here Crater I) approximately 34 km in diameter in the Chryse Planitia region near latitude +26.5°, longitude 38.8°W. Viking Mosaic 211-5382. (c) Ejecta facies of an unnamed crater (MC-5, Zm; labeled here Crater II) approximately 45 km in diameter situated 600 km northeast of Arandas in the fractured plains near latitude +46.4°, longitude 359.7°W. Viking Mosaic 211-5031.

(c)

Gault, 1979; Mouginis-Mark, 1979a). Previous studies have focused on the existence/absence of secondaries (Mouginis-Mark, 1979b) or on the maximum secondary crater diameter (Allen, 1979). The present study is concerned with the distribution of secondary crater sizes and shapes and the relation of these distributions to inferred martian lithology and ballistic history.

Differences in secondary-crater size/range distributions around unmodified craters can be attributed to differences in gravitational potential, lithologic properties, and atmospheric interactions. All three variables can affect the ballistic range. Material ejected at a given velocity on Mercury and Mars obviously will travel a shorter distance than on the Moon owing to the larger gravitational attraction. Material ejected at high or low angles due to unusual lithologic properties will similarly travel shorter ballistic ranges than material ejected at nominal angles. Finally, sufficiently small ejecta can be aerodynamically decelerated in an atmospheric environment. Consequently, we first compare the size/range distributions of secondaries around selected lunar, mercurian, and martian craters in order to further establish the gravitational effects discussed by Gault *et al.* (1975). Second, we compare the size/range distributions of secondaries around martian craters in different geologic terrains in order to establish possible lithologic effects. These distributions are complemented by measures of circularity that could indicate high or low ejection angles. Third, we make first-order corrections for gravity and ejection angles in order to evaluate whether or not the last variable (atmospheric effects) must be considered. Finally, we consider complementary morphologic evidence for separation of the variables.

DATA SET

Primary craters were selected on the basis of crater size, inferred target lithology, available resolution, and completeness of coverage. As previous studies have shown (Carr, *et al.,* 1977; Schultz and Gault, 1979; Mouginis-Mark, 1979b), the complexity of ejecta facies on Mars increases with increasing crater size. Ejecta facies of craters smaller than about 1 km–3 km in diameter typically resemble lunar craters. Ejecta facies of martian craters between 5 km and 50 km exhibit clear evidence for flow expressed as rampart-bordered and multi-lobed deposits with highly variable secondary crater populations. Ejecta facies of craters larger than 50 km typically are highly complex and composed of extensive secondary crater populations with multiple and overlapping ejecta flow lobes. The intermediate size range (5 km–50 km) was selected because of the diversity in emplacement style. Although this size range could be applied easily to the Moon, equivalent size craters on Mercury could not be selected, owing to resolutions that were insufficient for clear identification of secondaries.

Three martian craters smaller than 50 km in diameter were selected on the basis of contrasting inferred target lithologies and completeness of coverage at sufficient resolutions. The crater Arandas (24 km in diameter) occurs in the northern fractured plains and is surrounded by an extensive multi-lobed ejecta deposit (Fig. 1). The fractured plains are interpreted as regions of high water-ice content analogous to terrestrial patterned ground (Carr and Schaber, 1977). Viking coverage of this region permits resolving secondary craters as small as 0.2 km in diameter in two 180° sectors out to 6 crater radii from the primary-crater center and in two 30° sectors beyond 10 crater radii.

A 34-km diameter crater in Chryse Planitia displays markedly different ejecta facies (Fig. 1b) dominated by secondary craters in the outer deposits with multi-lobed ejecta flows in the inner deposits. The crater is designated as Crater Wx in MC-11 (see Batson *et al.,* 1979), but is denoted as Crater I in this paper. The Chryse plains are interpreted as basalt surfaces partly mantled by aeolian and outwash materials (Greeley, *et al.,* 1977; Theilig and Greeley, 1979). Contiguous Viking coverage of Crater I permits resolving 0.2 km diameter secondaries within a 120° sector out to 8 crater radii, and within a 30° sector beyond 12 crater radii.

An additional but larger crater (Fig. 1c) approximately 45 km in diameter was selected in the fractured plains region 600 km northeast of Arandas (Crater Zm in MC-5 but denoted here as Crater II). This additional crater was selected, since the large number of secondaries around Crater I relative to Arandas might be attributed to its slightly larger size rather than differences in target lithology. Contiguous Viking coverage of Crater II allows resolving 0.2 km-diameter secondaries within a 180° sector out to 10 crater radii, and within a 30° sector beyond 12 crater radii.

The craters Aristarchus (39 km; Fig. 2b) and Copernicus (96 km; Fig. 2a) comprise the lunar data set. Copernicus is significantly larger than the selected martian craters, but owing to the difference in gravity, the impact velocity of ejecta at a given *relative* range approximates the impact velocity of ejecta from a 50 km diameter martian crater. Ejecta velocities around Aristarchus approximate velocities from smaller martian craters comparable in size to Arandas.

Selection of an appropriate mercurian crater was hampered by incomplete coverage at sufficiently high resolutions. In order to assess the secondary crater populations to relative ranges of 7 crater radii, a primary crater significantly larger than the selected craters on Mars was considered. Nevertheless, these data provide a useful comparison. Mariner 10 images of the crater Alencar (110 km; Fig. 3) permit measurements nearly 180° in azimuth, but topography and another major crater in one sector limits useful data out to 5 crater radii within a 90° sector and 7 crater radii within a 30° sector.

Secondary craters were identified on the basis of occurrence as chains or clusters or as members of a crater ray. By lunar analogy, other characteristics included elongate or irregular plan, shallow profile, subdued morphology, and the herringbone pattern. These characteristics eliminate certain types of secondary craters, but represent consistent criteria. Because of the wide range in available resolutions, only secondaries larger than 1–2% of the primary crater diameter were recorded. The distinguishing characteristics of secondaries become less pronounced at large distances from the primary (>12 R) and can be confused with secondaries from other craters. Consequently, limiting resolutions also prevented confident identification of secondaries for a given primary at large ranges.

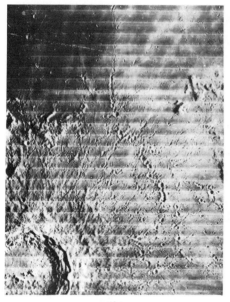 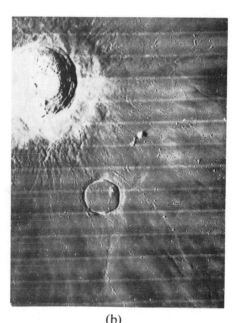

(a)   (b)

Fig. 2. Comparison of ejecta facies of selected lunar craters. (a) The lunar crater Copernicus (96 km) near latitude +9.7°, 19.9°W. Lunar Orbiter IV-121-H2. (b) The lunar crater Aristarchus (39 km) near latitude +23.7°, longitude 47.4°W. Lunar Orbiter IV-150-H3.

RESULTS

The distribution of secondary craters through a complete 360° azimuth could not be obtained for all craters owing to incomplete coverage, overlapping secondary crater fields, insufficient resolution, and contrasting resolutions dependent on terrain. Consequently, the collected data were analyzed within 30° arc sectors where the most consistent data could be obtained to the farthest distance from the primary. Comparison of different azimuth sectors of the same crater provides a check for any gross asymmetries in the secondary crater populations.

Figures 4 and 5 permit comparison of the raw data and include the approximate limits of the continuous ejecta deposits. The continuous ejecta facies (CEF) of the mercurian crater Alencar (Fig. 4) extends to about 1.8 R from the crater center in contrast to 2.3 R and 3 R for the lunar craters Copernicus and Aristarchus, respectively. The minimum extent of the ejecta flow lobe of Crater I (Chryse Planitia, Fig. 5) approximates the limit of the mercurian ejecta facies (2 R), and the maximum extent of both the inner lobe and the outer lobe of Crater I extends to about 3 R (2.8 R and 3.1 R, respectively). Data from Gault *et al.* (1975) as well as Fig. 4, indicate that the extent of the continuous ejecta facies increases with decreasing crater size. Consequently, the extent of the multiple lobes of Crater I on Mars is not significantly different from the limit of the

Fig. 3. The mercurian crater Alencar (110 km) near latitude −63.3°, longitude 104.0°W.
FDS 166727.

continuous ejecta facies for a crater on Mercury. Similarly, the extent of the
thick, inner ejecta lobes of Arandas approximate the continuous ejecta facies of
a mercurian crater of comparable size. The inner lobes of Crater II are not easily
delineated, as is characteristic of larger multi-lobed martian craters (Schultz and
Gault, 1979; Mouginis-Mark, 1979a). In contrast, the outer ejecta lobes of both
Crater II and Arandas extend to about 5 R.

Secondary craters become prominent beyond the limit of the continuous ejecta
facies for Alencar, Copernicus, and Aristarchus. Although subdued depressions
within these deposits are also probable secondaries, they are not always identifiable if lighting conditions are not favorable. Figure 4b shows that the large
number of secondaries around these lunar/mercurian craters are matched by
martian Crater I (Chryse Planitia). Crater II and Arandas, however, are generally
devoid of secondaries. Secondaries occur within the ejecta lobes of Crater II, but
generally appear to be overridden or surrounded by the relatively thin ejecta lobe.
Arandas secondaries larger than 2% of the primary diameter are relatively rare,
even beyond the outer limit of the ejecta flow lobes. This paucity of secondaries
is confirmed in other azimuth sectors.

Figure 5 allows comparison of secondary crater frequency within the same
relative areas (annuli) at different ranges from the crater center. This representation normalizes the secondary crater population within each annulus to the
crater size and permits direct comparison of the number of secondary craters
produced by different size primary craters. Figure 5 shows that the *relative* areal

density of secondary craters for Copernicus and Aristarchus are nearly the same. It should be remembered, however, that the *absolute* number of secondaries around Aristarchus is significantly less than the number around Copernicus. The absence of secondaries between the rim and 2 R from the rim of Aristarchus is largely due to the larger extent of its continuous ejecta facies, as noted in Fig. 4a. There are significantly fewer secondaries between 3 R and 4 R for Copernicus than for Aristarchus. Shoemaker (1962) noted a similar deficiency and attributed it to confusion with Eratosthenian secondaries and lighting conditions for the terrestrial view used. Although lighting limitations are not a problem with Lunar

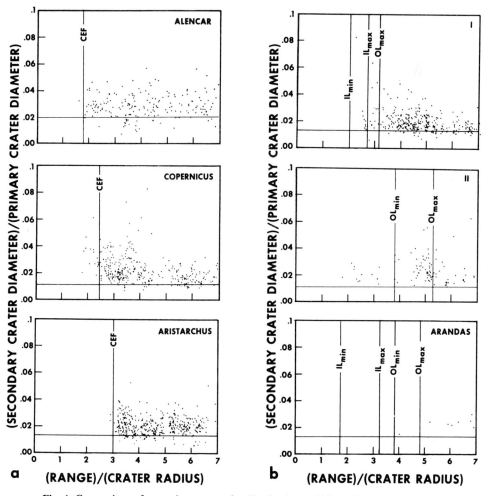

Fig. 4. Comparison of secondary crater size distributions within a 30° sector. (a) Craters Alencar (Mercury), Copernicus (Moon), and Aristarchus (Moon). The limit of the continuous ejecta facies is indicated by CEF. (b) The martian craters, Crater I, Crater II, and Arandas (see Fig. 2). The minimum and maximum extents of the inner (IL) and outer (OL) flow lobes are indicated.

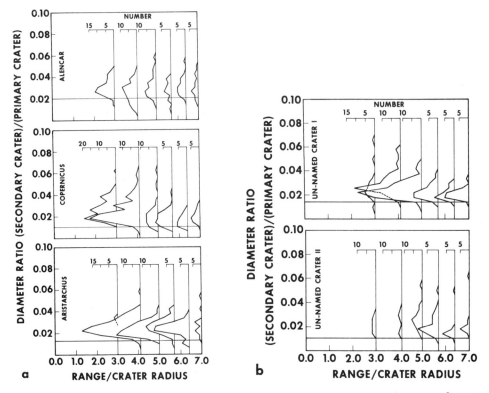

Fig. 5. The size-frequency distribution of secondaries within the same relative areas of different craters within 30° sectors. The horizontal lines indicate the approximate resolution cut-off of the collected data. The crater Arandas is not illustrated owing to the few secondaries counted above the resolution cut-off.

Orbiter photography of this region, secondaries from Eratosthenes complicate consistent identification of Copernican secondaries. The reduced number of secondaries within 5 R for the mercurian crater Alencar reflects the difficulty in clearly identifying secondaries in the more rugged terrains. Although the theoretical resolution is normally quoted as 2.2 pixels (Saunders *et al.*, 1975), practical resolution is dependent on terrain and is significantly poorer. In rugged terrains, as many as 90% of the craters 2.2 pixels in diameter are unidentified (Schultz, 1977).

Figure 5b confirms the similarity in the secondary crater distributions between Crater I and the lunar and mercurian craters, whereas Crater II displays a notable deficiency. The crater Arandas (not included in Fig. 5b) has even fewer secondaries, as shown in Fig. 4b.

Short and long axes of each secondary crater were measured and rectified, thereby permitting a measure of circularity. Figure 6 reveals that Alencar, Copernicus, Aristarchus, and Crater I have secondaries with wide ranges in circularity: values of short/long axes typically near 0.78, 0.8, 0.7, and 0.6, respectively.

In contrast, the identified secondaries of Arandas generally are highly circular (0.9) within 8 R and show lower circularity beyond 8 R. Smaller secondaries (not included in the data), however, have more elongate plans. Secondaries associated with Crater II are also more circular (0.8) than those around Crater I, but clearly more elongate or irregular than secondaries associated with Arandas.

DISCUSSION

Five parameters affect the size distribution of secondary craters: gravitational potential, ejection angle, ejection velocity, ejecta size, and ejecta state (solid, dispersed). Gault, *et al.* (1975) showed that the secondary crater population is not appreciably different beyond the region of continuous ejecta for the Moon and Mercury. Figure 5 indicates that Crater I in Chryse Planitia on Mars is also similar. The distribution of secondaries around Copernicus and Aristarchus can be adjusted simply for the gravity of Mars and compared with Crater I. Figure 7 reveals that such a correction most notably affects the near-rim distribution of secondaries, as observed by Gault *et al.* (1975). Secondaries at large ranges are pulled closer to the rim and increase this population without drastically modifying the overall distribution. Crater I exhibits a distribution very similar to the gravity-

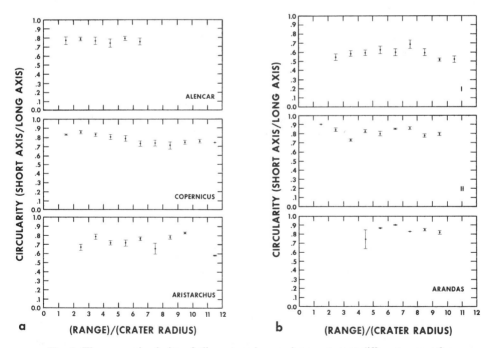

Fig. 6. The mean circularity of all measured secondary craters at different ranges for different craters. The error bars represent a one standard-deviation statistical spread of the data. The large deviation for Arandas near 4 R indicates the paucity rather than statistical spread of the data.

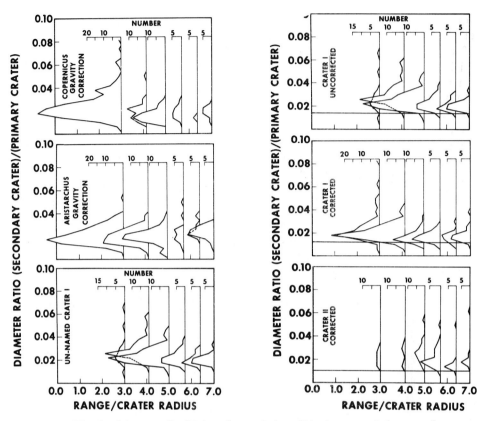

Fig. 7. The size-frequency distribution of secondaries within the same relative areas for Copernicus and Aristarchus after the observed range of secondaries (Fig. 4a) was reduced to match conditions of martian gravity. The large number of secondaries within 2 R of the rim would be represented by the continuous ejecta facies. Crater I in Chryse Planitia on Mars shows a similar size-frequency distribution beyond the multi-lobed ejecta flows.

Fig. 8. Comparison of the size-frequency distribution of secondaries within the same relative areas for martian Crater I before (top) and after (middle) reduction of range due to increasing the assumed ejection angle from 45° to 75°. As in Fig. 7, the most pronounced change occurs near the rim and may contribute to the massive ejecta flows developed around certain martian craters. Increased ejection angles may not account completely for the few craters beyond the multiple ejecta lobes of Crater II and are even less likely to account for the relative absence of secondaries around Arandas (Fig. 4b).

adjusted distributions for the lunar craters, except within the ejecta flow lobes, which incorporated or buried the near-rim secondaries. The paucity of secondaries around Crater II and Arandas, therefore, suggests the influence of parameters other than gravity.

Gault and Greeley (1978) and Greeley *et al.* (1980) demonstrated that impacts in low-viscosity targets eject material at appreciably higher ejection angles from the horizontal than impacts in high-viscosity targets. Thomsen *et al.* (1979; 1980) confirmed this for different material of low viscosity and calculated a similar crater flow field. Different ejection angles theoretically affect both the distribution of secondaries with distance from the primary and the size of the secondary. High (or very low) ejection angles will reduce the ballistic range for the same ejection velocity, thereby influencing the spatial distribution of secondaries. Gault (1974) notes that crater size increases as the impact angle increases according to $(\sin \theta)^{1/3}$. Consequently, an increased impact angle (and, at the same range, an increased velocity) would also be expected to increase secondary crater size, an effect that should increase the observed population of secondaries as they are raised above the resolution cut-off. For example, at 6 R from Arandas an ejection angle of 75° rather than 45° would increase the size 66%: 37% due to the increased ejection angle and 21% due to the increased velocity required to achieve the same ballistic range.

If the observed secondaries around Crater I were produced by ejecta with an ejection angle of 45° (producing the maximum ballistic range), then an ejection angle of 75° would reduce the range by a factor of two for the same ejection velocity. Such a range reduction applied to Crater I produces the distribution shown in Fig. 8. This reduction is insufficient to produce the deficiency of secondaries associated with Crater II and even less effective for secondaries of Arandas. Therefore, high ejection angles alone perhaps cannot account for the few secondaries associated with Arandas and Crater II. It should be noted, however, that the velocity flow fields in two contrasting media are only beginning to be addressed (Greeley *et al.*, 1980; Thomsen *et al.*, 1980; Austin *et al.*, 1980), and the assumption that the vector velocities of ejecta are the same is most likely in error. More detailed theoretical and experimental studies are clearly required.

The greater circularity for secondaries of Arandas (Fig. 6) supports the contention that ejection angles were higher than those for Crater I, Aristarchus, Copernicus, and Alencar. Consequently, suggestions that the target material in this region of Mars exhibited a low yield strength at the time of formation (Carr *et al.*, 1976; Gault and Greeley, 1978; Boyce, 1979; Schultz and Gault, 1979; Greeley *et al.*, 1980) gain further support. However, the few Arandas secondaries of sufficient size to be included in this study require further discussion.

Ejecta size and physical state also can affect the formation of secondary craters. Ejecta below a critical size can be decelerated by aerodynamic drag, thereby reducing the ballistic range and incorporating this fraction in a late-stage, near-rim ejecta deposit (Schultz and Gault, 1979). For Arandas-size craters this critical size is about 10 cm. If the pre-impact target lithology contains large amounts of water or ice, shock vaporization of this component may lead to increased comminution. Moreover, the dynamic forces should disperse any liquid ejecta and result in a low-density cloud of atomized liquid and solids traveling in close proximity, rather than a single ejecta block. If massive enough, such an impacting system may form detached fluidized ejecta deposits (Greeley *et al.*, 1980). If the velocity is high enough, craters with mounded floors may be formed (Schultz *et al.*, 1980). Highly dispersed ejecta, however, will not penetrate the surface to great depths and will scour the surface without producing large secondary craters.

The few secondaries around Arandas illustrated in Fig. 4b, therefore, may have three causes: little mass arriving beyond the ejecta flow lobes; dispersed small-size ejecta (but above the critical particle size) not forming identifiable secondary craters; and/or ejecta configurations producing deposits and secondaries that do not fit within the selection criteria used in this study. The first possibility results from high ejection angles, lower relative velocities, and aerodynamic effects on small-size ejecta. The combination of these factors can result in late-arriving ejecta that override early-arriving ejecta, which comprise the massive overturned flap of smaller craters (Schultz and Gault, 1979). Figure 9 illustrates a 20 km-diameter crater in Elysium Planitia region of Mars (+35°, 211°) where grooves and ridges continuously cross the inner lobe of ejecta deposits and extend through the outer lobe. This morphology is not unique and is consistent with a scouring action by late-arriving ejecta materials, perhaps analogous to a base surge. A similar sequence has been proposed for Arandas (Mouginis-Mark, 1980).

The second possible cause of missing secondaries involves small-size but undecelerated ejecta. Faint ray systems extend from Arandas (Fig. 10a) and other martian craters. Occurrence of ray systems indicates that high-velocity, small-

Fig. 9. Martian crater (MC-7, Nd; 20 km diameter) near latitude +35°, longitude 211°W that exhibits radial grooves (arrow) and ridges extending across inner, thick ejecta lobe onto the outer ejecta lobe. This sequence of emplacement suggests a late-stage flow of ejecta overriding the inner lobe. Viking Frame 86A14.

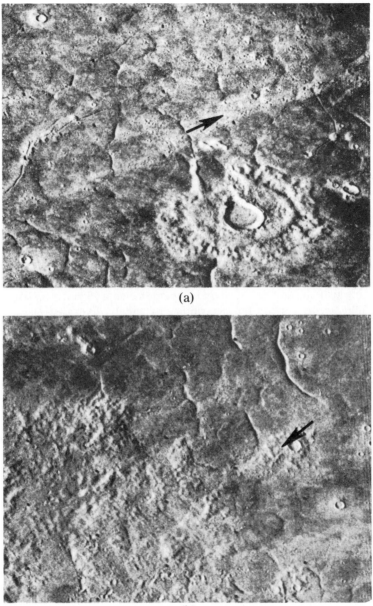

(a)

(b)

Fig. 10. The martian crater Arandas. (a) Ray system composed of small (<0.5 km) secondary craters below the size limit used in this study. Such rays were formed by ejecta large enough to be unaffected by aerodynamic drag, but small enough to produce barely recognizeable secondary craters. Area shown is approximately 30 km across. Viking Frame 32A21. (b) Hummocky units probably associated with Arandas and representing detached segments of ejecta deposits. Such units also are not included in the statistics but may represent low velocity, viscous ejecta or recocheted pieces of the ejecta flow lobes from the continuous facies. Viking Frame 32A32.

size ejecta extend to significant ranges without forming secondaries large enough to be included in this study. In addition, numerous secondaries can be identified beyond the ejecta lobes of Arandas, but are below our size-selection criterion. Many of these secondaries appear to be more elongate than the larger secondaries; therefore, they may indicate a high-velocity, low ejection-angle component of the ejecta.

The third cause is best illustrated by hillocky deposits radial to Arandas, but separated from the ejecta flow lobes (Fig. 10b). Secondary craters commonly are not associated with these deposits, yet the association with Arandas is unmistakable. They may be related to groups of dispersed fluidized ejecta that did not form secondary craters, but coalesced at impact to produce a limited flow separated from the main lobe as observed in the laboratory (Gault and Greeley, 1978; Greeley *et al.*, 1980). Alternatively, they may represent lower velocity fluidized segments recocheted and detached from the uprange ejecta lobe.

Figure 11a illustrates a type of crater not classed as a secondary crater, yet might be one. Such secondaries typically are circular with a diffuse halo and a central mound. A central mound has been recognized for certain lunar secondaries (Schultz and Mendenhall, 1979) and has been reproduced in the laboratory by impacting systems and fluidized and dispersed ejecta (Schultz and Mendenhall, 1979; Schultz *et al.*, 1980). The possibility of widely dispersed ejecta groups is illustrated by secondary crater clusters and chains. A more tightly clustered system perhaps associated with Arandas is shown in Fig. 11b.

CONCLUDING REMARKS

Although the number of primary craters used in this study is small, the results nevertheless have implications for interpreting the statistics of martian crater morphologies on a global scale. The following major points can be summarized:

1. The similarity in the distribution of secondaries and in the extent of the continuous ejecta facies between Crater I on Mars and craters of similar size on Mercury suggests that neither atmospheric nor target effects drastically modified the radial distribution of ejecta for certain martian craters.
2. Although numerous martian impact craters have major differences in the morphology and maximum extent of the ejecta flow lobes, the inner flow lobe of Arandas and other multi-lobed craters extends to about the same relative distance (2 R) as the continuous ejecta facies of similar-size mercurian craters.
3. Two craters with extensive flow lobes in the fractured plains region have very few secondaries larger than 2% the diameter of the primary.
4. Secondary craters associated with two multi-lobed craters in the fractured plains are significantly more circular than secondaries associated with a crater in the Chryse Planitia as well as secondaries around lunar and mercurian craters.

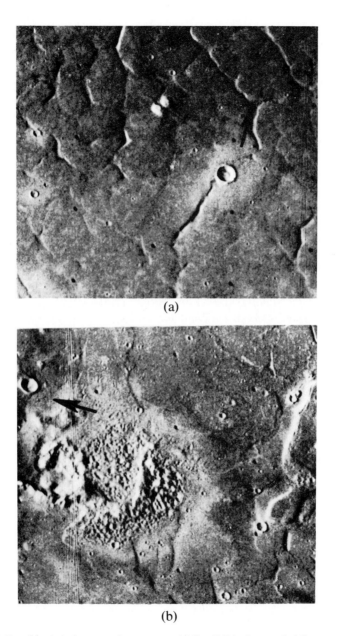

(a)

(b)

Fig. 11. Possible Arandas secondary craters. (a) Small (2 km) mounded-floor crater that could have developed by low velocity impacts (characteristic of secondary craters) by low viscosity projectiles. Such craters were not included in the statistics shown in Fig. 4. Viking Frame 32A35. (b) A tight cluster of small (<1 km) craters interpreted as secondaries from Arandas (direction indicated by arrow). Diffuse bright halo (15 km across) is slightly asymmetric, extending away from Arandas. The symmetry of this tertiary ejecta halo is in contrast to secondary craters on the Moon and further supports high angle ejection and impact of primary ejecta. Viking Frame 32A35.

5. Simple models correcting for possible high ejection angles do not appear to account for the paucity of secondaries beyond the outer ejecta flow lobes of Arandas.
6. The outer ejecta lobe of certain multi-lobe impact craters appear to have overridden the inner ejecta lobe.

The wide variations in secondary crater populations around medium-size (5–50 km) martian craters can be interpreted as effects of contrasting target lithologies. The extensive secondary cratering and elongate secondaries of Crater I in Chryse Planitia are consistent with a competent lithology of basaltic plains, as inferred from various photogeologic studies (e.g., Greeley *et al.*, 1977). Such targets would be expected to have relatively large ejecta, and experiments (Gault and Heitowit, 1963) indicate relatively low ejection angles (<50°). The relatively few large secondaries around Arandas, and their high circularity, are consistent with a lithology exhibiting low yield strength, as inferred for the fractured plains (e.g., Carr and Schaber, 1977). Here ejecta sizes probably are small owing to pre-impact grain size and enhanced comminution. The high ejection angles shown in experiments will contribute to a more massive near-rim arrival of ejecta (Gault and Greeley, 1978; Schultz and Gault, 1979; Greeley *et al.*, 1980). Aerodynamically decelerated fine-size ejecta will enhance near-rim arrival and could result in a late-stage flow overriding the inner ejecta flow lobes and forming or enhancing the outer flow lobes (Schultz and Gault, 1979; Mouginis-Mark, 1980).

Acknowledgments—The authors appreciate the detailed reviews by D. Wilhelms and C. Wood. Jill Singer gratefully acknowledges participation in the LPI Summer Intern Program. We also greatly appreciate the computer programming of John Harvey and Kin Leung. The Lunar and Planetary Institute is operated by the Universities Space Research Association under Contract No. NSR 09-051-001 with the National Aeronautics and Space Administration. This paper constitutes the Lunar and Planetary Institute Contribution No. 418.

REFERENCES

Allen C. C. (1979) Large lunar secondary craters: Size-range relationships. *Geophys. Res. Lett.* **6**, 51–54.
Austin M. G., Thomsen J. M., Ruhl S. F., Orphal D. L., and Schultz P. H. (1980 Calculational investigation of impact cratering dynamics: Material motions during the crater growth period. *Proc. Lunar Planet. Sci. Conf. 11th.* This volume.
Batson R. M., Bridges P. M., and Inge J. L. (1979) *Atlas of Mars.* NASA SP-48. 146 pp.
Boyce J. M. (1979) A method for measuring heat flow in the martian crust using impact crater morphology. Reports of Planetary Geology Program 1978–1979, p. 114–118. NASA TM-80339.
Carr M. H., Crumpler J. A., Cutts R., Greeley R., Guest J. E., and Masursky H. (1977) Martian impact craters and emplacement of ejecta by surface flow. *J. Geophys. Res.* **82**, 4055–4065.
Carr M. H. and Schaber G. G. (1977) Martian permafrost features. *J. Geophys. Res.* **82**, 4039–4054.
Gault D. E. (1974) Impact cratering. In *A Primer in Lunar Geology* (R. Greeley and P. Schultz, eds.), p. 137–176. TM X-62359.
Gault D. E. and Greeley R. (1978) Exploratory experiments of impact craters formed in viscous-liquid targets: analogs for rampart craters? *Icarus* **34**, 386–495.

Gault D. E., Guest J. E., Murray J. B., Dzurisin D., and Malin M. C. (1975) Some comparisons of impact craters on Mercury and the Moon. *J. Geophys. Res.* **80,** 2444–2460.

Gault D. E. and E. D. Heitowit (1963) The partitioning of energy for hypervelocity impact craters formed in rock. *Proc. 6th Symposium on Hypervelocity Impact,* p. 419–456.

Greeley R., Fink J., Gault D. E., Snyder D. B., Guest J. E., and Schultz P. H. (1980). Impact cratering in viscous targets: Laboratory experiments. *Proc. Lunar Planet. Sci. Conf. 11th.* This volume.

Greeley R. Theilig E., Guest J. E., Carr M. H., Masursky H., and Cutts J. A. (1977) Geology of Chryse Planitia. *J. Geophys. Res.* **82,** 4093–4109.

Mouginis-Mark P. (1979a) Ejecta emplacement of the martian impact crater Bamburg. *Proc. Lunar Planet. Sci. Conf. 10th,* p. 2651–2668.

Mouginis-Mark P. (1979b) Martian fluidized crater morphology: variations with crater size, latitude, altitude, and target material. *J. Geophys. Res.* **82,** 8011–8022.

Mouginis-Mark P. (1980) An emplacement sequence for martian fluidized ejecta craters (abstract). In *Lunar and Planetary Science XI,* p. 753–755. Lunar and Planetary Institute, Houston.

Saunders R. S., Mutch T. A., and Jones K. L. (1975) Guide to the use of Mariner images. *JPL Tech. Mem. 33–723.* 27 pp.

Schultz P. H. (1977) Endogenic modification of impact craters on Mercury. *Phys. Earth Planet. Inter.* **15,** 202–219.

Schultz P. H. and Gault D. E. (1979) Atmospheric effects on martian ejecta emplacement. *J. Geophys. Res.* **84,** 7669–7687.

Schultz P. H., Gault D. E., and Mendenhall M. H. (1980) Multiple-body impacts: implications for secondary impact processes (abstract). In *Lunar and Planetary Science XI,* p. 1006–1008. Lunar and Planetary Institute, Houston.

Schultz P. H. and Mendenhall M. H. (1979) On the formation of basin secondary craters by ejecta complexes (abstract). In *Lunar and Planetary Science X,* p. 1078–1080. Lunar and Planetary Institute, Houston.

Shoemaker E. M. (1962) Interpretation of lunar craters. In *Physics and Astronomy of the Moon* (Z. Kopal, ed.), p. 283–359. Academic, N.Y.

Theilig E. and Greeley R. (1979) Plains and channels in the Lunae Planum-Chryse Planitia region of Mars. *J. Geophys. Res.* **84,** 7994–8010.

Thomsen J. M., Austin M. G., Ruhl S. F., Schultz P. H. and Orphal D. L. (1979) Calculational investigation of impact cratering dynamics: early time material motions. *Proc. Lunar Planet. Sci. Conf. 10th,* p. 2741–2756.

Thomsen J. M., Austin M. G., Schultz P. H. (1980) The development of the ejecta plume in a laboratory-scale impact cratering event (abstract). In *Lunar and Planetary Science XI,* p. 1146–1148. Lunar and Planetary Institute, Houston.

Proc. Lunar Planet. Sci. Conf. 11th (1980), p. 2261–2273.
Printed in the United States of America

Hypervelocity impacts on Skylab IV/Apollo windows

Uel S. Clanton,[1] Herbert A. Zook,[1] and Richard A. Schultz[2]

[1]Geology Branch, NASA Johnson Space Center, Houston, Texas 77058
[2]Lunar and Planetary Institute, Houston, Texas 77058

Abstract—The three largest Skylab IV Command Module windows that were exposed for 84 days to space were optically scanned for impact features as small as 30 μm in diameter. This scanning effort, which was carried out at an optical magnification of 35×, detected features approximately three times smaller than were found in the original 5× scanning effort over the entire window surface by Cour-Palais (1979). Some 289 features were recorded from the 35× scan for later detailed analyses. Sixty of the largest and most promising features were cored from the windows for SEM and EDS analysis. Twenty-six of the cores contained craters with glassy pits, and of these, fourteen were found to contain strikingly obvious liners coating the interior of the glassy pit. The six largest features cored from the windows do not have a central glassy pit which leaves their previously reported hypervelocity origin in some doubt.

The remaining twenty-eight features that were cored from the windows show no clear evidence for a hypervelocity origin and evidence available at this time is insufficient to identify an origin in earth orbit or as ground damage. EDS analysis of six of the seven liners that have been examined show detectable aluminum in the liner or lip of the glassy pit. The source of aluminum is most probably an earth orbiting population of aluminum oxide spherules, exhaust effluent from solid rocket motors.

INTRODUCTION

The role of microparticle impacts in space and on planetary bodies without an atmosphere has been the focus of considerable research in support of the space program. Laboratory studies and analysis of lunar samples provide the bulk of reference data. This research outlines some new and unique observations of crater morphology on the Skylab IV/Apollo windows, features that have not been observed previously. These unusual data have renewed interest and resurrected certain questions about the nature and origin of some of the impacting materials in space. Our preliminary results are presented in support of this renewed interest.

The Command Module (CM) windows of the Skylab III and IV missions recorded the near-earth impacting meteoroid flux for periods of 59-1/2 and 84 days, respectively. Cour-Palais (1979) examined these windows for meteoroid impact craters and obtained an impact flux in very satisfactory agreement with his previous analyses of windows from the earlier Apollo missions.

These data have several important applications including: (1) obtaining absolute lunar regolith evolution rates; (2) establishing the current absolute erosion rates of lunar rocks; (3) establishing the surface exposure duration for certain lunar rocks still in crater production (if the cratering rate is assumed constant in time); (4) providing a foundation for deducing the space survival time of meteoroids against the collisional destruction of other meteoroids. However, questions had arisen about the origin of the impacting flux. These questions were derived largely from the investigations of Hallgren and Hemenway (1976) and Nagel *et al.* (1976) who detected abundant aluminum in some of the impact craters analyzed from Skylab experiment (S-149). Hallgren and Hemenway showed from field-of-view considerations that some of the craters with aluminum were produced by hyper-velocity impacts and were not derived from secondary ejecta from the adjacent orbital workshop. As there are no expectations of meteoroids with only aluminum and no other elements (with $Z>11$ and thereby detectable by energy dispersive X-ray analysis), these results gave rise to a suspicion of an earth-orbiting cloud of debris.

The above considerations, in part, prompted us to undertake a careful reex-amination of the Skylab IV CM windows for meteoroid impacts. We rescanned these windows optically at a magnification of $35\times$. This compares with the orig-inal $5\times$ scan of the entire window surface and a $20\times$ scan of 224 cm^2 of surface area (Cour-Palais, 1979). With our detection threshold set for a 30 μm impact spall diameter which corresponds to a pit diameter of about 7 μm, we had hoped to detect the inflection point (where the graph curvature changes from convex to concave) in the cumulative pit diameter distribution seen by Morrison and Zinner (1977) in lunar data. With the increase in meteoroid impact velocity largely due to the gravitational field of the earth, we anticipated that this inflection point should move to about 10 μm assuming the fused silica windows reacted similarly to lunar materials. We should, therefore, have had some chance of detecting this inflection point with our improved optical resolution.

WINDOW EXAMINATION AND CORING

The spacecraft windows are held in place by a gasket and a frame that restricts the area of exposure. The exposed area of each window was determined by cutting out sheets of paper to fit snugly into the recessed area of the window frame and then measuring the areas of these paper sheets with a planimeter. We measured areas of 940 cm^2 each for the right and left windows and 685 cm^2 for the hatch window. The last number differs somewhat from Cour-Palais' (1979) determination of 740 cm^2 for the hatch window. We believe our procedure yielded a more precise result ($\pm 1\%$ error) than did his approximate procedure. We then scribed the outline of these exposed areas on the Skylab IV CM windows with a diamond point pen in order to minimize the amount of area we needed to search for hypervelocity impact craters.

After cleaning the surfaces with a detergent (Alconox), the windows were optically scanned for candidate impact craters at a magnification of 35×. A small area of about 30 cm² was scanned at 50×. We feel that we were able, at 35×, to detect and examine essentially all craters with a spall diameter larger than 40 μm. Our threshold was set at 30 μm but it is probable that we failed to detect something like 10% of the crater population at that threshold. For the 50× scan we set a threshold of 15 μm.

To do the optical scanning we used a System C-7200 COSCAN optical comparator made by Optronics International (Chelmsford, Mass.). This system has an optically transparent table riding on an air bearing with 25 cm of travel in both x and y directions. We used it with transmitted light so that the craters would show up as shadows in an otherwise clear view. The table was motor driven with a logarithmic x-y controller and proved to be very satisfactory for our purposes.

Each window was scribed into four quadrants which were then scanned separately. When a crater was found, we searched it for features (such as a glassy pit) that would indicate a possible hypervelocity impact origin for it. The optical comparator had a 2× zoom capability that was useful for more detailed viewing. All those craters that were considered to be of possible hypervelocity impact origin were then recorded by a penciled dot on a sheet of white paper precisely positioned by retaining tabs glued on the sides of the window. Each such dot was numbered for later reference. We recorded 25 craters on the hatch window, 140 on the right window and 124 on the left window for a total of 289 craters. Approximately two hundred other features were dismissed during the optical scanning as pits clearly resulting from glass polishing processes or from some other non-hypervelocity impact origin.

The sixty craters that were optically judged to be the best candidates for a hypervelocity origin were then cored from the three windows, 25 from the right window, 22 from the left window and 13 from the hatch window. Four other craters were accidentally destroyed during the coring process. A one millimeter thick wafer containing the crater was then sawed from each core and ultrasonically cleaned in acetone, methonal, liquid freon, and occasionally, triply distilled water. Samples were then sputter coated with about 25Å of Au40:Pd60 alloy to produce an artifact free surface using the technique of Morrison and Clanton (1979). The coated samples were examined with a JEOL SEM-100CX TEM-SCAN which is capable of better than 30Å resolution point-to-point at 100,000× in the SEM mode. Chemical analysis data were obtained with a Princeton Gamma-Tech 1000 EDS.

CRATER MORPHOLOGY AND CHEMISTRY

Although all of the cores contained features whose origin was suspected to be from hypervelocity impacts, the magnification, resolution and depth of field of the optical microscope was inadequate to characterize or classify the features. Based on SEM studies, the impact features can be grouped into three major types: (1) Glassy pit craters, microcraters with a central glass lined pit surrounded by a raised lip of impact-fluidized glassy material, (2) pitless craters, microcraters without a central glass-lined pit and raised lip but with well developed zones of radial and concentric fracture, (3) damage craters, shallow features without fluidized glass but with poorly developed radial and concentric fracture.

Glassy pit craters

Twenty-six examples of hypervelocity impacts that produced impact-fluidized glassy pits and raised rims have been documented. The morphology in general is similar to features observed on lunar samples but the spall zones often appear to be more shallow (Fig. 1). However, twelve of the glassy pit craters, unlike

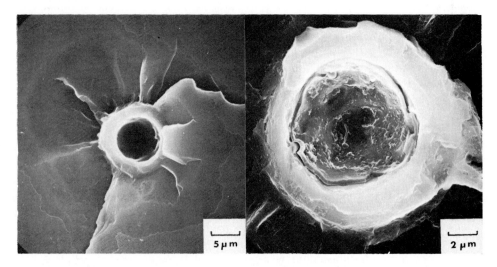

Fig. 1. SEM micrograph of a hypervelocity impact pit of probable micrometeorite origin. Fractures radiate from the glassy rim that surrounds the deep central pit. Most of the original surface near the point of impact has spalled away leaving conchoidal fracture scars. Four of the radial fractures extend beyond the conchoidal spall zone. A small portion of the original surface extends out over the glassy rim and serves to illustrate how the impact feature forms below the original surface. Surrounding the spall area is a thin white line that marks the edge of the magnesium flouride antireflection coating on the window that has been torn away by the impact event.

Fig. 2. SEM micrograph of a liner coating the interior of a hypervelocity impact pit. Based on the morphological relationships of the liner and the pit, a scenario during formation can be outlined. Some rather restricted conditions for the impact event are indicated and in particular, low impact velocities (<10 km/sec) are required because of the large amount of projectile material that survives the impact.

Initially, sufficient impact energy is available to form the classic glassy pit which chills rapidly. The projectile which is shock melted is sufficiently plastic to deform to the shape of the host pit yet not so fluid as to mix with and become an integral part of the glassy pit wall. The liner cools and contracts forming a cast of the pit wall. The separation of liner from the pit wall suggests materials with differing physical and chemical properties.

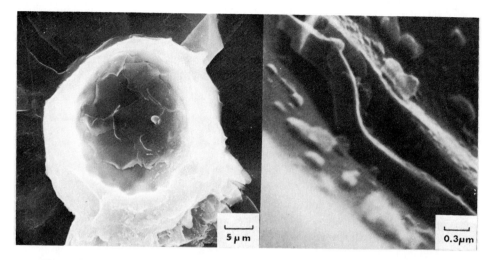

Fig. 3. SEM micrograph of a hypervelocity impact crater with two liners coating the interior of the glassy pit. Both the liners, although now incomplete and broken, appear to conform to the irregular interior of the pit yet are clearly separate from the wall and each other. The outer liner is crossed by several tension cracks; droplets and debris partially cover the exposed surface. The inner liner, exposed where the outer liner has broken away, is generally smooth with a surface that is almost totally free of debris.

Fig. 4. A higher magnification view of the broken edges of the double liner shown in Fig. 3. Some of the clues of the dynamics of crater and liner formation are illustrated in this SEM micrograph. The two layers were sufficiently plastic during emplacement to deform and coat the interior of the pit but because of physical or chemical differences, the layers do not mix with and become an integral part of the glassy wall. Further cooling and contraction tends to accentuate the separation of the liners from the host pit. The liners appear to have a fairly uniform thickness; the top liner is about 2000Å and the bottom liner is about 3000Å thick.

lunar samples, show clear evidence of a glassy liner in the pit (Fig. 2). Additionally, two other examples have been found that have a double liner (Figs. 3 and 4). Six elongate craters have also been documented and two of these have partially developed liners (Fig. 5). One crater which has a pit within a pit was also found (Fig. 6).

Pitless craters

This group contains the six largest features of possible hypervelocity impact origin observed on the windows (Fig. 7). One has a spall dimension of over 1 mm. The crater morphology is characterized by four distinct and concentric zones, (1) a large outer shallow spall, (2) a deeper spall with well developed radial fractures, (3) a depressed shatter zone of radial and concentric fractures and (4) a deep central shatter pit with well developed radial and concentric fractures. There is, however, no clear evidence impact-fluidized melt and no central pit of melted glass.

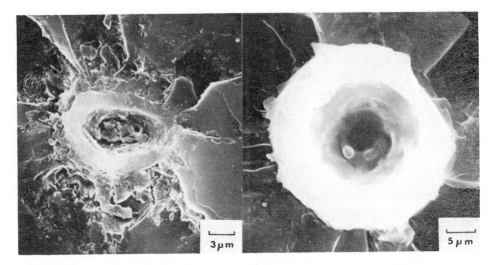

Fig. 5. Elongate hypervelocity impact craters on lunar samples are rare. This crater on the SL-4 window is not only elongate but also has a liner that is partially developed in the central pit. Although EDX analysis of the chemistry of the impact products is incomplete, Si and a small amount of Ti are the only elements detected in the rim materials of this crater.

Fig. 6. The SEM micrograph illustrates a hypervelocity impact pit of possible micrometeorite origin. Alhtough the general morphology resembles what has now come to be expected from a hypervelocity impact on the lunar surface, the interior morphology of the glassy pit is unusual. The normally concave floor is penetrated by a deeper and nearly concentric pit. An irregularly shaped projectile may explain the origin of the double pit.

Damage craters

These features have some morphological feature that suggest a hypervelocity impact origin when viewed under a binocular microscope. Further study with the SEM, however, dictates a low-velocity impact origin. The variety of fracture, shatter and spall morphology indicates that a wide range of particle sizes, densities and velocities contributed to the window damage. Figure 8 illustrates the damage from a low-velocity directional impact.

A higher energy origin for some of the damage craters may be argued on the basis of deep conchoidal spalls and well developed radial fractures. Some of these craters have what appears to be a fused aggregate of particles a few hundred angstroms in diameter that partially cover selected areas within the spall zone. These "popcorn" like features (Fig. 9) may represent incipient melting of projectile or target material, or perhaps some form of contamination that was not removed during the cleaning processes. Studies to date have not yet clearly identified an origin for these features.

Only a limited amount of chemistry has been attempted on the samples at this time. The data are limited to some qualitative EDS analysis and WDS analysis of some of the glassy pit craters with liners. The Skylab IV/Apollo window material is an optical grade of fused quartz and impurities do not exceed a few parts per thousand. The surface of the window is coated with about 200 angstroms of magnesium floride, an antireflection coating. EDS analysis of areas on the undamaged window easily detects the Mg-rich surface. In the spall areas, Si is typically the only element that can be detected. Six of the seven lined glassy pit craters that have been analyzed by EDS show detectable aluminum in the liner or rim material (Fig. 10). Additionally Ti has been detected in the rim of one of the elongate and lined glassy pit craters (Fig. 5).

Because much of the EDS analysis must be done on submicron thick features, considerable effort will be required to obtain more quantative data. The glassy pit crater shown in Fig. 2 was also subjected to extensive analysis on a Cambridge

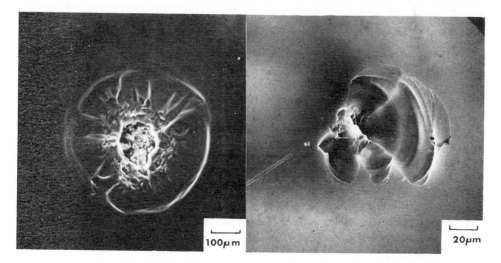

100μm   20μm

Fig. 7. The pitless craters are the largest damage features on the SL-4 windows and do not have a clear hypervelocity origin—no glassy central pit remains. There is evidence based on some laboratory tests, however, that the glassy pit may be dislodged by the violence of the impact event.

The damage typically forms four distinct zones (1) an outer very shallow spall about 1 mm in diameter, (2) a deeper spall about 500 μm in diameter that is characterized by large well developed radial fractures, (3) a depressed shatter zone about 230 μm in diameter of smaller radial and concentric fractures, and (4) a deep shatter pit about 150 μm in diamter with well developed radial and concentric fractures.

Fig. 8. A number of features too small to be clearly characterized with an optical microscope proved under SEM analysis not to have a hypervelocity impact origin; no central glassy pit had been developed. The variety of fracture, spall and shatter forms indicate a range in particles sizes or densities or velocities may have contributed to the window damage.

A zone of shatter marks the impact point of the projectile. The conchoidal spalls are unequally developed and asymmetrically arranged. An origin from a low-velocity directional impact is indicated.

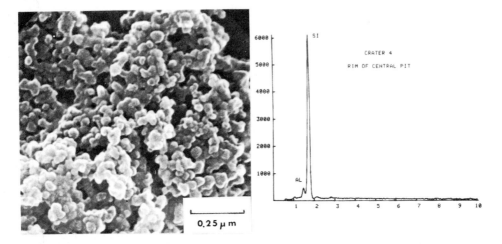

0.25 μm

Fig. 9. Some of the craters in the Skylab IV/Apollo windows have patches of what appear to be individual particles, some 400 to 600Å in diameter, that appear to have been fused together to form a "popcorn" morphology. This material may represent target material that has been partially sintered or may represent some form of contamination that was not removed during the cleaning process. EDS analysis of submicron particles is difficult and a definite composition has not yet been obtained.

Fig. 10. EDS spectra from the rim of a glassy pit crater. The verticle axis is count rate, the horizontal is energy (KeV). The dominant peak is silicon from the fused quartz window. The aluminum occurs only in the glassy rim material. The source of the aluminum is thought to be from aluminum oxide spherules, exhaust effluent from solid fuel rocket motors.

SEM equipped with wave length dispersive spectrometers (WDS). This WDS study confirmed the earlier EDS analysis; Al was the only foreign element that could be detected in the glassy pit/liner.

Crater morphology from hypervelocity impacts on lunar samples has been extensively documented since the return of the Apollo 11 samples (e.g., Carter and McGregor, 1970; Frondel *et al.*, 1970; Goldstein *et al.*, 1970; McKay *et al.*, 1970). Later, several research groups (e.g., Hörz *et al.*, 1971, and Fechtig *et al.*, 1976) carried out exhaustive surveys of microcratering on lunar rock surfaces. An extensive bibliography and review of the cratering literature is given by Ashworth (1978). Additionally, the work of Morrison and Clanton (1979) documents details of craters less than 1000Å in diameter on lunar samples.

Experimental studies of hypervelocity impacts under controlled laboratory conditions (e.g., Roy *et al.*, 1972; Roy and Slattery, 1973; Mandeville and Vedder, 1971; Vedder, 1971, 1976) provides an additional insight into such variables as projectile velocity, density, angle of incidence and the role of different target materials. A review of the equipment and the various techniques that have been used to produce hypervelocity impacts under laboratory conditions is given by Fechtig *et al.* (1978). Additionally, a brief description of the micrometeoroid detectors that have been flown in space is provided along with some of the problems associated with cross calibration of equipment/experiments.

The literature on lunar samples and laboratory simulations fails to document previous observations of liner morphology similar to those occurring on the Skylab IV/Apollo windows. The literature does, however, provide a precedent for hypervelocity craters with high aluminum. Hallgren and Hemenway (1976) analyzed 18 craters found on the S-149 Skylab experiment using a SEM with EDS capability and observed that most of the craters had high aluminum contents. The high aluminum contents are inconsistent with the observations of Anders *et al.* (1973) for the composition of meteorites in lunar soils. Although Nagel *et al.* (1976) and Hallgren and Hemenway (1976) could relate most of the hypervelocity pits to primary or secondary impacts from space debris, some craters appeared to have a true micrometeorite origin.

Our detection of high aluminum in six of the seven glassy pit craters with liners supports the findings of Hallgren and Hemenway (1976) and Nagel *et al.* (1976). A source for the aluminum may be inferred from indirect evidence; Brownlee *et al.* (1976) comment that 90 percent of the collected stratospheric particles in the 3 to 8 μm range are aluminum oxide spherules. Sampling flights through the exhaust plumes of Titan III rockets identified the source of these particles as exhaust effluent of solid rocket motors (Ferry and Lem, 1974). The review by Brownlee (1978) of stratospheric microparticle collection and analyses discusses the basis for identifying cosmic dust particles in this background of rocket exhaust effluent, terrestrial contamination and other man-induced space debris (titanium based paint flakes).

Our original goal was to determine the micrometeorite flux in near earth orbit. Our study has presented us with a much more complex problem than was first anticipated. We now find that we must separate a natural flux from a man-induced flux. At this time, the origin for two of the crater morphologies is not totally clear. The literature fails to provide a previous example of glassy liners in glassy pit craters in glass targets. However, Cour-Palais (pers. comm.) has observed an example of aluminum from a metallic projectile lining a hypervelocity pit in a copper target.

Additionally, the largest craters on the window do not have a clear hypervelocity origin. The radial and concentric fracture pattern of these features is characteristically associated with hypervelocity craters with glassy pits, yet no clear trace of a glassy pit remains. Some basis for arguing that the glassy central pit may have spalled from the surface can be developed. Cour-Palais (pers. comm.) has carried out a number of hypervelocity impact experiments that resulted in the ejection of the glass-lined pits from the fused silicate targets. These experiments were done with a light gas gun with projectile velocities in the range of 7 to 8 km/s. Also Carter and McKay (1971) noted an example of a pit that had nearly left its parent crater in a heated (750°C) fused silica target. The impact velocity in this case was 7.2 km/s. Notwithstanding these observations, however, we cannot yet be sure that the six largest craters did not result from processes occurring during manufacture, checkout or recovery of the Apollo spacecraft.

CRATER SIZE DISTRIBUTION

Cumulative crater size distribution based on SEM studies obtained from 32 of the 60 Skylab IV/Apollo window cores are shown in Fig. 11. The six largest features, the pitless craters, are plotted as closed symbols; the 26 glassy pit craters comprise the remaining plot of data. The 28 damage craters are not included. The 14 craters that have a liner which coats the central glassy pit are, in addition, replotted separately to the lower left in Fig. 11. In each case the total

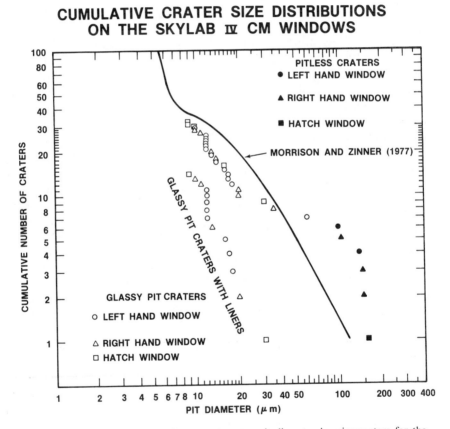

Fig. 11. Cumulative crater number versus crater pit diameter in micrometers for the three Apollo Command Module windows from the 84-day Skylab IV mission. Except for the six largest craters shown as filled symbols, only those craters were chosen that have a remelted glass-lined pit strongly indicative of hypervelocity impact. The data points to the upper right constitute a plot of all the larger candidate hypervelocity craters from the three windows, while the data points to the lower left are a subset in which the glass-lined pit has a separate inner liner of apparently foreign material. Morrison and Zinner's (1977) plot (renormalized to equal 7.5 at a 40 μm pit diameter) of impact pits on lunar rock 12054 is shown for comparison.

number of craters with a pit diameter larger than some chosen diameter are plotted versus that diameter. Both axes are scaled logarithmically.

Pit diameter was measured on SEM micrographs from rim center to rim center as presented by Morrison and Zinner (1977) but unlike Cour-Palais (1979) who measured the diameter of the interior of the rim. This difference in measurement becomes increasingly important in the smaller craters where the rim width is comparable to the rim diameter.

The shape of the cumulative curve formed by the 32 Skylab IV/Apollo datum points does not resemble the lunar impact pit size distribution curve obtained by Morrison and Zinner (1977) on lunar rock 12054, shown as a solid line on Fig. 11 (renormalized for easier comparison). These observations raise questions as to the origin of either the lunar impact pit data or the window impact pit data. One possible solution could be that neither the pitless craters nor the glassy pit craters with liners are due to meteoroid impact. A curve excluding both the pitless craters and the glassy pit craters with liners does give rise to a curve nearly parallel to the Morrison and Zinner (1977) lunar rock data. Another possible solution is to presume that many of the smaller lunar impact craters are formed by hypervelocity secondary ejecta. This latter solution seems less probable, however, because one would then also expect to see numerous low velocity impact features; but lunar samples are dominated by the glassy pit craters.

Because of our uncertainty as to the actual fraction of the impact craters that were due to meteoroid impacts, a flux vs. size curve is not now presented. However, to make a flux calculation in numbers of impacts per cm^2 per year, one merely divides the observed number of craters down to some chosen limiting diameter by 215. The number 215 is derived by reducing the window area (2565 cm^2) to an effective area that sees 2π steradians of space after accounting for Skylab and ATM shielding, window inset shielding and earth shielding (see Cour-Palais, 1979 for details) and then multiplying that effective area by the exposure duration in years (84 days = 0.23 years) of the windows. If one subtracts both the pitless craters and the glassy pit craters with liners from the data, a flux approximately 2.5 times lower than that given by Cour-Palais (1979) is obtained.

SUMMARY

Although more work clearly needs to be done to fully understand the origin and distribution of the microcraters on the Skylab IV/Apollo windows, the following observations seem pertinent:

1. Aluminum is detected as the only foreign component in six of the seven lined glassy pit craters so far examined by EDS analysis. The most probable source is from aluminum oxide spherules, exhaust effluent of solid fuel rocket motors. The seventh crater contains titanium which may have been derived from an impact of a chip of thermal paint.

2. The size distribution of the lined glassy pit craters appears to be compatible with an origin by hypervelocity impacts of aluminum oxide spherules. If the

aluminum oxide spherules are in earth orbit, impact velocities largely in the range of 7 to 10 km/s should be expected. Thus, the impact velocities are well below those expected for most impacts by micrometeorites and these lower velocities may be significant in the development of the lined glassy pit craters.

3. The documentation of hypervelocity impacts on Skylab IV/Apollo windows that contain aluminum support the observations of Hallgren and Hemenway (1976) and Nagle *et al.* (1976) and strongly indicate that there is a significant population of man-induced micro-debris in earth orbit.

4. The six largest craters observed on the windows are pitless craters. No impact fluidized glass is in evidence and an origin has not been clearly established. However, the conchoidal and radial fracture pattern of the pitless craters more closely resembles that observed with hypervelocity impacts than damage caused during polishing, window installation, ground operations, recovery, etc.

5. The shape of the curve of the Skylab IV/Apollo cumulative number versus pit diameter plot compares favorably with the Morrison and Zinner (1977) lunar curve only when the pitless craters and the lined glassy pit craters are excluded. We have not yet seen the inflection point expected at ~10 μm pit diameter.

Acknowledgments—We are deeply indebted to Fred Pearce of the Flight Equipment Section of NASA-JSC for the use of his laboratory facilities. His patience and understanding during the three month period while we used the optical comparator is sincerley appreciated.

A portion of this research was done while R. Schultz was an undergraduate Summer Intern at the Lunar and Planetary Institute, which is operated by the Universities Space Research Association under Contract No. NAS 9-3310 with the National Aeronautics and Space Administration. This paper constitutes Lunar and Planetary Institute Contribution No. 419.

REFERENCES

Anders E., Granapathy R., Krähenbühl U. R. S., and Morgan J. W. (1973). Meteoritic material on the Moon. *The Moon* **8**, 3–24.

Ashworth D. G. (1978) Lunar and planetary impact erosion. In *Cosmic Dust* (J. A. M. McDonnell, ed.), p. 427–526. Wiley, N.Y.

Brownlee D. E. (1978) Microparticle studies by sampling techniques. In *Cosmic Dust* (J. A. M. McDonnell, ed.), p. 295–336. Wiley, N.Y.

Brownlee D. E., Ferry G. V., and Tomandl D. (1976) Stratospheric aluminum oxide. *Science* **191**, 1270–1271.

Carter J. L. and MacGregor I. D. (1970) Mineralogy, petrology and surface features of some Apollo samples. *Proc. Apollo 11 Lunar Sci. Conf.*, p. 247–265.

Carter J. L. and McKay D. S. (1971) Influence of target temperature on crater morphology and implications on the origin of craters on lunar glass spheres. *Proc. Lunar Sci. Conf. 2nd*, p. 2653–2670.

Cour-Palais B. G. (1979) Results of the examination of the Skylab/Apollo windows for micrometeoroid impacts. *Proc. Lunar Planet. Sci. Conf. 10th*, p. 1665–1672.

Fechtig H., Gentner W., Hartung J. B., Nagel K., Neukum G., Schneider E., and Storzer D. (1976) Microcraters on lunar samples. In *The Soviet-American Conference on Cosmochemistry of the Moon and Planets* (J. H. Pomeroy and N. J. Hubbard, eds.), p. 585–604. NASA SP-370. Washington, D.C.

Fechtig H., Grün E., and Kissel J. (1978) Laboratory simulations. In *Cosmic Dust* (J. A. M. McDonnell, ed.), p. 607–669. Wiley, N.Y.

Ferry G. V. and Lem H. Y. (1974) Particulates in solid fuel rocket exhaust. *EOS (Trans. Amer. Geophys. Union)* **56**, 1123.

Frondel C., Klein C. Jr., Ito J., and Drake J. C. (1970) Mineralogical and chemical studies of Apollo 11 lunar fines and selected rocks. *Proc. Apollo 11 Lunar Sci. Conf.*, p. 445–474.

Goldstein J. I., Henderson E. P., and Yakowitz H. (1970) Investigation of lunar metal particles. *Proc. Apollo 11 Lunar Sci. Conf.*, p. 499–512.

Hallgren D. S. and Hemenway C. L. (1976) Analysis of impact craters from the S-149 Skylab experiment. In *Interplanetary Dust and Zodiacal Light,* Lecture Notes in Physics, **48** (H. Elsässer and H. Fechtig, eds.), p. 270–274. Springer-Verlag, N.Y.

Hörz F., Hartung J. B., and Gault D. E. (1971) Micrometeorite craters and lunar rock surfaces. *J. Geophys. Res.* **76**, 5770–5798.

Mandeville J.-C. and Vedder J. F. (1971) Microcraters formed in glass by low density projectiles. *Earth Planet. Sci. Lett.* **11**, 297–306.

McKay D. S., Greenwood W. R., and Morrison D. A. (1970) Origin of small lunar particles and breccia from the Apollo 11 site. *Proc. Apollo 11 Lunar Sci. Conf.*, p. 673–694.

Morrison D. A. and Clanton U. S. (1979) Properties of microcraters and cosmic dust of less than 1000Å dimensions. *Proc. Lunar Planet. Sci. Conf. 10th,* p. 1649–1663.

Morrison D. A. and Zinner E. (1977) 12054 and 76215: New measurements of interplanetary dust and solar flare fluxes. *Proc. Lunar Sci. Conf. 8th,* p. 841–863.

Nagle K., Fechtig H., Schneider E., and Neukam G. (1976) Micrometeorite impact craters on Skylab S-149. In *Interplanetary Dust and Zodiacal Light,* Lecture Notes in Physics, **48** (H. Elsässer and H. Fechtig, eds.), p. 275–278. Springer-Verlag, N.Y.

Roy N. L. and Slattery J. C. (1973) Study of impact cratering in lunar-like materials. TRW Final Report 17433–6002–RO–00 prepared under NASA contract No. NASW-2311. 95 pp.

Roy N. L., Slattery J. C., and Frichteniot J. F. (1972) Study for Apollo Window Meteoroid Experiment (S-176). TRW Final Report 209021-6001-RO-00 prepared under NASA contract No. NAS 9-12072. 95 pp.

Vedder J. F. (1971) Microcraters in glass and minerals. *Earth Planet. Sci. Lett.* **11**, 291–296.

Vedder J. F. (1976) Hollow lunar spherules and microcratering. *Meteoritics* **11**, 149–161.

Proc. Lunar Planet. Sci. Conf. 11th (1980), p. 2275–2308.
Printed in the United States of America

Computer code simulations of the formation of Meteor Crater, Arizona: Calculations MC-1 and MC-2

David J. Roddy[1], Sheldon H. Schuster[2], Kenneth N. Kreyenhagen[2] and
Dennis L. Orphal[3]

[1]U.S. Geological Survey, Branch of Astrogeologic Studies, Flagstaff, Arizona 86001, [2]California
Research and Technology, Incorporated, Woodland Hills, California 91367, [3]California Research
and Technology, Incorporated, Livermore, California 94550

Abstract—A series of numerical investigations of Meteor Crater, Arizona, is in progress to examine the formation of large bowl-shaped impact craters. The calculations are being performed by the two dimensional arbitrary Lagrangian-Eulerian computer code, CRALE, which uses finite difference methods to solve the continuum mechanics cratering models. Material modeling of the specific rock types and geologic conditions at Meteor Crater included effects of strength and fracturing, stratigraphic layering, porosity, water table and water-saturated strata. Two calculations, MC-1 and MC-2, have now been completed using 3.8 Megatons (1.6×10^{23} ergs) of impact energy and velocities of 25 km/sec and 15 km/sec, respectively. The results of MC-1 produced acceptable shock-induced flow fields, including particle velocity and dynamic pressure fields. Repeated ballistic extrapolations for MC-1 showed maximum crater growth and a terminal profile by ~272 msec. Permanent structural uplift of the upper rim strata and an overlying ejecta blanket were both shown in the solutions. Calculation of late-time rebound of a large region beneath the crater floor displayed a fractured and disorganized zone consistent with the location of the breccia lens observed at Meteor Crater. The calculated crater depth was approximately the same as the observed value for Meteor Crater, but the calculated crater diameter and volume were smaller for MC-1 than the observed values. The diameter was smaller by ~73% and the volume was smaller by a factor of ~3.8. The calculated profile was similar to a scaled-down profile of Meteor Crater, except in the deeper central region of the crater floor. The bowl-shaped profile, after scaling, matched 94% (by volume occupied) of the observed profile of Meteor Crater. To determine if a lower-velocity impact coupled more energy into deeper strata because of less surface heating and vaporization, a second calculation, MC-2, was computed using 3.8 Megatons impact energy and 15 km/sec impact velocity. Solution of MC-2 gave essentially the same results as those for MC-1. Present interpretation of both Calculations MC-1 and MC-2 suggests that the energy of formation of a bowl-shaped crater approximately one kilometer in diameter, using the initial conditions of Meteor Crater, is on the order of ~15 megatons (~6×10^{23} ergs). A new series of calculations is planned using a higher impact energy and modified material modeling to further examine this problem.

I. INTRODUCTION

Hypervelocity impact processes have been widely accepted as playing a major role in the evolution of the terrestrial planets and satellites. Recently, increased emphasis has been placed on developing quantitative methods to describe impact cratering, such as the use of continuum mechanics to numerically simulate impact events. Solutions of such complex equations commonly make use of two-dimensional, finite difference, computer codes. Applications of such code calculations to planetary impact problems were well documented in the first study of its kind by Bjork (1961), and more recently by Ahrens and O'Keefe (1978), Bjork *et al.* (1967), Bryan *et al.* (1978, 1980), Kreyenhagen and Schuster (1977), O'Keefe and Ahrens (1975, 1976, 1977, 1980), Orphal, (1977a, 1977b), Orphal *et al.* (1980), Roddy *et al.* (1980), Thomsen *et al.* (1979, 1980), and others. From these studies, it has become increasingly clear that the results of code calculations are greatly improved when initial impact conditions can be defined and when the numerical results can be tested against field and laboratory data.

In order to address this problem, a numerical code study of the formation of Meteor (Barringer) Crater, Arizona, has been undertaken with two goals in mind. The first goal is to complete a series of parametric calculations that quantitatively examine the formation of this crater under different impact conditions. Specifically, the objective is to determine the combinations of the impact velocities, impact energies, geologic, and material models that most accurately define the observed field and laboratory data for Meteor Crater. The second, longer term goal of this study is to calibrate the cratering code for impact studies by testing the results in detail against the Meteor Crater data, modifying the code input where necessary, and then generalizing it for use in other planetary environments.

This paper describes the major results from our first two code calculations, MC-1 and MC-2, that have been completed for Meteor Crater. Both calculations used an iron meteorite with a kinetic energy of 3.8 Megatons (1.6×10^{23} ergs). Calculation MC-1 had an impact velocity of 25 km/sec and MC-2 had an impact velocity of 15 km/sec. Material modeling of the specific rock types at Meteor Crater was used in both calculations. The calculations produced acceptable shock-induced flow fields as well as other major cratering aspects, such as structural uplift of the rim strata. Both calculations, however, indicate a final crater that is volumetrically smaller than Meteor Crater by a factor of ~3.8. Based on these first results, a second series of code calculations is planned to examine the effects of a larger impact energy and modified material models.

II. PREVIOUS NUMERICAL STUDIES

In the past, numerical simulation studies of the formation of large-scale natural impact craters have been highly limited because of the uncertainties in the initial conditions of the impacting body and the material responses of the target rocks. More recently, however, studies by a number of workers, such as Cooper (1977), Knowles and Brode (1977), Kreyenhagen and Schuster (1977), Maxwell (1977), Orphal (1977a, 1977b, 1980), Öpik (1976), Swift (1977), Trulio (1977), Shoemaker (1977), Shoemaker *et al.* (1979), Ullrich *et al.* (1977), Wetherill (1979), have helped place more reliable bounds on both the impacting bodies and material responses.

In 1960, Shoemaker completed the first numerical study outlining a specific natural impact event, i.e., the formation of Meteor Crater, Arizona. Shoemaker studied the geology of Meteor Crater and of two nuclear craters, Teapot ESS and Jangle U, at the Nevada Test Site. After comparing the types of structural deformation, he concluded that Meteor Crater was most similar to the Teapot ESS crater which had a 20.4 m depth of burial (DOB) for its nuclear device. Using three different diameters (hinge, apparent, and apparent before slumping) for Teapot ESS, Shoemaker calculated three scaled energies-of-formation for Meteor Crater of 1.4, 1.7, and 1.8 Megatons (~6 to ~8 $\times$ 10^{22} ergs). This scaling employed the empirical cube root scaling law of Lampson (Glasstone, 1957) for constant scaled depth of explosion, i.e., $d \propto W^{1/3}$, where d is diameter and W is nuclear yield. Choosing 1.7 Megatons (~7 $\times$ 10^{22} ergs) as an average and a velocity of 15 km/sec, the mass of the meteorite was calculated to have been 63,000 tons and to have had a diameter of 24.8 m if it was spherical. Shoemaker also used analytical calculations of one-dimensional shock to determine the peak pressures for several impact velocities (10, 15, 20 km/sec) and to outline the penetration history of the iron meteorite during compression and initial rarefaction. His results (for 15 km/sec) gave a maximum penetration of meteorite and compressed rock of about 164 m below ground, and a ". . . center of gravity of the energy released, or apparent origin of the shock . . . at roughly . . . 100 m to 122 m from the original surface along the path of penetration." Shoemaker concluded with an interpretation of the sequence of cratering events shown in a schematic series of geologic cross-sections which were based on both his geologic observations and calculations.

In 1961, Bjork, prompted by Shoemaker's request for a more quantitative description of the formation of Meteor Crater, completed the first application of a two-dimensional, finite difference, computer code calculation of a large-scale, natural, meteorite impact event. Bjork used the Particle-in-Cell (PIC) code to calculate the early stage of formation of Meteor Crater. For his numerical code solution, Bjork, following Shoemaker (1960), assumed an iron meteorite impacted vertically at a velocity of 30 km/sec, weighed 12,000 m tons, measured 12 m $\times$ 12 m (right cylinder), and had an impact energy of 1.1 Megatons (~5 $\times$ 10^{22} ergs). A homogeneous target consisting of tuff was ". . . used only because its equation of state was readily available to the author." Both the iron and the tuff were represented in the solution as purely hydrodynamic media without strength. Bjork assumed that a no-strength hydrodynamic calculation was valid at least for the early time because the very high pressures greatly exceeded the strength of the target rock and that the *early-time effects* should be similar for an impact into limestone during the initial penetration and shock compression phase. The solution was carried purely hydrodynamically to 61 msec, at which time the peak axial pressures were on the order of 20 kbar. At this time, however, material near the axis of penetration between depths of ~150 m to ~300 m (region of actual crater-floor depth to bottom of breccia lens) was still moving strongly downward, and calculated radial velocities extended only out to a maximum range of ~225 m (actual crater radius is ~593 m). Downward and outward velocities were still as high as 100 m/sec to 1,000 m/sec at 61 milliseconds. Consequently, the calculation used velocity vectors and pressure plots to describe only the dynamic *early-time history* of the iron penetration into the tuff mainly in the near-field cratered region. The iron meteorite, as well as part of the tuff, totally vaporized under these impact conditions. At 61 msec the shock wave had only expanded downward into the rock target to about 275 m depth and outward about 225 m along the ground surface.

The numerical uncertainties in such early code solutions, run to what were then considered very long times, precluded reasonable continuation of the calculation. Consequently, *no* calculation was made of the *final* crater profile, dimensions, or volume. Instead, a visual inspection of the field plots and scaling relationships of the impacting body with respect to a final crater size suggested a preliminary *estimate* of the final crater size to be about 150 m in depth and 500 m in radius (Bjork, 1961; pers. comm., 1980). Bjork noted that this underestimated the dimensions of the crater by about 20%, and suggested that scaling relationships between the crater and impacting iron indicated a larger meteorite ~14.4 m in diameter and weighing ~21,000 tons. Bjork considered the fact that the tuff was weaker than the actual limestone and sandstone at Meteor Crater, and said ". . . a rough estimate is that the crater calculated here is about 50 percent larger than would be obtained with these materials. This would indicate that a body of length and diameter 21.6 m would be required to produce the crater in limestone and sandstone, and such a meteorite would have a mass of about 71,000 tons." Since the initial impact energy and velocity are unknown, Bjork concluded that ". . . if the impact velocity were 11 km/sec, the required mass would be 194,000 tons; if it were 72 km/sec, about 30,000 tons." At constant momentum, the impact energies would be 2.8 and 18.5 Megatons (~1.2×10^{23} ergs and ~7.9×10^{23} ergs) respectively.

In a more recent study, Bryan *et al.* (1978) used the two-dimensional Eulerian, finite difference code called SOIL to calculate the "early-time or dynamic phase" of the formation of Meteor Crater. Bryan *et al.* (following Shoemaker, 1960 and Roddy, 1978), selected a set of initial conditions that included an iron meteorite impacting vertically at 15 km/sec into homogeneous limestone. The impact energy used by Bryan *et al.* (1978) was 4.5 Megatons (~2×10^{23} ergs). The meteorite would have weighed 167,000 tons and was given the shape of a right cylinder with dimensions of 30 m × 30 m. Both the iron and the target limestone were assumed to have non-hysteretic (return to same physical state after shock unloading) hydrodynamic properties, and the calculation was continued hydrodynamically to 500 msec. At this time, the material at the bottom of the transient cavity was still moving downward at about 200 m/sec or slightly greater, similar to Bjork's calculation in terms of velocities and ranges. This is a normal consequence of the hydrodynamic model. A standard ballistic extrapolation was then made in which upward moving material was allowed to pass through the original ground surface and leave as ejecta, and downward moving material was essentially fixed in place. This is conventional practice in the ballistic extrapolation technique. However, in a hydrodynamic model (no material rebound) deeper regions with high-velocity flow fields that are directed downward become fixed in place, thereby preventing any further expansion of the cavity beyond 500 msec. The termination of the calculation at this time was based upon the experience previously gained from numerous explosion code calculations (Bryan, pers. comm., 1980).

Bryan *et al.* (1978) indicated that they had also completed a second SOIL calculation which included material strengths for limestone with a shear modulus of 35 GPa (350 bars) and a von Mises yield strength of 0.02 GPa (0.2 bar). They stated that, "Preliminary analysis of results show similar behavior in the two [hydrodynamic vs. material strength] calculations." According to Bryan *et al.*, addition of elastic-plastic strength to the limestone reduced most of the crater dimensions and meteorite penetration depth by about 8 percent. They suggested that post-shock conditioned responses of the target rocks ". . . coupled with computational studies emphasizing times well beyond 0.5 seconds will be required to quantify the importance of material strengths in the late time formation of Meteor Crater."

To complete their calculations, Bryan *et al.* (1978) next adjusted the ballistic crater profile by calculating ". . . [a] final crater profile where a slope stability adjustment has been applied and the lower part of the crater was fit to a hyperbola." The ejecta on the flanks of the final crater profile also were repositioned in the upper part of the crater with a 1.2 bulking factor and 35 degrees for the slope stability angle. These final two adjustments, based upon techniques developed for subsurface explosion crater studies (Bryan, 1980, pers. comm.), gave their final crater profile and dimensions. Bryan *et al.* indicate that, "These calculational results are in good agreement with Meteor Crater in spite of simplifying assumptions." Depending on their choice of bulking factors (1.2 or 1.0), the final calculated dimensions of Bryan *et al.* had an apparent radius of 485 m or 505 m, apparent depth of 194 m, and apparent volume of $6.35 \times 10^7 m^3$ or $6.99 \times 10^7 m^3$. The estimated pre-erosion dimensions and present-day observed dimensions of Meteor Crater are shown in Fig. 1.

The code calculations MC-1 and MC-2 discussed in the remainder of this paper are similar to those of Bjork (1961) and Bryan *et al.* (1978) in that they also utilize two-dimensional, finite difference, computer codes. The information reported in this paper, however, results from a number of computational techniques not used in the earlier studies; the most prominent of these additions is material strength modeling with fracture properties.

III. PRE-IMPACT INITIAL, GEOLOGIC AND METEORITE CONDITIONS

The pre-impact conditions determined for the geology and those inferred for the meteorite are shown in Figs. 1 and 2. These conditions, used in our calculations of Meteor Crater, are described in detail in Roddy (1978). In summary, Meteor Crater is located in north-central Arizona near the southern edge of the Colorado Plateau. It was formed between 20,000 and 30,000 years ago by an iron meteorite impacting into nearly flat-lying sandstone, siltstone, and dolomitic limestone. The impact formed a bowl-shaped crater about 1.2 km across and about 200 m deep, measured at the rim crest. The crater is well-preserved and ~80 percent of its continuous ejecta blanket remains intact (Roddy *et al.*, 1975). The crater also still retains its general bowl-shape, except for a minor decrease in depth of ~30 m due to post-impact deposition of lake beds, alluvium and talus. The immediate post-impact dimensions estimated for the crater prior to erosion are shown in Fig. 1. The value for the Apparent Diameter/Apparent Depth was ~6.8 and the value for the Rim Crest Diameter/Rim Crest Depth was ~5.3, each estimated for the crater prior to erosion (Roddy, 1978). Detailed summaries of the geology and crater are given in Shoemaker (1960), Shoemaker and Kieffer (1974), and Roddy *et al.* (1975). Descriptions of the pre-impact geologic conditions, depth to water table, selected physical properties, impact energy calculations, estimated meteorite velocities and dimensions, crater initial dimensions and orientations of joints, faults and crater walls are given in Roddy (1978). The reader is referred to this companion paper for more complete discussions of the different initial conditions used in the code calculations reported here.

IV. EQUATIONS OF STATE AND MATERIAL MODELS

The basic initial conditions for the meteorite and target geology, including the geometry of the near-field computational grid, are shown in Fig. 2. The code calculations require symmetry about the axis of penetration, and therefore the impact is modeled with a vertical angle of incidence. Since the calculations are axisymmetric, only one-half of each velocity field plot is shown in the following figures. A standard 1 g gravitational field is used in both Calculations MC-1 and MC-2, but no atmosphere or overburden stresses are included because of their negligible effects on these calculations.

In Calculation MC-1, the meteorite was modeled as a right cylinder 20.32 m in length and diameter and is represented by 32 grid zones 2.54 m on a side. The Moenkopi sandstones and siltstones were modeled as a flat-lying unit 10 m thick with grid zones 2.54 m long in the radial direction and grid zones 2.5 m long in the vertical direction. The grid zoning for the finite difference calculations was increased geometrically with increasing distance from the impact point so that a region 1,700 m in radius and 2,000 m deep was finally included with 4,500 grid zones (50 lines radial by 90 lines vertical) at the time of conclusion of the calculation (272 msec). In calculation MC-2, the near-field grid was finer zoned by a factor of approximately two. No boundary effects in terms of shock reflections were experienced in the numerical solutions. All calculations were performed using the CRT two-dimensional arbitrary Lagrangian-Eularian finite difference code, CRALE.

The iron meteorite, as well as the sandstone and limestone strata, were modeled with updated versions of the Schuster and Isenberg (1972) and Isenberg (1972) equations of state (EOS) used extensively in nuclear and chemical explosion

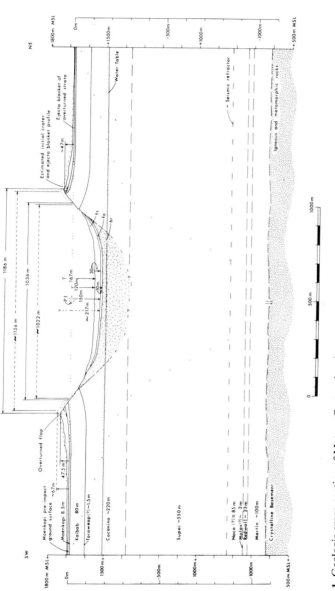

Fig. 1. Geologic cross-section of Meteor Crater showing a summary of stratigraphic thicknesses, present water table, pre-erosion and post-erosion crater dimensions and structure. The short-dashed information lines associated with the rim crest and apparent diameters, rim crest depth, and rim height are estimated initial pre-erosion values. All other crater dimensions are present post-erosion values. Two points of impact (PI) locations are shown to denote geometric center of crater and geometric center of this specific NE-SW cross-section. The symbols stand for: filling sediments = fs, fallout = fo, and breccia lens = br. The wide-spaced dotted line below the crater is an interpreted transitional zone enclosing fractured rock as inferred from seismic data (Ackermann *et al.*, 1975). Modified from Shoemaker (1960) and Roddy (1978).

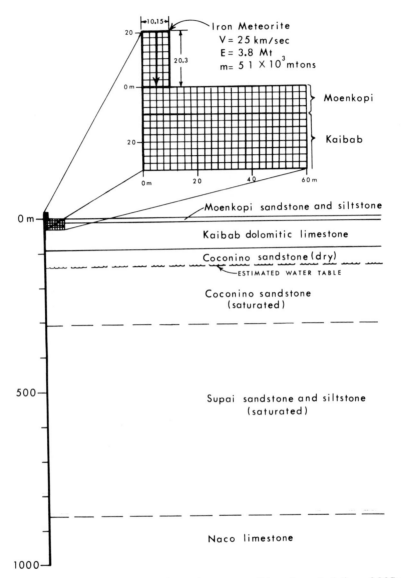

Fig. 2. Summary of initial meteorite and target conditions for calculation of MC-1. Conditions described in this paper and in Roddy (1978).

cratering studies. In these equations of state, the stress-strain-energy behavior is decomposed into a mean stress or pressure relationship plus the deviatoric stress tensor. The mean stress is further decomposed into two terms, i.e.,

$$P = P_s + P_v \tag{1}$$

where P_s represents the contribution to the mean stress of the solid and/or liquid phases and P_v represents the contribution of the vapor. Hysteresis, low-energy thermal effects, and reversible solid-solid phase changes are incorporated into the calculation of P_s. For a non-hysteretic material, such as the iron,

$$P_s = K_m\mu - (K_m - K_o)\mu^*(1 - e^{-\mu/\mu^*}) \tag{2}$$

where μ is the excess compression ($[\rho - \rho_o]/\rho_o$). The bulk moduli at zero and infinite compression, K_o and K_m, and μ^*, a free parameter, were selected (Table 1) to fit the Hugoniot data in Figs. 3a,c,d. The thermal energy dependence of the solid is incorporated by adding the effect of thermal expansion to μ to form an effective μ_{eff}

$$\mu + \beta E \rightarrow \mu_{eff} \tag{3}$$

where E = energy/unit mass and β = coefficient of thermal expansion/unit of energy. This is equivalent to the Grüneisen correction used in other models, with a variable Grüneisen gamma. At a solid-solid phase change, the effective μ is again altered to reflect the decrease in $dP/d\mu$. Hence, μ_{eff} is replaced by μ_Δ such that,

$$\mu_\Delta = \delta(\mu_{eff} - \mu_p) \tag{4}$$

where μ_p and δ are phase change parameters. Hysteresis is treated in a similar fashion (see Schuster and Isenberg, 1972) by a shift of the effective μ during loading. The vapor term, P_v, is computed using a variable gamma law gas,

$$P_v = (\gamma - 1)\rho E^* \tag{5}$$

where γ is ratio of specific heats, and ρ is density, and where

$$\gamma - 1 = .4 + .23 \log(\rho) + [.35 \log(E^*/\rho) - .464]^2 \tag{6}$$

and E^* is a non-negative effective energy density,

$$E^* = \max \begin{cases} (E - E_m)\left(1 - e^{\left(\frac{E_m - E}{E_m}\right)}\right), & \text{where } E > E_m \\ 0, & \text{where } E \leq E_m \end{cases} \tag{7}$$

where E is energy/gram and E_m is minimum energy for P_v. Thus, P_v is zero for materials with energy densities less than E_m.

Incremental deviatoric stresses, $d\sigma'_{ij}$, are computed from changes in the deviatoric strain tensor, $d\epsilon'_{ij}$, using the elastic equation,

$$d\sigma'_{ij} = -2G d\epsilon'_{ij}. \tag{8}$$

Similar to the variability of the bulk modulus in hysteretic materials, the shear modulus, G, is also a function of compression,

$$G = G_m - (G_m - G_o)e^{-\mu/\mu_g} \tag{9}$$

where G_o and G_m are the shear moduli at zero and infinite compression and μ_g is a free parameter determining the rate G increases from G_o to G_m. The second invariant of the deviatoric stress tensor, $\sqrt{J_2'}$, is then compared to a plastic yield surface, Y, where

$$Y = \min \begin{cases} C_o + \alpha p \\ Y_{vm} \end{cases} \qquad (10)$$

where C_o is the cohesion, α is the slope of the Mohr-Coulomb surface, and Y_{vm} is the limiting von Mises yield surface. If $\sqrt{J_2'}$ exceeds Y, the material has yielded and the deviatoric stresses are reduced by the standard Drucker-Prager flow rule, i.e., without volumetric plastic strain. Finally, the principal stresses are tested against the tensile limit, T. If a stress exceeds this limit the material is assumed to crack. The offending stress is set to zero and a crack volume computed. Subsequently, cracks can open perpendicular to this first crack and each can close upon the appropriate reloading of the material. Cracks do not heal so that if one closes it will re-open as soon as the stress perpendicular to it becomes tensile in nature; i.e., it does not have to reach the tensile limit again. Values of the input parameters for the materials used in this study are collected in Table 1.

The calculated Hugoniot for the iron is compared with experimental data up to 1 Mb in Fig. 3a. The Hugoniot to 5.0 Mb and several release adiabats shown in Fig. 3b illustrate the effect of energy on the release path. The release adiabats for an energy-independent equation of state would all lie on the Hugoniot. Although the elastic-plastic and failure properties of the iron are included for completeness, the initial pressures in the meteorite ($\sim$10Mb) vaporize the iron so that only the pressure-volume-energy relationship is significant for MC-1 at 25 km/sec.

The Kaibab limestone, between 10 m and 90 m depth, and the deeper limestones (below $\sim$850 m) were modeled separately because of the differences in their initial densities. The Kaibab, with 15% air voids, is highly hysteretic (experiences irreversible compaction after compression) under loading and unloading, thereby providing a significant mechanism to attenuate the shock. The Hugoniots for the two regions of limestone are compared to the experimental data in Fig. 3c. The lower initial densities in the *in situ* geologic materials generate more waste energy at higher pressures and cause the calculated Hugoniot curves to shift to the left above 0.5 Mb relative to the data which was obtained from small uniform samples near maximum crystal density ($\rho_o = 2.7$).

Similarly, three equations of state were necessary to model the various sandstones because of their differing initial properties. The Moenkopi sandstones and siltsones (0–10 m deep) have an initial density of $\sim$2.25 gm/cc and an air-filled porosity of $\sim$16%. The dry Coconino sandstone (90–140 m) has a reported average density of $\sim$2.03 and an air-filled porosity of $\sim$24%. The pre-impact level of the water table was placed at a depth estimated to have been $\sim$140 m (see Roddy, 1978). No distinction was made between the deeper saturated Coconino and underlying Supai sandstones since their *in situ* material properties are inferred (from petrologic examination) to be approximately the same. The saturated

Table 1. Summary of all physical and numerical constants used in calculation of MC-1 and MC-2.

Parameter	Symbol	Units	Meteorite Iron	Moenkopi Sandstone Siltstone Dry	Kaibab Limestone Dry	Coconino Sandstone Dry	Coconino Sandstone Saturated	Deeper limestone —
Initial bulk density	ρ_i	gm/cc	7.86	2.25	2.30	2.03	2.35	2.58
Reference grain density[1]	ρ_o	gm/cc	7.86	2.68	2.70	2.68	2.35	2.70
Air voids	—	%	0.0	16.0	14.8	24.3	0	4.4
Initial loading bulk mod.	K_i	Mb	2.0	0.06	0.1	0.06	0.225	0.5
Initial unloading bulk mod.	K_o	Mb	2.0	0.45	0.6	0.45	0.225	0.6
Maximum bulk modulus	K_m	Mb	2.0	0.70	0.8	0.70	0.58	0.8
Exponential factor	μ^*	—	0.0	0.25	0.4	0.25	0.30	0.4
Coef. of thermal expansion	β	cc/(cc−eu)[2]	7.86	3.0	3.0	3.0	3.0	3.0
Phase change	μ_P	—	0.04835	0.5996	0.05	0.5996	0.8243	0.05
parameters	δ	—	12.28	0.536	14.286	0.536	0.47	14.286
Minimum energy for P_v	E_m	eu	0.005	0.037	0.01	0.037	0.025	0.01
Initial shear modulus	G_o	Mb	0.508	0.05	0.15	0.05	0.045	0.15
Maximum shear modulus	G_m	Mb	0.508	0.12	0.15	0.12	0.11	0.15
Exponential factor	μ_g	—	—	0.0015	—	0.0015	0.0015	—
Cohesion	C_o	Mb	0.003312	5×10^{-5}	0.0002	5×10^{-5}	5×10^{-5}	0.0002
Tan Θ, where Θ = angle of internal friction	α	—	0.0	0.75	1.0	0.75	0.75	1.0
von Mises yield limit	Y_{vm}	Mb	0.003312	0.003	0.003	0.003	0.003	0.003
Tensile limit	T	Mb	−0.0052	-0.667×10^{-4}	−0.00015	-0.667×10^{-4}	-0.667×10^{-4}	−0.00015

(1) Zero air-filled porosity; (2) eu = Energy Unit = 10^{12} ergs/gm.

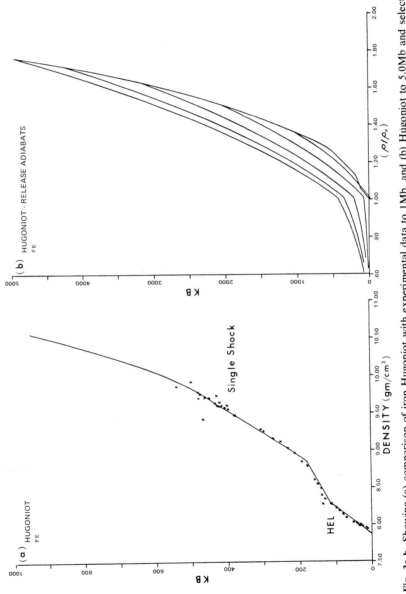

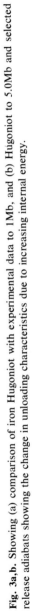

Fig. 3a,b. Showing (a) comparison of iron Hugoniot with experimental data to 1Mb, and (b) Hugoniot to 5.0Mb and selected release adiabats showing the change in unloading characteristics due to increasing internal energy.

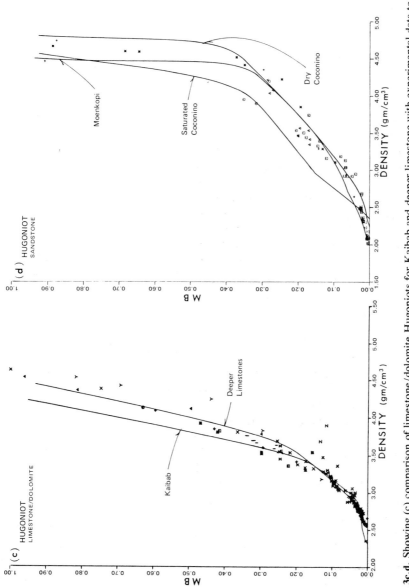

Fig. 3c,d. Showing (c) comparison of limestone/dolomite Hugoniots for Kaibab and deeper limestone with experimental data to 1Mb, and (d) Hugoniots for Moenkopi, dry Coconino, and saturated Coconino sandstones to 1Mb.

Coconino and Supai sandstone (140–850 m) were estimated to have a density of ~2.35 with no air voids. These different initial conditions lead to very dissimilar Hugoniots (Fig. 3a) due to the effect of the large differences in internal energy (PΔV/2) generated at high pressures.

V. RESULTS OF CALCULATIONS MC-1 AND MC-2

a) Introduction

This section summarizes the results of our first code calculation for Meteor Crater, designated MC-1, and briefly compares them with the preliminary results of the second code calculation, MC-2. The initial conditions for the meteorite used in both MC-1 and MC-2 are shown in Table 2. The results summarized here for MC-1 show the basic cratering flow fields, ballistic extrapolation of the final crater profile, crater dimensions and volume, determination of a subsurface fragmented zone co-located with the observed breccia lens and structural uplift in the rim. The difference in the size of the crater calculated in MC-1 vs. the observed size of the Meteor Crater is discussed in terms of an increase in the initial impact energy and possible revisions in material models. Other results from the calculations, such as those related to ejecta or shock metamorphism, will be presented later.

b) Particle velocity flow fields for MC-1

The numerical integration of Calculation MC-1 covered an interval from the time of initial impact to 272 msec after the impact. Figures 4 to 10 illustrate the material flow fields by particle velocity vectors in both the meteorite and target rocks. For MC-1 the calculated Hugoniot pressure is initially 10 Megabars for a coherent iron meteorite impacting sandstone at 25 km/sec. This high pressure drives the iron-sandstone interface downward at ~17 km/sec. A shock wave of this initial magnitude propagates into both the meteorite and into the Moenkopi and Kaibab rocks beneath the impact. As these shocks intersect free surfaces at the sides and back of the meteorite and in target media adjacent to the impact, pressure relief (rarefaction) propagates rapidly into the shocked region. The Moenkopi sandstones and siltstones adjacent to the penetration cavity are highly compressed and experience initial uplift at this time. The underlying Kaibab limestone is also highly compressed but is forced mainly downward during this early time. Peak pressures drop and the shock front in the target strongly diverges outward. Despite the rapid drop in shock pressures, the entire iron meteorite and adjacent rock experience sufficient peak pressures in the MC-1 calculation to produce vaporization upon relief. However, as noted by Moore *et al.* (1967), meteorite fragments that are now present around the crater exhibit limited to no shock metamorphic effects. Both Shoemaker (pers. comm., 1980) and Moore (pers.

Table 2. Summary of initial meteorite conditions used in Calculations MC-1 and MC-2 for Meteor Crater.

| Calculation | Impacting body | Impact energy | | Impact velocity (km/sec) | Meteorite mass (metric tons) | Meteorite size right cylinder (meters) | Momentum (tons-km/sec × 10^6) | Total calculation time (msec) |
		Megatons	Ergs					
MC-1	Iron meteorite	3.8	1.6×10^{23}	25	51,800	20.3	1.3	272
MC-2	Iron meteorite	3.8	1.6×10^{23}	15	144,000	28.6	2.2	70

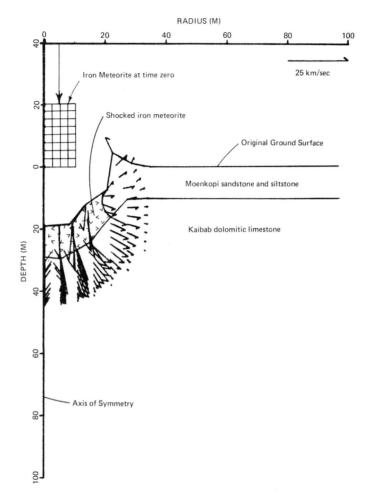

RADIUS (M)

25 km/sec

Iron Meteorite at time zero

Shocked iron meteorite

Original Ground Surface

Moenkopi sandstone and siltstone

Kaibab dolomitic limestone

DEPTH (M)

Axis of Symmetry

Fig. 4. Velocity vector field for Calculation MC-1 at 0.001770 seconds. Meteorite shown with grid zone and impact velocity of 25 km/sec at zero time. Particle velocities of vectors shown by scale.

comm., 1980) suggested these fragments may have spalled off the main meteorite during its atmospheric passage and were not directly involved in the impact.

By 1.77 msec (Fig. 4), the impacting body has penetrated below the original ground surface, become highly distorted, and expanded radially to about twice its initial diameter. By this time, for computational convenience, the thin Moenkopi unit beneath the meteorite has been merged into the Kaibab. It is retained in the calculation as a stratigraphic unit beyond the impact region to assist in ejecta tracing and definition of structural deformation in the rim. As noted by Orphal *et al.* (1980), the penetration of the meteorite deepens the cavity at a rate greater than the expansion of the radius during the early time.

The flow fields at 3.1 and 4.8 msec (Figs. 5 and 6) show that pressures in the meteorite and in Moenkopi strata near the surface have dropped to the point where large expansions associated with vaporization occur. The rear part of the meteorite, as well as Moenkopi strata, begins to expand into the transient cavity behind the meteorite. A similar sequence of phenomena has been observed in numerical analyses of the impact of iron bodies into aluminum at 20 and 72 km/sec (Bjork *et al.*, 1967).

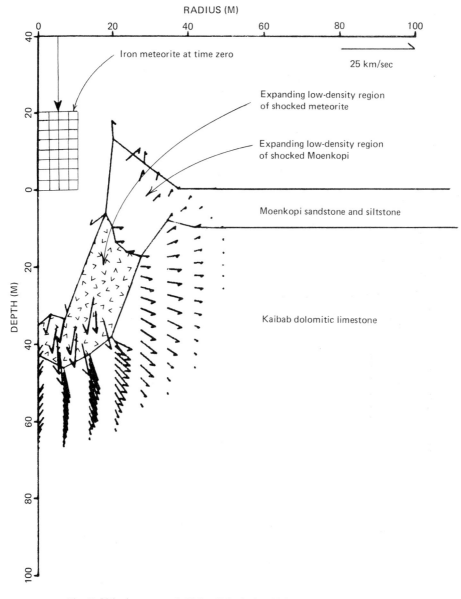

Fig. 5. Velocity vector field for Calculation MC-1 at 0.003137 seconds.

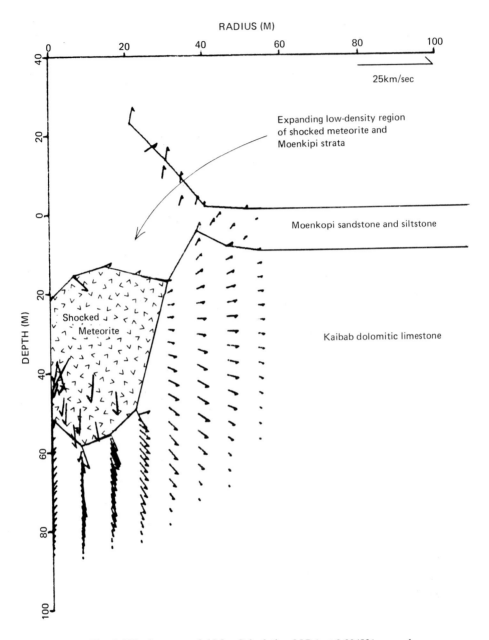

Fig. 6. Velocity vector field for Calculation MC-1 at 0.004831 seconds.

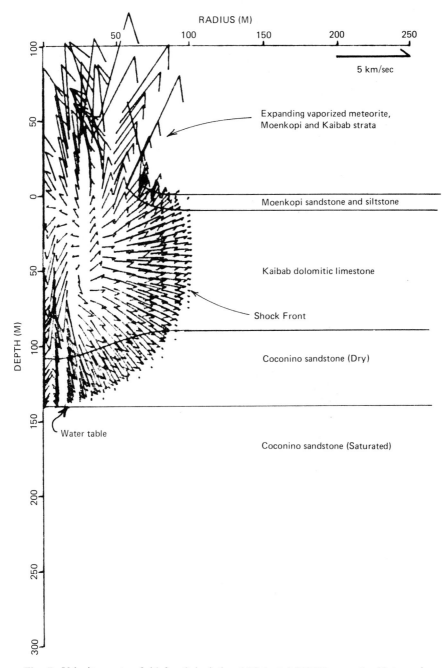

Fig. 7. Velocity vector field for Calculation MC-1 at 0.012626 seconds. Note scale change from Fig. 6.

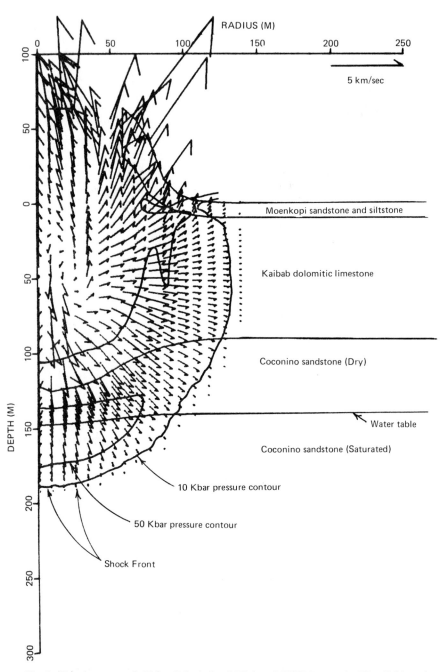

Fig. 8. Velocity vector field for Calculation MC-1 at 0.022396 seconds. The 10 kb and 50 kb pressure contours are superimposed on the particle velocity field.

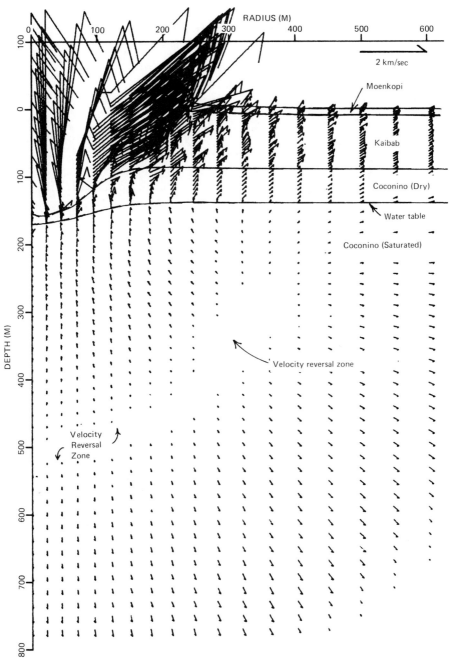

Fig. 9. Velocity vector field for Calculation MC-1 at 0.181428 seconds. Only every other velocity vector printed.

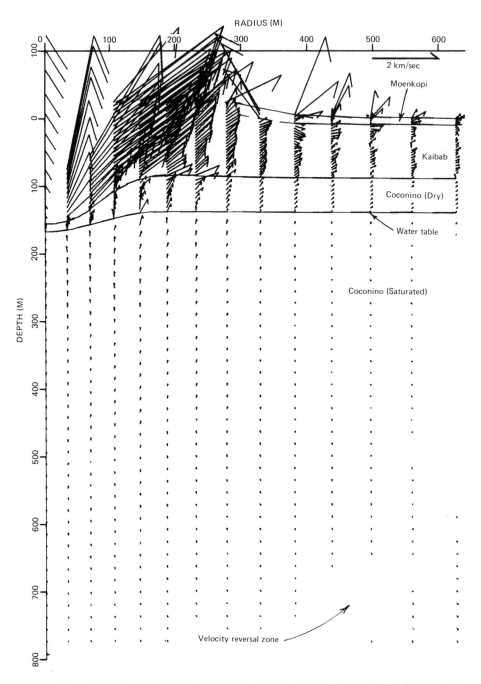

Fig. 10. Velocity vector field for Calculation MC-1 at 0.272403 seconds. Final calculated time for MC-1. Only every other velocity vector printed.

The continuing effects of vaporization of the iron meteorite and of Moenkopi and Kaibab strata are seen at 12.6 and 22.4 msec in Figs. 7 and 8. As pressure relief propagates into the strongly-shocked region, material still expands as a vapor and streams violently upward, thereby forming a cavity which is completely filled with vapor of continually diminishing density. For the remainder of the numerical solution, vapor escapes from this region at relatively high velocities.

Superimposed on Fig. 8, at 22.4 msec, are 10 kbar and 50 kbar (1 GPa and 5 GPa) pressure contours. Peak pressures in the strata are about 60 kbars (6 GPa) in the shocked rocks directly beneath the impact at ~130 to ~175 m depth, when peak pressures near the surface have dropped to about 10 kbars (1 GPa). Pressures on the order of 3 to 5 kbars (0.3 GPa to 0.5 GPa), which persist within the vapor-filled cavity, drop relatively slowly as the vapor continues to stream rapidly up and out.

By 181 msec (Fig. 9), sufficient vapor has escaped to reduce pressures in the cavity to the extent that material (rock) rebound occurs as deeper strata beneath the crater relax from their highly compressed state. A region of approximately zero velocity separates this rebound from the detached main shock, which continues to propagate into deeper layers (Fig. 9). An important point here is that elastic material rebound, which was not observed in earlier hydrodynamic solutions of Meteor Crater (Bjork, 1961; Bryan *et al.*, 1978), is critical to the final development of the breccia lens as discussed in a later section.

Integration of MC-1 was terminated at 272 msec at which time peak pressures in the crater field had dropped to ~200 bars to 300 bars (~0.02 GPa). The final velocity field, seen in Fig. 10, shows marked rebound in the rocks underlying the final crater region. There is continued blowout of vapor from the initially highly-shocked strata, although pressures have decayed to about 20 bars (~0.002 GPa). The finite difference solution was terminated at 272 msec.

c) Prediction of a final crater profile for MC-1 by ballistic extrapolation

Repeated ballistic extrapolations from about 250 msec to 272 msec showed that the crater volume and dimensions were changing less than about 1 percent for MC-1. Ballistic trajectories were computed for those cells of material having upward velocities sufficient to carry them up to the initial ground level or higher. The material in each of these cells was assumed to accumulate at the range where the terminal part of the trajectory would re-intersect the original ground level as the material landed and formed an ejecta blanket. To account for the lower density in the ejecta (bulking), such fallback was assumed to have a density of 2.0 g/cm³. Material having insufficient vertical velocity to reach the original ground level was not included in this ballistic extrapolation, but assumed to remain *in situ* at 272 msec.

The most important results of Calculation MC-1 are shown in Fig. 11 in which the crater profile, volume and dimensions (Fig. 11a) are compared with the observed field values for Meteor Crater (Fig. 11b). This solution, with detailed

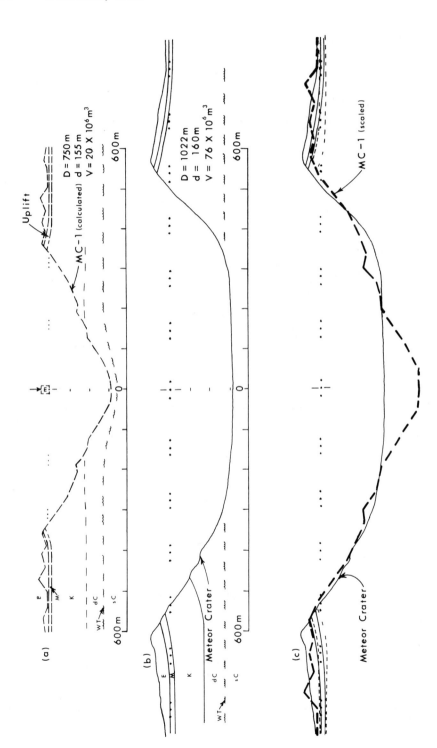

material modeling constructed specifically for these rocks, predicts that the impact of a 51,000-ton iron meteorite at 25 km/sec will produce a calculated crater volume which is only 26% of the observed volume. The calculated apparent volume is a factor of about 3.8 smaller than the observed apparent volume. The predicted diameter is ~72% of the present observed apparent diameter and ~73% of the inferred apparent diameter before erosion. The crater depth calculated for this impact, however, is nearly the same as that observed for the apparent crater floor of Meteor Crater.

The calculated profile of MC-1, scaled up by a factor of approximately 1.43 in diameter (approximately 3.8 in volume) is in reasonable agreement with the observed apparent profile of Meteor Crater, except for the lower central region (Fig. 11c). In particular, the calculated crater profile reproduces the shape of the observed middle and upper part of the crater walls relatively accurately (Fig. 11c), that is, those parts of the wall not covered by talus. It is important to note that the region within this part of the scaled calculated profile involves about 94% of the *total* crater volume. The calculated profile for the lower central part or floor of the crater (about 200 m in diameter) extends about 45 m below the observed apparent crater floor as shown in Fig. 11c. Although this central region appears relatively large as drawn in the cross-section of Fig. 11c, the departure of the scaled calculated profile from the observed profile actually involves a region comprising *only* about 6% of the total predicted crater volume.

The small central region that was calculated to extend downward to about 150 m below the original ground surface may have resulted from either the specific material modeling, positioning of the pre-impact water table, deep energy coupling, or some combination of these conditions. If such a region did actually exist in the cratering process, it would be expected to fill with fallback and talus flow from the walls of the crater. The possibility that the calculation "bottomed" on the water table, as is common in numerous explosion experiments (Roddy, 1976), is also a point yet to be considered. Although this central region involves only ~6% of the total crater volume, we expect to examine it further in other calculations. The *important point*, however, is that Calculation MC-1 reproduced

---

Fig. 11. The calculated crater profile and subsurface data for MC-1 are shown by dashed lines in (a), and the observed profile and subsurface data for Meteor Crater are shown by solid lines in (b). Both are drawn to the same horizontal and vertical scale. The estimated level of the water table is the same in both (a) and (b). In (c) the calculated profile of MC-1 is scaled up by a factor of approximately 1.43 in diameter (approximately 3.8 in volume) to fit the observed profile of Meteor Crater. The calculated profile *as scaled* in (c) matches approximately 94% of the total volume of the observed profile of Meteor Crater. The deeper central part of the scaled calculated profile below the crater floor is only approximately 6% of the total crater volume. All dimensions and volumes are apparent (original ground surface measured values) and the Meteor Crater values are estimated for a pre-erosion condition. The symbols include: E = ejecta, M = Moenkopi Formation, K = Kaibab Limestone, dC = dry Coconino Sandstone, sC = saturated Coconino Sandstone, WT = estimated level of water table and m = size of meteorite used in calculation. Section (c) modified after Roddy (1978) and Shoemaker (1960).

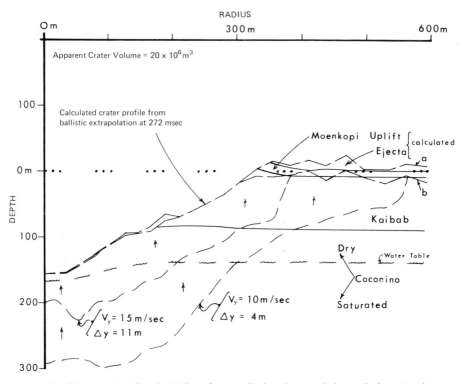

Fig. 12. Calculated profiles for MC-1 of strata displaced upwards beneath the crater in the general region observed as the breccia lens at Meteor Crater. Strata between the lowest dashed profile and the calculated crater profile will not be ejected from the crater, but instead will fall back in a disorganized and bulked state to form the breccia lens region. Material along the two dashed-line profiles has moved upward (↑) at a velocity (V_y) through a calculated distance (Δy) before falling back. Profile line marked (a) shows the top of the calculated ejecta. Profile line marked (b) shows the upper calculated limit of strata that did not rise above the original ground surface in the ballistic extrapolation. Material between these two lines (a and b) consists of ejecta, as well as strata that passed through the original ground surface and either remained *in situ* (uplifted) or fell back.

approximately 94% of the crater profile in terms of volume of the middle and upper parts of the apparent crater, thereby matching the upper and middle parts of the observed apparent profile of Meteor Crater reasonably well.

The most important structural feature calculated to exist in the rim occurs in the Moenkopi and upper Kaibab strata (Fig. 11). This structure consists of a permanent uplifted rim identical to that observed at Meteor Crater. The calculated value for the uplift is about 50% of the average measured rim uplift at the crater. A smaller value is to be expected, however, since the calculated crater is smaller than the actual observed crater. The calculated final crater profile has an Apparent Diameter/Apparent Depth of ~4.7 and a Rim Crest Diameter/Rim Crest Depth of ~4.4.

d) Calculation of a deep fragmented breccia lens zone in MC-1

The central part of the calculated profile in MC-1 represents the apparent crater floor observed at Meteor Crater (Fig. 11). Beneath this part of the profile is a large region of fragmented strata which was calculated to be moving upward at 272 msec due to rebound from the earlier compressed state. The velocities of these rocks, however, are not sufficient to reach pre-impact ground level. In the ballistic extrapolation shown in Fig. 11, such material did not contribute to the ejecta thrown from the crater. It was, in effect, fixed in place. While such material will not be excavated from the crater, it will nonetheless alter the crater floor profile by fragmenting and bulking upwards, and it will certainly contribute to the sub-crater structural deformation in a region commonly described as the breccia lens (Shoemaker, 1960; Roddy, 1978).

In order to examine the region of deep fragmentation and possible formation of a breccia lens, a modified set of ballistic-type contours was calculated. Figure 12 shows two contours of constant vertical velocity at 272 msec. Also shown is the maximum upward displacement that material along these contours experiences in the earth's gravitational field. For example, beneath the crater floor, these contours indicate that a region nearly 70 m thick will be displaced upward between 11 and 150 m before finally falling back in a disorganized, highly bulked state within the crater. As shown in Fig. 12, this involves a large volume of material that has experienced substantial upward displacement. The limited movement of this fragmented strata within the crater suggests that it would form fallback breccia mainly *in situ* as opposed to fragmented rock slumping off the walls and flowing into the crater. It would be expected that the degree of disorganization in the fallback, i.e., the breccia lens, would be related to the specific geometry of the paths of maximum upward and downward displacements, as well as other turbulent mixing caused during movement. Presumably material near the surface would be most severely disorganized and mixed, with the degree of disorganization decreasing with depth. This would be consistent with material mixing in breccia lenses in explosion craters (Roddy, 1976, 1977).

A more detailed analysis of these calculations, such as some type of dynamic mass flow adjustment, is necessary to determine quantitative changes in the final crater profile. Note, however, that no additional rock is ejected from the crater in these calculations, and consequently the total volume will not change. Instead, the basic effect of such late-stage material motions would only be to decrease the depth of the crater and increase its radius.

e) Calculation MC-2 and energy partitioning at a lower impact velocity

Calculation MC-1 predicted a reasonable bowl-shaped profile, but gave a diameter and volume smaller than that observed for Meteor Crater. One reason for the difference could be that an impact at 25 km/sec with 3.8 Megatons (1.6 $\times$ 10^{23} ergs) of kinetic energy does not couple sufficient energy to the deeper rock.

Simply stated, more energy is used to heat and vaporize the meteorite and near-surface target rocks in a high velocity impact as opposed to a low velocity impact.

To examine the possibility that a 25 km/sec impact is less efficient as compared to a lower velocity impact in forming crater volume, a second calculation, MC-2, was performed for a 15 km/sec impacting meteorite. The same amount of incident kinetic energy was used in MC-2 (3.8 Megatons) as was used in MC-1 by increasing the iron mass to 142,000 tons (a 28.6×28.6 m right cylinder). This solution was then run in order to compare energy partitioning and peak pressures with the corresponding quantities from the results of the higher velocity impact in MC-1.

Figure 13 compares the distribution of energy in the meteorite, and in the Moenkopi, Kaibab, and Coconino layers in solutions MC-1 and MC-2 up to 70 msec. The iron meteorite transfers much of its energy to the shallow Moenkopi and Kaibab layers earlier in the 25 km/sec impact, due to the higher shock pressure and faster shock speed generated in these layers. The critical point here is that after about 20 msec, the energy distributions become very similar for *both* MC-1 and MC-2. After this time the flow fields and energy distributions are essentially identical between the two calculations. In both impacts, about 90% of the incident kinetic energy is coupled into the Kaibab and Coconino layers, i.e., the region of strata in which over 95% of the crater is subsequently developed. More important, after about 20 msec the cratering flow fields appear nearly identical for a 3.8 Megaton event despite the two different impact velocities.

Figure 14 compares peak pressures in MC-1 and MC-2 along the axis of penetration for the two impacts. There are the expected differences in pressures near the surface, reflecting the higher initial Hugoniot pressures in the higher velocity impact for MC-1. At depths below about 60 m, however, the peak pressures nearly converge.

From these comparisons we tentatively conclude that for the same level of incident energy, a difference in impact velocity of 10 km/sec in the 15–25 km/sec range does not substantially affect the total energy coupled into the target media, and therefore does not substantially affect the crater volume.

VI. DISCUSSION

The first part of a new study of Meteor Crater, Arizona, has been completed using two-dimensional, finite difference, computer code techniques to simulate formation of this large bowl-shaped impact crater. Two code solutions, MC-1 and MC-2, have been computed using the most recent estimates of impact energy and range of impact velocities (Roddy, 1978). The energy was estimated by scaling nuclear explosion data from Cooper (1976), and the velocity range was taken from r.m.s. velocity values of Shoemaker (1977). Real physical properties for the observed rock types were included in the material modeling for the code calculations of both MC-1 and MC-2. Calculation of MC-1 (25 km/sec and 3.8 Megatons), after scaling, gave a reasonably accurate bowl-shaped profile for the

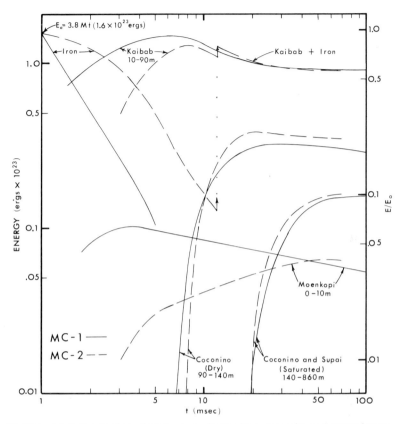

Fig. 13. Calculated distribution of total energy in the iron meteorite and target layers versus time for MC-1 (25 km/sec) and MC-2 (15 km/sec). The solid lines (—) are for MC-1 and the dashed lines (_ _) are for MC-2. The dotted line represents the merging of the iron meteorite and Kaibab rocks for computational convenience at about 10 msec.

middle and upper parts of the crater, encompassing ~94% of the total volume. Results from Calculation MC-1 also displayed certain types of structural deformation, such as a fragmented (brecciated) region beneath the crater and an uplifted rim, both of which are observed at Meteor Crater. The calculated depth of the crater was approximately the same as the observed value. The crater diameter and volume, however, were smaller for MC-1 than the observed values for Meteor Crater. We consider the differences in calculated versus observed values may be related in the case of Meteor Crater to 3 different potential causes, including a) excessive strength modeling of the target rocks, b) low energy coupling due to a high (25 km/sec) impact velocity, or simply c) an impact energy that is too low.

The first possible cause, material strength modeling, is always of concern in code calculations. However, the material parameters used in Table 1 were derived from measurements on laboratory samples of the actual rocks at Meteor Crater

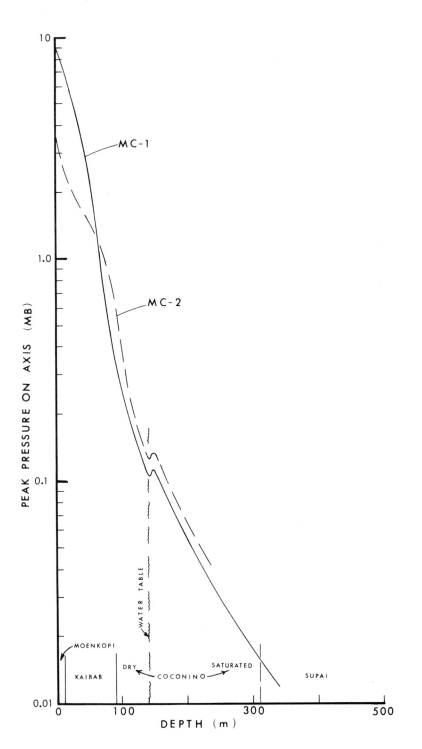

or from data reported for identical rock types. Values of *in situ* physical properties can be somewhat different than when measured in the laboratory, but our experience in comparable explosion code studies suggests that potential *in situ* differences in these rock types would not seriously affect the general cratering dynamics calculated for a Meteor Crater-type event.

Shock degradation of material strength properties by severe distortion and cracking is another potential problem in material modeling (Ullrich, 1976). Swift (1977) has shown that when physical properties are degraded to very low strengths, a larger crater can form. In his model, however, conditions for strength degradation by shock and the degree of degradation were specifically chosen to emphasize their effects on cratering. For example, the cohesive strength of material experiencing as little as a 100 bar (0.01 GPa) shock pressure was reduced to ~2 bar (~0.002 GPa) as soon as the pressure started to unload. Such strength changes are quoted for relatively unconsolidated materials. Such extreme strength degradation would probably be unrealistic for the well-consolidated sandstones and limestones at Meteor Crater.

In our Calculations MC-1 and MC-2, strength degradation was not explicitly modeled. Instead, a tensile cracking and subsequent distention material modeling was used as part of the CRALE code. In the Kaibab limestone, for example, such cracking is calculated to extend to ranges well beyond the calculated crater rim. This had the net effect of weakening (degrading) the resistance of rock to radial expansion of the crater in that region. While the effects of reduced or degraded strength on cratering certainly require further consideration, we do not presently believe this factor is the source of the large volume discrepancy between the calculated crater in MC-1 and the observed volume of Meteor Crater.

The effect of a lower water table on the material modeling would have been to deepen the crater, but not greatly increase the volume because of the small radius-squared effect at depth. A shallower water table would have decreased the crater depth, as seen in numerous explosion experiments and calculations, but that would have further decreased the calculated volume. The estimated location of the water table certainly would have affected the calculated crater volume, but it does not appear that it could have changed the volume by more than ~5 to 10%.

The second area of possible concern, variation of kinetic energy coupling with depth as a function of impact velocity, was tested by comparing MC-1 with MC-2. In these two calculations, the kinetic energy of impact was held constant at 3.8 Megatons (1.6×10^{23} ergs). MC-1 used 25 km/sec and MC-2 used 15 km/sec for their respective impact velocities. Our initial thought was that the higher-velocity impact conditions coupled too much energy into the near-surface rocks

---

Fig. 14. Calculated peak pressures along the vertical axis of penetration for MC-1 using 25 km/sec impact velocity and MC-2 using 15 km/sec. Shock pressures caused by an iron meteorite impacting normal to the target surface with 3.8 Megatons (1.6×10^{23} ergs) impact energy.

by extreme heating and vaporization very early in the impact at shallow depths. On the other hand, a lower-velocity impact that heated and vaporized very little near-surface material could conceivably be expected to couple more energy to deeper strata and thereby form a larger crater. The result of calculation MC-2, shown in Fig. 13, indicates that approximately the *same amount* of total energy was transferred to the deeper rocks for *both* MC-1 and MC-2 by 20 msec after impact. After that time the flow-fields and total energy distribution were essentially the same. The same is true for the peak pressures shown in Fig. 14. Our interpretation of the comparisons between MC-1 and MC-2 is that the final crater volume and dimensions would be approximately the same if run to comparable times. That is, under conditions of constant impact energy, variation of the impact velocity from 15 km/sec to 25 km/sec does not seriously change the cratering results.

The final area of concern, i.e., the impact energy is too low, appears to be the most likely explanation for the differences between the calculated volume and diameter and the observed volume and diameter of Meteor Crater. More specifically, the scaled profiles shown in Fig. 11 suggest the impact energy used in MC-1 is low by a factor of ~3.8. This would indicate that energy scaling from nuclear and HE explosion data, such as in Roddy (1978) and others, underestimates the energy required to form a large impact crater; i.e., nuclear and HE energy scaling for equivalent kinetic impact energy appear to require a calibration constant(s). A new set of calculations is planned using a higher impact energy and modified material modeling to further examine this problem.

Acknowledgments—Reviews and suggestions for this paper by J. Bryan, H. Moore, R. Pike, H. Swift, and W. Ullrich greatly assisted the writers, and we express our thanks to each for their time and help. The suggestions of R. Pike were especially helpful in improving our discussion of the scaling relationships, and we greatly appreciate the editorial patience of P. Criswell and R. Edwards of the LPI. We remain indebted to Dr. G. Sevin and Major R. Swedock of the Defense Nuclear Agency for their continued support of these studies. D. J. Roddy again expresses his continued appreciation to J. Roddy for her assistance in this study and to M., M. and M. Roddy for their help in the collection of field data. This work was supported through the cooperation of the National Aeronautics and Space Administration under Contract W-13,130 and the Defense Nuclear Agency, Department of Defense.

REFERENCES

Ackermann H. D., Godson R. H., and Watkins J. S. (1975) A seismic refraction technique used for subsurface investigations at Meteor Crater, Arizona. *J. Geophys. Res.* **80**, 765–775.

Ahrens T. J. and O'Keefe J. D. (1978) Energy and mass distributions of impact ejecta blankets on the moon and Mercury. *Proc. Lunar Planet. Sci. Conf. 9th*, p. 3787–3802.

Bjork R. L. (1961) Analysis of the formation of Meteor Crater, Arizona. *Proc. Acad. Nat. Sciences of Philadelphia* **76**, 275–278.

Bjork R. L., Kreyenhagen K. N., and Wagner M. H. (1967) Analytical studies of impact effect applied to the meteoroid hazard. NASA CR–757. 186 pp.

Bryan J. B., Burton D. E., Cunningham M. E., and Lettis L. A. Jr. (1978) A two-dimensional computer simulation of hypervelocity impact cratering: Some preliminary results from Meteor Crater, Arizona. *Proc. Lunar Planet. Sci. Conf. 9th*, p. 3931–3964.

Bryan J. B., Burton D. E., Lettis L. A. Jr., Morris L. K., and Johnson W. E. (1980) Calculations of impact crater size versus meteorite velocity (abstract). In *Lunar and Planetary Science XI*, p. 112–114. Lunar and Planetary Institute, Houston.

Cooper H. F. Jr. (1976) Estimates of crater dimensions for near-surface explosions of nuclear and high-explosive sources. RDA-TR-2604-001, sponsored by Lawrence Livermore Laboratory. 47 pp.

Cooper H. F. Jr. (1977) A summary of explosion cratering phenomena relevant to meteor impact events. In *Impact and Explosion Cratering* (D. J. Roddy, R. O. Pepin and R. B. Merrills, eds.), p. 11–14. Pergamon, N.Y.

Glasstone S. (1957) The effects of atomic weapons, U.S. Atomic Energy Commission, Washington, D.C. 579 pp.

Isenberg J. (1972) Part II Mechanical properties of earth materials, Nuclear Geoplosics. DNA 1285H2. Defense Nuclear Agency, Washington, D.C.

Knowles C. P. and Brode H. L. (1977) The theory of cratering phenomena, and overview. In *Impact and Explosion Cratering* (D. J. Roddy, R. O. Pepin and R. B. Merrill, eds.), p. 869–896. Pergamon, N.Y.

Kreyenhagen K. N. and Schuster S. H. (1977) Review and comparison of hypervelocity impact and explosion cratering calculations. In *Impact and Explosion Cratering* (D. J. Roddy, R. O. Pepin, and R. B. Merrill, eds.) p. 983–1002. Pergamon, N.Y.

Maxwell D. E. (1977) Simple *Z* model of cratering, ejection, and the overturned flap. In *Impact and Explosion Cratering* (D. J. Roddy, R. O. Pepin, and R. B. Merrill, eds.), p. 1003–1008. Pergamon, N.Y.

Moore C. B., Birrell P. J., and Lewis C. F. (1967) Variations in the chemical and mineralogical composition of rim and plains specimens of the Canyon Diablo meteorite. *Geochim.Cosmochim. Acta* **31**, 1885–1892.

O'Keefe J. D. and Ahrens T. J. (1975) Shock effects from a large impact on the Moon. *Proc. Lunar Sci. Conf. 6th*, p. 2831–2844.

O'Keefe J. D. and Ahrens T. J. (1976) Impact ejecta on the Moon. *Proc. Lunar Sci. Conf. 7th*, p. 3007–3025.

O'Keefe J. D. and Ahrens T. J. (1977) Impact-induced energy partitioning, melting, and vaporization on terrestrial planets *Proc. Lunar Sci. Conf. 8th*, p. 3357–3374.

O'Keefe J. D. and Ahrens T. J. (1980) Cometary impact calculations: flat-floors, multi-ring and central peaks (abstract). In *Lunar and Planetary Science XI*, p. 830–832. Lunar and Planetary Institute, Houston.

Öpik E. J. (1976) *Interplanetary Encounters*. Elsevier, N.Y. 155 p.

Orphal D. L. (1977a) Calculations of explosion cratering, Part I: the shallow-buried nuclear detonation JOHNNY BOY. In *Impact and Explosion Cratering* (D. J. Roddy, R. O. Pepin, and R. B. Merrill, eds.), p. 897–906. Pergamon, N.Y.

Orphal D. L. (1977b) Calculations of explosion cratering, Part II: the shallow-buried nuclear detonation JOHNNIE BOY. In *Impact and Explosion Cratering* (D. J. Roddy, R. O. Pepin, and R. B. Merrill, eds.), p. 907–918. Pergamon, N.Y.

Orphal D. L., Borden W. F., and Larson S. A. (1980) Calculations of impact generation and transport (abstract). In *Lunar and Planetary Science XI*, p. 833–835. Lunar and Planetary Institute, Houston.

Roddy D. J. (1976) High-explosive cratering analogs for bowl-shaped central uplift, and multiring impact craters. *Proc. Lunar Sci. Conf. 7th*, p. 3027–3056.

Roddy D. J. (1977) Large-scale impact and explosion craters: Comparisons of morphological and structural analogs. In *Impact and Explosion Cratering* (D. J. Roddy, R. O. Pepin, and R. B. Merrill, eds.), p. 185–246. Pergamon, N.Y.

Roddy D. J. (1978) Pre-impact geologic conditions, physical properties, energy calculations, meteorite and initial crater dimensions and orientations of joints, faults, and walls at Meteor Crater, Arizona. *Proc. Lunar Planet. Sci. Conf. 9th*, p. 3891–3930.

Roddy D. J., Boyce J. M., Colton G. W., and Dial A. L. (1975) Meteor Crater, Arizona, rim drilling with thickness, structural uplift, diameter, depth, volume, and mass balance calculations. *Proc. Lunar Sci. Conf. 6th*, p. 2621–2644.

Roddy D. J., Kreyenhagen K. N., Schuster S. H., and Orphal D. L. (1980) Theoretical and observational support for formation of flat-floored central uplift craters by low-density impacting bodies (abstract). In *Lunar and Planetary Science XI*, p. 943–945. Lunar and Planetary Institute, Houston.

Schmidt R. M. (1980) Meteor Crater—Implications of centrifuge scaling (abstract). In *Lunar and Planetary Science XI*, p. 984–986. Lunar and Planetary Institute, Houston.

Schuster S. H. and Isenberg J. (1972) Equations of state for geologic materials. DNA 2925Z Report. Defense Nuclear Agency, Washington, D.C.

Shoemaker E. M. (1960) Penetration mechanics of high velocity meteorites illustrated by Meteor Crater, Arizona. In *Structure of the Earth's Crusts and Deformation of Rocks*, Rept. 18, p. 418–434. International Geological Congress, XXI Session, Copenhagen.

Shoemaker E. M. (1977) Astronomically observable crater-forming projectiles. In *Impact and Explosion Cratering* (D. J. Roddy, R. O. Pepin, and R. B. Merrill, eds.), p. 617–728. Pergamon, N.Y.

Shoemaker E. M. and Kieffer S. (1974) Synopsis of the geology of Meteor Crater. In *Guide Book to the Geology of Meteor Crater*, 37th Ann. Met. Soc. Mtg., August 1974. 66 pp.

Shoemaker E. M., Williams J. G., Helin E. F., and Wolfe R. F. (1979) Earth-Crossing asteroids: orbital classes, collision rates with Earth and origin. In *Asteroids* (T. Gehrels, ed.), p. 253–282. Univ. Arizona Press, Tucson.

Swift R. P. (1977) Material strength degradation effect on cratering dynamics. In *Impact and Explosion Cratering* (D. J. Roddy, R. O. Pepin, and R. B. Merrill, eds.), p. 1025–1042. Pergamon, N.Y.

Thomsen J. M. and Austin M. G. (1980) The development of the ejecta plume in a laboratory-scale impact cratering event (abstract). In *Lunar and Planetary Science XI*, p. 1146–1148. Lunar and Planetary Institute, Houston.

Thomsen J. M., Austin M. G., Ruhl S. F., Schultz P. H., and Orphal D. L. (1979) Calculational investigation of impact cratering dynamics: Early time material motions. *Proc. Lunar Planet. Sci. Conf. 10th*, p. 2741–2756.

Trulio J. G. (1977) Ejecta formation: Calculated motion for a shallow buried nuclear burst, and its significance for high velocity impact cratering. In *Impact and Explosion Cratering* (D. J. Roddy, R. O. Pepin, and R. B. Merrill, eds.), p. 919–958. Pergamon, N.Y.

Ullrich G. W. (1976) The mechanics of central peak formation in shock wave cratering events. AFWL-TR-75-88, Air Force Weapons Laboratory, Kirtland AFB, New Mexico.

Ullrich G. W., Roddy D. J., and Simmons G. (1977) Numerical simulations of a 20-ton TNT detonation of the Earth's surface and implications concerning the mechanics of central uplift formation. In *Impact and Explosion Cratering* (D. J. Roddy, R. O. Pepin, and R. B. Merrill, eds.), p. 959–982. Pergamon, N.Y.

Wetherill G. W. (1979) Fragmentation of asteroids and delivery of fragments to Earth. In *Comets, Asteroids, Meteorites—Interrelations, Evolution and Origin* (A. H. Delsemme, ed.), p. 283–291. Univ. Toledo, Ohio.

Note added in proof: Recent centrifuge studies by R. M. Schmidt (This volume) also suggest, as does our calculational study, that the energy of formation for Meteor Crater is higher than the few-megaton range quoted in previous explosion scale studies. Schmidt interprets their experimental results to indicate energies in excess of 25 megatons for impact velocities in the 15 to 25 km/sec range.

Proc. Lunar Planet. Sci. Conf. 11th (1980), p. 2309–2323.
Printed in the United States of America

Impact melt generation and transport

D. L. Orphal[1], W. F. Borden[1], S. A. Larson[1], and P. H. Schultz[2]

[1]California Research and Technology, Incorporated, 4049 First Street—Suite 135, Livermore, California 94550, [2]Lunar and Planetary Institute, 3303 NASA Road 1, Houston, Texas 77058

Abstract—A series of continuum mechanics computer code calculations are planned to investigate the effects of variations in impactor mass and velocity on the generation and transport of impact melt. Results from the first two calculations are reported. In both of these calculations the impactor is modeled as a spherical iron projectile having a mass of 1×10^{12}g. The target is modeled as a gabbroic anorthosite (GA) half-space. In one calculation (NASA-1) the impact velocity is 5 km/sec and in the second calculation (NASA-2) it is 15.8 km/sec. In the NASA-1 calculation 6.8×10^{10}g (about 0.07 projectile masses) of GA is completely or partly melted. In NASA-2, 3.6×10^{12} g of GA are partially vaporized while 10.4×10^{12}g (about 10.4 projectile masses) of GA are completely or partly melted. The 5 km/sec and 15.8 km/sec calculations were performed to real times of 1.0 and 1.3 seconds, respectively. Ballistic extrapolations from these times give a crater diameter, D, of about 0.9 km and a volume, V, of about 6.5×10^7 m^3 for NASA-1. For NASA-2, D $\approx$ 2 km, V $\approx 6.8 \times 10^8$ m^3. Crater radii and volume scale approximately as $(KE)^{1/3}$. Ejection angles are nearly constant throughout most of the calculation at ~50–60° from the horizontal. Early-time ejection velocities are 1–2 km/sec in both cases. For NASA-1 all the melted GA is ejected from the crater. The maximum impact range for this ejected melted material is about 30 km. For NASA-2 about 50% of the melted GA is ejected from the crater to ranges up to about 130 km.

The attenuation of peak shock pressure with depth is reported for both calculations. Transient cavity dynamics are described and compared to that for surface and near-surface explosions.

INTRODUCTION

The characteristics and distribution of impact melt at terrestrial craters has often played an important role in the interpretation of the mechanics of crater formation (e.g., Phinney and Simonds, 1977 and Grieve *et al.*, 1977). However, there are a number of fundamental questions regarding not only the generation of impact melt (e.g., Ahrens and O'Keefe, 1972 and 1977) but, equally important, its subsequent dynamics and final disposition. In particular, the influence of variations in impact parameters on the generation *and transport* of impact melt is not well understood. An improved understanding of these processes is expected not only to enhance our understanding of cratering and ejection mechanics but also to aid in the interpretation of the lunar samples.

A matrix of theoretical continuum mechanics calculations has been initiated to address basic questions concerning the effects of variations in impact parameters on melt generation and transport as well as transient cavity and ejection dynamics more generally. The first two calculations of this matrix have recently been completed and some of the basic results are reported here. Analysis of the computations is continuing and additional results will be reported in future papers.

INITIAL CONDITIONS AND MATERIAL MODELS

The initial conditions, boundary conditions and material models for the two calculations reported here are identical in all respects except the impact velocity. In one calculation, denoted NASA-1, the impact velocity of the projectile was taken as 5 km/sec and in the second, NASA-2, it was 15.8 km/sec. Thus, the kinetic energy of the projectile in the NASA-2 calculation was ten times that in the NASA-1 calculation. The earth's gravitational field (980 cm/sec^2) was used in both calculations.

In both calculations the impact was modeled as normal (i.e., 90° to the target plane) and axial symmetry was assumed. The impact projectile was taken as an iron sphere of mass 10^{12} grams and a radius of about 31.2 meters. The equation of state used to model the iron projectile was that due to Tillotson (1962) and included both melting and vaporization. The equation-of-state constants for the iron are reported in Ahrens and O'Keefe (1977). The iron was modeled as a temperature-dependent von Mises material with a constant shear modulus, μ. For the projectile:

$$\mu = 0.625 \text{ Mb (62.5 GPa)}$$

and

$$\sqrt{3J_2} = Yo(1 - E/E_m)$$

where J_2 = second invariant of the stress deviator tensor.
    Yo = 3 kb (0.3 GPa)
    E_m = specific internal energy required for melting = 1.05×10^{10} erg/g.

The tensile strength for iron was taken as 667 bars (66.7 MPa).

In both calculations the iron projectile impacted a gabbroic anorthosite (GA) half-space. The equation of state used to model the GA was essentially identical to that reported by Ahrens and O'Keefe (1977) and O'Keefe and Ahrens (1978). The equation of state model for the GA included a solid-solid phase transition beginning at a pressure of about 150 kb (15 GPa) as well as melting and vaporization. The GA was also modeled as a temperature dependent von Mises material with a constant shear modulus. For the GA:

$$\mu = 0.325 \text{ Mb (32.5 GPa)}$$

and

$$\sqrt{3J_2} = Yo(1 - E/E_m)$$

where Yo = 3.464 kb (0.3464 GPa) and E_m = 1.76×10^{10} erg/g. The tensile strength of the GA was taken as 300 bars (30 MPa).

Table 1. Hugoniot shock pressures (Mb) inducing a change of state upon isentropic release to atmospheric pressure.

	Iron	Gabbroic anorthosite
Incipient melting	2.2	0.43
Complete melting	2.6	0.52
Incipient vaporization	4.2	1.02
Complete vaporization	16.8	5.9

For both the iron and the GA the shock pressures inducing a change of state upon isentropic release to atmospheric pressure are given in Table 1. The ideal one-dimensional Hugoniot pressure for the NASA-1 impact (5 km/sec) is about 950 kb (95 GPa) which is sufficient to completely melt, but not vaporize, the GA and is insufficient to melt the iron. The ideal Hugoniot impact pressure for NASA-2 (15.8 km/sec) is about 5.6 Mb (560 GPa) which is sufficient to begin vaporization in both the GA and the iron.

The computational geometry and initial impact conditions are summarized in Fig. 1. The initial finite-difference computational grid was the same for both calculations. Figure 2 shows this computational grid in the impact region. The finite difference zone on the axis of symmetry and immediately beneath the projectile was initially 3.5 m by 3.5 m. In the radial direction zone size increased with range geometrically with a geometric ratio of 1.03. Zone size increased with

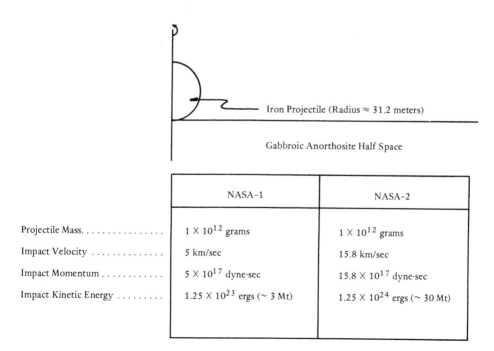

	NASA–1	NASA–2
Projectile Mass...............	1×10^{12} grams	1×10^{12} grams
Impact Velocity	5 km/sec	15.8 km/sec
Impact Momentum	5×10^{17} dyne-sec	15.8×10^{17} dyne-sec
Impact Kinetic Energy	1.25×10^{23} ergs (~ 3 Mt)	1.25×10^{24} ergs (~ 30 Mt)

Iron Projectile (Radius ≈ 31.2 meters)

Gabbroic Anorthosite Half Space

Fig. 1. Geometry and initial conditions for calculations.

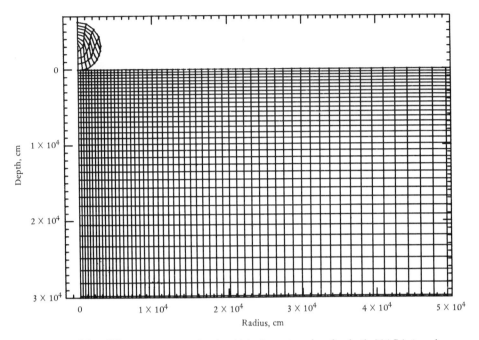

Fig. 2. Finite difference computational grid in impact region for both NASA-1 and NASA-2.

depth with a geometric ratio of 1.05. The interface between the projectile and the half-space is modeled as a frictionless sliding surface. The calculations were performed using the CRT two-dimensional Lagrangian finite difference code WAVE-L.

ENERGY PARTITIONING AND MELT GENERATION

Energy coupling and partitioning for the two calculations are shown in Fig. 3. Shown in Fig. 3 are the projectile kinetic and total energy and the GA kinetic, internal and total energy as a function of time for both calculations. Essentially all the initial projectile kinetic energy is coupled to the GA in both impacts. However, for the low velocity impact (NASA-1 at 5 km/sec) full coupling of the energy to the GA requires about 50 ms as compared to about 15 ms for the 15.8 km/sec impact (NASA-2). This is consistent with the factor of a little more than three difference in the impact velocities. In both calculations, the maximum kinetic energy in the GA is achieved at about the time energy coupling is complete. For NASA-1 the maximum GA kinetic energy is about 0.43 of the initial impact energy while for the higher velocity NASA-2 impact the maximum ratio of GA kinetic energy to initial impact energy is about 0.37.

Figure 4 shows the peak pressure on the axis of symmetry versus depth normalized to the initial projectile radius, d/R, for each of the calculations. For

NASA-1 the attenuation of peak pressure with depth is well represented over the range $0.5 \leq d/R \leq 4.82$ (89 kb $\leq P \leq$ 687 kb) by the equation:

$$\log P = -0.494 - 0.582 \log \frac{d}{R} - 0.313 \left[\log \frac{d}{R} \right]^2 - 0.047 \left[\log \frac{d}{R} \right]^3$$

where P is in Mbars.

The attenuation of peak pressure with depth for the NASA-2 calculation exhibits two distinct regimes and is well described by the equations:

$$\log P = 0.474 - 0.167 \log \frac{d}{R} \qquad \frac{d}{R} < 1$$

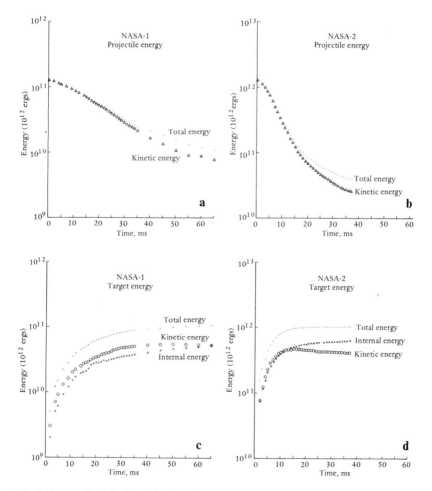

Fig. 3. Energy Partitioning. Kinetic and total energy versus time for NASA-1 projectile (a) and NASA-2 projectile (b) and kinetic, internal and total energy in GA versus time for NASA-1 (c) and NASA-2 (d).

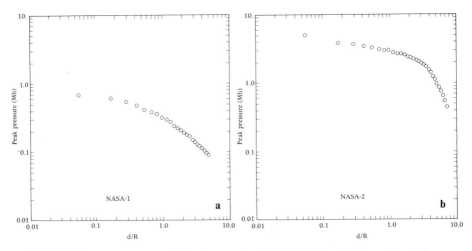

Fig. 4. Peak pressure versus depth, d, normalized to the initial projectile radius, R, for NASA-1 (left) and NASA-2 (right).

and

$$\log P = 0.476 - 0.430 \log \frac{d}{R} + 0.985 \left[\log \frac{d}{R}\right]^2 - 1.852 \left[\log \frac{d}{R}\right]^3, 1 \le \frac{d}{R} \le 7.22$$

In both calculations at depths below about $\frac{d}{R} \approx 3$, the attenuation of peak pressure with depth approximates an equation of the form:

$$P = a \left(\frac{d}{R}\right)^{-1.7}$$

where a is a constant.

In the NASA-1 calculation, GA extending to an original depth of about 10 m on axis is fully melted. Incipient melting on the axis of symmetry extends to an original depth of about 15 m. In NASA-1 all state changes in the GA are completed by about 3 ms after the impact.

The higher impact velocity NASA-2 calculation results in melting to much greater depths, as expected. For NASA-2 partial vaporization of the GA extends to an original depth of about 155 m. Complete melting in NASA-2 extends to an original depth of about 210 m while incipient melting extends to a depth of about 225 m. In NASA-2 all state changes in the GA are complete by about 21 ms after the impact. Figure 5 shows the spatial distribution of the partially vaporized, completely melted and partially melted GA for both of the calculations at about the time when changes of state in the GA are complete.

In NASA-1 the mass of material experiencing complete melting was about 3.6 $\times$ 10^{10}g or a mass equivalent to about 4% of the projectile mass. The mass of

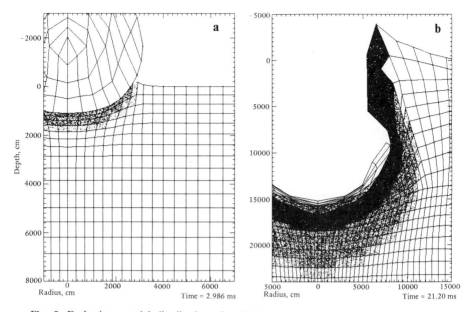

Fig. 5. Early-time spatial distribution of partially vaporized and melted GA for the NASA-1 calculation at a time of 3 ms (left) and the NASA-2 calculation at a time of 21 ms (right). Times correspond approximately to times when solid-liquid phase change in GA terminates. Stippling shows *dominant* material phase in each zone. Because of rezoning, zones may contain mixtures of partially vaporized, melted, partially melted and solid GA. Heavy, intermediate and light stippling denotes partially vaporized, fully melted and partially melted material, respectively. Note that in these plots the axis of symmetry is not coincident with the margin of the figure.

partially melted GA (i.e., at the melting temperature but not fully melted) was nearly the same, about 3.2×10^{10}g or 3% of the projectile mass.

The NASA-2 15.8 km/sec impact, which involved ten times the NASA-1 impact kinetic energy, resulted in over 150 times as much completely and partially melted GA as well as about 3.6 projectile masses of partially vaporized GA. For the NASA-2 impact about 7.6 projectile masses of GA were completely melted while about 2.8 projectile masses were partially melted. Table 2 summarizes the masses of GA undergoing state changes in each of the calculations.

Table 2. Gabbroic anorthosite melt and vapor generation.

	NASA-1 (5 km/sec)		NASA-2 (15.8 km/sec)	
	Mass (g)	Fraction of projectile mass	Mass (g)	Fraction of projectile mass
Fully vaporized	0	0	0	0
Partially vaporized	0	0	3.6×10^{12}	3.6
Fully melted	3.6×10^{10}	0.04	7.6×10^{12}	7.6
Partially melted	3.2×10^{10}	0.03	2.8×10^{12}	2.8

TRANSIENT CAVITY DYNAMICS

For surface and near-surface explosions, Orphal (1977a, 1977b) suggested that transient cavity formation may be conveniently viewed as a two-stage process: 1) nearly hemispherical growth of the transient cavity until maximum depth is achieved and 2) shearing of material along the transient cavity walls and continued radial growth following attainment of maximum cavity depth. Transient cavity dynamics are somewhat different for the impact conditions and velocity regime studied here in which there is little or no projectile vaporization. The early time cavity dynamics under these conditions is dominated by the penetration of the projectile. During this phase, the "depth" of the transient crater increases as the projectile penetrates the target while the "radius" of the transient cavity is nearly constant and of the same order of magnitude as the projectile radius. In short, during this early penetration phase the geometry of the transient cavity is approximately cylindrical. This early penetration phase is illustrated in Fig. 6 for the NASA-2 calculation.

During the penetration phase the projectile is decelerated and at some point in time the projectile velocity is less than the shock propagation speed in the target material in front of the projectile. At this point in time the shock begins to "break away" from the penetrating projectile and material flow begins to become more hemispherical about this point below the surface. For NASA-2 this "breakaway" occurs by 55 ms after impact as shown in Fig. 6.

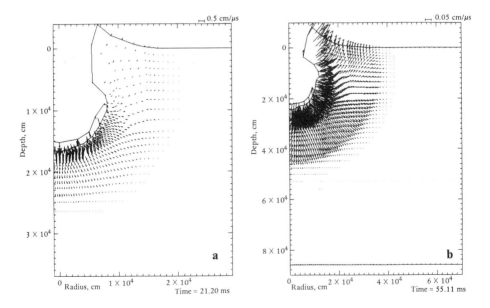

Fig. 6. Vector velocity plots of material flow field during early-time penetration phase of NASA-2 calculation showing cylindrical nature of early-time transient cavity and "break away" of shock front from projectile between 21 ms (left) and 55 ms (right).

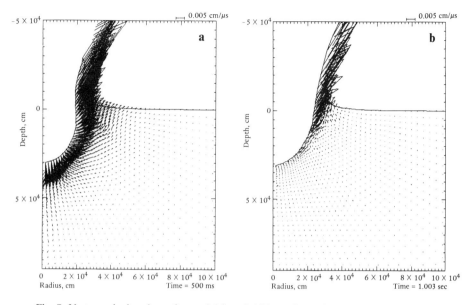

Fig. 7. Vector velocity plots of material flow field in cavity region for NASA-1 showing quasi-hemispherical flow at 500 ms and before attainment of maximum depth (left) and shear flow along cavity wall at 1 sec after maximum cavity depth is achieved (right).

Following the shock "breakaway" the transient cavity dynamics begins to more closely approach the two-state process described by Orphal (1977b) for near-surface explosions. At later times, the flow field near the transient cavity qualitatively resembles very much that for near-surface explosions as can be seen by comparing Fig. 7 with a similar figure (Fig. 1) in Orphal (1977b). There is a significant quantitative difference, however, between the transient cavity dynamics for the impacts reported here and near-surface explosions. For near-surface explosions the early growth of the transient cavity is nearly hemispherical with cavity radius and depth essentially equal until maximum depth is achieved. In contrast, for the impacts reported here the depth of the transient cavity is greater than the cavity radius until very late times in the cratering process. This can be seen in Fig. 8 which shows cavity radius and depth versus time for both the NASA-1 and the NASA-2 calculations. For NASA-1 the maximum cavity depth of about 325 m is achieved at a time of about 0.6 seconds. At this same time the radius of the transient cavity is only about 200 m or about ⅔ the depth. Even at a time of 1.0 second, when this calculation was terminated, the radius of the transient cavity is less than the depth. The same growth history is also seen for the NASA-2 crater.

Figure 9 shows the computed material flow fields for the NASA-1 and NASA-2 calculations at 1 second and 1.3 seconds, respectively. A ballistic extrapolation of the flow fields at these times results in the *estimated* final crater depths, radii and volumes given in Table 3.

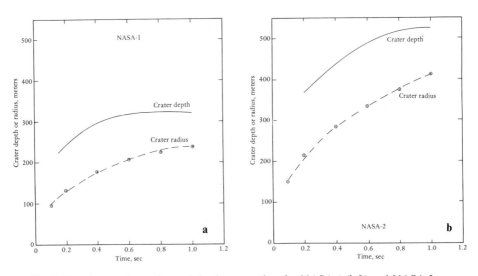

Fig. 8. Transient cavity radius and depth versus time for NASA-1 (left) and NASA-2 (right).

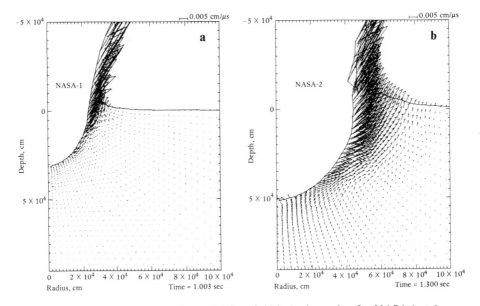

Fig. 9. Vector velocity plots of material flow field in cavity region for NASA-1 at 1 second (left) and NASA-2 at 1.3 seconds (right). Computations were terminated at these times.

Table 3. Final crater depth, radius and volume.*

	Depth	Radius	Volume
NASA-1 (5 km/sec)	315 m	~450	6.5×10^7 m³
NASA-2 (15.8 km/sec)	510 m	~1,000 m	6.8×10^8 m³

*Values *estimated* on basis of ballistic extrapolation of flow fields shown in Fig. 9.

SCALING

For the two calculations reported here the scaling of crater radius with the initial impact energy may be examined in a preliminary manner using the estimated crater radii given in Table 3. Then,

$$\frac{R_2}{R_1} = \left(\frac{KE_2}{KE_1}\right)^\alpha$$

and using $R_2 = 1,000$ m, $R_1 = 450$ m and $KE_2/KE_1 = 10$,

$$\alpha = 0.35 \approx 1/3.$$

This, of course, closely approximates the well known cube-root scaling with energy, indicating that for these two calculations crater radius is being dominated by hydrodynamic flow, as opposed to effects of gravity or the shear strength of the GA.

Similarly, using the estimated crater volumes given in Table 3,

$$\frac{V_2}{V_1} = \left(\frac{KE_2}{KE_1}\right)^\beta$$

and with $V_2 = 6.8 \times 10^8$ m³ and $V_1 = 6.5 \times 10^7$ m³,

$$\beta \doteq 1.02 \approx 1.$$

as required by hydrodynamic scaling.

On the other hand, crater depth or penetration depth does not scale as the cube-root of the energy, as can be verified by examining the crater depths given in Table 3. It is of some interest to examine the scaling of penetration depth or crater depth in terms of the impact velocity. It can be shown that penetration depth, p, scales as (Dienes and Walsh, 1970; Eichelberger and Gehring, 1962):

$$\frac{p}{d} \propto \left(\frac{\rho_p}{\rho_t}\right)^m \left(\frac{v}{a}\right)^n$$

where p = projectile penetration depth
      d = projectile diameter
      ρ_p = projectile density
      ρ_t = target density
      v = impact velocity
      a = sound speed in the target

For purely hydrodynamic behavior, n = ⅔, which is the equivalent of cube-root scaling of penetration depth with the initial kinetic energy. For momentum scaling, n = ⅓.

For the NASA-1 and NASA-2 calculations, the projectile diameters are identical as is the quantity ρ_p/ρ_t and the sound speed in the target. So:

$$\frac{p_1}{p_2} \propto \left(\frac{v_1}{v_2}\right)^n$$

Taking p_1 = 315 m, v_1 = 5 km/sec, $p_2 \cong$ 510 m and v_2 = 15.8 km/sec,

$$n \approx 2/5$$

or a value between those for momentum and energy scaling.

This apparent reduction in penetration efficiency is particularly intriguing since both the crater radii and the crater volumes scale very closely as required by hydrodynamic theory. With the results of only two calculations it is not possible at this time to determine with certainty the cause(s) of this apparent reduced penetration efficiency. However, because of the importance of penetration and excavation depths for interpretation of the origin of lunar samples, this aspect will be closely examined when the results of future calculations are available.

EJECTION DYNAMICS AND MELT TRANSPORT

In both the NASA-1 and NASA-2 calculations ejection angles are relatively constant at about 50–60° from the horizontal. Early-time ejection velocities are about 1–2 km/sec. Detailed analyses of ejection dynamics and ejecta deposition are planned and will be reported in a future paper.

The NASA-1 calculation was run to a real time of 1 second; NASA-2 was run to 1.3 seconds. The estimates of final crater dimensions and volumes of melt ejected from the crater are based on ballistic extrapolations from these times. Based on an analysis of the cratering flow fields at these times our experience suggests that these estimates will be very good approximations to the values that would be obtained if the full computation were continued until material motion has ceased.

For the NASA-1 impact, all the melted (and partially melted) GA is ejected from the crater. The melted GA is ejected up to a maximum range of about 30 km or about 60 crater radii.

In contrast, only about 53% of the completely melted GA and about 45% of

Table 4. Melt transport.

	Mass of fully melted GA ejected	Percentage of fully melted GA ejected	Mass of partially melted GA ejected	Percentage of partially melted GA ejected
NASA-1 (5 km/sec)	3.6×10^{10}g	~100%	3.2×10^{10}g	~100%
NASA-2 (15.8 km/sec)	4.0×10^{12}g	~53%	1.3×10^{12}g	~45%

From ballistic extrapolation at 1.0 sec.

the partially melted GA is ejected from the NASA-2 crater. The melted GA is ejected to a maximum range of about 130 km or about 130 crater radii.

Table 4 summarizes the masses of melted and partially melted GA ejected from the crater in both calculations.

DISCUSSION

The results reported here are from the first two of a planned matrix of calculations to investigate the effects of variations in impact parameters on melt generation and *transport*. With only two calculations completed it is not possible at this time to assess fully the scaling of parameters characterizing melt generation and transport and transient cavity dynamics with the variation of impact parameters included in the full matrix of calculations. However, several interesting observations are possible on the basis of the results available to date.

O'Keefe and Ahrens (1977) analyzed the scaling of the volume of impact melt generated with the initial impact kinetic energy in terms of the nondimensional parameter:

$$S = (\rho_p/\rho_t)(v/a)^2$$

where, as above, ρ_p = density of projectile, ρ_t = low pressure density of target, a = sound speed in the target and v = impact velocity. Taking ρ_p = 7.86 g/cc, v = 15.8 km/sec, and ρ_t = 2.936 g/cc and a = 7.4 km/sec for the high pressure phase of the GA, for the NASA-2 calculation S = 12.3. Then, for NASA-2, the volume of melt, V_m, as a ratio of the volume of the projectile, V_p, is:

$$\frac{V_m}{V_p} = 2.3S$$

This result is consistent with the value 2.5S reported by O'Keefe and Ahrens (1977).

For $S \lesssim 4$, O'Keefe and Ahrens report that V_m/V_p is not a linear function of the nondimensional parameter S. The NASA-1 results are consistent with this result. In fact, for the iron on GA impacts considered here, V_m approaches zero as the impact velocity approaches a value of about 3 km/sec.

Lange and Ahrens (1979), using results from O'Keefe and Ahrens (1977) and Oberbeck, *et al.* (1975) suggest that the volume of melt generated may be related to crater diameter, D, by:

$$V_m(km^3) = 2 \times 10^{-4} D^{3.4}$$

Taking $V_m = 3.5 \times 10^{-3}$ km^3 (from mass in Table 2 and assuming $\rho = 2.936$ g/cc for the low pressure phase of GA) and D = 2 km for NASA-2, $V_m/D^{3.4} \approx 3.3 \times 10^{-4}$, in reasonable agreement with the result of Lange and Ahrens (1979) and the terrestrial crater data reported in their paper.

Dence (1965) suggested that:

$$V_m(km^3) = 0.002D^3$$

on the basis of studies of terrestrial impact craters. The calculational results reported here agree only fairly with this scaling relationship with $V_m/D^3 \cong 3 \times 10^{-4}$ for the NASA-2 computation.

Comparing the volume of melted target rock with the volume of the crater, V_c, the NASA-2 calculation gives:

$$\frac{V_m}{V_c} \approx 4 \times 10^{-3}$$

Additional calculations are required to determine how this ratio scales with variations in impact conditions.

Acknowledgments—The work reported here was performed under NASA contract NASW 3276. The authors wish to express their appreciation to K. N. Kreyenhagen and S. H. Schuster for their informal reviews and advice during the course of the calculations and to Tom Ahrens and J. Dugan O'Keefe for reviewing the draft manuscript and offering a number of helpful suggestions and comments.

The Lunar and Planetary Institute is operated by the Universities Space Research Association under Contract No. NSR 09-051-001 with the National Aeronautics and Space Administration. This paper constitutes the Lunar and Planetary Institute Contribution No. 415.

REFERENCES

Ahrens T. J. and O'Keefe J. D. (1972) Shock melting and vaporization of lunar rocks and minerals. *The Moon* **4**, 214–249.

Ahrens T. J. and O'Keefe J. D. (1977) Equations of state and impact-induced shock wave attenuation of the moon. In *Impact and Explosion Cratering* (D. J. Roddy, R. O. Pepin, and R. B. Merrill, eds.), p. 639–656. Pergamon, N.Y.

Dence M. R. (1965) The extraterrestrial origin of Canadian craters. *Ann. N.Y. Acad. Sci.* **123**, 941–969.

Dienes J. K. and Walsh J. M. (1970) Theory of impact: Some general principles and the method of Eulerian codes. In *High Velocity Impact Phenomena* (Ray Kinslow, ed.), p. 45–104. Academic, N.Y.

Eichelberger R. J. and Gehring J. W. (1962) Effects of meteoroid impacts on space vehicles. *Amer. Rocket Soc. J.* **32**, No. 10, 1583.

Grieve R. A. F., Dence M. R., and Robertson P. B. (1977) Cratering processes: As interpreted from the occurrence of impact melts. In *Impact and Explosion Cratering* (D. J. Roddy, R. O. Pepin, and R. B. Merrill, eds.), p. 791–814. Pergamon, N.Y.

Lange M. A. and Ahrens T. J. (1979) Impact melting during the first 1.5 b.y. of lunar history (abstract). In *Lunar and Planetary Science X*, p. 700–702. Lunar and Planetary Institute, Houston.

Oberbeck V. R., Hörz F., Morrison R., Quaide W. L., and Gault D. E. (1975) On the origin of the lunar smooth plains. *The Moon* **12**, 19–54.

O'Keefe J. D. and Ahrens T. J. (1977) Impact-induced energy partitioning, melting, and vaporization on terrestrial planets. *Proc. Lunar Sci. Conf. 8th,* p. 3357–3374.

O'Keefe J. D. and Ahrens T. J. (1978) Impact flows and crater scaling on the moon. *Phys. Earth Planet. Inter.* **16**, 341–351.

Orphal D. L. (1977a) Calculations of explosion cratering I: The shallow-buried nuclear detonation JOHNIE BOY. In *Impact and Explosion Cratering* (D. J. Roddy, R. O. Pepin, and R. B. Merrill, eds.), p. 897–906. Pergamon, N.Y.

Orphal D. L. (1977b) Calculations of explosion cratering II: Cratering mechanics and phenomenology. *Impact and Explosion Cratering* (D. J. Roddy, R. O. Pepin, and R. B. Merrill, eds.), p. 907–917. Pergamon, N.Y.

Phinney W. C. and Simonds C. H. (1977) Dynamical implications of the petrology and distribution of impact melt rocks. *Impact and Explosion Cratering* (D. J. Roddy, R. O. Pepin, and R. B. Merrill, eds.), p. 771–790. Pergamon, N.Y.

Tillotson J. H. (1962) Metallic equations of state for hypervelocity impact. General Atomic Report. GA 3216.

Proc. Lunar Planet. Sci. Conf. 11th (1980), p. 2325–2345.
Printed in the United States of America

Calculational investigation of impact cratering dynamics: Material motions during the crater growth period

Michael G. Austin,[1] Jeffrey M. Thomsen,[1] Stephen F. Ruhl,[1] Dennis L. Orphal,[2] and Peter H. Schultz[3]

[1]Physics International Company, 2700 Merced Street, San Leandro, California 94577,   [2]California Research and Technology, Incorporated, 4049 First Street, Suite 135, Livermore, California 94550,   [3]Lunar and Planetary Institute, 3303 NASA Road 1, Houston, Texas 77058

Abstract—A two-dimensional finite difference calculation of a laboratory-scale hypervelocity (6 km/sec) impact of a 0.3 g spherical 2024 aluminum projectile into a homogeneous plasticene clay half-space has been carried out to a time of 600 μsec and the material motions during this crater growth period analyzed. Energy coupling from the projectile to the target is completed within the first 18 μsec after impact. The transient crater grows quickly to a depth of 4 cm in the first 100 μsec and then more slowly to a depth of 7.5 cm at 600 μsec while it maintains a radius-to-depth ratio of 0.85. The cratering flow field developed within the target has been analyzed in detail to find the differences and similarities to previously calculated flow fields for near-surface explosions. Application of Maxwell's analytical Z-Model (developed to interpret the flow fields of near-surface explosion calculations) to the flow field at 18 μsec showed a good fit to the Maxwell Z-Model if the flow field center was taken to be located at 0.65 cm beneath the original target surface rather than at the surface as had been previously done for near-surface explosion cratering calculations.

Between 18 μsec and 600 μsec, however, the analysis of the cratering flow field in terms of the Z-Model becomes more complicated. The two parameters of the Maxwell Z-Model are α, which characterizes the strength of the flow field, and Z, which characterizes its shape. For near-surface explosion cratering calculations, the steady-flow Maxwell Z-Model (i.e., α and Z almost constant with time during most of the crater growth period) gives good agreement with the calculated flow fields. This impact cratering calculation does not exhibit good agreement with the steady-flow feature of the Z-Model. Taking the center of the flow field to be fixed at 0.6 cm or at nearby depths results in α and Z which are time dependent during much of the crater growth process. If α and Z are to be almost constant, the flow field center is forced steadily deeper with time, thereby making the coordinate system itself time dependent. In either case, the Z-Model cannot be applied in a straightforward way to predict final cratering displacements based on the flow field shape and strength at very early times in the cratering process, as it can in some near-surface explosion cratering calculations.

Although α and Z are time dependent for a fixed flow field center for this calculation, at any given time they provide a good mathematical description of the cratering flow field. Only one calculation for one projectile material and one target material and with one set of initial conditions is analyzed here. Therefore it is premature to say in general to what degree the steady-flow Maxwell Z-Model is applicable to other impact conditions and other projectile and target materials.

I. INTRODUCTION

A coordinated program of calculations, experiments, and model construction is necessary for a better understanding of the detailed dynamics of impact cratering processes, which in turn is necessary for a better understanding of planetary surfaces. This present effort is a continuation of a study of impact dynamics in a uniform, non-geologic material at impact velocities achievable in laboratory-scale experiments (Thomsen *et al.*, 1979a). A calculation of a 6 km/sec impact of a 0.3 g spherical 2024 aluminum projectile into low strength (50 kPa) homogeneous plasticene clay has been continued from 18 μsec to past 600 μsec. The cratering flow field, defined as the material flow field in the target beyond the transient cavity but well behind the outgoing shock wave, has been analyzed in detail to see how applicable the Maxwell Z-Model, developed from analysis of near-surface explosion cratering calculations, is to impact cratering. A series of experiments was conducted at the NASA Ames Research Center's Vertical Gun Range (AVGR) using the same materials and impact conditions as the calculation. Some experimental results and comparisons between the experiments and the calculation have been reported by Thomsen *et al.* (1980), Ruhl and Thomsen (1980), and Austin *et al.* (1980). This paper concentrates on just the calculational effort and on the application of Maxwell's Z-Model to the calculated cratering flow field after the projectile has coupled its energy to the target but before final cratering displacements occur.

Laboratory-scale impact experiments have been performed by Gault *et al.* (1968), Oberbeck (1971), Gault and Wedekind (1977) and Moore (1976), but few calculations have been performed for this scale. Bjork (1961), Bryan *et al.* (1978), O'Keefe and Ahrens (1975, 1976, 1977, 1978, 1979), and Orphal *et al.* (1980) have concentrated on larger meteorite planetary-scale impacts. Attempts to summarize quantitatively the cratering process and build analytic predictive models for it have been limited to the Maxwell Z-Model (Maxwell, 1977; Orphal, 1977b) and to the semi-empirical model of Ivanov (1976).

This discussion divides the cratering process into the *compression stage* and the *excavation stage* as described empirically by Gault *et al.* (1968). The *compression stage* begins with the initial contact between projectile and target, includes the shock compression of the target and projectile, and ends when the coupling of the projectile energy and momentum to the target is essentially complete. The *excavation stage* then begins and lasts much longer. It describes the processing of the bulk of affected target material by the outgoing shock wave, the modification of the initially radial motions by the continuous fan of rarefactions generated at the target surface, and the orderly ejection of material from the transient cavity. The first part of the present effort covers the *compression stage* and the earliest part of the *excavation stage* during which an orderly material flow field is established in the target, and is reported by Thomsen *et al.* (1979a). This paper covers the calculation during the major portion of the *excavation stage* during which the crater achieves most of its growth.

II. CALCULATION OF CRATER GROWTH DURING THE EXCAVATION STAGE

The calculation initial conditions, materials, material models, and results during the *compression stage* have been previously reported (Thomsen *et al.* 1979a, 1979b), but will be briefly summarized here to provide a proper starting point for the discussion of the calculation results during the *excavation stage* of the crater growth.

Figure 1 summarizes the initial conditions for the calculation. The calculation simulates the impact of a 6 mm diameter 2024 aluminum projectile into a plasticene clay target having a von Mises strength envelope of 50 kPa. Axial symmetry, terrestrial gravity, and a vacuum environment are assumed. These initial conditions were chosen to allow directly comparable experiments to be performed at the AVGR (Thomsen *et al.* 1980; Ruhl and Thomsen, 1980). An infinite half-space of clay is used for the target boundaries to eliminate any possible reflections or rarefactions due to the outer boundaries. Thereby the analysis of the calculated flow field applies to the flow field itself and not to any extraneous waves. The plasticene clay, a uniform, nongeologic material containing no air voids, was chosen in particular because it was easily available for performing experiments and could be simply characterized for the calculation based on data from Maxwell and Reaugh (1972), Christensen *et al.* (1968), and Maxwell *et al.* (1972). The plasticene clay target has an initial density of 1.69 Mg/m^3, a compressional wave velocity of 1.4 m/msec, a shear wave velocity of 0.475 m/msec, and a Poisson's ratio of 0.435. The 2024 aluminum projectile has an initial density of 2.783 Mg/m^3, a compressional wave velocity of 5.343 m/msec, and a maximum strength of the Mohr-Coulomb failure envelope of 320 MPa which is much greater than the clay strength (van Thiel, 1977; McQueen *et al.* 1970). The two dimensional computer code PISCES 2DELK (Hancock, 1976) was used to perform the calculation. Further details of the material models and on the calculational technique and zoning used may be found in Thomsen *et al.*, (1979a, 1979b).

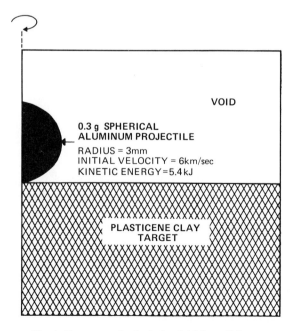

Fig. 1. Summary of calculation initial conditions.

During the *compression stage* of the calculation, the maximum calculated pressure of 50 GPa is less than the 60 GPa required to cause incipient melting of the aluminum (Gehring, 1970), but considerable vaporization of the clay occurs. In the first 1.5 μsec after impact 2.75 g of target material containing 1.5 kJ of internal energy (roughly 28 percent of the total impact energy) was vaporized. Jetting velocities about 1.8 times the initial impact velocity were calculated to occur during the first 1 μsec. The projectile quickly becomes severely deformed as it penetrated to a depth of about three times its original diameter in the first 18 μsec, at which time the projectile no longer contained any appreciable kinetic energy. Further details of the calculational results for this *compression stage* and the very early *excavation stage* of the crater growth may be found in Thomsen *et al.* (1979a, 1979b).

Figure 2 shows the developing crater lip and the material flow field at 18.0 μsec which is early in the *excavation stage*. The outgoing shock can clearly be seen in the pressure contours of Fig. 2. The shock wave has traveled almost 5 cm from the initial impact point and the transient cavity is a little more than 2 cm in depth. The ejection process is well underway in the region near the crater lip as the rarefactions from the free surface have caused the initial outwardly radial material velocity field to become more upwardly directed in the target region nearest the surface.

From 18 μsec to 600 μsec, the crater continues to grow and the crater lip grows

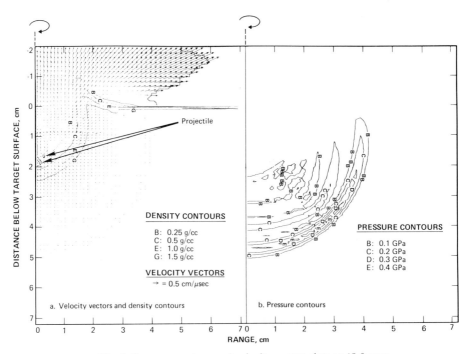

Fig. 2. Pressure contour and velocity vector plots at 18.0 μsec.

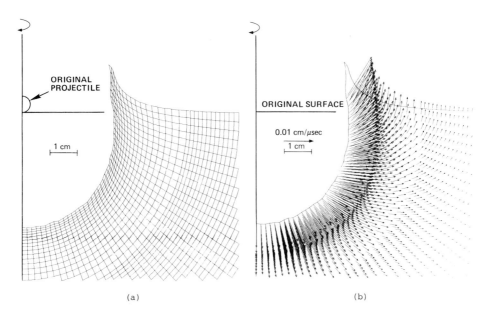

Fig. 3. (a) The Lagrangian grid in comparison to the original projectile at 97.7 μsec; (b) Velocity vector plot in the Lagrangian grid at 97.7 μsec.

and develops as the material velocity field slows more and more and as the pressure field in the cratering region decays to ambient levels. During this time almost all of the cratering flow field is entirely within the Lagrangian portion of the coupled Lagrangian-Eulerian grid, but the Eulerian grid is kept in the calculation at all times to provide the proper boundary condition for the rest of the grid. Figure 3 shows a portion of the Lagrangian grid at 97.7 μsec. Several zoning changes were made between 18 μsec and 192 μsec as required for computational economy while keeping adequate spatial resolution in all parts of the grid. The last zoning change at 192 μsec resulted in a grid approximately twice as coarsely zoned as the grid shown in Fig. 3. At all times sufficient zones were included in the outer portion of the grid so that the target effectively was an infinite half-space.

Figure 3 shows that the peak material velocity in the cratering flow field at 97.7 μsec is about 0.01 cm/μsec or about half the peak at 18 μsec. Figure 4 shows that at this time the peak shock has decayed to about 28 MPa and has propagated out to a distance of about 16 cm. The pressures in the cratering flow field region are mostly between 2 and 8 MPa as compared to pressures of about 100 MPa at 18 μsec.

At 600 μsec, Fig. 5 shows that the peak velocities in the cratering flow field have decayed further to about 0.003 cm/μsec as the crater has grown to a depth of approximately 7.5 cm. The tip of the Lagrangian grid is more than 10 cm above the original target surface which is outside the range plotted. The pressures in the cratering flow field region are mostly less than 2 MPa as shown in Fig. 6.

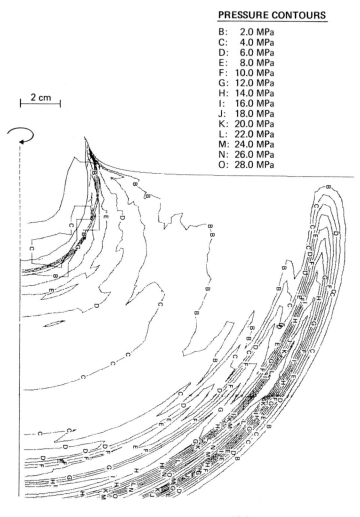

PRESSURE CONTOURS

B: 2.0 MPa
C: 4.0 MPa
D: 6.0 MPa
E: 8.0 MPa
F: 10.0 MPa
G: 12.0 MPa
H: 14.0 MPa
I: 16.0 MPa
J: 18.0 MPa
K: 20.0 MPa
L: 22.0 MPa
M: 24.0 MPa
N: 26.0 MPa
O: 28.0 MPa

2 cm

Fig. 4. Pressure contours at 98.8 μsec.

The flow field at 600 μsec is slowing rapidly and the crater is near but not yet at its final dimensions. The low material strength of 50 kPa is not enough to slow the crater growth to a complete stop at this calculated time.

III. IMPACT CRATERING PHENOMENOLOGY

The initial impact generated shock wave decays rapidly as it propagates outward from the point of impact. The highest shock levels in the target occur directly below the center of the projectile. For a given range from the point of impact, the peak pressures decrease gradually going circumferentially from the vertically downward direction to the target surface. Figure 7 shows the attenuation of peak pressure with range for $\theta = 0$ degrees (vertically downward) and $\theta = 45$ degrees. The magnitudes of the pressures differ, but the attenuation rates are the same. The peak pressures in the first few centimeters beyond the transient cavity are GPa's and tenths of GPa's, but these shock peaks occur early and quickly decay to much lower levels as shown in Figs. 4 and 6. For example, a peak pressure of 0.3 GPa occurs at a position 5 cm vertically downward from the impact point when the shock wave arrives there at about 20 μsec, but during the bulk of the *excavation stage*, the pressures are a few MPa at that position. After the initial formation of the transient cavity, most of the crater growth occurs when pressures

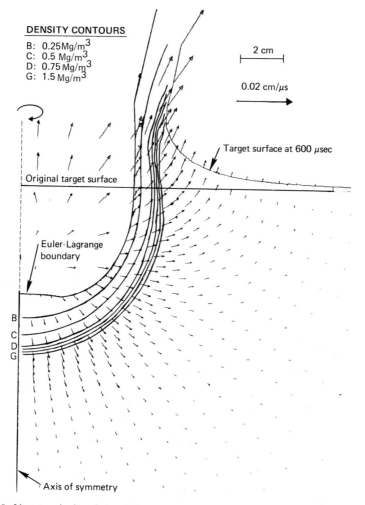

Fig. 5. Vector velocity plot at 600 μsec with overlaid density contours and Eulerian-Lagrangian boundary (the initial density of the clay is 1.69 Mg/m³).

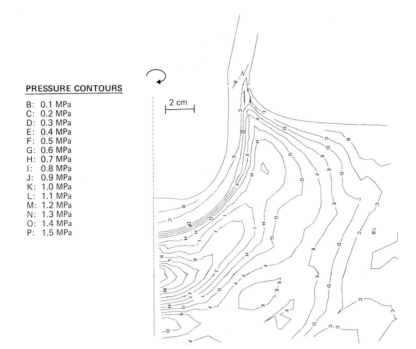

PRESSURE CONTOURS

B: 0.1 MPa
C: 0.2 MPa
D: 0.3 MPa
E: 0.4 MPa
F: 0.5 MPa
G: 0.6 MPa
H: 0.7 MPa
I: 0.8 MPa
J: 0.9 MPa
K: 1.0 MPa
L: 1.1 MPa
M: 1.2 MPa
N: 1.3 MPa
O: 1.4 MPa
P: 1.5 MPa

2 cm

Fig. 6. Pressure contours at 600 μsec.

in the cratering flow field have decayed to levels much lower than the initial peak shock pressures.

The shape of the transient cavity remains relatively constant as it grows. Measuring from the initial point to the 1.5 Mg/m^3 density level along the original target surface, and in the vertically downward direction gives a radius-to-depth (R/D) ratio of about 0.7 for times between 6 and 18 μsec. This ratio increases to a fairly constant 0.85 between 100 and 600 μsec. This contrasts with the more hemispherical (R/D $\gtrsim$ 1) transient cavity calculated for the JOHNIE BOY near-surface explosion crater (Orphal, 1977a). Orphal *et al.* (1980) also find that for a planetary-scale impact cratering calculation the transient cavity is less hemispherical at early times than is the case for near-surface explosion craters.

The coupling of the projectile's initial kinetic energy to the target proceeds rapidly in the first few μsec as shown in Fig. 8. By 18 μsec, the projectile's kinetic energy is less than 0.3 percent of its initial value. Less than 5 percent of the total impact energy still resides in the projectile in the form of internal energy. About 55 percent of the total impact energy has been coupled into the kinetic energy of the target material. The rest of the original energy is present in the form of internal energy of the target material. Although the total target kinetic energy remains relatively constant after the first few μsecs, its distribution within the target changes with time. Figure 9 shows the increase with time in the kinetic energy of the target material located initially beyond a cup-shaped region 2.5 cm in radius and 3.0 cm deep, centered about the original impact point. The target material initially inside the cup-shaped region has lost an equivalent amount of its kinetic energy.

IV. MAXWELL'S Z-MODEL

Maxwell's Z-Model is an analytical description of the cratering flow field developed from study of near-surface explosion cratering calculations. It provides a framework for describing the cratering flow field, defined as the target material velocity field beyond the transient crater wall but well behind the outgoing shock

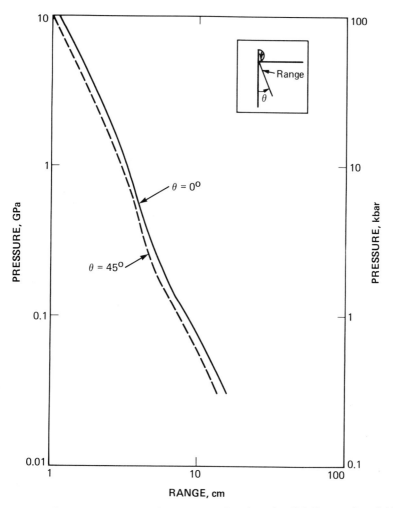

Fig. 7. Maximum shock pressure in target as a function of radial distance from initial impact point for $\theta = 0$ degrees (vertically downward) and 45 degrees.

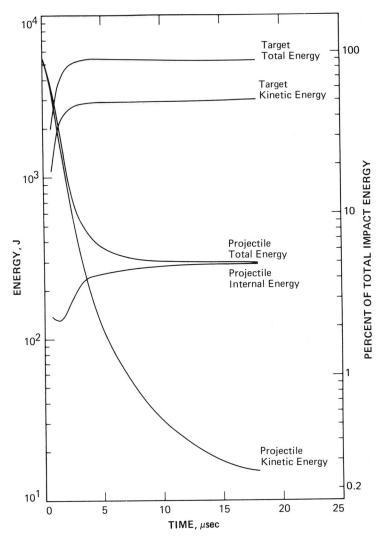

Fig. 8. Energy partitioning between projectile and target during the *compression stage*.

wave. Using the Z-Model equations and parameters characteristic of a specific flow field, it is possible to calculate ejection angles and velocities and final crater dimensions. In order to judge the applicability of this near-surface explosion cratering model to impact cratering, we examined in detail the cratering flow field at various times in our calculation.

Maxwell (1977), Orphal (1977b) and Thomsen *et al.* (1979a) have previously described Maxwell's Z-Model. Figure 10 defines the spherical polar coordinate system centered at the on-axis detonation point. R is the radial distance from this center to a given point in the cratering flow field and θ is the angle measured from the vertically downward direction. For near-surface explosion cratering calculations Maxwell (1977) found that at early times crater growth was nearly hemi-spherical and that after passage of the primary shock wave the cratering flow could be described as a very orderly process. The flow field could be described by:

$$\dot{R} = \alpha(t)R^{-z}, \tag{1}$$

where $\dot{R}$ is the radial velocity of the flow field, α is a time-dependent coupling term describing the flow field strength, and Z defines the rate of velocity decay with range, R. Maxwell (1977) also observed that the density in the cratering flow field region was approximately constant, yielding incompressible flow:

$$\nabla \cdot U(R,\theta) = 0 \tag{2}$$

where $U(R,\theta)$ is the vector velocity of the flow field. Combining Eqs. (1) and (2) permits derivation of the full equation of motion of a mass element along stationary streamlines within the cratering flow field:

$$U(R,\theta) = \dot{R}\hat{R} + \dot{R}(Z - 2)\tan(\theta/2)\hat{\theta}, \tag{3}$$

where $\hat{R}$ and $\hat{\theta}$ are unit vectors in spherical polar coordinates.

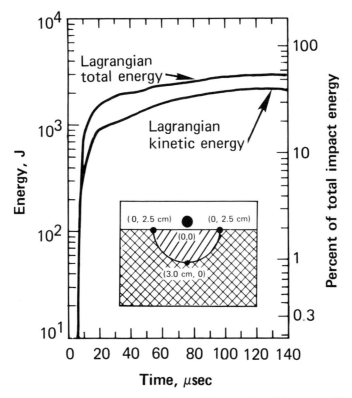

Fig. 9. Energy partitioning between the inner Eulerian portion of the target grid and the outer Lagrangian portion (cross-hatched in the insert) during the first part of the *excavation stage*.

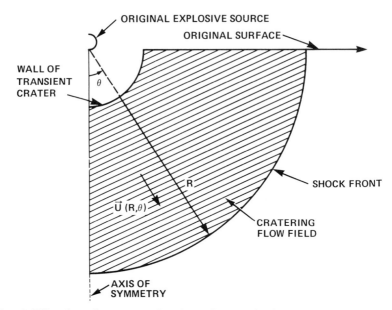

ORIGINAL EXPLOSIVE SOURCE

ORIGINAL SURFACE

WALL OF
TRANSIENT
CRATER

θ

R

$\vec{U}$ (R,θ)

SHOCK FRONT

CRATERING
FLOW FIELD

AXIS OF
SYMMETRY

Fig. 10. This schematic representation shows the cratering flow field region as defined for near-surface explosion cratering calculations with θ and R as used in the Maxwell-Z Model.

Z describes the shape of the cratering flow field. For Z = 2, U is radial at all points, giving irrotational flow. Values of Z > 2 describe flow fields which are rotational in the direction of the surface. Maxwell (1977) found for near-surface explosion cratering calculations that $Z \approx 2$ for $\theta = 0°$, that $Z \approx 2.7$ for $30° \lesssim \theta \lesssim 60°$; and that $Z \gtrsim 4$ for $\theta \gtrsim 75°$. An average value of $Z \approx 3$ was found to be representative of the entire cratering flow field produced by a near-surface explosion.

Maxwell (1977) and Orphal (1977b), for near-surface explosion cratering calculations, observed that α and Z varied somewhat for different streamlines (or angles θ), but to a good first approximation they were both constant with time. This steady-flow approximation, Eqs. (1), (2), (3), and consideration of material strength and gravity effects permitted a reasonable prediction of the displacement field, including ejecta and crater dimensions, at all subsequent times.

The special editing routine written to analyze our calculated flow fields performs a least squares fit to points in log R-log $\dot{R}$ space to get the parameters α and Z, as shown in Fig. 11. At various times, the calculated R's and $\dot{R}$'s are fit in this way for rays extending out every 5 degrees from possible coordinate centers on the axis every 0.1 cm from the initial impact point downward. The goodness of the least squares fit is evaluated by the percentage difference between the calculated and the fitted $\dot{R}$ for a given R (for a given time, center and angle). For a given time and center, various averages and standard deviations of α and Z for ranges of θ are also calculated.

V. ANALYSIS OF THE IMPACT CRATERING FLOW FIELD

The Z-Model editing routine was applied to the cratering flow field of the calculation at various times in order to judge the applicability of the Maxwell Z-Model by testing each of its assumptions and by making reasonable modification when necessary and feasible.

For the bulk of the *excavation stage,* the assumption of essentially incompressible material flow holds true, because the variations are small in the target material density throughout the cratering flow field region. At 18 μsec, the maximum density variation is about 3%. By 60 μsec, it is less than 1%, and at 600 μsec, it is less than 0.1%.

Thomsen *et al.* (1979a) reported that at 18 μsec the shock wave had detached itself from the immediate vicinity of the projectile and the cratering flow field had been established. Application of the editing routine to the flow field at that time gave the best overall fit and smallest standard deviations for an average Z = 2.11 and an average $\alpha = 0.084$ cm^{Z+1}/μsec for the flow field center taken to be located on-axis at a depth of 0.65 cm beneath the original impact point. At times less than 12 μsec, the initial directed momentum of the projectile caused Z to be less than 2 in the cratering flow field in the target material in the region directly beneath the projectile. These two basic results of our calculation at 18 μsec and before: a flow field center located beneath, not at, the target surface and the early time effect of the projectile on the flow field are basic differences between near-surface

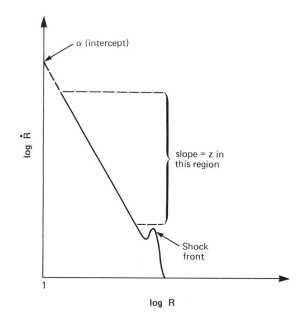

Fig. 11. Typical plot of $\dot{R}$ vs R in log-log space for one value of θ at one time in the calculation.

explosion cratering and impact cratering. However, the cratering flow field at 18 μsec was well described by the α and Z values given.

The requirement that the flow field center for the Z-Model be located some distance beneath the target surface is in qualitative agreement with work of Oberbeck (1971) and Bryan *et al.* (1979). Oberbeck (1971) found experimentally for impact of aluminum projectiles into non-cohesive quartz sand targets at 2 km/sec, that the resulting crater was the same as an equivalent energy explosion-generated crater in the same material when the explosive was buried at a depth of 6.3 ± 2 mm beneath the target surface. Bryan *et al.* (1979) found calculationally that for Meteor Crater, Arizona, a buried equivalent energy explosive source gave a similar crater to one produced by a meteor impact. It should be noted, however, that late time equivalence of crater shapes between impact and equivalent explosive events does not necessarily imply closely similar cratering flow fields at all times.

At times between 18 μsec and 600 μsec, possible on-axis flow field centers

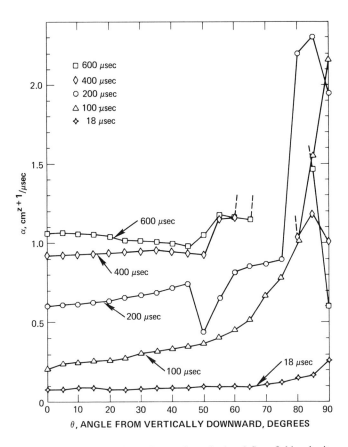

Fig. 12. α values from least squares fits to the calculated flow field velocity vectors' radial components with respect to a flow field center at 0.6 cm below the original impact point. (Dotted lines indicate that some out of range points were not plotted).

were examined every 0.1 cm from the target surface downward to find at each time which center gave the best fit of the calculated flow field to α and Z values for rays extending every 5° about the center. Before 200 μsec, the best fit was provided by the center located at 0.6 cm below the original target surface. The average difference between the calculated radial velocity component and the least squares fitted value was 3.5 percent for $\theta < 60°$. Between 200 and 600 μsec, the center providing the best fit becomes less pronounced and centers less than about 1 cm deep provide similarly good fits. The fits, however, become poorer later in time and go from about 5 percent average difference between calculated and fitted velocities at 200 μsec to about 15 percent at 600 μsec.

Figures 12 and 13 show the calculated α and Z values at various times for the center located at 0.6 cm below the original target surface. As previously observed for near-surface explosion calculations, the variation of α and Z is more regular for $\theta < 60°$ than it is for $\theta > 60°$. The interactions with the free surface make the flow field near the surface more complicated. The variations of α and Z with θ

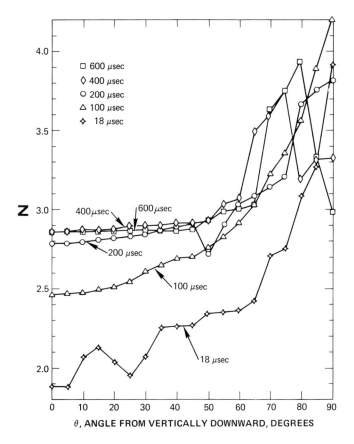

Fig. 13. Z values from least squares fits to the calculated flow field velocity vectors' radial components with respect to a flow field center at 0.6 cm below the original impact point.

for a given time as shown in Figures 12 and 13 are no greater than those observed in near-surface explosion cratering calculations.

The most striking feature in Figs. 12 and 13 is the strong variation of α and Z with time. α and Z fits to the cratering flow field at times between 18 and 100 μsec and between 100 and 200 μsec give values intermediate to the ones shown. The time of greatest change in α and Z is between 18 and 200 μsec, with Z fairly constant after 200 μsec for $\theta < 60°$ and α increasing much more slowly after about 400 μsec with a fairly constant value for $\theta < 60°$. But 200 μsec is already a large part of the *excavation stage*, and typical behavior of α and Z for near surface explosion cratering calculations is that α and Z remain fairly constant with time after the early part of the *excavation stage* at scaled times comparable to the 18 μsec time of this calculation.

The meaning of α and Z increasing with time can be seen in Fig. 14, where the least squares fits to radial velocities and ranges with respect to the flow field

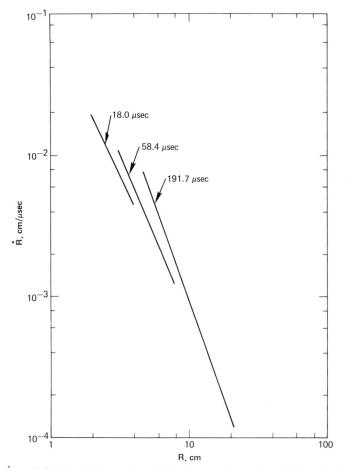

Fig. 14. $\dot{R}$ vs R for $\theta = 0$ degrees (vertically downward) at selected times during the *excavation stage*.

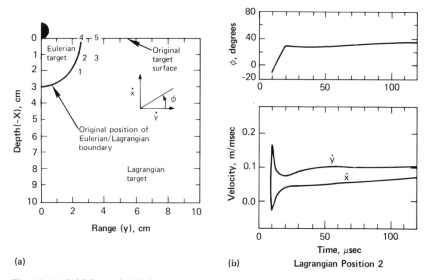

Fig. 15. (a) Initial coupled Eulerian-Lagrangian grid and initial location of selected Lagrangian points. (b) Velocity components and angle at Lagrangian point 2.

center at a depth of 0.6 cm are given for target material in the vertically downward direction in the cratering flow field at various times. If α and Z did not vary with time there would be one line instead of three. α increasing with time means that the fits at later times intercept the abscissa axis (R = 1 cm) at higher values of $\dot{R}$. Z increasing with time means that the fitted lines have steeper slopes at later times. Together these observations imply that the radial velocity components of the closer-in cratering flow field in the vertically downward direction are decaying more slowly than they should according to the steady-flow Maxwell Z-Model. In fact, particle velocities in the cratering flow field near the transient crater wall have increasing radial velocities over some period of time in the *excavation stage* during which the steady-flow Maxwell Z-Model would predict the radial component of the velocity should be decreasing according to Eq. (1). For example, Fig. 15 shows the particle velocity components for one Lagrangian point near the transient crater wall. With respect to the flow field center at 0.6 cm from 30 to 50 μsec the R of this Lagrangian point increases 6 percent, but the $\dot{R}$ increases by 12 percent instead of decreasing.

Modification of the steady-flow Maxwell Z-Model sufficient to keep the coupling constant α and the shape of the flow field as described by Z relatively constant with time during the *excavation stage* requires description of the flow field as having migrating flow field centers (slightly different for either constant α or constant Z) which follow the crater depth downward as shown in Figure 16. At 18 μsec the cratering flow field is well established and free from the earlier time effects of the projectile. For these migrating flow field centers which preserve at later times either the α or the Z values calculated for the vertically downward direction at 18 μsec, the α and Z values show slightly less variation with time or

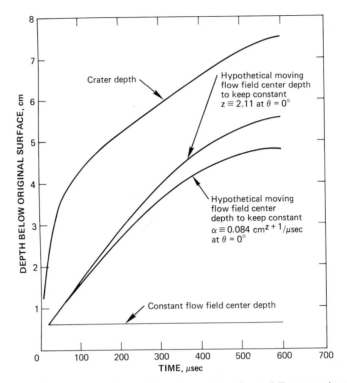

Fig. 16. Hypothetical flow field center location vs time if α and Z are restricted to be constant with time from the beginning of the excavation stage at $\theta = 0$ degrees (vertically downward).

with θ for $\theta \gtrsim 60°$ than those calculated with respect to the flow field center held fixed at 0.6 cm. However, the least squares fits for these deeper centers are only about half as good as the fits for the constant depth center in terms of the average percentage differences between the calculated and fitted radial velocities. Allowing the flow field center to migrate also shifts the time dependence of the flow field from the α and Z values to the coordinate system itself, so that the steady-flow Maxwell Z-Model with stationary streamlines still does not apply directly.

VI. DISCUSSION AND CONCLUSIONS

Using the Maxwell Z-Model of near-surface explosion cratering to analyze the finite difference calculation of a spherical aluminum projectile impact at 6 km/sec into a homogeneous plasticene clay target with a von Mises material yield strength of 50 kPa shows that in many ways the steady-flow Maxwell Z-Model applies directly or can be modified slightly, but that in one important way it does not apply directly. The basic assumption of incompressible flow holds for this calculation as shown by the small density variations in the cratering flow field during the *excavation stage*. At a given time and for a given angle, the regular power law decay of $\dot{R}$ with R holds within a few percent. However, the apparent flow field center is located beneath the surface, not at the surface as it is for near-surface explosion cratering. At a specific time, α and Z are about as spatially dependent (dependent on the angle θ) for this impact calculation as for previous near-surface explosion cratering calculations.

The major negative conclusion of the analysis of this impact calculation is that both α and Z are significantly time dependent during a major portion of the *excavation stage* of crater growth, or that the flow center itself migrates deeper with time. This is in contrast to observations of near surface explosion cratering calculations and to the steady-flow assumption underlying the most basic form of the Maxwell Z-Model. In the steady-flow form, the Maxwell Z-Model assumes that both α and Z are time independent and that the flow field center is stationary. In a modified form, the Z-Model allows α to decrease with time, but still assumes that the parameter Z, which characterizes the shape of the flow field, is time independent. Time dependence of Z or of the flow field center means that the incompressible flow occurs along time dependent streamlines so that much of the simplicity of the steady flow model is destroyed.

In practice, this means that α and Z calculated at early times in the *excavation stage* cannot be used directly in the machinery of the simple steady-flow Maxwell Z-Model to crank out the displacements, crater dimensions and ejecta angles and velocities at later times in the *excavation stage*.

This analysis has been applied to only one impact cratering calculation at one impact velocity (6 km/sec) at one scale (laboratory), at one gravity value (terrestrial), at one material strength value (50 kPa) for one target material (plasticene clay, which is very suitable for basic cratering phenomenology studies, but is not intended to be a direct simulant of materials existing on lunar or planetary surfaces). With only one calculation, what causes the departures from the simple steady-flow Z-Model is highly uncertain. Quantifying the degree to which these conclusions apply to other impact cratering flow fields with different initial conditions and different material models requires more calculations and analysis. In the meantime, the steady-flow Maxwell Z-Model should only be applied with caution to impact cratering.

Acknowledgments—This work was supported by the National Aeronautics and Space Administration under Contract No. NASW-3168. We appreciate the helpful review comments of R. Schmidt and T. Maxwell and the editorial comments of R. B. Schaal.

The Lunar and Planetary Institute is operated by the Universities Space Research Association under Contract No. NSR 09-051-001 with the National Aeronautics and Space Administration. This paper constitutes the Lunar and Planetary Institute Contribution No. 416.

REFERENCES

Austin M. G., Thomsen J. M., Ruhl S. F., Orphal D. L., and Schultz P. H. (1980) Calculational investigation of impact cratering dynamics: Material motions during the crater growth periods (abstract). In *Lunar and Planetary Science XI*, p. 46–48, Lunar and Planetary Institute, Houston.

Bjork R. L. (1961) Analysis of the formation of Meteor Crater, Arizona: A preliminary report. *J. Geophys. Res.* **66**, 3379–3387.

Bryan J. B., Burton D. E., Cunningham M. E., and Lettis L. A. Jr. (1978) A two-dimensional computer simulation of hyper-velocity impact cratering: Some preliminary results for Meteor Crater, Arizona. *Proc. Lunar Planet. Sci. Conf. 9th*, p. 3931–3964.

Bryan J. B., Burton D. E., and Lettis L. A. Jr. (1979) Calculational comparisons of explosion and impact cratering in two dimensions using barringer crater as a prototype (abstract). In *Lunar and Planetary Science X*, p. 159–161. Lunar and Planetary Institute, Houston.

Christensen D. M., Godfrey C. S., and Maxwell D. E. (1968) Calculations and model experiments to predict crater dimensions and free field motion. Report DASA-2360, Defense Atomic Support Agency, Washington, D.C. 78 pp.

Gault D. E., Quaide W. L., and Oberbeck V. R. (1968) Impact cratering mechanics and structures. In *Shock Metamorphism of Natural Materials*, (B. M. French and N. M. Short, eds.), p. 87–99. Mono, Baltimore.

Gault D. E. and Wedekind J. A. (1977) Experimental Hypervelocity Impact into Quartz Sand-II, effects of gravitational acceleration. In *Impact and Explosion Cratering* (D. J. Roddy, R. O. Pepin, and R. B. Merrill, eds.), p. 1231–1244. Pergamon, N.Y.

Gehring J. W. Jr. (1970) Theory of impact on thin targets and shields and correlation with experiment. In *High-Velocity Impact Phenomena* (R. Kinslow, ed.), p. 105–156. Academic, N.Y.

Hancock S. L. (1976) Finite difference equations for PISCES 2DELK, a coupled Euler Lagrange continuum mechanics computer program. Report TCAM 76-2, Physics International Company, San Leandro, Ca. 174 pp.

Ivanov B. A. (1976) The effect of gravity on crater formation: Thickness of ejecta and concentric basins. *Proc. Lunar Sci. Conf. 7th*, p. 2947–2965.

Maxwell D. E. (1977) Simple Z-model of cratering, ejection and the overturned flap. In *Impact and Explosion Cratering*, (D. J. Roddy, R. O. Pepin, and R. B. Merrill, eds.), p. 1003–1008. Pergamon, N.Y.

Maxwell D. E., McKay M. W., Reaugh J. E., and Seifert K. D. (1972) Advanced methods to predict the response of a site under nuclear attack. Report PIPR 388-2, Physics International Company, San Leandro, CA. 18 pp.

Maxwell D. E. and Reaugh J. E. (1972) Advanced methods to predict the response of a site under nuclear attack. Report PIPR-388-1, Physics International Company, San Leandro, CA. 35 pp.

McQueen R. G., Marsh S. P., Taylor J. W., Fritz J. N., and Carter W. J. (1970). The equation of state of solids from shock wave studies. In *High-Velocity Impact Phenomena* (R. Kinslow, ed.), p. 293–417. Academic, N.Y.

Moore H. J. (1976) Missile impact craters (White Sands Missile Range, New Mexico) and applications to lunar research. Geological Survey Professional Paper 812-B, U. S. Government Printing Office, Washington, D.C. 47 pp.

Oberbeck V. R. (1971) Laboratory simulation of impact cratering with high explosives. *J. Geophys. Res.* **76**, 5732–5749.

O'Keefe J. D. and Ahrens T. J. (1975). Shock effects from a large impact on the moon. *Proc. Lunar Sci. Conf. 6th*, p. 2831–2844.

O'Keefe J. D. and Ahrens T. J. (1976) Impact ejecta on the moon. *Proc. Lunar Sci. Conf. 7th*, p. 3007–3025.

O'Keefe J. D. and Ahrens T. J. (1977) Impact-induced energy partitioning, melting, and vaporization on terrestrial planets. *Proc. Lunar Sci. Conf. 8th*, p. 3357–3374.

O'Keefe J. D. and Ahrens T. J. (1978) Impact flows and crater scaling on the moon. *Phys. Earth Planet. Inter.* **16**, 341–351.

O'Keefe J. D. and Ahrens T. J. (1979) The effect of gravity on impact crater excavation time and maximum depth; comparison with experiment (abstract). In *Lunar and Planetary Science X*, p. 934–936. Lunar and Planetary Institute, Houston.

Orphal D. L. (1977a) Calculations of explosion cratering—I the shallow-buried nuclear detonation JOHNIE BOY: In *Impact and Explosion Cratering* (D. J. Roddy, R. O. Pepin, and R. B. Merrill, eds.), p. 897–906. Pergamon, N.Y.

Orphal D. L. (1977b) Calculations of explosion cratering—II cratering mechanics and phenomenology. In *Impact and Explosion Cratering* (D. J. Roddy, R. O. Pepin, and R. B. Merrill, eds.), p. 907–917. Pergamon, N.Y.

Orphal D. L., Borden W. F., Larson S. A., and Schultz P. H. (1980) Impact melt generation and transport. *Proc. Lunar Planet. Sci. Conf. 11th*. This volume.

Ruhl S. F. and Thomsen J. M. (1980) Impact jetting in plasticene clay: A computational and experimental comparison (abstract). In *Lunar and Planetary Science XI*, p. 958–960. Lunar and Planetary Institute, Houston.

van Thiel M. (1977) Compendium of shock wave data. Report UCRL 50108, Vol. 1, Lawrence Livermore Laboratory, Livermore, CA. 331 pp.

Thomsen J. M., Austin M. G., Ruhl S. F., Schultz P. H., and Orphal D. L. (1979a) Calculational investigation of impact cratering dynamics: Early time material motions. *Proc. Lunar Planet. Sci. Conf. 10th*, p. 2741–2756.

Thomsen J. M., Sauer F. M., Austin M. G., Ruhl S. F., Schultz P. H., and Orphal D. L. (1979b) Impact cratering calculations Part I: Early time results. Report PIFR-1220. Physics International Company, San Leandro. CA. 76 pp.

Thomsen J. M., Austin M. G., and Schultz P. H. (1980). The development of the ejecta plume in a laboratory-scale impact cratering event (abstract). In *Lunar and Planetary Science XI*, p. 1146–1148. Lunar and Planetary Institute, Houston.

Proc. Lunar Planet. Sci. Conf. 11th (1980), p. 2347–2378.
Printed in the United States of America

Cratering flow fields: Implications for the excavation and transient expansion stages of crater formation

Steven Kent Croft

Lunar and Planetary Institute, 3303 NASA Road 1, Houston, Texas 77058

Abstract—A Maxwell Z-model cratering flow field originating at non-zero depths-of-burst has been used to calculate theoretical depths and volumes of excavation, hinge radii, ejection angles, and transient structural rim uplifts for comparison with experimental and field data from impact and explosion craters. The model flow fields match the observed data well for values of Z between 2.5 and 3.0 for both explosion and impact craters, and effective depths-of-burst near one projectile diameter for impacts. The model flow field is therefore inferred to be a reasonable first-order quantitative approximation for several important crater structures, and to embody the important qualitative features of impact and explosion cratering flows. Formation of a hinge about which the coherent ejecta flap rotates at the rim of the transient crater divides material in the flow field into ejected and downward and outward-driven portions. Ejected material originates from an *excavation cavity* which has a geometry distinct from the *transient crater*. The excavation cavity and transient crater have the same diameter, D_{tc}, but the depth of the excavation cavity is ~ 0.1 D_{tc}, or about one-third of the transient crater depth, and, in the case of simple bowl-shaped craters, about one-half the depth of the final apparent crater. Down-driven material, including a central "cone" of shallow, highly-shocked material, moves downward and outward to form the walls of the transient crater and displaces an equivalent volume above the original ground surface to form the structural rim uplift.

The shallow depths of excavation both observed in impact and explosion craters and predicted by the Z-model flow fields imply that thickness estimates of lunar geologic units, such as the maria basalts, determined by assuming that excavation depths are similar to final or transient crater depths must be reduced by factors of two to three, respectively. In the Z-model flow field, streamline shapes are gravity independent and geometrically similar (except very near the center of flow). This implies that excavation cavities, whose shapes depend on hinge streamline geometry, are geometrically similar in craters of all sizes. Consequently, lunar basin excavation cavities are inferred to exhibit proportional growth and to have maximum depths of excavation near 0.1 the diameter of the basin transient crater. Thus, basin transient craters may attain diameters $\sim 10\text{X}$ the local crustal thickness before ejecting mantle material. The observed paucity of lunar mantle materals on the lunar surface around the Imbrium basin is compatible with the proportional growth of the excavation cavity if the diameter of the excavation cavity was ≥ 700 km, or near one of the innermost rings.

INTRODUCTION

Grieve (1979) has recently re-emphasized the necessity of a coherent model for the excavation stage of impact craters. Such a model is necessary to obtain structural information about the lunar crust from the lunar sample collection, and to interpret the geology of terrestrial impact structures and the petrography of their associated melt and ejecta deposits. Considerable effort has gone into the numerical simulation of impact and explosion crater formation based on the physical properties of shock waves, target and projectile materials (e.g., Bjork *et al.,* 1967; O'Keefe and Ahrens, 1978a; Orphal, 1977; Bryan *et al.,* 1978; Thomsen *et al.,* 1979; Roddy *et al.,* 1980; Orphal *et al.,* 1980; Austin *et al.,* 1980). Cratering calculations are extremely complex, but often limited to the earliest stages of crater formation so that there are few points of contact between phenomena indicated in the calculations and the crater examined by the geologist in the field. A few calculations carried to later stages of crater formation (e.g., Austin *et al.,* 1980; Orphal *et al.,* 1980) have yielded systematics in particle movements during the excavation stage that may help bridge the gap between the physics of cratering and geology of craters.

A simplified analytical description of systematic particle movements during cratering, i.e., the cratering flow field, has been described by Maxwell (1977) for explosion craters. In the following discussions, the general properties of cratering flow fields are first described. Second, a modified form of Maxwell's (1977) flow field model is derived that is proposed to describe both impact and explosion cratering flow fields. Third, specific predictions derivable from the modified flow field are compared with field observations to evaluate the applicability of the modified flow field model to actual craters. Last, implications of features of crater formation predicted by the flow field model for the depth of origin of lunar samples is discussed.

Cratering flow fields

Crater formation in explosions or impacts may be divided into three stages: a short high-pressure phase, a longer cratering flow phase, and a modification stage. The first two stages have been modeled numerically and are described in detail by Bjork *et al.* (1967) and Kreyenhagen and Schuster (1977), among others. Briefly, the high pressure phase is characterized by an expanding region of extremely high pressure behind the primary shock created by the explosion or impact. Rarefactions propagating from free surfaces quickly reduce pressures behind the primary shock to low levels, "detaching" the primary shock from the zone immediately around the explosion or impact. Virtually all material ultimately ejected from the crater is "shock processed" during the high-pressure phase, but comparatively little material movement or ejection occurs during this phase due to its short duration. In contrast, the cratering flow phase is characterized by a low-pressure, large-deformation inertial flow of target material fractured and heated by the passage of the primary shock. The preponderance of particle motion in crater formation occurs under the low-pressure conditions of the relatively lengthy cratering flow stage. The modification stage is characterized by (possibly) complex particle motions occurring very late in the cratering process in very large craters, or craters in weak materials.

The cratering flow field may be thought of as the aggregate of paths followed by particles set in motion by an impact or explosion that ultimately produces a crater. Consequently, the properties of the flow field determine how individual particles in the projectile and the cratered surface move with respect to each other and where they will be found in relation to the final crater. Therefore, if the nature of cratering flow fields can be deduced from theory or observation, a coherent model of crater formation can be constructed. A qualitative description of individual particle paths during impact cratering was given by Gault *et al.* (1968) on the basis of observations of the development and final structures of hypervelocity impact craters in sand in the laboratory. They found that during the cratering flow stage, which they called the excavation stage, particles traveled in concave-upward arcuate paths in response to rarefactions propagating downward from the free surface. A quantitative description of particle motion during the cratering flow stage of near-surface explosions, the so called Z-model, was developed and described by Maxwell and others (Maxwell, 1973, 1977; Maxwell and Seifert, 1974; Orphal, 1977). The Z-model is derived from three assumptions:

1. Flow below the ground plane is incompressible.

2. The radial velocity, $\dot{R}$, of particles below the ground plane is given by $\dot{R} = \alpha/R^Z$, where R is the radial distance from the effective origin of flow, α is a measure of the strength of the flow field, and Z determines the change of velocity with increasing radial distance.

3. Particles follow independent ballistic trajectories after spallation at or near the ground plane. The assumption of incompressibility and the expression for $\dot{R}$ lead to particle paths that are stationary streamlines similar to the empirical paths described by Gault *et al.* (1968). Values of Z for realistic computed flow fields for near-surface explosions vary from $Z \simeq 2$ near the vertical downward axis to $Z \simeq 4$ near the ground surface, with an average for the whole flow below the ground plane of $Z \simeq 3$. If α and Z are assumed to be constant, the flow field at all times can be explicitly evaluated, and quantitative descriptions of several features of near-surface explosion crater formation are obtained, including early hemispheric growth of the transient cavity (Orphal, 1977), angle of ejection, and development and deposition of an inverted ejecta flap (Maxwell, 1977). The assumption of constant α and Z is not consistent with the conservation of energy (Maxwell, 1977), but provides a good first order approximation to a real flow field (Orphal, 1977).

The Z-model, however, was generalized from numerical simulations of near-surface explosions, thus the application of the Z-model to impact craters was problematical, despite its great utility. Thomsen *et al.* (1979) investigated the ability of a constant α, constant Z-model to represent the impact cratering flow field of a 6 km/sec impact of aluminum on clay, and found that the Z-model was applicable at very early times provided: 1) the origin of the flow field is at some depth below the ground surface, and 2) the model is applied late enough in time that the projectile's momentum has dissipated. The depth of the effective center of Z-model flow ($\equiv$EDOZ) found by Thomsen *et al.* (1979) was equivalent to ~1.08 projectile diameter (D_p). Subsequent calculations by Thomsen *et al.* (1980) and Austin *et al.* (1980) of the aluminum on clay impact carried out to much later times, and to nearly final crater dimensions, showed that the evolution of the flow field could be viewed in two ways: 1) for EDOZ fixed at ~1.0 D_p, α generally increased throughout the entire calculation, and Z generally increased until the crater reached ~½ final dimensions, after which Z remained nearly constant; or, 2) holding α and Z fixed at early time values forced EDOZ to move steadily deeper into the target. Both viewpoints imply a departure from a constant α,Z flow field for impact craters. However, Austin *et al.* (1980) find that the assumption of incompressibility still holds during the cratering flow stage, hence the concept of streamlines is still valid. The result that α changes continuously for constant EDOZ in impact crater flow fields implies that particle movements in *time* along streamlines cannot be well represented by a constant α, Z, EDOZ model. The Z parameter, however, determines the shape of the streamlines. Therefore, the observation that Z is approximately constant for EDOZ $\simeq$ 1.0 D_p while the crater grows from ~½ to nearly final dimensions implies that the concept of stationary streamlines is still a reasonable quantitative approximation for particle movements in *space* for ~70–90% of the ejected volume. The variation of α and Z at early times, particularly the probably unphysical increase in α, imply that a constant α, Z, EDOZ flow field may be an oversimplified description of the earliest stages of crater flow. Accordingly, quantitative predictions may be suspect. However, because α and Z vary smoothly during early times, it is suggested that the dominant qualitative spatial features of the early-time impact flow field are embodied in a constant Z, EDOZ flow field.

There is also evidence that an effective center of flow near one projectile diameter as found by Thomsen *et al.* (1979) may have physical meaning for impact flow fields in general. Oberbeck (1971) produced a series of equal energy explosion craters in dry sand at varying depths-of-burst to determine the equivalent depth-of-burst (EDOB) of impact craters produced by aluminum projectiles impacting identical dry sand targets at ~2 km/sec, with the same energy as the explosives. Oberbeck's criteria of equivalent craters included crater dimensions, ejecta plume characteristics, and subsurface deformation. It can be shown from projectile data given in his paper that the cylindrical projectiles used were ~0.68 cm long. By comparison, the experimentally determined EDOB of 0.63 ± 0.02 cm is very nearly equal to one equivalent spherical projectile diameter. Thus the EDOZ of Thomsen *et al.* (1979) bears nearly the same relation as the EDOB of Oberbeck (1971) to the projectile diameter, despite significant differences in impact velocity and target composition. Holsapple (1980) used both theory and experimental explosions and impacts to determine the EDOB's of impact craters in quartz sand using crater volume as an equivalence criterion. Holsapple's results imply that the EDOB of impact craters lies between 0.5 and 2 projectile diameters for a large range of energies, impact velocities, and target to projectile density ratios. These results imply that explosion flow fields are most like impact flow fields when the depth of the explosive center is near one projectile diameter. Oberbeck's (1971) results, in particular, show significant differences in ejecta plume development and subsurface deformation (both direct results of the flow field) between impact and explosion craters when the explosive center was moved away from ~1.0 D_p in depth.

It is suggested that the parallels between the results of the studies of Thomsen *et al.* (1979), Austin *et al.* (1980), Oberbeck (1971), and Holsapple (1980) provide physical justification for the use of a presumably explosive flow field (the constant Z model) centered at ~1.0 D_p beneath the target surface to approximate an impact cratering flow. The ultimate utility of any approximate model of the impact cratering flow field is determined by the insights it provides by simplification of a very complex phenomenon, and by comparison of model predictions with observation. Appendix A gives mathematical expressions for particle streamlines, ejection angles, ejected volumes, maximum depths of excavation, and structural rim uplifts derived from a Maxwell constant Z-model centered at arbitrary EDOZ. These derived expressions describe primarily spatial relationships for two reasons: 1) as discussed above, in the spirit of a first-order approximation, a constant Z, EDOZ flow field may be expected to yield coarse quantitative information about spatial relations, especially at distances over half the apparent crater radius from the point of impact, but not temporal relations; and 2) most observations of explosion craters and laboratory impact craters are spatial in nature. Only spatial relations are available for large terrestrial impact craters.

Qualitative flow field features

The transient crater and the final apparent crater are two important stages in the process of crater formation. The final apparent crater is the observed void of a freshly formed crater after all cratering-related modifications (slumping, rebound, etc.) have taken place. The ratio of the apparent depth (d_a) to the apparent diameter (D_a) for simple (bowl-shaped) craters is typically 1:5 (Pike, 1977; Croft 1978, 1979b). The terms *transient crater* and *transient cavity* are frequently used interchangeably. In this paper, however, the transient crater is defined as the maximum extension of the transient cavity, where the transient cavity is defined as the growing void at the center of the flow field. This usage follows the convention suggested by Dence *et al.* (1977). Dence (1973) and Dence *et al.* (1977) infer from geologic evidence that the d_a:D_a ratio of the transient craters of both simple and complex terrestrial impact craters is ~1:3. Figure 1 shows the superposition of a constant Z, EDOZ flow field on a 1:5 simple crater profile and on a 1:3 parabolic transient crater profile. Even though the apparent radius of the final apparent crater is probably slightly larger than the apparent radius of the transient crater (e.g., see Dence *et al.*, 1977; Settle and Head, 1979), they are drawn equal in Fig. 1 in the spirit of a first-order approximation. To understand the significance of the volumes V_e, V_u and V_t indicated in Fig. 1, consider for a moment the velocity field of a growing crater. At the original ground surface, ejection velocities are highest near the incipient crater center, and decay rapidly with increasing range. The final ranges of particles passing through the original ground surface are the sums of their initial ranges at the surface and their ballistic ranges. Ballistic ranges are directly proportional to the square of the ejection velocity. Therefore, because ejection velocities decrease with increasing distance from the crater center, the ballistic ranges of particles at increasingly large initial

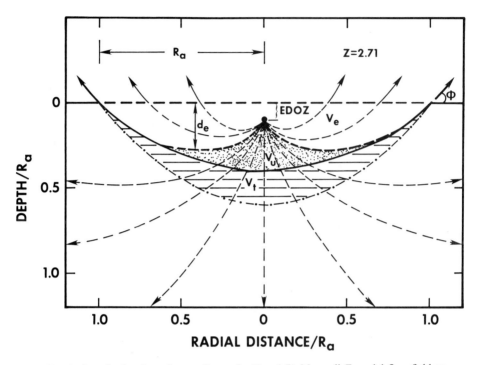

Fig. 1. A scale drawing of streamlines of a Z = 2.71 Maxwell Z-model flow field superposed on a parabolic 1:3.3 depth/diameter ratio transient cavity and a typical 1:5 simple crater profile. V_e is the ejected volume; V_u (stippled zone) is the volume of the permanent structural rim uplift, and V_t (horizontal hatched zone) is the volume difference between the transient and final craters.

ranges are monotonically decreasing. Consequently, between the surface particles initially near the crater center whose final ranges are large due to high ejection velocities, and surface particles initially far from the crater center whose final ranges are large due to their large initial ranges, there is a surface particle with a minimum final range (Ivanov, 1976; Killian and Germain, 1977). From Fig. 1, it is apparent that particles ejected at a given surface range from the crater center originate along a streamline that extends back to the EDOZ. The uppermost streamline of the four shown in Fig. 2 passes through the initial position of the surface particle with the minimum final range. This particle is defined here as the *hinge particle* because, as is seen in Fig. 2, it is translated by the flow field to the innermost position of the uplifted original surface, which position becomes a hinge about which the ejecta flap rotates (Maxwell, 1977). The streamline passing through the initial position of the hinge particle is defined as the *hinge streamline*. The hinge streamlines (heavy dashed lines in Fig. 1) represent a fundamental boundary in the flow field; particles initially on streamlines above the hinge streamlines (the clear field labeled V_e in Fig. 1) are thrown out of the crater as ejecta; particles initially on streamlines below the hinge streamlines are driven primarily downward and outward into the walls and floor of the transient crater. The initial position of the hinge particle becomes the apparent radius of the transient crater, R_a, because (as seen in Fig. 2) all material along streamlines passing through the ground surface inside R_a is ejected.

The fundamental division of the flow field into ejected and down-driven portions implies that all material ejected during crater formation originated from a volume of space bounded above by the original ground surface, and bounded below by the surface defined by axisymmetric rotation of the hinge streamline. As seen in Fig. 1, this volume of space is distinct from both the transient and final apparent craters. Therefore, following Dence *et al.* (1977), who recognized that the volume of the transient crater is formed in part by excavation and in part by displacement of target materials, this volume of space is defined as the *excavation cavity*. The excavation cavity is purely a spatial construct which is never seen, even momentarily, during the cratering process, because particles are moving simultaneously along all streamlines. But the excavation cavity is important because no material is excavated or uncovered from depths greater than the depth of the excavation cavity, d_e (with the possible exception of central peak materials uncovered during the modification phase of complex craters), which in Fig. 1 is seen to be significantly shallower than the transient crater and even the final apparent crater.

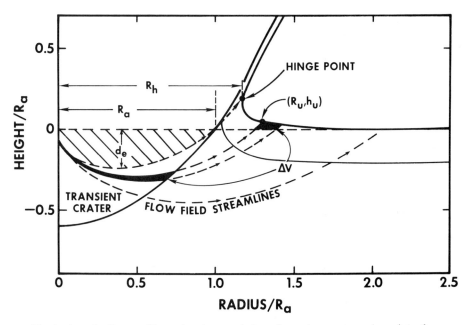

Fig. 2. A scale diagram illustrating the translation of transient crater volume into the structural rim uplift. Material in the hatched zone is ejected. Material below the ejected zone is driven along streamlines into the transient crater wall. Under the assumptions of incompressibility and steady streamlines, an equal volume is pushed into a structural rim uplift as indicated. The structure of the hinge zone is also indicated.

Orderly flow of material along the streamlines and ballistic paths indicated in Fig. 1 implies important secondary spatial characteristics of the ejected and down-driven material:

Ejected material
Particles are primarily thrown beyond the hinge zone to form the continuous and discontinuous portions of the ejecta blanket. On planets with atmospheres, smaller ejected particles experience drag and form a fallout layer which blankets the entire cratered area, including the interior of the final apparent crater (Schultz and Gault, 1979; Settle, 1980). Pre-cratering spatial positions of ejected particles are reversed: Particles initially near the crater center are thrown farther than particles initially farther from the crater center, and initially deep particles are ejected later and land on top of initially shallower particles. This reversal produces the inverted stratigraphy observed in ejecta blankets (e.g., Shoemaker, 1960; Stöffler *et al.*, 1975). Particles in the fallout layer defined by Shoemaker and Kieffer (1974) and Roddy (1978) are thoroughly mixed.

Down-driven material
Particles remain inside the crater and retain their unexcavated direct stratigraphy relative to the center of flow. In particular, this will be true of the "cone" of material immediately below the EDOZ (peak of the stippled field in Fig. 1), which consists of a small volume of material from shallow layers that are otherwise absent within the transient crater due to ejection. Material in the cone moves along streamlines that diverge and intercept the whole inner face of the transient crater. Consequently, the cone materials are "smeared out" to line the transient crater in an exceedingly thin layer that nevertheless retains the original direct stratigraphy. Relative movement between neighboring streamlines, particularly between streamlines immediately below the hinge streamline, will lead to considerable shearing in the crater floor. The volume of down-driven material $(V_u + V_t)$ translates directly into a structural rim uplift. This is illustrated in Fig. 2: The volume of material, ΔV, denoted by the shaded area bounded by the two closely spaced streamlines, the center of flow, and the transient crater wall, is carried by the flow into the transient crater wall. Under the assumptions of incompressibility and streamline flow, an equivalent volume, ΔV, is pushed above the original ground level at the range where the streamlines intersect the ground surface.

Comparison of flow field predictions with field data

Excavation Cavity Shape
Direct evidence for the shape of the excavation cavity is provided by Snowball, a 500 ton TNT explosion crater with a flat floor and central peak (Roddy, 1976). Prior to detonation, vertical columns containing numbered marker cans were placed at various distances radial to ground zero (GZ). Figure 3, a north-south cross-section through the crater, shows the *pre-shot* positions of marker cans

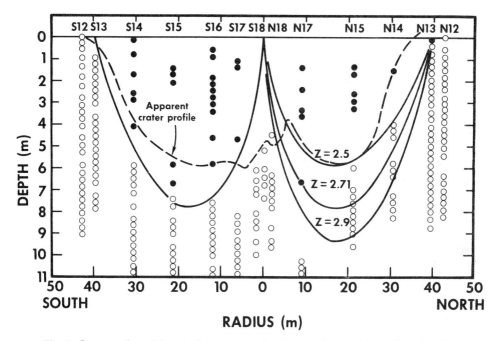

Fig. 3. Cross-section of Snowball target area showing pre-shot positions of numbered marker cans. Cans represented by solid circles were recovered after the shot in the ejecta; cans represented by open circles remained in the crater floor and walls. Data (from Jones, 1976) provide direct measure of *in situ* excavation cavity. Curves are excavation cavity profiles predicted by the Z-model.

that were recovered after the detonation. Marker cans were originally placed at regular intervals from the surface down in each column, thus gaps in the columns represent cans never recovered. The data are from Jones (1976). Filled circles designate cans found in the ejecta blanket after the detonation, thus defining points in the *in situ* excavation cavity. Open circles designate cans recovered in the post-shot crater floor and walls, thus defining points below the excavation cavity. Streamlines for $Z = 2.5$, 2.71, and 2.9 passing through the measured apparent crater radius of ~41 m are shown for comparison. Though there is some asymmetry, the observed excavation cavity corresponds well with that predicted by a $Z = 2.71$ flow field. In particular, a number of cans in the central three columns at depths considerably shallower than the depths of cans excavated on either side provide evidence of an unexcavated central cone of material. The profile of the final apparent crater is also shown in Fig. 3. The final crater is also slightly asymmetric and is shallower than the excavation cavity. The final positions of the marker cans in the floor and walls in the Snowball crater are extremely complex, especially in the region of the central peak (see Jones, 1976), presumably reflecting complex late-stage modifications not directly related to the cratering flow during the excavation stage.

Less direct evidence of a central down-driven cone of shallow material in

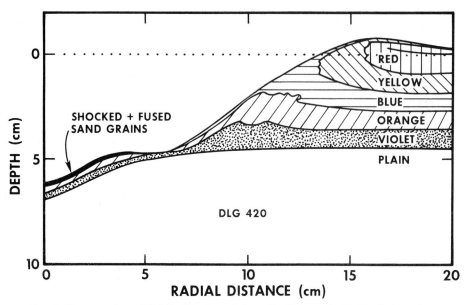

Fig. 4. Cross-section of DLG 420 adapted from Stöffler *et al.* (1965). Only grains from the top three layers have been ejected. The shaded layer labeled "shocked and fused sand grains" consists of highly shocked and thinned remnants of the top three layers in pre-impact stratigraphy. Deepest exposed material is on the upraised rim.

impact craters is provided by DLG 420, an experimental impact crater in sand with an apparent diameter of ~28 cm (Stöffler *et al.*, 1975). Figure 4 is a cross-section of DLG 420. As noted by Stöffler *et al.* (1975), the floor and walls of the crater are lined with extremely thin remnants of the top three layers of the target oriented in the original (pre-impact) stratigraphic sequence of red-yellow-blue sand going from surface to interior. The remnants are too thin to be differentiated in Fig. 4, but are represented by the (exaggeratedly thick) layer designated as shocked and fused sand grains. Because grains in the remnants are highly shock comminuted and fused, the remnants must have originated near the point of impact. The remnants are also well separated from other ejected and displaced portions of their respective colored layers, implying a distinct mode of origin. The shallow origin of these remnants near the point of impact, the preserved pre-impact stratigraphy, thinness, and position as lining of the inner wall of the crater are characteristics parallel to those predicted for down-driven central cone materials by the flow field model. Hence the remnants are inferred to be the down-driven cone material.

Kilometer-sized simple terrestrial impacts also appear to have deep seated layers that may be interpreted as down-driven cone material. It has been found by deep drilling that the relatively unshocked breccia lens of Brent Crater (rim diameter ~3.8 km) is underlain by a thin layer of melt, which in turn is underlain by thin layers of progressively less shocked material (Robertson and Grieve,

1977; Grieve *et al.*, 1977; Dence *et al.*, 1977). Crustal rocks at Brent lack stratigraphic markers, so positive identification of pre-impact positions of these deep layers cannot be made. However, the decrease of shock pressure with increasing distance from the point of impact permits the use of assumed pressure decay rates to reconstruct initial ranges of rocks of uniform shock deformation from the point of impact in a manner analogous to replacing layers of uniform composition in a layered target. Robertson and Grieve (1977) note that extrapolation of the *in situ* apparent shock pressure attenutation rate of the layers back to the point of impact imply improbably high impact pressures, suggesting that the layers were originally much closer to the point of impact. The qualitative lines of material movement deduced by Grieve *et al.* (1977) to explain the origin of the shocked layers are very similar to those in Fig. 1. The thinness, highly shocked nature, and monotonic decrease of shock features with increasing depth of the layers are analogous to the properties of the remnant layers in DLG 420, and suggest a similar origin as down-driven cone material. Dence *et al.* (1977) note that similar layers of highly shocked material at the base of the breccia lens have been reported for West Hawk Lake, Canada, and Lonar Lake, India.

Drilling at Meteor Crater gives evidence for a similar highly-shocked layer at this crater. Barringer (as quoted and discussed by Hager, 1953) reports finding meteoritic iron fragments between depths of 137–207 m in 14 of the 28 holes drilled in the central portion of the crater. Shoemaker (1960) and Brett (1968) report the fragments to be nickel-iron spherules surrounded by glass (from cuttings 168 to 198 m below the present crater floor). Barringer (Hager, 1953) also notes a layer of very white and fine "silica flour" immediately below the meteorite bearing layer, in and below which no meteoritic material was found. Below the layer of silica flour, less and less fractured sandstone was found until unaltered bedrock was finally reached. The description of material in the silica flour layer closely corresponds to the description of Kieffer's (1971) class lb shocked Coconino sandstone, and suggests the silica flour to consist of similarly shocked material. Thus at Meteor Crater there appears to be a crude series of shocked layers: glass plus meteorite spherules (highly shocked), silica flour (intermediate shock), and fractured bedrock (low shock), that appear in the same stratigraphic position as the melted and shocked layers at Brent. Figure 5 is a scale drawing of the breccia lens of Meteor Crater (Shoemaker, 1960) including the shocked layer. The exact lateral extent of the fused rock and meteorite and silica flour is not precisely known, but drill holes in the crater center extend to roughly the radial position of the question marks in Fig. 5 (Shoemaker and Kieffer, 1974), which is similar to the radial extent of the shocked material at Brent (Robertson and Grieve, 1977).

Volumes of excavation

The total volume predicted for the excavation cavity by the constant Z, EDOZ model is given by Eq. 3e in Appendix A. Accurate measurements of the volume of ejecta only (excluding rim uplift and correcting for slump, etc.) are almost non-existant for large impact craters, because either erosion has removed significant

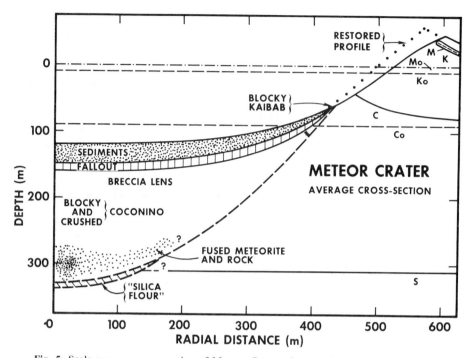

Fig. 5. Scale average cross-section of Meteor Crater. Current profile is top solid line. Restored pre-erosion profile is dotted line. Pre-impact stratigraphic horizons are labeled M_o (Moenkopi), K_o (Kaibab), and C_o (Coconino). The tight wavy line in the extreme upper right extension of the breccia lens is an estimated contact between Kaibab and Coconino in the brecciated material. Drawing compiled from data in Hager (1953), Shoemaker (1960), Brett (1968), Roddy et al. (1975) and Roddy (1978).

amounts of the ejecta, or sufficient probing of the upraised rim to correct for uplift, bulking, etc., has not been carried out. Consequently, only the ejecta volumes of Meteor Crater (Table 1), DLG 420 (Table 2), and the Prairie Flat explosion crater (Table 3) are used here. If the indication discussed in the flow fields section that a constant Z, EDOZ model quantitatively represents the steady state flow field for most of the ejected volume, an estimate of the Z of the flow field can be made using Eq. 3e, the ejecta volume (V_e), the apparent radius (R_a), and EDOZ. The values of EDOZ for Meteor Crater and DLG 420 were assumed to be at depths of approximately one projectile diameter, which for the apparent diameters given in the tables, and projectile diameters of ~30 m for Meteor Crater (Bryan et al., 1978; Roddy et al., 1980) and ~0.8 cm for DLG 420 (Stöffler et al., 1975), yields EDOZ $\simeq 0.031\ D_a$ for both craters. EDOZ was chosen to be 0.0 for Prairie Flat, which was formed by a spherical charge of TNT resting on the surface. Normalized ejecta volumes of the three craters are plotted on curves of appropriate EDOZ in Fig. 6c. The implied values of Z for the flow fields forming these craters is ~2.8 for Meteor Crater and Prairie Flat, and ~2.5 for

Table 1. Meteor Crater data summary[a]

			A. Individual Layers			
Layer	Depth (m)	Depth/R_a	V_d (displaced)	V_u (uplift)	V_e (ejected)	$\Sigma V/V_e$
Moenkopi	0	0.0	6.4 Mm³[b]	~ 0 Mm³	6.4 Mm³	0.102
Kaibab	8.5	0.017	48.7	~ 3.1	45.6	0.825
Toroweap	88.5 ± 2	0.181	0.7	~ 0.1	0.6	0.835
	90 ± 2	0.184	21.7	~ 11.3	10.4	1.000
Coconino	310	0.633				
Supai						
		Total	77.5	~ 14.5	63.0	

B. Whole Crater

R_a(m)	d_a	d_e	R(Hinge)/R_a	V_e/R_a^3	Ve/Vu
495 m	170 m[c]	120 m	1.14	0.519	5.24

[a] Layer depths adopted from Roddy (1978). Displaced volumes (V_d) are defined equal to the sums of the permanently uplifted (V_u) and ejected volumes (V_e). Values of V_d, V_u, and V_e are derived from a new model of the restored crater dimensions (Croft, 1980) and differ slightly from ejected volumes in Roddy *et al.* (1975).

[b] Symbol Mm³ = million cubic meters.

[c] Apparent depth of displaced crater volume, equals present depth (~120 m) plus thickness of sediment layer (~30 m), plus thickness of fallout layer (~10 m, Roddy, 1978), plus depth acquired by removing bulking in breccia lens (~10 m, Croft, 1980).

DLG 420. As an example of the uncertainty in Z encountered using this method, the curve for EDOZ = 0.084 D_a is included in Fig. 6c. This curve approximates the EDOZ for Meteor Crater assuming the center of flow is near the 85 m (~2.5 D_p) estimated by Bryan *et al.* (1978). For this EDOZ, the estimated value of Z for the Meteor Crater flow field would be ~2.5.

Table 2. DLG 420 data summary[a].

			A. Individual Layers			
Layer	Depth (cm)	Depth/R_a	V_d (cm³)	~V_u (cm³)	V_e (cm³)	$\Sigma V/V_e$
Red	0.0	0.0	652	115	537	0.54
Yellow	0.9	0.065	473	105	368	0.37
	1.8	0.131	127	35	92	0.09
Blue	2.7	0.196				
		Total	1252	~255	~997	

B. Whole Crater

D_a (cm)	d_a (cm)	d_e (cm)	Ve/R_a^3	Ve/Vu	R(Hinge)/Ra
27.5	5.55	2.5	0.384	~2.31[b]	1.20

[a] Data are taken directly or estimated from data and cross-sections of DLG 420 in Stöffler *et al.* (1975).

[b] V_u includes volumes of uplifted sand in rim (~255 cm³) plus volume in crater due to compaction (~176 cm³).

Table 3. Prairie Flat crater data summary.

A. Dimensions and Volumes

D_a	=	61 m[a]	V_e (ejecta)	=	13,000 m[3a]
d_a	=	4.4 m[a]	V_{tu} (transient uplift)	=	9,400 m[3b]
d_e	$\approx$	6.25 m	V_s (settled)	=	6,800 m[3b]
R(Hinge)/R_a	=	1.18	V_{tc} ($V_a + V_{tu} + V_s$)	=	27,300 m[3]
V_a	=	11,100 m[3c]	V_e/R_a^3	=	0.46

B. Transient Uplift

Average maximum[d]		Normalized to apparent radius (R_a)[a]	
Range (R_u), m	Height (h_u), m	R_u/R_a	h_u/R_a
33.3 ± 0.75	4.67 ± 0.70	1.191 ± 0.027	0.167 ± 0.025
34.8 ± 0.69	2.57 ± 0.21	1.242 ± 0.025	0.092 ± 0.008
43.8 ± 0.28	0.90 ± 0.09	1.565 ± 0.010	0.032 ± 0.003
52.5 ± 0.08	0.36 ± 0.04	1.874 ± 0.003	0.013 ± 0.001

[a] Roddy (1977a).
[b] Jones (1970).
[c] Rooke et al. (1972).
[d] Averages calculated from data given by Sauer (1970).
[e] Normalized to R_a = 28 m, the apparent radius for the profile along which accelerometers were located. This radius is slightly smaller than the average apparent crater radius of 30.5 m.

Hinge radii

Numerical analysis of rim uplifts of the type illustrated in Fig. 2 using the model in section 4 of Appendix A indicated that the ratio of the hinge radius (R_h in Fig. 2) to the apparent radius was a function of Z and EDOZ. Figure 6a shows R_h/R_a as a function of Z for the same values of EDOZ in Fig. 6c. Hinge radii are estimated for DLG 420 (Fig. 4), Meteor Crater (Fig. 5) and Prairie Flat (Fig. 9) by measuring the radius of the innermost point of the original ground surface, where the orientation of stratigraphic horizons changes from direct in the uplifted rim to inverted in the ejecta flap. Because the hinge point shown in Fig. 2 is still in motion, a hinge radius measured from the final apparent crater is an approximation of R_h in Fig. 2. Ratios of the hinge radius to the apparent radius of each of the three craters are given in the appropriate table and plotted on the appropriate EDOZ curves in Fig. 6a. The implied values of Z are similar to those predicted by the V_e/R_a^3 vs. Z relations.

Depths of excavation

Limits on the depth of excavation can be set on craters formed in targets consisting of layers that can be distinguished by composition, color, texture, etc. If, as illustrated in Fig. 7, the limit of excavation extends into layer 3 as shown, then samples from layer 3 will be present in the ejecta blanket and fallout layer (if any). The mere presence of samples from layers 1, 2, and 3, but not from layer

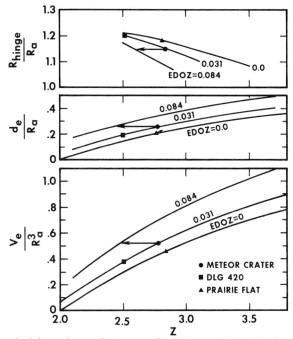

Fig. 6. Theoretical dependence of (A) normalized hinge radii, (B) depths of excavation and (C) ejecta volumes on Z and EDOZ. Data shown from craters indicate Z values between 2.5 and 2.9 for flows producing the craters.

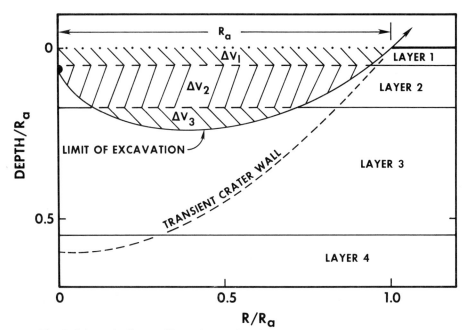

Fig. 7. Schematic diagram illustrating spatial relations used in technique for estimating depths of excavation for craters in layered targets.

4 sets upper and lower limits on the depth of excavation if the depths of the lower and upper boundaries, respectively, of layer 3 are known. The transient (or final apparent) crater may extend into layer 4 as shown, but no samples from layer 4 will be found on the surface unless the excavation cavity extended that deep. If the thickness of individual layers is small, limits on excavation depths will be narrow. If the individual layers are thick, such as layer 3 in Fig. 7, the limits on the excavation depth will be large.

An estimate of the depth of excavation into a thick layer can be made if the ejected volumes of individual layers (ΔV_1, ΔV_2 and ΔV_3 in Fig. 7) are known and the shape of the excavation cavity can be estimated. If the shape of the excavation cavity is given by the hinge streamline, then the volume of material, ΔV, ejected from each layer is a function of Z, EDOZ, R_a, and the thickness of each layer. The predicted cumulative volume ejected down to a given depth for any set of model parameters may be calculated and plotted as shown for EDOZ = 0.031 D_a in Fig. 8. If the total ejected volume, V_e, is known for the case shown in Fig. 7, then $\Delta V_1/V_e$ would plot at the normalized depth of the bottom of layer 1,

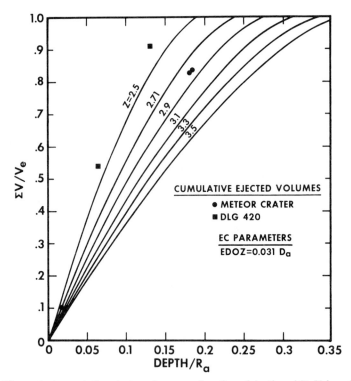

Fig. 8. Theoretical cumulative ejecta volume as a function of depth and Z. Value of $\Sigma V/V_T$ at a given depth is the fraction of the total ejecta volume originating at shallower levels. Depth where $\Sigma V/V_T = 1$ is the excavation depth. Volume data from DLG 420 and Meteor Crater are reproduced by flow fields of $Z \cong 2.4$ and 2.8, respectively. The difference may be indicative of the effect of material properties on streamline shape.

$(\Delta V_1 + \Delta V_2)/V_e$ at the normalized depth of the bottom of layer 2, etc. The depth of excavation may be estimated by passing a curve through the available depth vs. $\Sigma V/V_e$ points and extrapolating to 1.0 V_e. If different excavation cavity shapes are assumed, the $\Sigma V/V_e$ vs. depth curves will be different. For example, a cylindrical-shaped excavation cavity produces a family of straight lines through the origin in Fig. 8. A conic excavation cavity produces a family of curves much more sharply bent to the upper left corner of Fig. 8 than the Z-derived curves. The data points shown in Fig. 8 are normalized depth vs. normalized ejecta volumes for discrete layers in Meteor Crater (Table 1) and DLG 420 (Table 2). Both the number of craters (2) and the number of data points for each crater are admittedly small, but they are the only data available. The data points are not inconsistent with the calculated Z curves, implying that a constant Z, EDOZ excavation cavity shape is not inappropriate. The extrapolated depth of excavation for Meteor Crater is ~0.25 R_a, or ~120 m. This implies d_e to be approximately 30 m shallower than the top of the fallout layer, the apparent bottom of the pristine Meteor Crater, and 40 m above the top of the breccia lens, which presumably would correspond to the apparent bottom of Meteor Crater in the absence of an atmosphere (Settle, 1979). The extrapolated depth of excavation of DLG 420 is ~0.18 R_a (d ~ 2.5 cm), or slightly less than half the apparent crater depth. This represents excavation to nearly the bottom of the blue layer. That this was indeed the case is evidenced by the lack of any orange sand outside the crater, and the apparent "split" of the blue layer seen in Fig. 4; i.e., most of the blue layer was driven upward, but a thin layer of blue sand was driven downward, maintaining continuous contact under the crater floor.

The depth of excavation of Prairie Flat crater was estimated by examination of pre-shot *in situ* depths of material found in the ejecta blanket. A pre-shot bore hole through the Prairie Flat GZ (Jones, 1970) showed the following stratigraphy: Silty clay with layers of silt to 2 m depth, underlain by layers of fine to medium sand to ~6.25 m depth, underlain by a layer of brown and blue clay extending to a depth of ~20.9 m. Excavation of north-south trenches through the post-shot crater rims confirm that a similar stratigraphy underlay the entire crater. The excavated cross-section of the ejecta flaps of Prairie Flat (Jones, 1970, Figs. 36, 38) show that the flaps consist of nearly complete stratigraphic sections of the upper layers of silt and sand. However, none of the clay layers below 6.25 m was ejected (Roddy, pers. comm.), therefore, the depth of excavation is ~6.25 m (the value given in Table 3), or $d_e \simeq 0.20 R_a$. The apparent depths, apparent diameters, and normalized depth of excavations for DLG 420, Prairie Flat, Snowball (from Fig. 3), and Meteor Crater are collected together in Table 4 for comparison.

A unique confluence of different sources of lunar data provide a possible observational constraint on the depth of excavation of multi-kilometer sized lunar craters. Head *et al.* (1978) used various remote sensing techniques to define basalt flow units in Mare Crisium. A magnesium rich basalt (Head *et al.*'s type 1) was found to occur on a structural shelf in SE quadrant of the mare, and in the ejecta blankets of four craters: Picard, Peirce, Greaves and Cleomedes F. Head *et al.* interpret this to mean that these four craters have penetrated a surface layer of chemically distinct basalts and excavated portions of a subsurface magnesium-rich basalt layer, which apparently underlies the surface basalts of most of Mare Crisium. Andre *et al.* (1978) came to the same conclusion, and postulate further that the magnesium-rich layer corresponds to the 1.4 km deep reflection surface found by the lunar radar sounder experiment (May *et al.*, 1976). Peeples *et al.* (1978) published a detailed cross-section of the subsurface structure of Mare Crisium derived from the radar sounder that extends across the marginal shelf and central floor of the basin and passes through the crater Peirce. The subsurface layering appears remarkably regular. Lunar Topographic Orthophoto (LTO) coverage is available for the same region. If it is assumed (1) that the 1.4 km discontinuity does represent the top of the magnesium-rich layer, and (2) that the regular basin-wide layering found by the sounder experiment is approximately valid for the rest of the western end of Mare Crisium, then the presence of excavated magnesium-rich material in the ejecta blankets of some craters but not in others sets lower and upper limits, respectively, on the ratio d_e/R_a. Table 4 gives apparent diameter, apparent depth, and estimated depth to the top of the mg-rich layer beneath the four craters that penetrated the layer, and Swift, the largest crater on the mare that did not penetrate the layer. Peirce and Swift provide the strongest constraints on d_e, because they lie closest to the observed sounder profile. Note that the bottom of Swift not only lies below that of Peirce, but is actually ~200 m deeper than the postulated magnesium-rich layer. This is similar to DLG 420 (Fig. 4) where d_a is significantly deeper than the depths of layers that were not excavated. Greaves and Cleomedes F, only slightly larger than Swift, did penetrate the magnesium-rich layer because they are on the mare shelf, where the depth to the layer is significantly less.

By this analysis, the depths of excavation for these lunar craters lie between depths of ~0.09 D_a and ~0.17 D_a, which bracket the values of d_e found for the terrestrial impact and explosion craters. The data suggest that $d_e \simeq 0.1 D_a$ may be a general characteristic of large-scale impact craters formed by single, low-porosity projectiles. In comparison, the depths of excavation predicted by constant Z, EDOZ flow fields (Eq. 2i in Appendix A) are ~0.1 D_a (= 0.2 R_a) for values of Z between ~2.5 and ~3.0 as seen in Fig. 6b. Normalized depths of excavation for Meteor Crater, DLG 420 and Prairie Flat are plotted on curves of appropriate EDOZ, as was done for the ejecta volumes and hinge radii, with similar results: observed depths of excavation can be accounted for by flow fields of $Z \simeq 2.5$ for DLG 420, and $Z \simeq 2.8$ for Meteor Crater and Prairie Flat. From Fig. 3, the depth of excavation of Snowball is compatible with a flow field of $Z \simeq 2.7$.

Transient rim uplift

The field data discussed to this point have related primarily to the excavation cavity. Data are presented in this section which indicate the role played by the down-driven portion of the flow field in the formation of the transient crater. As suggested in the discussion of Fig. 2, under the assumptions of incompressibility and steady streamlines, the down-driven volume between the bottom of the excavation cavity and the transient crater wall pushes an equal volume of rim material into a transient rim uplift. Assuming a constant Z, EDOZ flow field, expressions are derived in section 4 of Appendix A for the uplifted range (R_u) and height (h_u) of points on the original ground surface as functions of the initial surface range of the point, Z, EDOZ, and the shape of the transient crater. Sufficient rim uplift data are available for the Prairie Flat explosion crater (Sauer, 1970; Jones, 1970; Roddy, 1976; Roddy *et al.*, 1977) to compare model predictions with observations, provided estimates of the transient crater can be made. Figure 9 is a drawing of the east profile of Prairie Flat showing the present surface and simplified geology of the crater rim region. The maximum transient uplift of the original ground surface (Sauer, 1970) is indicated by a dashed line. Velocity gauges and accelerometers were placed at radii between ~20 and ~100 m from ground zero at depths between 0 and 8 m. Roddy *et al.* (1977) integrated the observed velocities and accelerations to determine material movements in both time and space. They found that the entire instrumented rim region was uplifted as a coherent unit; i.e., material to a depth of 8 m, including all layers of silt and sand and the top two meters of clay, responded to crater formation as an unseparated unit. There is no reason to suspect that material separation occurred below 8 m in the clay, so material motions were probably continuous over the entire half-space. It is suggested that the measured and implied continuous particle motions in the cratered half space satisfies the assumption of incompressibility and streamline flow, and indicate that the volume of the transient crater was transmitted through the cratered medium into the transient uplift. Conservation

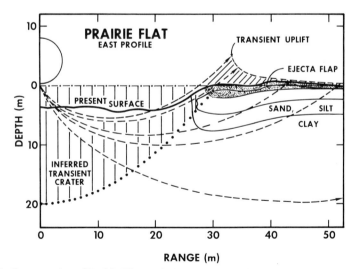

Fig. 9. Cross-section of Prairie Flat explosion crater compiled from data given by Jones (1970), Sauer (1970), and Roddy (1976). Half circle at left is size and location of high-explosive. Stippled zone is ejecta. Heavy solid line is post-shot surface. Dashed "dynamic uplift" line is position of maximum displacement of original ground surface now underlying ejecta. Dashed curved lines are Z = 2.7 streamlines through pre-shot surface positions of accelerometers. Dotted line is inferred transient crater achieved at maximum displacement of ground surface as discussed in the text.

of volume in the rim uplift also suggests that the total volume of the transient crater (V_{tc}) can be estimated from present apparent crater volume (V_a) plus the volume of the transient uplift (V_{tu}), plus a "settled" volume (V_s) equal to the volume between the original ground surface and its present down-slumped position. Estimates of these volumes made from geologic cross-sections and uplift measurements are given in Table 3. The estimated transient crater volume is ~27,300 m³. (There may be considerable uncertainty in this figure because estimates for V_s are based on only two profiles and V_{tu} on uplift measurements along a single radius.) If, following Dence (1973), a parabolic transient crater profile is assumed, then the implied depth of the transient crater, d_{tc}, is:

$$d_{tc} = 8V_{tc}/(\pi D_a^2) \simeq 19 \text{ m}$$

for the Prairie Flat crater. The inferred transient crater is indicated by the dotted line in Fig. 9.

Transient structural rim uplifts calculated from the inferred Prairie Flat transient crater (apparent transient crater dimensions: $D_a = 61$ m, $d_a = 19$ m) for various values of Z are shown in Fig. 10. The curves represent the maximum

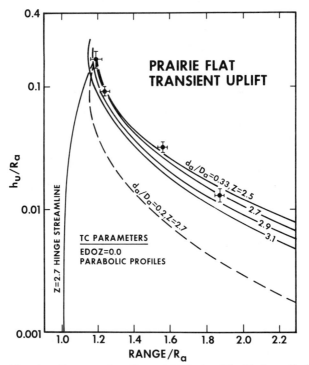

Fig. 10. Semi-log plot of measured maximum transient uplift with theoretical transient rim uplifts (solid curves) derived from the transient crater in Fig. 9 and an uplift model derived for a constant Z, EDOZ cratering flow field. Dashed curve is theoretical rim uplift for transient crater with same diameter as transient crater indicated in Fig. 10, but with a depth/diameter ratio of only 0.2.

uplifted positions of the original ground surface. The four data points are averages of the measured maximum displaced ranges and vertical displacements of velocity gauges initially located at different depths at four different ranges from ground zero (Sauer, 1970; given in Table 3). The observed uplift is reasonably well reproduced by the calculated uplifts. The close spacing of the calculated uplifts for different Z implies that the results are not very sensitive to Z. Departures of the transient crater profile from parabolic while maintaining constant V_{tc} changes the shapes of the calculated curves somewhat, and may be the explanation for the "high" observed uplift near Range$/R_a \simeq 1.6$ relative to the calculated curves. However, changes in the calculated rim uplift curves for reasonable variations in the transient crater profiles are of the same order of magnitude as the variations due to changing Z. The dashed curve shown is the uplift calculated for a parabolic transient crater with $D_a = 61$ m, $d_e = 12$ m (0.2 D_a), and $V_{tc} \simeq 18,000$ m^3. The large change in the calculated uplifts for a 30% decrease in V_{tc} implies that the most important parameter for rim uplift calculations is the volume of the transient crater; i.e., the transient crater volume of Prairie Flat crater could not have been much different from 27,000 m^3 and still match the rim uplift data. This conclusion correlates with the finding of Cooper and Sauer (1977) that transient rim uplifts are proportional to $V_a^{1/3}$ for explosion craters over a wide range of energies and target materials. Consequently the flow-field model provides a quantitative physical explanation for Cooper and Sauer's (1977) observations.

DISCUSSION

The observations cited in the preceding section appear to corroborate several important predictions of a constant Z, EDOZ flow field for both impact and explosion craters. Observed values of depths and volumes of excavation, and hinge radii were plotted on theoretical curves in Fig. 6 to estimate the approximate values of Z characterizing flows that could account for each feature. The consistency of the estimated value of Z predicted for each crater's flow field by each of these features implies that flow fields with values of Z and EDOZ unique to each crater can successfully predict values of hinge radius, depth and volume of excavation that are also unique to each crater. Further, in the individual cases where data are available, flow fields of unique Z and EDOZ also successfully predict excavation cavity shape (Snowball) and transient rim uplift (Prairie Flat). Qualitative evidence for a down-driven cone of shallow target material predicted by the constant Z, EDOZ flow field is found for Snowball, DLG 420, Brent and Meteor Crater. Likewise, deformations in the floor and walls of DLG 420 (Fig. 4) are consistent with motions predicted by the Z, EDOZ flow field. These corroborations imply that a constant Z, EDOZ flow field is a reasonable first-order quantitative approximation for several important crater structures, and does indeed embody the important qualitative features of impact and explosion cratering flows. This result has several important implications for planetary cratering mechanics, of which two will be discussed in detail here: 1) the nature of the excavation cavity, and 2) the depths of excavation of lunar samples.

The excavation cavity

In addition to the transient and final apparent craters, a distinct excavation cavity may be defined for any crater. The excavation cavity differs from the transient and final apparent craters in shape, depth/diameter ratios, and field measurements relevant to its structure.

Field measurements relevant to the final crater are the morphology and structure of the crater as it appears in the field after all cratering and readjustment processes have occurred. The transient crater is characterized directly by transient rim uplift, and indirectly by geological features discussed by Dence *et al.* (1977). The excavation cavity is characterized by volumes and depths of origin of ejecta and fallout breccias. The origin of the hinge zone at the surface radius of division between ejected and displaced material (see Fig. 2) implies that the diameters of the transient crater and excavation cavity coincide at the original ground surface. Because of its temporary nature, direct estimates of the transient crater are most difficult to obtain. In particular, the geologic indicators of transient cavity depth are in reality lower bounds because of the (unknown) extent of rebound or re-adjustment of displaced floor materials.

The peculiar shape of the excavation cavity predicted by the constant Z, EDOZ flow field model has been pointed out in Figs. 1, 2, 3, and 7. Figure 11 compares the relation between the excavation cavity and transient crater defined in this paper (Fig. 11d) with those suggested by Dence *et al.* (1977). The excavation cavities of Dence *et al.* were reconstructed assuming:

1) the highly shocked layers currently at the base of the breccia lens represents the downward-displaced floor of the glass-lined excavation cavity;

2) the pre-displacement position of the center of the excavation cavity floor could be estimated using the observed spread of shock pressures in the rock and the shock wave attenuation rates shown by each figure; and (implicitly)

3) the pre-displacement position of the floor center was the *bottom*, or lowest point, of an excavation cavity roughly parabolic in cross-section. Excavation cavity A in Fig. 11 uses a shock attenuation rate similar to experimentally determined rates (Croft, in preparation), but implies that the ratio of ejected volume to uplifted volume (apparently assumed by Dence *et al.* to be equal to the displaced volume) is $V_e/V_u \sim 0.3$, much smaller than the observed V_e/V_u ratios for Meteor Crater (Table 1) and DLG 420 (Table 2). Excavation cavity C in Fig. 11 predicts $V_e/V_u \sim 3$, in accord with observed values, but requires shock attenuation rates that are probably too great. The center point of the excavation cavity predicted by the flow field (Fig. 11d), however, is not the lowest point of the profile, in contrast to Dence *et al.*'s third assumption. Consequently, the floor center can be placed high in the down-driven cone, accommodating both shock attenuation rate of $P \propto R^{-2}$, and a large value for V_e/V_u. It must be noted that the values of V_e/V_u observed for the final apparent crater are dependent on the assumed mode of crater modification (Croft, in preparation), but that the total rim volume of both the transient and final apparent craters is equal to the *sum* of the ejecta and uplift volumes. Thus the use of Pike's (1967) relation for crater

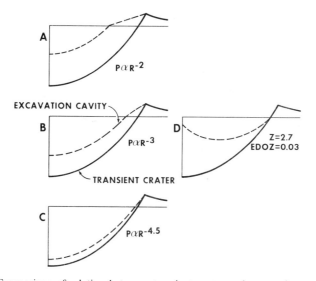

Fig. 11. Comparison of relation between transient crater and excavation cavities predicted by Dence *et al.* (1977) for shock wave attenuation rates shown (Figs. 12A, B, and
C), and excavation cavity predicted by constant Z, EDOZ flow field (Fig. 12d). The
primary difference between the models is the prediction of a central cone of shallow
material in the crater center that is driven downward and retained in the crater during
formation of the transient cavity, so that the deepest point of the excavation cavity is
off-center. Dence *et al.* (1977) assume the deepest point in the excavation cavity to be
in the center, implicitly requiring material to be "scooped" out of the excavation cavity,
rather than to flow away from the point of impact.

rim volumes to estimate the volume of displaced rock in the transient cavity, as
done by Dence *et al.* (1977), is inappropriate.

The differences between the respective apparent depth/apparent diameter ratios of the excavation cavity, the transient crater, and the final apparent crater
further illustrates the differences in their shapes. For simple craters, the apparent
diameters of the three cavities are approximately equal, but the apparent depths
are different. From the limited data in Table 4, the excavation depth appears to
be ~ 0.1 D_a while the depth of the final apparent crater is ~ 0.2 D_a (Pike, 1977).
Transient crater depths inferred from the highly shocked layers at the bases of
the breccia lenses of craters like Brent, Meteor Crater, and others, are near ~ 0.3
D_a (Dence, 1973; Dence *et al.*, 1977). Thus it appears that excavation depths
are significantly less than transient crater depths (at least for kilometer-sized
impact craters in rock), and, in the case of simple craters, even shallower than
the final apparent crater. For complex craters, the diameter of the final apparent
crater is usually considered larger than the transient crater diameter because of
the slumped nature of the crater walls. The apparent diameter of the transient
crater and excavation cavity are assumed to be the same based on the flow field
geometry illustrated in Fig. 2. The normalized final apparent depths of complex
craters are progressively smaller at increasing crater diameters (Pike, 1977).

However, field geological investigations by Dence (1973) and Dence et al. (1977) indicate that transient crater depths of complex craters are between 0.2 D_a and 0.3 D_a, not unlike the transient crater depths inferred for kilometer-sized simple craters. These geological investigations model reconstructions of transient craters by accounting for uplift and inward motions indicated by circumferential shortening of stratigraphic marker beds, and by re-orientation of beds containing shatter cones. The model transient crater depth of Prairie Flat near 0.3 D_a is consistent with these results. Again, from the limited data in Table 4, excavation depths of complex impact and explosion craters appear to be near the simple crater value of ~0.1 D_a.

Excavation depths predicted by constant Z, EDOZ flow fields for values of Z between 2.5 and 3, which apparently characterize real cratering flows, are also near 0.1 D_a (see Fig. 6b). Thus, based on the limited field data and the theoretical results of the Z-flow field, it appears that a ratio of the depth of excavation (d_e) to the apparent diameter of the transient crater of:

$$d_e/D_a \simeq 0.1$$

is valid for both simple and complex craters. In subsequent discussion, Eq. 1 is assumed to be valid.

It is noted that these results infer that the shallow excavation depth of DLG 420 is *not* anomalous, nor due solely to the compactibility of the target sand as has been suggested (Settle and Head, 1979), but a property common to most craters, and is a direct result of the dividing of the cratering flow field into ejected

Table 4. Observed depths of excavation.

Crater	D_a	d_a	d (layer)	d_e/D_a
		Terrestrial Craters		
DLG 420[a]	27.5 cm	5.5 cm	—	0.091
Prairie Flat[b]	~61 m	~4.4 m	—	~0.102
Snowball[c]	~83 m	~5.8 m	—	~0.094
Meteor Crater[d]	~990 m	170 m	—	~0.122
		Lunar Craters[e]		
Swift	8.35 km	1.59 km	1.4 km	<0.168
Cleomedes F	9.7 km	1.8 km	0.95 km	>0.098
Greaves	11.1 km	1.80 km	0.95 km	>0.086
Peirce[f]	15.5 km	1.37 km	1.4 km	>0.090
Picard	19.4 km	1.58 km	1.4 km	>0.072

[a] Stöffler *et al.* (1975)

[b] Roddy (1977a)

[c] Roddy (1977b)

[d] Croft (in preparation)

[e] Data from Croft (1979b). The first three craters are simple craters. Peirce and Picard are complex.

[f] Diameter adjusted for collapse using model of Croft (1979b).

and down-driven portions. The compressibility, even of sand, is not thought to be important during the *cratering flow* stage (i.e., in the material flow *behind* the primary shock; Orphal, 1977) because pressures are so low. Similarly, the depth of excavation at Snowball and Prairie Flat are suggested to not be solely related to the depth of the water table, but simply the consequence of a typical flow field. Material properties probably do affect the Z of the flow field—the apparent contrast of Z ~ 2.4 for DLG 420 and Z ~ 2.8 for Meteor Crater may be direct evidence of that—but d_e changes only slightly in the range $2.5 \leq Z \leq 3.0$.

Depth of origin of lunar samples

One of the most important, and disagreed on problems in lunar studies is the depth of origin of rocks on the lunar surface excavated during cratering events. Crater excavation depths are used in several ways, including thickness estimates of surface layers, such as the maria basalts, that have been penetrated by craters excavating subsurface layers distinct in chemistry and albedo, and depth estimates of origins of lunar samples used to interpret lunar crustal structure and history. In such studies it is usually assumed that excavation was by scooping out all material above a central nadir in a manner similar to Dence *et al.*'s (1977) model shown in Fig. 11c, such that the maximum excavation depth was virtually equivalent to the transient cavity depth, which in turn was equivalent to the final crater depth in simple craters (e.g., see Settle and Head, 1979; Whitford-Stark, 1980). Consequently, previous discussions of excavation depths have centered on an important problem in cratering mechanics: changes of shape (if any), and hence depth, of the transient crater with increasing diameter of the final apparent crater.

Two types of transient crater behavior at large diameters have been proposed: *proportional growth,* i.e., transient craters at all scales maintain a uniform 1:3 d_a/D_a ratio, and *non-proportional growth,* i.e., transient craters become progressively shallower with increasing crater diameter such that the transient craters of basins as large as Imbrium were only a few tens of kilometers deep. Evidence in favor of proportional growth includes geological evidence of terrestrial complex craters mentioned above, theoretical calculations that show penetrations of solid single projectiles to at least several projectile radii, nearly independent of scale (O'Keefe and Ahrens, 1978b; Roddy *et al.*, 1980), and, as discussed in this paper, transient rim uplift data. Evidence in favor of non-proportional growth includes complex crater restoration models (Settle and Head, 1979) and the general paucity of samples of the lunar mantle on the lunar surface in spite of the large lunar basins. Evidence cited in favor of either mode may be viewed as counter evidence against the other. For example, extrapolation of Settle and Head's (1979) depth/diameter relation to a transient crater dimension of ~970 km for the Imbrium Basin, as suggested by Head *et al.* (1975), yields a depth of ~50 km, or about one projectile diameter deep if the projectile were iron (O'Keefe and Ahrens, 1975), significantly less than a projectile diameter if the projectile were stony.

Certainly, the actual transient cavity diameter of multi-ringed basins is subject for dispute (see Croft, 1979b), and the composition, size, and impact velocity of the Imbrium projectile are not well known, but the physical reasons why transient craters should become so relatively shallow at basin dimensions are obscure. Conversely, if the transient craters are deep, then even for the relatively small transient crater diameter of ~570 km for Imbrium suggested by Dence (1976), there ought to be literally tens of millions of cubic kilometers of lunar mantle strewn around the lunar surface. Further, because the mantle material is stratigraphically deep in the original surface, it should be on top of everything else in the ejecta blanket, dominating our sample collections and orbital spectral maps. Thus the general lack of obvious mantle materials is difficult to explain in terms of sampling problems, or anything else, except the cratering model.

The nature of the excavation cavity proposed in this paper on the basis of analysis of the flow field has direct implications for both estimates of deposit thicknesses and maximum depths of origin of lunar samples. First, if one assumes excavation depths as derived from the constant Z, EDOZ model, then the excavation depth/apparent diameter ratio of impact craters is ~0.1. This value of d_e/D_a is ⅓ to ½ the depths of excavation usually assumed (e.g., Whitford-Stark, 1980). If this is true, a direct consequence is that the thickness estimates of lunar maria basalt deposits, for example, must be reduced by factors of ⅓ to ½.

Second, the distinction between the excavation cavity and the transient crater, and association of depths of origin of ejected material with the former, transfers the proportional vs. non-proportional growth controversy to evaluation of the excavation cavity. The apparent validity of Eq. 1 for craters like the lunar craters Peirce and Picard with diameters up to 20 km, well into the size range dominated by gravity scaling (Gault et al., 1975), implies proportional growth is valid for craters to at least 20 km in diameter. In the constant Z, EDOZ description of the flow field: 1) streamline shapes in the flow field are derived from the assumption of incompressibility, and thus are independent of gravity; 2) all streamlines have the same relative shape in a constant Z, EDOZ flow field except very near the EDOZ; and 3) the primary role played by gravity in the formation of craters controlled by gravity scaling is in determining which streamline becomes the hinge streamline. Therefore, as the absolute dimensions of the flow field increases with increasingly larger impacts, and gravity retains material in the crater along streamlines relatively nearer to EDOZ, the shape of the hinge streamline, and thus the shape of the excavation cavity, remains proportionally the same. These considerations provide theoretical support for the extrapolation of proportional growth of excavation cavities to impact basin sizes. If this line of reasoning is correct, then the depths of excavation of large basins are ~0.1 D_{tc}, permitting basins to possess proportional transient craters with diameters up to ~10X the local crust thickness before ejecting any mantle material, and somewhat larger before ejecting significant amounts. Therefore Imbrium may have possessed a proportional transient crater ~600 to 700 km in diameter without ejecting mantle material. The innermost rings of the Imbrium Basin lie within this diameter range.

Consequently, for basin formation models postulating transient crater rims near the innermost rings (e.g., Dence, 1976; Croft, 1979b), the general lack of lunar mantle samples on the lunar surface is no longer inconsistent with proportional growth of excavation cavities. It has also been pointed out in a previous study (Croft, 1979a) that a restoration model taking bulking of ejected material into account allows reconstruction of proportional transient craters for lunar craters up to basin dimensions. Thus a model of crater formation based on proportional growth of features associated with the flow field can be constructed that is consistent with all relevant observations.

Acknowledgments—I gratefully acknowledge the constructive reviews by Jon Bryan, Dave Roddy, and Jeff Thomsen, and extensive discussions with Peter Schultz. The encouragement offered by Fred Hörz is greatly appreciated, as are the near-Herculean efforts of Dennis Orphal in helping hammer the paper into publishable form. Efforts by Lila Mager and Gwen Stokes in manuscript preparation, and Terry Jackson in drawing the figures are also gratefully acknowledged.

The Lunar and Planetary Institute is operated by the Universities Space Research Association under Contract No. NSR 09-051-001 with the National Aeronautics and Space Administration. This paper constitutes the Lunar and Planetary Institute Contribution No. 421.

REFERENCES

Andre C. G., Wolfe R. W., and Adler I. (1978) Evidence for a high-magnesium subsurface basalt in Mare Crisium from orbital X-ray fluorescence data. In *Mare Crisium: The View from Luna 24* (R. B. Merrill and J. J. Papike, eds.), p. 1–12. Pergamon, N.Y.

Austin M. G., Thomsen J. M., Ruhl S. F., Orphal D. L., and Schultz P. H. (1980) Calculational investigation of impact cratering dynamics: Material motions during the crater growth period. *Proc. Lunar Planet. Sci. Conf. 11th.* This volume.

Bjork R. L., Kreyenhagen K. N., and Wagner N. H. (1967) *Analytical Study of Impact Effects Applied to the Meteoroid Hazard.* NASA CR-757. 186 pp.

Brett R. (1968) Opaque minerals in drill cuttings from Meteor Crater, Arizona. U.S. Geol. Survey Prof. Paper 600-D, p. 179–180.

Bryan J. B., Burton D. E., Cunningham M. E., and Lettis L. A. Jr. (1978) A two-dimensional computer simulation of hypervelocity impact cratering: Some preliminary results for Meteor Crater, Arizona. *Proc. Lunar Planet. Sci. Conf. 9th,* p. 3931–3964.

Cooper H. F. Jr. and Sauer F. M. (1977) Crater-related ground motions and implications for crater scaling. In *Impact and Explosion Cratering* (D. J. Roddy, R. O. Pepin and R. B. Merrill, eds.), p. 1133–1163. Pergamon, N.Y.

Croft S. K. (1978) Lunar crater volumes: Interpretation by models of impact cratering and upper crustal structure. *Proc. Lunar Planet. Sci. Conf. 9th,* p. 3711–3733.

Croft S. K. (1979a) Proportional vs. non-proportional growth of basin-sized excavation cavities: A reconciliation (abstract). In *Lunar and Planetary Science X,* p. 248–250. Lunar and Planetary Institute, Houston.

Croft S. K. (1979b) Impact Craters from Centimeters to Megameters. Ph.D. dissertation. University of California at Los Angeles. 264 pp.

Dence M. R. (1973) Dimensional analysis of impact structures (abstract). *Meteoritics* **8,** 343–344.

Dence M. R. (1976) Notes toward an impact model for the Imbrium Basin. In *Interdisciplinary Studies by the Imbrium Consortium,* Vol. 1 (J. A. Wood, ed.). LSI Contr. No. 267D, p. 147–155. Center for Astrophysics, Cambridge.

Dence M. R., Grieve R. A. F., and Robertson P. B. (1977) Terrestrial impact structures: Principle characteristics and energy considerations. In *Impact and Explosion Cratering* (D. J. Roddy, R. O. Pepin, and R. B. Merrill, eds.), p. 247–275. Pergamon, N.Y.

Gault D. E., Guest J. E., Murray J. B., Dzurisin D., and Malin M. C. (1975) Some comparisons of impact craters on Mercury and the Moon. *J. Geophys. Res.* **80**, 2444–2460.

Gault D. E., Quaide W. L., and Oberbeck V. R. (1968) Impact cratering mechanics and structures. In *Shock Metamorphism of Natural Materials* (B. M. French and N. M. Short, eds.), p. 87–99. Mono, Baltimore.

Grieve R. A. F. (1979) Cratering Record: Processes and Effects (abstract). In *Papers Presented to the Conference on the Lunar Highlands Crust*, p. 27–29. Lunar and Planetary Institute, Houston.

Grieve R. A. F., Dence M. R., and Robertson P. B. (1977) Cratering processes: As interpreted from the occurrence of impact melts. In *Impact and Explosion Cratering* (D. J. Roddy, R. O. Pepin, and R. B. Merrill, eds.), p. 791–814. Pergamon, N.Y.

Hager D. (1953) Crater mound (Meteor Crater), Arizona, a geologic feature. *Bull. Amer. Assoc. Petroleum Geologists* **37**, 821–857.

Head J. W., Adams J. B., McCord T. B., Pieters C., and Zisk S. (1978) Regional stratigraphy and geologic history of Mare Crisium. In *Mare Crisium: The View from Luna 24* (R. B. Merrill and J. J. Papike, eds.), p. 43–74. Pergamon, N.Y.

Head J. W., Settle M., and Stein R. S. (1975) Volume of material ejected from major lunar basins and implications for the depth of excavation of lunar samples. *Proc. Lunar Sci. Conf. 6th*, p. 2805–2829.

Holsapple K. A. (1980) The equivalent depth of burial for impact cratering. *Proc. Lunar Planet. Sci. Conf. 11th*. This volume.

Ivanov B. A. (1976) The effect of gravity on crater formation: Thickness of ejecta and concentric basins. *Proc. Lunar Sci. Conf. 7th*, p. 2947–2965.

Jones G. H. S. (1970) *Prairie Flat Crater and Ejecta Study*. DASA Report POR 2115 (WT 2115). Defense Atomic Support Agency, Washington, DC. 276 pp.

Jones G. H. S. (1976) *The Morphology of Central Uplift Craters*. Suffield Report 281, Defense Research Establishment Suffield, Alberta, Canada. 207 pp.

Kieffer S. W. (1971) Shock metamorphism of the Coconino sandstone at Meteor Crater, Arizona. *J. Geophys. Res.* **76**, 5449–5473.

Killian B. G. and Germain L. S. (1977) Scaling of cratering experiments—an analytical and hueristic approach to the phenomenology. In *Impact and Explosion Cratering* (D. J. Roddy, R. O. Pepin, and R. B. Merrill, eds.), p. 1165–1190. Pergamon, N.Y.

Kreyenhagen K. N. and Schuster S. H. (1977) Review and comparison of hypervelocity impact and explosion cratering calculations. In *Impact and Explosion Cratering* (D. J. Roddy, R. O. Pepin, and R. B. Merrill, eds.), p. 983–1002. Pergamon, N.Y.

Maxwell D. E. (1973) *Cratering Flow and Crater Prediction Methods*. Tech. Memo TCAM 73-17, Physics International, Calif. 50 pp.

Maxwell D. E. (1977) Simple Z model of cratering, ejection, and the overturned flap. In *Impact and Explosion Cratering* (D. J. Roddy, R. O. Pepin and R. B. Merrill, eds.), p. 1003–1008. Pergamon, N.Y.

Maxwell D. and Seifert K. (1974) *Modeling of Cratering, Close-in Displacements, and Ejecta*. Report DNA 3628F. Defense Nuclear Agency, Washington, DC. 110 pp.

May T. W., Peeples W. J., Maxwell T., Sill W. R., Ward S. H., Phillips R. J., Jordan R. L., and Abbott E. A. (1976) Subsurface layering in Maria Serenitatis and Crisium: Apollo Lunar Sounder Results (abstract). In *Lunar Science VII*, p. 540–542. Lunar Science Institute, Houston.

Oberbeck V. R. (1971) Laboratory simulation of impact cratering with high explosives. *J. Geophys. Res.* **76**, 5732–5749.

O'Keefe J. D. and Ahrens T. J. (1975) Shock effects from a large impact on the moon. *Proc. Lunar Sci. Conf. 6th*, p. 2831–2844.

O'Keefe J. D. and Ahrens T. J. (1978a) Impact flows and crater scaling on the moon. *Phys. Earth Planet. Inter.* **16**, 341–351.

O'Keefe J. D. and Ahrens T. J. (1978b) Late stage crater flows and the effect of strength on transient crater depth (abstract). In *Lunar and Planetary Science IX,* p. 823–825. Lunar and Planetary Institute, Houston.

Orphal D. L. (1977) Calculations of explosion cratering—II Cratering mechanics and phenomenology. In *Impact and Explosion Cratering* (D. J. Roddy, R. O. Pepin, and R. B. Merrill, eds.), p. 907–917. Pergamon, N.Y.

Orphal D. L., Borden W. F., Larson S. A., and Schultz P. H. (1980) Calculations of impact melt generation and transport (abstract). In *Lunar and Planetary Science XI,* p. 833–835. Lunar and Planetary Institute, Houston.

Peeples W. J., Sill W. R., May T. W., Ward S. H., Phillips R. J., Jordan R. L., Abbott E. A., and Killpack T. J. (1978) Orbital radar evidence for lunar subsurface layering in Maria Serenitatis and Crisium. *J. Geophys. Res.* **83,** 3459–3468.

Pike R. J. (1967) Schroeter's Rule and the modification of lunar crater impact morphology. *J. Geophys. Res.* **72,** 2099–2106.

Pike R. J. (1977) Apparent depth/apparent diameter relation for lunar craters. *Proc. Lunar Sci. Conf. 8th,* p. 3427–3436.

Robertson P. B. and Grieve R. A. F. (1977) Shock attenuation at terrestrial impact structures. In *Impact and Explosion Cratering* (D. J. Roddy, R. O. Pepin, and R. B. Merrill, eds.), p. 678–702. Pergamon, N.Y.

Roddy D. J. (1976) High-explosive cratering analogs for bowl-shaped, central uplift, and multi-ring impact craters. *Proc. Lunar Sci. Conf. 7th,* p. 3027–3056.

Roddy D. J. (1977a) Large-scale impact and explosion craters: Comparisons of morphological and structural analogs. In *Impact and Explosion Cratering* (D. J. Roddy, R. O. Pepin, and R. B. Merrill, eds.), p. 185–246. Pergamon, N.Y.

Roddy D. J. (1977b) Tabular comparisons of the Flynn Creek impact crater, United States, Steinheim impact crater, Germany and Snowball explosion crater, Canada. In *Impact and Explosion Cratering* (D. J. Roddy, R. O. Pepin, and R. B. Merrill, eds.), p. 125–162. Pergamon, N.Y.

Roddy D. J. (1978) Pre-impact geologic conditions, physical properties, energy calculations, meteorite and initial crater dimensions and orientations of joints, faults and walls at Meteor Crater, Arizona. *Proc. Lunar Planet. Sci. Conf. 9th,* p. 3891–3930.

Roddy D. J., Boyce J. M., Colton G. W., and Dial A. L. Jr. (1975) Meteor Crater, Arizona, rim-drilling with thickness, structural uplift, diameter, depth, volume, and mass-balance calculations. *Proc. Lunar Sci. Conf. 6th,* p. 2621–2644.

Roddy D. J., Schuster S., Kreyenhagen K., and Orphal D. (1980) Calculations of impact cratering mechanics at Meteor Crater, Arizona (abstract). In *Lunar and Planetary Science XI,* p. 946–948. Lunar and Planetary Institute, Houston.

Roddy D. J., Ullrich G. W., Sauer F. M., and Jones G. H. S. (1977) Cratering motions and structural deformation in the rim of the Prairie Flat multiring explosion crater. *Proc. Lunar Sci. Conf. 8th,* p. 3389–3407.

Rooke A. D., Meyer J. W., and Conway J. A. (1972) *Dial Pack: Crater and ejecta measurements from a surface-tangent detonation on a layered medium.* U.S. Army Engineer Waterways Experiment Station, Vicksburg, Mississippi. Misc. Paper N-72-9. 58 pp.

Sauer F. M. (1970) Summary of cratering and ground motion experiments. In *Operation Prairie Flat Symposium Report* (M. J. Dudash, ed.), p. 284–300. DASA 2377-1 **1.** DASA Information and Analysis Center, Santa Barbara, Calif.

Schultz P. H. and Gault D. E. (1979) Atmospheric effects on martian ejecta emplacement. *J. Geophys. Res.* **84,** 7669–7687.

Settle M. (1980) The roll of fallback ejecta in the modification of impact craters. *Icarus* **42,** 1–19.

Settle M. and Head J. W. (1979) The role of rim slumping in the modification of lunar impact craters. *J. Geophys. Res.* **84,** 3081–3096.

Shoemaker E. M. (1960) Penetration of high velocity meteorites, illustrated by Meteor Crater, Arizona. *International Geological Congress, XXI Session,* Part 18, p. 418–434.

Shoemaker E. M. and Kieffer (1974) *Guidebook to the Geology of Meteor Crater.* 37th Ann. Met. Soc. Mtg., Aug. 1974. 66 pp.

Stöffler D., Gault D. E., Wedekind J., and Polkowski G. (1975) Experimental hypervelocity impact into quartz sand: Distribution and shock metamorphism of ejecta. *J. Geophys. Res.* **80**, 4062–4077.

Thomsen J. M., Austin M. G., Ruhl S. F., Schultz P. H., and Orphal D. L. (1979) Calculational investigation of impact cratering dynamics: Early time material motions. *Proc. Lunar Planet. Sci. Conf. 10th*, p. 2741–2756.

Thomsen J. M., Austin M. G., and Schultz P. H. (1980) The development of the ejecta plume in a laboratory-scale impact cratering event (abstract). In *Lunar and Planetary Science XI*, p. 1146–1148. Lunar and Planetary Institute, Houston.

Whitford-Stark J. (1980) The craters of Mare Imbrium (abstract). In *Lunar and Planetary Science XI*, p. 1242–1244. Lunar and Planetary Institute, Houston.

APPENDIX A: MATHEMATICAL DESCRIPTION OF SELECTED SPATIAL FEATURES OF A CONSTANT Z, EDOZ FLOW FIELD

In this appendix, quantitative formulae are derived in outline for selected spatial features of a constant Z cratering flow field with an effective crater of flow (EDOZ) at constant depth. Such a field is termed a *constant Z, EDOZ* flow field in the main text. The geometry and symbols used are illustrated and defined in Fig. A1. In the derivations, it is assumed that: a) flow streamlines are defined by the expressions derived by Maxwell (1973, 1977) and Maxwell and Seifert (1974) for near-surface explosions (Eqs. 1a, 1b, and 1c below); b) streamlines originate at an EDOZ located at depth d; c) flow continues to follow the Z streamlines until spallation occurs at, or slightly above, the ground level. Assumptions b and c were suggested by Thomsen *et al.* (1979, 1980). The derivations of angles of ejection at the ground level (section 2) and total ejecta volumes (section 3) parallel derivations of the same quantities by Maxwell and Seifert (1974) for surface bursts (d = 0). The derived formulae are

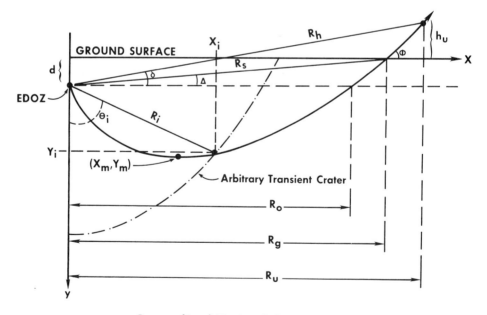

Generalized Z—Model Geometry

Fig. A1. Schematic drawing defining symbols used in derivations in Appendix.

shown in both cases to reduce to the expressions given by Maxwell and Seifert (1974) for d = 0. Two overlapping sets of coordinates are used: spherical polar (R = radial distance from EDOZ, θ = angle from vertical downward axis), and rectangular (X = horizontal distance from vertical axis through EDOZ, Y = distance below original ground plane).

1. Streamline properties:

Streamline equation (polar)	$R = R_0 (1-\cos\theta)^{1/(Z-2)}$	(a)
Streamline equation (rect.)	$X = R_0 \sin\theta$	(b)
	$Y = R_0 \cos\theta(1-\cos\theta)^{1/(Z-2)} + d$	(c)
Maximum streamline depth:	$X_m = R_0 \sin\theta_m(1-\cos\theta_m)^{1/(Z-2)}$	(d)
	$Y_m = R_0 (Z-2)[Z-1]^{(1-Z)/(Z-2)} + d$	(e)

where $\theta_m = \cos^{-1}[(Z-2)/(Z-1)]$.

Maximum depth of excavation, d_e: By definition, $d_e = Y_m$ for the streamline for which $R_g = R_a$, the apparent crater radius. From Fig. A1,

$$\tan\Delta = d/R_g \tag{f}$$

and

$$\cos\Delta = R_g/R_s. \tag{g}$$

From Eq. 1a: $R_s = R_0 (1 + \sin\Delta)^{1/(Z-2)}$.
Eliminating R_s, $R_g(\equiv R_a)$ and solving for R_0 yields:

$$R_0 = R_a/[\cos\Delta (1 + \sin\Delta)^{1/(Z-2)}]. \tag{h}$$

Substituting for R_0 and Y_m in Eq. 1e yields:

$$d_e = R_a (Z-2)[Z-1]^{(1-Z)/(Z-2)}/[\cos\Delta (1 + \sin\Delta)^{1/(Z-2)}] + d. \tag{i}$$

2. Angle of ejection at ground level:

The polar components of velocity along a streamline are (Orphal, 1977):

$$U_r = \alpha R^{-Z} \tag{a}$$
$$U_\theta = \alpha R^{-Z} (Z-2)\sin\theta/(1 + \cos\theta) \tag{b}$$

The rectangular components are found by orthogonal transformation to be:

$$U_x = U_r \cos\Delta - U_\theta \sin\Delta \tag{c}$$
$$U_y = U_r \sin\Delta + U_\theta \cos\Delta. \tag{d}$$

The angle ϕ (see Fig. 1) is given by:

$$\phi = \tan^{-1} (U_y/U_x). \tag{e}$$

Substitution of Eq. 2a, 2b, 2c, and 2d into Eq. 2e, and recognizing that $\theta = 90° + \Delta$ at the ground surface yields (after some algebra):

$$\phi = \tan^{-1}\left[\frac{\dfrac{\tan\Delta}{\cos\Delta} - \tan^2\Delta + (Z-2)}{\dfrac{1}{\cos\Delta} - \tan\Delta (Z-1)}\right]. \tag{f}$$

For d = 0, $\Delta = 0°$, Eq. 2f reduces to

$$\phi = \tan^{-1} (Z-2).$$

This is the result given by Maxwell and Seifert (1974) and Orphal (1977).

3. Ejecta volumes: in spherical coordinates:

$$V = \iiint r^2 \sin\theta \, d\theta \, dr \, d\phi. \tag{a}$$

For axial symmetry and Eq. 1a relating R and θ, the volumes of rotation of the area in Fig. 1A bounded by the line R_s and the streamline is given by:

$$V_{SL} = 2\pi \int_0^{Ro} \int_0^{90+\Delta} r^2 \sin\theta \, dr \, d\theta = \frac{2\pi R_o^3}{3} \int_0^{90+\Delta} (1-\cos\theta)^{\frac{3}{Z-2}} d\theta \qquad (b)$$

$$= \frac{2\pi}{3} R_o^3 \left[\frac{(Z-2)}{(Z+1)}\right] [1 + \sin\Delta]^{\frac{Z+1}{Z-2}}. \qquad (c)$$

The excavation cavity volume is seen in Fig. 1A to be the sum of V_{SL} and the volume of rotation of the triangle bounded by line RS, the ground surface, and the line bracketed by d. This latter volume, V_c, is that of a right circular cone of base $2R_a$ and height d. Therefore:

$$V_e = V_{SL} + V_c \qquad (d)$$

$$= \frac{2\pi}{3} R_o^3 \left[\frac{Z-2}{Z+1}\right] [1 + \sin\Delta]^{\frac{Z+1}{Z-2}} + \pi/3 \, d \, R_a^2.$$

Substituting Eq. 1h for R_o and rearranging yields:

$$V_e = (\pi/3 \frac{d}{R_a} + \frac{2\pi(Z-2)}{3(Z+1)} \frac{1 + \sin\Delta}{\cos^3\Delta}) R_a^3. \qquad (e)$$

For $d = 0$, $\Delta = 0$, Eq. 3e reduces to

$$V_e = \frac{2\pi}{3} R_a^3 \frac{(Z-2)}{(Z+1)} = \frac{2\pi}{3} R_a^3 \left(1 - \frac{3}{Z+1}\right). \qquad (f)$$

This is the result found by Maxwell and Seifert (1974).

4. *Structural rim uplift:* To obtain the structural rim uplift, it is assumed that material along a given streamline between EDOZ and the intersection point (X_i, Y_i) of the streamline with the transient cavity wall is pushed into the wall, displacing an equal volume (assuming no bulking or compression) above the original ground level at the range, R_g, where the streamline emerges. It is further assumed that the material continues along the streamline during uplift as indicated in Fig. 1A. The shape of the transient cavity is arbitrary. Given the intersection (X_i, Y_i); we have:

$$\theta_i = \tan^{-1} [X_i/(Y_i-d)] \qquad (a)$$
$$R_i = X_i/\sin\theta_i. \qquad (b)$$

Substituting into Eq. 1a and solving for R_o yields:

$$R_o = X_i/[\sin\theta_i (1-\cos\theta_i)^{1/(Z-2)}]. \qquad (c)$$

From Eqs. 1a, 1f, and 1g we find:

$$R_g = \cos\Delta \, R_o(1 + \sin\Delta)^{1/(Z-2)} \qquad (d)$$

which in practice must be interated to find Δ and R_g. The volumes along the streamline between $\theta = 0$ and θ_i, and $\theta = 90 + \Delta$ and $90 + \delta$ are equal by assumption. Solving Eq. 3b for each range of θ and equating yields (after rearranging):

$$\sin\delta = [(1-\cos\theta_i)^{\frac{Z+1}{Z-2}} + (1 + \sin\Delta)^{\frac{Z+1}{Z-2}}]^{\frac{Z-2}{Z+1}} - 1. \qquad (e)$$

Now from Eq. 1a

$$R_h = R_o (1 + \sin\delta)^{1/(Z-2)} \qquad (f)$$

where R_o is found from Eq. 4c and δ from Eq. 4e. Finally, the range, R_u, and height of uplift, h_u, are given by:

$$h_u = R_h \sin\delta - d \qquad (g)$$
$$R_u = R_h \cos\delta. \qquad (h)$$

Proc. Lunar Planet. Sci. Conf. 11th (1980), p. 2379–2401.
Printed in the United States of America

The equivalent depth of burst for impact cratering

K. A. Holsapple

Department of Aeronautics and Astronautics, University of Washington, Seattle,
Washington 98195, Consultant, Boeing Aerospace Company

Abstract—The concept of modeling an impact cratering event with an explosive event with the explosive buried at some equivalent depth of burst (d.o.b.) is discussed. Various and different ways to define this equivalent d.o.b. are identified. Recent experimental results for a dense quartz sand are used to determine the equivalent d.o.b. for various conditions of charge type, event size, and impact conditions.

The results show a decrease in equivalent d.o.b. with increasing energy for fixed impact velocity and a decrease in equivalent d.o.b. with increasing velocity for fixed energy. The values for an iron projectile are on the order of 2–3 projectile radii for energy equal to one ton of TNT, decreasing to about 1.5 radii at a megaton of TNT. The dependence on projectile and target mass density matches that included in common jet-penetration formulas for projectile densities greater than target densities and for the higher energies.

The values obtained in this paper are compared to other existing estimates.

1. INTRODUCTION

Large-scale impact cratering phenomenon studies are hindered by the fact that there are no cases where the impacting conditions are known. This is in contrast to the analogous case of explosively-formed craters where there do exist craters for a range of diameters into the hundreds of meters with known initial conditions. It is generally accepted that, while not identical, the processes in impact versus explosive cratering are similar in terms of crater morphologies and structural deformations (Baldwin, 1963; Oberbeck, 1971; Roddy, 1977; Piekutowski, 1980). Consequently, it is common to rely on comparisons with explosive craters in studies of impact cratering. A listing of the independent variables that determine the final crater in each of these types of cratering events reveals one fundamental parameter that is an independent variable in the case of explosive craters, but is not present in the case of impact craters. This parameter is the so-called depth of burst (d.o.b.), the distance from the center of energy release to the initial ground surface. If an impact crater is to be modeled by an explosive event, the "equivalent" d.o.b. for the impact event must be determined by the conditions of each of (1) impact event to be modeled and (2) the explosive event to do the modeling. This model generates a simplistic picture where the impact process is

2379

thought of in two parts; a penetration to some equivalent d.o.b., followed by an explosive energy release. Of course, the actual impact event is not this simple, but such a model is implicit in the concept of an equivalent d.o.b.

Quantitative or qualitative measures of the equivalent d.o.b. can be determined by either direct observation, calculation or by experimentation. Experiments designed to explore the equivalence must by necessity be on a small scale. There seems to exist only one such experimental study reported in the literature (Oberbeck, 1971) that is directly designed to determine this equivalence. In that study, an impact of 2 Km/sec of a 0.435-gm cylindrical projectile into a quartz-sand target material was compared to explosive events with a 0.150-gm low-density PETN cylindrical tamped-charge, buried at different depths of burst. Crater size and shape, ejection, and subsurface deformation were compared. It was concluded that the impact cratering event could be simulated by an explosive event when the explosive was placed 6.3 ± 2 mm below the surface.

Oberbeck cautions against the use of this result to the case of large impact craters. He notes a possible dependence on projectile velocity. In addition, questions might be raised about the effect of many of the other fixed conditions such as target material, charge type, size, and properties, projectile type and shape, etc. In spite of these questions, Oberbeck's results have been applied in a variety of cases simply because there was no other data.

Baldwin (1963) compares the shape of existing large-scale terrestrial and lunar craters to deduce an equivalent d.o.b. He suggests the formula

$$\frac{d}{W^{1/3}} \stackrel{\circ}{=} 0.1,$$

where d is in feet and W the equivalent weight of TNT in pounds. He states that the equivalent d.o.b. probably decreases below this value for larger craters and may even approach zero. He also notes that if the penetration is to remain low for any impact velocity, then the scaled d.o.b. must decrease as the velocity increases. Thus, while he suggests a velocity and energy dependence, his technique does not allow a quantitative determination of those effects.

The present paper uses results of recent experiments by Schmidt and Holsapple (1978a, 1979, 1980a, 1980b), Holsapple (1979), Schmidt (1980), together with experiments by Piekutowski (1974, 1975), Gault and Wedekind (1977), and Oberbeck (1977), of both impact and explosive events in quartz sands to determine equivalent d.o.b. for a variety of conditions. The experiments by Schmidt and Holsapple were based upon the observation that large-scale experiments with length scale L_1 and at gravity G_1 are similar to small-scale experiments with length scale L_2 and at gravity G_2, in the same material, whenever $G_1L_1 = G_2L_2$, as long as the material has no rate or size dependences (Schmidt and Holsapple, 1980a). This result has been exploited by performing small-scale tests on a centrifuge, with G_2 over 500 times earth's gravity, as a means of directly simulating tests 500 times larger in each linear dimension. The resulting equivalent or scaled energy is $(500)^3$: a factor of $1.25 \ 10^8$. Both explosive tests, and more recently light-gas-gun impact tests, have been performed and are reported in the listed references. While various materials have been used, the bulk of the data is for a dense Ottawa sand. The experiments confirm the expected similarity. Two explosive types, various heights and depths of burst, and a range of scaled energy, from grams to kilotons of equivalent TNT, have been tested in the explosive program. The impact program has investigated a range of impact velocities, projectile types, and scaled masses and energies. These data have been found to be extremely consistent with the previous experiments of Piekutowski (1974, 1975), Gault and Wedekind (1977) and Oberbeck (1977). All of the data has been fitted in terms of certain dimensionless parameters, for the entirety of conditions tested. The comparison of these results, between the explosive events and the impact events, allows a determination of an equivalent d.o.b. and its dependence on a variety of conditions. It is this comparison that is presented in this paper.

2. THE BASIS FOR AN EQUIVALENT D.O.B.

Preliminary to a study of the experimental data, it is worthwhile to present some general ideas about the concept of an equivalent d.o.b. The fundamental idea is to obtain, in some sense or other, the "same" resulting craters from an explosive and from an impact event. Various choices of this criteria for "sameness" can be imagined, ranging from the simplest choice: equal values for a single measure of the resulting craters (such as the volume); to very detailed comparisons of the entire final deformation pattern. Clearly the simpler criteria will lead to a larger number of ways to model the impact event, while a complex criteria may totally rule out *any* equivalent explosive event. The ultimate choice of equivalence criteria will depend in part upon how the equivalence is to be used. In this paper the simplest choice will be made: the determination of an equivalent d.o.b. that gives the same volume of resulting crater. The imposition of additional criteria will have as its solution some subset of the combinations for equivalence to be found here.

These general ideas can be presented in a mathematical form. For explosive craters the apparent crater volume V can be represented by

$$V = f[E, Q_e, \delta, P, g, m, d] \qquad (2.1)$$

where the explosive has energy E, specific energy Q_e, mass density δ; the event has atmospheric pressure P and gravity g; the symbol m denotes any set of target material properties; and finally d is the d.o.b. The explosive mass $W = E/Q_e$, and the radius of a spherical charge is given by $a = [3W/4\pi\delta]^{1/3}$.

Similarly an impact crater volume V* is represented by

$$V^* = f^*[E^*, U, \delta^*, P^*, g^*, m^*] \qquad (2.2)$$

where the projectile has kinetic energy E*, impact velocity U, and mass density δ^*; the target properties are represented by the single symbol m^*, and the event has atmospheric pressure P* and gravity g*. The equivalent energy density Q_e^*

of the projectile is given by ½ U^2, its total mass $W^* = 2E^*/U^2$ and its radius a^* (assuming a spherical shape) by $a^* = [3W/4\pi\delta^*]^{1/3}$. Note that Eq. (2.2) is comparable in all respects to Eq. (2.1) except for the one additional independent variable d in Eq. (2.1).

It has already been stated that the criteria for equivalence will be taken to be $V = V^*$. No comparisons of crater shape will be made. In addition, certain further conditions are commonly imposed. The cratered material is the same in both cases. Thus $m = m^*$. It will be assumed that both events have the same pressure, $P = P^*$, and the same gravity $g = g^*$. Finally, it is common (but not necessary) to explore only the case when the energy E of the explosive is equal to the kinetic energy E^* of the projectile. Making these substitutions in Eqs. (2.1) and (2.2), and equating the two, gives

$$f[E, Q_e, \delta, P, g, m, d] = f^*[E, U, \delta^*, P, g, m] \qquad (2.3)$$

which is assumed to be solvable for the depth of burst d as a function of the remaining variables:

$$d = g[E, Q_e, U, \delta, \delta^*, P, g, m] \qquad (2.4)$$

which is the solution for the equivalent d.o.b. It remains a function of the (equal) energies E, pressure P, gravity g and material type; as well as of the independent choices for explosive type characterized by Q_e and δ; and the projectile characterized by U and δ^*. In principle, explicit forms for the Eqs. (2.1) and (2.2) will give an explicit form for Eq. (2.4). In fact, for a given crater volume and other conditions fixed, there may be two burial depths that give that volume: one below and one above that so-called "optimum d.o.b." at which the crater volume is a maximum. The functions used in this paper for Eq. (2.1) are only valid for the smaller depths of burst below the optimum, and it is only that branch of the curve that is of interest here.

The general form of Eq. (2.4) gives the depth d to bury a given explosive in order to match the crater formed by a given projectile. However, the criteria imposed is only that of equal volume. Oberbeck (1971) chose an additional condition that he thought appropriate to obtain a match of the entire flow field. This condition, based upon a criteria of equal pressure transmitted in the initial shock wave, resulted in (almost) equal source energies per unit volume (internal for the explosive and kinetic for the projectile):

$$Q_e\delta = \tfrac{1}{2}U^2\delta^*. \qquad (2.5)$$

He arbitrarily chose an aluminum projectile, with $\delta^* = 2.8$ gm/cm³. In order to obtain the condition of Eq. (2.5) with a given explosive type (PETN) he was constrained to choose a form of PETN with $\delta = 1.0$ gm/cm³. The above discussion should emphasize that, at least insofar as the chosen criteria is equal volume, and for given projectile conditions, other explosive types are also possible and furthermore, these various explosive types may all have a different equivalent d.o.b.

If one imagines the impact process as a penetration to the depth d, followed by an explosion, then the equivalent explosive would have the same total energy and same total mass in the same total volume. This requires the stronger conditions $\delta = \delta^*$ and $Q_e = \frac{1}{2}U^2$, which were not separately matched by Oberbeck.

Various of these possibilities are pursued in numerical results to be given here.

3. EXPLOSIVE CRATER RESULTS

The various explosive experimental results reported elsewhere are now presented in the form of Eq. (2.1). These results are for dry quartz sand only, but since there was some difference in the initial mass density ρ, this density is retained as the one property of the target material included in the set m in Eq. (2.1). The volume V of the crater then depends on the seven independent variables E, Q_e, δ, P, g, ρ and d. These seven variables can be combined into five dimensionless "pi-groups" in a variety of ways. Those that have been found to be the most useful in the present application are defined by

$$\pi_v = \frac{V_\rho Q_e}{E} = \frac{V_\rho}{W}$$

$$\pi_2 = \frac{g}{Q_e^{4/3}}\left(\frac{E}{\delta}\right)^{1/3} = \frac{1.61 \, ga}{Q_e} \tag{3.1}$$

$$\pi_3 = \frac{\rho}{\delta}$$

$$\pi_4 = \frac{d}{a}$$

$$\pi_5 = \frac{P}{\rho Q_e}$$

where again $a = [3E/4\pi\delta Q_e]^{1/3}$ is the charge radius. (See Schmidt and Holsapple 1980a, for other possibilities.)

These pi-groups include those that have been used previously (Schmidt and Holsapple, 1980a), in addition to the last one listed, π_5, which includes the dependence on atmospheric pressure.

The experimental results are presentable in terms of a functional relation between these dimensionless groups, thus

$$\pi_v = F[\pi_2, \pi_3, \pi_4, \pi_5] \tag{3.2a}$$

or, written out:

$$\frac{V_\rho Q_e}{E} = F\left[\frac{g}{Q_e^{4/3}}\left(\frac{E}{\delta}\right)^{1/3}, \frac{\rho}{\delta}, \frac{d}{a}, \frac{P}{\rho Q_e}\right]. \tag{3.2b}$$

A plot of the experimental results, for atmospheric pressure only, is shown in Fig. 1 as a plot of π_v versus the gravity-scaled energy parameter π_2. Three curves are shown. One is for zero d.o.b., $\pi_4 = 0$. In this case, all points fall (to within

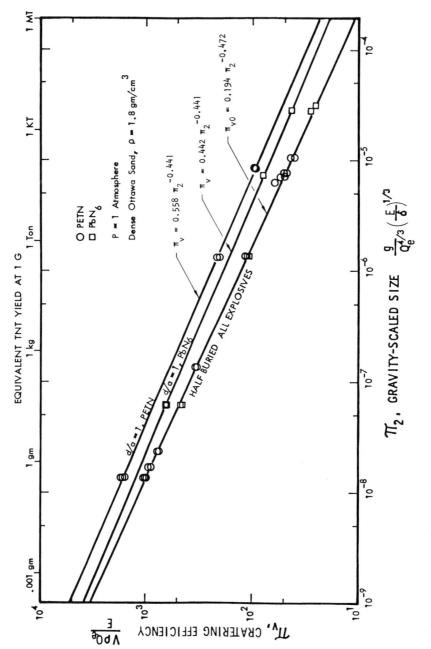

Fig. 1. Experimental results for crater volume in explosive cratering events in a dense Ottawa sand, in terms of the gravity-scaled size parameter π_2. Equivalent yield of TNT at 1-G is shown on the top scale.

about $\pm 5\%$) on the same curve, and are independent of the charge type (which affects the value of π_3 and π_5). Thus, at zero d.o.b. the π_v vs. π_2 relation is independent of both the explosive mass density δ and atmospheric pressure P. The independence from atmospheric pressure at zero d.o.b. was reported previously Herr (1971). It is this independence of the results on π_3 and π_5 that makes the above choice of dimensionless groups more useful than other possibilities.

For "tangent-below" buried shots, the explosive center is buried one charge radius a below the surface. Consequently, $\pi_4 = d/a = 1.0$. Two curves are shown with $d/a = 1$, one for each explosive type tested. The fact that these two curves are definitely distinct indicates a dependence on either π_3 and/or π_5.

In order to sort out the dependence on these two variables, and develop some confidence in the results, independent variations in charge properties δ and Q_e, or variations in pressure P would be needed. Since only two explosive types were tested, there were related changes in both δ and Q_e simultaneously when changing explosive type, so that the separate dependences on these variables could not be determined from the reported tests only. No variable pressure tests were conducted in the present explosive experiments. Herr (1971) has reported tests at variable ambient pressure, although his experiments concentrated on large d.o.b. and are of poor resolution at small d.o.b. However, with some reservations, Herr's data was used to determine the pressure dependence, and then the remaining dependences could be determined from the results shown in Fig. 1 for one-atmosphere events. The application of the results to other than one-atmosphere conditions will have to remain rather tentative at present.

The empirical fits to the explosive data have been reported elsewhere (Schmidt and Holsapple 1978a, 1979, 1980a, Holsapple and Schmidt, 1979). The zero d.o.b. results are expressed as the straight line shown on the log-log plot of Fig. 1, given as:

$$\pi_{vo} = \frac{V_o \rho Q_e}{E} = C_o \pi_2^{-\alpha} = C_o \left[\frac{g}{Q_e^{4/3}} \left(\frac{E}{\delta} \right)^{1/3} \right]^{-\alpha_0} \tag{3.3a}$$

where

$$C_o = 0.194 \pm 0.014 \tag{3.3b}$$

and

$$\alpha_o = 0.472 \pm 0.005, \tag{3.3b}$$

where the uncertainty shown for the constants is the standard deviation. The additional subscript "o" denotes the zero d.o.b. result. An equivalent form is

$$\log \pi_{vo} = \log C_o - \alpha_o \log \pi_2. \tag{3.3c}$$

For buried events, the zero d.o.b. curve (Eq. 3.3c) is modified to include an additional term:

$$\log \pi_v = \log \pi_{vo} + \frac{d}{a}(f_1 + f_2 \log \pi_2) \tag{3.4a}$$

where f_1 and f_2 depend upon explosive type via the parameters π_3 and π_5. As discussed previously, Herr's data is used to determine the pressure dependence (and therefore the π_5-dependence) and the data of Fig. 1 then gives the π_3-dependence. The final form chosen is given by

$$f_1 = C_1\, \pi_3^{\alpha_1}(\pi_3\pi_5)^{\alpha_2} \tag{3.4b}$$

$$f_2 = C_2 \tag{3.4c}$$

where

$$
\begin{aligned}
C_1 &= 0.151 \\
C_2 &= 0.031 \\
\alpha_1 &= 0.376 \\
\alpha_2 &= -0.095.
\end{aligned}
\tag{3.4d}
$$

The form of Eq. (3.4a) was based upon the reduction back to the case of $d/a = 1$, in which case it becomes:

$$
\begin{aligned}
\log \pi_v &= (\log C_o - \alpha_o \log \pi_2) + (1)(f_1 + f_2 \log \pi_2) \\
&= (\log C_o + f_1) - (\alpha_o - f_2) \log \pi_2
\end{aligned}
\tag{3.5}
$$

as a straight-line fit again on a log-log plot. For PETN at one-atmosphere pressure, for Ottawa sand

$$
\begin{aligned}
Q_e &= 5.5\ 10^{10}\ \text{erg/gm} \\
P &= 10^6\ \text{dyne/cm}^2 \\
\delta &= 1.7\ \text{gm/cm}^3 \\
\rho &= 1.8\ \text{gm/cm}^3
\end{aligned}
$$

so that

$$
\begin{aligned}
\pi_3 &= 1.059 \\
\pi_5 &= 1.010\ 10^{-5}
\end{aligned}
$$

which gives

$$
\begin{aligned}
f_1 &= 0.458 \\
f_2 &= 0.031
\end{aligned}
$$

so that

$$\log \pi_v = -0.254 - 0.441 \log \pi_2$$

or

$$\pi_v = 0.558\ \pi_2^{-0.441}$$

which is shown on Fig. 1.

Similarly, for lead-azide at one atmosphere and for Ottawa sand

$$
\begin{aligned}
Q_e &= 1.32\ 10^{10}\ \text{erg/gm} \\
P &= 10^6\ \text{dyne/cm}^2 \\
\delta &= 3.10\ \text{gm/cm}^3 \\
\rho &= 1.8\ \text{gm/cm}^3
\end{aligned}
$$

so that

$$\pi_3 = 0.581$$
$$\pi_5 = 4.20\ 10^{-5}$$

which gives

$$f_1 = 0.338$$
$$f_2 = 0.031$$

to obtain

$$\pi_v = 0.422\ \pi_2^{-0.441}$$

which is also shown in Fig. 1 to compare with the data.

Equation (3.4a) indicates a dependence of the d.o.b. dependence on scaled-size (the π_2 dependence) which has been taken to be the same for the two explosives. In Fig. 1 this is represented by the almost equal slope of the two curves for the two explosives for $d/a = 1$; and the divergence of each of these curves from the zero d.o.b. curve as scaled size increases. This result implies an increasing dependence on d/a for increasing event size, which is attributed to increasing confinement at a given value of d/a, due to increased lithostatic pressure.

The result that the two explosive types have different curves at the same d/a is reflected by the dependence of f_1 on the explosive properties in Eq. (3.4b). This difference is related to the fact that a higher energy-density explosive (the PETN) generates higher shock pressure and converts less of its total energy into the mechanical work of excavation (Burton *et al.*, 1975). However, some of these additional loss mechanisms are reduced by the extra confinement as the explosives are buried. Consequently, the higher energy-density explosive has a greater change in the cratering efficiency π_v as the d.o.b. variable d/a is changed.

The data shown in Fig. 1 is for values of d/a of zero and 1.0 only. At the higher scaled energies (larger π_2) no data is presently available for larger d.o.b. However, at the lower values, Piekutowski (1975) has data for Ottawa sand to large d.o.b. The form of Eq. (3.4) has been chosen to fit that data also, to values of d/a in excess of 4–5. (See Schmidt and Holsapple, 1980b.) It does not however, extend to depths approaching optimum d.o.b. and does not reflect the decreasing effect of d.o.b. for even deeper burial.

4. IMPACT CRATER RESULTS

As stated in the introduction, a series of impact tests was also performed into Ottawa sand. These consisted of spherical projectiles of various materials, fired with powder and light-gas guns, with a range of velocities of 2 to 6 Km/sec. Some of these experiments were done on a centrifuge with gravity in excess of 500 G's. The details of these experiments, as well as the complete data table, is given elsewhere in these proceedings (Schmidt, 1980).

A general dependence of the volume V^* in an impact event has been listed in Eq. (2.2). Compared to the explosive case, the d.o.b. variable d is missing, and the explosive specific energy Q_e is replaced by $\frac{1}{2} U^2$ where U is the projectile velocity. The resulting dimensionless groups are given by

$$\pi_v^* = \frac{V^* \rho U^2}{2E} = \frac{V^* \rho}{W^*}$$

$$\pi_2^* = \frac{g}{\left(\frac{1}{2} U^2\right)^{4/3}} \left(\frac{E}{\delta^*}\right)^{1/3} = \frac{3.22 \, g \, a^*}{U^2} \tag{4.1}$$

$$\pi_3^* = \frac{\rho}{\delta^*}$$

$$\pi_5^* = \frac{2P}{\rho U^2}$$

where the star superscript again distinguishes an impact event from an explosive event. Since all comparisons are at equal E, g, P and ρ, no superscripts are necessary on these parameters.

The functional relationship between these four dimensionless groups has the form

$$\pi_v^* = F^*[\pi_2^*, \pi_3^*, \pi_5^*]. \tag{4.2}$$

For these impact events, a range of each of π_2^*, π_3^* and π_5^* has been experimentally investigated. In particular a dependence on the pressure P, and consequently on π_5^* has been noted and reported (Holsapple, 1979). This was determined by a comparison of events at atmospheric pressure to events at about 1 mm ambient pressure.

The experimental points that have been used to determine the form of the function F* of Eq. (4.2) are shown in Fig. 2, on a log-log plot of π_v versus π_2. While an inspection of the data reveals a consistent difference between the previous experiments (Gault and Wedekind, 1977, Oberbeck, 1977), compared to the present data, this difference has been shown to be due simply to the packing density of the sand (Schmidt and Holsapple, 1978b; Schmidt et al., 1979). Only the data for the dense ($\rho = 1.8$ gm/cm^3) sand was used for the present purpose. The points shown for this case were at two pressures and a variety of velocities, and consequently have a variety of values for the pressure parameter π_5^*.

A two-parameter least-squares bilinear fit of this data, in log-log space, yielded the following result:

$$\pi_v^* = K_1 (\pi_2^*)^{-\beta_0} (\pi_5^*)^{-\beta_1} \tag{4.3}$$

where

$$K_1 = 0.257 \pm 0.023$$
$$\beta_0 = 0.474 \pm 0.006$$
$$\beta_1 = 0.027 \pm 0.008.$$

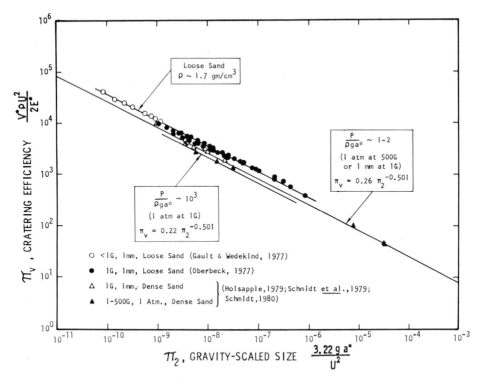

Fig. 2. Experimental results for crater volume in impact cratering events into two sand types: $\rho = 1.8$ gm/cm^3 and $\rho \simeq 1.7$ gm/cm^3, at various ambient pressure and impact conditions, plotted as a function of the gravity-scaled size parameter π_2.

Written out this becomes

$$\frac{V^*\rho U^2}{2E} = 0.257 \left[\frac{3.22\ g\ a^*}{U^2}\right]^{-0.474} \left[\frac{2P}{\rho U^2}\right]^{-0.027}, \qquad (4.4a)$$

which can be rewritten as

$$\frac{V^*\rho U^2}{E} = 0.520 \left[\frac{3.22\ ga^*}{U^2}\right]^{-0.501} \left[\frac{P}{\rho ga^*}\right]^{-0.027} \qquad (4.4b)$$

as previously given (Holsapple, 1980). These curves are valid for 1 mm $< P < 1$ atmosphere, velocity from about 2–7 Km/sec.

The experimental points fall into two categories. For P = 1 mm, 1 G, and projectile radii ranging around ½ cm, the value of the parameter $P/\rho ga^*$ is in the range from 1 to 2. For one-atmosphere tests at 500 G, with the projectile radius of 0.7 cm as used, this parameter has essentially the same value. Consequently, on a π_v versus π_2 plot these two sets of data should fall on the same line which, from Eq. (4.4b) is given by

$$\pi_v = 0.256 \, \pi_2^{-0.501}. \tag{4.5a}$$

The second category of data is obtained by the atmospheric tests at 1 G. In this case the parameter $p/\rho ga^* \sim 10^3$ and the Eq. (4.4b) gives

$$\pi_v = 0.215 \, \pi_2^{-0.501}. \tag{4.5b}$$

These two curves are shown on Fig. 2 to compare to the data. Note that it only applies to the dense ($\rho = 1.8\,gm/cm^3$) sand results.

5. THE EQUIVALENT D.O.B.

The explosive results, compared to the impact results, form the basis for the determination of an equivalent d.o.b. As previously discussed, there is a variety of ways to make this comparison. The comparison here is based upon equality of volume, $V = V^*$ between an explosive and an impact event, at equal energy $E = E^*$, and in the same soil, but considers a variety of ways to achieve that result. The equivalent d.o.b. so determined will then depend upon the choice of the other variables.

It is useful to derive the general result, based upon Eqs. (3.4) and (4.3) with arbitrary constants, in order that the dependence of the final result on these constants is apparent. Uncertainties in these constants will then translate into uncertainties in the value of the equivalent d.o.b.

The general result can be written in the form

$$\frac{d}{a} = \frac{A}{f_1 + f_2 \log \pi_2} \tag{5.1}$$

where f_1 and f_2 and π_2 are determined by the explosive event as given in Eqs. (3.4b), (3.4c) and (3.1); and the numerator A is given as the length sum

$$A = \left[\frac{1}{3}(\alpha_0 - \beta_0)\right] \log E + [\beta_1] \log \rho + \left[\frac{8}{3}\beta_0 + 2\beta_1 - 2\right] \log U$$

$$- [\beta_1] \log P + [\alpha_0 - \beta_0] \log g + \left[1 - \frac{4}{3}\alpha_0\right] \log Q_e + \left[\frac{1}{3}\beta_0\right] \log \delta^*$$

$$- \left[\frac{1}{3}\alpha_0\right] \log \delta + [\log 2 \, K_1 - \log C_0 - \left(\beta_1 + \frac{4}{3}\beta_0\right) \log 2]. \tag{5.1a}$$

This general result can then be applied to various special cases.

As a first application, it is assumed that an impact event, with any projectile type and velocity, at one atmosphere pressure, is to be modeled by an explosive event with a PETN charge of the type used in the explosives experiments reported. In this case the numerical values to be inserted are as follows:

$$\rho = 1.8 \text{ gm/cm}^3$$
$$g = 981 \text{ cm/sec}^2$$
$$P = 10^6 \text{ dyne/cm}^2$$
$$\delta = 1.7 \text{ gm/cm}^3$$
$$Q_e = 5.5 \times 10^{10} \text{ erg/gm.}$$

In addition, numerical values for the various constants in Eq. (5.1) are used. The result is given by

$$\frac{d}{a} = \frac{4.01 - 0.00067 \log E - 0.682 \log U + 0.158 \log \delta^*}{0.104 + 0.0103 \log E} \qquad (5.2)$$

as a function of the projectile (and explosive) energy E, projectile velocity U and mass density δ^*. Assuming further an aluminum projectile:

$$\delta^* = 2.8 \text{ gm/cm}^3$$

then

$$\frac{d}{a} = \frac{4.08 - 0.00067 \log E - 0.682 \log U}{0.104 + 0.0103 \log E}. \qquad (5.3)$$

The results of this formula are shown in Fig. 3 as a plot of equivalent d/a versus event energy E for various projectile velocities. It is again emphasized that this curve gives the depth to bury a PETN explosive in order that the crater volume be the same as an impact event. This depth is normalized to the explosive charge radius a, which is not the same as the projectile radius a*.

Some obvious trends are apparent on this figure. The equivalent d/a decreases some with increasing energy. More marked is the decrease in d/a with increasing velocity of impact. In fact, the formula Eq. (5.3) gives the equivalent d/a going to zero at velocities of about 10 Km/sec. The experimental results were for maximum velocities of about 6 Km/sec. Consequently, an extrapolation of the empirical results much above this velocity is uncertain, particularly in the present case when the specific energy of the explosive has been fixed, and is very different from that of the projectile.

Another interesting case to consider arises from the concept that the projectile penetrates to some given depth and then is equivalent to an explosive event. In this case the explosive is to have the same total energy and mass in the same volume as the projectile. Consequently, it is appropriate to choose equal specific energies and mass densities for the two processes:

$$\delta = \delta^*$$
$$Q_e = \frac{1}{2} U^2.$$

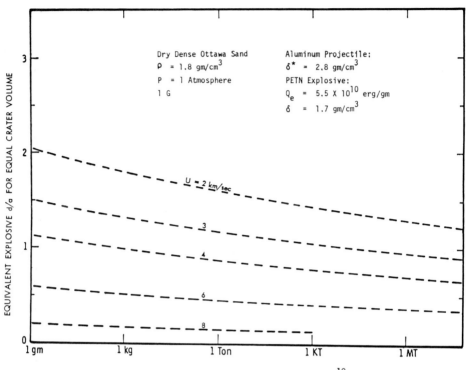

Fig. 3. The equivalent depth of burst normalized to charge radius as a function of event energy E and impact velocity U for a fixed explosive.

With the choice of

$$\rho = 1.8 \text{ gm/cm}^3$$
$$g = 981 \text{ cm/sec}^2$$
$$P = 10^6 \text{ dyne/cm}^2,$$

one obtains the rather complex equation:

$$\frac{d}{a^*} = \frac{-0.0479 - 0.00067 \log E + 0.0594 \log U + 0.0007 \log \delta}{0.105 - 0.0827 \log U + 0.0103 \log E - 0.0103 \log \delta + 0.0475 \, U^{0.190}\delta^{-0.281}}.$$

For an aluminum projectile $\delta = \delta^* = 2.8 \text{ gm/cm}^3$ which leaves the independent variables U and E. In this case the equivalent d.o.b. is normalized to the projectile radius a^*. The results are presented in Fig. 4. The equivalent d.o.b. in this case can be more nearly related to a projectile penetration. Again it shows a decrease as the velocity increases, but much less than on the previous figure, even when

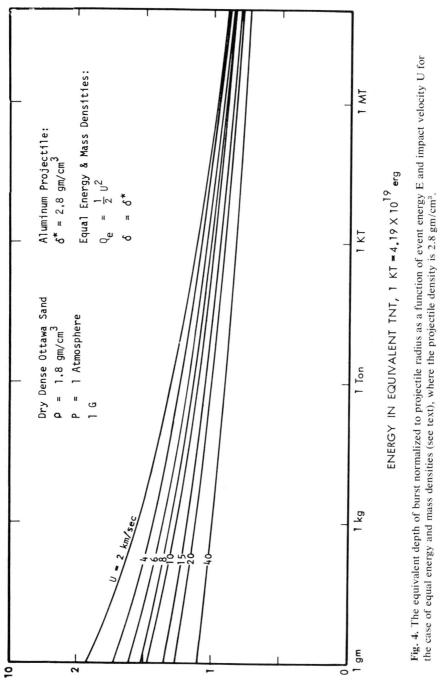

Fig. 4. The equivalent depth of burst normalized to projectile radius as a function of event energy E and impact velocity U for the case of equal energy and mass densities (see text), where the projectile density is 2.8 gm/cm³.

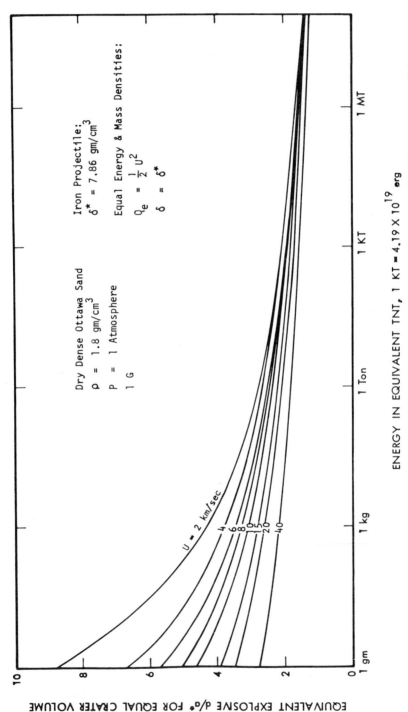

Fig. 5. The equivalent depth of burst, normalized to projectile radius as a function of event energy E and impact velocity U for the case of equal energy and mass densities (see text), where the projectile density is 7.86 gm/cm³.

extrapolated to extremely high velocities. As before, a definite dependence on energy is also apparent.

A similar case, but with an iron projectile, $\delta = \delta^* = 7.86$ gm/cm³ is shown in Fig. 5. All equivalent d/a* are larger, particularly for the small energy regime.

Interesting cross-plots of this case are shown in Figs. 6 and 7. Figure 6 shows the dependence on velocity for two extreme energy values. Figure 7 shows the effect of the density ratio ρ/δ of the soil to the projectile for this case of equal specific energy and mass density, for various combinations of velocity and energy. This figure shows a decrease in d/a* with ρ/δ, with a dependence of about $(\rho/\delta)^{-\frac{1}{2}}$ for the larger energies and at lower values of ρ/δ. For projectiles less dense than the target, the present results show a decreasing dependence in ρ/δ, approaching $(\rho/\delta)^{-1/3}$.

In principle, estimates of the uncertainty in the numerical results could be made from the uncertainties in the various empirical constants and the general form of the results, Eqs. (5.1) and (5.1a). However, at the present time insufficient data

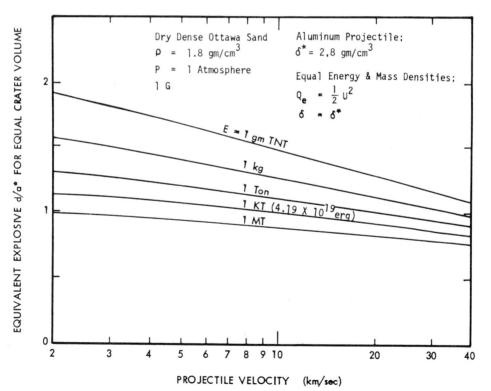

Fig. 6. The equivalent depth of burst as a function of impact velocity.

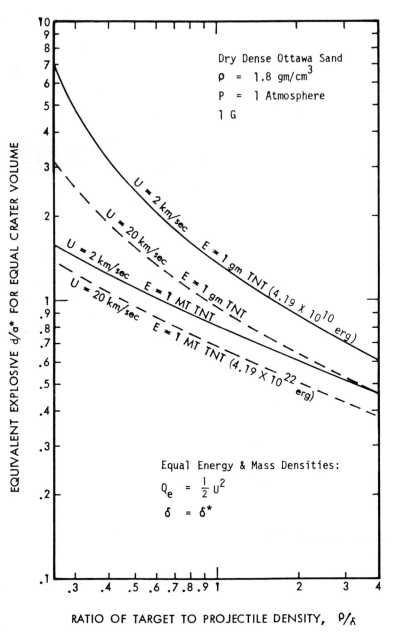

Fig. 7. The equivalent depth of burst as a function of the ratio of target to mass density.

is available to do this in a meaningful way. For example, no values are available at present for the uncertainty in the pressure dependence of the explosive results. Furthermore, while uncertainties have been given for the coefficients and exponents of the explosive and impact events at one-atmosphere pressure, these uncertainties are not independent and consequently the uncertainty in an algebraic combination of these constants such as appears in Eq. (5.1a) cannot be determined. For these reasons, it will have to suffice at present to give a qualitative discussion of the uncertainty in the results, and indicate the regimes of most confidence.

The explosive and impact events which served as the data base for the empirical results spanned a range of specific energies, mass densities and (scaled) energies. However, the explosive results have not been tested for specific energies in excess of about 6×10^{10} erg/gm. Extrapolations to 10^{11} are plausible, but caution should be exercised for further extrapolations. Similarly, the projectile maximum velocities were about 7×10^5 cm/sec, with a corresponding specific energy of about 2.5×10^{11} erg/gm. Extrapolations beyond, perhaps, 10^6 cm/sec are subject to question. Energies on the order of 10^{20} ergs have at least been simulated via the gravity-scaling technique. This is in the kiloton energy regime. Extrapolations to tens of kilotons are probably acceptable, extrapolations to megatons become more doubtful, but the results seem to indicate a smooth and reasonable behavior in the extrapolation.

Finally, there is the dependence on atmospheric pressure. As stated, the pressure dependence of the explosive results has to be considered tenuous at present. Thus, even though the results have been carried through with a pressure dependence, in order to serve as a framework when better results become available, any application to other than one-atmosphere pressure conditions is presently inappropriate.

6. DISCUSSION

The numerical and graphical results have been given in the preceding section. While various meaningful ways to define an equivalent d.o.b. have been identified, all results show similar trends with changes in energy and projectile velocity. These results are for fixed one-atmosphere pressure, dense Ottawa sand, and 1-G conditions.

It was stated in the introduction that another experimental determination of an equivalent d.o.b. has been reported. (Oberbeck, 1971). This was for a small (0.15 gm) explosive of a low-density PETN ($\delta = 1.0$ gm/cm^3). His experiment was conducted with near vacuum conditions (P = 1 mm).

The present results are thought to be valid only for atmospheric pressure. However, it is interesting to compare to Oberbecks' result. When the numerical values corresponding to Oberbecks' experiments, but with P = 1 atmosphere, are inserted into the formulas given above, an equivalent $d/a = 1.8$ is obtained. Oberbeck reports a d.o.b. of 0.15 gm, and a charge density of 1.0 gm/cm^3, this gives $d/a = 1.9 \pm 0.6$, which compares well.

The indications in the present work are that, for an explosive at a fixed burst depth, the volume increases more rapidly with a reduction in atmospheric pressure than for impact events, and therefore the equivalent d/a decreases as atmospheric pressure is reduced. Indeed, a direct application of Eq. (5.1) to the pressure of 1 mm gives a value for d/a = 0.8, which falls considerably below Oberbecks' result. However, it is again emphasized that there is no reason to believe that Eq. (5.1) is particularly accurate for reduced pressure. On the other hand, the material used by Oberbeck was a lower density sand (ρ = 1.7 gm/cm^3) than that used to derive the formulas (ρ = 1.8 gm/cm^3) which has a significant effect on cratering phenomena (Schmidt *et al.*, 1979) and perhaps on equivalent d.o.b. as well. Furthermore, the charge used by Oberbeck (1971) was similar to a detonator cap, with a cylindrical aluminum shell and an explosive tamped from one end, rather than a bare spherical explosive. A similar charge configuration was used by Johnson *et al.* (1969) in a series of crater experiments conducted at reduced gravity. A comparison of the d.o.b. effects measured in those experiments with that obtained by Piekutowski (1974, 1975) for spherical charges indicates a definite directional effect for the detonator caps, and an effective depth either above or below the charge center, depending upon orientation.

Various calculations have also been reported with various measures of an equivalent d.o.b. Bryan *et al.* (1978, 1980) in a recent calculation have estimated an equivalent d.o.b. by fitting a hemisphere to iso-density contours of the expanding cavity at a relatively early time in the calculation (a fixed ½ sec. for a range of energies from 0.08 to 12 megatons of equivalent TNT). Their values, written in terms of d/a*, equal 2.9 at the lower velocity of 2 Km/sec, jump to 5.2 at 10 Km/sec, and then decrease smoothly to 4.4 at 25 Km/sec. The values from the present results are in the range of about 1.5 at these energy levels with a decrease with velocity (Fig. 5). Consequently, with the exception of the 2 Km/sec point, the trends agree, but the values are substantially different.

One difference in this comparison is that the calculations by Bryan *et al.* (1978, 1980) are for a strengthless material and should not be expected to directly compare to the present results. In Bryan *et al.* (1978), a calculation was reported with yield strength included in the material model. At an early time in the calculation (0.5 secs with 4.5 megatons energy) there was reported to be an 8% reduction in penetration depth, and presumably then in the effective d.o.b. The effect of this strength for later times is not reported.

A probably more significant difference between these calculations and the experimental results reported here is that the definition of equivalent d.o.b. is very different. Piekutowski (1980) reports that in visual observations of tangent-below explosive shots, the effective flow center is nearer the charge bottom than its center. This observation suggests a possible factor of two difference between the equivalent d.o.b. based upon charge center and that based upon the flow-field center.

Many estimates of an effective d.o.b. are based upon formulas for the depth of penetration of a high-velocity jet. Birkhoff *et al.* (1948) gives the depth

$$d = L (\rho/\delta)^{-\frac{1}{2}} \qquad (6.1)$$

for the penetration of a jet of length L. While noting that the application of this formula to a spherical projectile is questionable, Shoemaker (1963) uses this formula to obtain an estimate of the effective d.o.b. of Meteor Crater. His result, 8–10 projectile radii, is considerably above any value in the present work, except for dense projectiles at relatively low velocities (2 Km/sec).

While the values obtained from Eq. (6.1) may not agree, the form of the dependence on the mass density ratio agrees surprisingly well. As previously noted in regard to Fig. 7, for values of ρ/δ less than about unity, and for large energies, the dependence on ρ/δ is about a negative one-half power.

Dienes and Walsh (1970) report a formula for penetration based upon the development of a "Late-Stage Equivalence" concept and computer calculations. They give the formula

$$\frac{d}{a} = K \left(\frac{\rho}{\delta}\right)^{-1/3} \left(\frac{U}{c}\right)^{0.58}$$

where c is the target sound speed and K depends upon the target strength. The ρ/δ dependence of this formula agrees with the present results for larger values of ρ/δ only. The increase with impact velocity is not observed.

Baldwin (1963) has obtained an estimate of the equivalent d.o.b. by a comparison of the shapes of existing large-scale explosive and impact craters. His formula, rewritten in c.g.s. units and using the notation of this paper is

$$\frac{d}{a^*} \doteq (1.46 \ 10^{-4})\delta^{1/3}U^{2/3}. \tag{6.2}$$

For iron projectiles at a velocity of 5×10^5 cm/sec, the formula gives

$$\frac{d}{a^*} = 1.83$$

independent of projectile size. This value agrees reasonably well with the larger energy values shown on Fig. 5, but does so only for velocities near this same value. The increasing depth with velocity predicted by Eq. (6.2) is not observed, and Baldwin specifically notes that the assumption that the effective d.o.b. remains near the surface would require that d/a^* decrease with increasing velocity rather than increase as in Eq. (6.2). On the other hand, the dependence or projectile density is in the range observed here, particularly for larger values of ρ/δ or smaller values of energy. Baldwin also indicates an expected decrease in the coefficient with increasing energy, which is the trend observed here.

A different definition of an equivalent d.o.b. has been introduced by Thomson *et al.* (1979) and continued in the paper by Austin *et al.* (1980). In these calculations a "Z-model" was fitted to early-time flow fields. The fit was optimized by choosing a point below the surface for the center. They chose a point with a value of $d/a^* = 2.0$, and noted the correspondence to the results of Oberbeck (1971). If atmospheric pressure is assumed, the present formulas applied to their conditions gives $d/a^* = 1.7$, which is somewhat smaller. However, Austin *et al.* note that fits at later times require a Z-model center migrating downward. This makes clear the difference in the method of defining the equivalent d.o.b. Furthermore, their calculations are for an entirely different material (a plasticene clay). It would be interesting to fit a Z-model to a buried explosive event calculation to compare the Z-model center to the actual charge center in that case, to see if the same trends develop and clarify the relation between the equivalent d.o.b. and the Z-model center.

In closing, it is interesting to note an implication of the results of this paper in application to the estimates of the size and energy of the projectile that formed Meteor Crater. (See Bryan *et al.,* 1978 for a summary). Many of these estimates were based on some determination of the equivalent d.o.b., followed by the use of results from existing explosive craters. For example, Baldwin (1963), using the equivalent d.o.b. discussed above, compares to empirical explosive crater results to obtain the estimate of a required energy of 8.1 megatons. Shoemaker (1963), as discussed above, uses an estimated equivalent d.o.b. of 8–10 projectile radii, based upon a penetration formula, compares to the nuclear TEAPOT-ESS crater, which has about the right d.o.b. (based upon his estimates) and obtains an energy of 1.7 MT.

The experimental results here all support an equivalent d.o.b. much less than that assumed by these authors. Consequently, the required energy becomes much larger. Schmidt (1980) in these proceedings formulates a new energy estimate based directly upon the experimental impact results, without the intermediate step of determination of the equivalent d.o.b. As to be expected, his estimates range considerably higher than these previous estimates.

Acknowledgments—The author is indebted to Dr. R. M. Schmidt for continuing discussions and collaboration, and to C. R. Wauchope for his continuing help. This work was performed for the National Aeronautics and Space Administration under contract NASW-3291 to the Boeing Aerospace Company. This contract is under the direction of Dr. W. L. Quaide.

REFERENCES

Austin M. G., Thomsen J. M., and Ruhl S. F. (1980) Calculational investigation of impact cratering dynamics: Material motions during the crater growth period (abstract). In *Lunar and Planetary Science XI*, p. 46–48. Lunar and Planetary Institute, Houston.

Baldwin R. B. (1963) *The Measure of the Moon*. Univ. Chicago Press, Chicago. 488 pp.

Birkhoff G., MacDougall D. P., Pugh E. M., and Taylor G. (1948) Explosives with lined cavities. *J. Appl. Phys.* **19**, 563–582.

Bryan J. B., Burton D. E., Cunningham M. E., and Lettis L. A. Jr. (1978) A Two-Dimensional Computer Simulation of Hypervelocity Impact Cratering: Some Preliminary Results for Meteor Crater, Arizona. *Proc. Lunar Planet. Sci. Conf. 9th*, p. 3931–3964.

Bryan J. B., Burton D. E., Lettis L. A. Jr., Morris L. K., and Johnson W. E. (1980) Calculations of impact crater size versus meteorite velocity (abstract). In *Lunar and Planetary Science XI*, p. 112–114. Lunar and Planetary Institute, Houston.

Burton D. E., Snell C. M., and Bryan J. B. (1975) Computer design of high-explosive experiments to simulate subsurface nuclear detonations. *Nucl. Technol.* **26**, 65–87.

Dienes J. K. and Walsh J. M. (1970) Theory of impact: Some general principles and the method of Eulerian codes. In *High-Velocity Impact Phenomena* (R. Kinslow, ed.), p. 46–104. Academic, N.Y.

Gault D. E. and Wedekind J. A. (1977) Experimental hypervelocity impact into quartz sand—II, effects of gravitational acceleration. In *Impact and Explosion Cratering,* (D. J. Roddy, R. O. Pepin, and R. B. Merrill, eds.), p. 1231–1260. Pergamon, N.Y.

Herr R. W. (1971) Effects of the atmospheric-lithostatic pressure ratio on explosive craters in dry soil. NASA TR R-366.

Holsapple K. A. (1979) Impact experiments with ambient atmospheric pressure (abstract). *EOS (Trans. Amer. Geophys. Union)* **60,** 871.

Holsapple K. A. (1980) The equivalent depth of burst for impact cratering (abstract). In *Lunar and Planetary Science XI,* p. 456–458. Lunar and Planetary Institute, Houston.

Holsapple K. A. and Schmidt R. M. (1979) A material-strength model for apparent crater volume. *Proc. Lunar Planet Sci. Conf. 10th,* p. 2757–2777.

Holsapple K. A. and Schmidt R. M. (1980) On the scaling of crater dimensions I: Explosive processes. *J. Geophys. Res.* In press.

Johnson S. W., Smith J. A., Franklin E. G., Moraski L. K., and Teal D. J. (1969) Gravity and atmospheric pressure effects on crater formation in sand. *J. Geophys. Res.* **74,** 4838–4850.

Oberbeck V. R. (1971) Laboratory simulation of impact cratering with high explosives. *J. Geophys. Res.* **76,** 5732–5749.

Oberbeck V. R. (1977) Application of high explosion cratering data to planetary problems. In *Impact and Explosion Cratering* (D. J. Roddy, R. O. Pepin, and R. B. Merrill, eds.), p. 45–65. Pergamon, N.Y.

Piekutowski A. J. (1974) Laboratory-scale high-explosive cratering and ejecta phenomenology studies. Air Force Weapons Laboratory report AFWL-TR-72-155, Albuquerque, N.M. 328 pp.

Piekutowski A. J. (1975) A comparison of crater effects for lead azide and PETN explosive charges. Air Force Weapons Laboratory Report AFWL-TR-74-182, Albuquerque, N.M. 140 pp.

Piekutowski A. J. (1980) Formation of bowl-shaped craters. *Proc. Lunar Planet. Sci. Conf. 11th.* This volume.

Roddy D. J. (1977) Large-scale impact and explosive craters: Comparisons of morphological and structural analogs. In *Impact and Explosion Cratering* (D. J. Roddy, R. O. Pepin and R. B. Merrill, eds.) p. 185–246. Pergamon, N.Y.

Schmidt R. M. (1980) Meteor Crater: Energy of Formation—Implications of Centrifuge Scaling. *Proc. Lunar Planet. Sci. Conf. 11th.* This volume.

Schmidt R. M. and Holsapple K. A. (1978a) Centrifuge crater scaling experiments I: Dry granular soils. Defense Nuclear Agency Report 4568F, Washington, D.C. 177 pp.

Schmidt R. M. and Holsapple K. A. (1978b) A gravity-scaled energy parameter relating impact and explosive crater size (abstract). *EOS (Trans. Amer. Geophys. Union)* **59,** 1121.

Schmidt R. M. and Holsapple K. A. (1979) Centrifuge crater scaling experiments II: Material strength effects. Defense Nuclear Agency Report 4999Z. Washington, D.C. 111 pp.

Schmidt R. M. and Holsapple K. A. (1980a) Theory and experiments on centrifuge cratering. *J. Geophys. Res.* **85,** 235–252.

Schmidt R. M. and Holsapple K. A. (1980b) Centrifuge crater scaling experiments III: HOB/DOB effects. Defense Nuclear Agency Report, Contract DNA 001-78-C-0149, Washington, D.C.

Schmidt R. M., Watson H. E., and Wauchope C. R. (1979) Projectile density/target density correlation for impact cratering (abstract). *EOS (Trans. Amer. Geophys. Union)* **60,** 871.

Shoemaker E. M. (1963) Impact mechanics at Meteor Crater, Arizona. In *The Moon, Meteorites, and Comets* (B. M. Middlehurst and G. P. Kuiper, eds.), p. 301–306. Univ. Chicago Press. Chicago.

Thomsen J. M., Austin M. G., Ruhl S. F., Schultz P. H., and Orphal D. L. (1979) Calculational Investigation of Impact Cratering Dynamics: Early Time Material Motions. *Proc. Lunar Planet. Sci. Conf. 10th,* p. 2741–2756.

Proc. Lunar Planet. Sci. Conf. 11th (1980), p. 2403–2421.
Printed in the United States of America

Mars Tharsis region: Volcanotectonic events in the stratigraphic record

David H. Scott and Kenneth L. Tanaka

U.S. Geological Survey, Flagstaff, Arizona 86001

Abstract—Detailed geologic mapping and stratigraphic studies using Viking images have provided new information on the evolutionary history of the Tharsis region of Mars. In this report we describe some results of mapping more than 20 geologic units of this region in relation to Tharsis volcanism and tectonism. Most of the map units are lava flows whose relative ages are based on stratigraphic relations, but substantiated and extended in places by crater counts. Nine periods of major volcanism have been inferred in the Tharsis region, beginning with the resurfacing of basement rocks. Most of the lava flows originated from Olympus Mons and the three large shield volcanoes aligned along the crest of Tharsis Montes. Major eruptions also occurred at Alba Patera, Syria Planum, and around the present site of Olympus Mons, whose aureoles may consist of pyroclastic deposits. Volcanism diminished after the earlier eruptions from Tharsis Montes but continued with surges of activity until late in the history of the planet. Olympus Mons, the largest volcanic structure on Mars, formed during the last stages in the evolution of the region over a relatively brief interval. Faults and fractures radiate outward from the central part of the Tharsis region and form complex patterns in the basement terra units. The density and orientation of fault systems recorded on individual lava flows allow a sequence of tectonic events to be determined in conjunction with major volcanic episodes. We compare the estimated volumes of lava extruded in seven Tharsis flow units with those of a few of the large better known terrestrial basalt and pyroclastic deposits.

INTRODUCTION

The most recent volcanism and much of the oldest tectonic activity on Mars apparently have occurred within the Tharsis region. Broadly defined, this region extends from the northern lowland plains southward to Solis Planum, and from Arcadia Planitia eastward to Lunae Planum. Physiographically, the Tharsis region is a somewhat elongate and irregular regional dome with three large shield volcanoes aligned along its crest to form Tharsis Montes. The geographic localities referred to in this report are shown on the "Topographic map of Mars" (U.S. Geological Survey, 1976).

Detailed geologic mapping (Schaber *et al.*, 1978; Scott *et al.*, 1980) from Viking images has provided new information on the evolutionary history of the region. More than 20 geologic units have been mapped, mostly lava flows whose eruptive sequences have been established by overlap, embayment, and transection relations. The relative ages of the flows are based primarily on stratigraphy but have been substantiated and extended in places by crater-density determinations (Table 1). Broad relatively flat sheet flows with smooth-appearing surfaces are common on the extensive plains surrounding the Tharsis volcanoes and on their gentle

2403

Table 1. Description of lava flow units (Nc = Cumulative number of craters >1 km dia./km^2)

Flow Unit	Nc	Occurrence and Morphology	Stratigraphic Relations
tm_7	8×10^{-5}	Forms smooth, fan-shaped mantles on northeast and southwest upper slopes of Tharsis Montes volcanoes; few channels with levees; few fissures; distal margins feather edges	Overlaps units tm_5, s and vu
op	$7.0–9.0 \times 10^{-5}$	Occurs around basal scarp of Olympus Mons; surface relatively smooth with many narrow, tongue-shaped to broad lobate overlapping flows; channels rare; few fissures or faults	Overlaps units om_2, tm_4, cf, oa, ap_2, ap_1, yt, and ot
tm_6	$<1.0 \times 10^{-4}$	Occurs on lower northwest flank of Pavonis Mons as a thin series of smooth, relatively featureless flows; few fractures and faults	Overlaps unit tm_5 but is partly covered by unit s
om_2	$1.5–2.0 \times 10^{-4}$	Forms complex, finely textured, interfingering tongues and lobes with channels and levees extending down flanks of Olympus Mons; collapse pits common; few faults	Overlaps basal scarp in places and units om_1, oa, and yt
om_1		Buried unit; small patches inferred in places along upturned edges of basal scarp around Olympus Mons. Flows extruded before formation of basal scarp around Olympus Mons	
tm_5	$4.0–6.0 \times 10^{-4}$	Occurs as relatively light-colored flows with dark streaks in places around the upper parts of Tharsis Montes; flows narrow, tongue-shaped to broad and sheetlike; few faults	Overlaps units tm_4, cf, tm_3, tm_2, vu, yt, and ot
tm_4	$6.0–8.0 \times 10^{-4}$	Occurs along northeast and southwest sides of Tharsis Montes; consists of numerous overlapping light and dark flows, elongate on upper slopes, broad on lower; few channels. Very high resolution frames show ridges concentric with lobate frontal scarps; minor faulting	Overlaps units cf, tm_3, ap_1, vu, yt, and ot
ap_3	7.0×10^{-4} 1.0×10^{-3}	Occurs around and within central caldera of Alba Patera; buries some ring and radial fault systems but is cut by others; channels with levees common along crests of narrow elongate flows	Overlaps units ap_2 and yt

Unit	Age	Description	Stratigraphic relations
cf	8.0×10^{-4} 1.2×10^{-3}	Flows extend southwest from highly faulted region of Ceraunius Fossae; relatively smooth, even-toned to mottled and streaked surface; leveed channels in places on crest of long narrow flows; moderately faulted	Overlaps units ap_2, ap_1, vu, and yt
tm_3	$1.3–1.5 \times 10^{-3}$	Exposed on west and southeast flanks of Arsia Mons; flows generally moderately dark, light in places; channels and levees rare; moderately faulted	Overlaps units tm_2, tm_1, yt, and ot
tm_2	$1.4–1.6 \times 10^{-3}$	Exposed on south flank of Arsia Mons; dark, relatively smooth flows with broad frontal lobes; fractures and faults common; appears subdued, probably by eolian mantle	Overlaps units tm_1, and ot
tm_1	$1.7–2.0 \times 10^{-3}$	Forms rough, hummocky surface on south flank of Arsia Mons; marelike ridges in places; faults and fractures common	Overlaps unit ot
oa		Occurs around base of Olympus Mons as a series of broad, flat, corrugated, overlapping semicircular lobes; faults and fractures common	Overlaps units ap_1, yt, and ot
sp_2	$1.8–2.4 \times 10^{-3}$	Occurs around crestal area of Syria Planum; long, narrow flows and sheet-type flows with prominent lobes; mottled light and dark; pit craters common; channels rare	Partly buries most fracture and fault systems of Claritas Fossae; cut by fractures of Labyrinthus Noctis; overlaps units sp_1, yt, and ot
ap_2	$1.8–2.4 \times 10^{-3}$	Forms broad plains around Alba Patera; narrow flows with channels and levees common on upper slopes, sheet-type flows on lower slopes; many faults and grabens	Overlaps units ap_1, yt, and ot
sp_1	$2.4–3.2 \times 10^{-3}$	Occurs around base of Syria Planum; flows are mottled light and dark and form narrow tongues and broad lobes with some channels; surface smooth to hummocky; transected by many faults of Claritas Fossae	Overlaps unit ot
ap_1	$2.4–3.2 \times 10^{-3}$	Similar to unit ap_2; forms broad, smooth to hummocky plains around Alba Patera; boundaries between other units (ap_2 and plains material of northern lowlands) not clearly defined; numerous faults and fractures	Overlaps units yt and ot
vu	$1.0–4.0 \times 10^{-3}$	Forms central prominences of volcanoes in and around Tharsis Montes	Stratigraphic relations and crater counts indicate a wide age range

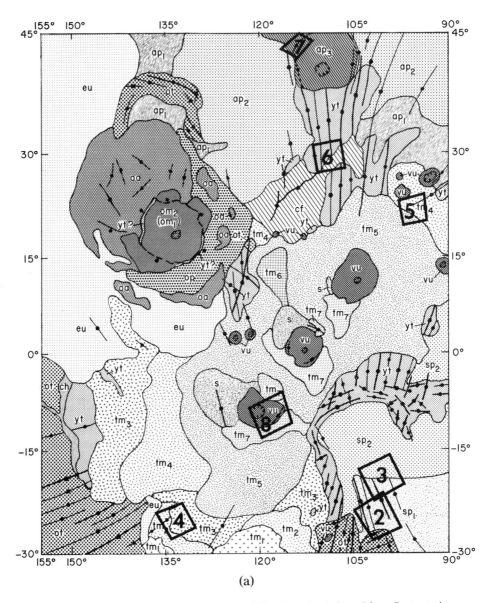

(a)

Fig. 1a. Generalized geologic-tectonic map of Tharsis region (adapted from Scott *et al.*, 1980). Numbered rectangles show locations of Viking images in Figs. 2–8.

lower slopes. Channel and tube-fed flows (Carr *et al.*, 1977) are more common on the steeper slopes higher up the sides of the volcanoes and have rougher textures. Most exposed flow units originated from two of the largest volcanoes, Olympus Mons and Arsia Mons.

Faults generally radiate outward from the central part of the Tharsis region (Carr, 1974; Masson, 1977; Wise *et al.*, 1979) and, in places, form intricate patterns whose complexity increases with the age of the faulted unit. The density and orientation of fault and fracture systems recorded on the lava flows allow a sequence of tectonic events to be determined in conjunction with major volcanic episodes.

FLOW STRATIGRAPHY AND VOLCANIC ACTIVITY

The systematic mapping of individual lava-flow units in the Tharsis region has been compiled in a series of maps (Scott *et al.*, 1980) that show the areal extent, eruptive sequence, and probable sources of the flows. The major flow units are

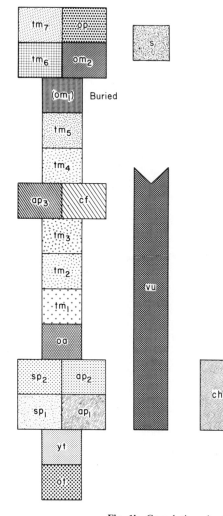

CORRELATION OF MAP UNITS

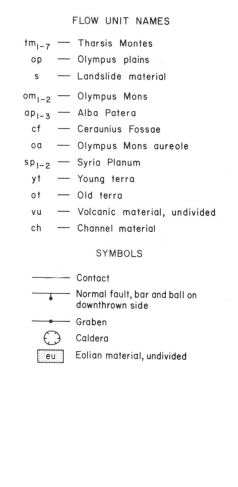

FLOW UNIT NAMES

tm_{1-7}	—	Tharsis Montes
op	—	Olympus plains
s	—	Landslide material
om_{1-2}	—	Olympus Mons
ap_{1-3}	—	Alba Patera
cf	—	Ceraunius Fossae
oa	—	Olympus Mons aureole
sp_{1-2}	—	Syria Planum
yt	—	Young terra
ot	—	Old terra
vu	—	Volcanic material, undivided
ch	—	Channel material

SYMBOLS

Contact

Normal fault, bar and ball on downthrown side

Graben

Caldera

eu   Eolian material, undivided

(b)

Fig. 1b. Correlation chart of map units shown in Fig. 1a.

shown on the generalized map and explanation (Fig. 1), along with basement rocks, channel and flood-plain deposits, and landslide and eolian materials. The area covered by each major flow unit can be calculated from the geologic map, although a volume determination relies on more speculative estimates of thickness. We noted while mapping the region, however, that many craters larger than about 5 km in diameter were incompletely buried by flow units. This relation suggested a thickness of some 200 m for each flow unit, if crater diameter-to-rim height relations measured on the moon (Pike, 1977) are applicable to martian craters of the same size. We have used volumes based on this estimated thickness in our calculations, except for the aureole deposits around Olympus Mons.

Nine periods of major volcanic activity have been inferred in the Tharsis region, beginning with the resurfacing of older basement rocks. Although much of the resurfacing was done by lava flows, eolian and fluvial processes were active as well; together they have created a relatively smooth, low relief topography on the terra around such areas as Memnonia, Noctis Labyrinthus, Lunae Planum, and parts of Ceraunius and Claritas Fossae. The sources of the flows are unknown but probably were fissures, subsequently buried by the flows that issued from them. The basement rocks may also consist of volcanic materials (and breccia) that have been so disrupted by faults, fractures, and impact craters that the morphologic characteristics of any original flows can no longer be recognized (Fig. 2).

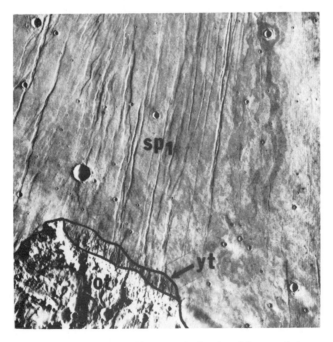

Fig. 2. Highly faulted and eroded older terra (unit ot) and less eroded, younger terra (unit yt) embayed by Syria Planum flows (unit sp₁). Flows, in turn, are cut by younger faults. North towards top; picture width about 280 km.

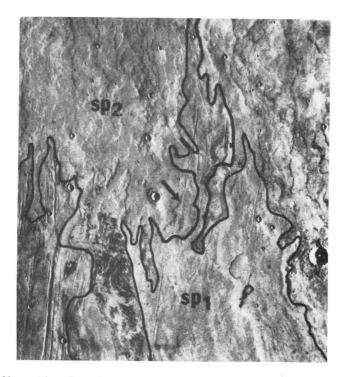

Fig. 3. Younger lava flows from Syria Planum (unit sp$_2$) covering faults in older flows (unit sp$_1$). Note that some flows have been extruded from and partly bury fissures (arrow). North towards top; picture width about 300 km.

After the volcanic-resurfacing episode, large volumes of lava were erupted from Alba Patera and the crestal area of Syria Planum onto the northern lowland plains and the highland plateau around the present Tharsis Montes (Fig. 3). These lava flows, the oldest, exposed in the Tharsis region, have crater densities (for diameters > 1 km) of about 1.8×10^{-3} to 3.2×10^{-3}/km for both the younger and older units at each locality, although a still-later stage of volcanism occurred around the summit of Alba Patera. Volcanic activity at these two centers extended over a considerable time interval as indicated by the wide range of crater counts obtained in their flow units.

As eruptions decreased toward the end of this period, new volcanic centers appeared around the present site of Olympus Mons. Volcanic materials extruded from several sources at different times formed the aureole deposits that cover approximately 2.3×10^6 km^2 of flat to gently rolling terrain. They form a succession of large overlapping plates with level, corrugated surfaces that resemble terrestrial deposits of ash-flow tuffs. The thickness of these plates probably exceeds 1000 m (E. C. Morris, pers. comm., 1980), and their volume is comparable to that of the larger lava flows in the Tharsis region. The relative ages of these aureole deposits around Olympus Mons can only be broadly established with respect to the other flow units; the deposits are overlapped in places by flows of Olympus Mons and the Olympus plains. Current studies by the authors suggest that the aureoles may also be overlain in places by even older flows from Ceraunius Fossae and Tharsis Montes. The aureole materials overlie basement rocks and the oldest flow unit from Alba Patera. Morphologically, they seem to be older than most lava flows in the Tharsis region, but this might result if they consist in part of poorly-welded pyroclastic deposits that are readily susceptible to erosion. Crater counts are not reliable age determinants on the aureole material because the ridged and grooved surfaces promote rapid degradation and obliteration of crater forms by mass wasting. On somewhat subjective evidence therefore, we provisionally place the aureole deposits just above the flows from Alba Patera in the stratigraphic sequence (Fig. 1).

Nearly contemporaneous with the voluminous floods of aureole material, eruptions in the ancestral Tharsis Montes started the growth of shield volcanoes along the axis of the structure. The main source vents were concentrated in three places equidistant along a probable northeast trending rift zone (Carr *et al.*, 1977). Regional uplift of Tharsis Montes, however, began at an earlier time when the fault

Fig. 4. Lava flows on southwest flank of Arsia Mons, about 1200 km from center of shield. Lighter-toned flow (unit tm$_3$) overlaps faults transecting older darker flow (unit tm$_2$). North towards upper right; picture width about 230 km.

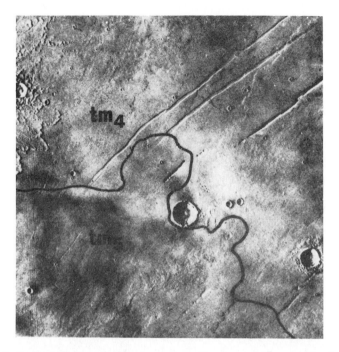

Fig. 5. Faults in older flows (unit tm₄) overlapped by younger flows (unit tm₅). North towards top; picture width about 220 km.

systems expressed in the terra of Memnonia and Tempe Fossae began to form (Wise *et al.*, 1979). The lava flows spread farther than 1200 km from the volcanic centers, especially the flows from Arsia Mons (Figs. 4, 5). During this period of massive volcanism, about 5×10^6 km^3 of lava were extruded to form the shield volcanoes and early flows of the Tharsis Montes. Volcanism sharply declined after this period but continued with several peaks of activity on a diminishing scale until late in the history of the planet.

As volcanism waned in the central part of the Tharsis region, eruptive centers shifted northward to the area around Ceraunius Fossae and the broad summit of Alba Patera. The flows at Ceraunius Fossae appear to have issued from fissure vents that followed the preexisting north-south-trending fault systems in the terra (Fig. 6). The last flows from Alba Patera were extruded around the crest of this volcano where they buried earlier sets of faults (Fig. 7). These flows, in turn, also have been displaced by more recent, and probably continuing, tectonic movements.

Olympus Mons, the largest single structure in the Tharsis region, formed late in the volcanic evolution of the region, probably over a relatively brief geologic interval. Stratigraphic relations are clearly defined between the flows from Olympus Mons and those on the surrounding plain, which partly encircle the volcano on its east side. The eruptions that built Olympus Mons may have occurred in

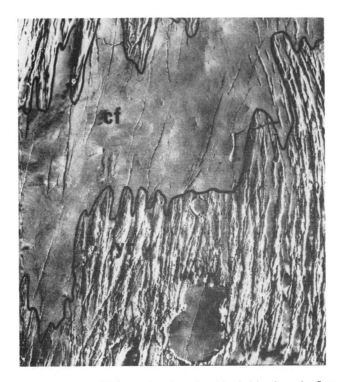

Fig. 6. Younger terra material (unit yt) embayed and buried in places by flows of Ceraunius Fossae (unit cf). A few faults and fractures transect lava flows. North towards top; picture width about 220 km.

two stages, separated by an episode of faulting that formed the prominent basal scarp around the volcano. Crater counts indicate that flows of two slightly different age groups may be present on Olympus Mons, although these data are limited by inherent errors in distinguishing between small impact and parasitic craters on the sides of the volcano. Most flows override the scarp and completely bury it in places. In some areas, however, an older surface is exposed along the raised edge of the scarp, and flow lines appear to be sharply truncated at its upper margin. No prescarp flows have been recognized around the base of Olympus Mons, although, they may have been covered by postscarp flows and more recent landslide material, which occur sporadically around the base of the volcano. Vertical displacements creating the basal scarp may have been induced by withdrawal of large volumes of the lava and pyroclastic materials that built the volcanic pile and the older aureoles. The raised edge along the upper part of the scarp indicates some subsidence of the central shield, probably as a consequence of the great uncompensated pressures (Sjogren, 1979) exerted at the base of Olympus Mons.

The youngest flows in the Tharsis region, high on the northeast and southwest

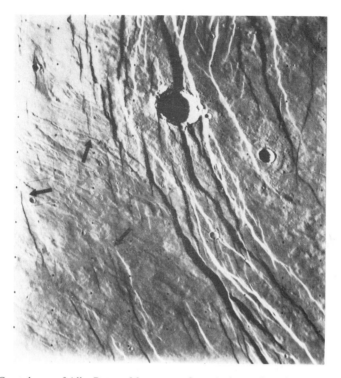

Fig. 7. Crestal area of Alba Patera. Most recent flows (unit ap₃) from fissures on volcano bury older sets of faults (arrow) but are cut in places by younger faults. North towards upper right; picture width about 120 km.

flanks of Arsia Mons, Pavonis Mons, and Ascraeus Mons (Fig. 8), originated from recent fissures along the ancient postulated rift zone transecting Tharsis Montes. The flows are relatively small in areal extent and appear to be thin because they do not obscure some topographic and structural features in the underlying terrain. The presence of these flows indicates some resurgence of volcanic activity along an old structural system.

Some measure of appreciation for the magnitude of volcanic activity in the Tharsis region is provided by comparisons with several well known localities of terrestrial volcanism. Most of these localities date from the more recent periods of earth history, although we recognize that great floods of lava have occurred throughout geologic time. The absolute timespan for volcanic events at Tharsis cannot be reliably determined without a time-frequency distribution scale for martian impact craters, corroborated by radiometrically dated rock samples, such as was done on the moon. Some constraints on the extent and duration of martian events, however, can be inferred from the available geologic data. Crater-frequency distributions, for example, indicate that volcanism progressed at Tharsis without any large gaps in the impact record (Fig. 9). Crater densities are the

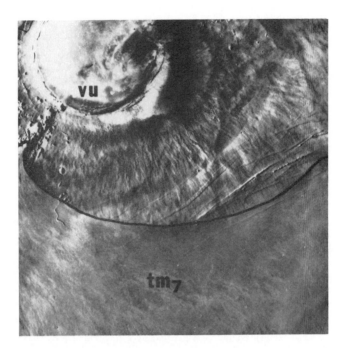

Fig. 8. Youngest flows (unit tm₇) in Tharsis region overlapping rough older surface of Arsia Mons shield volcano (unit vu).

cumulative numbers of caters with diameters larger than 1 km per km² on each unit. Textural details associated with such features as lava tubes and channels with levees, pressure ridges, and talus cones appear fresh on high-resolution images of flows considered to be young to intermediate in age (Scott *et al.*, 1980). Moreover, even some older flows retain morphologic features characteristic of the younger units, including tongue-shaped individual members with lobate fronts. These observations, together with the relative paucity of faults and fractures in the lava-flow units in comparison with the terra, suggests that Tharsis volcanism occurred late in martian history. Thus, we prefer the time scale of Soderblom *et al.* (1974) over that of Neukum and Wise (1976): Soderblom *et al.*'s scale restricts the volcanic activity to a period of several hundred million years, rather than to several billion years obtained from Neukum and Wise's scale.

Figure 10 shows volume estimates for several of the Tharsis flow units and shields in comparison with large lava and pyroclastic deposits on earth. Volumes were estimated from the areal extent of individual flow units and assuming an average thickness of 200 m for each unit. The areas covered by partly buried units were determined from paleostratigraphic maps (work in progress by the authors) showing the reconstruction of older surfaces. The logarithmic scale is somewhat deceiving; the Columbia River Plateau basalts, for example, have only one-fifth the volume of the lava flows on Syria Planum, and the aureole deposits of Olympus Mons are 20 times larger, in area and volume, than the ignimbrites in the Great Basin province of Nevada and Utah. These terrestrial examples, however, are middle Tertiary and younger in age, and their eruptive periods may have been much shorter (<30 m.y.) than those at Tharsis. Possibly some of the extensive oceanic basaltic plains are comparable to or even exceed the flows of the Tharsis region. The suboceanic volcanic fields, however, may more closely resemble those covering the vast lowland plains on Mars.

TECTONIC HISTORY

Most of the faulting in the Tharsis region is associated with large-scale tectonic activity, including the doming and uplift of Tharsis Montes, Syria Planum, and Alba Patera. Fault trends recorded by other investigators show general or local age relations by their respective superposition relations (Carr, 1974; Masson, 1977; Masursky *et al.*, 1978; McGill, 1978; Wise *et al.*, 1979). Using the detailed stratigraphy obtained from our lava-flow studies (Schaber *et al.*, 1978; Scott *et*

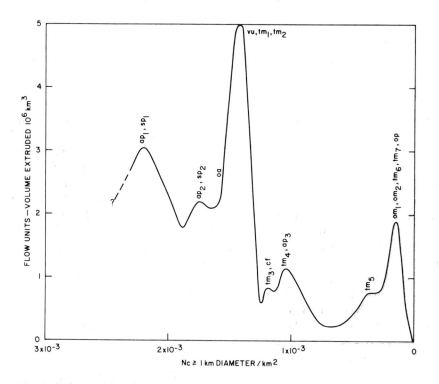

Fig. 9. Volume of lava flow units (individuals or groups) vs. total number of craters (Nc), larger than 1.0 km in diameter, normalized to an area of 1.0 km².

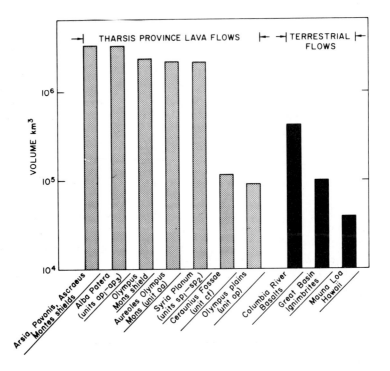

Fig. 10. Volumes of several Tharsis lava flows and volcanic structures in comparison with some similar features on earth.

al., 1980), we have related tectonic events to the major lava-flow units in the Tharsis region (Table 2). The regional geologic-tectonic map (Fig. 1) shows fault patterns for the different mapped units. Fault-density counts for areas considered to represent discrete tectonic episodes show several peaks at Alba Patera with minor activity around the other volcanic centers (Fig. 11). Faults and grabens in many of the older terra localities nearly cover the entire surface, thus these areas have maximum fault density. In general, graben widths are on the order of a few kilometers, although around Alba Patera they tend to be wider. With fairly uniform fault widths, therefore, a semi-quantitative degree of faulting relative to saturation can be estimated for a given area.

Only normal-type faults with vertical separation have been observed in the Tharsis region. These faults generally are long linear or arcuate features that in places extend more than 500 km and in many places are paired to form grabens as wide as 10 km or wider. Also, linear structurally controlled chains of pit craters commonly appear among the fault sets. Concentric faults and fractures around calderas are associated with the largest martian volcanoes, and a basal scarp encircles Olympus Mons. The long prominent escarpment in the terra forming Claritas Fossae may be a fault block (Masursky *et al.*, 1978) that represents a tectonic mountain range similar to those on earth.

Overall, the intensity of faulting increases with age (Fig. 11). Major tectonic episodes include initial fracturing due to the uplift of Syria Planum and Tharsis Montes, and two stages of concentric fracturing around Alba Patera (much of the faulting in unit ap_1 probably was contemporaneous with that in unit ap_2). Comparison of Figs. 9 and 11 shows no strong correlation between the volume of extruded lava and the degree of faulting, and indicates that most tectonic activity occurred before the extrusion of recognizable lavas in the region, except for the later episodes of faulting south of Syria Planum and around Alba Patera.

SUMMARY AND CONCLUSIONS

The eruptive sequence and areal extent of lava flows in the Tharsis region have been defined by stratigraphic studies and crater counts made on individual flow units. Faults and fractures transecting these units provide a record of changing tectonic intensity during this period of high volcanic activity.

Volcanism began with the resurfacing of basement rocks early in the history of the region and continued without large interruptions through nine major erup-

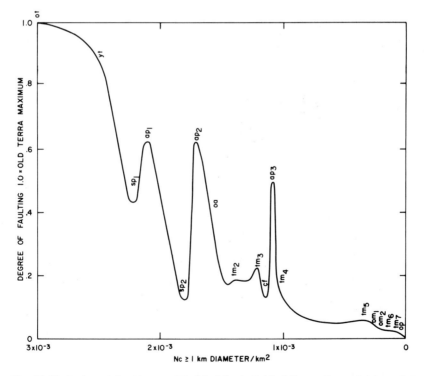

Fig. 11. Tectonic activity (degree of faulting) for individual flow units vs. total number of craters (Nc), as in Fig. 9.

Table 2. Faulting sequence (oldest to youngest)

Order	Terrain Description	Fault Trends	Faulted units
1	Basement block south of Syria Planum. Faults form broad, linear valleys	ENE-WSW (Masursky et al., 1978)	ot
2a	Syria Planum, radial and concentric faults around crest		
	i) Radial grabens south of Syria Planum	Fan NNE-SSW to NNW-SSE (McGill, 1978)	ot
	ii) Claritas Fossae; radial and concentric faults embayed to west by unit tm_3 and to east by unit sp_2	NNE-SSW and NE-SE	yt
	iii) Radial and concentric grabens north of Noctis Labyrinthus. Concentric faults mainly adjacent to dome, radial faults predominant to north. Grabens often enlarged by pit collapse and erosion. Overlain by flow units tm_5 and sp_2 to north and unit sp_2 to south	NNE-SSW and ENE-WSW to NW-SE (Masson, 1977)	yt
2b	Faults radial to Tharsis Montes		
	i) In Memmonia, low to moderate density, cut by Mangala Vallis channeling. Overlain by units tm_{1-4} to east	ENE-WSW (in the West) to NE-SE (in the east) (Mutch and Morris, 1979)	ot, yt
	ii) Faults in Tempe Fossae, embayed by unit ap_1 on south and unit ap_2 on west	NE-SW (Wise, 1979)	yt
	iii) Fissures of Ceraunius Fossae south of Alba Patera (Wise, 1979) and between Olympus Mons and Pavonis Mons. Overlain by various lava units, beginning with ap_1	N-S and N-S and NE-SW superposed by NNE-SSW, respectively (Carr, 1975)	yt
3a	Radial faults south of Syria Planum, faults in unit sp_1 covered by flows of sp_2 (Fig. 2)	NNW-SSE	sp_1
3b	Intense concentric faults on east and west sides of Alba Patera; offset lava flows	NNW-SSE to NE-SW (Wise, 1979)	ap_1, ap_2
3c	Radial faults in Memmonia continue and fault density diminishes after deposition of unit tm_3 (Fig. 4)	NE-SW	tm_{1-3}

3d	Fractures of Olympus Mons aureole deposit; curvilinear ridges and grooves transected by long linear fractures and grabens	NW-SE, others (Morris, 1979)	oa
4	Sparse localized faults radial to Tharsis Montes		
a)	Faults in units cf (Fig. 6) and tm_5 south of Alba Patera	N-S	cf, tm_5
b)	Faults in unit tm_4; fault density diminishes in unit tm_5 northeast of Ascraeus Mons (Fig. 5)	NE-SW	tm_4, tm_5
c)	Faults in unit tm_5 west of Pavonis Mons and Arsia Mons	NNW-SSE	tm_5
5	Intense concentric faults around Alba Patera, coeval with deposition of unit ap_3, offset some flows (Fig. 7)		ap_2, ap_3
6	Concentric faults around large Tharsis volcanoes		
a)	Encirclement of Olympus Mons by a scarp about 5 km high		yt(?), oa, om_1, om_2
b)	Formation of concentric grabens around Arsia Mons (Fig. 8) and Pavonis Mons (Crumpler and Aubele, 1978)		vu, tm_7

tive episodes. Volcanic centers shifted from place to place but resurgent activity occurred at several volcanoes. Volcanism culminated during the formation of large shield volcanoes along Tharsis Montes and thereafter appears to have gradually declined. Volumes of lava extruded were very large compared with terrestrial flows of the Tertiary Period, but the duration of Tharsis volcanism may have been much longer.

The dense and complex systems of faults that cut basement rocks are not nearly so common on even the older Tharsis flow units. Counts made of the number of faults on flow units of different age indicate that tectonic activity decreased with time. It appears to have been somewhat episodic, however, and was concentrated in different localities at different times.

Acknowledgments—The authors thank D. U. Wise and J. B. Plescia for their helpful comments. This research was supported by NASA Work Order W-13,709.

REFERENCES

Carr M. H. (1974) Tectonism and volcanism of the Tharsis region of Mars. *J. Geophys. Res.* **79**, 3943–3949.

Carr M. H. (1975) Geologic map of the Tharsis quadrangle of Mars. U.S. Geol. Survey Misc. Geol. Inv. Map I-893.

Carr M. H., Greeley R., Blasius K. R., Guest J. E., and Murray J. B. (1977) Some Martian volcanic features as viewed from the Viking orbiters. *J. Geophys. Res.* **82**, 3985–4015.

Crumpler L. S. and Aubele J. C. (1978) Structural evolution of Arsia Mons, Pavonis Mons, and Ascraeus Mons: Tharsis region of Mars. *Icarus* **34**, 496–511.

Masson P. (1977) Structure pattern analysis of the Noctis Labyrinthus-Valles Marineris regions of Mars. *Icarus* **30**, 49–62.

Masursky H., Dial A. L., and Strobell M. E. (1978) Geologic map of the Phoenicus Lacus quadrangle of Mars. U.S. Geol. Survey Misc. Geol. Inv. Map I-896.

McGill G. E. (1978) Geologic map of the Thaumasia quadrangle of Mars. U.S. Geol. Survey Misc. Geol. Inv. Map I-1077.

Morris E. C. (1979) The aureole of Olympus Mons (abstract). In *Reports of Planetary Geology Program, 1978–1979*, p. 239–240. NASA TM 80339.

Mutch T. A. and Morris E. C. (1979) Geologic map of the Memnonia quadrangle of Mars. U.S. Geol. Survey Misc. Geol. Inv. Map I-1137.

Neukum G. and Wise D. U. (1976) Mars: A standard crater curve and possible new time scale. *Science* **194**, 1381–1387.

Pike R. J. (1977) Size-dependence in the shape of fresh impact craters on the moon. In *Impact and Explosion Cratering* (D. J. Roddy, R. O. Pepin, and R. B. Merrill, eds.), p. 489–509. Pergamon, N.Y.

Schaber G. G., Horstman K. C., and Dial A. L. Jr. (1978) Lava flow materials in the Tharsis region of Mars. *Proc. Lunar Planet. Sci. Conf. 9th*, p. 3433–3458.

Scott D. H., Schaber G. G., Tanaka K. L., Horstman K. C., and Dial A. L. Jr. (1980) Lava flow maps of the Tharsis region of Mars. U.S. Geol. Survey Misc. Geol. Inv. Maps I-1266 to I-1282. In press.

Sjogren W. L. (1979) Mars gravity: High resolution results from Viking orbiter 2. *Science* **203**, 1006–1010.

Soderblom L. A., Condit C. D., West R. A., Herman B. M., and Kreidler T. J. (1974) Martian planetwide crater distributions: Implications for geologic history and surface processes. *Icarus* **22**, 239–263.

U.S. Geological Survey (1976) Topographic map of Mars. U.S. Geol. Survey Misc. Geol. Inv. Map I-961.

Wise D. U. (1979) Geologic map of the Arcadia quadrangle of Mars. U.S. Geol. Survey Misc. Geol. Inv. Map I-1154.

Wise D. U., Golombek M. P., and McGill G. E. (1979) Tharsis province of Mars: Geologic sequence, geometry, and a deformation mechanism. *Icarus* **38**, 456–472.

Proc. Lunar Planet. Sci. Conf. 11th (1980), p. 2423–2436.
Printed in the United States of America

Estimation of the thickness of the Tharsis lava flows and implications for the nature of the topography of the Tharsis plateau

Jeffrey B. Plescia[1,2] and R. Stephen Saunders[2]

[1]Department of Geological Sciences, University of Southern California, Los Angeles, California 90007,   [2]Jet Propulsion Laboratory, California Institute of Technology, Pasadena, California 91103

Abstract—In order to gain some insight into the nature of the topography of the Tharsis plateau, we have attempted to estimate the thickness of the volcanics which cover the large part of the Tharsis region. Estimates for the thickness of the flows were made using the numerous occurrences of the partially buried craters throughout the region and a relation between the crater diameter and crater rim height. Estimates were made along the southern contact with the heavily cratered terrain in the Phoenicis Lacus and Memnonia quadrangles, and along the northeast contact with older plains units in the Tharsis and Lunae Palus quadrangles. In these distal regions the thickness estimates range from 0 to 600–700 meters of lava. This would indicate that only 20–30 percent of the topography is the result of volcanic construction. If the 20–30 percent estimate is representative of Tharsis as a whole then lava thickness would reach only 2–3 kilometers near the center of the region. Thus only a small fraction of the topography presently observed can be attributed to units readily identifiable as volcanic. This additional evidence indicates that the model for Tharsis of a pre-existing elevated region of heavily cratered crust, capped by a veneer of volcanics is correct.

INTRODUCTION

The Tharsis region of Mars is unique in terms of topography, volcanism and tectonism. The region has long been known to be a broad domal area with approximately 10 km of relief relative to the adjacent plains. A topographic map (Fig. 1) illustrates the area of interest. Tharsis can be seen to be a wide asymmetric topographic high. From the crest along the Tharsis Montes line the elevations drop steeply to the northwest, while dropping more gradually to the southeast. Superimposed upon the domal region are the large Tharsis shields, which attain heights of approximately 15 km above the surrounding plains.

A number of mechanisms have been proposed to account for the origin of the topography, and the reader is referred to these for details of the various models (Phillips *et al.*, 1973; Wise *et al.*, 1979a, 1979b; Solomon and Head, 1980; Sleep and Phillips, 1979; and Phillips, 1978). Each of these models has somewhat different implications for the global evolution of Mars. However, despite their differences, they can be broadly classed into two groups. The first group (uplift),

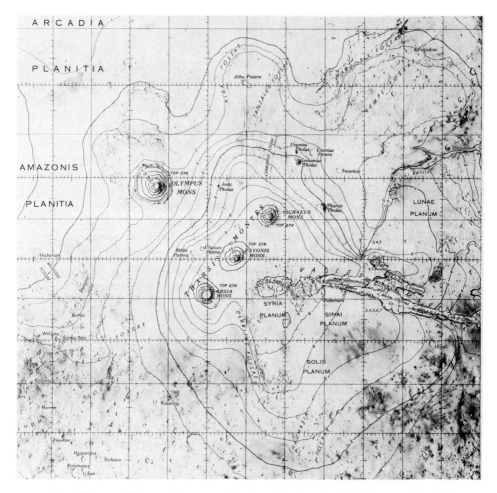

Fig. 1. Topographic map of the greater Tharsis region from the USGS Topographic Map of Mars. Contour interval is generally 1 km, except on the volcanic shields where it is 5 km. The map shows the roughly 10 km high plateau atop which sit the Tharsis Montes shields. The contour lines also indicate an asymmetry to the plateau with a steep slope along the western side relative to the east.

was originally proposed by Phillips *et al.* (1973), and proposes that the topography results from an uplift of the crust by an unspecified mechanism. The second group (constructional) proposes that Tharsis is the result of an accumulation of flows and coalescing shields, through time, to build the current topography. This model was recently reemphasized by Solomon and Head (1980).

These two hypotheses offer profoundly different views of the subject. Not only are the mechanisms different, but each carries a series of implications as to what the geological and geophysical history of the Tharsis region would have been.

Among these implications would be the nature and state of the crust and mantle with time, the timing of events and even the composition of the rocks exposed in the region.

The uplift model would entail an uplift of heavily cratered crust early in martian history. This uplift may have been continual or episodic in nature, but the result was a broad region of highly fractured elevated crust. At some time after this uplift the region would have been capped by a thin veneer of volcanic flows, and then finally the large shield volcanoes. The constructional model proposes that the topography of the region owes its origin to the piling up of material with time. The Tharsis area in cross section would reveal many kilometers of lavas and numerous coalesced shield volcanoes. All of the crust in the area would be volcanic in nature, including exposures of heavily cratered crust.

In order to gain insight into the question of the nature of the topography, we have attempted to determine the thickness of the lava flow units exposed in the region (units Apt, Aps and Apc from the Geologic Map of Mars, Scott and Carr, 1978).

The estimation of the thickness of those units will not resolve the argument as to the nature of the heavily cratered crust. However, it will allow a determination of the extent to which those lavas, formed over the last 3.0 billion years, have contributed to the topography. These thickness estimates coupled with crater counts should allow a clearer understanding of the volcanic history of the region.

METHODOLOGY

We have attempted to determine the thickness of the units using a method similar to that employed by DeHon (1979) in his study of the lunar western mare basalts. Scattered across the Tharsis plains are a number of large, generally 20–60 km diameter, craters which are partially to completely buried by volcanic cover. The assumption made here is that these partially buried craters are part of the heavily cratered crust which underlies the lava plains. An idealized cross section, with relevant dimensions, is illustrated in Fig. 2.

DeHon (1979) based his assumption, that these partially buried lunar craters were part of the basement, on a combination of spectral reflectance data and photogeologic interpretation. Spectral reflectance data are not of sufficient resolution on Mars to be applicable here. However, there are a few photogeologic observations that would tend to support our assumption. All of the volcanic flows are relatively young (Schaber *et al.*, 1978; Carr *et al.*, 1977), and hence post date the heavy bombardment. As a result, large craters are absent from these surfaces. There is, additionally, an extremely low probability that such large craters would form after the heavy bombardment ended. The size frequency distribution of these partially buried craters also reflects an ancient crust. The gentle slope exhibited by these curves parallels the ancient crust, rather than the steeper curves exhibited by the lava flows. We feel that these craters represent the heavily cratered crust, whatever it's origin.

DeHon (1979) used the relationship between the crater diameter and rim height as formulated by Pike (1974, 1977) to determine the thickness of the basalts on the mare. This relationship estimates the height of the rim (see Fig. 2) above the surrounding plains. This relationship ignores any radial variation around the rim, and hence represents an average. By estimating the exposed rim height (R_e) and subtracting it from the rim height (R_h), a thickness of the cover is obtained. This methodology ignores both the thickening within the crater itself, as well as the local thinning over the ejecta blanket around the crater. The thickness (T) should be viewed as a local average of the cover thickness away from the affect of the crater.

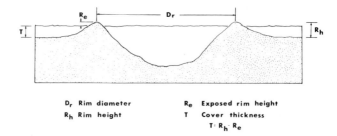

Fig. 2. Idealized cross section through a partially buried crater.

Crater rim height/diameter relationships have been determined for the moon by Pike (1974, 1977) and for Mercury by Cintala (1979). These studies indicate that despite the difference in g between the two bodies the rim height/diameter relationships are very similar. The difference between the values obtained for the moon and Mercury is approximately 10%, with the lunar rim heights being slightly lower than mercurian rim heights for craters of equal diameter.

As there have been no systematic studies of the martial crater rim height diameter relationship we have chosen to use the relationship for the moon based on the larger data base of Pike (162 craters) over that for Mercury from Cintala (30 craters). The equations used here, as determined for the moon, are:

$$R_h = 0.036 \ D_r^{1.104} \tag{1}$$
$$R_h = 0.236 \ D_r^{0.399} \tag{2}$$

where R_h equals rim height and D_r equals crater rim diameter (see Fig. 2). Equation (1) was used for craters with rim diameter of less than 17 km, while Eq. (2) was used for craters with rim diameters greater than 17 km.

Data from Schubert et al. (1977) indicate that very large martian craters (>100 km) have heights lower than corresponding lunar craters. That study used earth based radar to determine those heights. The resolution of that radar data was not sufficient to allow a determination of the rim height/diameter relationship for the craters used in this study.

The use of the lunar relationship will introduce an unknown amount of uncertainty into the calculations. There are additional complexities in the crater rim height/diameter relationship for Mars that do not affect the moon or Mercury. These complications will tend to affect the rim height with time. Isostatic rebound and various erosive agents, both aqueous and eolian, would tend to lower the rim height with time after formation. The presence of an atmosphere introduces a drag effect into the distance ejecta in thrown (Schultz and Gault, 1979). Hence, the presence of a much denser atmosphere in the past may have caused significantly larger accumulations of ejecta nearer the crater, raising the original rim height. All of these problems add to the uncertainty of the method. However, as none have been studied the degree of error introduced by each is unknown. Under the assumption that they are not inordinately large, the method should allow an order of magnitude estimation of the rim height. Since we are addressing the question of whether the lavas are nearer one km thick or 10 km thick, this seems a reasonable approach.

DATA BASE

Partially buried craters occur principally in two areas in the Tharsis region. The first is along the southern and southwestern contacts of the lava plains with the heavily cratered terrain in the Memnonia and Phoenicis Lacus quadrangles. The second area is to the northeast in the Tharsis and Lunae Palus quadrangles. Figures 3 and 4 illustrate parts of these areas and the location of the various partially buried craters.

The number listed on Figs. 3 and 4 indicate the calculated rim height based on the diameter of the crater. As the amount of exposed rim has not been subtracted, these numbers do not represent the thickness of the cover; however, they can be viewed as an upper limit. We have not attempted to subtract the rim height because of the error associated with measuring shadows on uneven terrain imaged at high altitude. Many of the images used were from the medium resolution mapping coverage with 200 to 300 meter/pixel resolution. With an uncertainty of ± 1 pixel, considerable error results. If, for example, an exposed rim height was 200 meters illuminated at an incidence angle of 55° it would produce a shadow of approximately 300 meters, approximately 1–2 pixels. An error of even 1 pixel would introduce an error, at these conditions, of ± 150 meters. The resulting errors are as large as the measured value. Again, the question of what is the order of magnitude of the lava thickness, can still be approached using the values obtained as upper limits.

MEMNONIA-PHOENICIS LACUS REGION

The Memnonia-Phoenicis Lacus region (Fig. 3) has the largest number of partially buried craters. These craters range from almost completely buried to almost

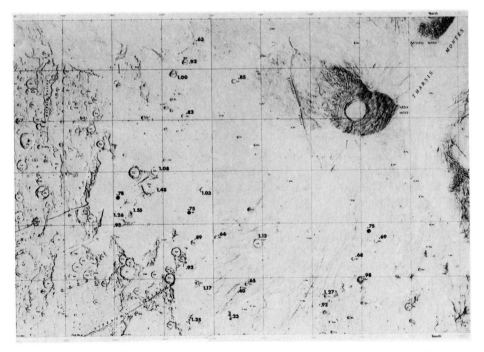

Fig. 3. Shaded relief map of part of the Memnonia and Phoenicis Lacus quadrangles. The numbers indicate the calculated rim heights for the various partially buried craters. The numbers do not take into account the amount of exposed rim height. The black dots indicate the location of those craters which were not on the shaded relief map.

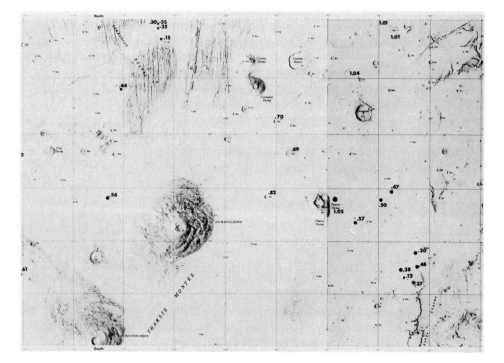

Fig. 4. Shaded relief map of parts of the Tharsis and Lunae Palus quadrangles. Again here the numbers indicate calculated rim height, not corrected for exposed rim height.

completely exposed. Figure 5 illustrates several examples of partially buried craters and the variation in burial. From Fig. 3 it can be seen that these craters have calculated rim heights of approximately 600–1600 meters and are randomly exposed across the area.

Earth based radar topography of the area south of Arsia Mons is illustrated in Fig. 6 and covers the region from 14°–21° south latitude (Roth *et al.,* 1980). The region south of Arsia Mons can be generally represented as a southward plunging arch. Between the contact of the volcanic plains and the cratered highlands (147° longitude) and the crest of the arch (120° longitude) the elevation rises approxiamtely 2.5 kilometers. Of importance here is the fact that despite an increase of 2.5 kilometers in elevation, craters with rim heights of approximately 700 meters are still exposed near the crest. This indicates that the lavas are less than 700 meters thick at the crest of the topography. This would thus imply that the majority of the elevation increase is due to the elevation of the basement. Because an uncertain amount of rim remains exposed, a reasonable estimate of the average thickness of the lavas in the areas would be 500–600 meters.

A second means of approximating the thickness of the lavas in the region is to look at the average local topographic variations in the heavily cratered terrain and compare it to that found on the volcanic plains. Figure 7 illustrates a single radar profile across 15.80° S from longitudes 170° to 125°. The contact between the two geologic units occurs at 147° and is marked by the arrow. Westward across the heavily cratered terrain the local topographic variations are on the order of 500 to 700 meters with local excursions to 1 kilometer. Conversely across the volcanic plains to the east, the local topographic variations are 100 to 200 meters. The large topographic excursions between 140° and 143° are the result of a cluster of three large craters located there (USGS designations, En, Cn and Dm). Hence, to bury the average topographic variations of the heavily cratered

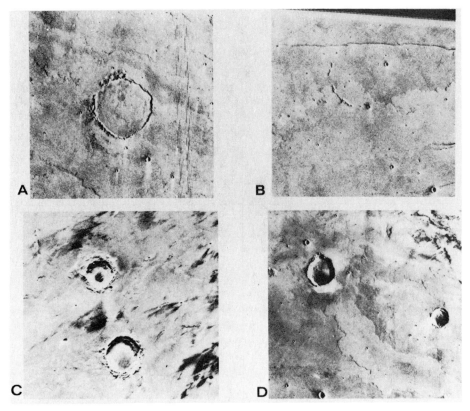

Fig. 5. Examples of partially buried craters from the Memnoia-Phoenicis Lacus quad-ranges. (A) This figure shows a section of 639 A 42 in which a 50 km crater has been significantly buried. Here only the upper reaches of the rim are exposed. To note is the fact that part of the rim is almost completely buried while the opposite side is more exposed. This probably results from a slope in the bedrock topography. (B) This section of 639 A 41 shows only an arcuate ridge, the remanents of an almost totally buried 50 km diameter crater. Again the absence of an exposed rim around the crater probably represents a slope in the bedrock topography. (C) Part of image 639 A 55 shows a couple of partially buried craters with exposed crater rims and partially exposed ejecta blankets. The larger of the two craters is approximately 40 km in diameter. (D) Part of image 639 A 39 shows two partially buried craters which have well exposed crater rims. Both craters have flooded floors and completely covered ejecta blankets. The larger crater is 28 km across.

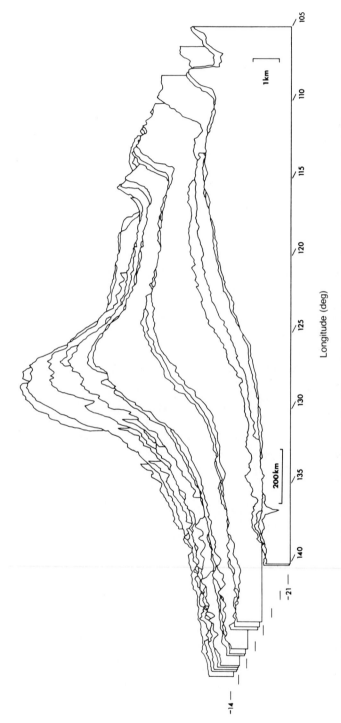

Fig. 6. Topography south of Arsia Mons from longitudes 140 to 105°. Vertical exaggeration 100 ×. Twenty-two Goldstone radar scans between latitudes −14.23° and −21.19°.

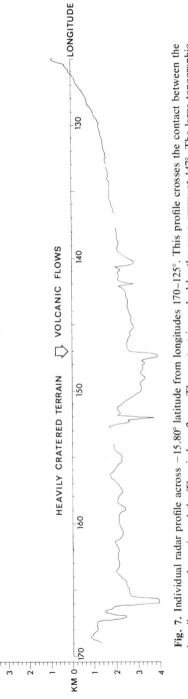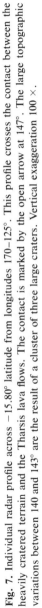

Fig. 7. Individual radar profile across −15.80° latitude from longitudes 170–125°. This profile crosses the contact between the heavily cratered terrain and the Tharsis lava flows. The contact is marked by the open arrow at 147°. The large topographic variations between 140 and 143° are the result of a cluster of three large craters. Vertical exaggeration 100 ×.

terrain would require a cover of approximately 500–600 meters. This is very similar to the result obtained from the crater rim height estimates. The larger excursions of topography would probably remain unburied and would be represented by the small knobs of bedrock exposed throughout the region. Examples of these knobs of bedrock are illustrated in Fig. 8.

From these two lines of evidence, local topographic relief, and crater rim heights, we would estimate that the lavas along the southern and southwestern regions are approximately 500 to 600 meters thick. This should be viewed as an average thickness as the lavas would be considerably thicker within a given crater and thinner over bedrock knolls. The point here is that the lavas comprise only about 25% of the observed topography. The remaining 75% is due to the elevation associated with the cratered basement, whatever its origin, beneath the flows.

LUNAE PALUS—THARSIS AREAS

The second area of partially buried craters (Fig. 4) occurs to the northeast in the Tharsis and Lunae Palus quadrangles. The density of partially buried craters here is considerably less than to the southwest. This probably results from a difference

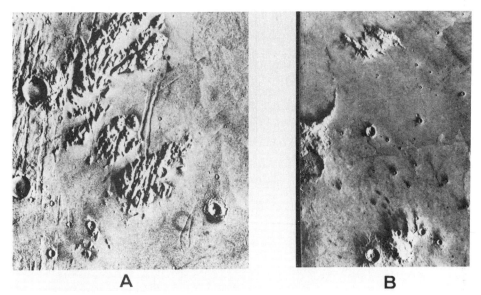

A   **B**

Fig. 8. Examples of bedrock knobs which protrude through the lava flows. These knobs are considered to be the result of local areas of large topographic variation in the bedrock. (A) Is part of frame 516 A 34 which shows an outcrop of faulted crust in the northern Tharsis plains. Note that the wide graben do not cut the surrounding plains, but can be traced between exposures of faulted crust. (B) This is part of frame 639 A 43 and shows the typical types of knobs exposed in the southern Tharsis plains. Most of the knobs are circular in plan and are furrowed in a radial manner.

in basement age. To the southwest the basement is heavily cratered terrain, while here the basement is of similar crater density and appears to be continuous with the Lunae Planum region further east. Depsite the limited data base some estimates for lava thickness can be made.

As shown by the topographic maps of the Tharsis area (Fig. 1) the elevation rises approximately 2 kilometers from the contact with older faulted crust (90° longitude 25°N latitude) to the region at 95° longitude and 20° N latitude. Here a few craters with rim heights of 700 to 800 meters are exposed. As in the south west part of the rim remains unburied, therefore a reasonable estimate of the cover thickness would be approximately 500 to 600 meters. The implication for this area is that only about 25% of the elevation is the result of construction by volcanic material. The remaining 75% is the result of the topography of the underlying basement. This result is similar to that determined for the southwest region.

DISCUSSION

If these regions, the Tharsis-Lunae Palus and the Phoenicis Lacus-Memnonia area, are representative of Tharsis as a whole, then it would appear that the readily mappable lavas make up only about 20 to 30% of the observed topography. The flows would naturally thin towards the contact and thicken towards the center of the Tharsis area. Considering that the center of the elevated region stands approximately at 10 kilometers elevation, and if the 20 to 30% figure is correct, then the average thickness of the lavas even here would be only 2 to 3 kilometers.

Because of the limited amount and nature of the data we have not attempted to prepare an isopach map of lava thickness. Such an attempt would require considerable more data than exists.

A relatively thin volcanic cover over a broad bedrock high also seems reasonable in light of the exposures of older crust within the area. Figure 9 is a sketch map of the distribution of the principal bedrock units. We have divided the terrain into three generalized units; the young units identified as volcanic flows, an older smooth plains units and the oldest heavily cratered terrain. Exposures of older crust occur well within the Tharsis region. Basement outcrops in a broad arc from the northeast corner to the southwest region. Clartias Fossae and the fractured terrain north of Noctis Labyrinthus form a prong of older terrain into the heart of Tharsis. Scattered across the north and northwest sides are outcrops of older and heavily fractured terrain. Additionally two very large craters, Fessenkov (22°N, 86.5° longitude: 85 km diameter) and an unnamed crater (USGS designation Ng; 8.5°N, 112.5° longitude: 77 km diameter) occur within the center of the area. These craters are relatively well preserved and have not been significantly buried, as is evident from the fact that ejecta near the crater rim can still be seen. They do however predate the volcanic flows and due to their size it could be argued from statistics that they date back to the terminal bombardment, and hence represent exposures of ancient crust.

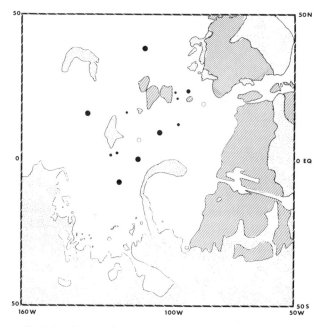

Fig. 9. Generalized sketch map of the pre-volcanic bedrock around the Tharsis region. The area has been divided into three general units. The white areas indicate cover of lava flows or other younger material of an unspecified nature. The striped areas are Lunae Planum plains and units with similar crater density. The stippled areas are regions of heavily cratered crust. The black dots indicate the location of the large Tharsis volcanoes, while the open circles indicate the location of large impact craters.

Solomon and Head (1980) have suggested that the heavily cratered terrain, as well as the lava plains, are volcanic in nature. The heavily cratered crust, they feel, would represent the earliest outpourings of Tharsis related lavas. We reject this suggestion for a variety of reasons. Data from several sources indicate that the heavily cratered crust in the Tharsis region is similar to that found elsewhere on the planet. Color mapping by Soderblom *et al*. (1978) does not show any difference in color between the Memnonia crust and other areas of heavily cratered crust, while a pronounced difference exists between the heavily cratered crust and the volcanics of Tharsis. Thermal inertia and residual temperature maps of Zimbelman and Kieffer (1979) show major differences between the Memnonia crust and the volcanics to the east. However, little difference can be seen between the Memnonia crust and other areas of heavily cratered crust.

There is also a lack of evidence for a volcanic nature for the heavily cratered crust. Volcanic features such as flows, flow fronts or large volcanoes are absent from the area. These features commonly occur on other areas interpreted to be volcanic.

As the heavily cratered terrain dates back to an early period in martian history, perhaps some 3.5 to 4 billion years ago, and considering the fact that flows of Olympus Mons may be as young as a couple of hundred million years (Blasius,

1976; Carr *et al.*, 1977; Plescia and Saunders, 1979; Neukum and Hiller, 1980*),
if the older crust is volcanic it would require an extremely long-lived thermal
anomaly under the Tharsis region. From the topographic map (Fig. 1) it can be
seen that the Tharsis bulge extends well out into the cratered terrain. Hence, in
addition to such a long lived thermal regime, one would also have to postulate
a much broader, several times larger than the present, area of volcanism.

One of the implications of the Solomon and Head model is that there should
be a progression of surfaces with crater ages which range from the heavily cratered crust to the youngest, essentially, uncratered flow units. Detailed crater
counts and stratigraphic separation of individual lave flow units have been compiled by Schaber *et al.* (1978). They have defined a number of flow units in the
Tharsis area which range in crater density from approximately 90 to 3200 craters
greater than 1 km per 10^6 km^2. This would simply a time span, based on the
chronology of Soderblom (1977) of 300 million to slightly more than 1.5 billion
years. Hence, there is a distance bimodality of ages for the Tharsis region, an
ancient 3.5–4.0 billion year old heavily cratered crust and a set of younger volcanic units.

The absolute age of a surface, derived from crater chronologies, is model dependent. Neukum and Hiller (1980*) have recently modified the crater chronology
of Neukum and Wise (1976) and proposed a model whereby the cratering rate is
approxiamtely 1.5 times the lunar value. This model compares closely with that
of Soderblom (1977). A second model proposed by Neuum and Hiller (1980*) with
a cratering rate 1/3 of the lunar value would push the Tharsis volcanics back in
time, but would still indicate an age of less than 3 billion years. Both models
would thereby produce the bimodality of ages observed between the heavily
cratered terrain and the lava plains. This bimodality of ages is in contrast to what
would be expected by the Solomon and Head model, but it would be expected
by an uplift of ancient crust followed by a veneering of lava.

We do not disagree with Solomon and Head (1980) that volcanism of the Tharsis
region may have been long lived. Our data as well as a variety of other data
indicate that several of the small volcanoes are very old. However, these appear
to represent a different style of volcanism than is responsible for the bulk of the
broad scale basaltic volcanism which presently dominates Tharsis.

In conclusion, our data indicates that the volcanic units of the Tharsis area are
a relatively thin cover of an older topographically higher area. The lava flow
appears to make up only 20 to 30 percent of the present topography. The thickness
of the lava appears to range from 2 to 3 kilometers near the center of Tharsis. A
variety of evidence would tend to support the original model of Phillips *et al.*
(1973) that the topographic dome of the Tharsis region is tectonic, rather than
constructional, in nature.

Acknowledgments—This paper is JPL Planetology Publication Number 326-80-61 and represents the
results of one phase of research carried out at the Jet Propulsion Laboratory, California Institute of
Technology, under Contract NAS 7-100, sponsored by the Planetary Geology Program Office, Planetary Division, Office of Space Science, National Aeronautics and Space Administration.

---

* Martian ages. Submitted to *J. Geophys. Res.*

REFERENCES

Blasius K. R. (1976) The record of impact cratering on the great shields of the Tharsis region of Mars. *Icarus* **29**, 343–361.

Carr M. H., Greeley R., Blasius K. R., Guest J. E., and Murray J. B. (1977) Some martian volcanic features as viewed from the Viking orbiters. *J. Geophys. Res.* **82**, 3984–4015.

Cintala M. J. (1979) Mercurian crater rim heights and some interplanetary comparisons. *Proc. Lunar Planet. Sci. Conf. 10th,* p. 3635–3650.

DeHon R. A. (1979) Thickness of the western mare basalts. *Proc. Lunar Planet. Sci. Conf. 10th,* p. 2935–2955.

Neukum G. and Wise D. U. (1976) Mars: A standard crater curve and possible new time scale. *Science* **194**, 1381–1387.

Phillips R. J. (1978) Topical problems in martian geophysics. *Reports of the Planetary Geology Program 1977–1978,* p. 69–70. NASA TM 79729.

Phillips R. J., Saunders R. S., and Conel J. (1973) Mars: Crustal structure inferred from Bougeur gravity anomalies. *J. Geophys. Res.* **78**, 4815–4820.

Pike R. J. (1974) Depth/diameter relations of fresh lunar craters: Revision from spacecraft data. *Geophys. Res. Lett.* **1**, 291–294.

Pike R. J. (1977) Size dependence in the shape of fresh impact craters on the moon. In *Impact and Explosion Cratering* (D. J. Roddy, R. O. Pepin, and R. B. Merrill, eds.), p. 489–510. Pergamon, N. Y.

Plescia J. B. and Saunders R. S. (1979) The chronology of the martian volcanoes. *Proc. Lunar Planet. Sci. Conf. 10th,* p. 2841–2859.

Roth L. E., Downs G. S., Saunders R. S., and Schubert G. (1980) Radar altimetry of south Tharsis, Mars. *Icarus*. In press.

Schaber G. G., Horstman K. C., and Dial A. L. Jr. (1978) Lava flow materials in the Tharsis region of Mars. *Proc. Lunar Planet. Sci. Conf. 9th,* p. 3430–3458.

Schubert G., Lingenfelter R., and Terrile R. (1977) Crater evolutionary tracks. *Icarus* **32**, 131–146.

Schultz P. H. and Gault D. E. (1979) Atmospheric effects on martian ejecta emplacement. *J. Geophys. Res.* **84**, 7669–7687.

Scott D. H. and Carr M H. (1978) Geologic map of Mars. U. S. Geol. Survey, Misc. Geol. Inv. Map I–1083.

Sleep N. H. and Phillips R. J. (1979) An isotatic model for the Tharsis province, Mars. *Geophys. Res. Lett.* **6**, 803–806.

Soderblom L. A. (1977) Historical variations in the density and distribution of impacting debris in the inner solar system: Evidence from planetary imaging. In *Impact and Explosion Cratering* (D. J. Roddy, R. O. Pepin, and R. B. Merrill, eds.), p. 629–633. Pergamon, N. Y.

Soderblom L. A., Edwards K., Eliason E. M., Sanchez E. M., and Charette M. P. (1978) Global color variations on the martian surface. *Icarus* **34**, 446–464.

Solomon S. C. and Head J. W. (1980) Tharsis: An alternative explanation (abstract). In *Reports of Planetary Geology Program 1979–1980,* p. 71–73. NASA TM 81776.

Wise D. U., Golombek M. P., and McGill G. E. (1979a) Tharsis province of Mars: geologic sequence, geometry, and a deformation mechanism. *Icarus* **38**, 456–472.

Wise D. U., Golombek M. P., and McGill G. E. (1979b) Tectonic evolution of Mars. *J. Geophys. Res.* **84**, 7934–7939.

Zimbelman J. R. and Keiffer H. H. (1979) Thermal mapping of the northern equatorial and temperate latitudes of Mars. *J. Geophys. Res.* **84**, 8239–8251.

Proc. Lunar Planet. Sci. Conf. 11th (1980), p. 2437–2446.
Printed in the United States of America

The geology and morphology of Ina

P. L. Strain and F. El-Baz

Center for Earth and Planetary Studies, National Air and Space Museum, Washington, D.C. 20560

Abstract—Ina is a unique D-shaped depression located atop a 15-km diameter extrusive dome southeast of the Imbrium basin. It lies in a region that is dissected by lineations radial to both Imbrium and Serenitatis. The rim of Ina is characterized by discontinuous concentric fractures, flows, and a smooth raised border. The floor contains 4 separate units, including smooth, sparsely cratered protrusions or mounds ranging in height from 5 to 25 m. The mounds are compared to the terrestrial lava pillars called Dimmuborgir located in northern Iceland. These pillars are morphologically similar to the Ina mounds and may have formed by the collapse and subsidence of a partially solidified lava lake. However, the distribution of the mounds in Ina, the disparate heights of their summits, and the frequent presence of moats around their bases suggest that the mounds originated as discrete extrusive features.

The dome underlying Ina appears comparable to low terrestrial basaltic shields although it exhibits some morphological differences. The unusual morphology of the Ina structure may have been influenced by its location in a highly fractured region between two of the largest nearside impact basins, and by the composition of the magma.

INTRODUCTION

The lunar feature Ina, an unusual D-shaped depression, is located in Lacus Felicitatis north of Mare Vaporum (Fig. 1). Before being officially named by the International Astronomical Union, it was sometimes informally referred to as "D-Caldera" (El-Baz, 1973). Attention was drawn to the feature because of its strange "blistered" appearance. It was first noticed on Apollo 15 panoramic camera photographs by E. A. Whitaker (1972a) who pointed out its apparently unique character. El-Baz (1972; 1973) proposed that it was a collapse caldera displaying evidence of episodic volcanic activity. Using detailed topographic data available from the Lunar Topographic Orthophotomap (LTO) series, we can now depict the morphology and structure of Ina and the surrounding region more accurately than was previously possible. The maps used in this study are 41C3 and 41C4 at 1:250,000 scale, and 41C3S1 (Ina) at 1:10,000 scale. In addition, profile data were specially produced by the Defense Mapping Agency/Topographic Center.

REGIONAL SETTING

Ina lies in a region between the Apennine Mountains of the Imbrium basin and the Haemus Mountains of Serenitatis. The area is characterized by highland materials dominated by northwest-southeast trending lineations radial to Imbrium (El-Baz, 1973). A pattern radial to Serenitatis is also evident, although less pronounced. Irregular patches of maria (lacus) are present and are often bordered by scarps radial to the basins (Fig. 1). Ina is located atop a raised basalt plateau that bisects Lacus Felicitatis. The plateau is tilted towards the west. Its western scarp reaches heights of more than 200 m while the eastern scarp is about 600 m high. The scarps bordering the plateau have slopes of about 9°. The western scarp extends into the highlands to the south and appears to emerge as a mare ridge that crosses Mare Vaporum (see Fig. 30-11, El-Baz, 1973) and terminates northeast of Sinus Medii in a region marked by dark mantle material (Wilhelms and McCauley, 1971). In Mare Vaporum the ridge marks a change in elevation on the order of 100 m.

The surface of Lacus Odii, located northeast of Felicitatis, also slopes down to the west, exhibiting a drop in elevation of more than 400 m. A lobate flow front (Fig. 1) indicates that during a late episode of extrusion, lava in Odii flowed

Fig. 1. Apollo 17 metric view of the Ina region. Ina, the D-shaped depression in Lacus Felicitatis, is about 2.7 km along its straight edge. Arrow designates lobate flow front discussed in text (AS17-M-1517).

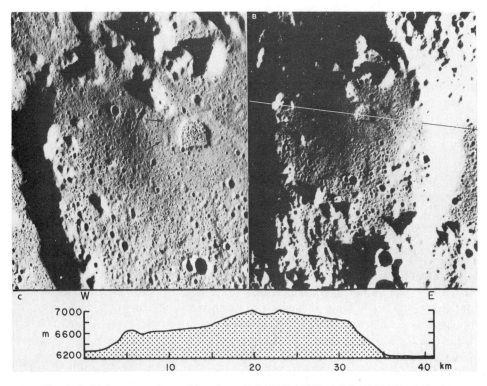

Fig. 2. (a, b) Low sun views of Ina dome (AS17-154-23672, AS17-159-23931). Arrows mark depressions along Ina's rim. (c) Profile across dome. Vertical exaggeration is about 6X. Trace marked in b.

from west to east. This suggests the original topographic slope was different from that of the present day. This evidence, coupled with the existence of the relatively raised Felicitatis plateau indicates a vertical displacement of several hundred meters for much of the maria in the region.

MORPHOLOGIC CHARACTERISTICS AND ORIGINS OF FEATURES

The dome. Topographic maps indicate that Ina is located on a dome which is about 15 km in diameter and which rises up to 300 m above the surrounding mare surface (Fig. 2). Its flanks slope at about 2–3°. Comparison with other analogous lunar domes, which range from 3–17 km in diameter (Head and Gifford, 1980), shows it to be among the largest on the moon. The dome's profile is somewhat asymmetric, perhaps the result of the tilting of the plateau. Atop the dome and along Ina's rim is a raised "collar". It is about 1.6 km wide and is bordered on the north and west by discontinuous graben-like depressions concentric to Ina (Fig. 2). Both on the collar and adjacent dome overlapping flow fronts indicate flow away from Ina.

Close examination of the dome reveals evidence for three lava surfaces of increasing age (i.e., increasing relative crater densities). The youngest is a very sparsely cratered unit found on the dome's summit. Another more densely cratered unit exists on the lower portions of the dome. A population of relatively large, flat-floored and apparently flooded craters suggests a third surface covered by thin younger flows. The dome as a whole exhibits a lower density of craters > 300 m than the surrounding mare units. Although the small areas involved and the flooded nature of many of the craters on the dome make age determinations by conventional crater counting methods unreliable and inconclusive, the relative crater densities indicate that the dome is younger than the mare material of the rest of the plateau, which has been mapped as Eratosthenian by Wilhelms and McCauley (1971).

The extrusive nature of the dome is supported both by its lower crater density and by its distinct color on the IR-UV photographs of Whitaker (1972b). These photographs are constructed by combining an IR positive and a UV negative. The process results in the enhancement of color differences. Color differences in the maria have been related to compositional variations (Whitaker, 1972b). On the IR-UV photos the dome correlates with a region that is bluer than the surrounding materials. This is in keeping with the observation made from orbit by the Apollo 17 astronauts that the Ina structure has a bluish tint (Evans and El-Baz, 1973).

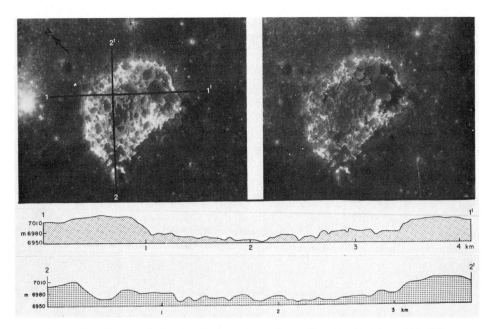

Fig. 3. Top. Stereo pair showing diverse units on floor of Ina (Apollo 15 pan 176, 181). Bottom. Profiles across Ina. Vertical exaggeration is about 3X. Data supplied by Defense Mapping Agency/Topographic Center.

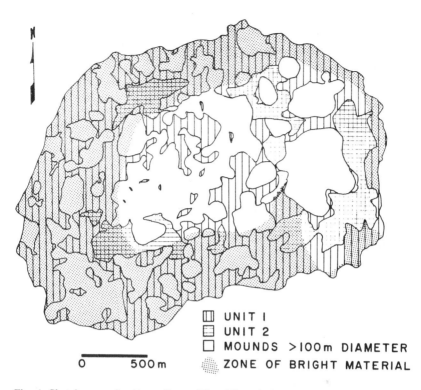

Fig. 4. Sketch map of units on floor of Ina. The stippled area marks the zone of the major occurrence of bright material.

Ina. Ina's diameter ranges from about 2.7 km along its straight edge to 2.9 km at its widest point. It is about 60 m deep. Its rim and walls are composed of mare basalts. Four units can be distinguished on the floor (Figs. 3 and 4): (1) A light-colored, low relief, roughly-textured unit covering much of the floor. This largely uncratered surface displays subdued interlocking polygonal hummocks and small sinuous scarps that may be flow features. (2) A dark-colored, hilly unit found predominantly along the eastern edge of the depression and in small patches on the floor. It is covered with rimless craters, often found at hill summits. (3) Smooth textured, extremely sparsely cratered protrusions, here referred to as mounds (Fig. 5). The mounds are often ringed by small depressions or "moats". (4) A unit found along most of the outer edge of the floor characterized by bright material around positive features and along the base of the wall. The zone in which the material occurs is bordered in places by patches of unit 2.

The morphologies of these units suggest that most or all are volcanic in origin. The possible flow fronts of unit 1 as well as its polygonal texture, which is reminiscent of the surfaces of terrestrial lava lakes dissected by cooling fractures, indicate an extrusive origin. The low albedo of unit 2 suggests a mantle of pyroclastic materials. The nature and origin of the mounds are discussed in detail below. The bright material of unit 4 is generally restricted to the outer edge of the depression. The variation between this outer portion of the floor and the central part may be compositional or physical. For example the bright material may mark a more highly fractured zone, which would be in keeping with its previous interpretation as sublimates deposited along cracks in the surface (Whitaker, 1972a). Alternatively, the bright unit may represent freshly exposed fine material eroded from high areas and collected in low-lying zones such as the moats around the mounds.

The Mounds. The material composing the mounds (unit 3) appears to be very similar to the mare basalts found on Ina's rim and walls. This similarity is substantiated by observations of the Apollo 17 astronauts that the mounds and the material surrounding Ina are the same color (Evans and El-Baz, 1973). Many of these structures are somewhat flat-topped while others are more rounded. The individual mounds have heights ranging from about 5 m to more than 25 m and areas from .003 km^2 to .26 km^2.

Some of the mounds appear to have summit craters (El-Baz, 1973). These craters resemble small vents as they are often roughly centrally located, rimless, and rayless. However, there is no consistent relation between summit crater diameter and mound diameter as has been noted by Head and Gifford (1980) for lunar mare domes in general. Some of the craters may, therefore, be fortuitous impacts. There is, however, a tendency for the craters to favor the tallest mounds. More than 80% of the mounds over 15 m high have the large summit craters, while less than 15% of those below 15 m do. This may result from the fact that craters on the smaller mounds may be too small to be resolved, or that they may be small enough to be easily filled.

Along the margins of Ina's floor the mounds are discrete, whereas in the center they coalesce to form one large complex exhibiting a concentric pattern (Fig. 5). It has been previously suggested by El-Baz (1973) that the mounds are extrusive features. Their preferential location in the lower central part of the depression and the concentric plan, which may reflect a fracture system, support this hypothesis.

Alternatively, however, the mounds may be analogous to the features called Dimmuborgir (Dark Castles) found near Mývatn, northern Iceland (Fig. 6), as was suggested by Wood *et al.* (1977). Although there are differences in morphology between the two sites (particularly, differences in scale and steepness of slopes), when photographs of similar resolution are compared, the features are quite similar in form. Despite the discrepancies, which might be explained by differences in the terrestrial and lunar environments, the Dark Castles are apparently the terrestrial features whose morphology most closely resembles that of the Ina mounds. As described by Barth (1950) the Dark Castles are "lava pillars" found in a circular depression at the center of a slightly raised area in a lava plain. The depression is about 15 m deep and the pillars are tall enough to reach the level of the surface surrounding the depression. Barth (1950) suggested the area was a lava lake that had drained leaving the pillars (areas that had already solidifed) standing. In support of this theory he cited vertical grooves similar to slickensides, and horizontal markings at different levels on the pillars as evidence

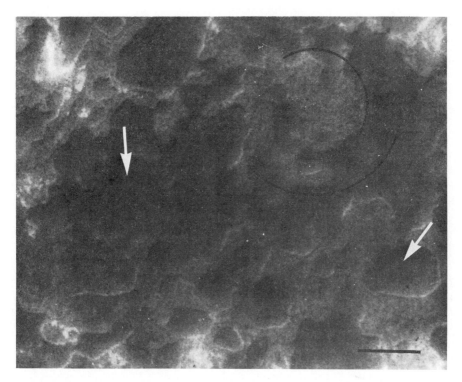

Fig. 5. Mounds on the floor of Ina. Note concentric pattern. Scale bar is about 200 m. Arrows indicate summit craters.

of the sinking of the lake's crust. Similarly, Wood (pers. comm.) suggests that the often flat tops and relatively steep slopes indicate that the Ina mounds may represent the remnants of a former surface which has partially collapsed. However, a difficulty exists in that the Ina mounds, unlike the Dark Castles, are not at similar levels. Topographic data indicate that their summits differ in elevation over a range of 25 m.

Another obstacle to this theory is the presence of moats around some of the mounds. Such features would not be expected in a simple collapse model. If the mounds were formed by extrusive flows, the loading of lava above and withdrawal below could cause the contemporaneous collapse of the surface and the formation of a depression in the vicinity of the mound. This is similar to a mechanism discussed by Schultz and Greeley (1976) for the formation of ring-moat structures on the lunar maria (see also McKee and Stradling, 1970, which describes the terrestrial analogue). We are not proposing that the Ina mounds are comparable to these ringed domes, nor that the mechanisms are identical. We merely suggest that there may be some similarities. The ring-moat structure theory involves the squeezing up of lava as a result of the sinking of the mare above. This process alone would not be sufficient to account for the considerable heights of the Ina mounds which might be better explained by continuous extrusions.

Fig. 6. (a) Arrows indicate location of Dimmuborgir on this aerial view. The large crater at the top of the photo is about 1 km in diameter. (Photo courtesy of C. A. Wood). (b) Lava pillars of Dimmuborgir (photo courtesy of C. A. Wood).

Although in some places the mounds appear connected to the mare unit on the wall and rim, they do not represent the same surface. The two units have different crater densities, the mounds appearing younger than the wall and rim material. There is no evidence, such as a high lava mark along the wall to indicate the former existence of a younger surface within the depression. Considering the distribution of the mounds, the heights of their summits, the presence of moats, the distribution of summit pits, and the lack of high lava marks, we favor the hypothesis that the mounds formed as discrete extrusive features.

CONCLUSION

Ina is an apparently unique volcanic feature. It rests on a dome displaying evidence of episodic extrusive activity. Its odd D-shape is probably related to a local fracture pattern, since the straight edge is aligned with other small lineaments in the area (El-Baz, 1973). Ina's floor exhibits several volcanic units including unusual mounds which we propose are small extrusive features. The exceedingly low crater density of the mounds (less even than that of the youngest unit on the dome) suggests that they may represent some of the youngest volcanism on the moon (El-Baz, 1972). Units on Ina's floor appear to be the culmination of a series of extrusive events spanning a considerable time period, as evidenced by the sequence of decreasing crater densities noted on the Ina structure and dome. The unusual morphology of the floor units may indicate changes in composition of the magma through this time span by differentiation in a localized magma source.

The extensive fracturing of the Ina region, caused by its proximity to two of the largest nearside impact basins, may have allowed abundant extrusion of lavas and the later collapse and subsidence of some mare units, as evidenced by their obvious vertical displacement. The unique morphology of Ina may have also been influenced by this structural setting, the numerous fractures perhaps serving as extrusive outlets. The diverse volcanic units on Ina's floor and the location of the depression on one of the largest lunar domes may reflect the ease with which the volcanic materials found their way to the surface. The correlation between dome formation and regions at the intersection of basins has been observed elsewhere on the moon. Head and Gifford (1980) noted that two of the three regions where lunar domes are found to be most highly concentrated (the areas around the craters Cauchy and Hortensius) lie near basin intersections. The combination of the possible existence of the differentiated magma source and the easy access of this magma to the surface may help to explain the unique nature of the Ina structure.

We believe the dome on which Ina rests is somewhat comparable to low terrestrial shields, the possibility of which was discussed by Pike (1978). Head and Gifford (1980) and Wood (1979) have proposed that lunar mare domes in general may be similar to these terrestrial structures. By comparing topographic dimensions, Pike (1979) showed that mare domes for which good Apollo topographic data exist do not closely resemble any terrestrial volcano class, but they have similarities to the low basaltic shields. Comparison with Pike's (1978) data shows that the Ina dome is morphologically similar to terrestrial shields. The diameter and height of the dome and the depth and depth/diameter ratio of the summit crater are all within the range of values measured for the terrestrial shields. Only the diameter of the summit crater exceeds the terrestrial range, although the diameter of the dome is also larger than most terrestrial examples. The ratio of summit crater diameter to dome diameter is higher than the average terrestrial case (Wood, 1979). However, Wood (1979) has suggested that differences between the diameters of such summit craters may be explained by the difference in gravitational acceleration for the earth and the moon, and consequently need not preclude similar origins for the features.

Ina's location at the summit of an extrusive dome, evidence of flows and fractures along the rim, and the morphology of the diverse floor units confirm its volcanic origin. It is questionable, however, if the term caldera, which has been previously proposed for the feature, is entirely appropriate. Since differences in the lunar and terrestrial environments may effect the morphologies of features formed by similar processes, it is not odd that it is difficult to place Ina strictly into any terrestrial classification scheme. However, it is a caldera in the sense that it is a volcanic crater larger than 1 mile in diameter, as defined by MacDonald (1972). As a shallow, single depression, however, it does not appear to have the scale or complexity often attributed to a true caldera.

Acknowledgments—The authors wish to thank C. A. Wood, T. A. Maxwell, J. Fink, J. L. Whitford-Stark, and P. Mouginis-Mark for their helpful comments. This work was performed under NASA Grant NSG-7188.

REFERENCES

Barth T. F. W. (1950) Volcanic geology, hot springs and geysers of Iceland. *Carnegie Inst. Wash., Publ. 587.* 174 pp.

El-Baz F. (1972) New geological findings in Apollo 15 photography. *Proc. Lunar Sci. Conf. 3rd,* p. 39–61.

El-Baz F. (1973) "D-Caldera": New photographs of a unique feature. *Apollo 17 Prelim. Sci. Rep.,* NASA SP-330, p. 30–13 to 30–16.

Evans R. E. and El-Baz F. (1973) Geological observations from lunar orbit. *Apollo 17 Prelim. Sci. Rep.,* NASA SP-330, p. 28–1 to 28–24.

Head J. W. and Gifford A. W. (1980) Lunar Mare Domes: Classification and modes of origin. *Moon and Planets* 22, 235–258.

MacDonald G. A. (1972) *Volcanoes.* Prentice-Hall, N.J. 510 pp.

McKee B. and Stradling D. (1970) The sag flowout: A newly described volcanic structure. *Bull. Geol. Soc. Amer.* 81, 2035–2043.

Pike R. J. (1978) Volcanoes on the inner planets: Some preliminary comparisons of gross topography. *Proc. Lunar Planet. Sci. Conf. 9th,* p. 3239–3273.

Schultz P. H. and Greeley R. (1976) Ring-moat structures: Preserved flow morphology on lunar maria (abstract). In *Lunar Science VII,* p. 788–790. The Lunar Science Institute, Houston.

Whitaker E. (1972a) An unusual mare feature. *Apollo 15 Prelim. Sci. Rep.,* NASA SP-289, p. 25–84 to 25–85.

Whitaker E. (1972b) Lunar color boundaries and their relationship to topographic features: A preliminary survey. *The Moon* 4, 348–355.

Wilhelms D. E. and McCauley J. F. (1971) Geologic map of the near side of the moon. U.S. Geol. Survey, Misc. Geol. Inv. Map I-703.

Wood C. A. (1979) Monogenetic volcanoes of the terrestrial planets. *Proc. Lunar Planet. Sci. Conf. 10th,* p. 2815–2840.

Wood C. A., Whitford-Stark J. L., and Head J. W. (1977) Iceland Field Itinerary. Basaltic Volcanism Study Project, Contrib. no. 6. Brown University, Providence.

Proc. Lunar Planet. Sci. Conf. 11th (1980), p. 2447–2462.
Printed in the United States of America

Ridge systems of Caloris: Comparison with lunar basins

Ted A. Maxwell and Ann W. Gifford

Center for Earth and Planetary Studies, National Air and Space Museum, Smithsonian Institution,
Washington, D.C. 20560

Abstract—Ridges within the Caloris basin on Mercury display several of the traits that are characteristic of mare-type ridges in multi-ring basins on the moon. Among these similarities are the predominance of basin-concentric and radial orientations, and the location of ridges inside topographic benches at the edge of the basin. The ratio of the most prominent ridge-ring diameter to basin rim diameter for Caloris is consistent with that of lunar multi-ring basins. However, if it is assumed that the major circumferential ridge systems formed directly over peak rings, then there does not seem to be a simple progression in the diameter of peak rings from unfilled lunar basins to multi-ring basins.

Based on comparable measurements of Caloris and the eastern portions of lunar multi-ring basins, both the orientation and frequency of ridges in lunar basins indicate the effect of global-scale, east-west compression. In the northern and southern sectors of lunar basins, globally induced stress dominated over that produced by basin downdropping, and caused ridge systems to deviate into the highlands, perhaps controlled by older fracture systems. In contrast to the ridge systems of lunar multi-ring basins, the strong basin-related trends within Caloris support previous studies that have suggested a post contraction/despinning(?) time of formation for structural features within the Caloris basin.

INTRODUCTION

The wrinkle ridge systems of the Caloris basin on Mercury display many of the traits that are characteristic of ridges in the mare-filled lunar multi-ring basins. Because of limitations of Mariner 10 image resolution, it is unknown whether small-scale features such as the crenulated crests and the dominant "ropy" type of morphology associated with lunar mare ridges are present within Caloris. Nonetheless, the broad, arch-like appearance, concentric and radial orientations, and predominant localization within the main ring of the Caloris basin suggest that these features are analogous to lunar mare ridges and arches. Among the several hypotheses that have been advanced to explain the origin of lunar mare ridges, a tectonic mode of formation has been preferred on the basis of several previous studies (Muehlberger, 1974; Maxwell *et al.*, 1975; Lucchitta, 1976, 1977; Maxwell, 1978). Assuming an origin by horizontal compression, Solomon and Head (1980) used the placement of mare ridges in lunar basins in order to model thickness variations of the elastic lunar lithosphere. Consequently, although the details of mare ridge formation are still not well known, the use of these features for tectonic models necessitates further study of their distribution and probable

2447

origin. In this study, we will concentrate on the ridge systems within the Caloris basin on Mercury, and implications for the origin of ridges within basins on both Mercury and the moon.

Previous studies of ridges and scarps on Mercury have been done primarily on a global scale, although several details of the Caloris ridges have not been overlooked. Strom *et al.* (1975) noted that ridges occur on both the hummocky and smooth plains surrounding Caloris, and that within Caloris, they are best developed in a zone 170 km wide, centered 120 to 140 km from the edge of the basin. Subsidence of the interior of Caloris was thought to be responsible for generating the compressive stress needed for ridge formation, although uplift and fracture of the central part of the basin may have occurred at a later time (Strom *et al.*, 1975). In an extensive study of ridges and scarps on Mercury, Dzurisin (1978) suggested that subsidence of Caloris may have been due to magma withdrawal from beneath the basin to form the later-emplaced smooth plains surrounding Caloris. Continuing isostatic adjustment to basin excavation was thought to be responsible for later uplift and the extensional fractures of the central part of the basin (Dzurisin, 1978).

The results of previous studies of mercurian scarps (Strom *et al.*, 1975; Cordell and Strom, 1977; Melosh and Dzurisin, 1978a; Dzurisin, 1978) all suggest that the tectonic features of the Caloris basin were formed *after* or during the waning stages of global thermal contraction or tidal despinning. Consequently, ridges and other tectonic features within Caloris should be primarily the result of basin-induced stress with a relatively minor component of a global stress field. Therefore, the purpose of this study is to compare the distribution and orientations of ridges within the Caloris basin to those of lunar basins in an attempt to provide photogeologic evidence for the relative effects of global versus basin-induced stress on both the Moon and Mercury.

RIDGE SYSTEMS OF THE CALORIS BASIN

Although broadly similar to the placement of ridges within lunar mare-filled basins, ridges in Caloris are more widely distributed than their lunar counterparts. Based on the 1:5,000,000 shaded relief map of the Caloris Planitia Area of Mercury (USGS, 1979), the main ring of the Caloris basin consists of a broken arc describing a basin diameter of 1420 km (centered at 30°N, 197°W; 40 km west of the center of the map projection used in Caloris map sheet). The main ring of ridges (inner-ring ridge system) occurs at a diameter of 1060 km (Fig. 1). In addition, a subtle ring in the central part of the basin (approximately 800 km in diameter) is represented by a highly-fractured, gentle topographic rise approximately 30 km wide.

The major morphological difference between ridges of Caloris and their lunar counterparts is that the narrow, sinuous ridges that occur on the crests of lunar mare arches appear to be absent on the structures within Caloris (Strom *et al.*, 1975). However, as pointed out by Malin (1978), the resolution of earth-based photographs of the moon is the most comparable to that of Mariner 10 images of

Fig. 1. Photomosaic of the Caloris Basin. Two inner-basin rings are indicated; innermost ring is composed of a gentle topographic arch 800 km in diameter. Prominent ring of ridges (second ring) is 1060 km in diameter.

Mercury. Examination of ridge systems on photographs from Kuiper *et al.* (1967) reveals that the narrow crestal ridges documented on Lunar Orbiter images and Apollo photographs appear severely degraded, but can still be recognized on properly oriented, sun-facing portions of the ridge. Similar narrow ridges are present in parts of the Caloris basin interior, where they generally follow the same orientations as the underlying arch although they deviate from side to side. Because of the varying resolution of Mariner 10 images, it is not possible to estimate the abundance of the smaller ridges within Caloris, but their existence in a few locations supports the analogy with lunar mare ridges.

In contrast to the occurrence of lunar mare ridges in single rings, the multiple concentric ridges of Caloris occur in a broad zone just within the main rim of the basin (Strom *et al.,* 1975; Dzurisin, 1978). They extend from a diameter of 1000 km to 1320 km (Fig. 2), and are accompanied by abundant fractures to a diameter range of 1300 km. The observation that these fractures crosscut the ridges led Strom *et al.* (1975) and Dzurisin (1978) to hypothesize later updoming of the center of the basin. Since no similar fractures are present in the central portions of lunar basins, it is possible that Caloris experienced a late-stage tectonic episode very different from the evolutionary trend of lunar basins. Nonetheless, an early episode of basin subsidence, consistent with the gross topography derived from isophotes (Hapke *et al.*, 1975), may have formed the Caloris ridges in a manner similar to that hypothesized for lunar ridges.

On the moon, the association of ridges with changes in mare level (Lucchitta,

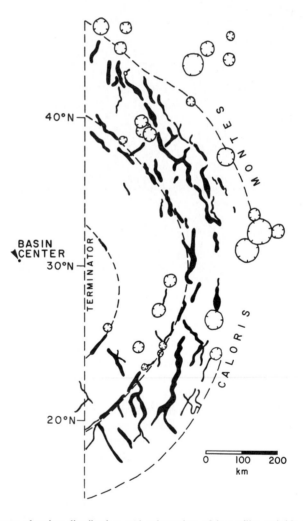

Fig. 2. Sketch map showing distribution and orientation of lunar-like wrinkle ridges within the Caloris Basin. Ridges are located predominantly between the second and third rings of the basin.

Fig. 3. Topographic benches and associated ridges in lunar basins (from earth-based photos) and Caloris (Mariner 10). A) Northwest Imbrium; Sinus Iridum (left side of picture) is topographically higher than central Mare Imbrium. B) Caloris; note bench (letter "B"), and graben-like trough similar to that of the Sulpicius Gallus region in southwestern Mare Serenitatis (C). D) Southeastern quadrant of the Humorum Basin; note basin-concentric systems of rilles and ridges and 80 km wide bench region.

1977), the edges of surface disc mascon models (Phillips *et al.*, 1972) and topographic benches surrounding the mascon basins all support an origin related to subsidence of the inner basin. Additional evidence for offsetting of subsurface mare contacts and probable normal faulting in Mare Crisium (Maxwell and Phillips, 1978), and subsidence of the inner part of Mare Serenitatis (Peeples *et al.*, 1978) indicates that downdropping took place before the latest period of mare basalt emplacement. Within Caloris, lunar-like topographic benches occur on both the northeastern and southeastern portions of the basin rim. In northeastern Caloris, a 70 km wide bench extends for 250 km along the arc of the basin, and is separated from the inner part of the basin by an upraised lip of small, discontinuous mare ridges (Fig. 3). A similar bench is present in southeastern Caloris, but here is only 30 km wide. Apparently, the later uplift of the central basin did not affect the bench regions, since the fractures occur basinward of the scarps.

As shown above, the morphology, basin-related setting and probable structural relationships with the benches in Caloris are similar to the setting of ridge systems of lunar mare basins. However, the use of Caloris ridges as an example of basin-induced stress is strongly dependent on the interpretation that these features formed after the period of mercurian global compression. Evidence for their relatively late-stage formation is based on both the global density of lobate scarps, and the relative age of the Caloris-related smooth plains. Cordell and Strom (1977) found that relatively few lobate scarps were found in the latitudes occupied by hilly and lineated terrain, which implied that any scarps that may have existed were destroyed at the time of the Caloris impact. They also noted, however, that several scarps are present on the post-Caloris smooth plains. Dzurisin (1978) also supported a post-contraction/despinning time of origin for Caloris, and attributed the scarps in the smooth plains surrounding Caloris to local gravitational adjustments. Separating the mercurian global lineaments from the arcuate scarps, Melosh and Dzurisin (1978a) found that very few lineaments and no arcuate scarps are present in the area affected by Caloris, and thus concluded that the Caloris basin formed after contraction ceased.

Additional evidence for the late-stage formation of Caloris-related structures is based on the time of emplacement of the Caloris smooth plains. Wood *et al.* (1977) noted that the smooth plains were emplaced after the beginning of the time of formation of Class 1 craters (LPL Classification; fresh craters). Thus, the material that is deformed by the ridges most likely was not present until well after the end of heavy bombardment.

In spite of these stratigraphic arguments for the formation of Caloris-related structures after global compression, it is still possible that the global lineament system may have affected the distribution and orientation of Caloris ridges. In addition to this possible inheritance factor, both Solomon (1977) and Strom (1979) note that Mercury should have experienced a relatively longer period of global compression than the other terrestrial planets. Consequently, the comparison of Caloris ridges with those of lunar basins should also allow an assessment of the relative importance of pre-basin fracture systems.

COMPARISON OF CALORIS WITH LUNAR BASINS

Comparison of Caloris basin structure with that of lunar basins is limited by three factors: 1) sun azimuth, which may hinder identification of east-west oriented structures; 2) low resolution of Mariner 10 images, when compared to Apollo metric and panoramic camera photographs and Lunar Orbiter images; and 3) images of only one-third of the Caloris basin are available for comparison with the more completely photographed lunar basins.

Due to the sinuous, segmented nature of both lunar and mercurian ridges, sun azimuth does not seem to be an important factor in biasing observations of ridge distribution and orientation. East-west oriented ridges can be seen on both the moon (in southern Mare Serenitatis and southern Mare Imbrium) and Mercury (Fig. 1), which suggests that their absence in other parts of the basins is real rather than an artifact of lighting. To compensate for the low resolution of Mariner 10 images, only those ridges and arches visible at the resolution of telescopic photographs (Kuiper *et al.*, 1967) were included for comparison with Caloris.

The most important limitation, the incomplete coverage of Caloris, is more difficult to contend with. In order to have a lunar data set comparable to that for Caloris, we have used only the eastern one-third of lunar multi-ring basins in our analysis of the distribution and orientation of ridges. Similar orientation data for entire lunar basins exist elsewhere in the literature (Strom, 1964; Elston *et al.*, 1971; Maxwell *et al.*, 1975; Fagin *et al.*, 1978), although there is no standardization for mapping or weighting procedures.

Ridge orientation

Two methods were used to study the orientation of ridges within Caloris and the lunar mare-filled basins of Crisium, Humorum, Serenitatis, and Nectaris. The orientation of each linear ridge segment was measured with respect to north, weighted according to its length (organized into number of kms of ridge per orientation), and plotted for the entire eastern one-third of the basins studied. In addition, these data were also broken down into 15° sectors, and plotted according

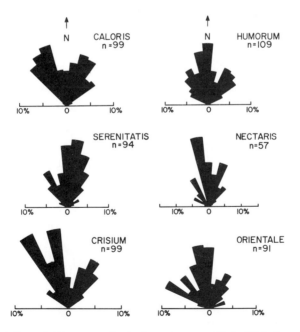

Fig. 4. Orientations of ridges occurring in the eastern one-third of Caloris and four lunar mare-filled basins. Hypothetical ridge orientations for Orientale are based on lineament trends and the few existing mare ridges that are present in the central part of the basin.

to the location of the sector. In order to provide some estimate on the expected subsurface structure within lunar basins, the orientations of lineaments, existing ridges, and areas of protruding topographic relief in the eastern part of the Orientale basin were measured in a manner similar to that used for Caloris and the mare-filled lunar basins. Because it is uncertain whether some of the smaller-scale ridges or lineaments might contribute to hypothetical Orientale ridge systems, these data were not broken down into 15° segments. Nonetheless, the major trends of Orientale are included for comparison with total (one-third) basin orientations.

As shown in the rose diagrams of ridge orientations for the eastern one-third of the basins (Fig. 4), Caloris shows much stronger maxima at N30°E and N30°W than do lunar basins. Although these orientations are consistent with fracture trends predicted by Melosh and Dzurisin (1978a) for a model of early tidal despinning of Mercury, the locations of these prominent trends indicate that they are imposed primarily by the geometry of the basin. In contrast, the orientations in Serenitatis, Humorum, Nectaris, and Orientale show maxima in a general northerly direction, while Crisium is strongly skewed to the northwest.

When plotted according to location within the basin, the reason for the NW and NE enhancement of ridge trends in Caloris is apparent (Fig. 5). The north-

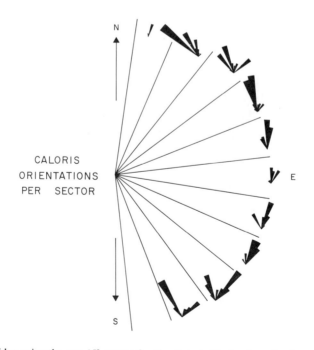

CALORIS
ORIENTATIONS
PER   SECTOR

Fig. 5. Ridge azimuths per 15° sector for the Caloris Basin. Length of rose diagram segments are normalized to total length of ridges in each sector. Note the prevalence of basin concentric and radial trends, and weak NS orientations in the eastern sectors of the basin.

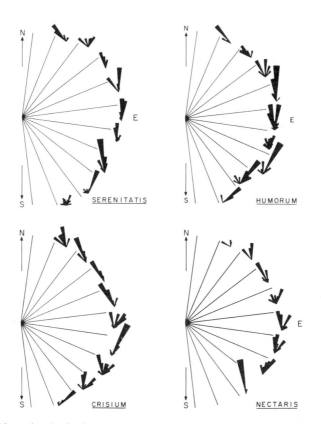

Fig. 6. Ridge azimuths for four lunar mare-filled basins; method of plotting is the same as used for Caloris (Fig. 5). Lunar ridges exhibit much greater variability in orientation than those of Caloris. Strong N-S trends are particularly well-developed in the Serenitatis and Humorum basins.

west trends are most strongly developed in the northeast where they are concentric to the basin, and in the southeast where they are radial to the basin. Similarly, the slightly weaker northeast mode is due to less prominent concentric ridges in the southeast, and radial ridges in the northeast. Both the low number of ridges in the eastern sector and the basinwide lack of north-trending ridges distinguish the Caloris orientation pattern from that of lunar basins (Fig. 6).

The orientations of ridges in the Caloris basin are more strongly concentric than those of either ridges in the flooded lunar basins studied here, or lineaments in the Orientale basin. In Caloris, 61% of all orientations measured are within 20° of being concentric to the center of the basin, and 19% are within 20° of being radial. For comparison with a flooded lunar basin, only 43% of the ridges in the eastern third of Serenitatis are within 20° of being concentric, and 16% are within 20° of being radial. In the relatively pristine Orientale basin, 43% of the lineaments, scarps and ridges are concentric, while 11% of these features form radial trends.

Consequently, the total orientations for the corresponding portions of Caloris and five lunar basins, and the orientations per sector of the basins both suggest that Caloris ridges are more closely related to basin-controlled radial and concentric orientations than are lunar basin ridge systems. On the basis of these data plotted for the entire eastern third of the basins studied here, the closest lunar analog to the Caloris ridge systems would be those of Crisium (Fig. 4). However, when broken down into sectors, it is apparent that the NW and NE modes of Crisium ridges are not created by strong radial components (Fig. 6).

Ridge distribution

A comparable treatment of the geographic distribution of Caloris and lunar basin ridges was done by: 1) comparing the diameter of the most prominent ring of ridges to that of the next outer ring of the basin, and 2) simply counting the number of ridges present along a line radial to the basin, in order to identify areas of maximum stress.

Relation to basin ring structure

Although much controversy surrounds both the mode of origin of lunar basin rings and the meaning of various ratios of ring diameters, the concentric pattern of ridge systems and their extension as highland fractures indicates involvement of the pre-mare surface, which was formed as a result of impact and later basin modification. In fact, the location of the lunar ridge systems has traditionally been used to define the inner-rings of mare-filled basins (Wilhelms and McCauley, 1971). Hartmann and Wood (1971) advocated a genetic relationship between peak rings and wrinkle ridges, although they attributed ridge formation to intrusion and extrusion of mare basalt. Nonetheless, they found a somewhat consistent ratio between the diameter of the prominent circular ridge system and the next largest basin rim (see Table VI in Hartmann and Wood, 1971). Brennan (1976), in a review of basin ring spacings, found no consistent ratio for outer rings, but did find an extremely high correlation between the diameter of the ridge ring and basin rim for five lunar mare-filled basins. Head (1977) believed that the position of the concentric mare ridge system marked the location of the central peak ring (termed "central peak ring" in Head, 1977) in lunar multi-ringed basins, and noted that Imbrium exhibits both the ridge system and numerous peak ring segments.

In order to examine the role of basin ring structure in controlling ridge location in both Caloris and lunar basins, we first extrapolated the best-fit curve of peak ring diameter (Dr, for peak-ring basins) vs. crater diameter (Db, for the next largest ring) to multi-ring basins. Based on the 17 lunar peak-ring basins documented by Wood and Head (1976), the best-fit linear relationship has a correlation coefficient of 0.97, but substantially underestimates the diameters of the ridge ring in multi-ring basins. If multi-ring basins are considered separately from peak-ring basins, two equally good fits can be made to both sets of data (Fig. 7). The

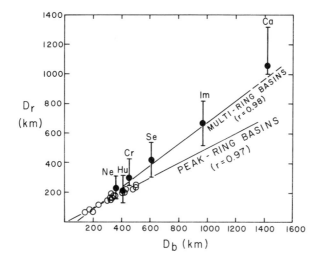

Fig. 7. Comparison of peak-ring and ridge-ring diameters (Dr; vertical axis) with crater or basin rim diameter (Db). Peak rings occur at smaller diameters than ridge rings in multi-ring basins, suggesting two distributions. Vertical line shown for multi-ring basins (name abbreviated above the line) indicates range of location of ridges.

slope of the line for peak-ring basins (0.51) is only slightly less than that found by Head (1978) for peak-ring basins on the Moon, Mars and Mercury. The relatively steeper slope of the regression line for multi-ring basins is consistent with their generally higher ridge-ring/crater rim ratios. A statistical basis for separating the two(?) ring distributions must await further studies detailing the variation in peak-ring location. It is possible that using the outer or inner edges of the peak ring may substantially change the best fit lines used here.

On the basis of ring geometry alone, therefore, ridge rings do not mark the predicted location of buried peak rings. This discrepancy in ring locations could result from: 1) peak rings forming at greater diameters in multi-ring basins as opposed to smaller basins and craters; 2) modification-stage effects during multi-ring basin formation which may act to retain the rim crest in its original position or move it inward; or 3) ridges forming only in near surface units, and having no relationship to the underlying structure. Although both basement influence and bending-stress solutions are consistent with the basin-related setting, the consistent ratio of ridge ring to basin rim would be fortuitous in light of the variations in load experienced by the mare-filled basins.

Several photogeologic observations support the first two hypotheses. According to Head (1977), the discrepancy in ridge ring/crater rim ratios may be due to the decreased role of terracing in the second ring of multi-ring basins, or to megaterrace formation outside the rim crest which acts to shift the rim crest inward. Both processes could explain the relatively larger ridge ring/crater rim ratios for multi-ring basins. In Mare Imbrium, isolated highland peaks protrude through the mare at a slightly larger diameter than the ridge systems, suggesting the influence of a buried peak ring. In several smaller craters on the moon (e.g., Letronne, the Flamsteed ring, Reiner R) the rim is defined both on the basis of ridges and isolated peaks of an original rim. In addition, mare ridges that deviate into the highlands are characterized by scarps and highland sculpture that suggests the influence of structure beneath the mare.

As shown in Fig. 7, the ridge ring of the Caloris basin lies close to the regression line for lunar multi-ring basins and their associated ridge systems. Despite the different tectonic history implied by the fracturing (and uplift?) of the central Caloris basin, this correlation provides further support for basin-control of the location of the ridge system. This result is also consistent with the conclusions of Wood and Head (1976) and Head (1978), who found no planet-specific variations between peak ring diameter and crater diameters.

Ridge density in azimuth

As shown in the histograms of number of ridges per azimuthal zone (Fig. 8), ridge systems of Caloris have a bimodal distribution resulting from relatively few ridges in the eastern sector. This region of relatively few ridges coincides with the area noted by Strom *et al.* (1975) where the rim of the Caloris basin is poorly developed. The lack of compressional stress in this sector could result from the absence of lateral confinement during subsidence of the basin.

Lunar basins, however, are either nonmodal (as in Crisium), or tend to approach a unimodal distribution with the greatest number of ridges in the eastern sectors. Even within the corresponding sectors of the coverage of Caloris, the decrease in the number of ridges in the northern and southern sectors is evident, but this tendency is even more pronounced if the coverage is increased to show the north and south sectors of the lunar basins (Fig. 8).

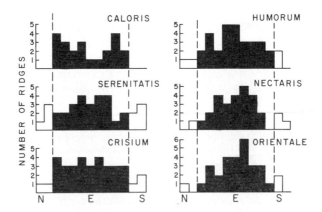

Fig. 8. Histograms showing the number of ridges per azimuth for Caloris and lunar multi-ring basins (represented in 15° bins). Vertical dashed line and filled-in portion of the histograms indicates the extent of Mariner 10 images of Caloris. As a result of deviations into neighboring highlands, relatively few ridges are present in the northern and southern sectors of lunar basins.

Fig. 9. Deviations of mare ridges into highlands bordering the northern edge of Imbrium (A), Nectaris (B), Serenitatis (C), and the southern edge of Crisium (D). Note the absence of a well-defined circumferential ridge system in the northern sectors of Imbrium, Nectaris and Serenitatis.

For reasons stated above, we believe that the low number of ridges in the northern and southern sectors of lunar basins is not an artifact of E-W lighting, but results from a combination of structural enhancement of N-S trending ridges that continue as highland lineaments (Fig. 9), and a lack of development of circumferential ridges in the northern and southern sectors. Stratigraphically, it is impossible to tell whether the highland scarps are older, reactivated structures, or were formed at the same time as the mare ridges themselves. Assuming a homogeneous tectonic response across the mare-highland contact, the ridges that continue into the highlands could be due to global stress combining with basin-induced hoop stress enhancing the N-S trends. However, since the observed ridge patterns are not truly radial to the basins, and their extensions as highland scarps and lineaments occur outside the range of significant load-induced compression (Solomon and Head, 1980), we believe that the contribution of basin-induced hoop stress was relatively minor compared to that of global compression localized by pre-existing fractures.

The lack of well-defined circumferential ridges and topographic benches in the northern and southern sectors of several lunar basins seems to contradict evidence presented above for control by pre-existing basin structure. Presumably, any impact-related ring structure would tend towards circular symmetry, so there is no *a priori* reason to expect a lack of ridges. Consequently, it is most likely that global stress dominated in these sectors and resulted in reactivation of old fracture trends.

CONCLUSIONS

Comparison of the ridge system of Caloris with those of lunar multi-ring basins indicates that the early evolution of Caloris was similar to that of lunar mascon basins. The morphology of ridges within Caloris compares favorably with lunar ridges when viewed on similar resolution earth-based lunar photographs. Ridges in Caloris occur from 1000 to 1320 km diameter range, and are situated within the boundary delineated by topographic benches in the northeastern and southeastern parts of the basin. The orientation of Caloris ridges is more dominantly concentric than ridge orientations in lunar basins, which supports an origin for Caloris ridges after the major effects of global contraction and/or despinning. In addition, the location of the most prominent ridge-ring in Caloris is similar to that of lunar basins, although the peak ring may not directly underlie the ridge ring. In contrast to lunar basins, however, ridges in Caloris are more abundant than in lunar basins, resulting from either material property differences, or the possibility that we are seeing an early stage of basin settling not covered by later mare basalts as on the moon.

In lunar basins, the effect of global east-west compression is most evident in the northern and southern sectors of the basins, where ridges are continuous with highland scarps and lineaments. Here, global-scale compression dominated over that induced by basin subsidence, resulting in the reactivation of pre-mare fractures. In order to account for the enhancement of the circumferential trends in the eastern and western sectors of lunar basins, and the lack of these trends in the northern and southern sectors, we believe that a significant component of global-scale compression must have been present at the time of ridge formation. Unfortunately, images of the key northern and southern sectors of Caloris do not exist, but based on the distribution and orientation of ridges in the available one-third of the basin, our results provide further evidence that the ridge systems of Caloris developed primarily by basin-induced stress.

The difference in tectonic style between Caloris and lunar basins is also expressed outside the basin rim. In lunar basins, circumferential graben most likely resulted from subsidence of the basin interior in response to basalt loading and thus producing extensional stress surrounding the basins (Solomon and Head, 1980). The lack of graben surrounding Caloris has been attributed to the overriding effect of global compression acting to inhibit extensional faulting (Cordell and Strom, 1977). However, it is also possible that Caloris did not receive the same magnitude of excess mass (volcanic fill?) as lunar basins (Melosh and Dzurisin, 1978b) or that the fill was not basin-wide in extent. The bulk of Caloris fill may have been confined within the zone represented by the most prominent ridge systems, centered around the second ring (McKinnon, 1979). Thus, the radial and concentric fractures of the central Caloris basin could have formed during the latest stages of subsidence, by extensional stress across the central basin created by subsidence of an annular load. Because of later filling, it is unknown whether lunar basins went through a similar history of loading and tectonic responses.

Acknowledgments—We thank J. W. Head, S. C. Solomon, J. L. Whitford-Stark and C. A. Wood for their helpful reviews. Mariner 10 images used in this study were provided by NSSDC. This research was supported by the National Air and Space Museum, Smithsonian Institution, and by NASA Grant NSG-7188.

REFERENCES

Brennan W. J. (1976) Multiple ring structures and the problem of correlation between lunar basins. *Proc. Lunar Sci. Conf. 7th,* p. 2833–2843.

Cordell B. M. and Strom R. G. (1977) Global tectonics of Mercury and the Moon. *Phys. Earth Planet. Inter.* **15,** 146–155.

Dzurisin D. (1978) The tectonic and volcanic history of Mercury as inferred from studies of scarps, ridges, troughs, and other lineaments. *J. Geophys. Res.* **83,** 4883–4906.

Elston W. E., Laughlin A. W., and Brower J. A. (1971) Lunar nearside tectonic patterns from Orbiter IV photographs. *J. Geophys. Res.* **76,** 5670–5674.

Fagin S. W., Worrall D. M., and Muehlberger W. R. (1978) Lunar mare ridge orientations: Implications for lunar tectonic models. *Proc. Lunar Planet. Sci. Conf. 9th,* p. 3473–3479.

Hapke B., Danielson G. E. Jr., Klaasen K., and Wilson L. (1975) Photometric observations of Mercury from Mariner 10. *J. Geophys. Res.* **80,** 2431–2443.

Hartmann W. K. and Wood C. A. (1971) Moon: Origin and evolution of multi-ring basins. *The Moon* **3,** 3–78.

Head J. W. (1977) Origin of outer rings in lunar multi-ringed basins: Evidence from morphology and ring spacing. In *Impact and Explosion Cratering* (D. J. Roddy, R. O. Pepin, and R. B. Merrill, eds.), p. 563–573. Pergamon, N.Y.

Head, J. W. (1978) Origin of central peaks and peak rings: Evidence from peak-ring basins on Moon, Mars, and Mercury (abstract). In *Lunar and Planetary Science IX,* p. 485–487. Lunar and Planetary Institute, Houston.

Kuiper G. P., Whitaker E. A., Strom R. G., Fountain J. W., and Larson S. M. (1967) Consolidated lunar atlas, Supp. nos. 3 and 4 to USAF photographic lunar atlas. *Contr. Lunar and Planetary Lab., Univ. Ariz.* **4,** 24 pp. 226 photographs.

Lucchitta B. K. (1976) Mare ridges and related highland scarps—Result of vertical tectonism? *Proc. Lunar Sci. Conf. 7th,* p. 2761–2782.

Lucchitta B. K. (1977) Topography, structure, and mare ridges in southern Mare Imbrium and northern Oceanus Procellarum. *Proc. Lunar Sci. Conf. 8th,* p. 2691–2703.

Malin M. C. (1978) Surfaces of Mercury and the moon: Effects of resolution and lighting conditions on the discrimination of volcanic features. *Proc. Lunar Planet. Sci. Conf. 9th,* p. 3395–3409.

Maxwell T. A. (1978) Origin of multi-ring basin ridge systems: An upper limit to elastic deformation based on a finite-element model. *Proc. Lunar Planet. Sci. Conf. 9th,* p. 3541–3559.

Maxwell T. A., El-Baz F., and Ward S. H. (1975) Distribution, morphology and origin of ridges and arches in Mare Serenitatis. *Bull. Geol. Soc. Amer.* **86,** 1273–1278.

Maxwell T. A. and Phillips R. J. (1978) Stratigraphic correlation of the radar-detected subsurface interface in Mare Crisium. *Geophys. Res. Lett.* **5**, 811–814.

McKinnon W. B. (1979) Caloris: Ring load on an elastic lithosphere (abstract). *EOS (Trans. Amer. Geophys. Union)* **60**, 871.

Melosh H. J. and Dzurisin D. (1978a) Mercurian global tectonics: A consequence of tidal despinning? *Icarus* **35**, 227–236.

Melosh H. J. and Dzurisin D. (1978b) Tectonic implications for the gravity structure of Caloris Basin, Mercury. *Icarus* **33**, 141–144.

Muehlberger W. R. (1974) Structural history of southeastern Mare Serenitatis and adjacent highlands. *Proc. Lunar Sci. Conf. 5th,* p. 101–110.

Peeples W. J., Sill W. R., May T. W., Ward S. H., Phillips R. J., Jordan R. L., Abbott E. A., and Killpack T. J. (1978) Orbital radar evidence for lunar subsurface layering in Mare Serenitatis and Crisium. *J. Geophys. Res.* **83**, 3459–3468.

Phillips R. J., Conel J. E., Abbott E. A., Sjogren W. L., and Morton J. B. (1972) Mascons: Progress toward a unique solution for mass distribution. *J. Geophys. Res.* **77**, 7106–7114.

Solomon S. C. (1977) The relationship between crustal tectonics and internal evolution in the Moon and Mercury. *Phys. Earth Planet. Inter.* **15**, 135–145.

Solomon S. C. and Head J. W. (1980) Lunar mascon basins: Lava filling, tectonics, and evolution of the lithosphere. *Rev. Geophys. Space Phys.* **18**, 107–141.

Strom R. G. (1964) Analysis of lunar lineaments, I: Tectonic maps of the Moon. *Contr. Lunar and Planetary Lab, Univ. Ariz.* **2**, 205–216.

Strom R. G. (1979) Mercury: A post-Mariner 10 assessment. *Space Sci. Rev.* **24**, 3–70.

Strom R. G., Trask N. J., and Guest J. E. (1975) Tectonism and volcanism on Mercury. *J. Geophys. Res.* **80**, 2478–2507.

U.S.G.S. (1979) Shaded relief map of the Caloris Planitia area of Mercury. U.S. Geol. Survey Misc. Geol. Inv. Map I-1172, 1:5,000,000.

Wilhelms D. E. and McCauley J. F. (1971) Geologic map of the near side of the Moon. U.S. Geol. Survey Misc. Geol. Inv. Map I-703, 1:5,000,000.

Wood C. A. and Head J. W. (1976) Comparison of impact basins on Mercury, Mars and the Moon. *Proc. Lunar Sci. Conf. 7th,* p. 3629–3651.

Wood C. A., Head J. W., and Cintala M. J. (1977) Crater degradation on Mercury and the Moon: Clues to surface evolution. *Proc. Lunar Sci. Conf. 7th,* p. 3503–3520.

Proc. Lunar Planet. Sci. Conf. 11th (1980), p. 2463–2477.
Printed in the United States of America

Lunar cold traps and their influence on argon-40

R. Richard Hodges, Jr.

The University of Texas at Dallas, P. O. Box 688, Richardson, Texas 75080

Abstract—In polar areas of the moon the maximum temperatures reached in some permanently shaded areas are well below the temperature required to retain water ice for billions of years, and cold enough to hold other volatiles for shorter periods. Aside from water, the most significant lunar volatiles are the radiogenic gases, of which argon-40 is the most easily detected, both *in situ* and as retrapped ions in rocks returned from the surface on the moon. Argon-40 escapes from the moon at a surprisingly high rate that is between 3% and 6% of its total production. Its brief lifetime in the lunar exosphere is marked by numerous adsorption/desorption events. Collisions with the lunar surface in cold, permanently shaded areas lead to long term storage, forming reservoirs of trapped gas that may be disturbed occasionally to produce sudden increases in atmospheric argon. It is postulated that this may explain at least part of the time variations in Apollo 17 mass spectrometer measurements of argon that were previously attributed to internal processes associated with the release of radiogenic gases from the moon.

INTRODUCTION

The first evidence of an atmosphere on the moon was the excess argon-40 discovered in Apollo 11 soils. Heymann and Yaniv (1970) noted that the apparent K-Ar ages of these samples, about 18 b.y., requires a source other than the decay of potassium within the soil grains. They, along with Manka and Michel (1970), argued further that the only reasonable source of the excess argon-40 is the lunar atmosphere, where ions, formed by solar radiation or charge exchange with solar wind ions, are accelerated by induced electric fields in the solar wind. Some of the atmospheric ions are driven into the surface of the moon with enough energy to be implanted in rocks.

Heymann and Yaniv (1970) calculated that the rate of trapping of argon-40 in lunar rocks is roughly 4×10^2 atoms cm^{-2} sec^{-1}, implying a total argon trapping rate in excess of 1.5×10^{20} atoms sec^{-1}. Neutral argon-40 lunar exosphere models fitted to the *in situ* measurements of argon made by the mass spectrometer at the Apollo 17 site suggest a total exospheric argon loss rate due to ionization of about $1.4 \cdot \times 10^{21}$ atoms sec^{-1}. The ion trajectory calculations of Manka and Michel (1971) show that almost half of the atmospheric argon ions should strike the lunar surface. Hence the mass spectrometer data require that the present flux of argon-40 ions to the lunar surface be roughly half the total loss rate, i.e., on the order of 7×10^{20} sec^{-1}. Because the fraction of these ions that attain sufficient energy to be implanted in rocks on the surface of the moon must be somewhat less than unity, the present rate of entrapment of argon-40 is probably on the order of 10^{20} atoms sec^{-1}, and hence similar in magnitude to the long term trapping rate determined from analyses of returned samples.

The argon-40 that forms a significant part of the lunar exosphere arises as the result of radioactive decay of potassium within the moon. If the average potassium abundance in the moon is assumed to be 100 ppm as suggested by Taylor and Jakeš (1974), the present rate of production of ^{40}Ar within the moon is 2.5 $\times$ 10^{22} atom sec^{-1}. Thus the present rate of loss of argon ions from the lunar atmosphere, as determined from *in situ* measurements, is 6% of the total production in the moon. Because the acceleration of ions by the induced electric fields in the solar wind causes at least half of them to escape, it is necessary that more than 3% of the all argon atoms effuse from the interior of the moon to supply the net exospheric loss of argon-40.

To sustain argon loss from the interior of the moon at a rate of 3% (or greater) of its production over a long time it is necessary that an efficient mechanism release argon, and other radiogenic gases, from solid rocks as they are produced. Speculation concerning the origin of the lunar atmospheric argon can be found in Hodges and Hoffman (1974) and Hodges (1975, 1977). Briefly, it has been argued that the source region is not near the lunar surface. Certainly, if the regolith were the source it would be depleted of argon rather than having an excess.

Models of the lunar interior generally place most of the potassium at shallow depths as a result of early differentiation (cf. Taylor and Jakeš, 1974). Assuming a potassium abundance of 600 ppm in the depth range of 0 to 25 km, the escaping argon amounts to 12% of production. On earth, about 10% of the argon is released, presumably by crustal weathering and tectonic processes that are essentially nonexistent on the moon. However, it has been noted in Hodges (1977) that there are apparent correlations of temporal increases in total atmospheric argon on the moon with the shallow moonquakes detected by the ALSEP seismic network (Nakamura *et al.*, 1974).

The most reasonable hypothesis on the origin of the atmospheric argon is that it diffuses from localized hot source regions that hold at least a few percent of the moon's potassium. Time variations in the argon abundance seem to require that the source region be compact. A strong argument has been made that these criteria place important constraints on lunar evolution models (cf. Hodges, 1977). In particular, the argon data favors a model in which accretional heating differentiated the outer part of the moon, leaving a central asthenosphere composed of primitive lunar materials, including K, U and Th, as proposed by Taylor and Jakeš (1974).

While the true nature of the origin of the escaping argon remains uncertain, there are theoretical reasons to suggest that adsorption processes on the lunar surface may be responsible for some of the time variations of argon-40 detected by the Apollo 17 mass spectrometer. If so, some of the constraints on the source identification problem may be removed, although the average rate of argon supply cannot change.

Previous attempts to relate the atmospheric argon measurements to source and escape rates have used a smooth moon temperature model that, in effect, averaged out the enormous lateral temperature gradients caused by shadows of large surface features. This approach is a satisfactory approximation except at high latitudes where permanently shaded regions may act as cold traps, holding adsorbed argon atoms for long periods, and possibly releasing them sporadically in response to surface movements caused by shallow seismic activity or slumping, or to seasonal changes in solar illumination. The idea of accumulating volatiles in polar cold traps was addressed by Watson *et al.* (1961), who found that a large fraction of the lunar surface (about 0.5%) is never exposed to sunlight, and that most of this area is always at temperatures below 120 K. These conclusions have been confirmed by Arnold (1979) using terminator photographs obtained from lunar orbiting spacecraft. Arnold has suggested that 10^{16} to 10^{17} grams of water, accreted from meteor and comet impacts, formed by solar wind reduction of Fe in the soil, and possibly degassed from primitive materials, have accumulated in cold traps as solid ice.

If there is reason to believe that water may be trapped in the lunar regolith for billions of years, then it is essential to consider the effects of the cold traps on argon-40. In subsequent discussion it is shown that the maximum temperature allowable for efficient argon trapping is around 40 K on surfaces contaminated by water, and nearly 70 K on clean rocks. This makes the area available for argon storage much less than that for water, necessitating a careful examination of the distribution of surface temperature in the shaded areas produced by orographic features on the lunar surface.

ADSORPTION ON THE LUNAR SURFACE

There are two important parameters that characterize an encounter of a volatile atom with a regolith surface: the time that the atom resides on each soil grain it strikes (i.e., the time between adsorption and desorption), and the number of grains that it encounters before reemerging from the regolith on an exospheric (ballistic) trajectory. The usual representation of the mean sticking time for atoms on a surface at temperature T is

$$\langle t \rangle = t_0 e^{Q/RT} \tag{1}$$

where Q is activation energy, R is the gas constant, and t_0 is a characteristic time (which may be a function of temperature). The probability of desorption after time t obeys an extinction relation of the form $\exp(-t/\langle t \rangle)$.

The number of collisions an atom makes with soil grains during one regolith encounter has a probability distribution that is similar to the first return to zero of a one dimensional random walk. If z is depth measured in random walk step lengths, and n is the number of collisions (a necessarily odd number), the probability of emerging (z > 0) after exactly n collisions is

$$p_n = \frac{2^{-n}(n-1)!}{\left(\frac{n+1}{2}\right)!\left(\frac{n-1}{2}\right)!}$$ (2)

(c.f. Feller, 1960). This function decreases gradually with increasing n, becoming approximately

$$P_n = \sqrt{2/\pi n^3}$$ (3)

as n becomes large. It is obvious that because P_n decreases as $n^{-3/2}$, the first moment of the distribution of n is a divergent sum, and hence there is no definable mean number of collisions.

A major deficiency of the one dimensional random walk approximation is that it fails to account for impinging ballistic atoms that pass between surface layer soil grains, making first collisions at finite depths, so that subsequent upward departures do not automatically lead to emergences. Hence Eq. (2) must overestimate true probabilities for small n.

To determine the possible implications of initial surface penetration, two approximate soil diffusion models were devised. Each was a Monte Carlo simulation in which the free path through the soil was given by an exponential extinction function. In one model the direction of departure of the atom from a soil grain was completely random, a condition that would arise if an adsorbed atom were mobile and could migrate arbitrarily over the grain surface. In the following discussion this model is denoted by M (for mobile). The other model placed a geometric restriction on the direction of departure equivalent to what would occur for immobile adsorbed atoms; and the model is accordingly designated I. Probability functions analogous to (2) were obtained by tracing many atoms in these models. The results are displayed in Fig. 1 in terms of the summation

$$f = \sum_{i=n+1}^{\infty} P_i$$ (4)

which is the probability that an atom has survived n collisions without emerging from the regolith. The immobile (I) atoms are more likely to emerge after one collision (25%) than the mobile (M) atoms (20%). At large values of n the apparent differences may not be significant because the data have statistical uncertainties the order of ±10%. It should be noted that both models closely follow the form of the random walk distribution (offset by a factor of about 1.6), and thus have similar attributes, including the $n^{-3/2}$ asymptotic behavior.

The important point to be made regarding the multiple collisions involved in the interactions of free exospheric atoms with a regolith is that the process cannot be characterized by a mean number of adsorption/desorption events. However, for Monte Carlo simulation of the lunar exosphere it is possible to generate a random deviate, n, for the number of collisions at each regolith encounter. Then

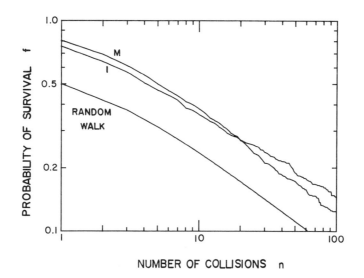

Fig. 1. Probability that an exospheric atom survives n collisions with soil grains before emerging from the regolith. Curves M and I are from soil models in which adsorbed atoms are mobile and immobile, respectively.

the total adsorption time for that encounter is the sum of n random deviates of the single collision sticking time characterized by Eq. (1). Because desorption probability obeys an exponential extinction law, the sum of sticking times is statistically equivalent to one random time deviate for desorption with time constant equal to $n\langle t\rangle$. The time variation of temperature, and hence $\langle t\rangle$, with rotation of the moon must be accounted for in calculating this deviate.

The above method of treating regolith adsorption and desorption was applied to the exosphere simulator used in earlier lunar atmosphere model work (cf. Hodges, 1975 and 1977). Random deviate values for the number of collisions were chosen from the M regolith encounter model probability distribution. A value of $t_0 = 1.6 \times 10^{-13}$ sec, as suggested by Frisillo *et al.* (1974), was used. The best fit of the calculated synodic variation of the argon-40 surface flux to the Apollo 17 mass spectrometer measurements was obtained with Q = 6000 cal/mole. This exceeds the activation energy for argon on glass of 3800 cal/mole measured by Clausing (see de Boer, 1968). The high activation energy on the moon seems to be a verification of an intuitively attractive hypothesis, that the degree of cleanliness of soil grains on the lunar surface is unattainable in the laboratory, and that laboratory activation energy data necessarily depict properties of surfaces contaminated with water vapor and possibly other volatiles. This idea seems to be supported by the extremely low argon activation energy (500–700 cal/mole) deduced by Frisillo *et al.* (1974) from measurements of BET "C" constants for returned lunar soils reported by Holmes *et al.* (1973). However, the BET "C" constants usually give low heats of adsorption because they are determined by the last part of monolayer formation, which occurs at sites where activation energies are low (Holmes *et al.*, 1973, and Frisillo *et al.*, 1974). Thus the low heats of adsorption of argon deduced from "C" constants may not be analogous to the adsorption of argon on clean rocks on the lunar surface. The difference between laboratory results for argon adsorption on glass and on returned lunar samples is more puzzling.

COLD TRAP TEMPERATURE REQUIREMENTS

The trapping of volatiles in cold, permanently shaded areas on the moon is certain; but what is uncertain is the time spent by volatiles in the traps. These time constants must be inferred from available data, which are sparse.

Arnold (1979) has noted that water has existed on the moon, and that much of this water must have migrated to polar cold traps, forming deposits of ice billions of years old. Areas cold enough to trap argon must be saturated with adsorbed water, and possibly coated with solid ice. Exceptions are areas very near the poles which experience seasonal temperature changes due to the slight tilt of the axis of rotation of the moon.

Figure 2 shows time constants for desorption as functions of temperature for argon and water. The curve for argon on glass corresponds to the aforementioned laboratory measurements of Clausing (see de Boer, 1968) which gave an activation energy of 3800 cal/mole, while the argon on lunar regolith curve is the result of model exosphere synthesis of the Apollo 17 mass spectrometer argon-40 data, and corresponds to 6000 cal/mole.

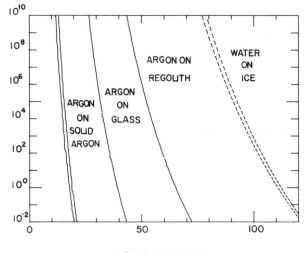

Fig. 2. Desorption time constants as functions of temperature.

Also shown in Fig. 2 are curves for argon on solid argon and water vapor on water ice that were derived by fitting vapor pressure (P_v) data found in the American Institute of Physics Handbook (1963) to the equation

$$P_v = P_o(T)e^{-Q/RT} \tag{5}$$

where P_o is a slowly varying function of T that is usually treated as a constant, but which must be proportional to $T^{5/2}$ as T approaches zero (cf. Kennard, 1938). Surface residence time $\langle t \rangle$ is related to the flux of evaporating atoms (or molecules), and hence to P_v by the relationships

$$\phi_{evap} = \frac{\langle v \rangle P_v}{4kT} = \frac{(\rho/m)^{2/3}}{\langle t \rangle} \tag{6}$$

where $\langle v \rangle$ is the mean thermal speed, k is Boltzmann's constant, ρ is density, and m is atomic (or molecular) mass. Curves for P_o proportional to $T^{5/2}$ and P_o = constant are shown for both argon and water, the difference in surface residence time in either case being inconsequential.

In Fig. 2 the temperature required to trap water ice for billions of years appears to be about 80 K. However the permanent retention of a monolayer of adsorbed water on soil grains should occur at a somewhat higher temperature, and the large ratio of total grain surface area to that of the spherical moon allows for considerable amounts of trapped water without forming ice. In addition, water accumulated as clathrate or amorphous ice as suggested by Arnold (1979) may have a vapor pressure that differs from the extrapolation of handbook data.

Owing to the average lunar radioactive heat flux of about 1.8×10^{-6} W cm^{-2} (Langseth *et al.*, 1976), the minimum surface temperature on the moon is about 25 K, precluding the formation of solid argon. As was noted in the previous section, argon adsorbed on surfaces that are clean by laboratory standards tends to have a lower activation energy than on the clean rocks in the equatorial region of the moon which influenced the Apollo 17 mass spectrometer data. Thus the argon on glass curve in Fig. 2 is probably a good estimator of the temperatures required to trap argon in lunar cold traps.

It can be seen in Fig. 1 that the median number of adsorption/desorption events per regolith encounter is about 5, so that the mean retention time for at least half of the ballistic argon flux exceeds $5\langle t \rangle$. Thus an estimate of the minimum value of $\langle t \rangle$ needed to trap argon for .5 y or more is .1 y. According to Fig. 2 this time constant is exceeded for temperatures less than 68 K on clean rocks and less than 40 K on the water-contaminated surfaces expected in cold areas.

SHADOW TEMPERATURES PRODUCED BY LUNAR SURFACE FEATURES

In a shadowed area the heat lost by radiation must be balanced by the inflow of conducted heat and the absorbed infrared radiation from nearby orographic features. Arnold (1979) has estimated that a shaded area of diameter 5–10 m that is shielded from the reradiated infrared flux should have a temperature less than 100 K. It is shown below that the size scales of shadows having temperatures much below 100 K are so large that lateral heat flow through the moon is negligible. However, the upward heat flow due to radioactivity is important.

The net reradiated infrared flux ϕ_r incident on point r must satisfy the integral equation

$$\phi_r = \int_A d^2r' \, \frac{\psi(r')\{\hat{n} \cdot (r' - r)\} \{\hat{n}' \cdot (r - r')\}}{\pi|r - r'|^4} \tag{7}$$

where r' is a variable of integration, $\hat{n}$ and $\hat{n}'$ are unit vectors normal to the surface at r and r', respectively, and $\psi(r')$ is the total heat flux radiated from r'. Support of the integral is the area A, which includes only points where the line of sight from r to r' is unobstructed. The total radiated flux is

$$\psi = \epsilon\{\phi_r + \phi_f \hat{r}' \cdot \hat{n}' + (1 - \alpha)U\phi_s \hat{s} \cdot \hat{n}'\} \tag{8}$$

where ϕ_f is the net vertical heat flux through the soil, ϕ_s is the solar constant (0.1352 W cm^{-2}), α is the albedo of the lunar surface (~0.08), ϵ is the infrared emissivity (~1.0), $\hat{r}'$ is the local vertical unit vector, $\hat{s}$ is a unit vector directed toward the sun, and U is unity in sunlit areas and zero in shadow. An interesting property of Eq. (7) is that by neglecting lateral heat conduction its solution is dependent only on orographic shape, and independent of physical size.

Owing to the poor conductivity of the upper layer of the lunar soil, the heat conducted downward during daytime is a small fraction of the influx. This is evidenced by the daytime surface temperature data of Keihm and Langseth (1973) which closely approximates radiative equilibrium. In the following the heat conducted downward is neglected. The upward heat flow due to radioactivity ($\langle\langle \phi_f \rangle \simeq 1.8 \times 10^{-6}$ W cm^{-2} according to Langseth *et al.* 1976) has been added to the reradiated influx after solving Eq. (7). This assures that an upper bound for the surface temperature results from the radiative equilibrium equation

$$T = \{(\phi_r + \langle\phi_f\rangle)/\epsilon\sigma\}^{1/4} \tag{9}$$

where σ is the Stefan-Boltzmann constant.

The surface features that produce most of the permanent shade on the moon are the bowl shaped craters, which have maximum wall slopes of about 40° (cf. the data of Wood and Anderson, 1978), and hence have permanently shaded areas at latitudes greater than 50°. The shape of a fresh bowl shaped crater is fairly accurately represented by a section of a sphere, a geometry that is special because the kernel of Eq. (7) simplifies to $1/4\pi R^2$, where R is the radius of curvature of the crater surface. The integral equation becomes simply

$$\phi_r(r) = \frac{1}{4\pi R^2} \int_A d^2r' \, \epsilon\{\phi_r + (1 - \alpha)U\phi_s \hat{s} \cdot \hat{r}'\} \tag{10}$$

and has as its solution ϕ_r = constant over the entire crater surface. That constant is

$$\phi_r = \frac{\phi_s \sin \zeta}{1 + D^2/4} \tag{11}$$

where ζ is the elevation angle of the sun, and D is the ratio of diameter to depth of the crater.

The morphometric data for craters on the moon of Wood and Anderson (1978) shows that bowl shaped craters with smooth rims range in size from quite small to diameters of about 20 km, and have a nearly constant diameter to depth ratio of about 5.4. Erosion processes tend to destroy rims and decrease wall slopes, increasing the effective value of D, the diameter to depth ratio, in Eq. (11). In Fig. 3 the equilibrium shadow temperatures in bowl shaped craters of several geometries are plotted as functions of solar elevation. The graph can also be interpreted as maximum noontime temperatures by referring to the latitude scale on the upper abscissa. It is important to note that this graph gives average shadow temperatures, and that surface irregularities may provide small areas that are shielded from both direct solar heating and its immediate reradiation, a subject that is discussed later. What is clear in Fig. 3 is that the fresh bowl shaped craters (D = 5.4) that are the source of much of the permanent shadow on the moon, do not provide a significant amount of surface area cold enough to retain water ice.

Shadow temperatures in bowl shaped craters decrease slowly with increasing

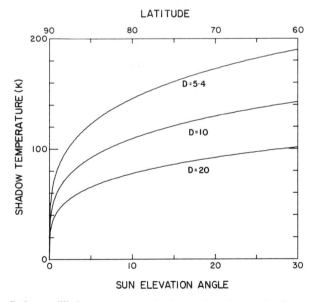

Fig. 3. Radiative equilibrium temperature in the shadowed part of a bowl shaped crater as a function of solar elevation for several values of the diameter to depth ratio (D).

Table 1. Flat floored crater classifications and characteristics (from Wood and Anderson, 1978).

Crater Type	Diameter (km.)	Diameter/Depth	Wall Slope
BIO (Biot)	<30	5.5	22°–26°
SOS (Sosigenes)	5–30	8–10	20°–26°
TYC (Tycho)	30–175	10–30	13°–15°

D, but craters with large diameter to depth ratios (D > 10) are usually not spherical in shape. Other morphologic categories of craters proposed by Wood and Anderson (1978) include three types of flat floored craters which produce large amounts of permanent shade on the moon. Table I gives a summary of the properties of these crater types.

The flat floored crater surface has been approximated by a truncated circular cone. To solve Eq. (7) for the reradiated flux in a flat floored crater it is necessary to use numerical methods, the simplest being to solve the equation at a fixed grid

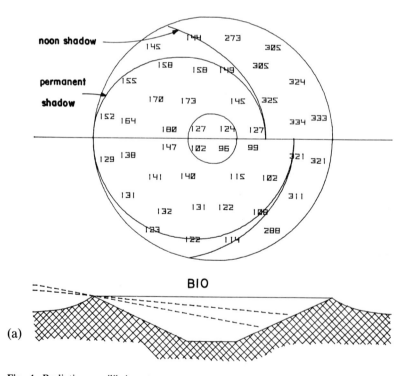

Fig. 4. Radiative equilibrium temperatures at representative points in three common types of flat floored craters. The upper half of each plan view corresponds to a latitude of 80° and the lower half to 85°. Noon and permanent shadow boundaries are marked on the plan views, and noon solar rays at each latitude are shown as dashed lines in the cross section views.

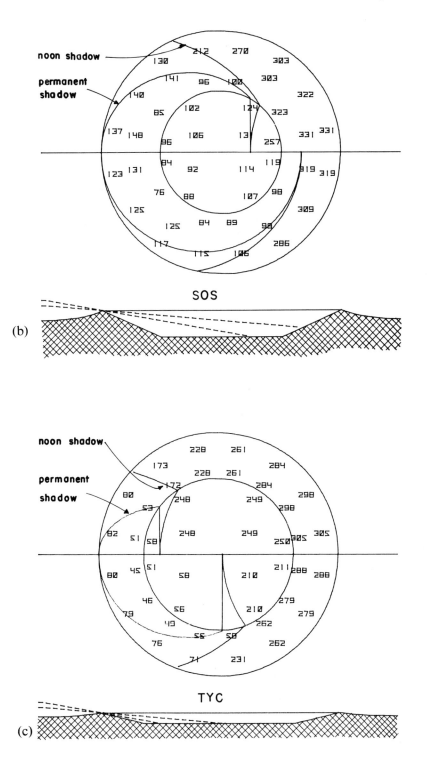

SOS

(b)

TYC

(c)

of points (96 points were used in these calculations). Then the integral in Eq. (9) becomes a matrix multiplication, and the reradiation flux is found by relatively simple matrix manipulations. Noontime temperatures at various points in each of the morphologic types of flat floored craters are shown in Fig. 4 for latitudes of 80° (upper half) and 85° (lower half). Arbitrary crater diameters of 10, 20, and 100 km were used for the BIO, SOS, and TYC types, respectively, and the respective values of D = 5.5, 9, and 22, were determined from regression formulas of Wood and Anderson (1978).

As in bowl shaped craters, there is a decreasing trend of minimum temperatures in flat floored craters with increasing diameter to depth ratio. It appears that the deep BIO type craters have temperatures similar to the bowl shaped craters. However, at latitudes above 80° water ice should exist on the floors of the large TYC type craters and on the equatorward walls of both TYC and SOS type craters. Large areas cold enough to trap argon on contaminated rocks exist only in TYC craters at latitudes greater than 85°.

All of the above temperature calculations have presumed smooth walls and floors. Actually these surfaces are pocked with smaller craters, and the walls of the larger craters are generally covered by tiers of terraces. These features provide many small depressions that are permanently shaded, not only from direct

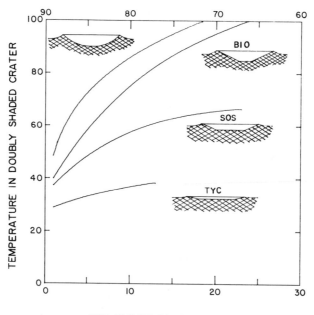

Fig. 5. Radiative equilibrium temperatures in small craters located in the permanent shadow on the floors of larger craters as functions of the elevation of the sun.

sunlight, but also from the first reradiation of the solar influx. The simplest approximation of such a doubly shaded area is a small crater on the floor of a larger crater. Fig. 5 gives the latitude dependence of the noontime temperatures in the doubly shielded part of a small bowl shaped crater located in the permanent shadow on the floor of each of the classes of large craters discussed above. It is evident that water should be retained as solid ice in bowl shaped craters at latitudes above 85°, in BIO type craters at latitudes greater than 80°, and in all permanently shaded areas of the larger craters. Temperatures cold enough to trap argon on clean, water free surfaces for periods up to a year (i.e., below 70 K) exist in SOS and TYC craters, but only in the very large TYC craters are there temperatures low enough (<40 K) to retain argon on surfaces contaminated by other volatiles.

CONCLUSION

The total surface area cold enough for water ice retention on the moon appears to be much less than that adopted by Watson *et al.* (1961) and by Arnold (1979), i.e., the total area in permanent shade. However, the immense surface area of soil grains in moderately cold permanent shadows can hold copious amounts of water in tightly bonded monolayer coatings. Furthermore, the vapor pressure data used in producing the ice curves in Fig. 2 may have been extrapolated unfairly, because it is possible that water has accumulated in some form of crystalline, clathrate or amorphous ice having a very low vapor pressure as suggested by Arnold (1979). Therefore there is no present reason to question Arnold's conclusion that a large amount of water must be stored on the moon.

The area available for argon retention for periods the order of a year or more on water contaminated rocks is mainly in doubly shielded places in large, flat floored craters located at latitudes greater than 75°. A rough estimate of this area can be made by assuming that these structures occupy about 10% of the surface above 75° latitude, that the average area in permanent shade is about 25% of the total crater area, and that about 20% of the permanent shade area is shielded from first reradiation of the solar influx. Hence the area available for argon adsorption is about 0.5% of the lunar surface at latitudes greater than 75°, or about 0.05% of the total lunar surface area.

Exospheric lateral migration tends to be from warm to cold areas, resulting in the equilibrium concentration being approximately proportional to $T^{-5/2}$ if the average desorption time is short (Hodges and Johnson, 1968). The smallness of the area available for argon trapping is compensated by the presence of a large fraction of the exospheric argon over the polar regions. It is therefore expected that argon is removed from the exosphere at a significant rate, and held in polar cold traps for reasonably long periods, forming a continually recycled reservoir that is occasionally disturbed, producing a sudden release.

Slight seasonal changes in surface illumination occur in lunar polar regions due to an axis tilt of 1.5° with respect to the ecliptic. This offsets the latitude scale of Fig. 5 by as much as 1.5° in the winter hemisphere, so that all of the bowl shaped craters within about 2.5° of the pole, and all of the small flat floored craters within about 3.5° have temperatures cold enough to trap argon for a few months. For a rough estimate of the area involved, assume that all of the area in the polar cap and 20% of that extending 1.5° from the polar cap is so affected. Then the area available for seasonal trapping of argon is about 0.03% of the total lunar surface area, and nearly the same as the estimated long term trapping area in the large flat floored craters. Further exosphere model work is needed to determine the amount of argon that cycles through these cold traps.

Acknowledgments—I am deeply indebted to J. R. Arnold for valuable discussions. This research was supported by NASA grant NSG-7034.

REFERENCES

American Institute of Physics Handbook (1963), Amer. Inst. of Physics (Y. E. Grayde, ed.), p. 4–218 and 4–274. McGraw-Hill, N.Y.

Arnold J. R. (1979) Ice in the lunar polar regions. *J. Geophys. Res.* **84**, 5659–5668.

de Boer J. H. (1968) *The Dynamical Character of Adsorption, 2nd ed*. Oxford Clarendon, N.Y., 240 pp.

Feller W. (1960) *An Introduction to Probability Theory and its Applications, 2nd ed*. Wiley, N.Y. 461 pp.

Frisillo A. L., Winkler J., and Strangway D. W. (1974) Molecular flow of gases through lunar and terrestrial soils. *Proc. Lunar Sci. Conf. 5th*, p. 2963–2973.

Heymann D. and Yaniv A. (1970) Ar^{40} anomaly in lunar samples from Apollo 11. *Proc. Apollo 11 Lunar Sci. Conf.*, p. 1261–1267.

Hodges R. R. (1975) Formation of the lunar atmosphere. *The Moon* **14**, 139–157.

Hodges R. R. Jr. (1977) Release of radiogenic gases from the moon. *Phys. Earth Planet. Inter.* **14**, 282–288.

Hodges R. R. and Hoffman J. H. (1974) Episodic release of ^{40}Ar from the interior of the moon. *Proc. Lunar Sci. Conf. 5th*, p. 2955–2961.

Hodges R. R. and Johnson F. S. (1968) Lateral transport in planetary exospheres. *J. Geophys. Res.* **73**, 7307–7317.

Holmes H. F., Fuller E. L. Jr., and Gammage R. B. (1973) Interaction of gases with lunar materials: Apollo 12, 14, and 16 samples. *Proc. Lunar Sci. Conf. 4th*, p. 2413–2423.

Keihm S. J. and Langseth M. G. Jr. (1973) Surface brightness temperatures at the Apollo 17 heat flow site: Thermal conductivity of the upper 15 cm of regolith. *Proc. Lunar Sci. Conf. 4th*, p. 2503–2513.

Kennard E. H. (1938) *Kinetic Theory of Gases*, McGraw-Hill, N.Y. 483 pp.

Langseth M. G. Jr., Keihm S. J., and Peters K. (1976) Revised lunar heat-flow values. *Proc. Lunar Sci. Conf. 7th*, p. 3143–3171.

Manka R. H. and Michel F. C. (1970) Lunar atmosphere as a source of argon-40 and other lunar surface elements. *Science* **169**, 278–280.

Manka R. H. and Michel F. C. (1971) Lunar atmosphere as a source of lunar surface elements. *Proc. Lunar Sci. Conf. 2nd*, p. 1717–1728.

Nakamura Y., Dorman J., Duennebier F., Ewing M., Lammlein D. and Latham G. (1974) High frequency lunar teleseismic events. *Proc. Lunar Sci. Conf. 5th*, p. 2883–2890.

Taylor S. R. and Jakeš P. (1974) The geochemical evolution of the moon. *Proc. Lunar Sci. Conf.* *5th,* p. 1287–1305.

Watson K., Murray B. C., and Brown H. (1961) The behavior of volatiles on the lunar surface. *J. Geophys. Res.* **66,** 3303–3313.

Wood C. A. and Anderson L. (1978) New morphometric data for fresh lunar craters. *Proc. Lunar Planet. Sci. Conf. 9th,* p. 3669–3689.

Proc. Lunar Planet. Sci. Conf. 11th (1980), p. 2479–2502.
Printed in the United States of America

Solar wind sputtering effects in the atmospheres of Mars and Venus

C. C. Watson*, P. K. Haff*, and T. A. Tombrello

W. K. Kellogg Radiation Laboratory, California Institute of Technology, Pasadena,
California 91125

Abstract—We have investigated the consequences of the direct collisional interaction of an energetic particle flux with the neutral components of a planetary atmosphere. A combination of Monte Carlo simulations and analytical analysis suggests that solar wind sputtering could provide an important exospheric mass sink on both Mars and Venus under appropriate conditions. We find that the efficient sputtering of the venusian atmosphere may result in a loss of He at the rate of $\sim 10^5$ atoms/cm²-sec, which could be a significant factor in the noble gas budget of that planet. Solar wind induced sputtering of the martian atmosphere could remove carbon, nitrogen and oxygen at the rates of 1×10^6 C atoms/ cm²-sec, 5×10^5 N atoms/cm²-sec and 3×10^6 O atoms/cm²-sec. These rates imply that sputtering would dominate over chemical and photo-chemical loss processes as a sink for carbon and nitrogen. Because of diffusive separation of lighter elements and isotopes and because the gravitational binding energy is proportional to the mass, the erosion process described here preferentially removes the lighter components of the atmosphere. Calculations based on a model martian atmosphere suggest that 99% of the N_2 and 43% of the CO_2 originally present could have been sputtered away over 4.5 $\times 10^9$ yr. In the same length of time the $^{15}N/^{14}N$ isotopic ratio for the bulk atmosphere would have increased by a factor of 1.97. Solar wind sputtering could thus compete favorably with other erosion mechanisms in generating substantial fractionation effects.

I. INTRODUCTION

Among the various possible exospheric sinks for planetary atmospheric mass, there is one which is unique to those bodies whose atmospheres are exposed to direct interaction with an energetic particle flux such as the solar wind (SW). We refer to the ejection of the ambient atomic and molecular species as a result of collisional energy transfer from the impinging ions. These ejectiles are said to have been sputtered. The sputtering mechanism is feasible when the intrinsic planetary magnetic field is not strong enough to exclude the SW from the atmospheric volume. This condition is satisfied when the magnetic induction is such that $B^2/8\pi \leq \rho v^2$, where ρ is the mass density and v is the bulk velocity of the SW. For typical ρ and v in the vicinity of the earth, the critical value of B is on the order of a milligauss. Measured by this criterion, both Venus and Mars possess at most only weak magnetic fields.

---

*also A. W. Wright Nuclear Structure Laboratory, Yale University, New Haven, Connecticut 06520

But a planetary body which does not possess an appreciable magnetic field may nevertheless at least partially divert the streaming solar wind plasma around itself if it possesses a sufficiently dense ionosphere (Spreiter *et al.*, 1970). The Mariner, Viking and Pioneer spacecraft probes have obtained substantial evidence for this latter type of interaction for Venus and, with less certainty, for Mars. Assuming a nonabsorptive flow pattern, the location of the ionopause, which is the boundary between the SW and the ionosphere, is fixed by the pressure balance between the flowing and static plasmas. The maximum SW pressure ($\approx \rho v^2$) occurs at the nose of the ionopause, i.e., the subsolar point. Unless there is absorption, the SW does not penetrate below the altitude of the nose. Upstream from the iono-pause, a standing bow shock wave is formed which acts to deflect the supersonic flow. Such a bow shock has been observed for both Venus and Mars. A basic difference between the bow shocks for nonmagnetic and magnetic planets such as the earth is that in the former case the shock surface lies much closer to the planet due to the much smaller apparent obstacle size offered by the ionosphere, compared to an Earth-like magnetosphere.

Between the bow wave and the ionopause the shocked SW forms a plasma sheath of ions flowing with reduced bulk velocity but greatly increased temper-ature and dayside density. This circumferential plasma flow about the ionosphere will interact directly with that portion of the neutral atmosphere which extends above the ionopause. In addition to sputtering, atmospheric mass loss may occur under these circumstances due to the mass loading of the SW by photo-ions produced in this region, if their density is not too great (Michel, 1971).

The preceding remarks describe a completely nonabsorptive SW flow. How-ever, there is growing evidence, particularly in the case of Venus, that the SW is absorbed through the ionopause to a significant extent in the vicinity of the subsolar point (Russell, 1977; Russell *et al.*, 1979; Wolfe *et al.*, 1979; Taylor *et al.*, 1979). The mechanism(s) responsible for this absorption are at present poorly known. Substantial fluctuations of the SW flux above its mean value may fre-quently result in pressures sufficient to overcome the ionospheric shielding (Sprei-ter *et al.*, 1970). A favorable alignment of the interplanetary magnetic field with the SW flow might also act to reduce the effectiveness of this shielding (Taylor *et al.*, 1979). Mass loading due to photo-ion pickup and the production of fast neutrals via change exchange can also lead to enhanced penetration of the ion-opause (Russell, 1977).

Whatever the mechanisms involved, data gathered by Mariner 5 and 10, Venera 4, 6, and 9 and Pioneer Venus at Venus indicate that the bow shock wave lies much closer to the planetary surface than would be anticipated simply from scal-ing the (nonabsorptive) flow pattern around the earth's magnetosphere. Russell (1977) has interpreted this to mean that a substantial fraction of the incident SW is directly absorbed in the subsolar region. Indeed, the absorption may on oc-casion be strong enough to result in a bow shock attached to the atmosphere. On the average, Russell (1977) estimates that as much as 29% of the SW flux incident on the planetary cross section may be absorbed.

To the extent that SW ions of sufficient energy penetrate a planetary atmosphere to its exobase, sputtering of the atmosphere can occur in a manner similar to the sputtering of a solid surface by an ion beam, i.e., through the mechanism of well developed collisional cascades (Haff *et al.*, 1978). This may or may not require that the SW penetrate the ionopause, depending on its altitude relative to the exobase. In sections II and III of this paper we present estimates of mass loss rates to be expected from this cascade type sputtering process in the atmospheres of Mars and Venus. The preliminary nature of these calculations, which derives chiefly from our incomplete understanding of the characteristics of the SW flow past these non-magnetic planets, scarcely needs to be emphasized. Nevertheless, we find that under favorable, though not unreasonable circumstances this sputtering mechanism can result in significant atmospheric mass loss.

If the SW does not penetrate to the bottom of the exosphere but flows around the planet at some higher altitude, sputtering will still occur since any collision between a SW ion and an atmospheric atom which transfers an energy to the atom greater than its gravitational potential energy has essentially a probability of 1/2 of removing that atom from the atmosphere. But generally speaking, this direct ejection mechanism, operating in the upper reaches of an exosphere whose density falls off exponentially with altitude, would not be as efficient as cascade type sputtering in the exobase region at removing atmospheric mass. Consequently, we shall limit ourselves to a brief qualitative description of direct sputtering in section II.B.

Perhaps more important than the total mass loss rate is the fact that the stoichiometry of the sputtered material may differ substantially from that of the bulk atmosphere, due primarily to the diffusive separation of the lighter components above the turbopause and to their lower gravitational binding energy. We shall explore in some detail the implications of such a mass fractionation effect for the compositional evolution of an atmosphere in section IV. We focus particularly on the anomalous $^{15}N/^{14}N$ isotopic ratio observed in the martian atmosphere.

II. MARS

On the basis of data from Mariners 4, 6, and 7, together with their model for nonabsorptive SW flow about Mars, Spreiter *et al.* (1970) have estimated the altitude of the nose of the martian ionopause to lie between 155 and 175 km. The neutral martian atmosphere as seen by the Viking I lander has its exobase at ~176 km. These figures imply a substantial direct interaction between the SW and the neutral martian atmosphere. We shall consider first the possibility of essentially normal (radial) flow down through the exobase region.

A. Cascade-type sputtering

The number density of a unimolecular atmosphere which is in thermal equilibrium above some reference altitude z_R (the turbopause) varies as $n(z) = n_R \exp(z -$

z_R)/H, where $H = kT/mg$ is the scale height, T is the absolute temperature, m is the molecular mass and g is the gravitational acceleration. (We shall neglect the variation of g with altitude.) The exobase, or critical height h_c, of such an atmosphere is defined as that altitude at which the mean free path of an atmospheric molecule in the horizontal direction equals the scale height. Thus

$$h_c = H\ln(H\sigma n_o)$$

where σ is the molecular cross section and $n_o \equiv n_R \exp(z_R/H)$. A fast SW ion passing down through the exosphere may collide with an atmospheric molecule. Recoiling atoms or molecules moving at altitudes $z \gg h_c$ suffer frequent collisions and are quickly thermalized with little chance for escape. A primary collision occurring in the vicinity of h_c, however, may generate a cascade of energetic secondary recoils, each of which has a substantial probability for escape if its velocity exceeds the escape velocity and is directed into the upper hemisphere. Thus, for normally incident ions, we expect sputtering to occur in a "critical layer" extending a few scale heights on either side of h_c.

An analogy may be drawn here to the sputtering of a solid surface. The exobase plays the role of the surface of the atmosphere as far as sputtering is concerned. The surface binding energy, U, of a molecule is just its gravitational potential energy—typically a few electron volts. Standard sputtering theory (Sigmund, 1969) treats an amorphous solid essentially as if it were a dense gas, ignoring any bulk, or lattice, binding energy. The density, ρ, of the target enters the calculation of the sputtering yield, which is the number of particles ejected per incident ion, in two mutually off-setting ways. On one hand, the number of secondaries formed per unit volume per unit time is directly proportional to ρ, while on the other hand the probability that a secondary of given energy will reach the target's surface without collision, integrated over the target volume, is inversely proportional to ρ. Since these two factors cancel, the sputtering yield is found to be independent of ρ, at least for a target of uniform density. Consequently, neglecting bulk binding energies and target non-uniformities, one expects similar yields within the framework of this model for solids and gases. This picture is valid for low-energy collisions as long as the separation of the atomic centers exceeds some critical distance whose order of magnitude is an atomic diameter. For solids this criterion is difficult to establish and the model described in this paper may in some cases be inadequate. Put another way, the conventional sputtering theory of solids is more appropriate when applied to the problem of the dilute gas target considered here than when applied to the problem for which it was constructed.

We have explored this model by making a Monte Carlo calculation of sputtering in a single component exponential atmosphere. Since the martian atmosphere is ~95% CO_2 (Owen and Biemann, 1976; Owen et al., 1977) we have chosen to consider a pure CO_2 atmosphere whose density in the critical layer was taken to be $n_{CO_2}(z) = (1.2 \times 10^{16} \text{ cm}^{-3}) \exp(-z/10 \text{ km})$, with reference to the initial Viking 1 data (Nier and McElroy, 1976). But note that according to the above remarks, our results should not be sensitive to the choice of $n_{CO_2}(z)$. We have based our model of the solar wind on its observed quiescent properties in the vicinity of the earth (Brandt, 1970). This yields a 1 keV proton flux of $9.5 \times 10^7 \text{ cm}^{-2} \text{ sec}^{-1}$ at Mars and a corresponding 4 keV α-particle flux of $5 \times 10^6 \text{ cm}^{-2} \text{ sec}^{-1}$. Of course, the energy of SW ions approaching the ionopause may be considerably reduced from these undisturbed values. Indeed, Cloutier et al. (1969) have estimated that the maximum interpenetration velocity of the SW at the stagnation point does not exceed 1 km/sec. According to the model calculations of Spreiter et al. (1970), however, the SW velocity increases rapidly away from the nose of the ionopause. In addition, it is expected that the shocked plasma would exhibit a large temperature increase so that over a large fraction of the dayside exosphere mean proton energies could possibly exceed 50 eV. This is roughly the minimum incident energy which will result in sputtering in our model. It has furthermore been postulated (Wallis and Ong, 1975) that charge exchange reactions may play an important role in the interaction between SW ions and atmospheric molecules. Such reactions would lead to the production of fast neutral H and He particles penetrating the exosphere.

It seems reasonable then that a substantial fraction of the SW crossing the bow shock could pass through the exosphere with sufficient energy to produce sputtering. In this energy range (50 eV $\leq$ E $\leq$ 1 keV for H^+, 20 eV $\leq$ E $\leq$ 4 keV for α) the nuclear stopping powers for H^+ and α on CO_2 exhibit broad peaks. The stopping power of a particle traversing some medium is defined as its energy loss per unit path length. Since this is just the energy which goes into the production of secondary recoils, it is not surprising that the sputtering yield, to the extent that it is mediated by a collisional cascade, is expected to be proportional to the stopping power. However, a deviation from this proportionality occurs for low-energy incident ions because much of the deposited energy is in the form of primary recoils having velocities less than the gravitational escape velocity. The energy at which the sputtering yield of a given species begins to differ markedly from its maximal value depends principally upon the ratio E_m/U where U is the gravitational binding energy of the species of interest and the energy E_m is characteristic of the most energetic primary recoils. E_m will depend not only on the incident ion energy but also on the composition of the exobase region. A useful rule of thumb appropriate for proton and α-particle sputtering is that the yield will be within a tenth of its maximum value for E_m $\gtrsim$ 3U (Watson, 1980). In the context of CO_2 sputtering on Mars, this estimate translates to E_{H+} $\gtrsim$ 100 eV and E_α $\gtrsim$ 35 eV. The situation is more favorable for the sputtering of lighter components such as N_2 and elemental C, N, and O.

Within these rather rough limits, the energies of the impinging ions are not of primary importance in the estimation of the cascade type sputtering process. The basic quantity we shall use to characterize the SW interaction with the critical layer, then, is the sputtering yield for ions with 1 keV/amu energy passing normally through the exobase. The calculation proceeds as follows. First, the ion-atom interactions are modeled by an r^{-2} screened Coulomb potential (Lindhard *et al.*, 1968). Dissociation of the CO_2 molecule is assumed if the energy transfer to the C or O atom exceeds some minimum value taken equal to the sum of its binding energy in the molecule and its gravitation potential energy. These energies are respectively 5.5 eV and 2.0 eV for O, and 11 eV and 1.5 eV for C atoms. Although smaller energy transfers might well result in substantial molecular dissociation, the resulting low energy recoils could not contribute to sputtering, and so will be neglected. At the other extreme, the most energetic atomic recoils are 216 eV O and 273 eV C.

These primary particles generate a cascade of recoiling molecules through subsequent collisions which are modeled in terms of a hard-sphere interaction. It is assumed that no molecular breakup occurs in these secondary collisions, and we furthermore neglect ionization and excitation processes. Experimental support for this nondissociative low-energy molecular collision picture is found in the recent study by Sheridan *et al.* (1979) of the dissociation of 6 to 12 keV N_2^+ ions in collision with O_2. It is found that the dissociation cross section is only 12–18% of the total cross section at these energies, and appears to be falling toward lower energies. In addition, the measured magnitude of the total molecular cross section is commensurate with our assumed CO_2 + CO_2 hard-sphere cross section, namely 22.5 Å^2. As a further check on the model, we have investigated the simpler case of 1 keV protons sputtering an O_2 gas through analytical calculations based on various assumptions for the collisional energetics (Watson, 1980). We find that the mass loss rate is quite insensitive to whether one assumes that molecular dissociation occurs in only the primary collision or in each secondary collision. The calculated sputtering yield varies by only a few percent as the translational energy assumed to be lost per collision in the model changes by an amount on the order of the dissociation energy.

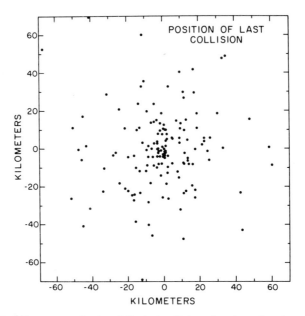

Fig. 1. Plot of the x-y coordinates of the last collision of each sputtered particle before leaving the atmosphere.

Each recoiling secondary particle is followed until either its energy falls below its escape energy (5.5 eV for CO_2), or until it moves above an altitude of 300 km. In the latter case it is recorded as a sputtered particle if its kinetic energy exceeds its escape energy. The proton and α particle sputtering yields as determined by our calculations are given in Table 1.

Figures 1–4 present some results of the proton sputtering calculation. Figure 1 shows the lateral position of the last collision suffered by a sputtered particle prior to leaving the atmosphere. The protons are all incident at the point (0,0). Nearly all sputtered molecules originate within ~50 km of the initial H^+-molecule impact, in the lateral dimension, and the width of a typical cascade is perhaps 60 km. The reason for this width may be understood by comparison with the case of a target of uniform density.

If we define a cascade to include only those secondaries having energies greater than the minimum energy U necessary for an atom to escape the target, and if $\overline{T}$ is the average energy of the primary initiating the cascade, then a typical cascade radius is $l(\overline{T}/U)^{\frac{1}{2}}$, where l is the mean free path of the target atoms. For CO_2 cascades in the critical layer U = 5.5 eV, l = H = 10 km and $\overline{T}$ = 40 eV, which is the average of the C and O primary energies. Thus we expect the radius of the cascade in the horizontal plane to be on the order of 30 km, as observed.

Figure 2 provides similar information on the vertical extent of the cascade, but in histogram form. The distribution is centered around the critical height h_c, as expected, with a full width of perhaps 40 km. The distribution is not symmetrical, as in Fig. 1, because of the variation in the atmospheric

Table 1. H^+ and α sputtering yields for a pure CO_2 atmosphere.

	$H^+ + CO_2$	$\alpha + CO_2$
S_{CO_2}	0.014	0.26
S_C	0.0064	0.010
S_O	0.0068	0.025

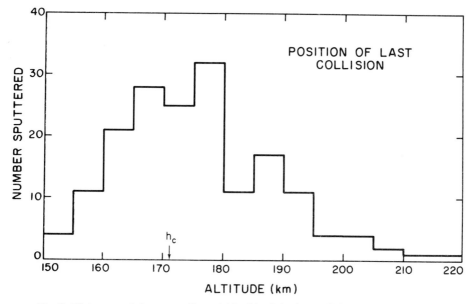

Fig. 2. Histogram of the z-coordinate (altitude) of the last collision of each sputtered particle before leaving the atmosphere.

density with altitude. The distribution in this figure would appear quite sharp were it not for the suppressed zeros. This points up the fact that as far as sputtering is concerned, the top of the atmosphere provides a reasonably sharp boundary.

It is interesting to compare the energy and angle distributions of Figs. 3 and 4 with those derived from the sputtering theory of solids. We assume (Thompson, 1968; Watson, 1980) that the energy spectrum within the solid is proportional to $1/E'^2$ and that the velocity distribution of recoiling atoms is isotropic. Then the flux of secondaries with energy in dE' about E' and velocity vector $\vec{v}'$ in the solid angle interval $d\Omega'$ about Ω', through a surface whose normal makes an angle θ' with $\vec{v}'$, has the form

$$\Phi(E',\Omega')d\Omega'dE' \propto \frac{\cos\theta'}{E'^2}\frac{d\Omega'}{4\pi}dE'.$$

Since an atom with energy E' greater than some minimum value U will escape the surface if its velocity is directed into the outer hemisphere, the conversion to an external sputtered flux is made by taking $E' = E + U$ and $\Omega = \Omega'$, so that

$$\Phi(E,\Omega)d\Omega dE \propto \frac{\cos\theta}{(E+U)^2}\frac{d\Omega}{4\pi}dE.$$

Here θ is the polar angle with respect to the normal to the target's surface. We find that the spectrum of Fig. 3 follows $1/(E+U)^2$ fairly well with U between 3.5 and 4.0 eV. The average gravitational binding energy for all particles considered here is 3.7 eV.

The angular distribution of Fig. 4 indicates that there is a depletion of the flux at small angles when compared to the above $\cos\theta$ dependence. $\theta = 0$ is upward from the planetary surface or backward relative to the solar wind flux. This indicates that, in fact, the flux of secondaries in the cascades is not completely isotropic.

Having established the nature of normal sputtering of an exponential atmosphere, we may proceed to estimate loss rates due to this mechanism for the

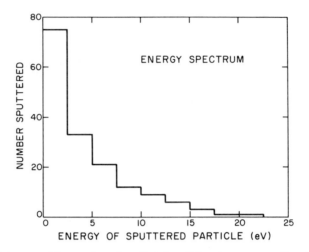

Fig. 3. Histogram of the number of sputtered particles at $z = +\infty$ as function of their energy (2.5 eV bins).

martian atmosphere. Sputtering will remove any component present in the critical layer. As a model for the martian upper atmosphere we adopt the composition and structure as determined by Viking 1 (Nier and McElroy, 1976; 1977) given in Fig. 5. We shall consider only the four largest components. The elemental oxygen component was not measured directly by VL1. Its structure has been estimated on the basis of its concentration at 130 km and an exospheric temperature of 169 °K (Nier and McElroy, 1977), assuming diffusive equilibrium. The critical height for a multicomponent atmosphere may be defined as that height at which the probability for a suitably energetic molecule, traveling radially outward, to escape the atmosphere without collision, averaged over all species present at that height, equals e^{-1}. For the atmosphere of Fig. 5 we find $h_c = 176$ km. At this height the atmosphere is 69.3% CO_2, 16.5% O, 10.7% N_2, and 3.6% CO. However, since cascade recoil kinematics depend on the recoil-particle mass, the

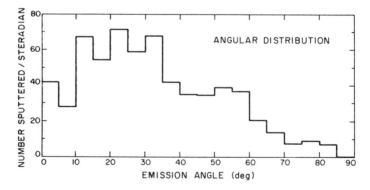

Fig. 4. Histogram of the number of sputtered particles per unit solid angle emitted in each 5° angle interval. $\theta = 0$ corresponds to upward emission.

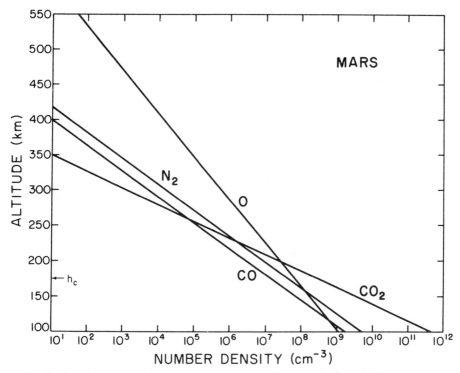

Fig. 5. Model for the martian atmosphere, based on Nier and McElroy (1977).

stoichiometry of the material sputtered is not necessarily identical to that of the critical layer. An analytical study (Watson and Haff, 1980; Watson, 1980) indicates, though, that the differences in the energy spectra of the recoiling species are likely to result in only small departures from the stoichiometry of the critical layer. The largest effect is due instead to differences between the gravitational potential energies, U_i, of the species. If S_i is the sputtering yield, n_i^c the density in the critical layer, and m_i the mass of species i, we find that to a good approximation,

$$S_i/S_j = U_j n_i^c / U_i n_j^c = m_j n_i^c / m_i n_j^c. \tag{1}$$

This relation applies in particular to those species whose recoil distributions evolve in well developed collisional cascades. It is a poorer approximation for atomic dissociation products, as may be seen from the yield values given in Table 1. A more accurate estimate may be made along the lines of Haff and Watson (1980) or Watson (1980), but since we are not interested here in the individual atomic yields per se it is sufficient for our purposes to scale the molecular equivalent yields of the various components from our calculated CO_2 yield according to Eq. (1). The molecular equivalent yield of a species is defined as the total mass of that species lost per incident ion, whether in atomic or molecular form, divided by the molecular mass. Averaging over a 95% proton and 5% α-particle SW flux we find the sputtering yields given in Table 2.

Table 2. Estimated sputtering yields in the exobase region of the Martian atmosphere.

Molecular	Total elemental
$S_{CO_2} = 0.021$	$S_O = 0.058$
$S_O = 0.014$	$S_C = 0.023$
$S_{N_2} = 0.0051$	$S_N = 0.010$
$S_{CO} = 0.0017$	

The magnitude of the exospheric mass sink generated by the sputtering process must be determined by coupling the above results with an analysis of the structure of the SW flow about Mars. The paucity of data relating to the details of this flow limit us at present to offering some general estimates which will be subject to refinement as the actual character of the SW-planetary interaction emerges.

Let the phase space density of the incidental plasma ions be denoted by $f_{SW}(\vec{r}, \vec{v}, t)$. We define a particle current density:

$$\vec{j}(\vec{r}, \vec{v}, t) = \vec{v}\, f_{SW}(\vec{r}, \vec{v}, t),$$

so that the flux of ions through the exobase at a point $\vec{r}$ and a time t, with velocities in d^3v about $\vec{v}$, is

$$|\hat{r} \cdot \vec{j}(\vec{r}, \vec{v}, t)|\, d^3v,$$

where $\hat{r}$ is the unit radius vector. Now because the mean free path of a solar wind ion in the exobase region is much greater than the width of the critical layer, the ith partial sputtering yield of such a particle, which traverses the exobase region at an angle θ with respect to $\hat{r}$, is expected to be enhanced by a factor of $|\cos\theta|^{-1}$ over its value for normal incidence, for angles θ not too near 90°. This results simply from the ion's increased path length through, and energy deposition in, the critical layer. Thus the partial yield of an ion impinging the exobase at point $\vec{r}$ is

$$S_i(\vec{r}, t) = \frac{S_i}{|\hat{r} \cdot \vec{j}(\vec{r}, \vec{v}, t)|/|\vec{j}(\vec{r}, \vec{v}, t)|}, \tag{2}$$

with S_i being the normal yield given in Table 2. We neglect here the dependence of the yield upon the ion's kinetic energy so long as it is within the limits of effectiveness noted above, e.g., 100 eV $\leq$ E $\leq$ 1 keV for a proton.

Assuming the exobase to have a more or less constant planetocentric radius r_{ex}, the total loss rate of species i per unit area, averaged over its entire surface may be written as

$$R_i(t) = \frac{1}{4\pi r_{ex}^2} \int_{exobase} d^2r \int d^3v\, |\hat{r} \cdot \vec{j}(\vec{r}, \vec{v}, t)|\, S_i(\vec{r}, t). \tag{3}$$

As a practical matter, the areal integration extends only over the exposed hem-

isphere. The integration over the ions' velocities should include only those consistent with the above energy limits.

With Eq. (2),

$$R_i(t) = \tfrac{1}{2} S_i \left[\frac{1}{2\pi} \int_{\text{exobase}} d\Omega_r \int d^3v \, |\vec{j}(\vec{r},\vec{v},t)| \right]. \tag{4}$$

If there were no planetary ionosphere and the SW flowed unperturbed directly
into the atmosphere, then on the dayside we would have $\int d^3v \, |\vec{j}(\vec{r},\vec{v},t)| = \phi(t)$,
where $\phi(t)$ is the magnitude of the interplanetary SW flux, and consequently

$$R_i(t) = \tfrac{1}{2} S_i \phi(t).$$

In the general case, $R_i(t)$ is proportional to the average magnitude of the SW
particle current density intercepting the exobase, as indicated by the bracketed
factor in Eq. (4). We are thus led to introduce a structure factor $\alpha(t)$ which directly
measures the ionospheric deviation of the SW. Recalling our definition of $\vec{j}(\vec{r},\vec{v},t)$
we define

$$\alpha(t) = \frac{1}{2\pi\phi(t)} \int_{\text{exobase}} d\Omega_r \int d^3v \, v \, f_{SW}(\vec{r},\vec{v},t), \tag{5}$$

so that

$$R_i(t) = \tfrac{1}{2} S_i \phi(t) \alpha(t). \tag{6}$$

The details of the SW plasma flow are therefore entirely incorporated in the
multiplicative factor $\alpha(t)$. It is important to note that $\alpha(t)$ may be greater than
unity even if all the flux incident on the planetary cross section does not penetrate
the critical layer. Nor is it necessary that the flow maintain a nonzero mean
velocity; a sufficiently high plasma temperature ($\sim 10^6 \, °K$) can also result in efficient sputtering.

Two distinct factors influencing the loss rates are isolated in expression (6).
$R_i(t)$ scales linearly with the magnitude of the interplanetary flux $\phi(t)$, while its
dependence upon the distribution of this flux in the exosphere is reflected in $\alpha(t)$.
A third factor, namely variation in the composition of the critical layer itself, will
be discussed in section IV below. As far as loss rates today are concerned, the
greatest uncertainty must be attached to the value of α. But until such time as
$\vec{j}(\vec{r},\vec{v},t)$ can be accurately specified, the most useful estimate is $R_i = \tfrac{1}{2} S_i \phi$, with
$\phi = 10^8 \, \text{cm}^{-2} \, \text{sec}^{-1}$, its average contemporary value. Summing over the molecular
yields of Table 2, we find the following erosion rates for the various elemental
species:

$$R_O = 3 \times 10^6 \, \text{O atoms/cm}^2 \, \text{sec},$$

$$R_C = 1 \times 10^6 \, \text{C atoms/cm}^2 \, \text{sec},$$

$$R_N = 5 \times 10^5 \, \text{N atoms/cm}^2 \, \text{sec}.$$

These rates would imply substantial mass loss from the planet over geological periods if we neglect the possible time variations discussed above. Integrated over a period of 4.5×10^9 yr, the total loss is on the order of the present mass of the martian atmosphere.

Although there is no information available which would allow us to establish the time dependence of $\alpha(t)$, the magnitude of the interplanetary flux $\phi(t)$ is a more tractable quantity. A recent study of SW nitrogen deposition in the lunar surface has suggested that the present-day SW flux is atypically low (Clayton and Thiemens, 1980). It is inferred that the average solar wind intensity over the entire lunar history has been greater by at least a factor of three than it is in the present epoch. Because such a sizeable increase in the SW intensity could substantially enhance the penetration of the SW ions into the neutral atmosphere, $\alpha(t)$ might be expected to be greater under such conditions if other factors, such as the structure of the ionosphere, remain the same. Consequently, on a geological time scale, sputter induced mass loss may have on the average considerably exceeded its present rate. The estimates given above would be amplified by a factor of three even if $\alpha(t)$ has remained constant over such periods.

An important measure of the significance of the above figures is to compare them with the loss rates estimated for other mechanisms. McElroy and others (McElroy, 1972; McElroy *et al.*, 1977) have calculated loss rates for C, O and N atoms due to chemical and photochemical processes in the martian atmosphere. We list below the important reactions and associated loss rates.

Oxygen:
$$CO_2^+ + O \rightarrow O_2^+ + CO$$
$$O_2^+ + e \rightarrow O + O \qquad R_O = 6 \times 10^7/cm^2\ sec$$
$$CO_2^+ + e \rightarrow CO + O$$

Carbon:
$$CO^+ + e \rightarrow C + O \qquad R_C = 1.5 \times 10^5/cm^2\ sec$$

$$h\nu + CO_2 \rightarrow C + O + O$$
$$h\nu + CO \rightarrow C + O$$
$$e + CO_2 \rightarrow e + C + O + O \qquad R_C = 6 \times 10^5/cm^2\ sec$$
$$e + CO \rightarrow e + C + O$$

$$e + CO_2^+ \rightarrow C + O_2 \qquad R_C \leq 4 \times 10^5/cm^2\ sec$$

Nitrogen:
$$N_2^+ + e \rightarrow N + N \qquad R_N = 3 \times 10^5/cm^2\ sec.$$

According to these numbers, the dominant mass loss mechanism is chemical ejection of oxygen atoms. The total mass loss due to sputtering of the critical layer with a structure factor equal to unity would amount to about 7% of the loss due to these other mechanisms. However, the single most important exospheric sink for carbon and nitrogen would be solar wind sputtering, especially in light of a long-term enhancement of the SW flux. The existence of this previously unstudied process may have a bearing on our understanding of the martian CO_2/H_2O ratio (Mutch *et al.*, 1976).

It is also instructive to compare these loss rates with those due to the SW sweeping of photo-ions in the upper atmosphere (Cloutier *et al.*, 1969; Michel, 1971). According to this mechanism, photo-ions produced in the upper atmosphere are carried away by their drift motion in the magnetized SW plasma. Such mass loading of the SW can proceed up to some critical mass addition rate before a substantial modification of the flow pattern must occur. It is not clear to what altitude the SW must penetrate in order for this optimal loss to be realized; however, maximum loss rates for CO_2 and N have been estimated to be (Mc-Elroy, 1972):

$$R_{CO_2} = 5 \times 10^5 \text{ molecules/cm}^2 \text{ sec}$$
$$R_N = 1 \times 10^6 \text{ atoms/cm}^2 \text{ sec.}$$

It would appear from this that the entrainment of photo-ions is a potentially significant atmospheric mass sink, particularly for nitrogen.

B. Direct ejection

The preceding analysis is valid to the extent that the SW does in fact penetrate the exobase region. If, on the other hand, the ionopause lies above the exobase, a substantial fraction of the SW may flow around the planet above the critical layer. In this event there can still be considerable interaction between the SW plasma and the lighter components of the atmosphere, which dominate at these higher altitudes. For the atmosphere of Fig. 5, the most important component in this connection is elemental oxygen.

A solar wind ion passing horizontally through the exosphere may collide with one of the ambient atoms, transferring to it an energy greater than its escape energy. If this primary atomic recoil is scattered into the upper hemisphere, it will almost certainly escape the planet. The recoiling ion, scattered downward, may lead to additional sputtering in the critical layer, as discussed for normal incidence. Based on the results of our CO_2 calculation, the mass loss due to this secondary ionic sputtering process would be on the order of 10% of that due to the direct ejection of the primary recoil. If, on the other hand, it is the primary atomic recoil, instead of the incident ion, which is scattered downward to the exobase, sputtering may still occur via the collisional cascade mechanism. A Monte Carlo calculation for protons and α particles on both N_2 and He atmospheres (cf. our subsequent discussion of Venus) indicates that the mass loss due to this type of secondary atomic sputtering should be roughly comparable to that resulting from direct ejection—for SW ions of inter-planetary energy. At lower energies such secondary sputtering effects are reduced. It should also be borne in mind that both the ionic and the atomic secondary sputtering mechanisms remove material in accordance with the composition of the critical layer, although there is a non-negligible probability in the latter case that the primary atomic particle will be reflected from the atmosphere and eventually escape.

The direct ejection sputtering mechanism has been discussed quantitatively in Watson (1980). It is shown there that the expected mass loss rate is directly proportional to the density of the atmosphere in those regions traversed by the SW and thus increases monotonically with the depth of the SW penetration. This result is in contrast to the insensitivity of the cascade type mechanism to the density of the critical layer. Calculations indicate that on Mars the contribution to atmospheric mass loss from the direct ejection process would be somewhat smaller than that due to cascade type sputtering if the SW flow does in fact reach the exobase. It has been suggested (Vaisberg *et al.*, 1976), however, that the nose of the martian ionopause lies at an altitude of about 400 km, instead of in the 155–175 km interval estimated by Spreiter *et al.* (1970). This value was obtained simply by scaling near-Earth data on the basis of the bow shock wave altitude at the subsolar point of 1500 km, as observed by the Mars 2, 3 and 5 spacecraft. In this event, and if there is no absorption of the SW through the ionopause, we would conclude (Watson, 1980) that sputtering does not play a substantial role in overall atmospheric mass loss from Mars.

Although we shall not pursue the direct ejection sputtering mechanism further with reference to the martian atmosphere, we would like to suggest that this mechanism might be important in generating mass loss from smaller planetary bodies with hot, extended exospheres. We have in mind specifically the sputter-induced erosion of the SO_2 atmosphere of the Jovian satellite Io (Haff and Yung, in preparation).

III. VENUS

The recent Pioneer Venus mission has revealed a dynamic, highly variable interaction between the SW and the ionosphere of Venus. A well-defined standing bow shock wave has been observed (Wolfe *et al.*, 1979), the nose of which lies at a radius of 1.23 R_V (altitude = 1400 km) (Russell *et al.*, 1979). Relative to the planetary radius, the Venusian bow shock thus lies much closer to Venus than the martian shock does to Mars. A distinct ionopause is seen whose altitude varies widely on a time scale of 24 hours in apparent response to varying SW pressure (Brace *et al.*, 1979; Taylor *et al.*, 1979; Kliore *et al.*, 1979; Knudsen *et al.*, 1979). The dayside ionopause has been observed at altitudes varying from 250 km to 1950 km (Taylor *et al.*, 1979; Brace *et al.*, 1979; Wolfe *et al.*, 1979). These observations, however, were made at fairly large solar zenith angles ($\gtrsim 60°$). It is to be expected that the altitude of the ionopause at the subsolar point would be substantially lower. Between the ionopause and the bow wave the shocked solar wind forms a plasma sheath. The mean, or bulk velocity of this plasma is expected to vary from near zero at the subsolar point on the ionopause to almost its undisturbed value near the planet's limbs (Spreiter, 1976). One must also expect that for a largely nondissipative flow, the energy loss represented by this reduction in bulk velocity should be compensated for by the compression and heating of the SW as it crosses the bow shock. The Pioneer Venus data (Taylor *et al.*, 1979; Wolfe *et al.*, 1979) offer support for this view. Indeed, it appears that the ionosheath temperature may frequently approach 10^6 °K (kT = 86 eV). But in order for a proton to eject, for instance, a helium atom from the venusian atmosphere, its required minimum energy is only about 3 eV. Thus the sputtering mechanism appears energetically quite tenable, at least for that portion of the flow above the exobase region.

The position of the ionopause with respect to the exobase in the subsolar region is still a matter of conjecture. But even if the ionopause lies above the critical layer, it may nevertheless be substantially penetrated by the SW. Russell (1977) has suggested that the relatively low position of the Venus bow shock indicates that on average perhaps 29% of the SW flux impinging Venus is absorbed through the ionopause. [For a less optimistic analysis, however, see Cloutier (1976).] If this flux is of sufficient energy, efficient sputtering of the critical layer can occur. The requisite ion energies are somewhat different here than those estimates given for Mars, due to the difference in gravitational binding energies. The criteria for the sputtering of elemental oxygen, which is the dominant constituent of the exobase region (see below), are $E_{H+} \gtrsim 120$ eV and $E_\alpha \gtrsim 41$ eV. The atmospheric component in which we shall have the most interest, however, is helium. Due to its low mass, cascade type sputtering of helium may be important for $E_{H+} \gtrsim 30$ eV and $E_\alpha \gtrsim 10$ eV. Of course, the energy available to these SW ions would depend critically on the mechanism responsible for their penetration to the exobase. But in any case, the quite low altitudes frequently attained by the ionopause

indicate a substantial interaction between the SW and the neutral atmosphere.

Our discussion of sputtering in the martian atmosphere is qualitatively equally valid for Venus. The total atmospheric mass of Venus, however, is a factor of 10^4 greater than that of Mars. This means that unless the SW structure factor is exceedingly large, the total mass loss due to sputtering would have little consequence for the evolution of the atmosphere as a whole. On the other hand, the selective sputtering of a minor component of the atmosphere could be of significance for the evolution of that component. Specifically, this is the case for helium. Although He dominates the neutral atmosphere of Venus at high altitudes, its bulk mixing ratio is probably at the 130 ppm level or less (von Zahn *et al.*, 1979).

In order to make our analysis quantitative we adopt the model atmosphere of Fig. 6, which derives from Pioneer Venus observations at a solar zenith angle of 88°, as the best available approximation to the average dayside composition (Niemann *et al.*, 1979). The critical height of this atmosphere lies at about 160 km. The sputter-induced mass loss due to passage of the SW through the exobase region may be estimated on the basis of Eq. (1) and the Monte Carlo yield for our model CO_2 martian atmosphere, with adjustment for the difference in gravitational binding energies. Recall that the atmospheric density is immaterial in a cascade type sputtering process. We shall again set $\alpha(t) = 1$, and assume a contemporary SW flux in the vicinity of Venus equal to 10^9 ions cm^{-2} sec^{-1}. This

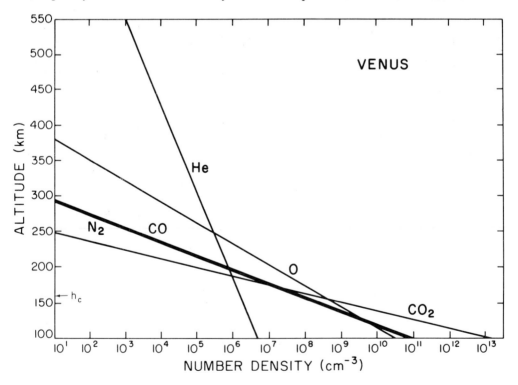

Fig. 6. Model for the atmosphere of Venus, based on Niemann *et al.* (1979).

value for $\phi(t)$ is consistent with the Pioneer Venus observations of the quiescent interplanetary flux (Wolfe *et al.*, 1979). As such it likely underestimates the mean flux intensity. In this respect, then, our results should be considered conservative.

From Eq. (6) we derive a total mass loss of

$$R = 2.7 \times 10^{-16} \text{ g cm}^{-2} \text{ sec}^{-1}.$$

Over a period of $T = 4.5 \times 10^9$ yr, with a SW flux enhanced by a factor of three, this rate would result in a total erosion of only 12 g cm^{-2}. This is to be compared with the present mass of the venusian atmosphere, 9.3×10^4 g cm^{-2}. Thus it is not likely that the sputtering process, at least as we have outlined it, has produced substantial erosion.

The situation with respect to the He component is quite different. Sputtering of the critical layer with a unitary structure factor is estimated to result in a contemporary loss rate of

$$R_{\text{He}} = 1.2 \times 10^5 \text{ atoms cm}^{-2} \text{ sec}^{-1}.$$

Integrating this value over 4.5×10^9 yr and including the usual correction factor for the SW flux, the implied total loss is 4.9×10^{22} atoms cm^{-2}. If the value of 130 ppm for the mixing ratio of He in the atmosphere of Venus is adopted, the present abundance of He is 1.7×10^{23} atoms cm^{-2}. By this measure then, SW sputtering of the exobase region would provide a significant sink for He.

The direct sputtering mechanism may also be operative in removing He, particularly since, according to Fig. 6, He dominates the venusian atmosphere above ~250 km. A quantitative analysis indicates that here, as was the case for the martian atmosphere, mass loss due to direct ejection alone begins to approach the above cascade type sputtering loss rate only if the SW penetrates to altitudes near the exobase. The details concerning the contribution of this direct sputtering process may be found in Watson (1980).

Another measure of the significance of the above loss rate is to compare it with possible sources of He in the atmosphere. One such source is outgassing. If we assume with Knudsen and Anderson (1969) that the production rate of He on Venus is the same as for the Earth, then the outgassing rate on Venus is 2×10^6 He atoms cm^{-2} sec^{-1}. A second source of He is the SW itself. Assuming the SW ion flux is 5% He, and that an average of 29% of the flux incident on Venus is absorbed (Russell, 1977), one finds a present deposition rate of $R = 3.6 \times 10^6$ He atoms cm^{-2} sec^{-1}. Thus, one would expect SW deposition to be at least as important a source of He in the venusian atmosphere as outgassing.

A comparison between these source strengths and the sputtering loss rate given above would imply that unless the SW structure factor is fairly large, it is likely that He accumulates in the atmosphere more rapidly than it is sputtered. But if there is in fact a substantial SW-planetary interaction, such an imbalance between the deposition and loss rates could not continue indefinitely. Since the deposition of He by the above two mechanisms is independent of the abundance of He in the atmosphere, while the sputter erosion rate increases roughly linearly with that abundance, the sputtering process tends to drive the concentration of He to some stable value. This value would not depend on the initial abundance of He,

but only on the efficiency of the sputtering mechanism relative to the He sources.

There is at least one other possible exospheric sink for He, namely, the entrainment of photo-ions by the SW. Michel (1971) estimates that the upper limit for the present He loss rate which could result from this sweeping process is 7.5×10^5 He atoms cm^{-2} sec^{-1}. This value is comparable to our exobase sputtering estimate, but in contrast to the latter, the SW sweeping rate would not increase above the quoted value in proportion to the concentration of He in the exosphere. Thus the conclusions reached in the last paragraph are not altered.

There is further evidence that the imbalance between the sources and sinks for He suggested by the above figures could not have persisted over the entire history of the planet. If the outgassing and SW sources of He had been operative at these strengths for 4.5×10^9 yr, then the total deposition of He would have been 1.8×10^{24} atoms cm^{-2}. This would imply a mixing ratio of He to CO_2 in the present atmosphere of 1.4×10^{-3}, which is an order of magnitude greater than the observed limit on this ratio of about 130 ppm. It is clear then that either the above deposition rate estimates are considerably too large when applied over geological periods, or else there is, in fact, an efficient sink for atmospheric He. In either case it is quite possible that sputtering due to solar wind impact could have played a substantial role in the evolution of the venusian atmosphere.

IV. FRACTIONATION

To the extent that solar wind sputtering occurs in the critical layer of the upper atmosphere, the loss rates for the various species should reflect their concentrations in this region and not their total abundances. For instance, since the Viking data (Nier and McElroy, 1976, 1977; Nier *et al.*, 1976a; Owen and Biemann, 1976; Owen *et al.*, 1977) show that the mixing ratio of N_2 to CO_2 is much greater at altitudes near 176 km than for the bulk Martian atmosphere we should expect the N_2 to be sputtered preferentially. Furthermore, an additional fractionation effect arises from the difference in the gravitational potential energies of the various species, as given in Eq. (1). These considerations apply equally well to the case of isotopes.

The implications of these effects for the history of the martian atmosphere have been investigated by considering a simple model atmosphere, with a present composition of 2.5% N_2 and 97.5% CO_2, assumed to have been rapidly outgassed 4.5×10^9 years ago. We assume a subsequent passive role for the surface and consider only that mass loss which is due to sputtering. Let $n_i(t)$ be the column density (molecules cm^{-2}) of species i for the bulk atmosphere and $n_i^c(t)$ be its number density in the critical layer at a time t (the atmosphere being formed at $t = 0$). The corresponding mixing ratios of N_2 to CO_2 are denoted by

$$f(t) = n_{N_2}(t)/n_{CO_2}(t),$$

and

$$f_c(t) = n_{N_2}^c(t)/n_{CO_2}^c(t).$$

The parameter $R \equiv f_c(t)/f(t)$ (Nier *et al.*, 1976b) is then a measure of the enrich-

ment of the critical layer in the lighter species due to diffusive separation. The dynamics of the atmosphere are assumed to be such as to maintain R essentially constant in time.

The sputter-induced fractionation of the atmosphere is determined by the relative magnitude of the time-integrated molecular loss rates. In addition to their obvious dependence on the intensity and structure of the SW flow, these rates will vary in response to the changing composition of the critical layer. In accordance with our proposal that energy sharing effects result in only small departures of the secondary recoil fluxes from the stoichiometry of the target medium, the sputtering yield of a normally incident ion in the exobase region of our model atmosphere may be written as

$$S_{CO_2}(t) = S \, n^c_{CO_2}(t)/[n^c_{CO_2}(t) + n^c_{N_2}(t)] = S/[1 + Rf(t)]. \qquad (7)$$

Here S is the yield for an unimolecular CO_2 atmosphere as determined, for instance, by our Monte Carlo calculation, and is constant in time. Using Eq. (1),

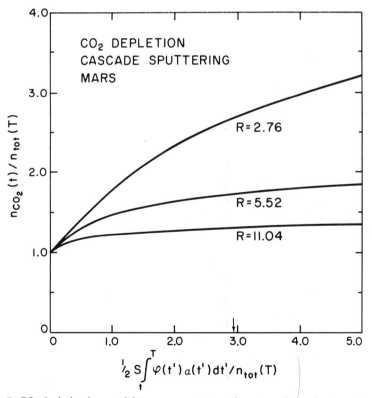

Fig. 7. CO_2 depletion in a model two component martian atmosphere, due to sputtering in the exobase region, for various diffusive enrichment parameters R. The abscissa is a time parametrization which in effect represents the integrated mass loss in a pure CO_2 atmosphere (see text).

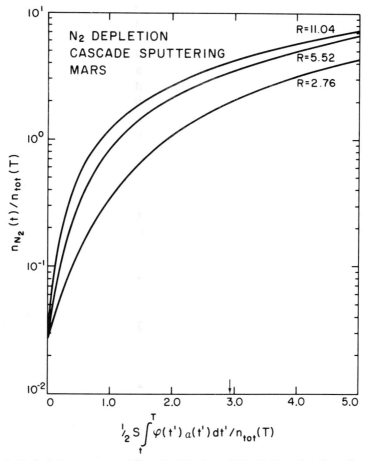

Fig. 8. N_2 depletion corresponding to the CO_2 loss of Fig. 7. Note that the ordinate in this case is logarithmic.

the N_2 yield is then given by

$$S_{N_2}(t)/S_{CO_2}(t) = m_{CO_2}n^c_{N_2}(t)/m_{N_2}n^c_{CO_2}(t) = Rf(t)\, m_{CO_2}/m_{N_2}. \tag{8}$$

The equations governing the evolution of the sputtered atmosphere are

$$dn_{N_2}(t)/dt = -S_{N_2}(t)\, \phi(t)\, \alpha(t)/2,$$

and $\qquad\qquad\qquad\qquad\qquad\qquad\qquad\qquad\qquad\qquad\qquad\qquad$ (9)

$$dn_{CO_2}(t)/dt = -S_{CO_2}(t)\, \phi(t)\, \alpha(t)/2.$$

Integrating these with the use of Eqs. (7) and (8), one finds

$$n_{N_2}(T)/n_{N_2}(t) = \left[n_{CO_2}(T)/n_{CO_2}(t) \right]^{R(m_{CO_2}/m_{N_2})} \tag{10}$$

and

$$n_{CO_2}(t) - n_{CO_2}(T) + [n_{N_2}(t) - n_{N_2}(T)](m_{N_2}/m_{CO_2}) = (S/2) \int_t^T \phi(t')\alpha(t')dt'. \quad (11)$$

The reference time T is the age of the atmosphere. Given the present abundance of N_2 and CO_2, Eqs. (10) and (11) may be solved to determine their abundances at any previous time t as a function of $(S/2) \int_t^T \phi(t')\alpha(t')dt'$. The latter quantity is a measure of the total amount of material lost to SW sputtering in the interval (t,T). The results of such a calculation for $n_{CO_2}(t)$ and $n_{N_2}(t)$ are given in Figs. 7 and 8 respectively for three different values of the diffusive separation parameter R. The value R = 5.52 is in the best agreement with the Viking data for the present martian atmosphere. In each figure, both scales are in units of the present total column density $n_{tot}(T) = n_{CO_2}(T) + n_{N_2}(T) = 2.25 \times 10^{23}$ cm^{-2}. The indicated point on the abcissa marks our nominally best estimate for the erosion parameter at t = 0. It results from taking $\alpha(t) = 1$, $\phi(t) = 3 \times 10^8$ cm^{-2} sec^{-1}, and S = 0.031. At this point, with R = 5.52, the model implies that tne initial martian atmosphere was 67% N_2 and 33% CO_2. Over a period of 4.5×10^9 yr 75% of the initial atmospheric mass could have been removed by sputtering, with 43% of the CO_2 and 99% of the N_2 being lost. An N_2 depletion of 99% is as great or greater than mass loss estimates based on other mechanisms, e.g., photochemical reactions (Nier et al., 1976b; McElroy et al., 1976; Brinkmann, 1971), so it would appear that sputtering of the critical layer could be significant for atmospheric evolution.

Given a value for $n_{N_2}(0)$, we may make an analysis similar to the above to determine the enrichment expected among nitrogen isotopes if the initial ratio of ^{15}N to ^{14}N equalled its present terrestrial value, $^{15}N/^{14}N = 0.00368$. Bearing in mind that nitrogen is lost mainly in molecular form, we have

$$n_{28}(T)/n_{28}(0) = [n_{29}(T)/n_{29}(0)]^{R_N(m_{29}/m_{28})} \quad (12)$$

which is analogous to Eq. (10). Assuming that the turbopause of the martian atmosphere is well defined and lies at an altitude of 125 km (Nier and McElroy, 1977), the $^{28}N_2/^{29}N_2$ diffusive separation parameter R_N may be determined from the observed $^{28}N_2$ scale height (Fig. 5), with the result that $R_N = 1.12$.

Figure 9 shows the behavior of the enrichment parameter

$$\epsilon_N(T) \equiv n_{15}(T)n_{14}(0)/[n_{14}(T)n_{15}(0)] \quad (13)$$

as a function of the integrated CO_2 equivalent molecular loss over the planet's history under the assumption that the N_2/CO_2 diffusive enrichment parameter R = 5.52. The indicated value is again our best estimation, viz., $\epsilon_N = 1.97$. This is to be compared to the enrichment observed by the Viking landers, $\epsilon_N = 1.62 \pm 0.2$ (Nier and McElroy, 1977). Clearly, fractionation due to efficient sputtering is a viable explanation for the observed isotopic ratio.

Other sets of atmospheric isotopes may also suffer fractionation as a result of preferential mass sinks. One such possibility is $^{13}C/^{12}C$. The carbon isotopic enrichment in the martian atmosphere relative to the terrestrial ratio is less than 5% (Nier and McElroy, 1977). The smallness of this value when compared to the nitrogen isotopic enrichment suggests that much less CO_2 than N_2 has been processed through the atmosphere, relative to their present abundances. This situation does, in fact, occur in the sputtering model developed here. Applying the above analysis to the relative ejection of $^{12}CO_2$ and $^{13}CO_2$ we obtain the second curve shown in Fig. 9. The indicated value of ϵ_c is about 1.07. Thus, the effects expected from SW sputtering of the exobase region are consistent with the observed scale of isotopic enrichments.

Fractionation may also result when sputtering occurs at higher altitudes. Indeed, the higher the altitude, the greater the diffusive separation which will exist between components. But on the other hand, the amount of material processed by sputtering decreases with increasing height. Although under favorable conditions sputtering of the upper exosphere could by itself produce significant isotopic enrichments, on the whole the direct ejection mechanism is not as effective in this regard as is sputtering of the exobase region (Watson, 1980).

V. SUMMARY

Our analysis has shown that, to the extent which an energetic particle flux such as the solar wind impinges upon a planetary atmosphere, the collisional ejection

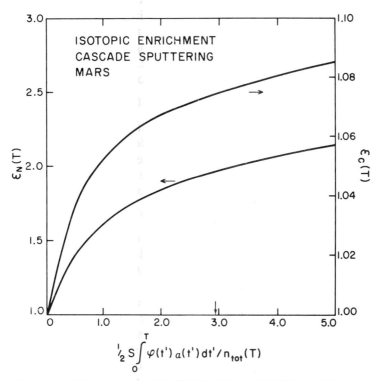

Fig. 9. Isotopic enrichment at time T for $^{15}N/^{14}N$ (left) and $^{13}C/^{12}C$ (right) in the model atmosphere of Figs. 7 and 8 (with R = 5.52), assuming initial terrestrial abundances.

of neutral species will be a potent force driving that atmosphere's evolution. The total mass loss to be expected for a given intensity of the incident plasma increases with the depth of penetration of the flow in proportion to the atmospheric density. A limiting sputtering rate is reached at the exobase at which point the direct ejection mechanism is superceded by a cascade type process similar in many respects to the mechanism which is thought to be operative in the sputter erosion of solid surfaces. Both because this cascade mechanism is independent of the target density, and because in any case the densities in the exobase regions of most atmospheres are similar, the importance of the total sputter-generated mass loss will be greater in less massive atmospheres. We have seen this to be the case in the comparison between Mars and Venus.

On the other hand, sputtering can lead to a significant loss of the less massive atmospheric components even when the total erosion is small. Such preferential ejection is due in part to the diffusive enrichment of the upper atmosphere in the lighter species. Although this enrichment increases with altitude, it is in competition with a decreasing sputtering rate so that the largest fractionation effects are experienced when loss from the exobase region is dominant. Needless to say, this diffusive separation factor will be operative for practically any exospheric mass sink. But sputter-induced loss rates involve an additional departure from stoichiometry in that they are inversely proportional to the masses of the sputtered species. In the case of cascade-type sputtering this is a consequence of the inverse proportionality of the yield to the gravitational binding energy, while for the direct ejection mechanism we have shown that the loss rate of a species depends directly on the scale height.

Preferential sputtering can lead to both elemental and isotopic mass fractionation of an atmosphere. The latter effect may be exemplified by the anomalous $^{15}N/^{14}N$ ratio on Mars. Not only can the sputtering mechanism account for a large part of the effect observed for this pair, but it is also consistent with the quite small fractionations found for $^{13}C/^{12}C$ (and $^{18}O/^{16}O$). On Venus, the elemental fractionation of He is of primary importance. The extent to which the present abundance of He in the venusian atmosphere has been controlled by interaction with the solar wind, as opposed to outgassing, is a question of fundamental significance in the context of those models for planetary formation which postulate a hot primordial solar nebula followed by condensation and grain accretion (Pollack and Black, 1979).

Our discussion of sputtering phenomena has necessarily been phrased in terms of contingencies due to the uncertainty in the details of the SW-planetary interaction. It seems probable that the SW does, in fact, impact the venusian atmosphere to a considerable extent, but the nature of the interaction at Mars is more equivocal. Fortunately, the information needed in order to evaluate the collisional processes explored here is actually rather limited. Basically what is desired is the average of the ion particle current density magnitude over either the exobase, or else over the inner boundary of the ionosheath. Even if a determination of this quantity implies only a marginal importance for SW induced sputtering in the present epoch, the fact that the SW wind intensity has probably been much greater on the average in the past could signify a substantial geological role for the phenomena we have delineated. It is our further hope that the utility of the present discussion will extend to other astrophysical environments not directly addressed here; for instance, to the interaction of the Jovian magnetosphere with the Galilean satellite Io.

Acknowledgments—P. K. H. and C. C. W. wish to acknowledge the encouragement and support of Professor D. A. Bromley. This work was supported in part by the National Aeronautics and Space Administration [NGR 05-002-333] and the National Science Foundation [PHY79-23638] at Caltech, and by the U.S. Department of Energy [EY-76-C-02-3074] at Yale.

REFERENCES

Brace L. H., Theis R. F., Krehbiel J. P., Nagy A. F., Donahue T. M., McElroy M. B., and Pedersen A. (1979) Electron temperatures and densities in the Venus ionosphere: Pioneer Venus orbiter electron temperature probe results. *Science* **203**, 763–765.

Brandt J. C. (1970) *Introduction to the Solar Wind*. Freeman, San Francisco. 199 pp.

Brinkmann R. T. (1971) Mars: has Nitrogen escaped? *Science* **174**, 944–945.

Clayton R. N. and Thiemens M. H. (1980) Lunar nitrogen: Evidence for secular change in the solar wind. In *Proc. Conf. Ancient Sun* (R. O. Pepin, J. H. Eddy, and R. B. Merrill, eds.), p. 463–474. Pergamon, N.Y.

Cloutier P. A., McElroy M. B. and Michel F. C. (1969) Modification of the Martian ionosphere by the solar wind. *J. Geophys. Res.* **74**, 6215–6228.

Cloutier P. A. (1976) Solar-wind interaction with planetary ionospheres. In NASA SP-397 (N. F. Ness, ed.), p. 111–119.

Haff P. K., Switkowski Z. E., and Tombrello T. A. (1978) Solar-wind sputtering of the Martian atmosphere. *Nature* **272**, 803–804.

Haff P. K. and Watson C. C. (1980) The erosion of planetary and satellite atmospheres by energetic atomic particles. *J. Geophys. Res.* **82**, 8436–8442.

Kliore A. J., Woo R., Armstrong J. W., Patel I. R., and Croft T. A. (1979) The polar ionosphere of Venus near the terminator from early Pioneer Venus orbiter radio occultations. *Science* **203**, 765–768.

Knudsen W. C. and Anderson A. D. (1969) Estimate of radiogenic He^4 and Ar^{40} concentration in the Cytherean atmosphere. *J. Geophys. Res.* **74**, 5629–5632.

Knudsen W. C., Spenner K., Whitten R. C., Spreiter J. R., Miller K. L., and Novak V. (1979) Thermal structure and major ion composition of the Venus ionosphere: first RPA results from Venus orbiter. *Science* **203**, 757–763.

Lindhard J., Nielsen V., and Scharff M. (1968) Approximation methods in classical scattering by screened Coulomb fields. *Mat. Fys. Medd. Dan Videnskab. Selskab* **36**, (10).

McElroy M. B. (1972) Mars, an evolving atmosphere. *Science* **175**, 443–445.

McElroy M. B., Kong T. Y., and Yung Y. L. (1977) Photochemistry and evolution of Mars' atmosphere: a Viking perspective. *J. Geophys. Res.* **82**, 4379–4388.

McElroy M. B., Yung Y. L., and Nier A. O. (1976) Isotopic composition of Nitrogen: implications for the past history of Mars' atmosphere. *Science* **194**, 70–72.

Michel F. C. (1971) Solar-wind-induced mass loss from magnetic field-free planets. *Planet. Space Sci.* **19**, 1580–1583.

Mutch T. A., Arvidson R. E., Head J. W. III, Jones K. L., and Saunders R. S. (1976) *The Geology of Mars*. Princeton Univ. Press, Princeton, New Jersey. 400 pp.

Niemann H. B., Hartle R. E., Kasprzak W. T., Spencer N. W., Hunten D. M. and Carignan G. R. (1979) Venus upper atmosphere neutral composition: preliminary results from the Pioneer Venus orbiter. *Science* **203**, 770–772.

Nier A. O., Hanson W. B., Seiff A., McElroy M. B., Spencer N. W., Duckett R. J., Knight T. C. D., and Cook W. S. (1976a) Composition and structure of the Martian atmosphere: preliminary results from Viking 1. *Science* **193**, 786–788.

Nier A. O., McElroy M. B., and Yung Y. L. (1976b) Isotopic composition of the Martian atmosphere. *Science* **194**, 68–70.

Nier A. O. and McElroy M. B. (1976) Structure of the neutral upper atmosphere of Mars: results from Viking 1 and Viking 2. *Science* **194**, 1298–1300.

Nier A. O. and McElroy M. B. (1977) Composition and structure of Mars' upper atmosphere: results from the neutral mass spectrometers on Viking 1 and 2. *J. Geophys. Res.* **82**, 4341–4349.

Owen T. and Biemann K. (1976). Composition of the atmosphere at the surface of Mars: detection of argon-36 and preliminary analysis. *Science* **193**, 801–803.

Owen T., Biemann K., Rushneck D. R., Biller J. E., Howarth D. W., and Lafleur A. L. (1977) The composition of the atmosphere at the surface of Mars. *J. Geophys. Res.* **82**, 4635–4639.

Pollack J. B. and Black D. C. (1979) Implications of the gas compositional measurements of Pioneer Venus for the origin of planetary atmospheres. *Science* **205**, 56–59.

Russell C. T. (1977) The Venus bow shock: detached or attached? *J. Geophys. Res.* **82**, 625–631.

Russell C. T., Elphic R. C., and Slavin J. A. (1979) Initial Pioneer Venus magnetic field results: dayside observations. *Science* **203**, 745–748.

Sheridan J. R., Merla T., and Enzweiler J. A. (1979) Cross sections for asymmetric charge transfer and collisional dissociation reactions of atmospheric ions at keV energies. *J. Geophys. Res.* **84**, 7302–7306.

Sigmund P. (1969) Theory of sputtering. I. Sputtering yield of amorphous and polycrystalline targets. *Phys. Rev.* **184**, 383–416.

Spreiter J. R., Summers A. L., and Rizzi A. W. (1970) Solar wind flow past nonmagnetic planets— Venus and Mars. *Planet Space Sci.* **18**, 1281–1299.

Spreiter J. R. (1976) Magnetohydrodynamic and gasdynamic aspects of solarwind flow around terrestrial planets a critical review. In NASA SP-397 (N. F. Ness, ed.), p. 135–149.

Taylor H. A., Brinton H. C., Bauer S. J., Hartle R. E., Cloutier P. A., Michel F. C., Daniell R. E. Jr., Donahue T. M., and Maehl R. C. (1979) Ionosphere of Venus: first observations of the effects of dynamics on the dayside ion composition. *Science* **203**, 755–757.

Thompson M. W. (1968) II. The energy spectrum of ejected atoms during the high energy sputtering of gold. *Phil. Mag.* **18**, 377–414.

Vaisberg O. L., Bagdanov A. V., Smirnov V. N., and Ramanov S. A. (1976) On the nature of the solar-wind-Mars interaction. In NASA SP-397 (N. F. Ness, ed), p. 21–40.

Wallis M. K. and Ong R. S. B. (1975) Strongly-cooled ionizing plasma flows with application to Venus. *Planet. Space Sci.* **23**, 713–721.

Watson C. C. (1980) Topics in Classical Kinetic Transport Theory with Applications to the Sputtering and Sputter-Induced Mass Fractionation of Solid Surfaces and Planetary Atmospheres. Ph.D. Thesis, Yale Univ., New Haven. 234 pp.

Watson C. C. and Haff P. K. (1980) Sputter-induced isotopic fractionation at solid surfaces. *J. Appl. Phys.* **51**, 691–699.

Wolfe J., Intriligator D. S., Mihalov J., Collard H., McKibben D., Whitten R., and Barnes A. (1979) Initial observations of the Pioneer Venus orbiter solar wind plasma experiment. *Science* **203**, 750–752.

von Zahn U., Krankowsky D., Mauersberger K., Nier A. O., and Hunten D. M. (1979) Venus thermosphere: *in situ* composition measurements, the temperature profile, and the homopause altitude. *Science* **203**, 768–770.

Errata

Note to Future Contributors: As you detect *important* typographical misprints, drafting errors, miscalculations, fallacious logic, or other mistakes in papers published in current or past *Proceedings* volumes please send corrections to the following address:

Publications Office
Lunar and Planetary Institute
3303 NASA Road 1
Houston, Texas 77058

Please keep notices of *errata* brief. Contributions greater in length than half of a printed page will be assessed page charges at the then current rate.

Proceedings of the Tenth Lunar and Planetary Science Conference
Geochimica et Cosmochimica Acta, Supplement 11

Volume 1

Norman M. D. and Ryder G. A summary of the petrology and geochemistry of pristine highlands rocks, p. 531–559.

Page 535: In Figure 2, cosmic abundance values of Ir and Re are ppm, not ppb.

Page 548: The text beginning on page 548 and ending at the bottom of page 549 is printed out of order, and should follow the text on page 541.

Page 547: In paragraph 1, line 9, change the age of the 73215 spinel troctolite clast from 4.6 ± 0.04 b.y. to 4.46 ± 0.04 b.y.

Watters T. R. and Prinz M. Aubrites: Their origin and relationship to enstatite chondrites, p. 1073–1093.

Page 1081: Data in Table 7, under the column heading Cumb. Falls should be changed as follows:

From	To
95.4	92.9
0.12	0.46
0.25	0.48
3.7	6.1
<0.02	0.18
99.5	100.1

Volume 2

Rao M. N., Venkatesan T. R., Goswami J. N., Nautiyal C. M., and Padia J. T.
Noble gas based solar flare exposure history of lunar rocks and soils, p. 1547–1564.

Page 1549: Table 1 data should be changed as follows:

From:

	Temperature (°C)	Helium ¾	^{4}He (10^{-5}) $^{++}$
R1	600	0.000446 ±0.000008	2.76 ±0.02
	1600	0.000591 ±0.000023	1.29 ±0.01
	Total	0.000495 ±0.000009	4.05 ±0.02

To:

	Temperature (°C)	Helium ¾	^{4}He $(10^{-5}$cc STP/g)
R1	600	0.000653 ±0.000013	2.25 ±0.02
	1600	0.002466 ±0.000100	0.37 ±0.01
	Total	0.00091 ±0.00001	2.62 ±0.02

Volume 3

Holsapple K. A. and Schmidt R. M. A material-strength model for apparent crater volume, p. 2757–2777.

Page 2769: Eq. (4.27), which reads

$$\pi_v = 0.194 \frac{\overline{\pi}_2}{\tan\phi + k_1}^{-0.472}$$

should be changed to read

$$\pi_v = 0.194 \left(\frac{\overline{\pi}_2}{\tan\phi + k_1} \right)^{-0.472}$$

Page 2771: Equation of Fig. 3 should be changed from $\pi_v \pi_2^{0.472} = 0.174$ to $\pi_v \overline{\pi}_2^{0.472} = 0.174$.

Lunar Sample Index

Pages   1– 838: Volume 1, Igneous Processes and Remote Sensing
Pages 839–1776: Volume 2, Meteorite and Regolith Studies
Pages 1777–2502: Volume 3, Physical Processes

Index entries were compiled from information supplied by the authors, and refer only to opening pages of articles.

Heavenly Body Index

Index entries were compiled from information supplied by the authors, and refer only to opening pages of articles.

Subject Index

Index entries were compiled from key words supplied by the authors, and refer only to opening pages of articles.

Author Index